Springer Series in Statistics

Springer Series in Statistics

Alho/Spencer: Statistical Demography and Forecasting
Andersen/Borgan/Gill/Keiding: Statistical Models Based on Counting Processes
Atkinson/Riani: Robust Diagnostic Regression Analysis
Atkinson/Riani/Ceriloi: Exploring Multivariate Data with the Forward Search
Berger: Statistical Decision Theory and Bayesian Analysis, 2nd edition
Borg/Groenen: Modern Multidimensional Scaling: Theory and Applications, 2nd edition
Brockwell/Davis: Time Series: Theory and Methods, 2nd edition
Bucklew: Introduction to Rare Event Simulation
Cappé/Moulines/Rydén: Inference in Hidden Markov Models
Chan/Tong: Chaos: A Statistical Perspective
Chen/Shao/Ibrahim: Monte Carlo Methods in Bayesian Computation
Coles: An Introduction to Statistical Modeling of Extreme Values
Devroye/Lugosi: Combinatorial Methods in Density Estimation
Diggle/Ribeiro: Model-based Geostatistics
Dudoit/van der Laan: Multiple Testing Procedures with Applications to Genomics
Efromovich: Nonparametric Curve Estimation: Methods, Theory, and Applications
Eggermont/LaRiccia: Maximum Penalized Likelihood Estimation, Volume I: Density Estimation
Fahrmeir/Tutz: Multivariate Statistical Modeling Based on Generalized Linear Models, 2nd edition
Fan/Yao: Nonlinear Time Series: Nonparametric and Parametric Methods
Ferraty/Vieu: Nonparametric Functional Data Analysis: Theory and Practice
Ferreira/Lee: Multiscale Modeling: A Bayesian Perspective
Fienberg/Hoaglin: Selected Papers of Frederick Mosteller
Frühwirth-Schnatter: Finite Mixture and Markov Switching Models
Ghosh/Ramamoorthi: Bayesian Nonparametrics
Glaz/Naus/Wallenstein: Scan Statistics
Good: Permutation Tests: Parametric and Bootstrap Tests of Hypotheses, 3rd edition
Gouriéroux: ARCH Models and Financial Applications
Gu: Smoothing Spline ANOVA Models
Györfi/Kohler/Krzyźak/Walk: A Distribution-Free Theory of Nonparametric Regression
Haberman: Advanced Statistics, Volume I: Description of Populations
Hall: The Bootstrap and Edgeworth Expansion
Härdle: Smoothing Techniques: With Implementation in S
Harrell: Regression Modeling Strategies: With Applications to Linear Models, Logistic Regression, and Survival Analysis
Hart: Nonparametric Smoothing and Lack-of-Fit Tests
Hastie/Tibshirani/Friedman: The Elements of Statistical Learning: Data Mining, Inference, and Prediction
Hedayat/Sloane/Stufken: Orthogonal Arrays: Theory and Applications
Heyde: Quasi-Likelihood and its Application: A General Approach to Optimal Parameter Estimation
Huet/Bouvier/Poursat/Jolivet: Statistical Tools for Nonlinear Regression: A Practical Guide with S-PLUS and R Examples, 2nd edition
Ibrahim/Chen/Sinha: Bayesian Survival Analysis
Jiang: Linear and Generalized Linear Mixed Models and Their Applications
Jolliffe: Principal Component Analysis, 2nd edition

(continued on p. 589)

Sandrine Dudoit
Mark J. van der Laan

Multiple Testing Procedures with Applications to Genomics

With 61 illustrations

Springer

Sandrine Dudoit
Division of Biostatistics
and Department of Statistics
University of California, Berkeley
101 Haviland Hall, #7358
Berkeley, CA 94720-7358
USA
sandrine@stat.berkeley.edu

Mark J. van der Laan
Division of Biostatistics
and Department of Statistics
University of California, Berkeley
101 Haviland Hall, #7358
Berkeley, CA 94720-7358
USA
laan@stat.berkeley.edu

Library of Congress Control Number: 2007927647

ISBN 978-0-387-49316-9 ISBN 978-0-387-49317-6 (eBook)

Printed on acid-free paper.

9 8 7 6 5 4 3 2 1

springer.com

À mes parents, Michèle Sutto-Dudoit et Alain Dudoit

To Martine, Laura, Lars, and Robin

Preface

Current statistical inference problems in areas such as astronomy, genomics, and marketing routinely involve the simultaneous test of thousands, or even millions, of null hypotheses. These hypotheses concern a wide range of parameters, for high-dimensional multivariate distributions, with complex and unknown dependence structures among variables.

Motivated by these applications and the limitations of existing multiple testing methods, we have developed and implemented resampling-based single-step and stepwise multiple testing procedures (MTP) for controlling a broad class of Type I error rates, defined as tail probabilities and expected values for arbitrary functions $g(V_n, R_n)$ of the numbers of Type I errors V_n and rejected hypotheses R_n (Birkner et al., 2005a,b,c, 2006, 2007; Dudoit et al., 2004a,b, 2006; Keleş et al., 2006; van der Laan et al., 2004a,b, 2005; van der Laan and Hubbard, 2006; Pollard et al., 2005a,b; Pollard and van der Laan, 2004; Rubin et al., 2006). Our proposed procedures take into account the joint distribution of the test statistics and provide Type I error control in testing problems involving general data generating distributions (with arbitrary dependence structures among variables), null hypotheses (defined in terms of submodels for the data generating distribution), and test statistics (e.g., t-statistics, χ^2-statistics, F-statistics). A key ingredient of the procedures is the null distribution used in place of the unknown joint distribution of the test statistics. The results of a given MTP are reported in terms of rejection regions (i.e., cut-offs) for the test statistics, confidence regions for the parameters of interest, and adjusted p-values.

This book provides a detailed account of the theoretical foundations of our multiple testing methodology and discusses its software implementation in R (multtest package; Gentleman et al. (2004); Pollard et al. (2005b); R Development Core Team (2006); `www.bioconductor.org`; `www.r-project.org`) and SAS (`www.sas.com`). The proposed methods are applied to a range of testing problems in biomedical and genomic research, including: the identification of differentially expressed and co-expressed genes in high-throughput gene expression experiments, such as microarray experiments; tests of association between gene expression measures and biological annotation metadata, such as Gene Ontology (GO, `www.geneontology.org`) annotation; protein sequence analysis; and the genetic mapping of complex traits using single nucleotide polymorphisms (SNP).

Intended readership

Methodological Chapters 4–7 are intended for readers with advanced undergraduate or graduate statistical training, whereas introductory Chapters 1–3 and applications Chapters 8–13 are also aimed at readers with biological background.

Some of the material discussed in this book was taught in the Division of Biostatistics at the University of California, Berkeley: upper division undergraduate course *Introduction to Statistical Methods in Computational and Genomic Biology* (PB HLTH 143); MA/PhD graduate course *Biostatistical Methods: Applications of Statistics to Genetics and Molecular Biology* (PB HLTH 240D); and MA/PhD graduate course *Multiple Testing and Loss Function Based Estimation: Applications in Biological Sciences* (PB HLTH 246C).

Overview

Chapter 1 introduces a general statistical framework for *multiple hypothesis testing* and discusses in turn the main ingredients of a multiple testing problem, including: the data generating distribution; the parameters of interest; the null and alternative hypotheses; the test statistics; multiple testing procedures; rejection regions for the test statistics; errors in multiple hypothesis testing: Type I, Type II, and Type III errors; Type I error rates; power; unadjusted and adjusted p-values; and stepwise multiple testing procedures.

Chapter 2 concerns a key feature of our proposed multiple testing methodology: the *test statistics null distribution* used to obtain rejection regions for the test statistics, confidence regions for the parameters of interest, and adjusted p-values. Indeed, whether testing single or multiple hypotheses, one needs the (joint) distribution of the test statistics in order to derive a procedure that probabilistically controls Type I errors. In practice, however, the true distribution of the test statistics is unknown and replaced by a null distribution. The choice of a proper null distribution is crucial in order to ensure that (finite sample or asymptotic) control of the Type I error rate under the assumed null distribution does indeed provide the desired control under the true distribution. This issue is particularly relevant for large-scale testing problems, such as those described above in biomedical and genomic research, which concern high-dimensional multivariate distributions, with complex and unknown dependence structures among variables.

Common approaches use a data generating distribution, such as a permutation distribution, that satisfies the complete null hypothesis that all null hypotheses are true. Procedures based on such a data generating null distribution typically rely on the subset pivotality assumption, stated in Westfall and Young (1993, p. 42–43), to ensure that Type I error control under the data generating null distribution leads to the desired control under the true data generating distribution. However, subset pivotality is violated in many important testing problems, because a data generating null distribution may result in a joint distribution for the test statistics that has a different dependence structure than their true distribution. In fact, in most problems, there does not exist a data generating null distribution that correctly specifies the joint distribution of the test statistics corresponding to the true null hypotheses. Indeed, subset pivotality fails for two types of testing problems that are highly relevant in biomedical and genomic data analysis: tests concerning correlation coefficients and tests concerning regression coefficients (Chapter 8; Pollard et al. (2005a); Pollard and van der Laan (2004)).

To address the shortcomings of existing approaches, we have formulated a general characterization of a test statistics null distribution for which the multiple testing procedures of Chapters 3–7 provide proper Type I error control (Section 2.2). Our general characterization is based on the intuitive notion of *null domination*, whereby the number of Type I errors is stochastically greater under the test statistics' null distribution than under their true distribution. Null domination conditions lead to the explicit construction of two main types of test statistics null distributions. The first original proposal of Dudoit et al. (2004b), van der Laan et al. (2004a), and Pollard and van der Laan (2004), defines the null distribution as the asymptotic distribution of a vector of *null shift and scale-transformed test statistics*, based on user-supplied upper bounds for the means and variances of the test statistics for the true null hypotheses (Section 2.3). The second and most recent proposal of van der Laan and Hubbard (2006) defines the null distribution as the asymptotic distribution of a vector of *null quantile-transformed test statistics*, based on user-supplied marginal test statistics null distributions (Section 2.4).

Either test statistics null distribution (or consistent estimators thereof) may be used in any of the multiple testing procedures proposed in Chapters 3–7, as they both satisfy the key property of joint null domination for the test statistics corresponding to the true null hypotheses. The latest proposal of van der Laan and Hubbard (2006) has the additional advantage that the marginal test statistics null distributions may be set to the optimal (i.e., most powerful) null distributions one would use in single hypothesis testing (e.g., permutation marginal null distributions, Gaussian or other parametric marginal null distributions). Resampling procedures (e.g., non-parametric or model-based bootstrap) are provided to conveniently obtain consistent estimators of the null distribution and of the corresponding test statistic cut-offs, parameter confidence regions, and adjusted p-values.

We stress the generality of our proposed test statistics null distributions: Type I error control does not rely on restrictive assumptions such as subset pivotality and holds for general data generating distributions, null hypotheses, and test statistics. In particular, the proposed null distributions allow one to address testing problems that cannot be handled by existing approaches, such as tests concerning correlation coefficients and parameters in general regression models (e.g., linear regression models where the covariates and error terms are allowed to be dependent, logistic regression models, Cox proportional hazards models).

As detailed in Section 2.8, the following two important points distinguish our approach from existing approaches to Type I error control and the choice of a null distribution. Firstly, we are only concerned with Type I error control under the true data generating distribution. The notions of weak and strong control (and associated subset pivotality) are therefore irrelevant for our methods. Secondly, we propose a null distribution for the test statistics, and not a data generating null distribution. The latter practice does not necessarily provide proper Type I error control, as a data generating null distribution may result in a joint distribution for the test statistics that has a different dependence structure than their true distribution.

The simulation studies of van der Laan and Hubbard (2006), Pollard et al. (2005a), and Pollard and van der Laan (2004), demonstrate that the choice of null distribution can have a substantial impact on the Type I error and power properties of a given multiple testing procedure (Chapter 8). In particular, Pollard et al. (2005a) show that procedures based on our general non-parametric bootstrap null shift and scale-transformed test statistics null distribution typically control the Type I error rate "on target" at the nominal level. In contrast, comparable procedures, based on parameter-specific bootstrap data generating null distributions, can be severely anti-conservative (bootstrapping residuals for testing regression coefficients) or conservative (independent bootstrap for testing correlation coefficients). van der Laan and Hubbard (2006) further illustrate that, for finite samples, the new null quantile-transformed test statistics null distribution provides more accurate Type I error control and is more powerful than the original null shift and scale-transformed null distribution.

Chapter 3 presents an overview of basic multiple testing procedures for controlling the number of Type I errors (family-wise error rate and generalized family-wise error rate, in Sections 3.2 and 3.3, respectively) and the proportion of Type I errors among the rejected hypotheses (false discovery rate and tail probabilities for the proportion of false positives, in Sections 3.4 and 3.5, respectively). The different procedures are stated in terms of adjusted p-values as well as cut-offs for individual test statistics or unadjusted p-values. Summary tables are provided in Appendix A.

Chapter 4 proposes general *joint single-step common-cut-off* and *common-quantile procedures* for controlling Type I error rates $\Theta(F_{V_n})$, defined as arbitrary parameters of the distribution of the number of Type I errors V_n (Section 4.2; Dudoit et al. (2004b); Pollard and van der Laan (2004)). Such error rates include the generalized family-wise error rate (gFWER), $gFWER(k) = 1 - F_{V_n}(k) = \Pr(V_n > k)$, i.e., the chance of at least $(k+1)$ Type I errors, and, in particular, the usual family-wise error rate (FWER), $FWER = gFWER(0) = 1 - F_{V_n}(0) = \Pr(V_n > 0)$. In the special case of $gFWER(k)$ control, the procedures are based on the $(k+1)$st largest test statistic and $(k+1)$st smallest unadjusted p-value, respectively. For control of the FWER, the procedures reduce to the single-step maxT and minP procedures, based on the maximum test statistic and minimum unadjusted p-value, respectively. Adjusted p-values are derived in Section 4.3. Single-step common-cut-off and common-quantile procedures, based on consistent estimators of the test statistics null distribution, are shown to provide asymptotic control of the Type I error rate $\Theta(F_{V_n})$. General bootstrap procedures are supplied to conveniently obtain consistent estimators of the single-step common cut-offs and common-quantile cut-offs and of the corresponding adjusted p-values (Section 4.4). This chapter also establishes equivalence results between Θ-specific single-step multiple testing procedures and parameter confidence regions (Section 4.6) and addresses the issue of test optimality, i.e., the maximization of power subject to a Type I error constraint (Section 4.7; Rubin et al. (2006)).

Chapter 5 focuses on control of the family-wise error rate, $FWER = 1 - F_{V_n}(0)$, and provides *joint step-down common-cut-off maxT* and *common-quantile minP procedures*, based on maxima of test statistics and minima of unadjusted p-values, respectively (Sections 5.2 and 5.3; van der Laan et al. (2004a)). Two main types of results are derived concerning asymptotic control of the FWER. The more general theorems prove that the step-down maxT and minP procedures provide asymptotic control of the FWER, under general asymptotic null domination assumptions for the test statistics null distribution. Exact asymptotic control results are obtained by making additional asymptotic separation assumptions for the test statistics for the true and false null hypotheses. Step-up procedures are discussed in Section 5.4. Step-down maxT and minP procedures, based on consistent estimators of the test statistics null distribution, are shown to provide asymptotic control of the FWER. General bootstrap procedures are supplied to conveniently obtain consistent estimators of the step-down maxT and minP cut-offs and of the corresponding adjusted p-values (Section 5.5).

Chapter 6 proposes a new general and flexible approach to multiple hypothesis testing, the augmentation method, whereby a set of suitably chosen null hypotheses are added to the set of hypotheses already rejected by an initial MTP, in order to control a second target Type I error rate (Dudoit et al., 2004a; van der Laan et al., 2004b). Specifically, given an initial gFWER-controlling procedure, this chapter provides (marginal/joint single-step/stepwise) *augmentation multiple testing procedures* (AMTP) for controlling generalized tail probability (gTP) error rates, $gTP(q,g) = \Pr(g(V_n, R_n) > q)$, for arbitrary functions $g(V_n, R_n)$ of the numbers of false positives V_n and rejected hypotheses R_n (Section 6.5). Simple augmentations of FWER-controlling procedures are treated in detail, for controlling tail probabilities for the number of false positives (gFWER), with $g(v,r) = v$, and tail probabilities for the proportion of false positives (TPPFP) among the rejected hypotheses, with $g(v,r) = v/r$ (Sections 6.2 and 6.5.3 for gFWER; Sections 6.3 and 6.5.4 for TPPFP). As shown in Section 6.5.2, the adjusted p-values for an augmentation multiple testing procedure are simply shifted versions of the ordered adjusted p-values for the initial MTP. Section 6.6 demonstrates that one can readily derive (conservative) procedures controlling generalized expected value (gEV) error rates, $gEV(g) = \mathrm{E}[g(V_n, R_n)]$, based on procedures controlling generalized tail probability error rates, $gTP(q,g) = \Pr(g(V_n, R_n) > q)$. Control of the false discovery rate (FDR), based on a TPPFP-controlling MTP, corresponds to the special case $g(v,r) = v/r$ (Section 6.4).

We stress the generality and important practical implications of the augmentation approach to multiple testing: any gFWER-controlling MTP immediately and trivially provides multiple testing procedures that control a wide variety of error rates, defined as tail probabilities $\Pr(g(V_n, R_n) > q)$ for arbitrary functions $g(V_n, R_n)$ of the numbers of false positives V_n and rejected hypotheses R_n. While existing approaches for controlling the proportion of false positives (e.g., TPPFP and FDR) typically assume either independence or specific dependence structures for the joint distribution of the test statistics, augmentation procedures can be derived for general data generating distributions (i.e., arbitrary joint distributions for the test statistics), null hypotheses, and test statistics. One can therefore build on the large pool of available FWER-controlling procedures to greatly expand the class of Type I error rates one can control (e.g., single-step and step-down maxT and minP procedures, summarized in overview Chapter 3 and discussed in detail in Chapters 4 and 5).

Chapter 7 builds on van der Laan et al. (2005) and proposes new *joint resampling-based empirical Bayes procedures* for controlling generalized tail probability error rates, $gTP(q,g) = \Pr(g(V_n, R_n) > q)$. The approach involves specifying: (i) a null distribution for vectors of null test statistics and (ii) a distribution for random guessed sets of true null hypotheses. By randomly sampling null test statistics and guessed sets of true null hypotheses, one obtains a distribution for a guessed g-specific function of the numbers of false positives and rejected hypotheses, for any given vector of cut-offs for the test statistics. Cut-offs can then be chosen to control tail probabilities for this distribution at a user-supplied level. This chapter also discusses empirical Bayes

q-value-based approaches to FDR control and connections to the frequentist step-up Benjamini and Hochberg (1995) procedure.

Chapter 8 presents simulation studies assessing the performance of the multiple testing procedures described in Chapters 1–7. The simulation studies focus on the choice of a test statistics null distribution in testing problems concerning correlation coefficients and regression coefficients in models where the covariates and error terms are allowed to be dependent (Pollard et al., 2005a).

Chapters 9–12 apply the proposed methodology to the following multiple testing problems in biomedical and genomic research: the identification of differentially expressed and co-expressed genes in high-throughput gene expression experiments (Chapter 9); tests of association between gene expression measures and biological annotation metadata, e.g., Gene Ontology annotation (Chapter 10); the identification of HIV-1 codon positions associated with viral replication capacity (Chapter 11); the genetic mapping of human obesity, based on tests of association between multilocus composite SNP genotypes and obesity-related phenotypes (Chapter 12).

The above testing problems share the following general characteristics: inference for high-dimensional multivariate distributions, with complex and unknown dependence structures among variables; broad range of parameters of interest, such as coefficients in general regression models relating possibly censored biological and clinical covariates and outcomes to genome-wide expression measures and genotypes; many null hypotheses, in the thousands or even millions; complex and unknown dependence structures among test statistics (e.g., directed acyclic graph (DAG) structure of GO terms in Chapter 10, Galois lattice for multilocus composite SNP genotypes in Chapter 12).

Due to their generality and flexibility, the multiple testing procedures of Chapters 1–7 are well-suited to address these and other high-dimensional testing problems arising in different areas of application of statistics. In particular, recall that the proposed procedures are designed to control a broad range of Type I error rates, for: general multivariate data generating distributions, with arbitrary dependence structures among variables; general null hypotheses, defined in terms of submodels for the data generating distribution; general test statistics, such as, t-statistics for tests of means, correlation coefficients, and coefficients in general regression models, and F-statistics for testing multiple-parameter null hypotheses.

Chapter 13 discusses the software implementation of the proposed multiple testing procedures in the R package multtest, released as part of the Bioconductor Project, an open-source software project for the analysis of biomedical and genomic data (Section 13.1; Gentleman et al. (2004); Pollard et al. (2005b); R Development Core Team (2006); `www.bioconductor.org`; `www.r-project.org`). This chapter also illustrates the implementation in SAS of a bootstrap-based single-step maxT procedure and gFWER- and

TPPFP-controlling augmentation multiple testing procedures (Section 13.2; Birkner et al. (2005b); SAS, Version 9, `www.sas.com`).

Appendix A contains summaries of basic definitions, notation, and multiple testing procedures.

Appendix B provides miscellaneous mathematical and statistical results used repeatedly throughout the book.

Appendix C supplies SAS code for some of the proposed multiple testing procedures.

Supplements

The book's website provides supplementary materials, such as, additional analyses, tables, and figures, articles, lecture notes, software, datasets, links, and errata (`www.stat.berkeley.edu/~sandrine/MTBook`; `www.springer-ny.com`).

The reader is referred to the National Center for Biotechnology Information (NCBI) website for online tutorials and other educational resources on genome biology (`www.ncbi.nlm.nih.gov/Education`). The supplements to *Nature Genetics* provide an overview of the biology, technology, and applications of microarray experiments (Phimister and Cohen (1999); Packer (2002); Packer and Axton (2005); `www.nature.com/ng/supplements`). The book edited by Speed (2003) discusses statistical methods for the analysis of microarray data.

Software packages (e.g., R package multtest), datasets, short course materials (e.g., lecture notes, computer labs), and documentation may be downloaded from the Bioconductor Project (Gentleman et al. (2004); `www.bioconductor.org`) and R Project (R Development Core Team (2006); `www.r-project.org`) websites. The monograph edited by Gentleman et al. (2005a) provides a survey of Bioconductor software packages and their applications to a range of problems in computational biology (`www.bioconductor.org/pub/docs/mogr`).

Technical reports are available from the *UC Berkeley Division of Biostatistics Working Paper Series* website (`www.bepress.com/ucbbiostat`).

Finally, our personal websites provide additional resources on multiple hypothesis testing (SD: `www.stat.berkeley.edu/~sandrine`; MJvdL: `www.stat.berkeley.edu/~laan`).

Acknowledgments

We warmly thank our former students and colleagues, Merrill D. Birkner, Sündüz Keleş, and Katherine S. Pollard, for most pleasant collaborations and their constructive comments on the following portions of the book: Chapters

8, 11, and 12, Section 13.2 (MDB); Chapter 10 (SK); Sections 9.3 and 13.1 (KSP).

We would also like to acknowledge the following colleagues and students for many stimulating discussions on multiple hypothesis testing: Simon E. Cawley, Yongchao Ge, Robert C. Gentleman, Torsten Hothorn, Jason C. Hsu, Alan E. Hubbard, Nicholas P. Jewell, Daniel Rubin, Juliet P. Shaffer, Phil Spector, and Terence P. Speed.

Much of our methodological work is motivated by statistical problems in biomedical and genomic research. We are most grateful to our collaborators in biology, computer science, and epidemiology for introducing us to fascinating new questions and for inspiring our research on multiple hypothesis testing: Alain Barrier, Jennifer C. Boldrick, Patrick O. Brown, Patricia A. Buffler, Matthew J. Callow, Karine Clément, Mélanie Courtine, Martyn T. Smith, and Jean-Daniel Zucker.

Last, but not least, we would like to thank students at UC Berkeley (PB HLTH 143, Spring 2004; PB HLTH 240D, Spring 2003 and 2005; PB HLTH 246C (formerly PB HLTH 243A), Fall 2003 and 2005) and participants in Bioconductor workshops for their valuable feedback on multiple testing methods and their software implementation.

Berkeley, CA
May 2007

Sandrine Dudoit
Mark J. van der Laan

Contents

List of Figures

* See color plates p. 321–344.

List of Tables

1

Multiple Hypothesis Testing

1.1 Introduction

1.1.1 Motivation

Current statistical inference problems in areas such as astronomy, genomics, and marketing routinely involve the simultaneous test of thousands, or even millions, of null hypotheses. These hypotheses concern a wide range of parameters, for high-dimensional multivariate distributions, with complex and unknown dependence structures among variables. Examples of testing problems in biomedical and genomic research include the following.

- The identification of differentially expressed genes in high-throughput gene expression experiments such as microarray experiments, i.e., genes whose expression measures are associated with possibly censored biological and clinical covariates and outcomes (Chapter 9).
- The identification of co-expressed genes in high-throughput gene expression experiments, i.e., pairs or sets of genes with correlated expression measures across biological samples (Chapter 9).
- Tests of association between gene expression measures and biological annotation metadata, e.g., Gene Ontology (GO, `www.geneontology.org`) annotation (Chapter 10).
- Tests of association between phenotypes and codon/amino acid mutations, e.g., association between viral replication capacity and HIV-1 sequence variation (Chapter 11).
- The genetic mapping of complex traits, based on tests of association between phenotypes and genotypes, e.g., individual single nucleotide polymorphisms (SNP), SNP haplotypes, microsatellite marker genotypes, and identity by descent status (Chapter 12).

The above testing problems share the following general characteristics.

- Inference for *high-dimensional multivariate distributions*, with complex and unknown dependence structures among variables.

- *Broad range of parameters* of interest, such as: regression coefficients in non-linear models relating patient survival data to genome-wide transcript (i.e., mRNA) levels, DNA copy numbers, or SNP genotypes; measures of association between GO annotation and parameters of the distribution of microarray expression measures; pairwise correlation coefficients between transcript levels.
- *Many null hypotheses*, in the thousands or even millions.
- *Complex and unknown dependence structures among test statistics*, e.g., directed acyclic graph (DAG) structure of GO terms in Chapter 10, Galois lattice for multilocus composite SNP genotypes in Chapter 12.

Motivated by these applications and the limitations of existing multiple testing methods, we have developed and implemented (in R and SAS) resampling-based single-step and stepwise multiple testing procedures (MTP) for controlling a broad class of Type I error rates, defined as tail probabilities and expected values for arbitrary functions $g(V_n, R_n)$ of the numbers of Type I errors V_n and rejected hypotheses R_n (Birkner et al., 2005a,b,c, 2006, 2007; Dudoit et al., 2004a,b, 2006; Keleş et al., 2006; van der Laan et al., 2004a,b, 2005; van der Laan and Hubbard, 2006; Pollard et al., 2005a,b; Pollard and van der Laan, 2004; Rubin et al., 2006). Our proposed procedures take into account the joint distribution of the test statistics and provide Type I error control in testing problems involving general data generating distributions (with arbitrary dependence structures among variables), null hypotheses (defined in terms of submodels for the data generating distribution), and test statistics (e.g., t-statistics, χ^2-statistics, F-statistics). A key ingredient of the procedures is the null distribution used in place of the unknown joint distribution of the test statistics. The results of a given MTP are reported in terms of rejection regions (i.e., cut-offs) for the test statistics, confidence regions for the parameters of interest, and adjusted p-values.

The different components of our multiple testing methodology are treated in detail in the collection of articles summarized below.

1.1.2 Bibliography for proposed multiple testing methodology

The early articles of **Dudoit et al. (2004b)** and **Pollard and van der Laan (2004)** establish a general statistical framework for multiple hypothesis testing. A key feature of the proposed MTPs is the *test statistics null distribution* (rather than a data generating null distribution) used to obtain rejection regions for the test statistics, confidence regions for the parameters of interest, and adjusted p-values. Indeed, whether testing single or multiple hypotheses, one needs the (joint) distribution of the test statistics in order to derive a procedure that probabilistically controls Type I errors. In practice, however, the true distribution of the test statistics is unknown and replaced by a null distribution. The choice of a proper null distribution is crucial in order to ensure that (finite sample or asymptotic) control of the Type I error rate under the assumed null distribution does indeed provide the desired control under the true distribution. This issue is particularly relevant for large-scale testing problems, such as those described above in biomedical and genomic research, which concern high-dimensional multivariate distributions, with complex and unknown dependence structures among variables. As illustrated in the simulation studies of van der Laan and Hubbard (2006), Pollard et al. (2005a), and Pollard and van der Laan (2004), the choice of null distribution can have a substantial impact on the Type I error and power properties of a given MTP.

Dudoit et al. (2004b) provide a general characterization for a proper test statistics null distribution, which leads to the explicit construction of two main types of test statistics null distributions. The first original proposal of Dudoit et al. (2004b), van der Laan et al. (2004a), and Pollard and van der Laan (2004), defines the null distribution as the asymptotic distribution of a vector of *null shift and scale-transformed test statistics*, based on user-supplied upper bounds for the means and variances of the test statistics for the true null hypotheses. The second and most recent proposal of **van der Laan and Hubbard (2006)** defines the null distribution as the asymptotic distribution of a vector of *null quantile-transformed test statistics*, based on user-supplied marginal test statistics null distributions. Resampling procedures (e.g., non-parametric or model-based bootstrap) are provided to conveniently obtain consistent estimators of the null distribution and of the corresponding test statistic cut-offs, parameter confidence regions, and adjusted p-values.

Dudoit et al. (2004b) and **Pollard and van der Laan (2004)** derive *joint single-step common-cut-off* and *common-quantile procedures* for controlling Type I error rates defined as arbitrary parameters $\Theta(F_{V_n})$ of the distribution F_{V_n} of the number of Type I errors V_n. Such error rates include the generalized family-wise error rate (gFWER), $gFWER(k) = 1 - F_{V_n}(k) = \Pr(V_n > k)$, i.e., the chance of at least $(k+1)$ Type I errors, and, in particular, the usual family-wise error rate (FWER), $FWER = gFWER(0) = 1 - F_{V_n}(0) = \Pr(V_n > 0)$. In the special case of $gFWER(k)$ control, such procedures are based on the $(k+1)$st largest test statistic and $(k+1)$st smallest unadjusted p-value, respectively.

The recent manuscript by **Rubin et al. (2006)** concerns *optimal multiple testing procedures*, i.e., MTPs that maximize power subject to a Type I error constraint.

van der Laan et al. (2004a) focus on control of the family-wise error rate, $FWER = 1 - F_{V_n}(0)$, and provide *joint step-down common-cut-off* and *common-quantile procedures*, based on maxima of test statistics (maxT) and minima of unadjusted p-values (minP), respectively.

van der Laan et al. (2004b) propose (marginal/joint single-step/stepwise) *augmentation multiple testing procedures* (AMTP) for controlling tail probabilities for the number of false positives (gFWER) and tail probabilities for the proportion of false positives (TPPFP) among the rejected hypotheses. The AMTPs are obtained by adding suitably chosen null hypotheses to the set of hypotheses already rejected by an initial FWER-controlling MTP. The results of a simulation study comparing augmentation procedures to existing gFWER- and TPPFP-controlling MTPs are reported in **Dudoit et al. (2004a)**. Dudoit et al. (2004a) further propose AMTPs for controlling generalized tail probability (gTP) error rates, $gTP(q,g) = \Pr(g(V_n, R_n) > q)$, for arbitrary functions $g(V_n, R_n)$ of the numbers of false positives V_n and rejected hypotheses R_n.

van der Laan et al. (2005) propose a TPPFP-controlling *joint resampling-based empirical Bayes procedure*, relying on a non-parametric mixture model for the test statistics.

Pollard et al. (2005a) address the choice of a test statistics null distribution in the context of tests for correlation coefficients and tests for regression coefficients in models where the covariates and error terms are allowed to be dependent. Simulation studies, comparing the Type I error control and power properties of MTPs based on our proposed bootstrap test statistics null distribution and existing bootstrap-based procedures, illustrate limitations of the latter approaches (which can be severely anti-conservative) and the importance of selecting a proper null distribution.

Dudoit et al. (2006) propose a general and formal statistical framework for multiple tests of association with biological annotation metadata. The methods are illustrated using the acute lymphoblastic leukemia microarray dataset of Chiaretti et al. (2004), with the aim of relating tumor differential gene expression to GO annotation.

Pollard et al. (2005b) discuss the software implementation of the aforementioned MTPs in the Bioconductor R package `multtest` (Gentleman et al. (2004); `www.bioconductor.org`).

Birkner et al. (2005b) demonstrate an implementation in SAS of a bootstrap-based single-step maxT procedure and gFWER- and TPPFP-controlling augmentation multiple testing procedures (SAS, Version 9, `www.sas.com`).

1.1.3 Overview of applications to biomedical and genomic research

The novel multiple testing procedures introduced above have been applied to a number of testing problems in biomedical and genomic research. Many of these applications concern *microarray experiments*, which are popular high-throughput assays for measuring the abundance of *deoxyribonucleic acids* (DNA) and *ribonucleic acids* (RNA) in different types of cell samples for thousands of sequences simultaneously (Phimister and Cohen, 1999; Packer, 2002; Packer and Axton, 2005; Speed, 2003).

- The identification of differentially expressed genes in microarray experiments, i.e., genes whose expression measures are associated with possibly censored biological and clinical covariates and outcomes (Chapter 9; Dudoit et al. (2002, 2003, 2006); Ge et al. (2003); Pollard et al. (2005a,b); Pollard and van der Laan (2004)):

  - lymphoma dataset of Alizadeh et al. (2000);
  - colon cancer datasets of Barrier et al. (2005a,b, 2006);
  - liver cancer datasets of Barrier et al. (2005c) and Chiappini et al. (2006);
  - bacteria dataset from the Boldrick et al. (2002) study of host (human peripheral blood mononuclear cells) genomic responses to bacterial (*Bordetella pertussis, Staphylococcus aureus*) infection;
  - Apo AI knock-out dataset from the Callow et al. (2000) study of metabolism and artherosclerosis susceptibility in mice;
  - acute lymphoblastic leukemia (ALL) dataset of Chiaretti et al. (2004);
  - acute lymphoblastic leukemia and acute myeloid leukemia (AML) dataset of Golub et al. (1999).
- Tests of association between the microarray expression measures of genes within the same *Pyrobaculum aerophilum* operon (Personal communication, Katherine S. Pollard, `www.docpollard.com`).
- The identification of differentially expressed and co-expressed microRNAs (miRNA) in cancerous and non-cancerous tissues (Chapter 9; Lu et al. (2005); Pollard et al. (2005a)).
- The identification of differentially expressed gene isoforms using alternative splicing microarrays, with probes in exons and in alternative and constitutive splice junctions (Blanchette et al., 2005).
- Monitoring spatial and temporal bacterial differential abundance in air samples from various US cities, based on measures from a 16s small-subunit ribosomal RNA (rRNA) microarray (Birkner et al., 2005a; DeSantis et al., 2005).
- Tests of association between microarray gene expression measures and Gene Ontology annotation (Chapter 10; Dudoit et al. (2006)).
- The identification of transcription factor (TF) binding sites in ChIP-Chip experiments, where chromatin immunoprecipitation (ChIP) of transcription factor-bound DNA is followed by microarray (Chip) hybridization of the IP-enriched DNA. E.g. Identification of binding sites for p53 transcription factor using high-density oligonucleotide chips for human chromosomes 21 and 22 (Cawley et al., 2004; Keleş et al., 2006).
- Tests of association between viral replication capacity and HIV-1 sequence variation (Chapter 11; Birkner et al. (2005b,c); Segal et al. (2004)).
- The genetic mapping of human obesity, based on tests of association between multilocus composite SNP genotypes and obesity-related phenotypes (Chapter 12; ObeLinks Project, `www.obelinks.org`; Birkner et al. (2007)).
- Tests of association between non-Hodgkin lymphoma subclass and SNPs in the ghrelin and neuropeptide Y genes (van der Laan and Hubbard, 2006).
- Tests of association between phenotypes and protein mass-spectroscopy measures. E.g. Identification of mass-to-charge ratios associated with leukemia (ALL vs. AML) class (Birkner et al., 2005a, 2006).

1.1.4 Road map

Scope of the book

This book focuses primarily on the above recently developed multiple hypothesis testing methodology.

For an account of classical approaches to multiple testing, the reader is referred to books by Hochberg and Tamhane (1987), Hsu (1996), and Westfall and Young (1993), and to overview articles by Dudoit et al. (2003) and Shaffer (1995).

References from the growing literature on procedures controlling the false discovery rate (FDR), i.e., the expected value $E[V_n/R_n]$ of the proportion of false positives among the rejected hypotheses, may be obtained from the following individuals' websites: Yoav Benjamini (`www.math.tau.ac.il/~ybenja`), Brad Efron (`www-stat.stanford.edu/~brad`), Christopher Genovese (`ib.stat.cmu.edu/~genovese`), and John Storey (`faculty.washington.edu/~jstorey`).

Note that a number of (empirical) Bayesian approaches have been proposed recently to address multiple testing problems in microarray data analysis (Efron, 2005; Efron et al., 2001a,b; Manduchi et al., 2000; Newton et al., 2001). Although such methods constitute an important alternative to frequentist approaches, their thorough treatment is beyond the scope of this book.

Outline of the book

This book provides a detailed account of the theoretical foundations of the multiple hypothesis testing methodology introduced in Section 1.1.2 and discusses its software implementation and application to a variety of testing problems in biomedical and genomic research.

The present chapter introduces a general statistical framework for *multiple hypothesis testing* and motivates the methods developed in Chapters 2–7. These methodological chapters provide specific multiple testing procedures for controlling a range of Type I error rates that are broadly defined as parameters $\Theta(F_{V_n,R_n})$ of the joint distribution F_{V_n,R_n} of the numbers of Type I errors V_n and rejected hypotheses R_n.

Chapter 2 discusses a key feature of our proposed multiple testing procedures, namely, the *test statistics null distribution* (rather than a data generating null distribution) used to obtain rejection regions for the test statistics, confidence regions for the parameters of interest, and adjusted p-values (Dudoit et al., 2004b; van der Laan et al., 2004a; van der Laan and Hubbard, 2006; Pollard and van der Laan, 2004).

Chapter 3 presents an overview of multiple testing procedures for controlling the number of Type I errors (FWER and gFWER, in Sections 3.2 and 3.3, respectively) and the proportion of Type I errors among the rejected hypotheses (FDR and TPPFP, in Sections 3.4 and 3.5, respectively).

Chapter 4 proposes general *joint single-step common-cut-off* and *common-quantile procedures* for controlling Type I error rates $\Theta(F_{V_n})$, defined as arbitrary parameters of the distribution of the number of Type I errors V_n (Dudoit et al., 2004b; Pollard and van der Laan, 2004). Such error rates include the generalized family-wise error rate, $gFWER(k) = 1 - F_{V_n}(k) = \Pr(V_n > k)$, i.e., the chance of at least $(k+1)$ Type I errors. In the special case of $gFWER(k)$ control, the procedures are based on the $(k+1)$st largest test statistic and $(k+1)$st smallest unadjusted p-value, respectively. This chapter also establishes equivalence results between Θ-specific single-step multiple testing procedures and parameter *confidence regions* and addresses the issue of test *optimality*, i.e., the maximization of power subject to a Type I error constraint (Rubin et al., 2006).

Chapter 5 focuses on control of the family-wise error rate, $FWER = gFWER(0) = 1 - F_{V_n}(0) = \Pr(V_n > 0)$, and provides *joint step-down common-cut-off* and *common-quantile procedures*, based on maxima of test statistics (maxT) and minima of unadjusted p-values (minP), respectively (van der Laan et al., 2004a).

Chapter 6 proposes a new general and flexible approach to multiple hypothesis testing, the augmentation method, whereby a set of suitably chosen null hypotheses are added to the set of hypotheses already rejected by an initial MTP, in order to control a second target Type I error rate (Dudoit et al., 2004a; van der Laan et al., 2004b). Specifically, given an initial gFWER-controlling procedure, this chapter provides (marginal/joint single-step/stepwise) *augmentation multiple testing procedures* (AMTP) for controlling generalized tail probability (gTP) error rates, $gTP(q,g) = \Pr(g(V_n, R_n) > q)$, for arbitrary functions $g(V_n, R_n)$ of the numbers of false positives V_n and rejected hypotheses R_n. Simple augmentations of FWER-controlling procedures are treated in detail, for controlling tail probabilities for the number of false positives (gFWER), with $g(v,r) = v$, and tail probabilities for the proportion of false positives (TPPFP) among the rejected hypotheses, with $g(v,r) = v/r$. This chapter also demonstrates that one can readily derive (conservative) procedures controlling generalized expected value (gEV) error rates, $gEV(g) = \mathrm{E}[g(V_n, R_n)]$, based on procedures controlling generalized tail probability error rates, $gTP(q,g) = \Pr(g(V_n, R_n) > q)$. Control of the FDR, based on a TPPFP-controlling MTP, corresponds to the special case $g(v,r) = v/r$.

Chapter 7 builds on van der Laan et al. (2005) and proposes new *joint resampling-based empirical Bayes procedures* for controlling generalized tail probability error rates, $gTP(q,g) = \Pr(g(V_n, R_n) > q)$. The approach involves specifying: (i) a null distribution for vectors of null test statistics and (ii) a distribution for random guessed sets of true null hypotheses. By randomly sampling null test statistics and guessed sets of true null hypotheses, one obtains a distribution for a guessed g-specific function of the numbers of false positives and rejected hypotheses, for any given vector of cut-offs for the test statistics. Cut-offs can then be chosen to control tail probabilities for this distribution at a user-supplied level. This chapter also discusses empirical Bayes q-value-based approaches to FDR control and connections to the frequentist step-up Benjamini and Hochberg (1995) procedure.

Chapter 8 presents simulation studies assessing the performance of the multiple testing procedures described in Chapters 1–7. The simulation studies focus on the choice of a test statistics null distribution in testing problems concerning correlation coefficients and regression coefficients in models where the covariates and error terms are allowed to be dependent (Pollard et al., 2005a).

Chapters 9–12 apply the proposed methodology to the following multiple testing problems in biomedical and genomic research.

- The identification of differentially expressed and co-expressed genes in high-throughput gene expression experiments (Chapter 9): Apo AI dataset of Callow et al. (2000) and cancer miRNA dataset of Lu et al. (2005).
- Tests of association between gene expression measures and biological annotation metadata, e.g., Gene Ontology annotation (Chapter 10; Dudoit et al. (2006)).
- The identification of HIV-1 codon positions associated with viral replication capacity (Chapter 11; Birkner et al. (2005b,c); Segal et al. (2004)).
- The genetic mapping of human obesity, based on tests of association between multilocus composite SNP genotypes and obesity-related phenotypes (Chapter 12; ObeLinks Project, `www.obelinks.org`; Birkner et al. (2007)).

Chapter 13 discusses the software implementation of the proposed multiple testing procedures in the R package multtest, released as part of the Bioconductor Project, an open-source software project for the analysis of biomedical and genomic data (Gentleman et al. (2004); Pollard et al. (2005b); R Development Core Team (2006); `www.bioconductor.org`; `www.r-project.org`). This chapter also illustrates the implementation in SAS of a bootstrap-based single-step maxT procedure and gFWER- and TPPFP-controlling augmentation multiple testing procedures (Birkner et al. (2005b); SAS, Version 9, `www.sas.com`).

Appendix A contains summaries of basic definitions, notation, and multiple testing procedures. **Appendix B** provides miscellaneous mathematical and statistical results used repeatedly throughout the book. **Appendix C** supplies SAS code for some of the proposed multiple testing procedures.

The book's website provides supplementary materials, such as, additional analyses, tables, and figures, articles, lecture notes, software, datasets, links, and errata (`www.stat.berkeley.edu/~sandrine/MTBook`; `www.springer-ny.com`).

Outline of Chapter 1

This first chapter is organized as follows. Section 1.2 introduces a general statistical framework for multiple hypothesis testing and discusses in turn the main ingredients of a multiple testing problem, including: the data generating distribution (Section 1.2.2); the parameters of interest (Section 1.2.3); the null and alternative hypotheses (Section 1.2.4); the test statistics (Section 1.2.5); multiple testing procedures (Section 1.2.6); rejection regions (i.e., cut-offs) for the test statistics (Section 1.2.7); errors in multiple hypothesis testing: Type I, Type II, and Type III errors (Section 1.2.8); Type I error rates (Section 1.2.9); power (Section 1.2.10); unadjusted and adjusted p-values (Section 1.2.12); stepwise multiple testing procedures (Section 1.2.13).

The key issues of Type I error control and the choice of a test statistics null distribution are treated in Chapter 2.

1.2 Multiple hypothesis testing framework

1.2.1 Overview

Hypothesis testing is concerned with using observed data to make decisions regarding properties of (i.e., hypotheses for) the unknown data generating distribution. A null hypothesis states that the data generating distribution belongs to a particular submodel, i.e., a set of possibly non-parametric distributions. Null hypotheses are often expressed in terms of parameters, defined as functions of the data generating distribution. Parameters of interest include, for example, the mean vector or correlation matrix of a multivariate distribution of microarray expression measures and regression coefficients in linear or non-linear models relating phenotypes to multilocus SNP genotypes.

A testing procedure is a data-driven rule for deciding which null hypotheses should be rejected. The decisions to reject or not the null hypotheses are based on test statistics, defined as functions of the data (e.g., t-statistics, χ^2-statistics, F-statistics, and likelihood ratio statistics). A multiple testing procedure (MTP), for the simultaneous test of $M \geq 1$ null hypotheses, provides rejection regions for each of the M hypotheses, i.e., sets of values for each of M test statistics that lead to the decision to reject the corresponding null hypotheses. In other words, a MTP produces a random (i.e., data-dependent) set of rejected hypotheses that estimates the set of false null hypotheses.

In any testing problem, two types of errors can be committed. A Type I error, or false positive, is committed by rejecting a true null hypothesis. A Type II error, or false negative, is committed by failing to reject a false null hypothesis. Ideally, one would like to simultaneously minimize both the number of Type I errors and the number of Type II errors. Unfortunately, this is not feasible and one seeks a trade-off between the two types of errors. This trade-off typically involves the minimization of Type II errors, i.e., the maximization of power, subject to a Type I error constraint. A multiple testing procedure is then a data-driven rule specifying which of the M null hypotheses to reject, while controlling a suitably defined Type I error rate.

Whether testing single or multiple hypotheses, one needs the (joint) distribution of the test statistics in order to derive rejection regions for procedures that probabilistically control Type I errors. In practice, however, the true distribution of the test statistics is unknown and replaced by a null distribution. As discussed in Chapter 2, the choice of a proper null distribution is crucial in order to ensure that (finite sample or asymptotic) control of the Type I error rate under the assumed null distribution does indeed provide the desired control under the true distribution. Resampling procedures (e.g., bootstrap and permutation) are particularly useful in this context.

As in the case of single hypothesis testing, one can report the results of a multiple testing procedure in terms of the following quantities: rejection regions for the test statistics, confidence regions for the parameters of interest, and adjusted p-values. Adjusted p-values, for the test of multiple hypotheses, are defined as straightforward extensions of unadjusted p-values, for the test of individual hypotheses: the adjusted p-value for a particular null hypothesis is the smallest nominal Type I error level (for the multiple test of all M hypotheses) at which one would reject this null hypothesis. The smaller the adjusted p-value, the stronger the evidence against the corresponding null hypothesis. The main components of a multiple testing procedure are discussed next and summarized in the flowchart of Table A.2, Appendix A.

1.2.2 Data generating distribution

Let $\mathcal{X}_n \equiv \{X_i : i = 1, \ldots, n\}$ denote a *random sample* of n independent and identically distributed (IID) random variables from a *data generating distribution* P, i.e., $X_i \overset{IID}{\sim} P$, $i = 1, \ldots, n$. Let P_n denote the corresponding *empirical distribution*, which places probability $1/n$ on each realization of X.

When needed, the cumulative distribution function (CDF), survivor function, and probability density function (PDF) corresponding to P may be denoted by F, $\bar{F}$, and f, respectively.

Suppose that the data generating distribution P is an element of a particular *statistical model* $\mathcal{M}$, i.e., a set of possibly non-parametric distributions.

In many testing problems of interest, the data structure consists of J-dimensional random vectors or, in short, J-vectors, $X = (X(j) : j = 1, \ldots, J) \sim P$, where the individual elements $X(j)$ correspond to possibly censored covariates/genotypes (e.g., microarray expression measures, amino acids, single nucleotide polymorphisms) and outcomes/phenotypes (e.g., tumor class, survival time, viral replication capacity, insulin level). In this context, a non-parametric model might be the set of all J-variate continuous distributions, whereas a parametric model might be the set of all J-variate Gaussian distributions with diagonal covariance matrix.

1.2.3 Parameters

Define *parameters* as arbitrary functions of the data generating distribution P: $\Psi(P) = \psi = (\psi(m) : m = 1, \ldots, M) \in I\!R^M$, where $\psi(m) = \Psi(P)(m) \in I\!R$.

Parameters of interest include (functions of) means, quantiles, covariances, correlation coefficients, and regression coefficients.

Example 1.1. Parameters of interest in microarray data analysis. In microarray data analysis, $X \sim P$ may denote a random J-vector consisting of two components: a G-vector of genome-wide expression measures, $X(1:G) = (X(j) : j = 1, \ldots, G)$, and a $(J-G)$-vector of possibly censored biological and clinical covariates and outcomes, $X((G+1):J) = (X(j) : j = G+1, \ldots, J)$. The dimension J of the data vector X is usually much greater than the sample size n, e.g., one may have thousands of gene expression measures for fewer than one hundred patients. Testing problems of interest may concern the following parameters of the (unknown) data generating distribution P.

Location parameters. Location parameters, measuring differential gene expression, include (functions of) means and quantiles.

- $\psi = \Psi(P) = \mathrm{E}[X(1:G)]$: G-vector of mean expression measures, $\psi(g) = \mathrm{E}[X(g)]$, for each gene g, $g = 1, \ldots, G$, in a population of yeast cells.
- $\psi = \Psi(P) = \mathrm{E}[X(1:G)|Y = 1] - \mathrm{E}[X(1:G)|Y = 0]$: G-vector of differences in (conditional) mean expression measures, $\psi(g) = \mathrm{E}[X(g)|Y = 1] - \mathrm{E}[X(g)|Y = 0]$, for each gene g, $g = 1, \ldots, G$, in two different populations of cancer cell samples, where $Y = X(G+1) \in \{0, 1\}$ is a cancer class indicator (e.g., survivor vs. non-survivor, ALL B-cell vs. ALL T-cell cancer).

Scale parameters. Scale parameters, measuring co-expression, include matrices of pairwise covariances and correlation coefficients for gene expression measures.

- $\sigma = \Sigma(P) = \mathrm{Cov}[X(1:G)]$: $G \times G$ covariance matrix of $X(1:G)$, with element $\sigma(g, g') = \mathrm{Cov}[X(g), X(g')]$ denoting the pairwise covariance for the expression measures of genes g and g', g, $g' = 1, \ldots, G$. We may adopt the shorter notation $\sigma^2(g) = \sigma(g, g) = \mathrm{Var}[X(g)]$ for the diagonal elements of σ, i.e., the variances.
- $\sigma^* = \Sigma^*(P) = \mathrm{Cor}[X(1:G)]$: $G \times G$ correlation matrix of $X(1:G)$, with element $\sigma^*(g, g') = \mathrm{Cor}[X(g), X(g')] = \sigma(g, g')/\sigma(g)\sigma(g')$ denoting the pairwise correlation coefficient for the expression measures of genes g and g', g, $g' = 1, \ldots, G$.

Regression parameters. Regression parameters, measuring the association of gene expression measures with possibly censored biological and clinical covariates and outcomes, include the following.

- $\psi = \Psi(P)$: G-vector of regression parameters $\psi(g)$, for univariate Cox proportional hazards models relating a survival time $T = X(G+1)$ to the expression measures $X(g)$ of each gene g, $g = 1, \ldots, G$.

- $\psi = \Psi(P)$: G-vector of interaction effects $\psi(g)$ of two drugs on the expression measures $X(g)$ of each gene g, $g = 1, \ldots, G$.
- $\psi = \Psi(P)$: G-vector of linear combinations $\psi(g) = a^\top \lambda(g)$, where $\lambda(g)$ denotes a $(J-G)$-dimensional regression parameter vector in a linear model relating the expression measure $X(g)$ of gene g to a $(J-G)$-dimensional covariate vector $Z = X((G+1):J) = (X(j) : j = G+1, \ldots, J)$, $\mathrm{E}[X(g)|Z] = Z^\top \lambda(g)$, $g = 1, \ldots, G$.

1.2.4 Null and alternative hypotheses

General submodel hypotheses

In order to cover a broad class of testing problems, define M pairs of null and alternative hypotheses in terms of a collection of M *submodels*, $\mathcal{M}(m) \subseteq \mathcal{M}$, $m = 1, \ldots, M$, for the data generating distribution P. Specifically, the M *null hypotheses* and corresponding *alternative hypotheses* are defined as

$$H_0(m) \equiv \mathrm{I}\left(P \in \mathcal{M}(m)\right) \qquad \text{and} \qquad H_1(m) \equiv \mathrm{I}\left(P \notin \mathcal{M}(m)\right), \tag{1.1}$$

respectively. Here, $\mathrm{I}(\cdot)$ is the indicator function, equal to one if the condition in parentheses is true and zero otherwise. Thus, $H_0(m)$ is true (i.e., $H_0(m) = 1$) if the data generating distribution P belongs to submodel $\mathcal{M}(m)$; $H_0(m)$ is false otherwise (i.e., $H_0(m) = 0$).

This general submodel representation covers tests of means, quantiles, covariances, correlation coefficients, and regression coefficients in linear and non-linear models (e.g., logistic, survival, time-series models).

For instance, for a random J-vector $X \sim P$, the full model $\mathcal{M}$ might refer to the set of all continuous J-variate distributions. The submodel $\mathcal{M}(m)$, corresponding to the mth null hypothesis, might be the subset of $\mathcal{M}$ for which the mth element of the mean vector $\mathrm{E}[X]$ is non-negative, i.e., $\mathcal{M}(m) = \{P \in \mathcal{M} : \mathrm{E}[X(m)] \geq 0\}$, $m = 1, \ldots, M = J$. Other submodels of interest may be of the form $\mathcal{M}(m) = \{P \in \mathcal{M} : X(m) \sim \mathrm{N}(0,1)\}$, $m = 1, \ldots, M = J$, where $\mathrm{N}(0,1)$ denotes the standard normal distribution, with mean 0 and variance 1. One could also consider the $M = J(J-1)/2$ submodels $\mathcal{M}(j, j') = \{P \in \mathcal{M} : X(j) \perp X(j')\}$, $j, j' = 1, \ldots, J$, $j < j'$, corresponding to pairwise independence of the elements of X.

Parametric hypotheses

In many testing problems, the submodels concern parameters, i.e., functions $\Psi(P) = \psi = (\psi(m) : m = 1, \ldots, M) \in I\!R^M$ of the data generating distribution P, and each null hypothesis $H_0(m)$ refers to a single parameter, $\psi(m) = \Psi(P)(m) \in I\!R$.

One distinguishes between two types of testing problems for such parametric hypotheses, one-sided and two-sided tests.

$$\begin{aligned} \textit{One-sided tests} \quad & H_0(m) = \mathrm{I}\left(\psi(m) \leq \psi_0(m)\right) \\ \text{vs.} \quad & H_1(m) = \mathrm{I}\left(\psi(m) > \psi_0(m)\right), \quad m = 1, \ldots, M. \end{aligned} \tag{1.2}$$

$$\begin{aligned} \textit{Two-sided tests} \quad & H_0(m) = \mathrm{I}\left(\psi(m) = \psi_0(m)\right) \\ \text{vs.} \quad & H_1(m) = \mathrm{I}\left(\psi(m) \neq \psi_0(m)\right), \quad m = 1, \ldots, M. \end{aligned} \tag{1.3}$$

The hypothesized *null values*, $\psi_0(m)$, are frequently zero. For instance, in microarray data analysis, one may be interested in testing the null hypotheses $H_0(m)$ of no differences in mean gene expression measures between two populations of patients or of no pairwise correlations in gene expression measures.

Sets of true and false null hypotheses

Let

$$\mathcal{H}_0 = \mathcal{H}_0(P) \equiv \{m : H_0(m) = 1\} = \{m : P \in \mathcal{M}(m)\} \tag{1.4}$$

denote the set of $h_0 \equiv |\mathcal{H}_0|$ *true null hypotheses*, where the longer notation $\mathcal{H}_0(P)$ emphasizes the dependence of this set on the data generating distribution P. Likewise, let

$$\mathcal{H}_1 = \mathcal{H}_1(P) \equiv \{m : H_1(m) = 1\} = \{m : P \notin \mathcal{M}(m)\} = \mathcal{H}_0^c(P) \tag{1.5}$$

be the set of $h_1 \equiv |\mathcal{H}_1| = M - h_0$ *false null hypotheses.*

The goal of a multiple testing procedure is to accurately estimate, i.e., *reject*, the set $\mathcal{H}_1$, while probabilistically controlling false positives.

Complete null hypothesis

The *complete null hypothesis* H_0^C is defined as

$$H_0^C \equiv \prod_{m=1}^{M} H_0(m) = \prod_{m=1}^{M} \mathrm{I}\left(P \in \mathcal{M}(m)\right) = \mathrm{I}\left(P \in \cap_{m=1}^{M} \mathcal{M}(m)\right). \tag{1.6}$$

The complete null hypothesis is true if and only if all M individual null hypotheses $H_0(m)$ are true, i.e., if and only if the data generating distribution P belongs to the intersection $\cap_{m=1}^{M} \mathcal{M}(m)$ of the M submodels.

1.2.5 Test statistics

A *testing procedure* is a *random* or *data-driven rule* for deciding which null hypotheses should be rejected, i.e., which $H_0(m)$ should be declared false (zero), so that one concludes that $P \notin \mathcal{M}(m)$.

The decisions to reject or not the null hypotheses are based on an M-vector of *test statistics*, $T_n = (T_n(m) : m = 1, \ldots, M)$, that are functions $T_n(m) =$

$T(m; \mathcal{X}_n) = T(m; P_n)$ of the data $\mathcal{X}_n$, i.e., of the empirical distribution P_n. Denote the typically unknown (finite sample) joint distribution of the test statistics T_n by $Q_n = Q_n(P)$.

As in Dudoit et al. (2004b) and Pollard and van der Laan (2004), for the test of single-parameter null hypotheses of the form $H_0(m) = \mathrm{I}(\psi(m) \leq \psi_0(m))$ or $H_0(m) = \mathrm{I}(\psi(m) = \psi_0(m))$, $m = 1, \ldots, M$, consider two main types of test statistics, *difference statistics*,

$$T_n(m) \equiv \text{Estimator} - \text{Null value} = \sqrt{n}(\psi_n(m) - \psi_0(m)), \tag{1.7}$$

and *t-statistics* (i.e., standardized differences),

$$T_n(m) \equiv \frac{\text{Estimator} - \text{Null value}}{\text{Standard error}} = \sqrt{n}\frac{\psi_n(m) - \psi_0(m)}{\sigma_n(m)}. \tag{1.8}$$

Here, $\hat{\Psi}(P_n) = \psi_n = (\psi_n(m) : m = 1, \ldots, M)$ denotes an *estimator* for the parameter $\Psi(P) = \psi = (\psi(m) : m = 1, \ldots, M)$ and $(\sigma_n(m)/\sqrt{n} : m = 1, \ldots, M)$ denote the estimated *standard errors* for elements $\psi_n(m)$ of ψ_n.

Consider *asymptotically linear estimators* ψ_n of the parameter ψ, with M-dimensional vector *influence curve* (IC) $IC(X|P) = (IC(X|P)(m) : m = 1, \ldots, M)$, such that

$$\psi_n(m) - \psi(m) = \frac{1}{n}\sum_{i=1}^{n} IC(X_i|P)(m) + o_P(1/\sqrt{n}) \tag{1.9}$$

and $\mathrm{E}[IC(X|P)(m)] = 0$, for each $m = 1, \ldots, M$. Let $\Sigma(P) = \sigma = (\sigma(m, m') : m, m' = 1, \ldots, M)$ denote the $M \times M$ covariance matrix of the vector influence curve $IC(X|P)$, where $\sigma(m, m') = \mathrm{E}[IC(X|P)(m) IC(X|P)(m')]$ and we may adopt the shorter notation $\sigma^2(m) = \sigma(m, m) = \mathrm{E}[IC^2(X|P)(m)]$ for variances. Similarly, let $\Sigma^*(P) = \sigma^* = (\sigma^*(m, m') : m, m' = 1, \ldots, M)$ denote the $M \times M$ correlation matrix of the IC, where $\sigma^*(m, m') = \sigma(m, m')/\sigma(m)\sigma(m')$. Assume that $\sigma_n^2(m)$ are *consistent estimators* of the IC variances $\sigma^2(m)$.

The influence curve of a given estimator can be derived as its mean-zero-centered functional derivative (as a function of the empirical distribution P_n for the entire sample of size n), applied to the empirical distribution for a sample of size one (Gill, 1989; Gill et al., 1995).

As illustrated in Section 2.6, this general representation for the test statistics covers standard one-sample and two-sample t-statistics for testing hypotheses concerning mean parameters, but also test statistics for correlation coefficients and regression coefficients in linear and non-linear models. F-statistics for multiple-parameter null hypotheses are discussed in Section 2.7. Test statistics for other types of null hypotheses include χ^2-statistics and likelihood ratio statistics.

1.2.6 Multiple testing procedures

A *multiple testing procedure* (MTP) provides *rejection regions* $\mathcal{C}_n(m)$, i.e., sets of values for each test statistic $T_n(m)$ that lead to the decision to reject the corresponding null hypothesis $H_0(m)$ and declare that $P \notin \mathcal{M}(m)$, $m = 1, \ldots, M$. In other words, a MTP produces a random (i.e., data-dependent) subset $\mathcal{R}_n$ of rejected hypotheses that estimates the set $\mathcal{H}_1$ of false null hypotheses,

$$\mathcal{R}_n = \mathcal{R}(T_n, Q_{0n}, \alpha) \equiv \{m : T_n(m) \in \mathcal{C}_n(m)\} = \{m : H_0(m) \text{ is rejected}\}, \tag{1.10}$$

where $\mathcal{C}_n(m) = \mathcal{C}(m; T_n, Q_{0n}, \alpha)$, $m = 1, \ldots, M$, denote possibly random rejection regions.

The long notation $\mathcal{R}(T_n, Q_{0n}, \alpha)$ and $\mathcal{C}(m; T_n, Q_{0n}, \alpha)$ emphasizes that the MTP depends on the following three ingredients:

1. the *data*, $\mathcal{X}_n = \{X_i : i = 1, \ldots, n\}$, through the M-vector of *test statistics*, $T_n = (T_n(m) : m = 1, \ldots, M)$;
2. an (estimated) M-variate *test statistics null distribution*, Q_{0n}, for deriving rejection regions, confidence regions, and adjusted p-values (Chapter 2);
3. the *nominal Type I error level* α, i.e., a user-supplied upper bound for a suitably defined Type I error rate (Section 1.2.9 provides definitions of Type I error rates and Section 2.2.1 elaborates on the distinction between the actual and nominal Type I error levels of a MTP).

1.2.7 Rejection regions

Having selected a proper test statistics null distribution Q_0 (or estimator thereof, Q_{0n}) according to Chapter 2 guidelines, there remains the main task of specifying *rejection regions* for each null hypothesis, i.e., cut-offs for each test statistic.

We consider MTPs based on *nested* rejection regions, so that

$$\mathcal{C}(m; T_n, Q_{0n}, \alpha_1) \subseteq \mathcal{C}(m; T_n, Q_{0n}, \alpha_2), \qquad \text{whenever } \alpha_1 \leq \alpha_2. \tag{1.11}$$

Rejection regions are typically defined in terms of intervals, such as, $\mathcal{C}_n(m) = (u_n(m), +\infty)$, $\mathcal{C}_n(m) = (-\infty, l_n(m))$, or $\mathcal{C}_n(m) = (-\infty, l_n(m)) \cup (u_n(m), +\infty)$, where $l_n(m) = l(m; T_n, Q_{0n}, \alpha)$ and $u_n(m) = u(m; T_n, Q_{0n}, \alpha)$ are to-be-determined lower and upper *critical values*, or *cut-offs*, computed under the null distribution Q_{0n} for the test statistics T_n. Two-sided rejection regions of the form $\mathcal{C}_n(m) = (-\infty, l_n(m)) \cup (u_n(m), +\infty)$ allow the use of asymmetric cut-offs for two-sided tests.

Unless specified otherwise, we assume that large values of the test statistic $T_n(m)$ provide evidence against the corresponding null hypothesis $H_0(m)$, that is, we consider one-sided rejection regions of the form $\mathcal{C}_n(m) = (c_n(m), +\infty)$, where $c_n(m) = c(m; T_n, Q_{0n}, \alpha)$. For two-sided tests of single-parameter null hypotheses using difference or t-statistics, as in Equations (1.7) and (1.8), one could take absolute values of the test statistics (details in Sections 1.2.12, 2.5.2, and 4.5).

Among the different approaches for defining rejection regions, we distinguish the following.

Marginal vs. joint multiple testing procedures. *Marginal multiple testing procedures* are based solely on the marginal distributions of the test statistics (e.g., classical FWER-controlling single-step Bonferroni Procedure 3.1; gFWER-controlling Lehmann and Romano (2005) Procedures 3.15 and 3.17). In contrast, *joint multiple testing procedures* take into account the dependence structure of the test statistics (e.g., gFWER-controlling single-step common-cut-off Procedure 3.18 and common-quantile Procedure 3.19).
Joint MTPs tend to be more powerful than marginal MTPs.
Note that while a procedure may be marginal, proof of Type I error control by this MTP may require certain assumptions on the dependence structure of the test statistics (e.g., FWER-controlling step-up Hochberg Procedure 3.13; TPPFP-controlling step-down Lehmann and Romano (2005) Procedure 3.24).

Single-step vs. stepwise multiple testing procedures. In *single-step procedures*, each null hypothesis is tested using a rejection region that is independent of the results of the tests of other hypotheses. Improvement in power, while preserving Type I error control, may be achieved by *stepwise procedures*, in which the rejection region for a particular null hypothesis depends on the outcome of the tests of other hypotheses. As detailed in Section 1.2.13, the testing procedure is applied to a *sequence of successively smaller nested random subsets of null hypotheses*, defined by the *ordering* of the test statistics (common cut-offs) or unadjusted p-values (common-quantile cut-offs).

Common-cut-off vs. common-quantile multiple testing procedures. In *common-cut-off procedures*, the same cut-off c_0 is used for each test statistic (cf. FWER-controlling single-step and step-down maxT Procedures 3.5 and 3.11, based on maxima of test statistics). In contrast, in *common-quantile procedures*, the cut-offs are the δ_0-quantiles of the marginal null distributions of the test statistics (cf. FWER-controlling single-step and step-down minP Procedures 3.6 and 3.12, based on minima of unadjusted p-values).
The latter p-value-based procedures place the null hypotheses on an "equal footing", i.e., are more balanced than their common-cut-off counterparts, and may therefore be preferable. However, this comes at the expense of increased computational complexity (Section 4.2.4).

The choice of a proper test statistics null distribution is addressed in detail in Chapter 2. An overview of available MTPs is provided in Chapter 3. Core methodological Chapters 4–7 discuss the following main approaches for deriving rejection regions.

Chapter 4. *Joint single-step common-cut-off* and *common-quantile procedures* for controlling *general Type I error rates* $\Theta(F_{V_n})$, defined as arbitrary parameters of the distribution of the number of Type I errors V_n (Dudoit et al., 2004b; Pollard and van der Laan, 2004). Error rates of the form $\Theta(F_{V_n})$ include the generalized family-wise error rate (gFWER), $gFWER(k) = 1 - F_{V_n}(k) = \Pr(V_n > k)$, i.e., the chance of at least $(k+1)$ Type I errors.

Chapter 5. *Joint step-down common-cut-off (maxT)* and *common-quantile (minP) procedures* for controlling the *family-wise error rate* (FWER), $FWER = gFWER(0) = 1 - F_{V_n}(0) = \Pr(V_n > 0)$ (van der Laan et al., 2004a).

Chapter 6. (Marginal/joint single-step/stepwise) *augmentation multiple testing procedures* (AMTP) for controlling *generalized tail probability* (gTP) error rates, $gTP(q, g) = \Pr(g(V_n, R_n) > q)$, for arbitrary functions $g(V_n, R_n)$ of the numbers of false positives V_n and rejected hypotheses R_n, based on an initial gFWER-controlling procedure (Dudoit et al., 2004a; van der Laan et al., 2004b). Error rates treated in detail include the generalized family-wise error rate, with $g(v, r) = v$, and tail probabilities for the proportion of false positives (TPPFP) among the rejected hypotheses, with $g(v, r) = v/r$.

Chapter 7. *Joint resampling-based empirical Bayes procedures* for controlling *generalized tail probability* error rates. The special case of TPPFP control is discussed in detail in van der Laan et al. (2005).

1.2.8 Errors in multiple hypothesis testing: Type I, Type II, and Type III errors

In any testing problem, two types of errors can be committed. A *Type I error*, or *false positive*, is committed by rejecting a true null hypothesis ($\mathcal{R}_n \cap \mathcal{H}_0$). A *Type II error*, or *false negative*, is committed by failing to reject a false null hypothesis ($\mathcal{R}_n^c \cap \mathcal{H}_1$).

The situation can be summarized as in Table 1.1, where the number of rejected null hypotheses is

$$R_n \equiv |\mathcal{R}_n| = \sum_{m=1}^{M} \mathrm{I}\left(T_n(m) \in \mathcal{C}_n(m)\right), \tag{1.12}$$

the number of Type I errors or false positives is

$$V_n \equiv |\mathcal{R}_n \cap \mathcal{H}_0| = \sum_{m \in \mathcal{H}_0} \mathrm{I}\left(T_n(m) \in \mathcal{C}_n(m)\right), \tag{1.13}$$

the number of Type II errors or false negatives is

$$U_n \equiv |\mathcal{R}_n^c \cap \mathcal{H}_1| = \sum_{m \in \mathcal{H}_1} \mathrm{I}\left(T_n(m) \notin \mathcal{C}_n(m)\right), \tag{1.14}$$

the number of *true negatives* is

$$\begin{aligned} W_n \equiv |\mathcal{R}_n^c \cap \mathcal{H}_0| &= \sum_{m \in \mathcal{H}_0} \mathrm{I}\left(T_n(m) \notin \mathcal{C}_n(m)\right) \\ &= M - R_n - U_n = h_0 - V_n, \end{aligned} \tag{1.15}$$

and the number of *true positives* is

$$\begin{aligned} S_n \equiv |\mathcal{R}_n \cap \mathcal{H}_1| &= \sum_{m \in \mathcal{H}_1} \mathrm{I}\left(T_n(m) \in \mathcal{C}_n(m)\right) \\ &= R_n - V_n = h_1 - U_n. \end{aligned} \tag{1.16}$$

Note that S_n, U_n, V_n, and W_n each depend on the unknown data generating distribution P through the unknown set of true null hypotheses $\mathcal{H}_0 = \mathcal{H}_0(P)$. Therefore, the numbers $h_0 = |\mathcal{H}_0|$ and $h_1 = |\mathcal{H}_1| = M - h_0$ of true and false null hypotheses are *unknown parameters* (row margins of Table 1.1), the number of rejected hypotheses R_n is an *observable random variable* (column margins of Table 1.1), and S_n, U_n, V_n, and W_n are *unobservable random variables* (cells of Table 1.1).

Ideally, one would like to simultaneously minimize both the number of Type I errors and the number of Type II errors. Unfortunately, this is not feasible and one seeks a *trade-off* between the two types of errors. A standard approach is to specify an acceptable level α for a suitably defined Type I error rate and derive testing procedures (i.e., rejection regions) that aim to minimize a Type II error rate (i.e., maximize power) within the class of tests with Type I error level at most α.

For two-sided tests concerning single-parameter null hypotheses, one is often interested in determining the *direction of rejection* for the null hypotheses. For instance, in microarray experiments, one may wish to know whether genes are *over-* or *under-expressed* in, say, treated cells compared to untreated cells. In this setting, one can commit a *Type III error* by correctly rejecting a false null hypothesis $H_0(m) = \mathrm{I}\left(\psi(m) = \psi_0(m)\right)$, but incorrectly concluding that $\psi(m) < \psi_0(m)$ when in truth $\psi(m) > \psi_0(m)$ (or vice versa). Control of Type III errors, as well as Type I errors, brings in additional complexities (Finner, 1999; Shaffer, 2002) and is not considered here.

1.2.9 Type I error rates

When testing multiple hypotheses, there are many possible definitions for the Type I error rate and power of a test procedure. Accordingly, we define a *Type I error rate* as a parameter $\theta_n = \Theta(F_{V_n,R_n})$ of the joint distribution F_{V_n,R_n} of the numbers of Type I errors $V_n = |\mathcal{R}_n \cap \mathcal{H}_0|$ and rejected hypotheses $R_n = |\mathcal{R}_n|$. We focus primarily on Type I error rates such that $\Theta(F_{V_n,R_n}) \in [0,1]$.

Type I error rates based on the distribution of the number of Type I errors: $\Theta(F_{V_n})$

The general representation $\Theta(F_{V_n,R_n})$ covers the following commonly-used Type I error rates, that are parameters $\Theta(F_{V_n})$ of the distribution F_{V_n} of the *number* of Type I errors V_n.

- The *family-wise error rate* (FWER) is the probability of at least one Type I error,
$$FWER \equiv \Pr(V_n > 0) = 1 - F_{V_n}(0). \tag{1.17}$$
- The *generalized family-wise error rate* (gFWER), for a user-supplied integer $k \in \{0, \ldots, M\}$, is the probability of at least $(k+1)$ Type I errors. That is,
$$gFWER(k) \equiv \Pr(V_n > k) = 1 - F_{V_n}(k). \tag{1.18}$$
When $k = 0$, the gFWER reduces to the usual family-wise error rate, FWER.
- The *per-comparison error rate* (PCER) is the expected value of the proportion of Type I errors among the M tests,
$$PCER \equiv \frac{1}{M}\,\mathrm{E}[V_n] = \frac{1}{M}\int v dF_{V_n}(v). \tag{1.19}$$
- The *per-family error rate* (PFER) is the expected value of the number of Type I errors,
$$PFER \equiv \mathrm{E}[V_n] = \int v dF_{V_n}(v). \tag{1.20}$$
- The *median-based per-family error rate* (mPFER) is the median number of Type I errors,
$$mPFER \equiv \mathrm{Median}[V_n] = F_{V_n}^{-1}(1/2). \tag{1.21}$$
- The *quantile number of false positives* (QNFP), for a user-supplied constant $\delta \in (0,1)$, is the δ-quantile of the distribution F_{V_n} of the number of Type I errors. That is,
$$QNFP(\delta) \equiv F_{V_n}^{-1}(\delta). \tag{1.22}$$
When $\delta = 1/2$, the QNFP reduces to the median-based per-family error rate, mPFER.

Until recently, most multiple testing procedures focused on control of the FWER, e.g., classical Bonferroni Procedure 3.1. Existing procedures for controlling the number of Type I errors are reviewed in Sections 3.2 and 3.3. Chapter 4 proposes joint single-step common-cut-off and common-quantile procedures for controlling general Type I error rates $\Theta(F_{V_n})$. Chapter 5 focuses on control of the FWER and provides joint step-down common-cut-off and common-quantile procedures, based on maxima of test statistics (maxT) and minima of unadjusted p-values (minP), respectively. Chapters 6 and 7 propose, respectively, joint augmentation multiple testing procedures and resampling-based empirical Bayes procedures for controlling tail probabilities for arbitrary functions of the number of Type I errors.

For controlling general Type I error rates $\Theta(F_{V_n})$, our proposed multiple testing procedures rely on the following two assumptions concerning the mapping $\Theta : F \to \Theta(F)$, that defines the Type I error rate as a parameter of the distribution F_{V_n} of the number of Type I errors V_n.

Given two cumulative distribution functions F_1 and F_2 on $\{0, \ldots, M\}$, define a *distance measure* d by

$$d(F_1, F_2) \equiv \max_{x=0,\ldots,M} |F_1(x) - F_2(x)|. \tag{1.23}$$

Assumption MΘ. [Monotonicity of Θ] The mapping Θ is *non-decreasing*. That is, given two CDFs F_1 and F_2 on $\{0, \ldots, M\}$,

$$F_1 \geq F_2 \qquad \Longrightarrow \qquad \Theta(F_1) \leq \Theta(F_2). \tag{1.24}$$

In other words, for one-sided rejection regions of the form $\mathcal{C}_n(m) = (c_n(m), +\infty)$, if the test statistics T_{2n} are marginally stochastically greater than the test statistics T_{1n}, the Type I error rate of a MTP based on T_{2n} is greater than the Type I error rate of a corresponding MTP based on T_{1n}.

Assumption CΘ. [Continuity of Θ] The mapping Θ is *uniformly continuous*. That is, given two sequences $\{F_{1n}\}$ and $\{F_{2n}\}$ of CDFs on $\{0, \ldots, M\}$,

$$\lim_{n\to\infty} d(F_{1n}, F_{2n}) = 0 \qquad \Longrightarrow \qquad \lim_{n\to\infty} (\Theta(F_{2n}) - \Theta(F_{1n})) = 0. \tag{1.25}$$

In most cases, we only need continuity at a fixed CDF F_1, i.e., the special case $F_{1n} = F_1$.

Type I error rates based on the distribution of the proportion of Type I errors among the rejected hypotheses: $\Theta(F_{V_n/R_n})$

The general representation $\Theta(F_{V_n,R_n})$ also covers the following commonly-used Type I error rates, that are parameters $\Theta(F_{V_n/R_n})$ of the distribution F_{V_n/R_n} of the *proportion* V_n/R_n of Type I errors among the rejected hypotheses, with the convention that $V_n/R_n \equiv 0$ if $R_n = 0$.

- The *tail probability for the proportion of false positives* (TPPFP) *among the rejected hypotheses*, for a user-supplied constant $q \in (0, 1)$, is defined as

$$TPPFP(q) \equiv \Pr\left(\frac{V_n}{R_n} > q\right) = 1 - F_{V_n/R_n}(q). \tag{1.26}$$

- The *false discovery rate* (FDR) is the expected value of the proportion of Type I errors among the rejected hypotheses,

$$FDR \equiv \mathrm{E}\left[\frac{V_n}{R_n}\right] = \int q dF_{V_n/R_n}(q). \tag{1.27}$$

  The FDR may be rewritten as

$$\begin{aligned} FDR &= \mathrm{E}\left[\frac{V_n}{\max\{R_n, 1\}}\right] \\ &= \mathrm{E}\left[\frac{V_n}{R_n}\middle| V_n > 0\right] \Pr(V_n > 0) \\ &= \mathrm{E}\left[\frac{V_n}{R_n}\middle| R_n > 0\right] \Pr(R_n > 0). \end{aligned} \tag{1.28}$$

  Under the complete null hypothesis $H_0^C = \mathrm{I}\left(P \in \cap_{m=1}^M \mathcal{M}(m)\right)$, all R_n rejected hypotheses are Type I errors, hence $V_n/R_n = 1$ and $FDR = FWER = \Pr(V_n > 0)$. FDR-controlling procedures therefore also control the FWER in the weak sense (Section 2.8). In general, because $V_n/R_n \leq 1$, the FDR is less than or equal to the FWER for any given MTP.
- The *proportion of expected false positives* (PEFP) is the ratio of the expected values of the numbers of Type I errors and rejected hypotheses,

$$PEFP \equiv \frac{\mathrm{E}[V_n]}{\mathrm{E}[R_n]} = \frac{\int v dF_{V_n}(v)}{\int r dF_{R_n}(r)}. \tag{1.29}$$

  Note that, while the FDR is an *expected value of a ratio*, the PEFP is a *ratio of expected values* and is therefore more tractable than the FDR.
- The *quantile proportion of false positives* (QPFP), for a user-supplied constant $\delta \in (0, 1)$, is the δ-quantile of the distribution F_{V_n/R_n} of the proportion of Type I errors among the rejected hypotheses. That is,

$$QPFP(\delta) \equiv F_{V_n/R_n}^{-1}(\delta). \tag{1.30}$$

In the remainder of this book, we use the shorter phrase *proportion of false positives* (PFP) to refer to the proportion V_n/R_n of false positives *among the R_n rejected hypotheses*, rather than among the M null hypotheses. Controlling the latter proportion would reduce to controlling the number of false positives, i.e., error rates of the form $\Theta(F_{V_n})$.

Error rates $\Theta(F_{V_n/R_n})$, based on the *proportion* of false positives (e.g., TPPFP and FDR), are especially appealing for the large-scale testing problems encountered in genomics, compared to error rates $\Theta(F_{V_n})$, based on the *number* of false positives (e.g., gFWER and PFER), as they do not increase exponentially with the number M of tested hypotheses.

However, error rates $\Theta(F_{V_n/R_n})$ tend to be more difficult to control than error rates $\Theta(F_{V_n})$, as they are based on the *joint* distribution of V_n and

R_n, rather than only the marginal distribution of V_n. In particular, error rates $\Theta(F_{V_n/R_n})$ involve the distribution of test statistics for the false null hypotheses $\mathcal{H}_1$, via the number $S_n = |\mathcal{R}_n \cap \mathcal{H}_1| = R_n - V_n$ of true positives.

Existing procedures for controlling the proportion of Type I errors among the rejected hypotheses are reviewed in Sections 3.4 and 3.5.

Tail probability Type I error rates: $\Pr(g(V_n, R_n) > q)$

Chapters 6 and 7 provide, respectively, joint augmentation multiple testing procedures and resampling-based empirical Bayes procedures for controlling *generalized tail probability* (gTP) error rates,

$$gTP(q, g) \equiv \Pr(g(V_n, R_n) > q), \tag{1.31}$$

for arbitrary functions $g(V_n, R_n)$ of the numbers of Type I errors V_n and rejected hypotheses R_n. The special cases $g(v, r) = v$ and $g(v, r) = v/r$ correspond, respectively, to the gFWER and TPPFP.

Chapters 6 and 7 also discuss control of *generalized expected value* (gEV) error rates,

$$gEV(g) \equiv \mathrm{E}[g(V_n, R_n)]. \tag{1.32}$$

The special cases $g(v, r) = v$ and $g(v, r) = v/r$ correspond, respectively, to the PFER and FDR.

Notation

To emphasize the dependence of a MTP on parameters defining the Type I error rate mapping Θ, we may adopt a longer notation, whereby sets of null hypotheses and their cardinality are indexed by these parameters as well as by the nominal Type I error level α.

For instance, for a gFWER-controlling MTP with k allowed false positives, we may use $\mathcal{R}_n(k; \alpha)$ and $R_n(k; \alpha) = |\mathcal{R}_n(k; \alpha)|$, for the set of rejected hypotheses and its cardinality, respectively, and $V_n(k; \alpha) = |\mathcal{R}_n(k; \alpha) \cap \mathcal{H}_0|$, for the number of Type I errors.

1.2.10 Power

Within a class of multiple testing procedures that control a given Type I error rate $\theta_n = \Theta(F_{V_n,R_n})$ at an acceptable level α, one seeks procedures that maximize power, that is, minimize a suitably defined Type II error rate. As with Type I error rates, the concepts of Type II error rate and power can be extended in various ways when moving from single to multiple hypothesis testing. Accordingly, we define *power* as a parameter $\vartheta_n = \Theta(F_{U_n,R_n})$ of the joint distribution F_{U_n,R_n} of the numbers of Type II errors $U_n = |\mathcal{R}_n^c \cap \mathcal{H}_1|$ and rejected hypotheses $R_n = |\mathcal{R}_n|$. Recall that the numbers of true positives S_n and Type II errors U_n satisfy $S_n + U_n = h_1$ (Table 1.1). Parameters of interest include the following.

- The probability of rejecting *at least one* false null hypothesis, i.e., of at least one true positive,

$$AnyPwr \equiv \Pr(S_n \geq 1) = \Pr(U_n \leq h_1 - 1) = F_{U_n}(h_1 - 1). \quad (1.33)$$

- The probability of rejecting *all* false null hypotheses, i.e., of no Type II errors,

$$AllPwr \equiv \Pr(S_n = h_1) = \Pr(U_n = 0) = F_{U_n}(0). \quad (1.34)$$

- The *average power*, i.e., the expected value of the proportion of rejected hypotheses among the false null hypotheses,

$$AvgPwr \equiv \frac{1}{h_1}\,\mathrm{E}[S_n] = \frac{1}{h_1}\,\mathrm{E}[h_1 - U_n] = 1 - \frac{1}{h_1}\int u dF_{U_n}(u). \quad (1.35)$$

- The *true discovery rate* (TDR), i.e., the expected value of the proportion of true positives among the rejected hypotheses,

$$TDR \equiv \mathrm{E}\left[\frac{S_n}{R_n}\right] = \mathrm{E}\left[\frac{R_n - V_n}{R_n}\right], \quad (1.36)$$

with the convention that $(R_n - V_n)/R_n \equiv 0$ if $R_n = 0$. The TDR may be rewritten as

$$TDR = \mathrm{E}\left[\frac{R_n - V_n}{R_n}\middle| R_n > 0\right]\Pr(R_n > 0) = \Pr(R_n > 0) - FDR.$$

One can think of the TDR as a power analogue of the FDR. If all null hypotheses are false (i.e., $h_1 = M$), then TDR reduces to $AnyPwr$. For a family of tests consisting of pairwise mean comparisons, the parameters in Equations (1.33)–(1.35) have been termed, respectively, the any-pair power, the all-pairs power, and the per-pair power (Ramsey, 1978; Shaffer, 1995).

1.2.11 Type I error rates and power: Comparisons and examples

Comparison of Type I error rates

Classical approaches to multiple hypothesis testing call for controlling the FWER, as in well-known Bonferroni Procedure 3.1. In general, for a given multiple testing procedure, $PCER \leq FWER \leq PFER$ and $FDR \leq FWER$, with $FDR = FWER$ under the complete null hypothesis ($M = h_0$). Thus, procedures controlling the FWER are typically more conservative, i.e., lead to fewer rejected hypotheses, than those controlling either the FDR or the PCER. Procedures controlling the PCER are generally less conservative than those controlling either the FDR or the FWER, but do not really address the multiplicity problem.

Note that, by Markov's Inequality (Equation (B.2)),

$$gFWER(k) = \Pr(V_n \geq k+1) \leq \frac{1}{k+1}\,\mathrm{E}[V_n] = \frac{1}{k+1} PFER. \tag{1.37}$$

Thus, PFER control at level $\alpha \times (k+1)$ implies $gFWER(k)$ control at level α. In the special case $k = 0$, $FWER = gFWER(0) \leq PFER$, thus PFER control at level α implies FWER control at level α.

To illustrate properties of the different Type I error rates, consider the complete null hypothesis $H_0^C = \prod_{m=1}^{M} H_0(m)$, i.e., assume that $P \in \cap_{m=1}^{M} \mathcal{M}(m)$. Suppose each null hypothesis $H_0(m)$ is tested individually at exact actual Type I error rate α_m and the decision to reject or not reject this hypothesis is based solely on the corresponding test statistic $T_n(m)$. Then, the PCER is simply the average of the individual Type I error rates α_m and the PFER is the sum of these α_m. In contrast, the FWER and the FDR cannot be expressed in terms of the individual Type I error rates α_m alone, but involve the joint distribution of the test statistics $T_n(m)$. One has

$$\begin{aligned} PCER &= \frac{1}{M}(\alpha_1 + \cdots + \alpha_M) \\ &\leq \max\{\alpha_1, \ldots, \alpha_M\} \\ &\leq FWER \\ &\leq \alpha_1 + \cdots + \alpha_M \\ &= PFER. \end{aligned} \tag{1.38}$$

The following simple example describes the behavior of different Type I error rates and power as one varies the total number M of tested hypotheses and the proportion h_0/M of true null hypotheses.

A simple example

Consider Gaussian random M-vectors $X \sim \mathrm{N}(\psi, I_M)$, with mean vector $\mathrm{E}[X] = \psi = (\psi(m) : m = 1, \ldots, M)$ and identity covariance matrix $\mathrm{Cov}[X] = I_M$. Suppose one wishes to test simultaneously the M null hypotheses $H_0(m) = \mathrm{I}(\psi(m) = 0)$ against the two-sided alternative hypotheses $H_1(m) = \mathrm{I}(\psi(m) \neq 0)$. Assume without loss of generality that the h_0 true null hypotheses are indexed by $\mathcal{H}_0 = \{1, \ldots, h_0\}$.

Suppose one has a random sample $\mathcal{X}_n = \{X_i : i = 1, \ldots, n\}$, of n IID M-vectors $X_i \sim \mathrm{N}(\psi, I_M)$. Then, a simple multiple testing procedure is based on absolute values of one-sample z-statistics $T_n(m) = \sqrt{n}|\bar{X}_n(m)|$, where $\bar{X}_n = \sum_i X_i/n$ is the empirical mean vector. One rejects null hypothesis $H_0(m)$ if $T_n(m) \geq z_{\alpha/2}$, that is, the set of rejected hypotheses is $\mathcal{R}_n = \{m : \sqrt{n}|\bar{X}_n(m)| \geq z_{\alpha/2}\}$, where $z_{\alpha/2}$ is such that $\Phi(z_{\alpha/2}) = 1 - \alpha/2$ and Φ is the standard normal cumulative distribution function. Let $R_n(m) =$

$\mathrm{I}\left(|\bar{X}_n(m)| \geq z_{\alpha/2}/\sqrt{n}\right)$ denote a rejection indicator for null hypothesis $H_0(m)$ and let $\gamma_n(m) = \mathrm{E}[R_n(m)] = 1 - \Phi(z_{\alpha/2} - \psi(m)\sqrt{n}) + \Phi(-z_{\alpha/2} - \psi(m)\sqrt{n})$ denote the chance of rejecting null hypothesis $H_0(m)$.

The number of rejected hypotheses is $R_n = \sum_{m=1}^{M} R_n(m)$, the number of Type I errors is $V_n = \sum_{m=1}^{h_0} R_n(m)$, and the number of Type II errors is $U_n = h_1 - S_n = h_1 - \sum_{m=h_0+1}^{M} R_n(m)$.

Analytical formulae for commonly-used Type I error rates can easily be derived as

$$\begin{aligned} PFER &= \mathrm{E}[V_n] = \sum_{m=1}^{h_0} \gamma_n(m), \\ PCER &= \frac{1}{M}\mathrm{E}[V_n] = \frac{1}{M}\sum_{m=1}^{h_0} \gamma_n(m), \\ FWER &= \Pr(V_n > 0) = 1 - \prod_{m=1}^{h_0} (1 - \gamma_n(m)), \\ FDR &= \mathrm{E}\left[\frac{V_n}{R_n}\right] = \sum_{r_1=0}^{1} \cdots \sum_{r_M=0}^{1} \frac{\sum_{m=1}^{h_0} r_m}{\sum_{m=1}^{M} r_m} \prod_{m=1}^{M} \gamma_n(m)^{r_m}(1-\gamma_n(m))^{1-r_m}, \end{aligned} \tag{1.39}$$

with the FDR convention that $0/0 = 0$.

In this simple example, the single test Type I error rate is constant, that is, $\gamma_n(m) = \alpha$ for the true null hypotheses $m \in \mathcal{H}_0$. If one further assumes a common shift parameter d for the false null hypotheses, i.e., $\psi(m) = d/\sqrt{n}$ for $m \in \mathcal{H}_1$, then the single test power is also constant. That is, $\gamma_n(m) = 1 - \Phi(z_{\alpha/2} - d) + \Phi(-z_{\alpha/2} - d) = \beta$ for the false null hypotheses $m \in \mathcal{H}_1$. The above expressions for the Type I error rates simplify to

$$\begin{aligned} PFER &= h_0\alpha, \\ PCER &= \frac{h_0\alpha}{M}, \\ FWER &= 1 - (1-\alpha)^{h_0}, \\ FDR &= \sum_{v=1}^{h_0}\sum_{s=0}^{h_1} \frac{v}{v+s}\binom{h_0}{v}\alpha^v(1-\alpha)^{h_0-v}\binom{h_1}{s}\beta^s(1-\beta)^{h_1-s}. \end{aligned} \tag{1.40}$$

Note that unlike the PFER, PCER, and FWER, the FDR depends on the distribution of the test statistics for the false null hypotheses $\mathcal{H}_1$, through the number of true positives $S_n = R_n - V_n = |\mathcal{R}_n \cap \mathcal{H}_1|$. Indeed, in this simple example, the FDR is a function of β, the rejection probability for the false null hypotheses. Thus, it is in general more difficult to derive procedures controlling the proportion of false positives V_n/R_n (e.g., TPPFP, FDR), than procedures controlling the number of false positives V_n (e.g., FWER, PFER).

Regarding power, recall that the number of Type II errors is $U_n = h_1 - S_n = h_1 - \sum_{m=h_0+1}^{M} R_n(m)$, thus the average power is given by

$$AvgPwr = \frac{1}{h_1} \sum_{m=h_0+1}^{M} \mathrm{E}[R_n(m)] = \frac{1}{h_1} \sum_{m=h_0+1}^{M} \gamma_n(m) = \beta. \tag{1.41}$$

In this special case of independent test statistics, one can also obtain simple formulae for the so-called any-pair power,

$$\begin{aligned} AnyPwr &= \Pr\left(\sum_{m=h_0+1}^{M} R_n(m) > 0 \right) \\ &= 1 - \Pr\left(\bigcap_{m=h_0+1}^{M} \{R_n(m) = 0\} \right) \\ &= 1 - \prod_{m=h_0+1}^{M} (1 - \gamma_n(m)) \\ &= 1 - (1-\beta)^{h_1}, \end{aligned} \tag{1.42}$$

and all-pairs power,

$$\begin{aligned} AllPwr &= \Pr\left(\sum_{m=h_0+1}^{M} R_n(m) = h_1 \right) \\ &= \Pr\left(\bigcap_{m=h_0+1}^{M} \{R_n(m) = 1\} \right) \\ &= \prod_{m=h_0+1}^{M} \gamma_n(m) \\ &= \beta^{h_1}. \end{aligned} \tag{1.43}$$

Figure 1.1 displays plots of the FWER, PCER, and FDR vs. the number of tested hypotheses M, for different proportions h_0/M of true null hypotheses, $h_0/M = 1$, 0.9, 0.5, 0.1, and for single test actual Type I error rate $\alpha = 0.05$ and alternative shift parameter $d = 1$. In general, the FWER and PFER increase sharply with the number M of tested hypotheses, whereas the PCER remains constant (the PFER is not shown because it is on a different scale, i.e., it is not restricted to belong to the interval $[0, 1]$). Under the complete null hypothesis ($M = h_0$), the FDR is equal to the FWER and both error rates increase sharply with M. However, as the proportion of true null hypotheses h_0/M decreases, the FDR remains relatively stable as a function of M and approaches the PCER. Type I error rates are only displayed for moderate values of M, between 1 and 100, to provide more detail in regions where there are sharp changes. For greater values of M, in the thousands as in microarray experiments, the error rates tend to reach a plateau.

Figure 1.2 displays plots of the FWER, PCER, and FDR vs. the single test actual Type I error rate α, for different proportions h_0/M of true null hypotheses, $h_0/M = 1, 0.9, 0.5, 0.1$, and for $M = 100$ and $d = 1$. The FWER and FDR are generally much greater than the PCER, the greatest differences being under the complete null hypothesis ($M = h_0$). As the proportion of true null hypotheses decreases, the FDR again becomes closer to the PCER. The error rates display similar behavior for greater values of M, with a sharper increase of the FWER as α increases.

1.2.12 Unadjusted and adjusted p-values

Unadjusted p-values

Definition

Consider testing M null hypotheses $H_0(m)$, $m = 1, \ldots, M$, individually at level α, based on test statistics $T_n = (T_n(m) : m = 1, \ldots, M)$, with unknown true distribution $Q_n = Q_n(P)$ and assumed null distribution Q_0 (or estimator thereof, Q_{0n}). Then, null hypothesis $H_0(m)$ is rejected at single test nominal Type I error level α if $T_n(m) \in \mathcal{C}_n(m; \alpha)$. The rejection regions $\mathcal{C}_n(m; \alpha) = \mathcal{C}(Q_{0,m}, \alpha)$ are based solely on the marginal null distributions $Q_{0,m}$ (or estimators thereof, $Q_{0n,m}$) and are chosen such that the chance of a Type I error is at most α for each test,

$$\Pr_{Q_{0,m}}(T_n(m) \in \mathcal{C}_n(m; \alpha)) \leq \alpha, \qquad (1.44)$$

and the nestedness assumption of Equation (1.11) is satisfied.

The *unadjusted p-value* $P_{0n}(m) = P(T_n(m), Q_{0,m})$, for the single test of null hypothesis $H_0(m)$, is defined as

$$\begin{aligned} P_{0n}(m) &\equiv \inf\{\alpha \in [0,1] : \text{Reject } H_0(m) \text{ at single test nominal level } \alpha\} \\ &= \inf\{\alpha \in [0,1] : T_n(m) \in \mathcal{C}_n(m; \alpha)\}, \qquad m = 1, \ldots, M. \qquad (1.45) \end{aligned}$$

That is, the unadjusted p-value $P_{0n}(m)$, for null hypothesis $H_0(m)$, is the *smallest nominal Type I error level* of the *single hypothesis testing procedure* at which one would reject $H_0(m)$, given $T_n(m)$. The smaller the unadjusted p-value $P_{0n}(m)$, the stronger the evidence against the corresponding null hypothesis $H_0(m)$. Unadjusted p-values may also be referred to as *marginal* or *raw* p-values.

Unadjusted p-value representation of a single hypothesis testing procedure

Under the nestedness assumption of Equation (1.11), one has two equivalent representations for a single hypothesis testing procedure, in terms of rejection regions for the test statistics and in terms of unadjusted p-values. Specifically, null hypothesis $H_0(m)$ is rejected at single test nominal Type I error level α, if either $T_n(m) \in \mathcal{C}_n(m;\alpha)$ or $P_{0n}(m) \leq \alpha$. That is, the set of rejected null hypotheses at single test nominal Type I error level α is

$$\mathcal{R}_n(\alpha) = \{m : T_n(m) \in \mathcal{C}_n(m;\alpha)\} = \{m : P_{0n}(m) \leq \alpha\}. \tag{1.46}$$

Distributional properties of unadjusted p-values

The single test procedure for null hypothesis $H_0(m)$ controls the Type I error rate at (actual) level α, if the marginal null distribution $Q_{0,m}$ *dominates* the true marginal distribution $Q_{n,m}$, in the sense that

$$\Pr_{Q_{n,m}}(T_n(m) \in \mathcal{C}_n(m;\alpha)) \leq \Pr_{Q_{0,m}}(T_n(m) \in \mathcal{C}_n(m;\alpha)), \qquad \forall\, \alpha \in [0,1],$$

that is, the chance of rejecting $H_0(m)$ is greater under the null distribution than under the true distribution.

Proposition 1.2. [Distributions of unadjusted p-values] *Consider an M-vector of test statistics $T_n = (T_n(m) : m = 1, \ldots, M)$, with true joint distribution $Q_n = Q_n(P)$ and joint null distribution Q_0. Define a collection of M rejection regions $\{\mathcal{C}_n(m;\alpha) : m = 1, \ldots, M\}$ and an M-vector of unadjusted p-values $(P_{0n}(m) : m = 1, \ldots, M)$, as in Equations* (1.44) *and* (1.45), *respectively. That is, $\mathcal{C}_n(m;\alpha)$ and $P_{0n}(m)$ are such that*

$$\Pr_{Q_{0,m}}(T_n(m) \in \mathcal{C}_n(m;\alpha)) \leq \alpha, \tag{1.47}$$
$$\mathcal{C}_n(m;\alpha_1) \subseteq \mathcal{C}_n(m;\alpha_2), \qquad \text{whenever } \alpha_1 \leq \alpha_2,$$
$$P_{0n}(m) = \inf\{\alpha \in [0,1] : T_n(m) \in \mathcal{C}_n(m;\alpha)\}, \qquad m = 1, \ldots, M.$$

Then, the unadjusted p-values $P_{0n}(m)$ satisfy the following inequality with respect to the marginal null distributions $Q_{0,m}$,

$$\Pr_{Q_{0,m}}(P_{0n}(m) \leq z) \leq z, \qquad \forall\, z \in [0,1]. \tag{1.48}$$

For continuous marginal null distributions $Q_{0,m}$, the unadjusted p-values $P_{0n}(m)$ are uniformly distributed on the interval $[0,1]$, that is, $P_{0n}(m) \sim U(0,1)$ and

$$\Pr_{Q_{0,m}}(P_{0n}(m) \leq z) = z, \qquad \forall\, z \in [0,1]. \tag{1.49}$$

Furthermore, Equation (1.48) *holds for the true marginal distributions $Q_{n,m}$, under the following* marginal null domination *condition,*

$$\Pr_{Q_{n,m}}(T_n(m) \in \mathcal{C}_n(m;\alpha)) \leq \Pr_{Q_{0,m}}(T_n(m) \in \mathcal{C}_n(m;\alpha)), \qquad \forall\, \alpha \in [0,1]. \tag{1.50}$$

That is, under Equation (1.50), *the unadjusted p-values $P_{0n}(m)$ satisfy the following inequality with respect to the true marginal distributions $Q_{n,m}$,*

$$\Pr_{Q_{n,m}}(P_{0n}(m) \leq z) \leq z, \qquad \forall\, z \in [0,1]. \tag{1.51}$$

Proof of Proposition 1.2. The proof relies on the nestedness assumption for rejection regions in Equation (1.11), so that

$$\bigcup_{\alpha \leq z} \mathcal{C}_n(m;\alpha) = \mathcal{C}_n(m;z), \qquad \forall\, z \in [0,1].$$

Then,

$$\begin{aligned}
\Pr_{Q_{0,m}}\left(P_{0n}(m) \leq z\right) &= \Pr_{Q_{0,m}}\left(\inf\left\{\alpha \in [0,1] : T_n(m) \in \mathcal{C}_n(m;\alpha)\right\} \leq z\right) \\
&= \Pr_{Q_{0,m}}\left(T_n(m) \in \bigcup_{\alpha \leq z} \mathcal{C}_n(m;\alpha)\right) \\
&= \Pr_{Q_{0,m}}\left(T_n(m) \in \mathcal{C}_n(m;z)\right) \\
&\leq z,
\end{aligned}$$

where the inequality follows by definition of the rejection regions $\mathcal{C}_n(m;\alpha)$ in Equation (1.47). For continuous marginal null distributions $Q_{0,m}$, $\Pr_{Q_{0,m}}(T_n(m) \in \mathcal{C}_n(m;z)) = z$, hence $\Pr_{Q_{0,m}}(P_{0n}(m) \leq z) = z$.

Finally, for the true marginal distributions $Q_{n,m}$, the marginal null domination condition of Equation (1.50) implies that

$$\begin{aligned}
\Pr_{Q_{n,m}}\left(P_{0n}(m) \leq z\right) &= \Pr_{Q_{n,m}}\left(T_n(m) \in \mathcal{C}_n(m;z)\right) \\
&\leq \Pr_{Q_{0,m}}\left(T_n(m) \in \mathcal{C}_n(m;z)\right) \\
&\leq z.
\end{aligned}$$

□

Section 2.2 elaborates on Type I error control and the choice of a test statistics null distribution. In particular, the notion of null domination is extended to the test of multiple null hypotheses.

Unadjusted p-values for one-sided tests

Suppose the null hypotheses $H_0(m)$ are rejected for *large* values of the test statistics $T_n(m)$, i.e., consider *one-sided rejection regions* of the form $\mathcal{C}_n(m;\alpha) = (c_n(m;\alpha), +\infty)$. The cut-offs $c_n(m;\alpha) = c(Q_{0,m},\alpha)$ may be defined in terms of the marginal null survivor functions, $\bar{Q}_{0,m}(z) = 1 - Q_{0,m}(z) = \Pr_{Q_{0,m}}(T_n(m) > z)$ (or estimators thereof, $\bar{Q}_{0n,m}$), as

$$c_n(m;\alpha) = \bar{Q}_{0,m}^{-1}(\alpha) = \inf\left\{z \in I\!R : \bar{Q}_{0,m}(z) \leq \alpha\right\}. \tag{1.52}$$

That is, the cut-off $c_n(m;\alpha)$ is the *least conservative* (i.e., smallest) value, so that the nominal Type I error level does not exceed α for the single test of hypothesis $H_0(m)$. In other words, $c_n(m;\alpha)$ is the cut-off for which it is the "easiest to reject" $H_0(m)$, subject to the Type I error constraint.

The unadjusted p-values are given by

$$\begin{aligned} P_{0n}(m) &= \inf\left\{\alpha \in [0,1] : c_n(m;\alpha) < T_n(m)\right\} \\ &= \inf\left\{\alpha \in [0,1] : \bar{Q}_{0,m}^{-1}(\alpha) < T_n(m)\right\}, \qquad m = 1,\ldots,M. \end{aligned}$$

If one assumes that the marginal null distributions $Q_{0,m}$ are continuous and strictly increasing, then the unadjusted p-values become

$$P_{0n}(m) = c_n^{-1}(m; T_n(m)) = \bar{Q}_{0,m}(T_n(m)), \qquad m = 1,\ldots,M, \tag{1.53}$$

where $c_n^{-1}(m;\cdot)$ are the inverses of the non-increasing functions of α, $c_n(m;\cdot) : \alpha \to c_n(m;\alpha) = \bar{Q}_{0,m}^{-1}(\alpha)$.

Single test procedures, based on the above one-sided rejection regions and corresponding unadjusted p-values, provide Type I error control only for a proper choice of the marginal null distributions $Q_{0,m}$, i.e., for distributions $Q_{0,m}$ that satisfy the null domination condition of Equation (1.50) for true null hypotheses. In the special case of one-sided rejection regions of the form $\mathcal{C}_n(m;\alpha) = (c_n(m;\alpha), +\infty)$, the null domination condition can be expressed in terms of the marginal survivor functions, as

$$\bar{Q}_{n,m}(z) = \Pr_{Q_{n,m}}(T_n(m) > z) \le \Pr_{Q_{0,m}}(T_n(m) > z) = \bar{Q}_{0,m}(z), \quad \forall\ z \in I\!R,$$

and Type I error control follows from

$$\begin{aligned} \Pr_{Q_{n,m}}(P_{0n}(m) \le \alpha) &= \Pr_{Q_{n,m}}(T_n(m) > c_n(m;\alpha)) \\ &\le \Pr_{Q_{0,m}}(T_n(m) > c_n(m;\alpha)) \le \alpha. \end{aligned}$$

Unadjusted p-values for two-sided tests

Suppose the null hypotheses $H_0(m)$ are rejected for *large and small* values of the test statistics $T_n(m)$, i.e., consider *two-sided rejection regions* of the form $\mathcal{C}_n(m;\alpha) = (-\infty, l_n(m;\alpha)) \cup (u_n(m;\alpha), +\infty)$, $l_n(m;\alpha) \le u_n(m;\alpha)$. The lower and upper cut-offs, $l_n(m;\alpha) = l(Q_{0,m},\alpha)$ and $u_n(m;\alpha) = u(Q_{0,m},\alpha)$, defined in terms of the marginal null distributions $Q_{0,m}$ (or estimators thereof, $Q_{0n,m}$), are such that the following Type I error constraints are satisfied,

$$\Pr_{Q_{0,m}}(T_n(m) < l_n(m;\alpha) \text{ or } T_n(m) > u_n(m;\alpha)) \le \alpha,$$

and the rejection regions $\mathcal{C}_n(m;\alpha)$ satisfy the nestedness assumption of Equation (1.11), that is, $l_n(m;\alpha)$ and $u_n(m;\alpha)$ are, respectively, non-decreasing and non-increasing as functions of α.

If one assumes that the marginal null distributions $Q_{0,m}$ are continuous and strictly increasing, then the unadjusted p-values are given by

$$\begin{aligned} P_{0n}(m) &= \inf\left\{\alpha \in [0,1] : T_n(m) < l_n(m;\alpha) \text{ or } T_n(m) > u_n(m;\alpha)\right\} \quad (1.54) \\ &= \inf\left\{\alpha \in [0,1] : \alpha \ge l_n^{-1}(m;T_n(m)) \text{ or } \alpha \ge u_n^{-1}(m;T_n(m))\right\} \\ &= \min\left\{l_n^{-1}(m;T_n(m)), u_n^{-1}(m;T_n(m))\right\}, \qquad m = 1,\ldots,M, \end{aligned}$$

where $l_n^{-1}(m;\cdot)$ and $u_n^{-1}(m;\cdot)$ are, respectively, the inverses of the non-decreasing and non-increasing functions of α, $l_n(m;\cdot) : \alpha \to l_n(m;\alpha)$ and $u_n(m;\cdot) : \alpha \to u_n(m;\alpha)$.

In the special case of *symmetric two-sided rejection regions*, i.e., for $l_n(m;\alpha) = -c_n(m;\alpha)$ and $u_n(m;\alpha) = c_n(m;\alpha)$, $c_n(m;\alpha) \geq 0$, the unadjusted p-values are given by

$$\begin{aligned} P_{0n}(m) &= \min\left\{c_n^{-1}(m;-T_n(m)), c_n^{-1}(m;T_n(m))\right\} \\ &= c_n^{-1}(m;|T_n(m)|), \qquad m = 1,\dots,M, \end{aligned} \tag{1.55}$$

where $c_n^{-1}(m;\cdot)$ are the inverses of the non-increasing functions of α, $c_n(m;\cdot) : \alpha \to c_n(m;\alpha)$, and $l_n^{-1}(m;z) = c_n^{-1}(m;-z)$.

For *symmetric marginal null distributions* $Q_{0,m}$, such that $\bar{Q}_{0,m}(z) = 1 - Q_{0,m}(z) = Q_{0,m}(-z)$ for each $z \in I\!R$, one may select the cut-off $c_n(m;\alpha)$ as the *least conservative* (i.e., smallest) value, so that the nominal Type I error level does not exceed α for the single test of hypothesis $H_0(m)$. That is,

$$c_n(m;\alpha) = \bar{Q}_{0,m}^{-1}(\alpha/2) = \inf\left\{z \in I\!R : \bar{Q}_{0,m}(z) \leq \alpha/2\right\}, \tag{1.56}$$

and

$$\begin{aligned} \Pr_{Q_{0,m}}(T_n(m) \in \mathcal{C}_n(m;\alpha)) &= \Pr_{Q_{0,m}}(|T_n(m)| > c_n(m;\alpha)) \\ &= Q_{0,m}(-c_n(m;\alpha)) + \bar{Q}_{0,m}(c_n(m;\alpha)) \\ &= 2\bar{Q}_{0,m}(c_n(m;\alpha)) \leq \alpha. \end{aligned}$$

The unadjusted p-values are then given by

$$P_{0n}(m) = c_n^{-1}(m;|T_n(m)|) = 2\bar{Q}_{0,m}(|T_n(m)|), \qquad m = 1,\dots,M. \tag{1.57}$$

Single test procedures, based on the above two-sided rejection regions and corresponding unadjusted p-values, provide Type I error control only for a proper choice of the marginal null distributions $Q_{0,m}$, i.e., for distributions $Q_{0,m}$ that satisfy the null domination condition of Equation (1.50) for true null hypotheses. In the special case of symmetric test statistics marginal distributions $Q_{n,m}$ and $Q_{0,m}$ and symmetric two-sided rejection regions of the form $\mathcal{C}_n(m;\alpha) = (-\infty, -c_n(m;\alpha)) \cup (c_n(m;\alpha), +\infty)$, the null domination condition can be expressed in terms of the marginal survivor functions, as

$$\bar{Q}_{n,m}(z) \leq \bar{Q}_{0,m}(z), \qquad \forall\, z \in I\!R.$$

Indeed, under the above condition, for each $z \in I\!R$,

$$\Pr_{Q_{n,m}}(|T_n(m)| > z) = 2\bar{Q}_{n,m}(z) \leq 2\bar{Q}_{0,m}(z) = \Pr_{Q_{0,m}}(|T_n(m)| > z).$$

Type I error control then follows from

$$\begin{aligned} \Pr_{Q_{n,m}}(P_{0n}(m) \leq \alpha) &= \Pr_{Q_{n,m}}(|T_n(m)| > c_n(m;\alpha)) \\ &\leq \Pr_{Q_{0,m}}(|T_n(m)| > c_n(m;\alpha)) \leq \alpha. \end{aligned}$$

Adjusted p-values

Definition

The notion of p-value extends directly to multiple testing problems as follows. Consider any multiple testing procedure $\mathcal{R}_n(\alpha) = \mathcal{R}(T_n, \mathcal{Q}_0, \alpha)$, with rejection regions $\mathcal{C}_n(m;\alpha) = \mathcal{C}(m; T_n, \mathcal{Q}_0, \alpha)$. Then, one can define an M-vector of *adjusted p-values*, $\widetilde{P}_{0n} = (\widetilde{P}_{0n}(m) : m = 1, \ldots, M) = \widetilde{P}(T_n, \mathcal{Q}_0) = \widetilde{P}(\mathcal{R}(T_n, \mathcal{Q}_0, \alpha) : \alpha \in [0,1])$, as

$$\begin{aligned} \widetilde{P}_{0n}(m) &\equiv \inf\{\alpha \in [0,1] : \text{Reject } H_0(m) \text{ at nominal MTP level } \alpha\} \quad (1.58)\\ &= \inf\{\alpha \in [0,1] : m \in \mathcal{R}_n(\alpha)\}\\ &= \inf\{\alpha \in [0,1] : T_n(m) \in \mathcal{C}_n(m;\alpha)\}, \qquad m = 1, \ldots, M. \end{aligned}$$

That is, the adjusted p-value $\widetilde{P}_{0n}(m)$, for null hypothesis $H_0(m)$, is the *smallest nominal Type I error level* (e.g., gFWER, TPPFP, or FDR) of the *multiple hypothesis testing procedure* at which one would reject $H_0(m)$, given T_n. As in single hypothesis tests, the smaller the adjusted p-value $\widetilde{P}_{0n}(m)$, the stronger the evidence against the corresponding null hypothesis $H_0(m)$. Thus, one rejects $H_0(m)$ for small adjusted p-values $\widetilde{P}_{0n}(m)$. Note that the unadjusted p-value $P_{0n}(m)$, for the single test of a null hypothesis $H_0(m)$, corresponds to the special case $M = 1$.

As mentioned in Section 1.2.7, the particular mapping defining the rejection regions $\mathcal{C}(m; T_n, \mathcal{Q}_0, \alpha)$ depends on the choice of MTP (e.g., marginal vs. joint, single-step vs. stepwise, common-cut-off vs. common-quantile procedure). For instance, the adjusted p-values for classical FWER-controlling Bonferroni Procedure 3.1 are $\widetilde{P}_{0n}(m) = \min\{MP_{0n}(m), 1\}$.

Chapter 3 provides adjusted p-values for commonly-used procedures controlling the FWER, gFWER, FDR, and TPPFP. Adjusted p-values for joint single-step common-quantile Procedure 4.1 and common-cut-off Procedure 4.2, controlling an arbitrary parameter $\Theta(F_{V_n})$ of the distribution of the number of Type I errors, are derived in Section 4.3. Adjusted p-values for FWER-controlling joint step-down common-cut-off (maxT) Procedure 5.1 and common-quantile (minP) Procedure 5.6 are the subject of Sections 5.2.4 and 5.3.4, respectively. Adjusted p-values for gTP-controlling augmentation multiple testing procedures are derived in Section 6.5.2. Adjusted p-values for gTP-controlling resampling-based empirical Bayes procedures are treated in Section 7.3.

Note that the unadjusted p-values $P_{0n}(m)$ and adjusted p-values $\widetilde{P}_{0n}(m)$ are *random variables*, i.e., functions of the data $\mathcal{X}_n$ via the test statistics T_n. As usual, denote *realizations* of the test statistic, unadjusted p-value, and adjusted p-value for null hypothesis $H_0(m)$ by the lowercase letters $t_n(m)$, $p_{0n}(m)$, and $\widetilde{p}_{0n}(m)$, respectively.

Adjusted p-value representation of a multiple hypothesis testing procedure

Under the nestedness assumption of Equation (1.11), one has two equivalent representations for a MTP, in terms of rejection regions for the test statistics and in terms of adjusted p-values. Specifically, the set of rejected null hypotheses at multiple test nominal Type I error level α is

$$\mathcal{R}_n(\alpha) = \{m : T_n(m) \in \mathcal{C}_n(m;\alpha)\} = \left\{m : \widetilde{P}_{0n}(m) \leq \alpha\right\}. \tag{1.59}$$

Let $O_n(m)$ denote the indices for the *ordered adjusted p-values*, so that $\widetilde{P}_{0n}(O_n(1)) \leq \cdots \leq \widetilde{P}_{0n}(O_n(M))$. Then, the set of rejected hypotheses $\mathcal{R}_n(\alpha)$ consists of the indices for the $R_n(\alpha) = |\mathcal{R}_n(\alpha)|$ hypotheses with the smallest adjusted p-values, that is, $\mathcal{R}_n(\alpha) = \{O_n(m) : m = 1, \ldots, R_n(\alpha)\}$.

We wish to stress that a multiple testing procedure $\mathcal{R}_n(\alpha)$, defined as in Equation (1.59), provides Type I error control only for a proper choice of the test statistics null distribution Q_0, i.e., for a joint null distribution Q_0 that dominates the true joint distribution Q_n, in the sense that the number of Type I errors is stochastically greater under Q_0 than under Q_n. Section 2.2 elaborates on Type I error control and the choice of a test statistics null distribution.

Advantages of the adjusted p-value representation

As in the single hypothesis case, reporting the results of a MTP in terms of adjusted p-values, as opposed to only rejection or not of the null hypotheses, offers several advantages.

- Adjusted p-values can be defined for *any Type I error rate* (e.g., gFWER, TPPFP, or FDR).
- They reflect the strength of the evidence against each null hypothesis in terms of the *Type I error rate for the entire MTP.*
- They are *flexible summaries* of a MTP, in the sense that results are supplied for *all Type I error levels* α, i.e., the level α need not be chosen ahead of time.
- They provide convenient *benchmarks to compare different MTPs*, whereby smaller adjusted p-values indicate a less conservative procedure.
- Plots of sorted adjusted p-values allow investigators to examine sets of rejected hypotheses associated with various Type I error rates (e.g., gFWER, TPPFP, or FDR) and nominal levels α. Such plots provide tools to decide on an appropriate combination of the number of rejected hypotheses and tolerable false positive rate for a particular experiment and available resources.

Adjusted p-values for one-sided tests

Suppose the null hypotheses $H_0(m)$ are rejected for large values of the test statistics $T_n(m)$, i.e., consider one-sided rejection regions of the form $\mathcal{C}_n(m;\alpha) = (c_n(m;\alpha), +\infty)$, based on an M-vector of cut-offs, $c_n(\alpha) = c(T_n, \mathcal{Q}_0, \alpha) = (c_n(m;\alpha) = c(m; T_n, \mathcal{Q}_0, \alpha) : m = 1, \ldots, M)$. Then, for a continuous test statistics null distribution $\mathcal{Q}_0$, the adjusted p-values are given by

$$\begin{aligned} \widetilde{P}_{0n}(m) &= \inf\{\alpha \in [0,1] : c_n(m;\alpha) < T_n(m)\} \\ &= c_n^{-1}(m; T_n(m)), \qquad m = 1, \ldots, M, \end{aligned} \tag{1.60}$$

where $c_n^{-1}(m;\cdot)$ are the inverses of the non-increasing functions of α, $c_n(m;\cdot) : \alpha \to c_n(m;\alpha) = c(m; T_n, \mathcal{Q}_0, \alpha)$.

Adjusted p-values for two-sided tests

Suppose the null hypotheses $H_0(m)$ are rejected for large and small values of the test statistics $T_n(m)$, i.e., consider two-sided rejection regions of the form $\mathcal{C}_n(m;\alpha) = (-\infty, l_n(m;\alpha)) \cup (u_n(m;\alpha), +\infty)$, based on M-vectors of lower and upper cut-offs, $l_n(\alpha) = l(T_n, \mathcal{Q}_0, \alpha) = (l_n(m;\alpha) = l(m; T_n, \mathcal{Q}_0, \alpha) : m = 1, \ldots, M)$ and $u_n(\alpha) = u(T_n, \mathcal{Q}_0, \alpha) = (u_n(m;\alpha) = u(m; T_n, \mathcal{Q}_0, \alpha) : m = 1, \ldots, M)$, with $l_n(m;\alpha) \leq u_n(m;\alpha)$. Then, for a continuous test statistics null distribution $\mathcal{Q}_0$, the adjusted p-values are given by

$$\begin{aligned} \widetilde{P}_{0n}(m) &= \inf\{\alpha \in [0,1] : T_n(m) < l_n(m;\alpha) \text{ or } T_n(m) > u_n(m;\alpha)\} \\ &= \min\left\{l_n^{-1}(m; T_n(m)), u_n^{-1}(m; T_n(m))\right\}, \qquad m = 1, \ldots, M, \end{aligned} \tag{1.61}$$

where $l_n^{-1}(m;\cdot)$ and $u_n^{-1}(m;\cdot)$ are, respectively, the inverses of the non-decreasing and non-increasing functions of α, $l_n(m;\cdot) : \alpha \to l_n(m;\alpha)$ and $u_n(m;\cdot) : \alpha \to u_n(m;\alpha)$.

In the special case of symmetric two-sided rejection regions, i.e., for $l_n(m;\alpha) = -c_n(m;\alpha)$ and $u_n(m;\alpha) = c_n(m;\alpha)$, $c_n(m;\alpha) \geq 0$, the adjusted p-values are given by

$$\begin{aligned} \widetilde{P}_{0n}(m) &= \min\left\{c_n^{-1}(m; -T_n(m)), c_n^{-1}(m; T_n(m))\right\} \\ &= c_n^{-1}(m; |T_n(m)|), \qquad m = 1, \ldots, M, \end{aligned} \tag{1.62}$$

where $c_n^{-1}(m;\cdot)$ are the inverses of the non-increasing functions of α, $c_n(m;\cdot) : \alpha \to c_n(m;\alpha)$, and $l_n^{-1}(m;z) = c_n^{-1}(m;-z)$.

1.2.13 Stepwise multiple testing procedures

Basic definitions

One usually distinguishes between two main classes of multiple testing procedures, single-step and stepwise procedures, depending on whether the rejection regions for the test statistics are constant or random (given the test statistics null distribution Q_0 or an estimator thereof, Q_{0n}), i.e., are independent or not of the data $\mathcal{X}_n$.

Specifically, in *single-step* procedures, each null hypothesis $H_0(m)$ is tested using a rejection region that is independent of the results of the tests of other hypotheses and is not a function of the data $\mathcal{X}_n$ (unless these data are used to estimate the null distribution Q_0, as in Sections 2.3.2 and 2.4.2).

Improvement in power, while preserving Type I error control, may be achieved by *stepwise* procedures, in which the decision to reject a particular null hypothesis depends on the outcome of the tests of other hypotheses. That is, the (single-step) test procedure is applied to a *sequence of successively smaller nested random (i.e., data-dependent) subsets of null hypotheses*, defined by the *ordering* of the test statistics (common-cut-off MTPs) or unadjusted p-values (common-quantile MTPs). The rejection regions are therefore allowed to depend on the data $\mathcal{X}_n$ via the test statistics T_n. Stepwise procedures are of two main types, step-down and step-up procedures, depending on the order in which the null hypotheses are tested.

In *step-down* procedures, the *most significant* null hypotheses (i.e., the null hypotheses with the largest test statistics for common-cut-off MTPs or smallest unadjusted p-values for common-quantile MTPs) are considered successively, with further tests depending on the outcome of earlier ones. As soon as one fails to reject a null hypothesis, no further hypotheses are rejected.

In contrast, for *step-up* procedures, the *least significant* null hypotheses are considered successively, again with further tests depending on the outcome of earlier ones. As soon as one null hypothesis is rejected, all remaining more significant hypotheses are rejected.

Generic marginal step-down and step-up common-quantile multiple testing procedures

The next two procedures provide general statements of marginal step-down and step-up common-quantile MTPs, in terms of sequences of non-decreasing unadjusted p-value cut-offs.

Let $P_{0n}(m)$ denote the unadjusted p-value for null hypothesis $H_0(m)$, $m = 1, \ldots, M$, where $P_{0n}(m)$ is computed as described in Equation (1.45), under a null distribution Q_0 for the test statistics T_n.

Let $O_n(m)$ denote the indices for the *ordered unadjusted p-values*, $P^{\circ}_{0n}(m) \equiv P_{0n}(O_n(m))$, so that $P_{0n}(O_n(1)) \leq \cdots \leq P_{0n}(O_n(M))$. Here, for common-quantile MTPs, the *mth most significant* null hypothesis refers to the hypothesis $H_0(O_n(m))$ with the mth smallest unadjusted p-value $P^{\circ}_{0n}(m)$, i.e., to the hypothesis with p-value rank m.

Procedure 1.3. [Generic marginal step-down common-quantile procedure]
Given suitably chosen unadjusted p-value cut-offs, $a_1(\alpha) \leq \cdots \leq a_M(\alpha)$, a marginal step-down common-quantile multiple testing procedure can be expressed as

$$\mathcal{R}_n(\alpha) \equiv \{O_n(m) : P_{0n}(O_n(h)) \leq a_h(\alpha),\ \forall\ h \leq m\}. \tag{1.63}$$

The number of rejected null hypotheses $R_n(\alpha)$ is given by

$$R_n(\alpha) \equiv \begin{cases} M, & \text{if } P_{0n}(O_n(m)) \leq a_m(\alpha)\ \forall\ m \\ \min\{m : P_{0n}(O_n(m)) > a_m(\alpha)\} - 1, & \text{otherwise} \end{cases}. \tag{1.64}$$

The corresponding adjusted p-values are given by

$$\widetilde{P}_{0n}(O_n(m)) = \max_{h=1,\ldots,m} \left\{\min\left\{a_h^{-1}(P_{0n}(O_n(h))), 1\right\}\right\}, \qquad m = 1, \ldots, M, \tag{1.65}$$

where a_m^{-1} are the inverses of the cut-off mappings $a_m : \alpha \to a_m(\alpha)$, which are assumed, for simplicity, to be continuous and strictly increasing in α.

Procedure 1.4. [Generic marginal step-up common-quantile procedure]
Given suitably chosen unadjusted p-value cut-offs, $a_1(\alpha) \leq \cdots \leq a_M(\alpha)$, a marginal step-up common-quantile multiple testing procedure can be expressed as

$$\mathcal{R}_n(\alpha) \equiv \{O_n(m) : \exists\ h \geq m \text{ such that } P_{0n}(O_n(h)) \leq a_h(\alpha)\}. \tag{1.66}$$

The number of rejected null hypotheses $R_n(\alpha)$ is given by

$$R_n(\alpha) \equiv \begin{cases} 0, & \text{if } P_{0n}(O_n(m)) > a_m(\alpha)\ \forall\ m \\ \max\{m : P_{0n}(O_n(m)) \leq a_m(\alpha)\}, & \text{otherwise} \end{cases}. \tag{1.67}$$

The corresponding adjusted p-values are given by

$$\widetilde{P}_{0n}(O_n(m)) = \min_{h=m,\ldots,M} \left\{\min\left\{a_h^{-1}(P_{0n}(O_n(h))), 1\right\}\right\}, \qquad m = 1, \ldots, M, \tag{1.68}$$

where a_m^{-1} are the inverses of the cut-off mappings $a_m : \alpha \to a_m(\alpha)$, which are assumed, for simplicity, to be continuous and strictly increasing in α.

The adjusted p-values for Procedures 1.3 and 1.4 may be derived straightforwardly from the general definition in Equation (1.58). Specifically, for step-down Procedure 1.3,

$$\begin{aligned}
\widetilde{P}_{0n}(O_n(m)) &= \inf\left\{\alpha \in [0,1] : O_n(m) \in \mathcal{R}_n(\alpha)\right\} \\
&= \inf\left\{\alpha \in [0,1] : R_n(\alpha) \geq m\right\} \\
&= \inf\left\{\alpha \in [0,1] : \min\left\{h : P_{0n}(O_n(h)) > a_h(\alpha)\right\} > m\right\} \\
&= \inf\left\{\alpha \in [0,1] : \min\left\{h : a_h^{-1}(P_{0n}(O_n(h))) > \alpha\right\} > m\right\} \\
&= \inf\left\{\alpha \in [0,1] : \max_{h=1,\ldots,m} a_h^{-1}(P_{0n}(O_n(h))) \leq \alpha\right\} \\
&= \max_{h=1,\ldots,m}\left\{\min\left\{a_h^{-1}(P_{0n}(O_n(h))),\, 1\right\}\right\}.
\end{aligned}$$

Likewise, for step-up Procedure 1.4,

$$\begin{aligned}
\widetilde{P}_{0n}(O_n(m)) &= \inf\left\{\alpha \in [0,1] : O_n(m) \in \mathcal{R}_n(\alpha)\right\} \\
&= \inf\left\{\alpha \in [0,1] : R_n(\alpha) \geq m\right\} \\
&= \inf\left\{\alpha \in [0,1] : \max\left\{h : P_{0n}(O_n(h)) \leq a_h(\alpha)\right\} \geq m\right\} \\
&= \inf\left\{\alpha \in [0,1] : \max\left\{h : a_h^{-1}(P_{0n}(O_n(h))) \leq \alpha\right\} \geq m\right\} \\
&= \inf\left\{\alpha \in [0,1] : \min_{h=m,\ldots,M} a_h^{-1}(P_{0n}(O_n(h))) \leq \alpha\right\} \\
&= \min_{h=m,\ldots,M}\left\{\min\left\{a_h^{-1}(P_{0n}(O_n(h))),\, 1\right\}\right\}.
\end{aligned}$$

Note that taking maxima of the quantities $\min\left\{a_h^{-1}(P_{0n}(O_n(h))),\, 1\right\}$ over subsets $\{1,\ldots,m\}$ in Equation (1.65) enforces the step-down property and monotonicity of the adjusted p-values. That is, one can only reject a particular null hypothesis provided all hypotheses with smaller unadjusted p-values were rejected beforehand and $\widetilde{P}_{0n}(O_n(1)) \leq \cdots \leq \widetilde{P}_{0n}(O_n(M))$. Likewise, taking minima of the quantities $\min\left\{a_h^{-1}(P_{0n}(O_n(h))),\, 1\right\}$ over subsets $\{m,\ldots,M\}$ in Equation (1.68) enforces the step-up property and monotonicity of the adjusted p-values. That is, as soon as one null hypothesis is rejected, all hypotheses with smaller unadjusted p-values are rejected and $\widetilde{P}_{0n}(O_n(1)) \leq \cdots \leq \widetilde{P}_{0n}(O_n(M))$.

In contrast to marginal single-step procedures, for which the cut-offs $a_m(\alpha)$ are constant in m (e.g., single-step Bonferroni Procedure 3.1, with $a_m(\alpha) = \alpha/M$), marginal stepwise procedures gain power by allowing the cut-offs $a_m(\alpha)$ to increase with m (e.g., step-down Holm Procedure 3.7, with $a_m(\alpha) = \alpha/(M-m+1)$).

Finally, note that one could possibly use the same unadjusted p-value cut-offs $a_m(\alpha)$ for a step-down and a step-up procedure (e.g., step-down Holm Procedure 3.7 and step-up Hochberg Procedure 3.13, with $a_m(\alpha) =$

$\alpha/(M-m+1)$). The main difference between step-down and step-up procedures is the *order* in which null hypotheses are tested: from most significant to least significant for the step-down approach vs. from least significant to most significant for the step-up approach. This distinction is reflected by taking maxima and minima of p-values in Equations (1.65) and (1.68), respectively.

Alternative formulations for marginal step-down and step-up common-quantile multiple testing procedures

Given a subset of null hypotheses $\mathcal{J} \subseteq \{1,\ldots,M\}$, with cardinality $|\mathcal{J}| = J$, let

$$m_n^{\mathcal{J}}(j) \equiv \min\{m : |\{O_n(1),\ldots,O_n(m)\} \cap \mathcal{J}| = j\}, \qquad j = 1,\ldots,J, \quad (1.69)$$

denote the rank, among all M hypotheses, of the null hypothesis with the jth smallest unadjusted p-value among the J hypotheses in $\mathcal{J}$. That is, $O_n(m_n^{\mathcal{J}}(j)) \in \mathcal{J}$ and the unadjusted p-values $P_{0n}^{\circ}(m_n^{\mathcal{J}}(j)) = P_{0n}(O_n(m_n^{\mathcal{J}}(j)))$ are non-decreasing in j, $j = 1,\ldots,J$. By definition of the ranks $m_n^{\mathcal{J}}(j)$, one must have $j \leq m_n^{\mathcal{J}}(j) \leq M - J + j$.

In particular, $m_n^{\mathcal{J}}(1) = \min\{m : O_n(m) \in \mathcal{J}\}$ is the rank of the null hypothesis with the smallest unadjusted p-value, $P_{0n}^{\circ}(m_n^{\mathcal{J}}(1)) = P_{0n}(O_n(m_n^{\mathcal{J}}(1))) = \min_{j\in\mathcal{J}} P_{0n}(j)$, among the J hypotheses in $\mathcal{J}$.

Proposition 1.5. [Generic marginal step-down common-quantile procedure] *Consider suitably chosen unadjusted p-value cut-offs, $a_1(\alpha) \leq \cdots \leq a_M(\alpha)$, and a subset of null hypotheses $\mathcal{J} \subseteq \{1,\ldots,M\}$, with cardinality $|\mathcal{J}| = J$. If marginal step-down common-quantile Procedure 1.3 rejects at least one of the null hypotheses in $\mathcal{J}$, then*

$$\min_{j\in\mathcal{J}} P_{0n}(j) = P_{0n}^{\circ}(m_n^{\mathcal{J}}(1)) \leq a_{M-J+1}(\alpha). \quad (1.70)$$

That is, the unadjusted p-value $P_{0n}^{\circ}(m_n^{\mathcal{J}}(1))$ for the most significant null hypothesis in $\mathcal{J}$ does not exceed its corresponding cut-off $a_{M-J+1}(\alpha)$ for the test of only the J null hypotheses in $\mathcal{J}$.

Proof of Proposition 1.5. If Procedure 1.3 rejects at least one of the null hypotheses in $\mathcal{J}$, then, by the step-down property, it must reject all null hypotheses that are at least as significant as the most significant null hypothesis in $\mathcal{J}$. The most significant null hypothesis in $\mathcal{J}$ has overall rank $m_n^{\mathcal{J}}(1)$, hence one has

$$P_{0n}^{\circ}(m) \leq a_m(\alpha), \qquad \text{for each } m \in \{1,\ldots,m_n^{\mathcal{J}}(1)\}.$$

In particular, since $m_n^{\mathcal{J}}(1) \leq M - J + 1$, monotonicity in m of the cut-offs $a_m(\alpha)$ implies that

$$\min_{j\in\mathcal{J}} P_{0n}(j) = P_{0n}^{\circ}(m_n^{\mathcal{J}}(1)) \leq a_{m_n^{\mathcal{J}}(1)}(\alpha) \leq a_{M-J+1}(\alpha).$$

□

Proposition 1.6. [Generic marginal step-up common-quantile procedure] *Consider suitably chosen unadjusted p-value cut-offs, $a_1(\alpha) \leq \cdots \leq a_M(\alpha)$, and a subset of null hypotheses $\mathcal{J} \subseteq \{1, \ldots, M\}$, with cardinality $|\mathcal{J}| = J$. If marginal step-up common-quantile Procedure 1.4 rejects at least one of the null hypotheses in $\mathcal{J}$, then*

$$P_{0n}^{\circ}(m_n^{\mathcal{J}}(j^\star)) \leq a_{M-J+j^\star}(\alpha), \qquad \textit{for some } j^\star \in \{1, \ldots, J\}. \tag{1.71}$$

That is, at least one of the unadjusted p-values $P_{0n}^{\circ}(m_n^{\mathcal{J}}(j^\star))$ for the null hypotheses in $\mathcal{J}$ does not exceed its corresponding cut-off $a_{M-J+j^\star}(\alpha)$ for the test of only the J null hypotheses in $\mathcal{J}$.

Proof of Proposition 1.6. If Procedure 1.4 rejects at least one of the null hypotheses in $\mathcal{J}$, then, by the step-up property, it must reject at least one null hypothesis that is not more significant than the most significant null hypothesis in $\mathcal{J}$. The most significant null hypothesis in $\mathcal{J}$ has overall rank $m_n^{\mathcal{J}}(1)$, hence one has

$$P_{0n}^{\circ}(m^\star) \leq a_{m^\star}(\alpha), \qquad \text{for some } m^\star \in \{m_n^{\mathcal{J}}(1), \ldots, M\}.$$

Let $j^\star = \max\{j : m_n^{\mathcal{J}}(j) \leq m^\star\}$, i.e., $m_n^{\mathcal{J}}(j^\star)$ is the overall rank of the least significant null hypothesis in $\mathcal{J}$ that is not less significant than the overall $m^\star$th most significant null hypothesis. Then, at least $(J - j^\star)$ null hypotheses are less significant than the $m^\star$th most significant null hypothesis and $m^\star \leq M - J + j^\star$. Monotonicity in m of the cut-offs $a_m(\alpha)$ implies that

$$P_{0n}^{\circ}(m_n^{\mathcal{J}}(j^\star)) \leq P_{0n}^{\circ}(m^\star) \leq a_{m^\star}(\alpha) \leq a_{M-J+j^\star}(\alpha).$$

□

Propositions 1.5 and 1.6 are applied in Section 3.2 to establish FWER control for various marginal step-down and step-up MTPs.

In particular, for step-down Holm Procedure 3.7, with unadjusted p-value cut-offs $a_m(\alpha) = \alpha/(M - m + 1)$, Equation (1.70) becomes

$$\min_{j \in \mathcal{J}} P_{0n}(j) \leq \frac{1}{J}\alpha,$$

where α/J is the Holm cut-off corresponding to the most significant null hypothesis among J hypotheses of interest. Likewise, for step-up Hochberg Procedure 3.13, with unadjusted p-value cut-offs $a_m(\alpha) = \alpha/(M - m + 1)$, Equation (1.71) becomes

$$P_{0n}^{\circ}(m_n^{\mathcal{J}}(j^\star)) \leq \frac{1}{J - j^\star + 1}\alpha, \qquad \text{for some } j^\star \in \{1, \ldots, J\},$$

where $\alpha/(J - j^\star + 1)$ is the Hochberg cut-off corresponding to the $j^\star$th most significant null hypothesis among J hypotheses of interest.

Comparison of step-down and step-up adjusted p-values

For step-down/step-up procedure pairs, defined in terms of the same test statistic or unadjusted p-value cut-offs, one can express the adjusted p-values as

$$\begin{aligned} \widetilde{P}^{\downarrow}_{0n}(O_n(m)) &= \max_{h=1,\ldots,m} \widetilde{P}^{-}_{0n}(O_n(h)) \qquad \text{[step-down]} \\ \widetilde{P}^{\uparrow}_{0n}(O_n(m)) &= \min_{h=m,\ldots,M} \widetilde{P}^{-}_{0n}(O_n(h)) \qquad \text{[step-up]}, \end{aligned} \tag{1.72}$$

for suitably defined common "pre-adjusted p-values" $\widetilde{P}^{-}_{0n}(O_n(m))$.

This representation holds for generic marginal common-quantile Procedures 1.3 and 1.4 and, in particular, for the marginal step-down/step-up Holm/Hochberg procedure pair introduced below and discussed in detail in Chapter 3 (Procedures 3.7 and 3.13). It also holds for the joint step-down/step-up maxT and minP procedure pairs introduced in Chapter 3 (Procedures 3.11 and 3.12) and considered in detail in Chapter 5.

It immediately follows from Equation (1.72) that, for each $m = 1, \ldots, M$,

$$\begin{aligned} \widetilde{P}^{\downarrow}_{0n}(O_n(m)) &= \max_{h=1,\ldots,m} \widetilde{P}^{-}_{0n}(O_n(h)) \\ &\geq \widetilde{P}^{-}_{0n}(O_n(m)) \\ &\geq \min_{h=m,\ldots,M} \widetilde{P}^{-}_{0n}(O_n(h)) = \widetilde{P}^{\uparrow}_{0n}(O_n(m)). \end{aligned} \tag{1.73}$$

A step-up procedure therefore leads to at least as many rejected null hypotheses as its step-down counterpart based on the same cut-offs, that is,

$$\begin{aligned} \mathcal{R}^{\downarrow}_n(\alpha) &= \left\{O_n(m) : \widetilde{P}^{\downarrow}_{0n}(O_n(m)) \leq \alpha\right\} \\ &\subseteq \left\{O_n(m) : \widetilde{P}^{\uparrow}_{0n}(O_n(m)) \leq \alpha\right\} = \mathcal{R}^{\uparrow}_n(\alpha). \end{aligned}$$

Indeed, by starting with the least significant null hypotheses and rejecting all remaining more significant hypotheses as soon as one null hypothesis is rejected, a step-up procedure gives each null hypothesis "several chances at rejection".

Examples: Marginal single-step Bonferroni, step-down Holm, and step-up Hochberg procedures

As detailed in Section 3.2, for control of the FWER, stepwise analogues of classical single-step Bonferroni (1936) Procedure 3.1, are step-down Holm (1979) Procedure 3.7 and step-up Hochberg (1988) Procedure 3.13. In these stepwise procedures, based solely on the marginal distributions of the test statistics, the mth smallest unadjusted p-value $P^{\circ}_{0n}(m) = P_{0n}(O_n(m))$ is compared to the cut-off $\alpha/(M-m+1)$, rather than the smaller, more conservative Bonferroni cut-off α/M. The adjusted p-values for these three simple marginal procedures are

$$\widetilde{P}_{0n}(O_n(m)) = \begin{cases} \min\left\{M P_{0n}(O_n(m)), 1\right\} & \text{[Bonferroni]} \\ \max_{h=1,\ldots,m}\left\{\min\left\{(M-h+1)\, P_{0n}(O_n(h)), 1\right\}\right\} & \text{[Holm]} \\ \min_{h=m,\ldots,M}\left\{\min\left\{(M-h+1)\, P_{0n}(O_n(h)), 1\right\}\right\} & \text{[Hochberg]} \end{cases}. \tag{1.74}$$

Figures 1.3 and 1.4 compare the FWER-controlling Bonferroni, Holm, and Hochberg procedures in terms of sets of rejected null hypotheses and adjusted p-values. Although the Holm and Hochberg MTPs rely on the same unadjusted p-value cut-offs, the order in which null hypotheses are tested can lead to different sets of rejected hypotheses. As detailed above, a step-up procedure is generally less conservative than its step-down counterpart based on the same cut-offs. This step-down vs. step-up distinction is reflected by taking maxima ($\max_{h=1,\ldots,m}$) vs. minima ($\min_{h=m,\ldots,M}$) of p-values in Equation (1.74).

Huang and Hsu (2005) illustrate geometrically that step-up Hochberg Procedure 3.13 "cuts corners off the acceptance region" of its step-down analogue, Holm Procedure 3.7. The rejection regions for both procedures are contrasted in Figure 1.5, in the special case of $M=2$ null hypotheses. This graphical representation clearly shows that the step-up procedure is less conservative than the step-down procedure. The regions where only one of the null hypotheses is rejected coincide for the two procedures ($\mathcal{R}_n(\alpha)=\{1\}$ and $\mathcal{R}_n(\alpha)=\{2\}$). However, the region where both null hypotheses are rejected is larger for the step-up procedure ($\mathcal{R}_n(\alpha)=\{1,2\}$): specifically, the step-up procedure adds the rejection region $[\alpha/2,\alpha]\times[\alpha/2,\alpha]$.

It is important to note that step-up Hochberg Procedure 3.13 is based on Simes' Inequality (Equation (B.5); Simes (1986)) and therefore only guarantees FWER control for certain forms of dependence structures among the test statistics. Likewise, FDR-controlling step-up procedures, such as Benjamini and Hochberg (1995) Procedure 3.22, typically rely on a number of assumptions concerning the joint distribution of the test statistics (e.g., independence or positive regression dependence as in Benjamini and Yekutieli (2001)). In general, establishing Type I error control for step-up procedures is more difficult than for step-down procedures.

Commonly-used single-step and stepwise MTPs for controlling the FWER, gFWER, FDR, and TPPFP, are reviewed in Sections 3.2–3.5. Chapter 4 focuses on joint single-step procedures for controlling general Type I error rates $\Theta(F_{V_n})$, while Chapter 5 considers FWER-controlling joint step-down procedures.

Table 1.1. *Type I and Type II errors in multiple hypothesis testing.* This table summarizes the different types of decisions and errors in multiple hypothesis testing. The number of rejected null hypotheses is $R_n = |\mathcal{R}_n|$, the number of Type I errors or false positives is $V_n = |\mathcal{R}_n \cap \mathcal{H}_0|$, the number of Type II errors or false negatives is $U_n = |\mathcal{R}_n^c \cap \mathcal{H}_1|$, the number of true negatives is $W_n = |\mathcal{R}_n^c \cap \mathcal{H}_0|$, and the number of true positives is $S_n = |\mathcal{R}_n \cap \mathcal{H}_1|$. Cells corresponding to errors are enclosed in boxes.

		Null hypotheses		
		Non-rejected, $\mathcal{R}_n^c$	**Rejected, $\mathcal{R}_n$**	
Null hypotheses	**True, $\mathcal{H}_0$**	$W_n = \lvert\mathcal{R}_n^c \cap \mathcal{H}_0\rvert$	$\boxed{V_n = \lvert\mathcal{R}_n \cap \mathcal{H}_0\rvert}$	h_0
	False, $\mathcal{H}_1$	$\boxed{U_n = \lvert\mathcal{R}_n^c \cap \mathcal{H}_1\rvert}$	$S_n = \lvert\mathcal{R}_n \cap \mathcal{H}_1\rvert$	h_1
		$M - R_n$	R_n	M

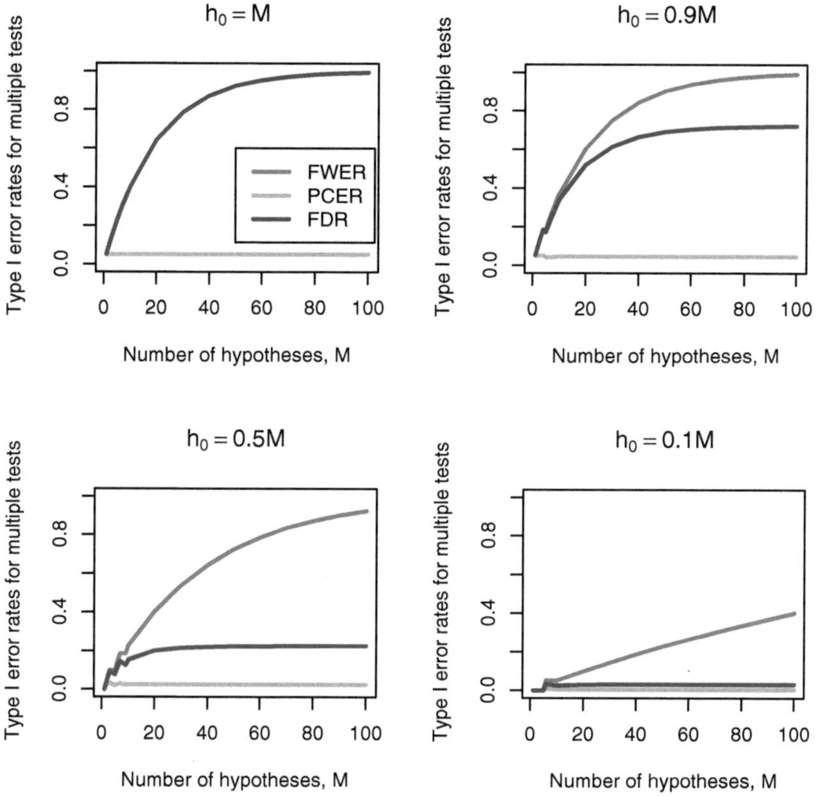

Figure 1.1. *Comparison of Type I error rates for a simple example.* Plots of Type I error rates for multiple tests (FWER: red curve; PCER: green curve; FDR: blue curve) vs. number of null hypotheses M, for different proportions h_0/M of true null hypotheses, $h_0/M = 1$, 0.9, 0.5, 0.1. The model and multiple testing procedures are described in Section 1.2.11. The single test actual Type I error rate is $\alpha = 0.05$ and the alternative shift parameter d is set to 1. The non smooth behavior for small M is due to the fact that it is not always possible to have exactly 90%, 50%, or 10% of true null hypotheses and rounding to the nearest integer may be necessary. (Color plate p. 321)

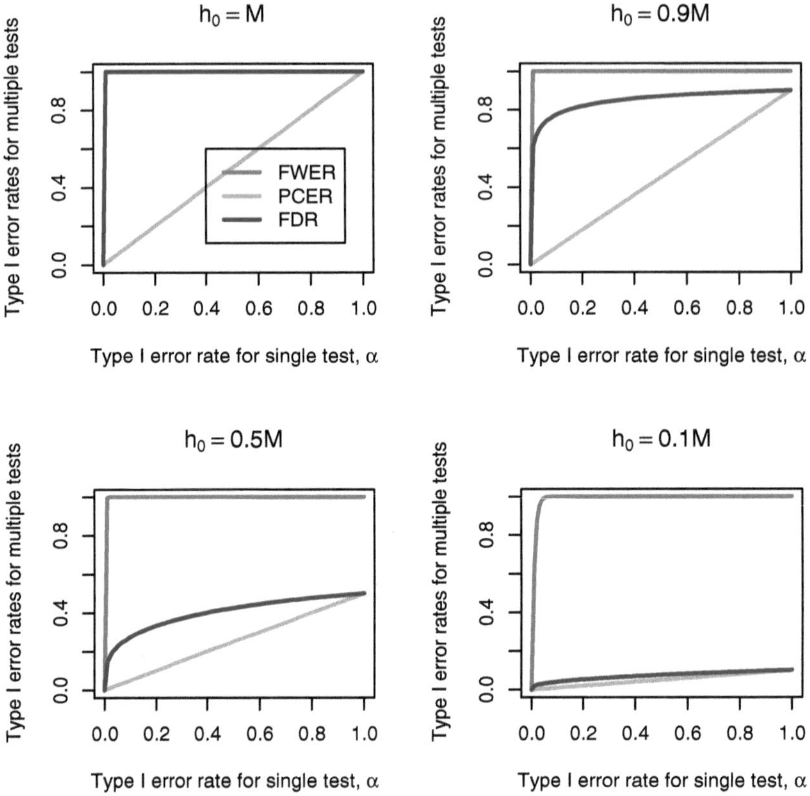

Figure 1.2. *Comparison of Type I error rates for a simple example.* Plots of Type I error rates for multiple tests (FWER: red curve; PCER: green curve; FDR: blue curve) vs. single test actual Type I error rate α, for different proportions h_0/M of true null hypotheses, $h_0/M = 1, 0.9, 0.5, 0.1$. The model and multiple testing procedures are described in Section 1.2.11. The number of hypotheses is $M = 100$ and the alternative shift parameter d is set to 1. (Color plate p. 322)

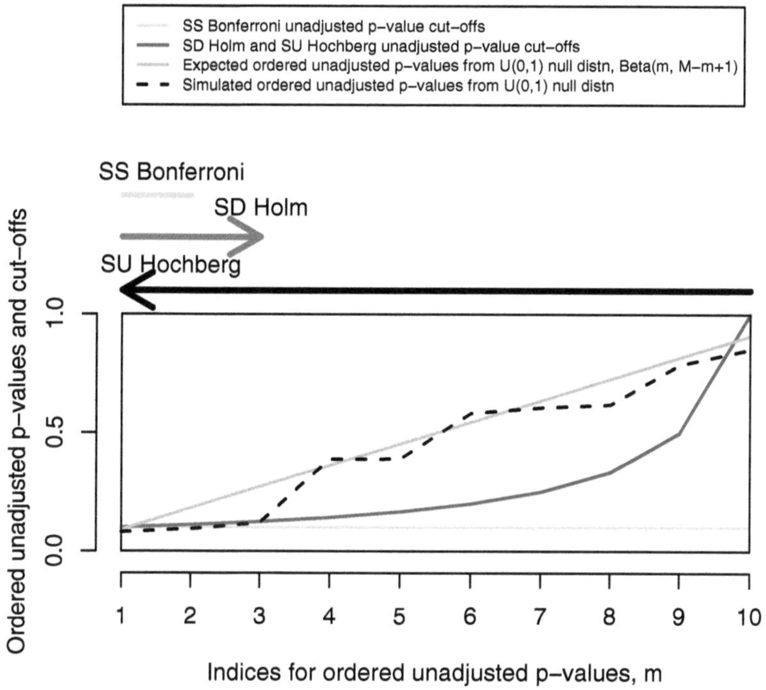

Figure 1.3. *Comparison of single-step, step-down, and step-up procedures: Cut-offs for FWER-controlling marginal Bonferroni, Holm, and Hochberg procedures.* The figure displays unadjusted p-value cut-offs for single-step Bonferroni Procedure 3.1 (SS Bonferroni; green curve), step-down Holm Procedure 3.7 (SD Holm; purple curve), and step-up Hochberg Procedure 3.13 (SU Hochberg; purple curve), for the test of $M = 10$ null hypotheses. Null unadjusted p-values are simulated as M independent realizations from the $\mathrm{U}(0,1)$ distribution. The simulated ordered unadjusted p-values and their expected values $m/(M+1)$ under the $\mathrm{Beta}(m, M-m+1)$ distribution are plotted as the dashed black curve and solid gray curve, respectively. The rejected null hypotheses are indicated by horizontal lines for each of the three procedures. An extreme nominal Type I error level $\alpha = 1$ is used for illustration purposes. (Color plate p. 323)

Ordered unadjusted p-values

$$P_{0n}(O_n(1)) \leq \quad P_{0n}(O_n(2)) \quad \leq \quad P_{0n}(O_n(3)) \quad \leq \cdots \leq P_{0n}(O_n(M))$$

Single-step Bonferroni adjusted p-values

$$MP_{0n}(O_n(1)) \leq \quad MP_{0n}(O_n(2)) \quad \leq \quad MP_{0n}(O_n(3)) \quad \leq \cdots \leq MP_{0n}(O_n(M))$$

$$\widetilde{P}_{0n}(O_n(m)) = \min\{MP_{0n}(O_n(m)), 1\}$$

Step-down Holm adjusted p-values

$$MP_{0n}(O_n(1))\,?\,(M-1)P_{0n}(O_n(2))\,?\,(M-2)P_{0n}(O_n(3))\,?\cdots\,?\;1P_{0n}(O_n(M))$$

$$\Longrightarrow$$

$$\widetilde{P}_{0n}(O_n(m)) = \max_{h=1,\ldots,m}\{\min\{(M-h+1)\,P_{0n}(O_n(h)), 1\}\}$$

Step-up Hochberg adjusted p-values

$$MP_{0n}(O_n(1))\,?\,(M-1)P_{0n}(O_n(2))\,?\,(M-2)P_{0n}(O_n(3))\,?\cdots\,?\;1P_{0n}(O_n(M))$$

$$\Longleftarrow$$

$$\widetilde{P}_{0n}(O_n(m)) = \min_{h=m,\ldots,M}\{\min\{(M-h+1)\,P_{0n}(O_n(h)), 1\}\}$$

Figure 1.4. *Comparison of single-step, step-down, and step-up procedures: Adjusted p-values for FWER-controlling marginal Bonferroni, Holm, and Hochberg procedures.* (Color plate p. 324)

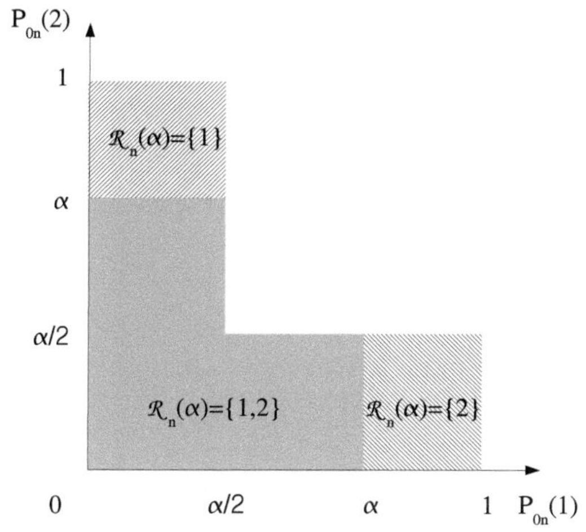

Panel (a): Step-down Holm Procedure 3.7

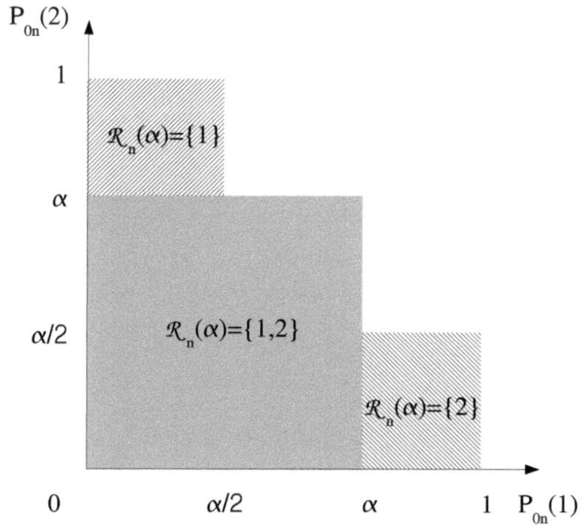

Panel (b): Step-up Hochberg Procedure 3.13

Figure 1.5. *Comparison of step-down and step-up procedures: Rejection regions for FWER-controlling marginal Holm and Hochberg procedures.* The figure displays rejection regions for step-down Holm Procedure 3.7 (Panel (a)) and step-up Hochberg Procedure 3.13 (Panel (b)), in the special case of $M = 2$ null hypotheses. The rejection regions are larger for the step-up procedure than for the step-down procedure, implying that the step-up procedure is less conservative than its step-down counterpart based on the same unadjusted p-value cut-offs.

2

Test Statistics Null Distribution

2.1 Introduction

2.1.1 Motivation

A key feature of our proposed multiple testing procedures (MTP) is the *test statistics null distribution* (rather than a data generating null distribution) used to obtain rejection regions (i.e., cut-offs) for the test statistics, confidence regions for the parameters of interest, and adjusted p-values. Indeed, whether testing single or multiple hypotheses, one needs the (joint) distribution of the test statistics in order to derive a procedure that probabilistically controls Type I errors. In practice, however, the true distribution of the test statistics is unknown and replaced by a null distribution. The choice of a proper null distribution is crucial in order to ensure that (finite sample or asymptotic) control of the Type I error rate under the *assumed null distribution* does indeed provide the desired control under the *true distribution*. This issue is particularly relevant for large-scale testing problems, such as those encountered in biomedical and genomic research (Chapters 9–12), which concern high-dimensional multivariate distributions, with complex and unknown dependence structures among variables.

Common approaches use a data generating distribution, such as a permutation distribution, that satisfies the *complete null hypothesis* that *all* null hypotheses are true. Procedures based on such a *data generating null distribution* typically rely on the *subset pivotality* assumption, stated in Westfall and Young (1993, p. 42–43), to ensure that Type I error control under the data generating null distribution leads to the desired control under the true data generating distribution. However, subset pivotality is violated in many important testing problems, because a data generating null distribution may result in a joint distribution for the test statistics that has a different dependence structure than their true distribution. In fact, in most problems, there does not exist a data generating null distribution that correctly specifies the joint distribution of the test statistics corresponding to the true null hypotheses.

Indeed, subset pivotality fails for two types of testing problems that are highly relevant in biomedical and genomic data analysis: tests concerning correlation coefficients and tests concerning regression coefficients (Chapter 8; Pollard et al. (2005a); Pollard and van der Laan (2004)).

We have formulated a general characterization of a test statistics null distribution for which the multiple testing procedures of Chapters 3–7 provide proper Type I error control. Our general characterization is based on the intuitive notion of *null domination*, whereby the number of Type I errors is stochastically greater under the test statistics' null distribution than under their true distribution. Null domination conditions lead us to the explicit construction of two main types of test statistics null distributions. The first original proposal of Dudoit et al. (2004b), van der Laan et al. (2004a), and Pollard and van der Laan (2004), defines the null distribution as the asymptotic distribution of a vector of *null shift and scale-transformed test statistics*, based on user-supplied upper bounds for the means and variances of the test statistics for the true null hypotheses. The second and most recent proposal of van der Laan and Hubbard (2006) defines the null distribution as the asymptotic distribution of a vector of *null quantile-transformed test statistics*, based on user-supplied marginal test statistics null distributions. Resampling procedures (e.g., non-parametric or model-based bootstrap) are provided to conveniently obtain consistent estimators of the null distribution and of the corresponding test statistic cut-offs, parameter confidence regions, and adjusted p-values.

We stress the generality of these two test statistics null distributions: Type I error control does not rely on restrictive assumptions such as subset pivotality and holds for general data generating distributions (with arbitrary dependence structures among variables), null hypotheses (defined in terms of submodels for the data generating distribution), and test statistics (e.g., t-statistics, χ^2-statistics, F-statistics). In particular, the proposed null distributions allow one to address testing problems that cannot be handled by existing approaches, such as tests concerning correlation coefficients and parameters in general regression models (e.g., linear regression models where the covariates and error terms are allowed to be dependent, logistic regression models, Cox proportional hazards models; Chapter 8; Pollard et al. (2005a)). The latest proposal of van der Laan and Hubbard (2006) has the additional advantage that the marginal test statistics null distributions may be set to the optimal (i.e., most powerful) null distributions one would use in single hypothesis testing (e.g., permutation marginal null distributions, Gaussian or other parametric marginal null distributions).

As illustrated in the simulation studies of Chapter 8 and articles by van der Laan and Hubbard (2006), Pollard et al. (2005a), and Pollard and van der Laan (2004), the choice of null distribution can have a substantial impact on the Type I error and power properties of a given multiple testing procedure. In particular, Pollard et al. (2005a) show that procedures based on our general non-parametric bootstrap null shift and scale-transformed test statistics null distribution typically control the Type I error rate "on target" at the nominal level. In contrast, comparable procedures, based on parameter-specific bootstrap data generating null distributions, can be severely anti-conservative (bootstrapping residuals for testing regression coefficients) or conservative (independent bootstrap for testing correlation coefficients). van der Laan and Hubbard (2006) further illustrate that, for finite samples, the new null quantile-transformed test statistics null distribution provides more accurate Type I error control and is more powerful than the original null shift and scale-transformed null distribution.

Finally, note that the null shift and scale-transformed and null quantile-transformed test statistics null distributions are only two among a family of null distributions that satisfy null domination conditions for a given testing problem. The explicit construction of null distributions with good Type I error control and power properties still represents an open and important research avenue.

2.1.2 Outline

Section 2.2 outlines the main features of our approach to Type I error control and the key choice of a test statistics null distribution based on the notion of null domination. Section 2.3 discusses in detail our first proposal of a null shift and scale-transformed test statistics null distribution. Section 2.4 introduces our most recent null quantile-transformed test statistics null distribution. Section 2.5 considers the choice of a null distribution for transformations of the test statistics, such as the absolute value transformation. Sections 2.6 and 2.7 focus on two particular examples of testing problems covered by our framework: the test of single-parameter null hypotheses using t-statistics (e.g., tests of means, correlation coefficients, regression coefficients in linear and non-linear models) and the test of multiple-parameter null hypotheses using F-statistics. The last two sections are devoted to contrasting our proposed methodology with existing approaches. Specifically, Section 2.8 revisits the notions of weak and strong control of a Type I error rate and the related assumption of subset pivotality. We stress that such conditions are made irrelevant by our general approach, which is only concerned with control of the Type I error rate under the *true data generating distribution* and is based on a *test statistics null distribution* rather than a data generating null distribution. Finally, Section 2.9 examines test statistics null distributions based on bootstrap and permutation data generating distributions.

2.2 Type I error control and choice of a test statistics null distribution

2.2.1 Type I error control

As in Section 1.2, consider the simultaneous test of M null hypotheses $H_0(m)$, $m = 1, \ldots, M$, based on test statistics $T_n = (T_n(m) : m = 1, \ldots, M)$, with finite sample joint distribution $Q_n = Q_n(P)$, under the data generating distribution P. We wish to derive rejection regions for the test statistics $T_n(m)$, such that Type I errors are probabilistically controlled at a user-supplied level α (see Section 1.2.9 for definitions of Type I error rates). In practice, however, the true distribution $Q_n(P)$ of the test statistics is unknown and replaced by a null distribution Q_0 (or estimator thereof, Q_{0n}). As in Section 1.2.6, let $\mathcal{C}_n(m) = \mathcal{C}(m; T_n, Q_0, \alpha)$, $m = 1, \ldots, M$, and $\mathcal{R}_n = \mathcal{R}(T_n, Q_0, \alpha)$ denote, respectively, the M rejection regions and corresponding set of rejected null hypotheses, for a MTP with nominal Type I error level α. That is,

$$\mathcal{R}(T_n, Q_0, \alpha) = \{m : T_n(m) \in \mathcal{C}(m; T_n, Q_0, \alpha)\}. \tag{2.1}$$

Given a random M-vector $Z = (Z(m) : m = 1, \ldots, M)$, with joint distribution Q, and a collection of M rejection regions $\mathcal{C} = \{\mathcal{C}(m) : m = 1, \ldots, M\}$[1], denote the numbers of rejected hypotheses and Type I errors by

$$R(\mathcal{C}|Q) \equiv \sum_{m=1}^{M} \mathrm{I}\left(Z(m) \in \mathcal{C}(m)\right) \tag{2.2}$$

and

$$V(\mathcal{C}|Q) \equiv \sum_{m \in \mathcal{H}_0} \mathrm{I}\left(Z(m) \in \mathcal{C}(m)\right),$$

respectively. For given rejection regions $\mathcal{C}$, adopt the following shorthand notation for the special cases where Q corresponds to the test statistics true distribution Q_n and null distribution Q_0,

$$\begin{aligned} R_n &\equiv R(\mathcal{C}|Q_n), & R_0 &\equiv R(\mathcal{C}|Q_0), \\ V_n &\equiv V(\mathcal{C}|Q_n), & V_0 &\equiv V(\mathcal{C}|Q_0). \end{aligned} \tag{2.3}$$

For one-sided rejection regions of the form $\mathcal{C}(m) = (c(m), +\infty)$, based on an M-vector of cut-offs $c = (c(m) : m = 1, \ldots, M) \in I\!R^M$, further denote the numbers of rejected hypotheses and Type I errors by $R(c|Q)$ and $V(c|Q)$, respectively.

Rejection regions are typically derived so that the Type I error rate $\Theta(F_{V_0,R_0})$, *under the test statistics null distribution* Q_0, is controlled at *nominal level* $\alpha \in (0,1)$[2], that is,

[1] N.B. In stepwise procedures, the rejection regions $\mathcal{C}$ may be random, i.e., may depend on Z.

[2] N.B. Without loss of generality, we focus for simplicity on Type I error rates $\Theta(F_{V_n,R_n}) \in [0,1]$.

$$\Theta(F_{V_0,R_0}) \leq \alpha. \tag{2.4}$$

The multiple testing procedure $\mathcal{R}_n$ is said to control the Type I error rate $\Theta(F_{V_n,R_n})$, *under the test statistics true distribution* Q_n, at *actual level* $\alpha \in (0,1)$, if

$$\begin{aligned} \Theta(F_{V_n,R_n}) &\leq \alpha && \text{[finite sample control]} \\ \limsup_{n\to\infty} \Theta(F_{V_n,R_n}) &\leq \alpha && \text{[asymptotic control].} \end{aligned} \tag{2.5}$$

Note that the actual Type I error rate $\Theta(F_{V_n,R_n})$ of a multiple testing procedure typically differs from its nominal level α, i.e., the level at which it claims to control Type I errors. Discrepancies between actual and nominal Type I error levels can be attributed to a number of factors, including the choice of a test statistics null distribution Q_0 and the type of rejection regions for a given choice of Q_0. A testing procedure is said to be *conservative* if the nominal Type I error level α is greater than the actual Type I error rate and *anti-conservative* if the nominal Type I error level α is less than the actual Type I error rate, that is,

$$\begin{array}{ll} \text{Conservative} & \Theta(F_{V_n,R_n}) < \alpha \\ \text{Anti-conservative} & \Theta(F_{V_n,R_n}) > \alpha. \end{array} \tag{2.6}$$

The choice of a suitable test statistics null distribution Q_0 is crucial in order to ensure that (finite sample or asymptotic) control of the Type I error rate under this assumed null distribution does indeed provide the desired control under the true distribution Q_n. For proper control, the Type I error rate under the null distribution Q_0 must *dominate* the Type I error rate under the true distribution Q_n. That is, the null distribution Q_0 must satisfy

$$\begin{aligned} \Theta(F_{V_n,R_n}) &\leq \Theta(F_{V_0,R_0}) && \text{[finite sample control]} \\ \limsup_{n\to\infty} \Theta(F_{V_n,R_n}) &\leq \Theta(F_{V_0,R_0}) && \text{[asymptotic control].} \end{aligned} \tag{2.7}$$

Chapter 8 and articles by van der Laan and Hubbard (2006), Pollard et al. (2005a), and Pollard and van der Laan (2004), present simulation studies investigating the impact of the null distribution on the Type I error control and power properties of a MTP.

2.2.2 Sketch of proposed approach to Type I error control

The following discussion motivates our general approach to the problem of Type I error control and highlights important considerations in choosing a test statistics null distribution. We focus on Type I error rates defined as arbitrary parameters $\Theta(F_{V_n})$ of the distribution of the number of Type I errors V_n (Section 1.2.9).

Recall that the distribution F_{V_n}, for the number of Type I errors $V_n = |\mathcal{R}_n \cap \mathcal{H}_0| = |\mathcal{R}(T_n, Q_0, \alpha) \cap \mathcal{H}_0(P)|$, depends on the following: the true distribution $Q_n = Q_n(P)$ of the test statistics T_n; the test statistics null distribution Q_0, used to derive the rejection regions $\mathcal{C}_n(m) = \mathcal{C}(m; T_n, Q_0, \alpha)$; the nominal Type I error level α of the MTP; and the set $\mathcal{H}_0 = \mathcal{H}_0(P)$ of true null hypotheses. Type I error control is therefore a statement about the true, unknown data generating distribution P, via $Q_n(P)$ and $\mathcal{H}_0(P)$.

Control of Type I error rates of the form $\Theta(F_{V_n})$ can be achieved by the three-step road map of Procedure 2.1, below. This road map provides intuition behind the general characterization (Section 2.2.3) and explicit construction (Sections 2.3 and 2.4) of a proper test statistics null distribution Q_0. It also provides a template for Θ-controlling joint single-step common-quantile Procedure 4.1 and common-cut-off Procedure 4.2. The main idea is to substitute control of the *unknown parameter* $\Theta(F_{V_n})$, for the *true distribution* F_{V_n} of the number of Type I errors, by control of the corresponding *known parameter* $\Theta(F_{R_0})$, for the *null distribution* F_{R_0} of the number of rejected hypotheses.

Procedure 2.1. [Three-step road map for controlling Type I error rates $\boldsymbol{\Theta(F_{V_n})}$]

1. **Null domination conditions for the Type I error rates $\boldsymbol{\Theta(F_{V_n})}$ and $\boldsymbol{\Theta(F_{V_0})}$.** Select a test statistics null distribution Q_0 such that the Type I error rate $\Theta(F_{V_0})$, under this null distribution Q_0, dominates the Type I error rate $\Theta(F_{V_n})$, under the true distribution Q_n. That is, the following *null domination* assumption for the *Type I error rates* is satisfied.

$$\Theta(F_{V_n}) \leq \Theta(F_{V_0}) \qquad \text{[finite sample control]}$$
$$\limsup_{n \to \infty} \Theta(F_{V_n}) \leq \Theta(F_{V_0}) \qquad \text{[asymptotic control]}. \qquad (\mathsf{ND}\Theta)$$

2. **Monotonicity of the Type I error rate mapping $\boldsymbol{\Theta}$.** Note that the number of Type I errors is always less than or equal to the total number of rejected hypotheses (i.e., $V_0 \leq R_0$), so that $F_{V_0} \geq F_{R_0}$. Hence, under monotonicity Assumption $\mathsf{M}\Theta$ for the Type I error rate mapping Θ, one has

$$\Theta(F_{V_0}) \leq \Theta(F_{R_0}). \qquad (2.8)$$

3. **Control of $\boldsymbol{\Theta(F_{R_0})}$.** Select rejection regions $\mathcal{C}_n(m) = \mathcal{C}(m; T_n, Q_0, \alpha)$ so that the following Type I error constraint is satisfied,

$$\Theta(F_{R_0}) \leq \alpha. \qquad (2.9)$$

That is, control the known parameter $\Theta(F_{R_0})$, corresponding to the number of rejected hypotheses $R_0 = \sum_{m=1}^{M} \mathrm{I}(T_n(m) \in \mathcal{C}_n(m))$, under the null distribution Q_0, i.e., assuming $T_n \sim Q_0$.

Combining Steps 1–3 provides the desired control of the actual Type I error rate $\Theta(F_{V_n})$ at level $\alpha \in (0,1)$, that is,

$$\begin{aligned} \Theta(F_{V_n}) \leq \Theta(F_{V_0}) \leq \Theta(F_{R_0}) \leq \alpha \qquad & \text{[finite sample control]} \\ \limsup_{n\to\infty} \Theta(F_{V_n}) \leq \Theta(F_{V_0}) \leq \Theta(F_{R_0}) \leq \alpha \qquad & \text{[asymptotic control]}. \end{aligned} \tag{2.10}$$

Note that the road map of Procedure 2.1 is conservative in two ways: (i) from the null domination of the Type I error rate in Step 1, $\Theta(F_{V_n}) \leq \Theta(F_{V_0})$; (ii) from controlling $\Theta(F_{R_0}) \geq \Theta(F_{V_0})$ in Step 3. Step 1 is often the most problematic and requires a judicious choice for the test statistics null distribution Q_0.

2.2.3 Characterization of test statistics null distribution in terms of null domination conditions

For certain families of Type I error rate mappings Θ and rejection regions $\mathcal{C}_n$, Θ-specific Type I error rate null domination Assumption ND$\boldsymbol{\Theta}$, in Step 1 of the road map, can be shown to hold under the following alternate forms of null domination.

- Null domination for the distributions F_{V_n} and F_{V_0} of the number of Type I errors.
- Null domination for the joint distributions $Q_{n,\mathcal{H}_0}$ and $Q_{0,\mathcal{H}_0}$ of the $\mathcal{H}_0$-specific subvector $(T_n(m) : m \in \mathcal{H}_0)$ of test statistics for the true null hypotheses $\mathcal{H}_0$.

Null domination conditions for the numbers of Type I errors V_n and V_0

One can specify *null domination* conditions in terms of the distributions of the *numbers of Type I errors* V_n and V_0, as follows. For each $x \in \{0, \ldots, M\}$,

$$\begin{aligned} F_{V_n}(x) \geq F_{V_0}(x) \qquad & \text{[finite sample control]} \\ \liminf_{n\to\infty} F_{V_n}(x) \geq F_{V_0}(x) \qquad & \text{[asymptotic control]}. \end{aligned} \tag{NDV}$$

That is, the number of Type I errors V_0, under the null distribution Q_0, is stochastically greater than the number of Type I errors V_n, under the true distribution Q_n for the test statistics T_n.

For Type I error rate mappings Θ that satisfy monotonicity Assumption MΘ and continuity Assumption CΘ at F_{V_0}, null domination Assumption NDV for the number of Type I errors implies null domination Assumption ND$\boldsymbol{\Theta}$ for the Type I error rate.

Joint null domination conditions for the $\mathcal{H}_0$-specific test statistics $(T_n(m) : m \in \mathcal{H}_0)$

One can also specify multivariate null domination conditions in terms of the joint distribution of the test statistics $(T_n(m) : m \in \mathcal{H}_0)$ for the true null hypotheses $\mathcal{H}_0$, based on the notion of multivariate stochastic order (Kamae et al., 1977, p. 899). Below are three equivalent *joint null domination* conditions for the $\mathcal{H}_0$-*specific test statistics* $(T_n(m) : m \in \mathcal{H}_0)$.

The null distribution $Q_{0,\mathcal{H}_0}$, of the $\mathcal{H}_0$-specific subvector of test statistics $(T_n(m) : m \in \mathcal{H}_0)$, is said to be stochastically greater than the corresponding true distribution $Q_{n,\mathcal{H}_0} = Q_{n,\mathcal{H}_0}(P)$, if, for all bounded componentwise increasing functions $\ell : I\!R^{h_0} \rightarrow I\!R$,

$$\begin{aligned} \mathrm{E}_{Q_{n,\mathcal{H}_0}}\left[\ell((T_n(m) : m \in \mathcal{H}_0))\right] &\leq \mathrm{E}_{Q_{0,\mathcal{H}_0}}\left[\ell((Z(m) : m \in \mathcal{H}_0))\right] \\ \limsup_{n\rightarrow\infty} \mathrm{E}_{Q_{n,\mathcal{H}_0}}\left[\ell((T_n(m) : m \in \mathcal{H}_0))\right] &\leq \mathrm{E}_{Q_{0,\mathcal{H}_0}}\left[\ell((Z(m) : m \in \mathcal{H}_0))\right], \end{aligned} \tag{2.11}$$

where, for the asymptotic statement, the null distribution $Q_{0,\mathcal{H}_0}$ is further required to be continuous.

An alternate formulation of joint null domination is that, for all Borel sets $\mathcal{B} \subseteq I\!R^{h_0}$ with componentwise increasing indicator function $\mathrm{I}_{\mathcal{B}} : z \in I\!R^{h_0} \rightarrow \mathrm{I}(z \in \mathcal{B}) \in \{0, 1\}$,

$$\begin{aligned} \mathrm{Pr}_{Q_{n,\mathcal{H}_0}}\left((T_n(m) : m \in \mathcal{H}_0) \in \mathcal{B}\right) &\leq \mathrm{Pr}_{Q_{0,\mathcal{H}_0}}\left((Z(m) : m \in \mathcal{H}_0) \in \mathcal{B}\right) \\ \limsup_{n\rightarrow\infty} \mathrm{Pr}_{Q_{n,\mathcal{H}_0}}\left((T_n(m) : m \in \mathcal{H}_0) \in \mathcal{B}\right) &\leq \mathrm{Pr}_{Q_{0,\mathcal{H}_0}}\left((Z(m) : m \in \mathcal{H}_0) \in \mathcal{B}\right), \end{aligned} \tag{2.12}$$

where, for the asymptotic statement, the null distribution $Q_{0,\mathcal{H}_0}$ is further required to be continuous.

A third, more compact formulation of joint null domination, in terms of the joint cumulative distribution functions of the test statistics $(T_n(m) : m \in \mathcal{H}_0)$, is that, for all $z \in I\!R^{h_0}$,

$$\begin{aligned} Q_{n,\mathcal{H}_0}(z) &\geq Q_{0,\mathcal{H}_0}(z) \qquad \text{[finite sample control]} \\ \liminf_{n\rightarrow\infty} Q_{n,\mathcal{H}_0}(z) &\geq Q_{0,\mathcal{H}_0}(z) \qquad \text{[asymptotic control]}, \end{aligned} \tag{jtNDT}$$

where, for the asymptotic statement, the null distribution $Q_{0,\mathcal{H}_0}$ is further required to be continuous. Note that Assumption jtNDT corresponds to Equation (2.12), with sets $\mathcal{B} = (-\infty, z]^c$ defined in terms of h_0-dimensional rectangles $(-\infty, z] = \prod_{m=1}^{h_0}(-\infty, z(m)] \subseteq I\!R^{h_0}$.

For ease of notation, we may simply refer to the finite sample and asymptotic joint null domination conditions as $Q_{n,\mathcal{H}_0} \geq Q_{0,\mathcal{H}_0}$ and $\liminf_n Q_{n,\mathcal{H}_0} \geq Q_{0,\mathcal{H}_0}$, respectively.

Relationships between null domination Assumptions jtNDT, NDV, and NDΘ

For one-sided rejection regions of the form $\mathcal{C}(m) = (c(m), +\infty)$, joint null domination Assumption jtNDT for the test statistics implies null domination Assumption NDV for the number of Type I errors. Indeed, for given $c = (c(m) : m = 1, \dots, M) \in I\!R^M$ and $x \in \{0, \dots, M\}$, one may apply Equation (2.11), with the bounded componentwise increasing function $\ell : I\!R^{h_0} \rightarrow I\!R$ defined such that

$$\ell((Z(m) : m \in \mathcal{H}_0)) = \mathrm{I}\left(\sum_{m \in \mathcal{H}_0} \mathrm{I}\left(Z(m) > c(m)\right) > x \right) = \mathrm{I}\left(V(c|Q) > x\right),$$

where $Z = (Z(m) : m = 1, \dots, M) \sim Q$. Then,

$$\begin{aligned} \Pr\left(V(c|Q_n) > x\right) &\leq \Pr\left(V(c|Q_0) > x\right) \\ \limsup_{n \to \infty} \Pr\left(V(c|Q_n) > x\right) &\leq \Pr\left(V(c|Q_0) > x\right). \end{aligned}$$

Noting that $\Pr(V(c|Q_n) > x) = 1 - F_{V_n}(x)$ and $\Pr(V(c|Q_0) > x) = 1 - F_{V_0}(x)$ yields Assumption NDV.

Monotonicity Assumption MΘ and continuity Assumption CΘ at F_{V_0} then imply null domination Assumption NDΘ for the Type I error rate.

Note that, for the asymptotic versions of null domination in Equations (2.11), (2.12), and (jtNDT), one could relax the continuity assumption on Q_0, by requiring, for example, that the cut-offs c be continuity points of Q_0.

To summarize, one has the following relationships among the three types of null domination assumptions introduced thus far. Under these assumptions, the road map of Procedure 2.1 provides (finite sample or asymptotic) control of general Type I error rates of the form $\Theta(F_{V_n})$.

Assumption jtNDT: Joint null domination for $\mathcal{H}_0$-specific test statistics

$$Q_{n,\mathcal{H}_0} \geq Q_{0,\mathcal{H}_0}.$$

$$\Downarrow$$

Assumption NDV: Null domination for number of Type I errors, for one-sided rejection regions of the form $\mathcal{C}(m) = (c(m), +\infty)$,

$$F_{V_n} \geq F_{V_0}.$$

$$\Downarrow$$

Assumption NDΘ: Null domination for Type I error rate, under Assumptions MΘ and CΘ,

$$\Theta(F_{V_n}) \leq \Theta(F_{V_0}).$$

Note that null domination is only a statement about the joint distribution of the subvector of test statistics $(T_n(m) : m \in \mathcal{H}_0)$ for the true null hypotheses $\mathcal{H}_0$.

More specific (i.e., less stringent) forms of null domination may be derived for given definitions of the Type I error rate mapping Θ and rejection regions (e.g., null domination conditions for FWER-controlling step-down common-cut-off and common-quantile MTPs in Chapter 5 and van der Laan et al. (2004a)).

General joint null domination Assumption jtNDT, for the $\mathcal{H}_0$-specific test statistics, provides a template for deriving test statistics null distributions that lead to proper Type I error control: identify a collection of M functions, $\ell_m : I\!R \rightarrow I\!R$, such that the joint distribution of the transformed test statistics $(\ell_m(T_n(m)) : m \in \mathcal{H}_0)$ dominates the joint distribution of the original test statistics $(T_n(m) : m \in \mathcal{H}_0)$. Based on this general characterization, Sections 2.3 and 2.4, below, provide two explicit constructions for a proper test statistics null distribution Q_0: the asymptotic distribution of a vector of null shift and scale-transformed test statistics, based on user-supplied upper bounds for the means and variances of the $\mathcal{H}_0$-specific test statistics (Section 2.3; Dudoit et al. (2004b); van der Laan et al. (2004a); Pollard and van der Laan (2004)) and the asymptotic distribution of a vector of null quantile-transformed test statistics, based on user-supplied marginal test statistics null distributions (Section 2.4; van der Laan and Hubbard (2006)).

Either test statistics null distribution may be used in any of the multiple testing procedures proposed in Chapters 3–7 of this book, as they both satisfy the key property of joint null domination for the $\mathcal{H}_0$-specific test statistics (Assumption jtNDT). Specifically, the null shift and scale-transformed null distribution (or a consistent estimator thereof) provides Type I error control for: $\Theta(F_{V_n})$-controlling joint single-step common-cut-off and common-quantile procedures (Chapter 4; Dudoit et al. (2004b)); FWER-controlling joint step-down common-cut-off (maxT) and common-quantile (minP) procedures (Chapter 5; van der Laan et al. (2004a)); gTP-controlling (marginal/joint single-step/stepwise) augmentation multiple testing procedures (Chapter 6; Dudoit et al. (2004a); van der Laan et al. (2004b)); gTP-controlling joint resampling-based empirical Bayes procedures (Chapter 7; van der Laan et al. (2005)). van der Laan and Hubbard (2006) argue that the above results also hold for the new null quantile-transformed test statistics null distribution.

2.2.4 Contrast with other approaches

One of our main contributions is the general characterization (Section 2.2.3) and explicit construction (Sections 2.3 and 2.4) of proper null distributions Q_0 (or estimators thereof, Q_{0n}) for the test statistics T_n. As detailed in Section 2.8, the following two main points distinguish our approach from existing approaches to Type I error control and the choice of a test statistics null distribution (e.g., Hochberg and Tamhane (1987) and Westfall and Young (1993)).

Type I error control under the true data generating distribution

Firstly, we are only concerned with control of the Type I error rate under the *true data generating distribution* P, i.e., under the joint distribution $Q_n = Q_n(P)$, implied by P, for the test statistics T_n. The concepts of weak and strong control of a Type I error rate are therefore irrelevant in our context.

In particular, our *null domination* Assumptions jtNDT, NDV, and NDΘ, introduced in Section 2.2.3, differ from the standard *subset pivotality* assumption of Westfall and Young (1993, p. 42–43), in the following senses: (i) null domination is only concerned with the true data generating distribution P, i.e., the subset $\mathcal{H}_0(P)$ of true null hypotheses and not all possible 2^M subsets $\mathcal{J}_0 \subseteq \{1, \ldots, M\}$ of null hypotheses; (ii) null domination does not require equality of the joint distributions $Q_{0,\mathcal{H}_0}$ and $Q_{n,\mathcal{H}_0}(P)$, for the $\mathcal{H}_0$-specific test statistics $(T_n(m) : m \in \mathcal{H}_0)$, but the weaker domination of $Q_{n,\mathcal{H}_0}(P)$ by $Q_{0,\mathcal{H}_0}$.

Null distribution for the test statistics

Secondly, we propose a *null distribution for the test statistics* ($T_n \sim Q_0$) rather than a *data generating null distribution* ($X \sim P_0$). A common choice of data generating null distribution P_0 is one that satisfies the complete null hypothesis, $H_0^C = \prod_{m=1}^M H_0(m) = \prod_{m=1}^M \mathrm{I}(P \in \mathcal{M}(m)) = \mathrm{I}(P \in \cap_{m=1}^M \mathcal{M}(m))$, that all M null hypotheses are true, i.e., $P_0 \in \cap_{m=1}^M \mathcal{M}(m)$. The data generating null distribution P_0 then implies a null distribution $Q_n(P_0)$ for the test statistics.

As discussed in Pollard et al. (2005a) and Pollard and van der Laan (2004), procedures based on $Q_n(P_0)$ do not necessarily provide proper Type I error control under the true distribution P. Indeed, the assumed null distribution $Q_{n,\mathcal{H}_0}(P_0)$ and the true distribution $Q_{n,\mathcal{H}_0}(P)$, of the $\mathcal{H}_0$-specific test statistics $(T_n(m) : m \in \mathcal{H}_0)$, may have different dependence structures and, as a result, may violate the required null domination condition for the Type I error rate (Assumption NDΘ, in Step 1 of the road map of Procedure 2.1). For instance, for test statistics with Gaussian asymptotic distributions (Section 2.6), the asymptotic covariance matrix of the $\mathcal{H}_0$-specific test statistics $\Sigma_{\mathcal{H}_0}(P)$, under the true distribution P, may be different from the corresponding covariance matrix $\Sigma_{\mathcal{H}_0}(P_0)$, under the complete null distribution P_0. For the two-sample test of means, based on difference statistics and the commonly-used permutation data generating null distribution P_0, Pollard and van der Laan (2004) show that $\Sigma_{\mathcal{H}_0}(P) = \Sigma_{\mathcal{H}_0}(P_0)$ if and only if (i) the two populations have the same covariance matrices or (ii) the population frequencies are equal (Section 2.9).

Consequently, approaches based on permutation or other data generating null distributions P_0 (e.g., Korn et al. (2004), Troendle (1995, 1996), and Westfall and Young (1993)) are only valid under certain assumptions for the true data generating distribution P. In fact, in most testing problems, there does not exist a data generating null distribution $P_0 \in \cap_{m=1}^{M} \mathcal{M}(m)$ that correctly specifies a joint null distribution for the $\mathcal{H}_0$-specific test statistics $(T_n(m) : m \in \mathcal{H}_0)$, i.e., such that the required null domination condition for the Type I error rate is satisfied (Assumption NDΘ).

In summary, unlike current methods that can only be applied to a limited set of multiple testing problems, the general constructions of Sections 2.3 and 2.4 lead to joint single-step and stepwise procedures that provide the desired Type I error control for general data generating distributions, null hypotheses, and test statistics. Our proposed test statistics null distributions can be used in testing problems that cannot be handled by traditional approaches based on a data generating null distribution and the associated assumption of subset pivotality. Such problems include tests for correlation coefficients and regression coefficients in linear and non-linear models where covariates and error terms are allowed to be dependent (Chapter 8; Pollard et al. (2005a)).

2.3 Null shift and scale-transformed test statistics null distribution

2.3.1 Explicit construction for the test statistics null distribution

Following Dudoit et al. (2004b), van der Laan et al. (2004a), and Pollard and van der Laan (2004), our first proposal for a test statistics null distribution is the asymptotic distribution of a vector of null shift and scale-transformed test statistics, based on user-supplied upper bounds for the means and variances of the $\mathcal{H}_0$-specific test statistics.

Theorem 2.2. [Null shift and scale-transformed test statistics null distribution]
Asymptotic test statistics null distribution. *Suppose there exist known M-vectors $\lambda_0 \in I\!R^M$ and $\tau_0 \in I\!R^{+M}$ of null values, so that, for each $m \in \mathcal{H}_0$,*

$$\limsup_{n\to\infty} \mathrm{E}[T_n(m)] \leq \lambda_0(m) \tag{2.13}$$

and

$$\limsup_{n\to\infty} \mathrm{Var}[T_n(m)] \leq \tau_0(m).$$

Let

$$\nu_{0,n}(m) \equiv \sqrt{\min\left\{1, \frac{\tau_0(m)}{\mathrm{Var}[T_n(m)]}\right\}} \tag{2.14}$$

and define an M-vector of null shift and scale-transformed test statistics $Z_n = (Z_n(m) : m = 1, \ldots, M)$ *by*

$$Z_n(m) \equiv \nu_{0,n}(m)\,(T_n(m) - \mathrm{E}[T_n(m)]) + \lambda_0(m), \qquad m = 1, \ldots, M. \tag{2.15}$$

Suppose that the random M-vector Z_n converges weakly to a random M-vector Z, with continuous joint distribution $Q_0 = Q_0(P)$,

$$Z_n \overset{\mathcal{L}}{\Rightarrow} Z \sim Q_0(P). \tag{2.16}$$

Then, the asymptotic joint distribution Q_0 satisfies asymptotic joint null domination Assumption jtNDT for the $\mathcal{H}_0$-specific subvector of test statistics $(T_n(m) : m \in \mathcal{H}_0)$. That is, for all $z \in \mathbb{R}^{h_0}$,

$$\liminf_{n\to\infty} Q_{n,\mathcal{H}_0}(z) \geq Q_{0,\mathcal{H}_0}(z).$$

In addition, for all $c = (c(m) : m = 1, \ldots, M) \in \mathbb{R}^M$ and $x \in \{0, \ldots, M\}$,

$$\liminf_{n\to\infty} \Pr_{Q_n}\left(\sum_{m\in\mathcal{H}_0} \mathrm{I}\left(T_n(m) > c(m)\right) \leq x\right)$$
$$\geq \Pr_{Q_0}\left(\sum_{m\in\mathcal{H}_0} \mathrm{I}\left(Z(m) > c(m)\right) \leq x\right).$$

Thus, for one-sided rejection regions of the form $\mathcal{C}_n(m) = (c_n(m), +\infty)$, the null distribution Q_0 satisfies asymptotic null domination Assumption NDV for the number of Type I errors,

$$\liminf_{n\to\infty} F_{V_n}(x) \geq F_{V_0}(x), \qquad \forall\, x \in \{0, \ldots, M\}.$$

If one further assumes that the Type I error rate mapping Θ meets monotonicity Assumption MΘ and continuity Assumption CΘ at F_{V_0}, then the null distribution Q_0 also satisfies asymptotic null domination Assumption NDΘ for the Type I error rate,

$$\limsup_{n\to\infty} \Theta(F_{V_n}) \leq \Theta(F_{V_0}).$$

Finite sample test statistics null distribution. *Suppose there exists a known M-vector $\lambda_{0,n} \in \mathbb{R}^M$ of null values, so that, for each $m \in \mathcal{H}_0$,*

$$\mathrm{E}[T_n(m)] \leq \lambda_{0,n}(m). \tag{2.17}$$

Define an M-vector of null shift-transformed test statistics $Z_n = (Z_n(m)\colon m = 1, \ldots, M)$ *by*

$$Z_n(m) \equiv T_n(m) - \mathrm{E}[T_n(m)] + \lambda_{0,n}(m), \qquad m = 1, \ldots, M. \tag{2.18}$$

Then, the finite sample joint distribution $Q_{0,n} = Q_{0,n}(P)$ of Z_n satisfies the finite sample versions of null domination Assumptions **jtNDT**, **NDV**, *and* **NDΘ**.

The asymptotic distribution Q_0, of the null shift and scale-transformed test statistics Z_n, generalizes the null distribution proposed in Pollard and van der Laan (2004) for the test of single-parameter null hypotheses based on t-statistics. In this special case, the null distribution Q_0 turns out to be a Gaussian distribution with mean vector zero (Section 2.6).

Dudoit et al. (2004b) and van der Laan et al. (2004a) prove that joint single-step and step-down procedures based on the null distribution of Theorem 2.2 (or a consistent estimator thereof) do indeed provide the desired asymptotic control of the Type I error rate $\Theta(F_{V_n})$, for general data generating distributions (with arbitrary dependence structures among variables), null hypotheses (defined in terms of submodels for the data generating distribution), and test statistics (e.g., t-statistics, χ^2-statistics, F-statistics).

As seen in Sections 2.6 and 2.7, the null distribution Q_0 is continuous for a broad class of testing problems. Otherwise, one could relax the continuity assumption on Q_0, by requiring, for example, that the cut-offs c be continuity points of Q_0.

Proof of Theorem 2.2.
Asymptotic test statistics null distribution. The proof is straightforward and is based on an intermediate random vector $\widetilde{Z}_n = (\widetilde{Z}_n(m) : m = 1, \ldots, M)$, defined as

$$\widetilde{Z}_n(m) = T_n(m) + \max\{0, \lambda_0(m) - \mathrm{E}[T_n(m)]\}, \qquad m = 1, \ldots, M. \tag{2.19}$$

First, note that $T_n(m) \leq \widetilde{Z}_n(m)$ for each $m = 1, \ldots, M$. Next, for $m \in \mathcal{H}_0$, since $\limsup_n \mathrm{E}[T_n(m)] \leq \lambda_0(m)$ and $\limsup_n \mathrm{Var}[T_n(m)] \leq \tau_0(m)$, then $\lim_n \nu_{0,n}(m) = 1$ and the $\mathcal{H}_0$-specific subvectors $(\widetilde{Z}_n(m) : m \in \mathcal{H}_0)$ and $(Z_n(m) : m \in \mathcal{H}_0)$ have the same asymptotic joint distribution. That is,

$$(\widetilde{Z}_n(m) : m \in \mathcal{H}_0) \stackrel{\mathcal{L}}{\Rightarrow} (Z(m) : m \in \mathcal{H}_0) \sim Q_{0,\mathcal{H}_0}.$$

Thus, asymptotic joint null domination Assumption **jtNDT** follows from the definition of weak convergence to a continuous limit distribution Q_0 (Equation (B.7)). That is, for each $z \in I\!R^{h_0}$ and corresponding h_0-dimensional rectangle $(-\infty, z] \subseteq I\!R^{h_0}$,

$$\begin{aligned}
\liminf_{n\to\infty} Q_{n,\mathcal{H}_0}(z) &= \liminf_{n\to\infty} \Pr\left((T_n(m) : m \in \mathcal{H}_0) \in (-\infty, z]\right) \\
&\geq \liminf_{n\to\infty} \Pr\left((\widetilde{Z}_n(m) : m \in \mathcal{H}_0) \in (-\infty, z]\right) \\
&= \Pr\left((Z(m) : m \in \mathcal{H}_0) \in (-\infty, z]\right) \\
&= Q_{0,\mathcal{H}_0}(z).
\end{aligned}$$

In addition, for all $c = (c(m) : m = 1, \ldots, M) \in I\!R^M$, the Continuous Mapping Theorem (Theorem B.3), applied to the function $\ell((z(m) : m \in \mathcal{H}_0)) = \sum_{m\in\mathcal{H}_0} \mathrm{I}\left(z(m) > c(m)\right)$, implies that

$$\begin{aligned}
\ell((\widetilde{Z}_n(m) : m \in \mathcal{H}_0)) &= \sum_{m\in\mathcal{H}_0} \mathrm{I}\left(\widetilde{Z}_n(m) > c(m)\right) \\
&\overset{\mathcal{L}}{\Rightarrow} \sum_{m\in\mathcal{H}_0} \mathrm{I}\left(Z(m) > c(m)\right) = \ell((Z(m) : m \in \mathcal{H}_0)).
\end{aligned}$$

Asymptotic null domination Assumption NDV then follows from Proposition B.2. That is, for all $c = (c(m) : m = 1, \ldots, M) \in I\!R^M$ and $x \in \{0, \ldots, M\}$,

$$\begin{aligned}
\liminf_{n\to\infty} F_{V_n}(x) &= \liminf_{n\to\infty} \Pr\left(\sum_{m\in\mathcal{H}_0} \mathrm{I}\left(T_n(m) > c(m)\right) \leq x\right) \\
&\geq \liminf_{n\to\infty} \Pr\left(\sum_{m\in\mathcal{H}_0} \mathrm{I}\left(\widetilde{Z}_n(m) > c(m)\right) \leq x\right) \\
&= \Pr\left(\sum_{m\in\mathcal{H}_0} \mathrm{I}\left(Z(m) > c(m)\right) \leq x\right) \\
&= F_{V_0}(x).
\end{aligned}$$

Finite sample test statistics null distribution. The finite sample results follow immediately by noting that, under Equation (2.17), $Z_n(m) \geq T_n(m)$ for $m \in \mathcal{H}_0$.

□

Remarks

1. **Role of null shift values λ_0.** The construction of the null distribution Q_0 in Theorem 2.2 is inspired by joint null domination Assumption jtNDT, for the $\mathcal{H}_0$-specific subvector of test statistics $(T_n(m) : m \in \mathcal{H}_0)$. The purpose of the null shift values $\lambda_0(m)$ is to generate $\mathcal{H}_0$-specific statistics $(Z_n(m) : m \in \mathcal{H}_0)$ that are asymptotically stochastically greater than the original test statistics $(T_n(m) : m \in \mathcal{H}_0)$. Thus, for one-sided rejection regions of the form $\mathcal{C}_n(m) = (c_n(m), +\infty)$, the number of Type I errors V_0, under the null distribution Q_0, is asymptotically stochastically

greater than the number of Type I errors V_n, under the true distribution Q_n. The null distribution Q_0 therefore satisfies asymptotic null domination Assumption NDV, for the number of Type I errors, and also Θ-specific asymptotic null domination Assumption ND$\boldsymbol{\Theta}$, for any Type I error rate mapping Θ that meets monotonicity Assumption MΘ and continuity Assumption CΘ at F_{V_0}.

2. **Role of null scale values $\boldsymbol{\tau_0}$.** In contrast, the null scale values $\tau_0(m)$ are not needed for Type I error control. The purpose of $\tau_0(m)$ is to avoid a degenerate null distribution and infinite cut-offs for the false null hypotheses ($m \in \mathcal{H}_1$), an important property for power considerations. This scaling is needed, in particular, for F-statistics that have asymptotically infinite means and variances for non-local alternative hypotheses (Section 2.7).
3. **Estimation of null values $\boldsymbol{\lambda_0}$ and $\boldsymbol{\tau_0}$.** The null values $\lambda_0(m)$ and $\tau_0(m)$ only depend on the marginal distributions of the test statistics $T_n(m)$ for the true null hypotheses $\mathcal{H}_0$ and are generally known from single hypothesis testing. For instance, for the test of single-parameter null hypotheses using t-statistics, the null values are $\lambda_0(m) = 0$ and $\tau_0(m) = 1$ (Section 2.6). For testing the equality of K population mean vectors using F-statistics, the null values are $\lambda_0(m) = 1$ and $\tau_0(m) = 2/(K-1)$, under the assumption of equal variances in the different populations (Section 2.7). More generally, the null values $\lambda_0(m)$ and $\tau_0(m)$ may depend on the unknown data generating distribution P, as is the case for F-statistics when population variances are unequal (Equation (2.54)). In such a situation, one may replace the parameters $\lambda_0(m)$ and $\tau_0(m)$ by consistent estimators thereof.
4. ***t*-statistics: Gaussian null distribution.** For a broad class of testing problems, such as the test of single-parameter null hypotheses using t-statistics, the null distribution $Q_0 = Q_0(P)$ is an M-variate Gaussian distribution, with mean vector zero and covariance matrix $\sigma^* = \Sigma^*(P)$, that is, $Q_0 = \mathrm{N}(0, \sigma^*)$ (Section 2.6). For tests where the parameter of interest is the M-dimensional mean vector $\Psi(P) = \psi = \mathrm{E}[X]$, the estimator ψ_n is simply the M-vector of empirical means and $\sigma^* = \Sigma^*(P) = \mathrm{Cor}[X]$ is the correlation matrix of $X \sim P$, that is, $Q_0(P) = \mathrm{N}(0, \mathrm{Cor}[X])$. More generally, for an asymptotically linear estimator ψ_n, $\Sigma^*(P)$ is the correlation matrix of the vector influence curve. This situation covers standard one-sample and two-sample t-statistics for tests of means, but also test statistics for correlation coefficients and regression coefficients in linear and non-linear models.
5. ***F*-statistics: Gaussian quadratic form null distribution.** For testing the equality of K population mean vectors using F-statistics, an F-statistic-specific null distribution Q_0^F may be defined as the joint distribution of an M-vector of quadratic forms of Gaussian random variables (Section 2.7).

6. **Estimation of the test statistics null distribution.** In practice, the test statistics null distribution $Q_0 = Q_0(P)$ is unknown, as it depends on the unknown data generating distribution P. As detailed in Section 2.3.2, below, resampling procedures, such as the bootstrap procedures proposed in Dudoit et al. (2004b), van der Laan et al. (2004a), and Pollard and van der Laan (2004), may be used to conveniently obtain consistent estimators Q_{0n} of the null distribution Q_0 and of the corresponding test statistic cut-offs and adjusted p-values.

2.3.2 Bootstrap estimation of the test statistics null distribution

As noted above, the test statistics null distribution $Q_0 = Q_0(P)$ proposed in Theorem 2.2 depends on the typically unknown data generating distribution P. Although in some cases marginal test statistics null distributions may be known from single hypothesis testing, the dependence structure among the test statistics is usually unknown. In practice, one therefore needs to estimate the joint null distribution Q_0.

Consistent estimators Q_{0n} of the test statistics null distribution Q_0 and of the corresponding test statistic cut-offs and adjusted p-values may be obtained according to the following three main approaches: (i) general direct bootstrap estimation; (ii) test statistic-specific estimation (e.g., for t-statistics, χ^2-statistics, F-statistics); (iii) data generating null distribution estimation.

Given an estimator Q_{0n} of the null distribution Q_0, Procedures 4.20 and 4.21 provide algorithms for estimating cut-offs and adjusted p-values for $\Theta(F_{V_n})$-controlling joint single-step common-quantile Procedure 4.1 and common-cut-off Procedure 4.2, respectively. Similar algorithms are proposed in Procedure 5.15 for FWER-controlling joint step-down maxT Procedure 5.1 and minP Procedure 5.6.

General direct bootstrap estimation

As discussed below, bootstrap procedures provide a very general approach for obtaining consistent estimators of the test statistics null distribution Q_0 proposed in Theorem 2.2. The method may be summarized as follows and is illustrated in Figure 2.1.

1. Given B bootstrap samples of the data $\mathcal{X}_n$, obtain an $M \times B$ matrix of test statistics, $\mathbf{T}_n^B = \big(T_n^B(m,b)\big)$, with rows corresponding to the M null hypotheses and columns to the B bootstrap samples.
2. Estimate the expected values, $\mathrm{E}[T_n(m)]$, and variances, $\mathrm{Var}[T_n(m)]$, of the test statistics (under the true data generating distribution P) by taking row means and variances of the matrix $\mathbf{T}_n^B$.
3. Row-shift and scale the matrix of bootstrap test statistics $\mathbf{T}_n^B$, with the user-supplied null values $\lambda_0(m)$ and $\tau_0(m)$, to produce an $M \times B$ matrix $\mathbf{Z}_n^B = \big(Z_n^B(m,b)\big)$.

4. Estimate the null distribution Q_0 by the empirical distribution Q_{0n} of the B columns of matrix $\mathbf{Z}_n^B$.

The remainder of this section provides details on the (non-parametric or model-based) bootstrap estimation of the null distribution Q_0 of Theorem 2.2. Specifically, let $P_n^\star$ denote an estimator of the true data generating distribution P. For the *non-parametric bootstrap*, $P_n^\star$ is simply the empirical distribution P_n, that is, samples of size n are drawn at random, with replacement from the observed data $\mathcal{X}_n = \{X_i : i = 1, \ldots, n\}$. For the *model-based bootstrap*, $P_n^\star$ belongs to a model $\mathcal{M}$ for the data generating distribution P, such as a family of multivariate Gaussian distributions.

A *bootstrap sample* consists of n IID copies, $\mathcal{X}_n^\# \equiv \{X_i^\# : i = 1, \ldots, n\}$, of a random variable $X^\# \sim P_n^\star$. Denote the M-vector of test statistics computed from such a bootstrap sample by $T_n^\# = (T_n^\#(m) : m = 1, \ldots, M)$. The null distribution Q_0 proposed in Theorem 2.2 can be estimated by the distribution of the null shift and scale-transformed bootstrap test statistics,

$$Z_n^\#(m) \equiv \sqrt{\min\left\{1, \frac{\tau_0(m)}{\mathrm{Var}_{P_n^\star}[T_n{}^\#(m)]}\right\}}\,(T_n^\#(m) - \mathrm{E}_{P_n^\star}[T_n{}^\#(m)]) + \lambda_0(m). \tag{2.20}$$

In practice, one can only approximate the distribution of $Z_n^\# = (Z_n^\#(m) : m = 1, \ldots, M)$ by an empirical distribution over B bootstrap samples drawn from $P_n^\star$, as described next in Procedure 2.3.

Procedure 2.3. [Bootstrap estimation of the null shift and scale-transformed test statistics null distribution]

1. Generate B bootstrap samples, $\mathcal{X}_n^b \equiv \{X_i^b : i = 1, \ldots, n\}$, $b = 1, \ldots, B$. For the bth sample, the X_i^b, $i = 1, \ldots, n$, are n IID copies of a random variable $X^\# \sim P_n^\star$.
2. For each bootstrap sample $\mathcal{X}_n^b$, compute an M-vector of test statistics, $T_n^B(\cdot, b) = (T_n^B(m, b) : m = 1, \ldots, M)$, that can be arranged in an $M \times B$ matrix, $\mathbf{T}_n^B = \big(T_n^B(m, b) : m = 1, \ldots, M;\ b = 1, \ldots, B\big)$, with rows corresponding to the M null hypotheses and columns to the B bootstrap samples.
3. Compute row means and variances of the matrix $\mathbf{T}_n^B$, to yield estimators of the means, $\mathrm{E}[T_n(m)]$, and variances, $\mathrm{Var}[T_n(m)]$, of the test statistics under the true data generating distribution P. That is, compute

$$\mathrm{E}[T_n^B(m,\cdot)] \equiv \frac{1}{B}\sum_{b=1}^{B} T_n^B(m,b), \tag{2.21}$$

$$\mathrm{Var}[T_n^B(m,\cdot)] \equiv \frac{1}{B}\sum_{b=1}^{B}(T_n^B(m,b) - \mathrm{E}[T_n^B(m,\cdot)])^2.$$

4. Obtain an $M \times B$ matrix, $\mathbf{Z}_n^B = (Z_n^B(m,b) : m=1,\ldots,M; b=1,\ldots,B)$, of null shift and scale-transformed bootstrap test statistics $Z_n^B(m,b)$, as in Theorem 2.2, by row-shifting and scaling the matrix $\mathbf{T}_n^B$ using the bootstrap estimators of $\mathrm{E}[T_n(m)]$ and $\mathrm{Var}[T_n(m)]$ and the user-supplied null values $\lambda_0(m)$ and $\tau_0(m)$. That is, define

$$Z_n^B(m,b) \equiv \sqrt{\min\left\{1, \frac{\tau_0(m)}{\mathrm{Var}[T_n^B(m,\cdot)]}\right\}} \times \left(T_n^B(m,b) - \mathrm{E}[T_n^B(m,\cdot)]\right) + \lambda_0(m). \tag{2.22}$$

5. The bootstrap estimator Q_{0n} of the null distribution Q_0 from Theorem 2.2 is the empirical distribution of the B columns $\{Z_n^B(\cdot,b) : b = 1,\ldots,B\}$ of matrix $\mathbf{Z}_n^B$.

For one-sided rejection regions of the form $\mathcal{C}_n(m) = (c_n(m), +\infty)$, bootstrap estimators of the unadjusted p-values $P_{0n}(m)$ may be obtained from the matrix $\mathbf{Z}_n^B = \left(Z_n^B(m,b)\right)$ by recording, for each row m, the proportion of null shift and scale-transformed bootstrap test statistics $Z_n^B(m,b)$ that are greater than or equal to the observed test statistic $T_n(m)$ (Section 1.2.12). That is,

$$P_{0n}(m) = \frac{1}{B}\sum_{b=1}^{B} \mathrm{I}\left(Z_n^B(m,b) \geq T_n(m)\right), \qquad m = 1,\ldots,M. \tag{2.23}$$

Figures 2.1 and 2.2 provide, respectively, graphical summaries of the bootstrap estimation of the null distribution Q_0 and of the corresponding unadjusted p-values $P_{0n}(m)$.

There is no obvious general recommendation for the number of bootstrap samples B. However, note that bootstrap unadjusted p-values are discrete tail probabilities, with steps of size $1/B$. Thus, for estimating very small p-values (e.g., of the order of 10^{-9}), one clearly needs a very large B in order to get enough resolution in the tails. In addition, according to the definition in Equation (2.23), unadjusted p-values are often zero, even for moderate numbers of bootstrap samples B. In order to deal with the discreteness of the bootstrap distribution, the marginal null distributions $Q_{0n,m}$ obtained from the matrix $\mathbf{Z}_n^B$ may be replaced by Gaussian approximations or smoothed (e.g., using kernel density estimation methods). Specific algorithms for accurate estimation of tail probabilities are beyond the scope of this book. In general, the user needs to find a balance between estimation accuracy and computational cost.

Test statistic-specific estimation: *t*-statistics and *F*-statistics

For certain types of test statistics T_n (e.g., t-statistics, χ^2-statistics, F-statistics) one may exploit the known parametric form of the null distribution Q_0 of Theorem 2.2. An advantage of test statistic-specific parametric estimation approaches, such as those discussed in Sections 2.6 and 2.7, is that they yield continuous null distributions, which do not suffer from the discreteness of the non-parametric bootstrap null distribution described above.

t-statistics

As detailed in Section 2.6, for the test of single-parameter null hypotheses using t-statistics, a t-statistic-specific null distribution $Q_0^t = Q_0^t(P)$ is the M-variate Gaussian distribution $\mathrm{N}(0, \sigma^*)$, where $\sigma^* = \Sigma^*(P)$ is the correlation matrix of the M-dimensional vector influence curve, $IC(X|P) = (IC(X|P)(m) : m = 1, \ldots, M)$, for an asymptotically linear estimator ψ_n of the parameter M-vector ψ (Section 1.2.5).

In this case, one can estimate Q_0^t by $Q_{0n}^t = \mathrm{N}(0, \sigma_n^*)$, where $\sigma_n^* = \hat{\Sigma}^*(P_n)$ is a consistent estimator of the correlation matrix σ^*. For example, one could use the correlation matrix σ_n^* corresponding to the following estimator of the $M \times M$ influence curve covariance matrix,

$$\sigma_n = \hat{\Sigma}(P_n) = \frac{1}{n}\sum_{i=1}^{n} IC_n(X_i) IC_n^{\top}(X_i), \tag{2.24}$$

where $IC_n(X) = (IC_n(X)(m) : m = 1, \ldots, M)$ is an estimator of the M-vector influence curve $IC(X|P)$.

Influence curves can be derived straightforwardly for simple parameters such as means. For example, when estimating the mean vector $\psi = \mathrm{E}[X]$, for a random M-vector $X \sim P$, using the corresponding empirical mean vector $\psi_n = \bar{X}_n$, the influence curves are $IC(X|P)(m) = X(m) - \psi(m)$ and corresponding estimators are $IC_n(X)(m) = X(m) - \psi_n(m)$, where $\psi_n(m) = \bar{X}_n(m) = \sum_i X_i(m)/n$, $m = 1, \ldots, M$. Then, σ_n^* is simply the empirical correlation matrix. Influence curves for estimators of correlation coefficients and regression coefficients are given in Section 2.6.

In cases where the influence curves are not readily available, the correlation matrix σ^* may be estimated with the bootstrap.

F-statistics

As detailed in Section 2.7, for testing the equality of K population mean vectors using F-statistics, an F-statistic-specific null distribution

$Q_0^F = Q_0^F(P_1, \ldots, P_K)$ can be defined in terms of a simple quadratic function of K independent Gaussian M-vectors, $Y_k \sim \mathrm{N}(0, \sigma_k)$, where $\sigma_k = \Sigma(P_k)$ denotes the covariance matrix for the kth population, $k = 1, \ldots, K$.

An estimator Q_{0n}^F of the null distribution Q_0^F can be obtained by estimating each population covariance matrix σ_k by the corresponding empirical covariance matrix or by using the bootstrap.

Data generating null distribution estimation

In certain testing problems, one may define a test statistics null distribution $Q_n(P_0)$, in terms of a data generating distribution P_0 that satisfies the complete null hypothesis $H_0^C = \prod_{m=1}^M H_0(m)$ that all M null hypotheses are true. Such a null distribution may be estimated by $Q_{0n} = Q_n(P_{0n})$, where, for example, P_{0n} is a bootstrap- or permutation-based estimator of P_0.

Test statistics null distributions based on bootstrap and permutation data generating distributions are discussed in Section 2.9. Parameter-specific bootstrap data generating null distributions are described in Chapter 8 for tests concerning regression coefficients and correlation coefficients (Procedures 8.4 and 8.6, respectively).

However, as discussed in Pollard et al. (2005a) and Pollard and van der Laan (2004), approaches based on a data generating null distribution can fail in important testing problems, as the assumed null distribution $Q_{n,\mathcal{H}_0}(P_0)$ and the true distribution $Q_{n,\mathcal{H}_0}(P)$, of the $\mathcal{H}_0$-specific test statistics $(T_n(m) : m \in \mathcal{H}_0)$, may have different dependence structures and, as a result, may violate the required null domination condition for the Type I error rate (Assumption $\mathsf{ND}\Theta$, in Step 1 of the road map of Procedure 2.1).

Indeed, the simulation studies of Chapter 8 show that bootstrap data generating null distributions can lead to severely anti-conservative (bootstrapping residuals for testing regression coefficients) or conservative (independent bootstrap for testing correlation coefficients) procedures.

2.4 Null quantile-transformed test statistics null distribution

Following van der Laan and Hubbard (2006), our second proposal for a test statistics null distribution is the asymptotic distribution of a vector of null quantile-transformed test statistics, based on user-supplied marginal test statistics null distributions. Because this promising approach represents a very recent development in our ongoing research on multiple testing, this book only introduces the main features of the null quantile-transformed null distribution. The reader is referred to van der Laan and Hubbard (2006) for formal theorems and proofs, a detailed treatment of tests based on t-statistics and χ^2-statistics, simulation studies, and an application to tests of association between non-Hodgkin lymphoma (NHL) subclass and single nucleotide polymorphisms (SNP) in the ghrelin (GHRL) and neuropeptide Y (NPY) genes.

This latest construction has the advantage that the marginal test statistics null distributions may be set to the optimal (i.e., most powerful) null distributions one would use in single hypothesis testing (e.g., permutation marginal null distributions, Gaussian or other parametric marginal null distributions). The preliminary results in van der Laan and Hubbard (2006) indeed illustrate that, for finite samples, the new null quantile-transformed null distribution provides more accurate Type I error control and is more powerful than the null shift and scale-transformed null distribution of Section 2.3.

2.4.1 Explicit construction for the test statistics null distribution

Marginal null domination conditions for the $\mathcal{H}_0$-specific test statistics $(T_n(m) : m \in \mathcal{H}_0)$

The main ingredients of the new null quantile-transformed test statistics null distribution are user-supplied marginal test statistics null distributions $q_{0,m}$, $m = 1, \ldots, M$, that satisfy the following *marginal null domination* condition[3]. For each $m \in \mathcal{H}_0$ and $z \in I\!R$,

$$\begin{aligned} Q_{n,m}(z) &\geq q_{0,m}(z) \qquad \text{[finite sample control]} \\ \liminf_{n\to\infty} Q_{n,m}(z) &\geq q_{0,m}(z) \qquad \text{[asymptotic control].} \end{aligned} \tag{mgNDT}$$

That is, the test statistics $(T_n(m) : m \in \mathcal{H}_0)$, for the true null hypotheses $\mathcal{H}_0$, are marginally stochastically greater under the null distributions $q_{0,m}$ than under the true distributions $Q_{n,m}$. Note that the above marginal null domination Assumption mgNDT is implied by the stronger joint null domination Assumption jtNDT.

Finite sample test statistics null distribution

Given marginal null distributions $q_{0,m}$, $m = 1, \ldots, M$, that satisfy marginal null domination Assumption mgNDT, the proposed finite sample joint null distribution is based on the *generalized quantile-quantile function transformation* of Yu and van der Laan (2002). Specifically, let $\breve{Q}_{0,n} = \breve{Q}_{0,n}(P)$ denote the joint distribution of the M-vector of *null quantile-transformed test statistics* $\breve{Z}_n = (\breve{Z}_n(m) : m = 1, \ldots, M)$ defined as

$$\breve{Z}_n(m) \equiv q_{0,m}^{-1} Q_{n,m}^{\Delta}(T_n(m)), \qquad m = 1, \ldots, M, \tag{2.25}$$

[3] N.B. In practice, user-supplied marginal null distributions, such as permutation distributions, depend on the sample size n. However, for simplicity, references to the sample size n are omitted from the notation $q_{0,m}$.

where $Q^{\Delta}_{n,m}(z) \equiv \Delta Q_{n,m}(z) + (1-\Delta)Q_{n,m}(z^-)$ and the random variable Δ is uniformly distributed on the interval $[0,1]$, independently of the data $\mathcal{X}_n$.

One can easily verify that the marginal distributions $\breve{Q}_{0,n,m}$, corresponding to the proposed joint null distribution $\breve{Q}_{0,n}$, are indeed equal to the user-supplied marginal null distributions $q_{0,m}$. For continuous user-supplied marginal null distributions $q_{0,m}$ and continuous true marginal distributions $Q_{n,m}$, one has $Q^{\Delta}_{n,m}(z) = Q_{n,m}(z)$ for each $z \in I\!R$ and, hence,

$$\begin{aligned}\breve{Q}_{0,n,m}(z) &= \Pr\left(\breve{Z}_n(m) \le z\right)\\ &= \Pr\left(q^{-1}_{0,m}Q_{n,m}(T_n(m)) \le z\right)\\ &= \Pr\left(Q_{n,m}(T_n(m)) \le q_{0,m}(z)\right)\\ &= q_{0,m}(z),\end{aligned}$$

where the last equality follows from Proposition 1.2.

In cases where the marginal distributions $Q_{n,m}$ and $q_{0,m}$ are not necessarily continuous, Lemma 2.4 of Yu and van der Laan (2002) ensures that the marginal distributions $\breve{Q}_{0,n,m}$ are indeed equal to the user-supplied marginal null distributions $q_{0,m}$.

Result 1 in van der Laan and Hubbard (2006) establishes that the finite sample joint null distribution $\breve{Q}_{0,n}$ satisfies null domination Assumption NDV for the number of Type I errors. That is, for each $c = (c(m) : m = 1,\ldots,M) \in I\!R^M$ and $x \in \{0,\ldots,M\}$,

$$\Pr\left(V(c|\breve{Q}_{0,n}) \le x\right) - \Pr\left(V(c|Q_n) \le x\right) \le 0 \tag{2.26}$$

$$\limsup_{n\to\infty}\left(\Pr\left(V(c|\breve{Q}_{0,n}) \le x\right) - \Pr\left(V(c|Q_n) \le x\right)\right) \le 0.$$

In other words, the number of Type I errors $V_{0,n} = V(c|\breve{Q}_{0,n}) = \sum_{m\in\mathcal{H}_0} \mathrm{I}\left(\breve{Z}_n(m) > c(m)\right)$, under the null distribution $\breve{Q}_{0,n}$, is stochastically greater than the number of Type I errors $V_n = V(c|Q_n) = \sum_{m\in\mathcal{H}_0} \mathrm{I}\left(T_n(m) > c(m)\right)$, under the true distribution Q_n. Null domination Assumption NDΘ for the Type I error rate follows for mappings Θ that satisfy monotonicity Assumption MΘ and uniform continuity Assumption CΘ.

Asymptotic test statistics null distribution

As in van der Laan and Hubbard (2006), further assume that the finite sample joint null distribution $\breve{Q}_{0,n} = \breve{Q}_{0,n}(P)$ converges weakly to an asymptotic joint null distribution $\breve{Q}_0 = \breve{Q}_0(P)$.

Result 2 in van der Laan and Hubbard (2006) is an analogue for $\breve{Q}_0$ of Result 1 for $\breve{Q}_{0,n}$. That is, the asymptotic joint null distribution $\breve{Q}_0$ satisfies null domination Assumption NDV for the number of Type I errors.

In general, proofs of null domination properties for the new null quantile-transformed null distribution are similar to those for the original null shift and scale-transformed null distribution (e.g., Theorem 2.2).

2.4.2 Bootstrap estimation of the test statistics null distribution

As for our original null shift and scale-transformed test statistics null distribution $Q_0 = Q_0(P)$ (Section 2.3), neither the finite sample null distribution $\breve{Q}_{0,n} = \breve{Q}_{0,n}(P)$ nor the asymptotic null distribution $\breve{Q}_0 = \breve{Q}_0(P)$ is known, as they both depend on the true, unknown data generating distribution P. van der Laan and Hubbard (2006) propose in their Section 2 a bootstrap procedure, similar to Procedure 2.3, for estimating the asymptotic null distribution $\breve{Q}_0$.

Procedure 2.4. [Bootstrap estimation of the null quantile-transformed test statistics null distribution]

1. Generate B bootstrap samples, $\mathcal{X}_n^b \equiv \{X_i^b : i = 1, \ldots, n\}$, $b = 1, \ldots, B$. For the bth sample, the X_i^b, $i = 1, \ldots, n$, are n IID copies of a random variable $X^\# \sim P_n^\star$.
2. For each bootstrap sample $\mathcal{X}_n^b$, compute an M-vector of test statistics, $T_n^B(\cdot, b) = (T_n^B(m, b) : m = 1, \ldots, M)$, that can be arranged in an $M \times B$ matrix, $\mathbf{T}_n^B = (T_n^B(m, b) : m = 1, \ldots, M;\ b = 1, \ldots, B)$, with rows corresponding to the M null hypotheses and columns to the B bootstrap samples.
3. Define M bootstrap marginal cumulative distribution functions $Q_{n,m}^B$, as the empirical CDFs of the rows of matrix $\mathbf{T}_n^B$, that is,
$$Q_{n,m}^B(z) \equiv \frac{1}{B} \sum_{b=1}^{B} \mathrm{I}\left(T_n^B(m, b) \leq z\right). \tag{2.27}$$
4. Obtain an $M \times B$ matrix, $\mathbf{Z}_n^B = (Z_n^B(m, b) : m = 1, \ldots, M;\ b = 1, \ldots, B)$, of null quantile-transformed bootstrap test statistics $Z_n^B(m, b)$, defined as
$$Z_n^B(m, b) \equiv q_{0,m}^{-1} Q_{n,m}^{B,\Delta}(T_n^B(m, b)), \tag{2.28}$$
where $Q_{n,m}^{B,\Delta}(z) \equiv \Delta Q_{n,m}^B(z) + (1 - \Delta) Q_{n,m}^B(z^-)$ and the random variable Δ is uniformly distributed on the interval $[0, 1]$, independently of the data $\mathcal{X}_n$.
5. The bootstrap estimator $\breve{Q}_{0n}$ of the null distribution $\breve{Q}_0$ is the empirical distribution of the B columns $\{Z_n^B(\cdot, b) : b = 1, \ldots, B\}$ of matrix $\mathbf{Z}_n^B$.

From Lemma 2.4 in Yu and van der Laan (2002), the generalized quantile-quantile function transformation $q_{0,m}^{-1}Q_{n,m}^{B,\Delta}(z)$ ensures that the margins $\breve{Q}_{0n,m}$, of the estimator $\breve{Q}_{0n}$ based on a finite number B of bootstrap samples, are equal to the user-supplied marginal null distributions $q_{0,m}$.

As discussed in Section 2.3.2, in the context of the null shift and scale-transformed null distribution, one could also envisage estimation approaches that are test statistic-specific (e.g., for t-statistics, χ^2-statistics, F-statistics) or based on a data generating null distribution. The reader is referred to Section 4 in van der Laan and Hubbard (2006) for a detailed treatment of null distributions for tests based on t-statistics and χ^2-statistics.

2.4.3 Comparison of null shift and scale-transformed and null quantile-transformed null distributions

This section compares our two main constructions for a test statistics null distribution. Recall from Section 2.3 that the first null distribution $Q_0 = Q_0(P)$, proposed in Dudoit et al. (2004b) and van der Laan et al. (2004a), is defined as the asymptotic distribution of the M-vector $Z_n = (Z_n(m) : m = 1, \ldots, M)$ of null shift and scale-transformed test statistics. That is,

$$Z_n(m) = \sqrt{\min\left\{1, \frac{\tau_0(m)}{\mathrm{Var}[T_n(m)]}\right\}}\,(T_n(m) - \mathrm{E}[T_n(m)]) + \lambda_0(m),$$

where $\lambda_0(m)$ and $\tau_0(m)$ are, respectively, user-supplied upper bounds for the means and variances of the $\mathcal{H}_0$-specific test statistics.

In contrast, the new null distribution $\breve{Q}_0 = \breve{Q}_0(P)$ of van der Laan and Hubbard (2006) is defined as the asymptotic distribution of the M-vector $\breve{Z}_n = (\breve{Z}_n(m) : m = 1, \ldots, M)$ of null quantile-transformed test statistics. That is,

$$\breve{Z}_n(m) = q_{0,m}^{-1}Q_{n,m}^{\Delta}(T_n(m)),$$

where $q_{0,m}$ are user-supplied marginal test statistics null distributions.

1. **Main ingredients: Null shift and scale values and null quantiles.** While our first proposal requires M-vectors of null values $\lambda_0 \in I\!R^M$ and $\tau_0 \in I\!R^{+M}$, so that $\limsup_n \mathrm{E}[T_n(m)] \leq \lambda_0(m)$ and $\limsup_n \mathrm{Var}[T_n(m)] \leq \tau_0(m)$ for $m \in \mathcal{H}_0$, the new proposal of van der Laan and Hubbard (2006) relies on marginal null distributions $q_{0,m}$ that dominate the true marginal distributions $Q_{n,m}$, i.e., satisfy marginal null domination Assumption mgNDT.
2. **$\mathcal{H}_0$-specific joint null distributions.** If the true marginal distributions $Q_{n,m}$, of the test statistics $T_n(m)$ for the true null hypotheses $m \in \mathcal{H}_0$, converge weakly (up to a location shift) to the corresponding user-supplied marginal null distributions $q_{0,m}$, then the two $\mathcal{H}_0$-specific joint null distributions $Q_{0,\mathcal{H}_0}$ and $\breve{Q}_{0,\mathcal{H}_0}$ coincide.

3. **$\mathcal{H}_1$-specific joint null distributions.** In general, for the false null hypotheses $\mathcal{H}_1$, the null-transformed test statistics $Z_n(m)$ and $\breve{Z}_n(m)$ can have very different finite sample and asymptotic marginal distributions. In particular, whereas the marginal distributions of $\breve{Q}_0$ coincide with the user-supplied marginal null distributions (i.e., $\breve{Q}_{0,m} = q_{0,m}$), the marginal distributions of Q_0 do not necessarily have this property. Hence, the $\mathcal{H}_1$-specific joint null distributions $Q_{0,\mathcal{H}_1}$ and $\breve{Q}_{0,\mathcal{H}_1}$ could in principle be very different and thus lead to procedures with different power properties.
4. **Estimation of the test statistics null distributions.** In practice, both test statistics null distributions $Q_0 = Q_0(P)$ and $\breve{Q}_0 = \breve{Q}_0(P)$ are unknown, as they depend on the unknown data generating distribution P. Similar bootstrap procedures may be used to obtain consistent estimators Q_{0n} and $\breve{Q}_{0n}$, of Q_0 and $\breve{Q}_0$, respectively (Procedures 2.3 and 2.4). However, bootstrap estimators $\breve{Q}_{0n}$ of the null quantile-transformed null distribution $\breve{Q}_0$ are expected to be more efficient than bootstrap estimators Q_{0n} of the null shift and scale-transformed null distribution Q_0. To see this, suppose that the two $\mathcal{H}_0$-specific joint null distributions $Q_{0,\mathcal{H}_0}$ and $\breve{Q}_{0,\mathcal{H}_0}$ coincide. The bootstrap estimator of $\breve{Q}_0$ is based on a model where all marginal distributions are given, whereas the bootstrap estimator of Q_0 ignores this information and considers a larger model with unspecified marginal distributions. As a result, the bootstrap marginal distributions $Q_{0n,m}$ are subject to finite sample variability and typically differ from the user-supplied marginal distributions $q_{0,m}$.
5. **Known optimal marginal null distributions.** The new null quantile-transformed null distribution is particularly appealing when one has available optimal marginal null distributions $q_{0,m}$ for single hypothesis testing. For example, consider a data structure $X = (X(m) : m = 1, \ldots, M+1)$, where $(X(m) : m = 1, \ldots, M)$ is an M-dimensional covariate/genotype vector and $Y = X(M+1)$ is a univariate outcome/phenotype. The covariates/genotypes could correspond to M microarray gene expression measures and the outcome/phenotype to a (censored) survival time or a tumor class. Suppose one wishes to test the M null hypotheses $H_0(m)$ of independence between the covariates $X(m)$ and the outcome $Y = X(M+1)$, $m = 1, \ldots, M$, based on an M-vector of arbitrary test statistics $T_n = (T_n(m) : m = 1, \ldots, M)$. Then, one can set the marginal null distributions $q_{0,m}$ equal to the permutation distributions of the corresponding test statistics $T_n(m)$. One knows from single hypothesis testing that if the null hypothesis $H_0(m)$ is true, then the permutation distribution of $T_n(m)$ is (exactly) equal to the true conditional distribution of $T_n(m)$, given the marginal empirical distributions of $X(m)$ and Y. In the special case of the test of single-parameter null hypotheses based on t-statistics, one could use standard normal marginal null distributions, that is, set $q_{0,m} = \Phi$, where Φ is the N(0, 1) CDF.

van der Laan and Hubbard (2006) argue in their Section 3 that Type I error control results proved in our earlier articles for the original null shift and scale-transformed test statistics null distribution Q_0 and its bootstrap estimators Q_{0n} also hold for the new null quantile-transformed test statistics null distribution. Specifically, the null quantile-transformed test statistics null distribution $\breve{Q}_0$ and its bootstrap estimators $\breve{Q}_{0n}$ provide Type I error control for: $\Theta(F_{V_n})$-controlling joint single-step common-cut-off and common-quantile procedures (Chapter 4); FWER-controlling joint step-down common-cut-off (maxT) and common-quantile (minP) procedures (Chapter 5); gTP-controlling (marginal/joint single-step/stepwise) augmentation multiple testing procedures (Chapter 6); gTP-controlling joint resampling-based empirical Bayes procedures (Chapter 7). The main point is that both test statistics null distributions satisfy joint null domination Assumption jtNDT for the $\mathcal{H}_0$-specific test statistics.

Section 4 in van der Laan and Hubbard (2006) is analogous to Sections 2.6 and 2.7, below, in that it examines properties of the null quantile-transformed null distribution for two types of testing problems: the test of single-parameter null hypotheses using t-statistics (e.g., tests of means, correlation coefficients, regression coefficients) and the test of multiple-parameter null hypotheses using χ^2-statistics.

In summary, either test statistics null distribution Q_0 or $\breve{Q}_0$ (or consistent estimators thereof) may be used in any of the multiple testing procedures proposed in Chapters 3–7 of this book, as they both satisfy the key property of joint null domination for the $\mathcal{H}_0$-specific test statistics (Assumption jtNDT). In particular, Type I error control does not rely on restrictive assumptions such as subset pivotality and holds for general data generating distributions (with arbitrary dependence structures among variables), null hypotheses (defined in terms of submodels for the data generating distribution), and test statistics (e.g., t-statistics, χ^2-statistics, F-statistics). The newly proposed null quantile-transformed null distribution has the additional advantage that it allows the user to select optimal marginal null distributions and hence tends to outperform the original null shift and scale-transformed null distribution. Unless stated otherwise, the simpler notation Q_0 and Q_{0n} refers to either null distribution.

2.5 Null distribution for transformations of the test statistics

2.5.1 Null distribution for transformed test statistics

Suppose one is interested in deriving rejection regions for an M-vector of test statistics $T_n^\ell = (T_n^\ell(m) : m = 1, \ldots, M)$, defined as *transformations* of the original test statistics $T_n = (T_n(m) : m = 1, \ldots, M)$, by $T_n^\ell(m) \equiv \ell_m(T_n(m))$, in terms of a collection of M functions $\ell_m : I\!R \rightarrow I\!R$.

The special case of the absolute value function ($\ell(z) = |z|$) is discussed in general terms in Section 2.5.2 and also in Section 4.5, in the context of single-step common-cut-off and common-quantile procedures.

As in Equation (2.2), given a random M-vector $Z = (Z(m) : m = 1, \ldots, M)$, with joint distribution Q, and a collection of M rejection regions $\mathcal{C} = \{\mathcal{C}(m) : m = 1, \ldots, M\}$, denote the numbers of rejected hypotheses and Type I errors for the transformed test statistics $\ell_m(Z(m))$ by

$$R^\ell(\mathcal{C}|Q) \equiv \sum_{m=1}^{M} \mathrm{I}\left(\ell_m(Z(m)) \in \mathcal{C}(m)\right) \tag{2.29}$$

and

$$V^\ell(\mathcal{C}|Q) \equiv \sum_{m \in \mathcal{H}_0} \mathrm{I}\left(\ell_m(Z(m)) \in \mathcal{C}(m)\right),$$

respectively. Also adopt the shorthand notation of Equation (2.3), for the special cases where Q corresponds to the true distribution Q_n and null distribution Q_0 for the original test statistics T_n,

$$\begin{aligned} R_n^\ell &\equiv R^\ell(\mathcal{C}|Q_n), & R_0^\ell &\equiv R^\ell(\mathcal{C}|Q_0), \\ V_n^\ell &\equiv V^\ell(\mathcal{C}|Q_n), & V_0^\ell &\equiv V^\ell(\mathcal{C}|Q_0). \end{aligned} \tag{2.30}$$

The proposition below specifies conditions under which deriving rejection regions for the transformed test statistics T_n^ℓ, based on a null distribution Q_0 for the original test statistics T_n, leads to proper Type I error control.

Proposition 2.5. [Null distribution for transformed test statistics] *Consider the simultaneous test of M null hypotheses $H_0(m)$, $m = 1, \ldots, M$, based on an M-vector of test statistics $T_n^\ell = (T_n^\ell(m) : m = 1, \ldots, M)$, defined as transformations of the original test statistics $T_n = (T_n(m) : m = 1, \ldots, M)$, by $T_n^\ell(m) = \ell_m(T_n(m))$, in terms of a collection of M functions $\ell_m : I\!R \rightarrow I\!R$. Let $Q_n = Q_n(P)$ and Q_0 denote, respectively, the true finite sample joint distribution of T_n and a null distribution that satisfies joint null domination Assumption jtNDT for the $\mathcal{H}_0$-specific subvector of test statistics $(T_n(m) : m \in \mathcal{H}_0)$.*

Scenario 1. *If the functions ℓ_m are continuous and non-decreasing, then joint null domination Assumption jtNDT for the original test statistics $(T_n(m) : m \in \mathcal{H}_0)$ implies joint null domination Assumption jtNDT for the transformed test statistics $(T_n^\ell(m) : m \in \mathcal{H}_0)$. Hence, for one-sided rejection regions of the form $\mathcal{C}(m) = (c(m), +\infty)$ for the transformed test statistics $T_n^\ell(m)$, null domination Assumption NDV is satisfied by the numbers of Type I errors V_n^ℓ and V_0^ℓ. If one further assumes that the Type I error rate mapping Θ meets monotonicity Assumption MΘ and continuity Assumption CΘ at $F_{V_0^\ell}$, then null domination Assumption NDΘ is* *satisfied by the Type I error rates* $\Theta(F_{V_0^\ell})$ *and* $\Theta(F_{V_n^\ell})$. *This means that one-sided rejection regions for the transformed test statistics* T_n^ℓ *may be derived based on the null distribution* Q_0 *for the original test statistics* T_n.

Scenario 2. *If joint null domination Assumption* jtNDT *holds with equality for the original test statistics* $(T_n(m) : m \in \mathcal{H}_0)$, *then it also holds with equality for the transformed test statistics* $(T_n^\ell(m) : m \in \mathcal{H}_0)$, *for any continuous functions* ℓ_m. *Hence, for any type of rejection regions* $\mathcal{C}(m)$ *for the transformed test statistics* $T_n^\ell(m)$, *null domination Assumption* NDV *is satisfied with equality by the numbers of Type I errors* V_n^ℓ *and* V_0^ℓ. *If one further assumes that the Type I error rate mapping* Θ *meets monotonicity Assumption* MΘ *and continuity Assumption* CΘ *at* $F_{V_0^\ell}$, *then null domination Assumption* **NDΘ** *is satisfied with equality by the Type I error rates* $\Theta(F_{V_0^\ell})$ *and* $\Theta(F_{V_n^\ell})$. *This means that any type of rejection regions for the transformed test statistics* T_n^ℓ *may be derived based on the null distribution* Q_0 *for the original test statistics* T_n.

The proof of this proposition is straightforward and is therefore omitted.

An alternative and more general approach for obtaining rejection regions for transformed test statistics T_n^ℓ would be to derive a null distribution Q_0^ℓ directly for T_n^ℓ, using the general constructions of Sections 2.3 and 2.4.

There is, however, a trade-off between generality and simplicity. For instance, consider the test of single-parameter null hypotheses using t-statistics T_n (Section 2.6). For the null shift and scale-transformed approach of Section 2.3, the null values are $\lambda_0(m) = 0$ and $\tau_0(m) = 1$ and the null distribution Q_0 for T_n is an M-variate Gaussian distribution, with mean vector zero and covariance matrix $\sigma^* = \Sigma^*(P)$ equal to the correlation matrix of the vector influence curve. For the transformed test statistics T_n^ℓ, the null shift and scale values are no longer 0 and 1 and the null distribution Q_0^ℓ is no longer Gaussian.

2.5.2 Example: Absolute value transformation

A special case of interest is the *absolute value* function, $\ell(z) = |z|$, which corresponds to *symmetric two-sided rejection regions* for the original test statistics $T_n(m)$: $\mathcal{C}_n(m;\alpha) = (-\infty, -c_n(m;\alpha)) \cup (c_n(m;\alpha), +\infty)$, for an M-vector of non-negative cut-offs $c_n(\alpha) = (c_n(m;\alpha) : m = 1, \ldots, M) \in I\!R^{+M}$. That is, for a MTP with nominal Type I error level α, the set of rejected null hypotheses is given by

$$\begin{aligned}\mathcal{R}^{||}(T_n, Q_0, \alpha) &= \{m : T_n(m) < -c_n(m;\alpha) \text{ or } T_n(m) > c_n(m;\alpha)\} \quad (2.31)\\ &= \{m : |T_n(m)| > c_n(m;\alpha)\}.\end{aligned}$$

Specifically, consider the two-sided test of single-parameter null hypotheses $H_0(m) = \mathrm{I}(\psi(m) = \psi_0(m))$ against alternative hypotheses $H_1(m) =$

$\mathrm{I}(\psi(m) \neq \psi_0(m))$, based on an M-vector of t-statistics, defined as in Section 2.6 by

$$T_n(m) = \sqrt{n}\frac{\psi_n(m) - \psi_0(m)}{\sigma_n(m)},$$

where ψ_n is an asymptotically linear estimator of the parameter ψ.

A similar argument as in the proof of Theorem 2.6 shows that

$$(T_n(m) : m \in \mathcal{H}_0) \stackrel{\mathcal{L}}{\Rightarrow} Q_{0,\mathcal{H}_0} = \mathrm{N}(0, \sigma^*_{\mathcal{H}_0}).$$

Hence, asymptotic joint null domination Assumption **jtNDT** is satisfied with equality for the $\mathcal{H}_0$-specific absolute t-statistics $(|T_n(m)| : m \in \mathcal{H}_0)$ and the null distribution $Q_0 = \mathrm{N}(0, \sigma^*)$ of Theorem 2.6. It follows from Proposition 2.5, Scenario 2, that asymptotic null domination Assumptions **NDV** and **NDΘ**, for the number of Type I errors and Type I error rate, are also satisfied with equality for any type of rejection regions for $|T_n(m)|$. Hence, as dictated by the three-step road map of Procedure 2.1, one has

$$\lim_{n\to\infty} \Theta(F_{V_n^{||}}) = \Theta(F_{V_0^{||}}) \leq \Theta(F_{R_0^{||}}) \leq \alpha. \quad (2.32)$$

Thus, multiple testing procedures based on absolute t-statistics $|T_n(m)|$ and any type of rejection regions $\mathcal{C}_n(m) = \mathcal{C}(m; T_n, Q_0, \alpha)$, derived under the Gaussian null distribution $Q_0 = \mathrm{N}(0, \sigma^*)$ of Theorem 2.6, do indeed provide the desired Type I error control. The special cases of single-step common-quantile Procedure 4.1 and common-cut-off Procedure 4.2 are discussed in detail in Section 4.5.

Note that, for the absolute value function and two-sided rejection regions, the stronger requirement of asymptotic *equality* of the test statistics true distribution $Q_{n,\mathcal{H}_0}$ and null distribution $Q_{0,\mathcal{H}_0}$ is essential, as the weaker domination property would only guarantee Type I error control for one of the tails.

2.5.3 Example: Null shift and scale and null quantile transformations

The random M-vectors of *null-transformed test statistics* Z_n and $\breve{Z}_n$ (Equations (2.15) and (2.25)), defining the null distributions proposed in Sections 2.3 and 2.4, correspond, respectively, to the following transformations,

$$\ell_{0,m}(z) \equiv \nu_{0,n}(m)\,(z - \mathrm{E}[T_n(m)]) + \lambda_0(m) \quad (2.33)$$

and

$$\breve{\ell}_{0,m}(z) \equiv q_{0,m}^{-1} Q_{n,m}^{\Delta}(z).$$

The null shift and scale functions $\ell_{0,m}$ are continuous and non-decreasing. For continuous marginal distributions $Q_{n,m}$ and $q_{0,m}$, the null quantile functions $\breve{\ell}_{0,m}$ are also continuous and non-decreasing. Thus, Scenario 1 in Proposition 2.5 applies to a broad range of testing problems.

2.5.4 Bootstrap estimation of the null distribution for transformed test statistics

Regarding the bootstrap estimation of rejection regions and adjusted p-values for MTPs based on transformed test statistics T_n^ℓ, one could first use general Procedure 2.3 or 2.4 (or a related procedure from Section 2.6 or 2.7) to derive a matrix $\mathbf{Z}_n^B = \left(Z_n^B(m,b)\right)$, of null-transformed bootstrap test statistics $Z_n^B(m,b)$, based on the original test statistics T_n. The null distribution Q_0, for the original test statistics T_n, is estimated by the empirical distribution Q_{0n} of the columns of matrix $\mathbf{Z}_n^B$. An estimated null distribution Q_{0n}^ℓ, for the transformed test statistics T_n^ℓ, is given by the empirical distribution of the columns of the transformed matrix $\ell(\mathbf{Z}_n^B) = \left(\ell_m(Z_n^B(m,b))\right)$.

Using Q_{0n}^ℓ to obtain rejection regions for the transformed test statistics T_n^ℓ leads to procedures that control the Type I error rate $\Theta(F_{V_n^\ell})$ under the two scenarios considered in Proposition 2.5.

For instance, bootstrap versions of single-step common-quantile Procedure 4.1 and common-cut-off Procedure 4.2 may be implemented as in Procedures 4.20 and 4.21, respectively, using transformed test statistics T_n^ℓ and the estimated null distribution Q_{0n}^ℓ.

2.6 Testing single-parameter null hypotheses based on t-statistics

2.6.1 Set-up and assumptions

In this section, we consider the one-sided test of M single-parameter null hypotheses $H_0(m) = \mathrm{I}\left(\psi(m) \leq \psi_0(m)\right)$ against alternative hypotheses $H_1(m) = \mathrm{I}\left(\psi(m) > \psi_0(m)\right)$, where $\Psi(P) = \psi = (\psi(m) : m = 1, \ldots, M)$ is an M-vector of real-valued parameters $\Psi(P)(m) = \psi(m)$.

The null hypotheses can be tested using *t-statistics*, defined as in Section 1.2.5 by

$$T_n(m) \equiv \sqrt{n}\frac{\psi_n(m) - \psi_0(m)}{\sigma_n(m)}, \tag{2.34}$$

where $\hat{\Psi}(P_n) = \psi_n = (\psi_n(m) : m = 1, \ldots, M)$ is an *asymptotically linear estimator* of the parameter M-vector $\Psi(P) = \psi$, with M-dimensional vector *influence curve* (IC) $IC(X|P) = (IC(X|P)(m) : m = 1, \ldots, M)$, such that

$$\psi_n(m) - \psi(m) = \frac{1}{n}\sum_{i=1}^{n} IC(X_i|P)(m) + o_P(1/\sqrt{n}), \tag{2.35}$$

and $\sigma_n^2(m)$ are consistent estimators of the variances $\sigma^2(m) = \sigma(m,m) = \mathrm{E}[IC^2(X|P)(m)]$, $m = 1, \ldots, M$. Let $Q_n = Q_n(P)$ denote the finite sample joint distribution of T_n, under the true, unknown data generating distribution P. Large values of the t-statistic $T_n(m)$ are assumed to provide evidence against the corresponding null hypothesis $H_0(m) = \mathrm{I}(\psi(m) \leq \psi_0(m))$, that is, tests are based on one-sided rejection regions of the form $\mathcal{C}_n(m) = (c_n(m), +\infty)$.

Next, we propose a *t-statistic-specific null distribution* Q_0^t that leads to asymptotic control of Type I error rates $\Theta(F_{V_n})$, defined as arbitrary parameters of the distribution of the number of Type I errors V_n.

2.6.2 Test statistics null distribution

Theorem 2.6. [*t*-**statistic-specific null distribution**] *Consider t-statistics $T_n = (T_n(m) : m = 1, \ldots, M)$, defined as in Equation* (2.34), *and a test statistics null distribution $Q_0^t = Q_0^t(P) \equiv N(0, \Sigma^*(P))$, defined as the M-variate Gaussian distribution with covariance matrix $\sigma^* = \Sigma^*(P)$ equal to the correlation matrix of the vector influence curve $IC(X|P)$ of Equation* (2.35). *Then, asymptotic null domination Assumption* NDV, *for the number of Type I errors, is satisfied by the t-statistics T_n and the null distribution Q_0^t. That is, for all $c = (c(m) : m = 1, \ldots, M) \in I\!R^M$ and $x \in \{0, \ldots, M\}$,*

$$\liminf_{n \to \infty} \Pr_{Q_n}\left(\sum_{m \in \mathcal{H}_0} \mathrm{I}(T_n(m) > c(m)) \leq x \right)$$

$$\geq \Pr_{Q_0^t}\left(\sum_{m \in \mathcal{H}_0} \mathrm{I}\left(Z^t(m) > c(m)\right) \leq x \right).$$

Thus, according to the three-step road map of Procedure 2.1, multiple testing procedures based on t-statistics T_n and the t-statistic-specific null distribution Q_0^t provide asymptotic control of general Type I error rates $\Theta(F_{V_n})$, for the one-sided test of single-parameter null hypotheses $H_0(m) = \mathrm{I}(\psi(m) \leq \psi_0(m))$ against alternative hypotheses $H_1(m) = \mathrm{I}(\psi(m) > \psi_0(m))$.

Proof of Theorem 2.6. Let us verify asymptotic null domination Assumption NDV for the t-statistics T_n of Equation (2.34) and the null distribution $Q_0^t = \mathrm{N}(0, \sigma^*)$. Firstly, note that the t-statistics $T_n(m)$ can be rewritten as

$$T_n(m) = \sqrt{n}\frac{\psi_n(m) - \psi(m)}{\sigma_n(m)} + \frac{\sigma(m)}{\sigma_n(m)}\sqrt{n}\frac{\psi(m) - \psi_0(m)}{\sigma(m)} \tag{2.36}$$

$$= Z_n^t(m) + \frac{\sigma(m)}{\sigma_n(m)} d_n(m),$$

in terms of deterministic shifts, $d_n(m) \equiv \sqrt{n}(\psi(m) - \psi_0(m))/\sigma(m)$, and standardized statistics, $Z_n^t(m) \equiv \sqrt{n}(\psi_n(m) - \psi(m))/\sigma_n(m)$. By Equation (2.35) and the Central Limit Theorem (Theorem B.4), one has

$$Z_n^t \stackrel{\mathcal{L}}{\Rightarrow} Z^t \sim Q_0^t(P) = \mathrm{N}(0, \Sigma^*(P)), \tag{2.37}$$

where $\sigma^* = \Sigma^*(P) = \mathrm{Cov}[Z^t]$ is the correlation matrix of the M-vector influence curve $IC(X|P)$. For $m \in \mathcal{H}_0$, $d_n(m) \leq 0$, so that $T_n(m) \leq Z_n^t(m)$. Thus, from the Continuous Mapping Theorem (Theorem B.3) and Proposition B.2,

$$\begin{aligned}
&\liminf_{n\to\infty} \Pr\left(\sum_{m\in\mathcal{H}_0} \mathrm{I}\left(T_n(m) > c(m)\right) \leq x\right) \\
&\geq \liminf_{n\to\infty} \Pr\left(\sum_{m\in\mathcal{H}_0} \mathrm{I}\left(Z_n^t(m) > c(m)\right) \leq x\right) \\
&= \Pr\left(\sum_{m\in\mathcal{H}_0} \mathrm{I}\left(Z^t(m) > c(m)\right) \leq x\right),
\end{aligned}$$

for all $c = (c(m) : m = 1, \ldots, M) \in I\!R^M$ and $x \in \{0, \ldots, M\}$.

□

The above theorem proposes a test statistics null distribution Q_0^t derived specifically in terms of the t-statistics T_n of Equations (2.34) and (2.35). As described below, it turns out that this null distribution Q_0^t corresponds to the general proposals Q_0 and $\breve{Q}_0$ of Sections 2.3 and 2.4, respectively.

Comparison to null shift and scale-transformed null distribution

One can show, under mild regularity conditions, that the t-statistic-specific null distribution $Q_0^t = \mathrm{N}(0, \sigma^*)$ of Theorem 2.6 corresponds to the general null shift and scale-transformed null distribution Q_0 of Theorem 2.2, with null values $\lambda_0(m) = 0$ and $\tau_0(m) = 1$.

To see this, consider the simple known variance case, where $\sigma_n(m) = \sigma(m)$. Then, $\mathrm{E}[T_n] = d_n$ and $\mathrm{Cov}[T_n] = \mathrm{Cor}[T_n] = \sigma^*$. Hence, $T_n(m) = Z_n^t(m) + \mathrm{E}[T_n(m)]$. In addition, for null values $\lambda_0(m) = 0$ and $\tau_0(m) = 1$, the M-vector Z_n, defining the general null distribution Q_0 in Theorem 2.2, reduces to Z_n^t. Hence, $Q_0 = Q_0^t = \mathrm{N}(0, \sigma^*)$.

Comparison to null quantile-transformed null distribution

A similar equivalence result is provided for the null quantile-transformed null distribution in Section 4.1 of van der Laan and Hubbard (2006). Specifically, for standard normal marginal null distributions $q_{0,m} = \Phi$, it is argued that the asymptotic null quantile-transformed null distribution $\breve{Q}_0$ is equal to the t-statistic-specific null distribution $Q_0^t = \mathrm{N}(0, \sigma^*)$ of Theorem 2.6. That is,

$$\breve{Z}_n = (\Phi^{-1} Q_{n,m}^{\Delta}(T_n(m)) : m = 1, \ldots, M) \stackrel{\mathcal{L}}{\Rightarrow} Q_0^t.$$

Theorem 2 of van der Laan and Hubbard (2006) further shows that the bootstrap estimator $\breve{Q}_{0n}$ of Procedure 2.4 converges weakly to Q_0^t.

2.6.3 Estimation of the test statistics null distribution

One can exploit the specific form of the t-statistics defined in Equations (2.34) and (2.35), to derive consistent estimators of the null distribution $Q_0^t = \mathrm{N}(0, \sigma^*)$ of Theorem 2.6.

First, consider the case where one knows the form of the M-vector influence curve, $IC(X|P) = (IC(X|P)(m) : m = 1, \ldots, M)$, for the estimator $\psi_n = \hat{\Psi}(P_n)$ (e.g., tests for means, correlation coefficients, and regression coefficients, treated in Sections 2.6.4–2.6.6, below). Given an estimator $IC_n(X) = (IC_n(X)(m) : m = 1, \ldots, M)$ of $IC(X|P)$, one can obtain the following estimator of the $M \times M$ influence curve covariance matrix $\sigma = \Sigma(P)$,

$$\sigma_n = \hat{\Sigma}(P_n) = \frac{1}{n} \sum_{i=1}^{n} IC_n(X_i) IC_n^{\top}(X_i). \tag{2.38}$$

An estimator of Q_0^t is then given by the M-variate Gaussian distribution $Q_{0n}^t = \mathrm{N}(0, \sigma_n^*)$, where $\sigma_n^* = \hat{\Sigma}^*(P_n)$ is the correlation matrix corresponding to the estimated covariance matrix σ_n.

When the influence curve is not readily available, $\sigma^* = \Sigma^*(P)$ can be estimated with the bootstrap as follows. Given an estimator $P_n^\star$ of the true data generating distribution P, let $\mathcal{X}_n^\# = \{X_i^\# : i = 1, \ldots, n\}$ denote a random sample of n IID copies of a random variable $X^\# \sim P_n^\star$. For each bootstrap sample $\mathcal{X}_n^\#$, with empirical distribution $P_n^\#$, compute the estimator $\psi_n^\# = \hat{\Psi}(P_n^\#)$. A bootstrap estimator of the covariance (and correlation) matrix $\sigma^* = \mathrm{Cov}_P[Z^t]$ is given by the covariance (and correlation) matrix $\sigma_n^* = \mathrm{Cov}_{P_n^\star}[Z_n^{t,\#}]$, of standardized bootstrap test statistics $Z_n^{t,\#}$ defined as either

$$Z_n^{t,\#}(m) = \frac{(\psi_n^\#(m) - \mathrm{E}_{P_n^\star}[\psi_n{}^\#(m)])}{\sqrt{\mathrm{Var}_{P_n^\star}[\psi_n{}^\#(m)]}} \tag{2.39}$$

or

$$Z_n^{t,\#}(m) = \frac{(\psi_n^\#(m) - \psi_n(m))}{\sqrt{\mathrm{Var}_{P_n^\star}[\psi_n{}^\#(m)]}}, \qquad m = 1, \ldots, M.$$

A parametric bootstrap estimator of the null distribution Q_0^t is then given by $Q_{0n}^t = \mathrm{N}(0, \sigma_n^*)$; a non-parametric bootstrap estimator is also provided by the joint distribution of the M-vector of standardized statistics $Z_n^{t,\#}$.

Note that, when an estimator of the influence curve is available, using the bootstrap to estimate σ^* does not necessarily pay off over direct estimation based on the original sample $\mathcal{X}_n$. When the correlation matrix is sparse, shrinkage estimation methods may be beneficial.

Alternately, a consistent estimator of the null distribution Q_0^t can be obtained using general bootstrap Procedure 2.3, for the null shift and scale-transformed null distribution, with null values $\lambda_0(m) = 0$ and $\tau_0(m) = 1$. Likewise, one could apply general bootstrap Procedure 2.4, for the null quantile-transformed null distribution, with standard normal marginal null distributions $q_{0,m} = \Phi$.

As mentioned in Section 2.3.2, above, one of the main advantages of a parametric estimator $Q_{0n}^t = \mathrm{N}(0, \sigma_n^*)$ is that it is continuous and hence does not suffer from the discreteness of non-parametric bootstrap estimators. Similar issues arise for F-statistics, as discussed in Section 2.7, below.

2.6.4 Example: Tests for means

A familiar testing problem, that falls within our single-parameter hypothesis testing framework, is that where $X \sim P$ is a random J-vector and the parameter of interest is the *mean* vector of X, $\Psi(P) = \psi = (\psi(j) : j = 1, \ldots, J) = \mathrm{E}[X]$, with elements $\psi(j) = \Psi(P)(j) = \mathrm{E}[X(j)]$. The $M = J$ null hypotheses, $H_0(m) = \mathrm{I}(\psi(m) \le \psi_0(m))$, then refer to individual elements of the mean vector ψ.

Given a random sample $\mathcal{X}_n = \{X_i : i = 1, \ldots, n\}$, from the data generating distribution P, the test statistics $T_n(m)$ of Equation (2.34) are the usual *one-sample t-statistics*, where $\psi_n(m) = \hat{\Psi}(P_n)(m) = \bar{X}_n(m) = \sum_i X_i(m)/n$ and $\sigma_n^2(m) = \sum_i (X_i(m) - \bar{X}_n(m))^2/n$ are the empirical means and variances of the M elements of X, respectively.

In this simple case, the elements of the M-vector influence curve are $IC(X|P)(m) = X(m) - \psi(m)$ and can be estimated by $IC_n(X)(m) = X(m) - \bar{X}_n(m)$. Thus, a consistent estimator of the test statistics null distribution Q_0^t of Theorem 2.6 is the M-variate Gaussian distribution $Q_{0n}^t = \mathrm{N}(0, \sigma_n^*)$, where $\sigma_n^* = \hat{\Sigma}^*(P_n)$ is the $M \times M$ empirical correlation matrix $\mathrm{Cor}_{P_n}[X]$.

2.6.5 Example: Tests for correlation coefficients

Another common testing problem covered by Theorem 2.6 is that where the parameter of interest is the $J \times J$ *correlation* matrix for a random J-vector $X \sim P$, that is, $\Psi(P) = \psi = (\psi(j, j') : j,\ j' = 1, \ldots, J) = \mathrm{Cor}[X]$, with elements $\psi(j, j') = \Psi(P)(j, j') = \mathrm{Cor}[X(j), X(j')]$. Suppose one is interested in testing the $M = J(J-1)/2$ null hypotheses that the J elements of X are uncorrelated, that is, null hypotheses $H_0(j, j') = \mathrm{I}(\psi(j, j') = 0)$, $j = 1, \ldots, J-1$, $j' = j+1, \ldots, J$.

Commonly-used test statistics for this problem are $T_n(j, j') = \sqrt{n}\psi_n(j, j')$, where $\psi_n(j, j') = \hat{\Psi}(P_n)(j, j')$ are the *empirical correlation coefficients.* As discussed in Westfall and Young (1993, Example 2.2, p. 43), subset pivotality fails for this testing problem. To see this, consider the simple case where $J = 3$ (and $M = 3$) and assume that $H_0(1, 2)$ and $H_0(1, 3)$ are true, so that $\psi(1, 2) = \psi(1, 3) = 0$. Then, the joint distribution of $(T_n(1, 2), T_n(1, 3))$ is asymptotically Gaussian, with mean vector zero, unit variances, and correlation of $\psi(2, 3)$, and thus depends on the truth or falsity of the third hypothesis $H_0(2, 3)$. In other words, the covariance matrix of the vector influence curve for the empirical correlation coefficients differs under the true data generating distribution P and under a data generating null distribution P_0 for which $\psi(j,j') = 0, \forall\, j \neq j'$. Tests for correlation coefficients thus provide an example where standard procedures based on subset pivotality fail, whereas procedures based on the t-statistic-specific null distribution of Theorem 2.6 or the general null distributions of Sections 2.3 and 2.4 achieve the desired Type I error control (Pollard et al., 2005a; Pollard and van der Laan, 2004).

The influence curves for the empirical correlation coefficients $\psi_n(j,j')$ can be obtained by applying the Delta-method with the function

$$f(\xi(j,j')) = \psi(j,j') = \frac{\gamma(j,j') - \gamma(j)\gamma(j')}{\sqrt{\gamma(j,j) - \gamma^2(j)}\sqrt{\gamma(j',j') - \gamma^2(j')}}, \tag{2.40}$$

defined in terms of a 5×1 parameter column vector $\xi(j,j') = \Xi(P)(j,j') = [\gamma(j), \gamma(j'), \gamma(j,j), \gamma(j',j'), \gamma(j,j')]^\top$, with elements $\gamma(j) = \Gamma(P)(j) = \mathrm{E}[X(j)]$ and $\gamma(j,j') = \Gamma(P)(j,j') = \mathrm{E}[X(j)X(j')]$, $j,\, j' = 1,\ldots,J$. Let $f'(\xi)$ denote the 1×5 gradient row vector of $f(\xi)$. Then,

$$\psi_n(j,j') - \psi(j,j') = f'(\xi(j,j'))\,(\xi_n(j,j') - \xi(j,j')) + o_P(1/\sqrt{n}), \tag{2.41}$$

where $\xi_n(j,j') = \hat{\Xi}(P_n)(j,j') = [\gamma_n(j), \gamma_n(j'), \gamma_n(j,j), \gamma_n(j',j'), \gamma_n(j,j')]^\top$ is a 5×1 estimator column vector for $\xi(j,j')$, based on the empirical moments. Hence, the influence curve for the estimator $\psi_n(j,j')$ is

$$\begin{aligned} IC(X|P)(j,j') &= f'(\xi(j,j'))(\xi_1(j,j') - \xi(j,j')) \\ &= \frac{1}{\sqrt{\sigma(j,j)}\sqrt{\sigma(j',j')}} \begin{bmatrix} \gamma(j)\frac{\sigma(j,j')}{\sigma(j,j)} - \gamma(j') \\ \gamma(j')\frac{\sigma(j,j')}{\sigma(j',j')} - \gamma(j) \\ -\frac{1}{2}\frac{\sigma(j,j')}{\sigma(j,j)} \\ -\frac{1}{2}\frac{\sigma(j,j')}{\sigma(j',j')} \\ 1 \end{bmatrix}^\top \begin{bmatrix} X(j) - \gamma(j) \\ X(j') - \gamma(j') \\ X^2(j) - \gamma(j,j) \\ X^2(j') - \gamma(j',j') \\ X(j)X(j') - \gamma(j,j') \end{bmatrix}, \end{aligned} \tag{2.42}$$

where covariances are denoted by $\sigma(j,j') = \gamma(j,j') - \gamma(j)\gamma(j')$.

Section 8.4 examines the choice of a test statistics null distribution in testing problems concerning correlation coefficients. Section 9.3 considers the identification of co-expressed miRNAs based on tests for correlations coefficients.

2.6.6 Example: Tests for regression coefficients

Consider a random $J = (M+1)$-vector $X \sim P$, from a data generating distribution P, where $(X(m) : m = 1,\ldots,M)$ is an M-dimensional covariate/genotype vector and $Y = X(M+1)$ is a univariate outcome/phenotype. For instance, the covariates/genotypes could correspond to M microarray gene expression measures and the outcome/phenotype to a (censored) survival time or a tumor class.

Assume the following model for the conditional expected value of the outcome Y given individual covariates $X(m)$,

$$\mathrm{E}[Y|X(m)] = g(X(m);\gamma_m) = h(\gamma_m(1) + \gamma_m(2)X(m)), \qquad m = 1,\ldots,M, \tag{2.43}$$

where $\Gamma_m(P) = \gamma_m = (\gamma_m(1), \gamma_m(2))$ are *regression coefficients* for the mth covariate $X(m)$. The parameter of interest is the M-vector of *slope parameters*, $\Psi(P) = \psi = (\psi(m) = \gamma_m(2) : m = 1,\ldots,M)$.

Given a random sample $\mathcal{X}_n = \{X_i : i = 1,\ldots,n\}$, from the data generating distribution P, one can estimate the regression parameters γ_m for each covariate $X(m)$ using the method of least squares, that is, by seeking γ_m that minimizes the sum of squared residuals, $\sum_i (Y_i - g(X_i(m);\gamma_m))^2$. The *least squares estimator*, $\hat{\Gamma}_m(P_n) = \gamma_{m,n} = (\gamma_{m,n}(1), \gamma_{m,n}(2))$, is obtained by solving the following equation for γ,

$$0 = \frac{\partial}{\partial\gamma}\sum_{i=1}^{n}(Y_i - g(X_i(m);\gamma))^2,$$

that is,

$$0 = \sum_{i=1}^{n}\left(\frac{\partial}{\partial\gamma}g(X_i(m);\gamma)\right)(Y_i - g(X_i(m);\gamma)).$$

Let $IC_m(X|P) = (IC_m(X|P)(1), IC_m(X|P)(2))$ denote the two-dimensional vector influence curve for the least squares estimator $\gamma_{m,n}$ of the regression parameters γ_m corresponding to covariate $X(m)$. Under mild regularity conditions (Lemma 2.1, p. 105, van der Laan and Robins (2003)), one can show that

$$\gamma_{m,n} - \gamma_m = \frac{1}{n}\sum_{i=1}^{n} c_m^{-1}(\gamma_m)\left(\frac{\partial}{\partial\gamma}g(X_i(m);\gamma)\right)\Bigg|_{\gamma=\gamma_m}(Y_i - g(X_i(m);\gamma_m)) + o_P(1/\sqrt{n}), \tag{2.44}$$

where, for a given $\gamma \in I\!R^2$,

$$c_m(\gamma) = E\begin{bmatrix} \left(\frac{\partial}{\partial\gamma(1)}g(X(m);\gamma)\right)^2 & \left(\frac{\partial}{\partial\gamma(1)}g(X(m);\gamma)\right) \times \left(\frac{\partial}{\partial\gamma(2)}g(X(m);\gamma)\right) \\ \left(\frac{\partial}{\partial\gamma(1)}g(X(m);\gamma)\right) \times \left(\frac{\partial}{\partial\gamma(2)}g(X(m);\gamma)\right) & \left(\frac{\partial}{\partial\gamma(2)}g(X(m);\gamma)\right)^2 \end{bmatrix}.$$

From the above expression, the influence curves are

$$IC_m(X|P) = c_m^{-1}(\gamma_m)\left(\frac{\partial}{\partial\gamma}g(X(m);\gamma)\right)\bigg|_{\gamma=\gamma_m}(Y - g(X(m);\gamma_m)). \quad (2.45)$$

The M-dimensional vector influence curve for the least squares estimators $\hat{\Psi}(P_n) = \psi_n = (\psi_n(m) = \gamma_{m,n}(2) : m = 1, \ldots, M)$, of the M slope parameters ψ, is

$$IC(X|P) = (IC_m(X|P)(2) : m = 1, \ldots, M).$$

The covariance matrix of the vector influence curve $IC(X|P)$ is

$$\sigma = \Sigma(P) = \mathrm{E}\left[IC(X|P)IC^\top(X|P)\right],$$

and can be estimated as in Equation (2.38), using the empirical covariance matrix for an estimator $IC_n(X)$ of the vector influence curve.

Linear regression

A common model for a continuous outcome $Y \in I\!R$ is the *linear model*, corresponding to the identity function $h(z) = z$. That is,

$$\mathrm{E}[Y|X(m)] = g(X(m);\gamma_m) = \gamma_m(1) + \gamma_m(2)X(m). \quad (2.46)$$

In this case, the influence curves for the least squares estimators $\gamma_{m,n}$ of the regression coefficients γ_m are given by

$$IC_m(X|P) = \frac{1}{\mathrm{Var}[X(m)]}\begin{bmatrix} \mathrm{E}[X^2(m)] & -\mathrm{E}[X(m)] \\ -\mathrm{E}[X(m)] & 1 \end{bmatrix}\begin{bmatrix} 1 \\ X(m) \end{bmatrix} \times (Y - \gamma_m(1) - \gamma_m(2)X(m)). \quad (2.47)$$

Logistic regression

A common model for a binary outcome $Y \in \{0, 1\}$ is the *logistic model*, corresponding to the *softmax* or *inverse logit* function $h(z) = \exp(z)/(1 + \exp(z))$. That is,

$$\Pr(Y = 1|X(m)) = g(X(m);\gamma_m) = \frac{\exp(\gamma_m(1) + \gamma_m(2)X(m))}{1 + \exp(\gamma_m(1) + \gamma_m(2)X(m))}. \quad (2.48)$$

Here,

$$\left(\frac{\partial}{\partial\gamma}g(X(m);\gamma)\right)\bigg|_{\gamma=\gamma_m} = \frac{\exp(\gamma_m(1) + \gamma_m(2)X(m))}{(1 + \exp(\gamma_m(1) + \gamma_m(2)X(m)))^2}\begin{bmatrix} 1 \\ X(m) \end{bmatrix}, \quad (2.49)$$

and the influence curves for the least squares estimators $\gamma_{m,n}$ of the regression coefficients γ_m can be derived by substituting for $\partial g(X(m);\gamma)/\partial\gamma$ in Equation (2.45), above.

Section 8.3 examines the choice of a test statistics null distribution in testing problems concerning regression coefficients in linear models where the covariates and error terms are allowed to be dependent. Section 9.3 considers tests for regression coefficients in logistic models relating cancer status to miRNA expression measures and tissue type (Pollard et al., 2005a).

2.7 Testing multiple-parameter null hypotheses based on F-statistics

2.7.1 Set-up and assumptions

Consider random M-vectors $X_k = (X_k(m) : m = 1, \ldots, M) \sim P_k$, from K different populations, with respective data generating distributions P_k, $k = 1, \ldots, K$. Let $\psi_k = \Psi(P_k) = \mathrm{E}[X_k]$ and $\sigma_k = \Sigma(P_k) = \mathrm{Cov}[X_k]$ denote, respectively, the mean vector and covariance matrix for Population k. Denote the elements of the covariance matrix σ_k by $\sigma_k(m, m') = \mathrm{Cov}[X_k(m), X_k(m')]$ and adopt the shorter notation $\sigma_k^2(m) = \sigma_k(m, m)$ for the diagonal elements of σ_k, i.e., the variances. Consider testing the M null hypotheses $H_0(m) = \mathrm{I}(\psi_1(m) = \psi_2(m) = \cdots = \psi_K(m))$, that the elements of the mean vectors are constant across the K populations.

Suppose one observes a random sample $\mathcal{X}_{k,n_k} = \{X_{k,i} : i = 1, \ldots, n_k\}$, of size n_k, from Population k, $k = 1, \ldots, K$[4]. Let $n = \sum_k n_k$ denote the total sample size and $\eta_{k,n} = n_k/n$ the empirical frequency for Population k. Assume that $\lim_n \eta_{k,n} = \eta_k > 0$, $\forall\ k = 1, \ldots, K$. The null hypotheses can be tested using *F-statistics*,

$$T_n(m) \equiv \frac{\frac{1}{K-1}\sum_{k=1}^K n_k(\bar{X}_{k,n_k}(m) - \bar{X}_n(m))^2}{\frac{1}{n-K}\sum_{k=1}^K \sum_{i=1}^{n_k}(X_{k,i}(m) - \bar{X}_{k,n_k}(m))^2}, \qquad m = 1, \ldots, M, \tag{2.50}$$

where $\bar{X}_{k,n_k} = \sum_i X_{k,i}/n_k$ denotes the empirical mean vector for the sample $\mathcal{X}_{k,n_k}$ from Population k and $\bar{X}_n = \sum_k \eta_{k,n}\bar{X}_{k,n_k} = \sum_k \sum_i X_{k,i}/n$ denotes the empirical mean vector for the pooled sample of size n. Large values of the F-statistic $T_n(m)$ are assumed to provide evidence against the corresponding null hypothesis $H_0(m) = \mathrm{I}(\psi_1(m) = \psi_2(m) = \cdots = \psi_K(m))$, that is, tests are based on one-sided rejection regions of the form $\mathcal{C}_n(m) = (c_n(m), +\infty)$.

Next, we propose an *F-statistic-specific null distribution* Q_0^F that leads to asymptotic control of Type I error rates $\Theta(F_{V_n})$, defined as arbitrary parameters of the distribution of the number of Type I errors V_n.

[4] N.B. With proper care, one could allow random sample sizes n_k.

2.7.2 Test statistics null distribution

Theorem 2.7. **[F-statistic-specific null distribution]** *Consider F-statistics $T_n = (T_n(m) : m = 1, \ldots, M)$, defined as in Equation (2.50), and a test statistics null distribution $Q_0^F = Q_0^F(P_1, \ldots, P_K)$, defined as the joint distribution of a random M-vector $Z^F = (Z^F(m) : m = 1, \ldots, M)$ of quadratic forms*

$$Z^F(m) \equiv \frac{1}{(K-1)\sum_{k=1}^K \eta_k \sigma_k^2(m)} \times \left(\sum_{k=1}^K (1-\eta_k) Y_k^2(m) - \sum_{\substack{k=1 \\ k \neq k'}}^K \sum_{k'=1}^K \sqrt{\eta_k \eta_{k'}} Y_k(m) Y_{k'}(m) \right), \tag{2.51}$$

based on K independent Gaussian M-vectors $Y_k = (Y_k(m) : m = 1, \ldots, M) \sim N(0, \sigma_k)$. In matrix notation, the quadratic forms are defined by

$$Z^F(m) \equiv \widetilde{Y}_m^\top A_m \widetilde{Y}_m, \tag{2.52}$$

based on M dependent Gaussian K-vectors $\widetilde{Y}_m = (Y_k(m) : k = 1, \ldots, K) \sim N(0, \widetilde{\sigma}_m)$, with diagonal covariance matrices $\widetilde{\sigma}_m$ such that $\widetilde{\sigma}_m(k,k) = \sigma_k^2(m)$, and M symmetric $K \times K$ matrices A_m with elements

$$A_m(k,k') \equiv \frac{1}{(K-1)\sum_{k=1}^K \eta_k \sigma_k^2(m)} \begin{cases} (1-\eta_k), & \text{if } k = k' \\ -\sqrt{\eta_k \eta_{k'}}, & \text{if } k \neq k' \end{cases}. \tag{2.53}$$

Then, the F-statistics $(T_n(m) : m \in \mathcal{H}_0)$ for the true null hypotheses converge weakly to the $\mathcal{H}_0$-specific quadratic forms $(Z^F(m) : m \in \mathcal{H}_0)$, that is,

$$(T_n(m) : m \in \mathcal{H}_0) \stackrel{\mathcal{L}}{\Rightarrow} (Z^F(m) : m \in \mathcal{H}_0) \sim Q^F_{0,\mathcal{H}_0}.$$

It follows that asymptotic null domination Assumption NDV, for the number of Type I errors, is satisfied with equality *by the F-statistics T_n and the null distribution Q_0^F. That is, for all $c = (c(m) : m = 1, \ldots, M) \in I\!R^{+M}$ and $x \in \{0, \ldots, M\}$,*

$$\lim_{n \to \infty} \Pr_{Q_n} \left(\sum_{m \in \mathcal{H}_0} \mathrm{I}(T_n(m) > c(m)) \leq x \right) = \Pr_{Q_0^F} \left(\sum_{m \in \mathcal{H}_0} \mathrm{I}(Z^F(m) > c(m)) \leq x \right).$$

Thus, according to the three-step road map of Procedure 2.1, multiple testing procedures based on F-statistics T_n and the F-statistic-specific null distribution Q_0^F provide asymptotic control of general Type I error rates $\Theta(F_{V_n})$, *for the test of multiple-parameter null hypotheses of the form* $H_0(m) = \mathrm{I}\,(\psi_1(m) = \psi_2(m) = \cdots = \psi_K(m))$.

Furthermore, the quadratic forms $Z^F(m)$ *have means and variances given, respectively, by*

$$\mathrm{E}[Z^F(m)] = \frac{1}{(K-1)\sum_{k=1}^K \eta_k \sigma_k^2(m)} \sum_{k=1}^K (1-\eta_k)\sigma_k^2(m) \tag{2.54}$$

and

$$\mathrm{Var}[Z^F(m)] = \frac{2}{(K-1)^2(\sum_{k=1}^K \eta_k \sigma_k^2(m))^2} \times \left(\left(\sum_{k=1}^K (1-2\eta_k)\sigma_k^4(m) \right) + \left(\sum_{k=1}^K \eta_k \sigma_k^2(m) \right)^2 \right).$$

In the special case of constant variances across populations, i.e., $\sigma_k^2(m) = \sigma^2(m)$, *then* $\mathrm{E}[Z^F(m)] = 1$, $\mathrm{Var}[Z^F(m)] = 2/(K-1)$, *and the quadratic forms have marginal* χ^2*-distributions with* $(K-1)$ *degrees of freedom, that is,*

$$(K-1)Z^F(m) \sim \chi^2(K-1). \tag{2.55}$$

Proof of Theorem 2.7. Firstly, note that the denominators of the F-statistics can be rewritten as

$$D_n(m) \equiv \frac{n}{n-K} \sum_{k=1}^K \eta_{k,n} \sigma_{k,n_k}^2(m), \tag{2.56}$$

where the empirical frequencies $\eta_{k,n} = n_k/n$ converge to the population frequencies η_k and the empirical variances $\sigma_{k,n_k}^2(m) = \sum_i (X_{k,i}(m) - \bar{X}_{k,n_k}(m))^2/n_k$ are consistent estimators of the population variances $\sigma_k^2(m)$, i.e., $\eta_{k,n} \to \eta_k > 0$ and $\sigma_{k,n_k}^2(m) \xrightarrow{P} \sigma_k^2(m)$, $k = 1, \ldots, K$. Thus, as $n \to \infty$,

$$D_n(m) \xrightarrow{P} D(m) \equiv \sum_{k=1}^K \eta_k \sigma_k^2(m). \tag{2.57}$$

The numerators of the F-statistics can be rewritten as quadratic forms

$$
\begin{aligned}
N_n(m) &\equiv \frac{1}{K-1}\sum_{k=1}^{K}\left(Y_{k,n_k}(m) - \sqrt{\eta_{k,n}}\sum_{k'=1}^{K}\sqrt{\eta_{k',n}}Y_{k',n_{k'}}(m)\right)^2 \qquad (2.58)\\
&= \frac{1}{K-1}\left(\sum_{k=1}^{K}Y_{k,n_k}^2(m)\right.\\
&\qquad - 2\left(\sum_{k=1}^{K}\sqrt{\eta_{k,n}}Y_{k,n_k}(m)\right)\left(\sum_{k'=1}^{K}\sqrt{\eta_{k',n}}Y_{k',n_{k'}}(m)\right)\\
&\qquad \left. + \sum_{k=1}^{K}\eta_{k,n}\left(\sum_{k'=1}^{K}\sqrt{\eta_{k',n}}Y_{k',n_{k'}}(m)\right)^2\right)\\
&= \frac{1}{K-1}\left(\sum_{k=1}^{K}Y_{k,n_k}^2(m) - \left(\sum_{k=1}^{K}\sqrt{\eta_{k,n}}Y_{k,n_k}(m)\right)^2\right)\\
&= \frac{1}{K-1}\left(\sum_{k=1}^{K}(1-\eta_{k,n})Y_{k,n_k}^2(m)\right.\\
&\qquad \left. - \sum_{k=1}^{K}\sum_{\substack{k'=1\\ k\neq k'}}^{K}\sqrt{\eta_{k,n}\eta_{k',n}}Y_{k,n_k}(m)Y_{k',n_{k'}}(m)\right),
\end{aligned}
$$

where $Y_{k,n_k} = (Y_{k,n_k}(m) : m = 1,\ldots,M)$ are K independent M-vectors defined by $Y_{k,n_k}(m) = \sqrt{n_k}(\bar{X}_{k,n_k}(m) - \bar{\psi}(m))$ and $\bar{\psi}(m) = \sum_k \eta_k\psi_k(m)$, $k = 1,\ldots,K$.

Thus, asymptotically, one can approximate the F-statistics $T_n = (T_n(m) : m = 1,\ldots,M)$ by a random M-vector $Z_n^F = (Z_n^F(m) : m = 1,\ldots,M)$ of quadratic forms, as follows,

$$
\begin{aligned}
T_n(m) &\approxeq \frac{N_n(m)}{D(m)} \qquad (2.59)\\
&\approxeq \frac{1}{(K-1)\sum_{k=1}^{K}\eta_k\sigma_k^2(m)}\\
&\quad \times\left(\sum_{k=1}^{K}(1-\eta_k)Y_{k,n_k}^2(m) - \sum_{k=1}^{K}\sum_{\substack{k'=1\\ k\neq k'}}^{K}\sqrt{\eta_k\eta_{k'}}Y_{k,n_k}(m)Y_{k',n_{k'}}(m)\right)\\
&\equiv Z_n^F(m).
\end{aligned}
$$

That is, the mth element $Z_n^F(m)$ of the random M-vector Z_n^F is a simple quadratic function $f_m(Y_{1,n_1},\ldots,Y_{K,n_K})$ of the mth elements $Y_{k,n_k}(m)$ of the

K random M-vectors Y_{k,n_k}, $k = 1, \ldots, K$. The M-vector Z_n^F may be expressed as $Z_n^F = f(Y_{1,n_1}, \ldots, Y_{K,n_K}) = (f_m(Y_{1,n_1}, \ldots, Y_{K,n_K}) : m = 1, \ldots, M)$.

By the Central Limit Theorem (Theorem B.4),

$$(Y_{k,n_k}(m) : m \in \mathcal{H}_0) \overset{\mathcal{L}}{\Rightarrow} (Y_k(m) : m \in \mathcal{H}_0),$$

for independent Gaussian M-vectors $Y_k = (Y_k(m) : m = 1, \ldots, M) \sim \mathrm{N}(0, \sigma_k)$, $k = 1, \ldots, K$. By the Continuous Mapping Theorem (Theorem B.3), it then follows that

$$(T_n(m) : m \in \mathcal{H}_0) \overset{\mathcal{L}}{\Rightarrow} (Z^F(m) : m \in \mathcal{H}_0) \sim Q_{0,\mathcal{H}_0}^F,$$

where $Z^F = f(Y_1, \ldots, Y_K)$ is the random M-vector of quadratic forms with joint distribution Q_0^F, defined as in Equations (2.51)–(2.53).

Hence, asymptotic null domination Assumption NDV, for the number of Type I errors, is satisfied with equality by the F-statistics T_n and the null distribution Q_0^F.

Note that the F-statistics $(T_n(m) : m \in \mathcal{H}_1)$ for the false null hypotheses have infinite limits. Indeed, for $m \in \mathcal{H}_1$, $Y_{k,n_k}(m) = \sqrt{n_k}(\bar{X}_{k,n_k}(m) - \psi_k(m)) + \sqrt{n_k}(\psi_k(m) - \bar{\psi}(m))$ converges to either $+\infty$ or $-\infty$ for some k, hence $\lim_n T_n(m) = +\infty$.

The moments of $Z^F(m)$ are obtained from standard results on quadratic forms (Theorem 1, p. 55, and Corollary 1.3, p. 57, Searle (1971)). In the special case of constant variances across populations, i.e., $Diag(\sigma_k) = (\sigma^2(m) : m = 1, \ldots, M)$, the matrices $(K-1)A_m \operatorname{Cov}[\widetilde{Y}_m]$ are idempotent; hence, the quadratic forms $(K-1)Z^F(m)$ have marginal $\chi^2(K-1)$-distributions (Theorem 2, p. 57, Searle (1971)).

□

The above theorem proposes a test statistics null distribution Q_0^F derived specifically in terms of the F-statistics T_n of Equation (2.50). This null distribution is the joint distribution of an M-vector of quadratic forms of Gaussian random variables and is entirely specified by the population covariance matrices σ_k and frequencies η_k (via the matrices A_m and the random M-vectors $Y_k \sim \mathrm{N}(0, \sigma_k)$, defining the quadratic forms Z^F in Equations (2.51)–(2.53)). Although properties of the marginal distributions of the F-statistics follow from standard univariate results on quadratic forms, Theorem 2.7 provides as a main contribution a joint null distribution Q_0^F that takes into account the dependence structure of these test statistics. Specifically, the dependence structure of the null distribution Q_0^F is implied by the dependence structure of the data generating distributions P_k, as indicated by the presence of the covariance matrices σ_k in the definition of the quadratic forms Z^F.

Note that the F-statistics $(T_n(m) : m \in \mathcal{H}_0)$ for the true null hypotheses converge weakly to the $\mathcal{H}_0$-specific joint null distribution $Q_{0,\mathcal{H}_0}^F$. Asymptotic joint null domination Assumption jtNDT for the test statistics $(T_n(m) : m \in \mathcal{H}_0)$ is therefore satisfied with equality. In contrast, the F-statistics $(T_n(m) :$

$m \in \mathcal{H}_1$) for the false null hypotheses have infinite limits, i.e., $\lim_n T_n(m) = +\infty$, for $m \in \mathcal{H}_1$. Key Assumption NDV of asymptotic null domination for the number of Type I errors is nonetheless satisfied, as it only concerns the test statistics $(T_n(m) : m \in \mathcal{H}_0)$ corresponding to the true null hypotheses. In other words, neither convergence to nor the weaker domination by Q_0^F is needed for the false null hypotheses.

Gaussian data generating distributions with constant variances across populations

In the special case of Gaussian data generating distributions $P_k = \mathrm{N}(\psi_k, \sigma_k)$, with constant variances across populations, i.e., $Diag(\sigma_k) = (\sigma^2(m) : m = 1, \ldots, M)$, the test statistics T_n have marginal *non-central F-distributions* (Section 2.4, Searle (1971)). Specifically, $T_n(m) \sim \mathrm{F}(\nu_1, \nu_2, \upsilon_n(m))$, where the degrees of freedom are $\nu_1 = (K-1)$ and $\nu_2 = (n-K)$ and the non-centrality parameter is

$$\upsilon_n(m) = \frac{1}{\sigma^2(m)} \sum_{k=1}^{K} n_k(\psi_k(m) - \bar{\psi}(m))^2, \qquad \bar{\psi}(m) = \sum_{k=1}^{K} \eta_k \psi_k(m). \quad (2.60)$$

For the true null hypotheses (i.e., for $m \in \mathcal{H}_0$), $\upsilon_n(m) = 0$. For the false null hypotheses (i.e., for $m \in \mathcal{H}_1$) and non-local alternative mean parameters $\psi_k(m)$, $\lim_n \upsilon_n(m) = +\infty$. In addition, $\lim_n \nu_2 = +\infty$.

The means and variances of the F-statistics are given by, respectively,

$$\mathrm{E}[T_n(m)] = \frac{(\nu_1 + \upsilon_n(m))\nu_2}{\nu_1(\nu_2 - 2)} \to \begin{cases} 1, & \text{if } m \in \mathcal{H}_0 \\ +\infty, & \text{if } m \in \mathcal{H}_1 \end{cases} \quad (2.61)$$

and

$$\begin{aligned} \mathrm{Var}[T_n(m)] &= \frac{2\nu_2^2\left(\nu_1^2 + (2\upsilon_n(m) + \nu_2 - 2)\nu_1 + \upsilon_n(m)(\upsilon_n(m) + 2\nu_2 - 4)\right)}{\nu_1^2(\nu_2 - 4)(\nu_2 - 2)^2} \\ &\to \begin{cases} 2/(K-1), & \text{if } m \in \mathcal{H}_0 \\ +\infty, & \text{if } m \in \mathcal{H}_1 \end{cases}. \end{aligned} \quad (2.62)$$

Furthermore, the F-statistics $T_n(m)$ have asymptotic marginal *non-central χ^2-distributions*, with $(K-1)$ degrees of freedom and non-centrality parameter $\upsilon_n(m)$. That is,

$$(K-1)T_n(m) \stackrel{\mathcal{L}}{\Rightarrow} \chi^2(K-1, \upsilon_n(m)). \quad (2.63)$$

Comparison to null shift and scale-transformed null distribution

Instead of the F-statistic-specific null distribution Q_0^F proposed in Theorem 2.7, one could apply the general construction of Theorem 2.2, whereby the null distribution Q_0 is defined as the asymptotic distribution of the M-vector $Z_n = (Z_n(m) : m = 1, \ldots, M)$ of null shift and scale-transformed test statistics,

$$Z_n(m) = \sqrt{\min\left\{1, \frac{\tau_0(m)}{\mathrm{Var}[T_n(m)]}\right\}}\,(T_n(m) - \mathrm{E}[T_n(m)]) + \lambda_0(m).$$

For F-statistics, the null values $\lambda_0(m)$ and $\tau_0(m)$ are based on, respectively, the means and variances of the quadratic forms Z^F (Equation (2.54)). In the special case of constant variances across populations, i.e., $Diag(\sigma_k) = (\sigma^2(m) : m = 1, \ldots, M)$, the null values do not depend on the unknown data generating distributions P_k and are given by $\lambda_0(m) = 1$ and $\tau_0(m) = 2/(K-1)$. Otherwise, one needs to estimate the population frequencies η_k and variances $\sigma_k^2(m)$ in order to use Equation (2.54).

Note that, in the construction of $Z_n(m)$, it is important to scale the test statistics $T_n(m)$ by $\nu_{0,n}(m) = \sqrt{\min\{1, \tau_0(m)/\mathrm{Var}[T_n(m)]\}}$, as these F-statistics converge to infinity for non-local alternative hypotheses. Without this scaling, one could have asymptotically infinite test statistic cut-offs and hence no power against the alternative hypotheses.

The F-statistic-specific null distribution Q_0^F of Theorem 2.7 and the general null distribution Q_0 of Theorem 2.2 are the same for the true null hypotheses ($m \in \mathcal{H}_0$), but may differ for the false null hypotheses ($m \in \mathcal{H}_1$). Thus, in choosing between Q_0^F and Q_0, the main issue is power.

Comparison to null quantile-transformed null distribution

Section 4.2 of van der Laan and Hubbard (2006) addresses a similar testing problem using the new null quantile-transformed null distribution introduced in Section 2.4. Specifically, for χ^2-statistics T_n and marginal null distributions $q_{0,m} = \chi^2(K-1)$, Theorem 3 proves that the null quantile-transformed test statistics $(\breve{Z}_n(m) : m \in \mathcal{H}_0)$ for the true null hypotheses converge weakly to the $\mathcal{H}_0$-specific subdistribution $Q_{0,\mathcal{H}_0}^{\chi}$, of a joint null distribution Q_0^{χ} with marginal $\chi^2(K-1)$-distributions. Theorem 3 further provides conditions under which estimators of Q_0^{χ} lead to proper Type I error control.

As previously discussed, the ability to control marginal null distributions should confer greater power to this new approach.

2.7.3 Estimation of the test statistics null distribution

A consistent estimator Q_{0n}, of the general null shift and scale-transformed null distribution Q_0 of Theorem 2.2, can be obtained using bootstrap Procedure 2.3, with null values $\lambda_0(m)$ and $\tau_0(m)$ defined as in Equation (2.54). In the special case of constant variances across populations, the null values are $\lambda_0(m) = 1$ and $\tau_0(m) = 2/(K-1)$. Otherwise, one needs to estimate the null values, as they depend on the unknown population frequencies η_k and variances $\sigma_k^2(m)$.

Estimation approaches for the general null quantile-transformed null distribution of Section 2.4 are discussed in Section 4.2 of van der Laan and Hubbard (2006).

Alternately, one can exploit properties of F-statistics to derive a consistent estimator Q_{0n}^F of the F-statistic-specific null distribution Q_0^F of Theorem 2.7. Recall that this null distribution is the joint distribution of an M-vector of quadratic forms of Gaussian random variables and is entirely specified by the population covariance matrices σ_k and frequencies η_k (Equations (2.51)–(2.53)). The main task is therefore to derive estimators $\sigma_{k,n}$ and $\eta_{k,n}$ of these population covariance matrices and frequencies, based on the K random samples $\mathcal{X}_{k,n_k} = \{X_{k,i} : i = 1, \ldots, n_k\}$, $k = 1, \ldots, K$. The null distribution Q_0^F may then simply be estimated by the joint distribution Q_{0n}^F of an M-vector of quadratic forms, defined using the empirical analogues of Equations (2.51)–(2.53), in terms of independent Gaussian M-vectors $Y_k \sim \mathrm{N}(0, \sigma_{k,n})$. Unlike the general non-parametric bootstrap estimator of Procedure 2.3, for the null distribution Q_0 of Theorem 2.2, this F-statistic-specific estimator has the advantage of being continuous.

Finally, another F-statistic-specific approach involves bootstrapping the centered observations $X_{k,i} - \bar{X}_{k,n_k}$ and estimating the null distribution Q_0^F by the bootstrap distribution of the corresponding F-statistics. In this method, the estimated null distribution of the test statistics is based on a data generating null distribution.

The last two approaches both provide consistent estimators of the F-statistic-specific null distribution Q_0^F of Theorem 2.7.

2.8 Weak and strong Type I error control and subset pivotality

As mentioned in Section 2.2.4, the multiple testing methodology developed in this book differs in a number of fundamental aspects from existing approaches to Type I error control and the choice of a test statistics null distribution. Our proposed multiple testing procedures are: (i) only concerned with controlling the Type I error rate under the *true data generating distribution* P, i.e., under the joint distribution $Q_n = Q_n(P)$ of the test statistics T_n implied by P; (ii) based on a *test statistics null distribution* rather than a data generating null distribution.

In this regard, one of our main contributions is the general characterization (Section 2.2.3) and explicit construction (Sections 2.3 and 2.4) of proper null distributions Q_0 (and estimators thereof, Q_{0n}) for the test statistics T_n. Procedures based on the proposed null distributions provide Type I error control for general data generating distributions (with arbitrary dependence structures among variables), null hypotheses, and test statistics.

In our framework, the notions of weak and strong control of a Type I error rate become irrelevant and Type I error control does not involve associated restrictive assumptions such as subset pivotality. The present section attempts nonetheless to formalize these concepts and discusses how they relate to the approach introduced in Section 2.2.

2.8.1 Weak and strong control of a Type I error rate

Usual definitions of weak and strong Type I error control

As discussed in Hochberg and Tamhane (1987, p. 3) and Westfall and Young (1993, p. 9–10), the multiple testing literature commonly distinguishes between weak and strong control of a Type I error rate.

Weak control refers to control of the Type I error rate under a data generating distribution P_0 that satisfies the *complete null hypothesis*, $H_0^C = \prod_{m=1}^M H_0(m) = \prod_{m=1}^M \mathrm{I}\left(P \in \mathcal{M}(m)\right) = \mathrm{I}\left(P \in \cap_{m=1}^M \mathcal{M}(m)\right)$, that all M null hypotheses are true, i.e., under a distribution P_0 that belongs to the intersection $\cap_{m=1}^M \mathcal{M}(m)$ of all M submodels.

In contrast, *strong control*, as defined in Westfall and Young (1993), considers *all* 2^M *possible subsets of null hypotheses*, $\mathcal{J}_0 \subseteq \{1, \ldots, M\}$, and refers to control of the Type I error rate under each of 2^M distributions $P_{\mathcal{J}_0} \in \cap_{m \in \mathcal{J}_0} \mathcal{M}(m)$ that satisfy subsets of null hypotheses $\mathcal{J}_0$. In particular, strong control implies weak control for $\mathcal{J}_0 = \{1, \ldots, M\}$.

As detailed below, the definitions of weak and strong control implicitly assume the existence of a *mapping* $\mathcal{J}_0 \rightarrow P_{\mathcal{J}_0}$, from subsets $\mathcal{J}_0$ of null hypotheses to data generating distributions $P_{\mathcal{J}_0} \in \cap_{m \in \mathcal{J}_0} \mathcal{M}(m)$ that satisfy each of the null hypotheses in $\mathcal{J}_0$.

It is important to recognize that, although strong control does consider the subset $\mathcal{H}_0 = \mathcal{H}_0(P)$ of true null hypotheses corresponding to the true data generating distribution P, Type I error control under P is not guaranteed by strong control, unless the mapping $\mathcal{J}_0 \rightarrow P_{\mathcal{J}_0}$ results in $P_{\mathcal{H}_0} = P$.

Defining a data generating distribution that satisfies a given subset of null hypotheses

In much of the multiple testing literature, Type I error rates are defined loosely in terms of probabilities *given subsets of null hypotheses*, rather than probabilities *under distributions that satisfy subsets of null hypotheses*, i.e., *under distributions that belong to intersections of submodels*. For example, Westfall and Young (1993, p. 9) refer to the FWEP as the family-wise error rate "computed under the *partial null hypothesis* (meaning that some subcollection of nulls, say $H_{j_1}, \ldots, H_{j_t}$, is true)" and provide the following definition in their Equation (1.2),

$$FWEP = \Pr(\text{Reject at least one } H_i,\ i = j_1, \ldots, j_t | H_{j_1}, \ldots, H_{j_t} \text{ are true}).$$

As discussed in Dudoit et al. (2004b) and Pollard and van der Laan (2004), such a quantity is not well-defined, because Type I error rates are parameters of a distribution for the number of Type I errors (and possibly the number of rejected hypotheses, as for the FDR) and can only be defined meaningfully with respect to such a distribution (Section 1.2.9). A more precise definition would be that FWEP is the family-wise error rate *under a data generating distribution* $P_{\mathcal{J}_0}$ that satisfies a certain subset $\mathcal{J}_0 = \{j_1, \ldots, j_t\}$ of null hypotheses, i.e., defined such that $P_{\mathcal{J}_0} \in \cap_{m \in \mathcal{J}_0} \mathcal{M}(m)$.

This immediately raises the issue of how to map from a subset $\mathcal{J}_0$ of null hypotheses to a well-defined data generating distribution $P_{\mathcal{J}_0} \in \cap_{m \in \mathcal{J}_0} \mathcal{M}(m)$. Except in very simple situations (e.g., null hypotheses concerning the mean vector of a multivariate Gaussian data generating distribution), each subset $\mathcal{J}_0$ of null hypotheses corresponds to a family of possible distributions. One approach is to define the distribution $P_{\mathcal{J}_0}$ as a projection of the true data generating distribution P onto the submodel $\cap_{m \in \mathcal{J}_0} \mathcal{M}(m)$, selecting, for example, the distribution $P_{\mathcal{J}_0} \in \cap_{m \in \mathcal{J}_0} \mathcal{M}(m)$ with the smallest Kullback-Leibler divergence with P. That is,

$$\begin{aligned} P_{\mathcal{J}_0} &= \Pi_{KL}(P| \cap_{m \in \mathcal{J}_0} \mathcal{M}(m)) \qquad (2.64) \\ &\equiv \arg\max_{P' \in \cap_{m \in \mathcal{J}_0} \mathcal{M}(m)} \int \log \left(\frac{dP'(x)}{d\mu(x)} \right) dP(x), \end{aligned}$$

for a dominating measure μ. Another possibility is to select the distribution $P_{\mathcal{J}_0}$ on the conservative boundary of the submodel $\cap_{m \in \mathcal{J}_0} \mathcal{M}(m)$. The reader is referred to Pollard and van der Laan (2004) for a discussion of multivariate data generating null distributions and proposals for specifying such joint distributions based on projections of the true data generating distribution P onto submodels satisfying subsets of null hypotheses.

However, as discussed by these authors, in many testing problems of interest, one simply cannot identify a data generating null distribution $P_0 \in \cap_{m=1}^{M} \mathcal{M}(m)$ that provides proper control of the Type I error rate under the true data generating distribution P. That is, in many cases, the assumed *null distribution* $Q_{n,\mathcal{H}_0}(P_0)$ and the *true distribution* $Q_{n,\mathcal{H}_0}(P)$ of the $\mathcal{H}_0$-specific subvector $(T_n(m) : m \in \mathcal{H}_0)$ of test statistics have different limits and thus violate null domination Assumption $\mathsf{ND}\Theta$ for the Type I error rate, i.e., $\lim_n \Theta(F_{V_n}) > \Theta(F_{V_0}) = \alpha$. Instead, for the test of single-parameter null hypotheses using t-statistics (Section 2.6), Pollard and van der Laan (2004) recommend using a test statistics null distribution such as the Kullback-Leibler projection of $Q_n = Q_n(P)$ onto the space of multivariate Gaussian distributions with mean vector zero. The projection null distribution corresponds to the null distribution $Q_0^t(P) = \mathrm{N}(0, \Sigma^*(P))$ proposed in Theorem 2.6.

Revised definitions of weak and strong Type I error control

As usual, consider the simultaneous test of M null hypotheses $H_0(m)$, $m = 1, \ldots, M$, based on test statistics $T_n = (T_n(m) : m = 1, \ldots, M)$, with true finite sample joint distribution $Q_n = Q_n(P)$ and null distribution Q_0. Let $\mathcal{C}_n(m) = \mathcal{C}(m; T_n, Q_0, \alpha)$, $m = 1, \ldots, M$, and $\mathcal{R}_n = \mathcal{R}(T_n, Q_0, \alpha)$ denote, respectively, the M rejection regions and corresponding set of rejected null hypotheses, for a MTP with nominal Type I error level α. That is,

$$\mathcal{R}(T_n, Q_0, \alpha) = \{m : T_n(m) \in \mathcal{C}(m; T_n, Q_0, \alpha)\}.$$

Given a subset of null hypotheses $\mathcal{J}_0 \subseteq \{1, \ldots, M\}$, define a data generating distribution $P_{\mathcal{J}_0} \in \cap_{m \in \mathcal{J}_0} \mathcal{M}(m)$ and let $Q_n(P_{\mathcal{J}_0})$ denote the corresponding joint distribution for the test statistics T_n. Following the notation introduced in Equations (2.2) and (2.3), denote the numbers of rejected hypotheses and Type I errors by

$$R_n(\mathcal{J}_0) \equiv R(\mathcal{C}_n | Q_n(P_{\mathcal{J}_0})) = \sum_{m=1}^{M} \mathrm{I}\left(T_n(m) \in \mathcal{C}_n(m)\right) \tag{2.65}$$

and

$$V_n(\mathcal{J}_0) \equiv V(\mathcal{C}_n | Q_n(P_{\mathcal{J}_0})) = \sum_{m \in \mathcal{J}_0} \mathrm{I}\left(T_n(m) \in \mathcal{C}_n(m)\right),$$

respectively, under the assumption that $T_n \sim Q_n(P_{\mathcal{J}_0})$.

Strong control of a Type I error rate at level α requires that

$$\max_{\mathcal{J}_0 \subseteq \{1,\ldots,M\}} \Theta(F_{V_n(\mathcal{J}_0), R_n(\mathcal{J}_0)}) \leq \alpha \qquad \text{[finite sample strong control]}$$

$$\limsup_{n \to \infty} \max_{\mathcal{J}_0 \subseteq \{1,\ldots,M\}} \Theta(F_{V_n(\mathcal{J}_0), R_n(\mathcal{J}_0)}) \leq \alpha \qquad \text{[asymptotic strong control]}. \tag{2.66}$$

Thus, strong control involves considering 2^M distributions $P_{\mathcal{J}_0}$, each corresponding to a subset $\mathcal{J}_0$ of null hypotheses. Note also that this definition of strong control is completely dependent upon the choice of mapping $\mathcal{J}_0 \to P_{\mathcal{J}_0}$.

Weak control corresponds to $\mathcal{J}_0 = \{1, \ldots, M\}$ and $P_0 = P_{\{1,\ldots,M\}}$.

Type I error control under the true data generating distribution P does not necessarily follow from strong control, unless the mapping $\mathcal{J}_0 \to P_{\mathcal{J}_0}$ results in $P_{\mathcal{H}_0} = P$ for $\mathcal{J}_0 = \mathcal{H}_0$. In other words, control under the true P could fail under strong control when an improper mapping is used to define $P_{\mathcal{H}_0}$.

In contrast, as discussed in Section 2.2, the methodology proposed in this book is only concerned with Type I error control under the true data generating distribution P. That is, we only require that Equation (2.66) hold in the special case where $\mathcal{J}_0 = \mathcal{H}_0$ and $P_{\mathcal{H}_0} = P$.

2.8.2 Subset pivotality

In practice, it is not feasible to consider all 2^M possible subsets of null hypotheses and commonly-used single-step and stepwise multiple testing procedures are typically based on cut-offs derived under a data generating distribution P_0 that satisfies the complete null hypothesis $H_0^C = \prod_{m=1}^{M} H_0(m)$, i.e.,

$P_0 \in \cap_{m=1}^{M} \mathcal{M}(m)$. Strong control of a Type I error rate, and in particular control under the true data generating distribution P, are then claimed to follow from weak control under conditions such as subset pivotality.

As stated in Condition 2.1, p. 42, in Westfall and Young (1993), "The distribution of $\mathbf{P}$ has the *subset pivotality* property if the joint distribution of the subvector $\{P_i : i \in K\}$ is identical under the restrictions $\cap_{i \in K} H_{0i}$ and H_0^C, for all subsets $K = \{i_1, \ldots, i_j\}$ of true null hypotheses". In our notation, K is a subset $\mathcal{J}_0 \subseteq \{1, \ldots, M\}$ of null hypotheses and $\mathbf{P}$ refers to the vector $(P_{0n}(m) : m = 1, \ldots, M)$ of unadjusted p-values (Section 1.2.12).

As for the definitions of weak and strong control, subset pivotality implicitly assumes the existence of a mapping $\mathcal{J}_0 \to P_{\mathcal{J}_0}$, from subsets $\mathcal{J}_0$ of null hypotheses to data generating distributions $P_{\mathcal{J}_0} \in \cap_{m \in \mathcal{J}_0} \mathcal{M}(m)$ that satisfy each of the null hypotheses in $\mathcal{J}_0$. A (finite sample) subset pivotality condition for test statistics can then be stated as

$$Q_{n,\mathcal{J}_0}(P_{\mathcal{J}_0}) = Q_{n,\mathcal{J}_0}(P_0), \qquad \forall\ \mathcal{J}_0 \subseteq \{1, \ldots, M\}, \tag{2.67}$$

in terms of distributions $P_{\mathcal{J}_0}$ corresponding to subsets $\mathcal{J}_0$ of null hypotheses and where $P_0 = P_{\{1,\ldots,M\}}$.

Note that the subset pivotality condition considers all 2^M possible subsets of null hypotheses, and not simply the subset $\mathcal{J}_0 = \mathcal{H}_0(P)$ corresponding to the true data generating distribution P. In this sense, and provided $P_{\mathcal{H}_0} = P$, the assumption is stronger than needed, because it is only of interest to control Type I error rates under the true P, that is, the only relevant condition is $Q_{n,\mathcal{H}_0}(P) = Q_{n,\mathcal{H}_0}(P_0)$ for $\mathcal{J}_0 = \mathcal{H}_0$. In general, however, subset pivotality does not guarantee control under the true P, if an improper mapping $\mathcal{J}_0 \to P_{\mathcal{J}_0}$ is used and $P_{\mathcal{H}_0} \neq P$.

Finally, as discussed in Section 2.2.4, the subset pivotality assumption in Equation (2.67) differs from our (finite sample) joint null domination Assumption jtNDT which: (i) only considers the subset $\mathcal{J}_0 = \mathcal{H}_0$; (ii) does not require the test statistics null distribution $Q_{0,n}$ or Q_0 to be defined in terms of a data generating null distribution P_0, i.e., $Q_{0,n} = Q_n(P_0)$; (iii) does not require equality of the true and null test statistics distributions, but the weaker null domination, i.e., $Q_{n,\mathcal{H}_0}(P) \geq Q_{n,\mathcal{H}_0}(P_0)$.

2.9 Test statistics null distributions based on bootstrap and permutation data generating distributions

Permutation procedures are widely-used in multiple testing to obtain data generating null distributions P_0 and corresponding test statistics null distributions $Q_n(P_0)$ (Westfall and Young, 1993). This section builds on Pollard and van der Laan (2004) and compares bootstrap- and permutation-based test statistics null distributions.

2.9.1 The two-sample test of means problem

Consider a two-sample test of means problem, with data structure $(X, Y) \sim P \in \mathcal{M}$, where $X = (X(m) : m = 1, \ldots, M)$ is a random M-vector and $Y \in \{1, 2\}$ a binary population label. For Population k, $k = 1, 2$, let $\eta_k = \Pr(Y = k)$ denote the population frequency, let $P_{X|k}$ denote the conditional data generating distribution of X given $Y = k$ (i.e., $X|Y = k \sim P_{X|k}$), and let $\psi_k = (\psi_k(m) : m = 1, \ldots, M) = \mathrm{E}[X|Y = k]$ and $\sigma_k = \mathrm{Cov}[X|Y = k]$ denote, respectively, the conditional M-dimensional mean vector and $M \times M$ covariance matrix of X. Consider testing the following M null hypotheses concerning the differences $\psi(m) = \psi_1(m) - \psi_2(m)$ in conditional means,

$$H_0(m) = \mathrm{I}\left(\psi(m) = 0\right), \qquad m = 1, \ldots, M. \tag{2.68}$$

Suppose one has a random sample $\mathcal{XY}_n = \{(X_i, Y_i) : i = 1, \ldots, n\}$, of n IID copies of the pair $(X, Y) \sim P \in \mathcal{M}$. Denote the (random) sample size for Population k by $n_k = \sum_i \mathrm{I}\left(Y_i = k\right)$ and estimate the conditional mean vector ψ_k by the corresponding empirical mean vector $\psi_{k,n_k} = \bar{X}_{k,n_k}$, with elements $\psi_{k,n_k}(m) = \bar{X}_{k,n_k}(m) = \sum_i \mathrm{I}\left(Y_i = k\right) X_i(m)/n_k$. The null hypotheses can be tested using (unstandardized) difference statistics,

$$\begin{aligned} D_n(m) &\equiv \sqrt{n}(\psi_{2,n_2}(m) - \psi_{1,n_1}(m)) & (2.69)\\ &= \sqrt{n} \sum_{i=1}^{n} \left(\frac{\mathrm{I}\left(Y_i = 2\right) X_i(m)}{n_2} - \frac{\mathrm{I}\left(Y_i = 1\right) X_i(m)}{n_1} \right), \qquad m = 1, \ldots, M. \end{aligned}$$

Consider the following two models, $\mathcal{M}$ and $\mathcal{M}_= \subseteq \mathcal{M}$, corresponding, respectively, to general non-parametric and location-shifted conditional data generating distributions $P_{X|1}$ and $P_{X|2}$.

Non-parametric model, $\mathcal{M}$. For the *non-parametric model* $(X, Y) \sim P \in \mathcal{M}$, $X|Y = 1 \sim P_{X|1}$ and $X|Y = 2 \sim P_{X|2}$, where $P_{X|1}$ and $P_{X|2}$ are arbitrary conditional data generating distributions for Populations 1 and 2, respectively.

Location shift model, $\mathcal{M}_=$. For the *location shift model* $(X, Y) \sim P_= \in \mathcal{M}_=$, $X|Y = 1 \sim P_{X|1} = P_X(\cdot - \psi_1)$ and $X|Y = 2 \sim P_{X|2} = P_X(\cdot - \psi_2)$, where P_X is a common M-dimensional distribution with mean vector zero. That is, $P_{X|1}$ and $P_{X|2}$ are identical except for a location shift.

The implications of each model are investigated in terms of the choice of an appropriate null distribution for the test statistics D_n. Model $\mathcal{M}_= \subseteq \mathcal{M}$ makes the strong assumption that, under the complete null hypothesis $H_0^C = \prod_{m=1}^{M} H_0(m) = \mathrm{I}\left(\psi_1 = \psi_2\right)$, the random vector X has the same conditional distribution in the two populations ($P_{X|1} = P_{X|2}$), that is, X and Y are independent. If one were testing the null hypothesis $\mathrm{I}\left(P_{X|1} = P_{X|2}\right)$ that the conditional data generating distributions are identical for the two populations, then $\mathcal{M}_=$ would clearly be a good choice of model from which to select a data generating null distribution. However, model $\mathcal{M}_=$ may be a poor choice for testing the null hypotheses in Equation (2.68), which only concern differences in means between the two populations and allow, in particular, different covariance structures σ_k in each population.

2.9.2 Distribution of the test statistics under two different data generating distributions

By the Central Limit Theorem (Theorem B.4), the difference statistics D_n have a Gaussian asymptotic distribution. This distribution is fully specified by its mean vector (with elements equal to zero for the true null hypotheses) and its covariance matrix. In what follows, we therefore focus on properties and estimation of the covariance matrix of the test statistics.

For simplicity, and without loss of generality, consider only $M = 2$ null hypotheses, i.e., a bivariate random vector X.

Proposition 2.8, below, provides asymptotic variances and covariances for the difference statistics D_n under two different data generating distributions for (X, Y).

Proposition 2.8. [Asymptotic variances and covariances of difference statistics for two-sample test of means, under two different data generating distributions] *Consider a data structure $(X, Y) \sim P \in \mathcal{M}$, where $X = (X(1), X(2))$ is a bivariate random vector and $Y \in \{1, 2\}$ is a binary population label, with $\eta_k = \Pr(Y = k)$, $k = 1, 2$. Let $P_{X|k}$ denote a bivariate distribution, with mean vector $\psi_k = [\psi_k(1), \psi_k(2)]^\top$ and covariance matrix $\sigma_k = (\sigma_k(m, m') : m, m' = 1, 2)$, $k = 1, 2$. Specifically, consider the following two data generating distributions for (X, Y).*

Non-parametric data generating distribution, P. *For $(X, Y) \sim P$, the conditional distribution of X given $Y = k$ is $P_{X|k}$, that is, $X|Y = k \sim P_{X|k}$, $k = 1, 2$.*

Independence data generating distribution, $P_\perp$. *For $(X, Y) \sim P_\perp$, X and Y are independent and X has the mixture distribution $X \sim \eta_1 P_{X|1} + \eta_2 P_{X|2}$.*

Then, for a random sample $\mathcal{XY}_n = \{(X_i, Y_i) : i = 1, \ldots, n\}$, of n IID copies of the pair $(X, Y) \sim P$, the asymptotic covariance matrix of the difference statistics $D_n = (D_n(1), D_n(2))$ of Equation (2.69) is given by

$$\varsigma \equiv \lim_{n \to \infty} \mathrm{Cov}_P[D_n] = \begin{bmatrix} \frac{\sigma_1(1,1)}{\eta_1} + \frac{\sigma_2(1,1)}{\eta_2} & \frac{\sigma_1(1,2)}{\eta_1} + \frac{\sigma_2(1,2)}{\eta_2} \\ \frac{\sigma_1(1,2)}{\eta_1} + \frac{\sigma_2(1,2)}{\eta_2} & \frac{\sigma_1(2,2)}{\eta_1} + \frac{\sigma_2(2,2)}{\eta_2} \end{bmatrix}. \tag{2.70}$$

For $(X, Y) \sim P_\perp$,

$$\varsigma_\perp \equiv \lim_{n \to \infty} \mathrm{Cov}_{P_\perp}[D_n] = \begin{bmatrix} \frac{\sigma_1(1,1)}{\eta_2} + \frac{\sigma_2(1,1)}{\eta_1} & \frac{\sigma_1(1,2)}{\eta_2} + \frac{\sigma_2(1,2)}{\eta_1} \\ \frac{\sigma_1(1,2)}{\eta_2} + \frac{\sigma_2(1,2)}{\eta_1} & \frac{\sigma_1(2,2)}{\eta_2} + \frac{\sigma_2(2,2)}{\eta_1} \end{bmatrix}. \tag{2.71}$$

It is interesting to note that the asymptotic covariance matrices ς and $\varsigma_\perp$ of the difference statistics D_n are identical, except for the roles of population frequencies η_1 and η_2 being reversed.

The expressions for ς and $\varsigma_\perp$ illustrate that, for most values of the parameters σ_k and η_k, $k = 1, 2$, the difference statistics have different asymptotic distributions for data generating distributions P and $P_\perp$. If, however, either (i) $\eta_1 = \eta_2$ or (ii) $\sigma_1 = \sigma_2$, then the asymptotic distributions are the same for both scenarios, i.e., $\varsigma = \varsigma_\perp$.

As discussed below, the bootstrap estimator of the distribution of the test statistics D_n converges to the asymptotic distribution of D_n under P, while the permutation estimator of the distribution of D_n converges to the asymptotic distribution of D_n under $P_\perp$. Thus, under the reduced location shift model $\mathcal{M}_=$, for which $\sigma_1 = \sigma_2$, a permutation data generating distribution yields a sensible test statistics null distribution.

It is somewhat surprising that, even when the data generating distribution P is not an element of the reduced model $\mathcal{M}_=$ (e.g., $\sigma_1 \neq \sigma_2$), one still has $\varsigma = \varsigma_\perp$ when $\eta_1 = \eta_2$. Thus, in the case of equal population frequencies (i.e., $\eta_1 = \eta_2$), permutation distributions, corresponding to the independence data generating distribution $P_\perp$, yield valid test statistics null distributions.

In summary, Proposition 2.8 suggests that, unless either (i) $\eta_1 = \eta_2$ or (ii) $\sigma_1 = \sigma_2$, one should use the bootstrap (rather than permutation) to estimate the null distribution of the test statistics D_n, since the bootstrap preserves the covariance structure ς of these test statistics. However, for equal population frequencies (i.e., $\eta_1 = \eta_2$) or covariance structures (i.e., $\sigma_1 = \sigma_2$, as in model $\mathcal{M}_=$), one could use permutation estimators, because the asymptotic covariance matrix of the test statistics D_n is the same for both data generating distributions P and $P_\perp$ (i.e., $\varsigma = \varsigma_\perp$). Furthermore, permutation estimators of the covariance matrix tend to be more efficient than non-parametric bootstrap estimators, because they correspond to a smaller model and make use of all n observations (Pollard and van der Laan, 2004).

Similar conclusions apply to the usual (standardized) two-sample Welch t-statistics,

$$T_n(m) \equiv \frac{\psi_{2,n_2}(m) - \psi_{1,n_1}(m)}{\sqrt{\frac{\sigma^2_{1,n_1}(m)}{n_1} + \frac{\sigma^2_{2,n_2}(m)}{n_2}}}, \tag{2.72}$$

where n_k, $\psi_{k,n_k}(m)$, and $\sigma^2_{k,n_k}(m)$ denote, respectively, the sample size, empirical means, and empirical variances, for Population k, $k = 1, 2$.

Proof of Proposition 2.8. The derivations of variances and covariances for the difference statistics $D_n = (D_n(1), D_n(2))$ are similar for the two data generating distributions P and $P_\perp$ and make use of the Double Expectation Theorem. For simplicity, and without loss of generality, assume that both null hypotheses $H_0(1)$ and $H_0(2)$ are true and that the mean vectors for $P_{X|1}$ and $P_{X|2}$ are zero, i.e., $\psi_1 = \psi_2 = [0, 0]^\top$. Then,

$\mathrm{E}[D_n] = [0,0]^\top$ and $\mathrm{Cov}[D_n] = \mathrm{E}[D_n D_n^\top]$ under both distributions P and $P_\perp$. Let $\mathcal{Y}_n = \{Y_i : i = 1, \ldots, n\}$.

Variances. First, derive the asymptotic variances of the difference statistics $D_n(m)$.

$$\begin{aligned}
\lim_{n\to\infty} \mathrm{Var}[D_n(m)] &= \lim_{n\to\infty} \mathrm{E}[D_n^2(m)] \\
&= \lim_{n\to\infty} n\,\mathrm{E}\left[\mathrm{E}\left[\left(\sum_{i=1}^{n}\left(\frac{\mathrm{I}(Y_i=2)\,X_i(m)}{n_2} - \frac{\mathrm{I}(Y_i=1)\,X_i(m)}{n_1}\right)\right)^2 \middle| \mathcal{Y}_n\right]\right] \\
&= \lim_{n\to\infty} n\,\mathrm{E}\left[\mathrm{E}\left[\sum_{i=1}^{n}\left(\frac{\mathrm{I}(Y_i=2)\,X_i(m)}{n_2} - \frac{\mathrm{I}(Y_i=1)\,X_i(m)}{n_1}\right)^2 \middle| \mathcal{Y}_n\right]\right] \\
&= \lim_{n\to\infty} n\,\mathrm{E}\left[\mathrm{E}\left[\sum_{i=1}^{n}\left(\frac{\mathrm{I}(Y_i=2)\,X_i^2(m)}{n_2^2} + \frac{\mathrm{I}(Y_i=1)\,X_i^2(m)}{n_1^2}\right) \middle| \mathcal{Y}_n\right]\right] \\
&= \lim_{n\to\infty} n\,\mathrm{E}\left[\sum_{i=1}^{n}\left(\frac{\mathrm{I}(Y_i=2)\,\mathrm{E}[X_i^2(m)|Y_i=2]}{n_2^2}\right.\right. \\
&\qquad\qquad \left.\left. + \frac{\mathrm{I}(Y_i=1)\,\mathrm{E}[X_i^2(m)|Y_i=1]}{n_1^2}\right)\right] \\
&= \mathrm{E}[X^2(m)|Y=2] \lim_{n\to\infty} \mathrm{E}\left[\frac{n}{n_2}\right] + \mathrm{E}[X^2(m)|Y=1] \lim_{n\to\infty} \mathrm{E}\left[\frac{n}{n_1}\right] \\
&= \frac{\mathrm{E}[X^2(m)|Y=2]}{\eta_2} + \frac{\mathrm{E}[X^2(m)|Y=1]}{\eta_1}.
\end{aligned}$$

The third equality follows by noting that the (X_i, Y_i) are independent, with $\mathrm{E}[X_i(m)|Y_i = k] = 0$, $m = 1, 2$, $k = 1, 2$; the fourth from $\mathrm{I}(Y_i = 1)\,\mathrm{I}(Y_i = 2) = 0$; the sixth from the fact that the (X_i, Y_i) are identically distributed; and the seventh from $\lim_n n_k/n = \eta_k$ a.s., $k = 1, 2$.

When $(X, Y) \sim P$,

$$\begin{aligned}
\lim_{n\to\infty} \mathrm{Var}[D_n(m)] &= \frac{\mathrm{E}[X^2(m)|Y=1]}{\eta_1} + \frac{\mathrm{E}[X^2(m)|Y=2]}{\eta_2} \\
&= \frac{\sigma_1(m,m)}{\eta_1} + \frac{\sigma_2(m,m)}{\eta_2}, \qquad m = 1, 2.
\end{aligned}$$

Similarly, when $(X, Y) \sim P_\perp$, the asymptotic variance of the difference statistic $D_n(m)$ is as follows.

$$
\begin{aligned}
\lim_{n\to\infty} \mathrm{Var}[D_n(m)] &= \frac{\mathrm{E}[X^2(m)|Y=1]}{\eta_1} + \frac{\mathrm{E}[X^2(m)|Y=2]}{\eta_2} \\
&= \frac{\mathrm{E}[X^2(m)]}{\eta_1} + \frac{\mathrm{E}[X^2(m)]}{\eta_2} \\
&= \left(\frac{1}{\eta_1} + \frac{1}{\eta_2}\right)(\eta_1\sigma_1(m,m) + \eta_2\sigma_2(m,m)) \\
&= \frac{\sigma_1(m,m)}{\eta_2} + \frac{\sigma_2(m,m)}{\eta_1}, \qquad m = 1,\, 2.
\end{aligned}
$$

The second and third equalities follow by noting that X and Y are independent, with X having the mixture distribution $X \sim \eta_1 P_{X|1} + \eta_2 P_{X|2}$, so that $\mathrm{Var}[X(m)] = \mathrm{E}[X^2(m)] = \mathrm{E}[X^2(m)|Y=1] = \mathrm{E}[X^2(m)|Y=2] = \eta_1\sigma_1(m,m) + \eta_2\sigma_2(m,m)$, $m = 1,\, 2$.

Covariances. Now consider the asymptotic covariance between the difference statistics $D_n(1)$ and $D_n(2)$.

$$
\begin{aligned}
\lim_{n\to\infty} \mathrm{Cov}[D_n(1), D_n(2)] &= \lim_{n\to\infty} \mathrm{E}[D_n(1)D_n(2)] \\
&= \lim_{n\to\infty} n\,\mathrm{E}\Bigg[\mathrm{E}\Bigg[\sum_{i=1}^{n}\left(\frac{\mathrm{I}(Y_i=2)\,X_i(1)}{n_2} - \frac{\mathrm{I}(Y_i=1)\,X_i(1)}{n_1}\right) \\
&\qquad \times \sum_{i=1}^{n}\left(\frac{\mathrm{I}(Y_i=2)\,X_i(2)}{n_2} - \frac{\mathrm{I}(Y_i=1)\,X_i(2)}{n_1}\right)\Bigg|\,\mathcal{Y}_n\Bigg]\Bigg] \\
&= \lim_{n\to\infty} n\,\mathrm{E}\Bigg[\mathrm{E}\Bigg[\sum_{i=1}^{n}\left(\frac{\mathrm{I}(Y_i=2)\,X_i(1)}{n_2} - \frac{\mathrm{I}(Y_i=1)\,X_i(1)}{n_1}\right) \\
&\qquad \times \left(\frac{\mathrm{I}(Y_i=2)\,X_i(2)}{n_2} - \frac{\mathrm{I}(Y_i=1)\,X_i(2)}{n_1}\right)\Bigg|\,\mathcal{Y}_n\Bigg]\Bigg] \\
&= \lim_{n\to\infty} n\,\mathrm{E}\Bigg[\mathrm{E}\Bigg[\sum_{i=1}^{n}\Bigg(\frac{\mathrm{I}(Y_i=2)\,X_i(1)X_i(2)}{n_2^2} \\
&\qquad + \frac{\mathrm{I}(Y_i=1)\,X_i(1)X_i(2)}{n_1^2}\Bigg)\Bigg|\,\mathcal{Y}_n\Bigg]\Bigg] \\
&= \lim_{n\to\infty} n\,\mathrm{E}\Bigg[\sum_{i=1}^{n}\Bigg(\frac{\mathrm{I}(Y_i=2)\,\mathrm{E}[X_i(1)X_i(2)|Y_i=2]}{n_2^2} \\
&\qquad + \frac{\mathrm{I}(Y_i=1)\,\mathrm{E}[X_i(1)X_i(2)|Y_i=1]}{n_1^2}\Bigg)\Bigg] \\
&= \mathrm{E}[X(1)X(2)|Y=2] \lim_{n\to\infty} \mathrm{E}\left[\frac{n}{n_2}\right] + \mathrm{E}[X(1)X(2)|Y=1] \lim_{n\to\infty} \mathrm{E}\left[\frac{n}{n_1}\right] \\
&= \frac{\mathrm{E}[X(1)X(2)|Y=2]}{\eta_2} + \frac{\mathrm{E}[X(1)X(2)|Y=1]}{\eta_1}.
\end{aligned}
$$

The third equality follows by noting that the (X_i, Y_i) are independent, with $\mathrm{E}[X_i(m)|Y_i = k] = 0$, $m = 1, 2$, $k = 1, 2$; the fourth from $\mathrm{I}(Y_i = 1)\,\mathrm{I}(Y_i = 2) = 0$; the sixth from the fact that the (X_i, Y_i) are identically distributed; and the seventh from $\lim_n n_k/n = \eta_k$ a.s., $k = 1, 2$.

When $(X, Y) \sim P$,

$$\begin{aligned}\lim_{n\to\infty} \mathrm{Cov}[D_n(1), D_n(2)] &= \frac{\mathrm{E}[X(1)X(2)|Y=1]}{\eta_1} + \frac{\mathrm{E}[X(1)X(2)|Y=2]}{\eta_2} \\ &= \frac{\sigma_1(1,2)}{\eta_1} + \frac{\sigma_2(1,2)}{\eta_2}.\end{aligned}$$

Similarly, when $(X, Y) \sim P_\perp$, the asymptotic covariance of the difference statistics $D_n(1)$ and $D_n(2)$ is as follows.

$$\begin{aligned}\lim_{n\to\infty} \mathrm{Cov}[D_n(1), D_n(2)] &= \frac{\mathrm{E}[X(1)X(2)|Y=1]}{\eta_1} + \frac{\mathrm{E}[X(1)X(2)|Y=2]}{\eta_2} \\ &= \frac{\mathrm{E}[X(1)X(2)]}{\eta_1} + \frac{\mathrm{E}[X(1)X(2)]}{\eta_2} \\ &= \left(\frac{1}{\eta_1} + \frac{1}{\eta_2}\right)(\eta_1\sigma_1(1,2) + \eta_2\sigma_2(1,2)) \\ &= \frac{\sigma_1(1,2)}{\eta_2} + \frac{\sigma_2(1,2)}{\eta_1}.\end{aligned}$$

The second and third equalities follow by noting that X and Y are independent, with X having the mixture distribution $X \sim \eta_1 P_{X|1} + \eta_2 P_{X|2}$, so that $\mathrm{Cov}[X(1), X(2)] = \mathrm{E}[X(1)X(2)] = \mathrm{E}[X(1)X(2)|Y = 1] = \mathrm{E}[X(1)X(2)|Y = 2] = \eta_1\sigma_1(1, 2) + \eta_2\sigma_2(1, 2)$.

□

2.9.3 Bootstrap and permutation test statistics null distributions

As suggested by Proposition 2.8, the non-parametric model $\mathcal{M}$ and the smaller location shift model $\mathcal{M}_=$ imply different bootstrap sampling distributions for estimating the distribution of the test statistics $D_n = (D_n(m) : m = 1, \ldots, M)$. In particular, each model implies different data generating and test statistics null distributions. For non-parametric model $\mathcal{M}$, one samples from the joint empirical distribution of the pair (X, Y), whereas for reduced model $\mathcal{M}_=$, one samples from a model-based estimator of the data generating distribution.

Bootstrap test statistics null distribution for model $\mathcal{M}$

For the non-parametric model $\mathcal{M}$, the bootstrap estimator of the *joint data generating distribution* P of the pair (X, Y) is the *joint empirical distribution*

P_n of the $n = n_1 + n_2$ pairs of (X, Y)-observations, $\{(X_i, Y_i) : i = 1, \ldots, n\}$. One resamples n pairs of (X, Y)-observations at random, with replacement from P_n, to form a bootstrap sample $\{(X_i^\#, Y_i^\#) : i = 1, \ldots, n\}$. The bootstrap test statistics null distribution Q_{0n} is the empirical distribution of the M-vectors of centered difference statistics, $Z_n^\# = \sqrt{n}((\psi_{2,n_2}^\# - \psi_{1,n_1}^\#) - (\psi_{2,n_2} - \psi_{1,n_1}))$, where $\psi_{k,n_k}^\#$ denotes the bootstrap empirical mean vector for Population k, that is, $\psi_{k,n_k}^\# = \sum_i \mathrm{I}\left(Y_i^\# = k\right) X_i^\# / \sum_i \mathrm{I}\left(Y_i^\# = k\right)$, $k = 1, 2$.

Note that an asymptotically equivalent estimator could be obtained by sampling n_1 observations at random, with replacement from the Population 1 sample, $\{(X_i, Y_i) : Y_i = 1, i = 1, \ldots, n\}$, and n_2 observations at random, with replacement from the Population 2 sample, $\{(X_i, Y_i) : Y_i = 2, i = 1, \ldots, n\}$.

Bootstrap test statistics null distribution for model $\mathcal{M}_=$

For the reduced location shift model $\mathcal{M}_=$, the bootstrap estimator of the *common mean-zero marginal data generating distribution* P_X of X is the *centered marginal empirical distribution* $P_{X,n}$ of the $n = n_1 + n_2$ centered X-observations, $\{X_{=,i} : i = 1, \ldots, n\}$, where $X_{=,i} = X_i - \mathrm{I}(Y_i = 1)\, \psi_{1,n_1} - \mathrm{I}(Y_i = 2)\, \psi_{2,n_2}$. One resamples n centered X-observations, $\{X_{=,i}^\# : i = 1, \ldots, n\}$, at random, with replacement from $P_{X,n}$, and sets $X_i^\# = X_{=,i}^\# + \psi_{k,n_k}$ and $Y_i^\# = k$ for a random subset of n_k such observations, $k = 1, 2$, to form a bootstrap sample $\{(X_i^\#, Y_i^\#) : i = 1, \ldots, n\}$. Again, the bootstrap test statistics null distribution Q_{0n} is the empirical distribution of the M-vectors of centered difference statistics, $Z_n^\# = \sqrt{n}((\psi_{2,n_2}^\# - \psi_{1,n_1}^\#) - (\psi_{2,n_2} - \psi_{1,n_1}))$.

Note that the above bootstrap procedure for model $\mathcal{M}_=$ is equivalent to the following approach: form the *mixture marginal empirical distribution* $P_{X,n}$ of the $n = n_1 + n_2$ (uncentered) X-observations, $\{X_i : i = 1, \ldots, n\}$; resample n X-observations, $\{X_i^\# : i = 1, \ldots, n\}$, at random, with replacement from $P_{X,n}$; set $Y_i^\# = k$ for a random subset of n_k such observations, $k = 1, 2$; and define Q_{0n} as the empirical distribution of the M-vectors of (uncentered) difference statistics, $D_n^\# = \sqrt{n}(\psi_{2,n_2}^\# - \psi_{1,n_1}^\#)$. This yields the non-parametric bootstrap (sampling *with* replacement) analogue of the commonly-used permutation (sampling *without* replacement) test, corresponding to the independence data generating distribution $P_\perp$ in Proposition 2.8.

Permutation test statistics null distribution

Permutation tests are known to be exact (up to the discreteness of the permutation distribution) under the location shift model $\mathcal{M}_=$ and the complete null hypothesis (Theorem 6, p. 231, Lehmann (1986); Puri and Sen (1971)). Indeed, if X and Y are independent, then the permutation distribution is equal to the conditional joint distribution of the pair (X, Y), given the marginal empirical distributions of X and Y. In contrast, bootstrap procedures corresponding to non-parametric model $\mathcal{M}$ are only approximate. In other words, model $\mathcal{M}_= \subseteq \mathcal{M}$ implies a stronger null model restriction than $\mathcal{M}$, as needed for an exact test.

As remarked in Section 2.9.2, above, when the permutation approach is appropriate, it tends to provide less variable estimators of the test statistics null distribution than the non-parametric bootstrap. Indeed, it should come as no surprise that, for small sample sizes, one typically obtains more accurate test results using a model-based (permutation or other suitable) estimator of the null distribution than a non-parametric estimator.

In general, estimation of the test statistics null distribution involves a bias-variance trade-off and raises the interesting open question of model selection.

The reader is referred to Pollard et al. (2005a) and Pollard and van der Laan (2004) for a more detailed discussion of the relative merits of bootstrap- and permutation-based multiple testing procedures.

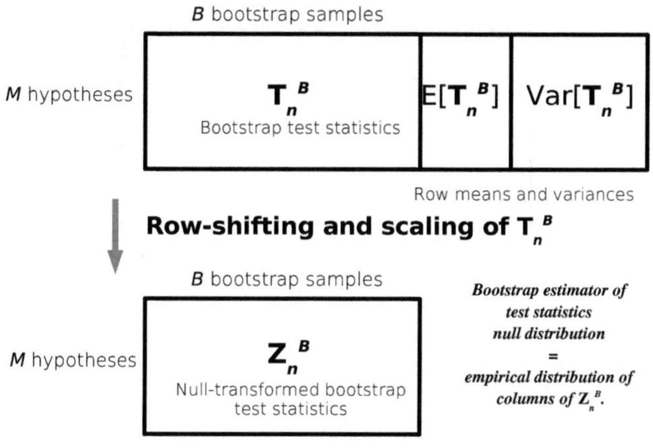

Figure 2.1. *Bootstrap estimation of the null shift and scale-transformed test statistics null distribution Q_0 (Procedure 2.3).* The bootstrap test statistics are stored in the $M \times B$ matrix $\mathbf{T}_n^B = \left(T_n^B(m,b)\right)$, with rows corresponding to the M null hypotheses and columns to the B bootstrap samples. Expected values, $\mathrm{E}[T_n(m)]$, and variances, $\mathrm{Var}[T_n(m)]$, of the test statistics are estimated by taking, respectively, row means and variances of $\mathbf{T}_n^B$. The matrix of test statistics $\mathbf{T}_n^B$ can then be row-shifted and scaled using the null values $\lambda_0(m)$ and $\tau_0(m)$, to produce an $M \times B$ matrix $\mathbf{Z}_n^B = \left(Z_n^B(m,b)\right)$. The null distribution Q_0 is estimated by the empirical distribution Q_{0n} of the columns of $\mathbf{Z}_n^B$.

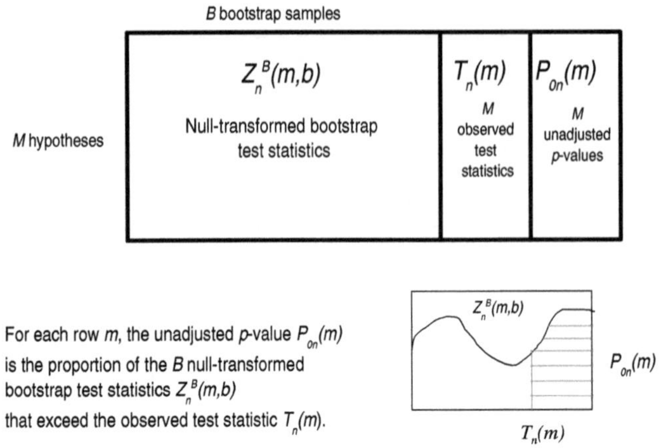

Figure 2.2. *Bootstrap estimation of the unadjusted p-values* $P_{0n}(m)$. Bootstrap estimators of the unadjusted p-values $P_{0n}(m)$ are obtained from the matrix $\mathbf{Z}_n^B = \big(Z_n^B(m,b)\big)$, of null-transformed bootstrap test statistics, by recording, for each row m, the proportion of $Z_n^B(m,b)$ that are greater than or equal to the observed test statistic $T_n(m)$.

3

Overview of Multiple Testing Procedures

3.1 Introduction

This chapter provides an overview of multiple testing procedures (MTP) for controlling the *number* of Type I errors (FWER and gFWER, in Sections 3.2 and 3.3, respectively) and the *proportion* of Type I errors among the rejected hypotheses (FDR and TPPFP, in Sections 3.4 and 3.5, respectively). The MTPs are summarized in Appendix A. Applications are described in Chapters 9–12 and software implementation is discussed in Chapter 13.

3.1.1 Set-up

As in Section 1.2, consider the simultaneous test of M null hypotheses $H_0(m)$, $m = 1, \ldots, M$, based on test statistics $T_n = (T_n(m) : m = 1, \ldots, M)$, with finite sample joint distribution $Q_n = Q_n(P)$, under the data generating distribution P. In practice, the true distribution $Q_n(P)$ of the test statistics is unknown and replaced by a null distribution $Q_0 = Q_0(P)$ (or estimator thereof, Q_{0n}), in order to derive rejection regions for the test statistics. Unless specified otherwise, assume that large values of the test statistic $T_n(m)$ provide evidence against the corresponding null hypothesis $H_0(m)$, that is, consider one-sided rejection regions of the form $\mathcal{C}_n(m;\alpha) = (c_n(m;\alpha), +\infty)$ and corresponding set of rejected null hypotheses $\mathcal{R}_n(\alpha) = \{m : T_n(m) \in \mathcal{C}_n(m;\alpha)\} = \{m : T_n(m) > c_n(m;\alpha)\}$.

Given a test statistics null distribution Q_0, define unadjusted p-values $P_{0n} = (P_{0n}(m) : m = 1, \ldots, M)$ as in Section 1.2.12. For continuous marginal null distributions $Q_{0,m}$ and one-sided rejection regions of the form $\mathcal{C}_n(m;\alpha) = (c_n(m;\alpha), +\infty)$, the unadjusted p-values are given by $P_{0n}(m) = \bar{Q}_{0,m}(T_n(m))$, where $\bar{Q}_{0,m}$ are the marginal survivor functions for Q_0.

Let $O_n(m)$ denote the indices for the ordered unadjusted p-values, $P^{\circ}_{0n}(m) = P_{0n}(O_n(m))$, so that $P_{0n}(O_n(1)) \leq \cdots \leq P_{0n}(O_n(M))$. For common-quantile MTPs, the mth most significant null hypothesis refers to the hypothesis $H_0(O_n(m))$ with the mth smallest unadjusted p-value $P^{\circ}_{0n}(m)$, i.e., to the hypothesis with p-value rank m; in contrast, for common-cut-off MTPs, the mth most significant null hypothesis is that with the mth largest test statistic.

The different multiple testing procedures presented in this chapter are stated in terms of their adjusted p-values,

$$\widetilde{P}_{0n}(m) = \inf\{\alpha \in [0,1] : m \in \mathcal{R}_n(\alpha)\} = \inf\{\alpha \in [0,1] : T_n(m) \in \mathcal{C}_n(m;\alpha)\}.$$

As discussed in Section 1.2.12, adjusted p-values provide a flexible representation of the results of a MTP and, in particular, are convenient benchmarks for comparing different MTPs: the smaller the adjusted p-values, the less conservative the procedure, i.e., the greater the number $R_n(\alpha) = |\mathcal{R}_n(\alpha)|$ of rejected null hypotheses at any nominal Type I error level α.

3.1.2 Type I error control and choice of a test statistics null distribution

Denote the number of Type I errors and the number of rejected null hypotheses, under the true distribution $Q_n = Q_n(P)$ for the test statistics T_n, by

$$V_n = V_n(\alpha) = \sum_{m \in \mathcal{H}_0} \mathrm{I}(T_n(m) \in \mathcal{C}_n(m;\alpha)) = \sum_{m \in \mathcal{H}_0} \mathrm{I}\left(\widetilde{P}_{0n}(m) \leq \alpha\right)$$

and

$$R_n = R_n(\alpha) = \sum_{m=1}^{M} \mathrm{I}(T_n(m) \in \mathcal{C}_n(m;\alpha)) = \sum_{m=1}^{M} \mathrm{I}\left(\widetilde{P}_{0n}(m) \leq \alpha\right),$$

respectively. Likewise, let V_0 and R_0 denote the corresponding quantities under the assumption that the test statistics T_n have the null distribution Q_0.

For Type I error rates $\Theta(F_{V_n})$, defined as arbitrary parameters of the distribution F_{V_n} of the number of Type I errors V_n, proofs of Type I error control are provided, for simplicity, under the assumption that the test statistics $(T_n(m) : m \in \mathcal{H}_0)$ for the true null hypotheses $\mathcal{H}_0$ have the null distribution $Q_{0,\mathcal{H}_0}$. In short, the proofs focus on control of the parameter $\Theta(F_{V_0})$, rather than $\Theta(F_{V_n})$.

Control of the actual Type I error rate $\Theta(F_{V_n})$, under the true distribution Q_n, follows only for a suitable choice of the null distribution Q_0. Section 2.2 provides a general characterization for a proper test statistics null distribution, in terms of null domination conditions for the joint distribution of the $\mathcal{H}_0$-specific test statistics $(T_n(m) : m \in \mathcal{H}_0)$ (Assumptions jtNDT, NDV, and NDΘ). This general characterization leads to the explicit proposal of two test statistics null distributions: the asymptotic distribution of a vector of null shift and scale-transformed test statistics (Section 2.3) and the asymptotic distribution of a vector of null quantile-transformed test statistics (Section 2.4).

The reader is referred to Sections 1.2.9 and 1.2.12 for further detail on Type I error rates and p-values, respectively, and to Chapter 2 for a more complete discussion of Type I error control and test statistics null distributions.

3.1.3 Marginal multiple testing procedures

We focus initially on *marginal multiple testing procedures*, that is, procedures based solely on the marginal distributions of the test statistics $T_n = (T_n(m) : m = 1, \ldots, M)$ (e.g., classical FWER-controlling single-step Bonferroni Procedure 3.1; gFWER-controlling single-step and step-down Lehmann and Romano (2005) Procedures 3.15 and 3.17). Such procedures typically test each null hypothesis $H_0(m)$ using cut-off rules based only on the marginal distribution of the corresponding individual test statistic $T_n(m)$ or unadjusted p-value $P_{0n}(m)$.

Rather than exploiting the dependence structure of the test statistics, the FWER-controlling marginal MTPs of Section 3.2 use explicit conservative upper bounds on tail probabilities $\Theta(F_{V_0}) = \Pr(V_0 > 0)$ for the number of Type I errors under a test statistics null distribution Q_0. The bounds typically rely on probability inequalities such as Boole's Inequality (Equation (B.1)) and are based on tail probabilities for the marginal null distributions $Q_{0,m}$. For instance, Bonferroni Procedure 3.1 relies on Boole's Inequality, $\Pr(\cup_{m=1}^{M} B_m) \leq \sum_{m=1}^{M} \Pr(B_m)$, for events B_m of the form $B_m = \{P_{0n}(m) \leq \alpha/M\}$. Other examples of FWER-controlling marginal methods are single-step and stepwise procedures based on parametric marginal distributions for the test statistics (e.g., standard normal distribution, $\mathrm{N}(0, 1)$) or on conservative finite sample marginal tail probabilities as obtained from Bernstein's, Boole's, Šidák's, or Simes' Inequality (Hochberg, 1988; Holland and Copenhaver, 1987; Holm, 1979; Hommel, 1988; Rom, 1990; Šidák, 1967; Simes, 1986).

While such marginal approaches circumvent the need for resampling techniques to estimate the joint distribution of the test statistics, they can result in very conservative multiple testing procedures.

Furthermore, although a procedure may be marginal, proof of Type I error control by this MTP may require certain assumptions on the dependence structure of the test statistics, i.e., on their joint distribution (e.g., FWER-controlling single-step Šidák Procedure 3.3; FWER-controlling step-up Hochberg Procedure 3.13; TPPFP-controlling step-down Lehmann and Romano (2005) Procedure 3.24). Failure to satisfy these assumptions can lead to anti-conservative procedures.

In summary, marginal MTPs (i) can be very conservative (e.g., Bonferroni-like adjustment), in order to provide Type I error control for general test statistics joint distributions, including the independence case, or (ii) provide Type I error control only under certain assumptions concerning the dependence structure of the test statistics (e.g., joint distribution that satisfies Šidák's or Simes' Inequality).

3.1.4 Joint multiple testing procedures

In contrast, the *joint multiple testing procedures* of Chapters 4–7 take into account the dependence structure of the test statistics and are generally more powerful than marginal procedures.

Chapter 4 provides two main classes of joint single-step procedures for controlling Type I error rates $\Theta(F_{V_n})$, defined as arbitrary parameters of the distribution F_{V_n} of the number of Type I errors V_n: common-cut-off Procedure 4.2 and common-quantile Procedure 4.1. The special cases for FWER control (i.e., $\Theta(F_{V_n}) = 1 - F_{V_n}(0)$) and gFWER control (i.e., $\Theta(F_{V_n}) = 1 - F_{V_n}(k)$) are introduced in Sections 3.2.2 and 3.3.2, respectively. For control of the gFWER, Procedures 4.2 and 4.1 are based, respectively, on the $(k+1)$st largest test statistic and $(k+1)$st smallest unadjusted p-value (Procedures 3.18 and 3.19). In particular, for FWER control, one recovers single-step maxT Procedure 3.5 and minP Procedure 3.6.

FWER-controlling joint step-down maxT and minP procedures are treated in detail in Chapter 5 and introduced in Section 3.2.3 (Procedures 3.11 and 3.12).

Chapter 6 demonstrates that any gFWER-controlling (marginal/joint single-step/stepwise) MTP can be straightforwardly augmented to control generalized tail probability (gTP) error rates, $gTP(q,g) = \Pr(g(V_n, R_n) > q)$, for arbitrary functions $g(V_n, R_n)$ of the numbers of false positives V_n and rejected hypotheses R_n. The special cases of gFWER- and TPPFP-controlling augmentation multiple testing procedures (AMTP) are introduced in Sections 3.3.3 and 3.5.3, respectively.

The gTP-controlling resampling-based empirical Bayes procedures of Chapter 7 also account for the dependence structure of the test statistics. The special cases of gFWER- and TPPFP-controlling empirical Bayes procedures are introduced in Sections 3.3.4 and 3.5.4, respectively.

3.2 Multiple testing procedures for controlling the number of Type I errors: FWER

3.2.1 Controlling the number of Type I errors

Classical approaches to multiple testing call for controlling the *family-wise error rate* (FWER), that is, the chance $FWER = \Theta(F_{V_n}) = 1 - F_{V_n}(0) = \Pr(V_n > 0)$ of committing at least one Type I error (Section 1.2.9).

Hochberg and Tamhane (1987) describe a variety of FWER-controlling single-step and stepwise methods, based on cut-off rules for ordered unadjusted p-values. Westfall and Young (1993) provide resampling-based multiple testing procedures for controlling the FWER.

As detailed in Chapter 4, Dudoit et al. (2004b) and Pollard and van der Laan (2004) propose a general class of joint single-step multiple testing procedures for controlling Type I error rates $\Theta(F_{V_n})$, defined as arbitrary parameters of the distribution F_{V_n} of the number of Type I errors V_n. In addition to the FWER, such error rates include the *generalized family-wise error rate* (gFWER), i.e., tail probabilities for the number of Type I errors, $gFWER(k) = \Theta(F_{V_n}) = 1 - F_{V_n}(k) = \Pr(V_n > k)$, $k \in \{0, \ldots, M\}$, and the *per-family error rate* (PFER), i.e., the mean number of Type I errors, $PFER = \Theta(F_{V_n}) = \int v dF_{V_n}(v) = \mathrm{E}[V_n]$. Pollard and van der Laan (2004) focus on single-parameter null hypotheses and test statistics that are asymptotically normally distributed. Dudoit et al. (2004b) extend the proposed approach to general null hypotheses (defined in terms of submodels for the data generating distribution) and test statistics (e.g., t-statistics, χ^2-statistics, F-statistics) and derive adjusted p-values for each procedure.

Chapter 5 discusses the FWER-controlling joint step-down maxT and minP procedures of van der Laan et al. (2004a).

As detailed in Chapters 6 and 7, respectively, the general augmentation approach of van der Laan et al. (2004b) and empirical Bayes approach of van der Laan et al. (2005), yield a variety of joint procedures for controlling Type I error rates of the form $\Theta(F_{V_n})$.

3.2.2 FWER-controlling single-step procedures

FWER-controlling single-step Bonferroni procedure

Bonferroni's (1936) classical procedure for FWER control is perhaps the best-known procedure in the multiple testing literature. It controls the FWER for arbitrary test statistics joint null distributions.

Procedure 3.1. [FWER-controlling single-step Bonferroni (1936) procedure]
For controlling the FWER at level α, the *single-step Bonferroni (1936) procedure* rejects any null hypothesis $H_0(m)$ with unadjusted p-value $P_{0n}(m)$ less than or equal to the common single-step cut-off $a_m(\alpha) \equiv \alpha/M$. That is, the set of rejected null hypotheses is

$$\mathcal{R}_n(\alpha) \equiv \left\{ m : P_{0n}(m) \leq \frac{1}{M}\alpha \right\}. \tag{3.1}$$

The corresponding adjusted p-values are thus given by

$$\widetilde{P}_{0n}(m) = \min\left\{ M\, P_{0n}(m), 1 \right\}, \qquad m = 1, \ldots, M. \tag{3.2}$$

Adjusted p-values are derived straightforwardly from the general definition in Equation (1.58). Indeed,

$$\begin{aligned}\widetilde{P}_{0n}(m) &= \inf\{\alpha \in [0,1] : m \in \mathcal{R}_n(\alpha)\}\\ &= \inf\left\{\alpha \in [0,1] : P_{0n}(m) \leq \frac{1}{M}\alpha\right\}\\ &= \min\{MP_{0n}(m), 1\}.\end{aligned}$$

Proposition 3.2. [FWER control for single-step Bonferroni (1936) Procedure 3.1] *Consider an M-vector of test statistics $T_n = (T_n(m) : m = 1, \ldots, M)$ and an associated M-vector of unadjusted p-values $P_{0n} = (P_{0n}(m) : m = 1, \ldots, M)$, defined as in Equation* (1.45)*, based on a test statistics null distribution Q_0. Then, single-step Bonferroni (1936) Procedure 3.1 controls the FWER under arbitrary null distributions Q_0. That is,*

$$\Pr(V_0 > 0) = \Pr_{Q_0}\left(\sum_{m \in \mathcal{H}_0} \mathrm{I}\left(P_{0n}(m) \leq \frac{1}{M}\alpha\right) > 0\right) \leq \alpha.$$

Proof of Proposition 3.2. Control of the FWER, $\Theta(F_{V_0}) = 1 - F_{V_0}(0) = \Pr(V_0 > 0)$, under an arbitrary null distribution $Q_{0,\mathcal{H}_0}$ for the test statistics $(T_n(m) : m \in \mathcal{H}_0)$ corresponding to the true null hypotheses, is established as follows.

$$\begin{aligned}\Pr(V_0 > 0) &= \Pr_{Q_0}\left(\sum_{m \in \mathcal{H}_0} \mathrm{I}\left(P_{0n}(m) \leq \frac{1}{M}\alpha\right) > 0\right)\\ &= \Pr_{Q_0}\left(\bigcup_{m \in \mathcal{H}_0} \left\{P_{0n}(m) \leq \frac{1}{M}\alpha\right\}\right)\\ &\leq \sum_{m \in \mathcal{H}_0} \Pr_{Q_0}\left(P_{0n}(m) \leq \frac{1}{M}\alpha\right)\\ &\leq \frac{h_0}{M}\alpha \leq \alpha,\end{aligned}$$

where the first inequality results from Boole's Inequality (Equation (B.1)) and the second from Proposition 1.2, whereby $\Pr_{Q_0}(P_{0n}(m) \leq z) \leq z$, $\forall\ z \in [0,1]$ and $m \in \mathcal{H}_0$.

Note that, the smaller the proportion h_0/M of true null hypotheses, the more conservative the procedure.

□

FWER-controlling single-step Šidák procedure

Closely related to the Bonferroni procedure is Šidák's (1967) single-step procedure, which controls the FWER for test statistics null distributions Q_0 that satisfy, for the true null hypotheses $\mathcal{H}_0$, an inequality known as Šidák's Inequality.

Šidák's Inequality. (Šidák, 1967)
Consider a random M-vector $Z = (Z(m) : m = 1, \ldots, M)$, with joint distribution Q_0, and an M-vector of constants $c = (c(m) : m = 1, \ldots, M) \in I\!R^M$. Then, *Šidák's Inequality* states that

$$\Pr_{Q_0}\left(\bigcap_{m=1}^{M} \{Z(m) \leq c(m)\}\right) \geq \prod_{m=1}^{M} \Pr_{Q_0}(Z(m) \leq c(m)). \tag{3.3}$$

For a random M-vector of unadjusted p-values $P_0 = (P_0(m) : m = 1, \ldots, M)$, defined as in Equation (1.45) based on Q_0, and an M-vector of constants $c = (c(m) : m = 1, \ldots, M) \in [0,1]^M$, the p-value version of Šidák's Inequality states that

$$\Pr_{Q_0}\left(\bigcap_{m=1}^{M} \{P_0(m) > c(m)\}\right) \geq \prod_{m=1}^{M} \Pr_{Q_0}(P_0(m) > c(m)). \tag{3.4}$$

Šidák's Inequality, also known as the *positive orthant dependence property*, holds for independent test statistics and for test statistics with certain parametric distributions. Specifically, the inequality was initially derived by Dunn (1958) for multivariate Gaussian distributions with mean vector zero and certain types of covariance matrices. Šidák (1967) extended the result to multivariate Gaussian distributions with arbitrary covariance matrices and Jogdeo (1977) showed that the inequality holds for a larger class of distributions, including some multivariate t- and F-distributions.

Procedure 3.3. [FWER-controlling single-step Šidák (1967) procedure]
For controlling the FWER at level α, the *single-step Šidák (1967) procedure* rejects any null hypothesis $H_0(m)$ with unadjusted p-value $P_{0n}(m)$ less than or equal to the common single-step cut-off $a_m(\alpha) \equiv 1 - (1-\alpha)^{1/M}$. That is, the set of rejected null hypotheses is

$$\mathcal{R}_n(\alpha) \equiv \left\{m : P_{0n}(m) \leq 1 - (1-\alpha)^{1/M}\right\}. \tag{3.5}$$

The corresponding adjusted p-values are thus given by

$$\widetilde{P}_{0n}(m) = 1 - (1 - P_{0n}(m))^M, \qquad m = 1, \ldots, M. \tag{3.6}$$

Adjusted p-values are derived straightforwardly from the general definition in Equation (1.58). Indeed,

$$\begin{aligned}\widetilde{P}_{0n}(m) &= \inf\{\alpha \in [0,1] : m \in \mathcal{R}_n(\alpha)\} \\ &= \inf\left\{\alpha \in [0,1] : P_{0n}(m) \leq 1-(1-\alpha)^{1/M}\right\} \\ &= \inf\left\{\alpha \in [0,1] : 1-(1-P_{0n}(m))^M \leq \alpha\right\} \\ &= 1-(1-P_{0n}(m))^M.\end{aligned}$$

Procedure 3.3 does not guarantee control of the FWER for arbitrary test statistics distributions. However, it can be shown to control the FWER for null distributions Q_0 that satisfy Šidák's Inequality (Equations (3.3) and (3.4)). In particular, under the complete null hypothesis ($h_0 = M$) and for independent U(0, 1) unadjusted p-values $P_{0n}(m)$, Procedure 3.3 provides exact FWER control, in the sense that $\Theta(F_{V_0}) = \Theta(F_{R_0}) = \alpha$. As with Bonferroni Procedure 3.1, the smaller the proportion h_0/M of true null hypotheses, the more conservative the procedure.

Proposition 3.4. [FWER control for single-step Šidák (1967) Procedure 3.3] *Consider an M-vector of test statistics $T_n = (T_n(m) : m = 1,\ldots,M)$ and an associated M-vector of unadjusted p-values $P_{0n} = (P_{0n}(m) : m = 1,\ldots,M)$, defined as in Equation* (1.45)*, based on a test statistics null distribution Q_0. Then, single-step Šidák (1967) Procedure 3.3 controls the FWER under null distributions Q_0 that satisfy Šidák's Inequality (Equations* (3.3) *and* (3.4)*) for the true null hypotheses $\mathcal{H}_0$. That is,*

$$\Pr(V_0 > 0) = \Pr_{Q_0}\left(\sum_{m\in\mathcal{H}_0} \mathrm{I}\left(P_{0n}(m) \leq 1-(1-\alpha)^{1/M}\right) > 0\right) \leq \alpha.$$

Proof of Proposition 3.4.
Case 1. Unadjusted p-values $(P_{0n}(m) : m \in \mathcal{H}_0)$ that are IID U(0, 1). Let us first prove that Procedure 3.3 controls the FWER under the assumption that the unadjusted p-values $(P_{0n}(m) : m \in \mathcal{H}_0)$ for the true null hypotheses are independently and identically distributed (IID) as U(0, 1) random variables under Q_0. Then, one has

$$\begin{aligned}\Pr(V_0 > 0) &= 1 - \Pr(V_0 = 0) \\ &= 1 - \Pr_{Q_0}\left(\bigcap_{m\in\mathcal{H}_0}\left\{P_{0n}(m) > 1-(1-\alpha)^{1/M}\right\}\right) \\ &= 1 - \prod_{m\in\mathcal{H}_0} \Pr_{Q_0}\left(P_{0n}(m) > 1-(1-\alpha)^{1/M}\right) \\ &= 1 - \prod_{m\in\mathcal{H}_0}\left(1-\left(1-(1-\alpha)^{1/M}\right)\right) \\ &= 1-(1-\alpha)^{h_0/M} \leq \alpha.\end{aligned}$$

Note that, under the complete null hypothesis ($h_0 = M$), the MTP provides exact control of the FWER at level α. Indeed,

$$\Pr(V_0 > 0) = 1 - \Pr(R_0 = 0) = 1 - (1-\alpha)^{M/M} = \alpha.$$

Case 2. Unadjusted p-values $(P_{0n}(m) : m \in \mathcal{H}_0)$ that satisfy Šidák's Inequality. More generally, when the unadjusted p-values $(P_{0n}(m) : m \in \mathcal{H}_0)$ for the true null hypotheses satisfy Šidák's Inequality under Q_0, then

$$\begin{aligned}
\Pr(V_0 > 0) &= 1 - \Pr_{Q_0}\left(\bigcap_{m \in \mathcal{H}_0} \left\{ P_{0n}(m) > 1 - (1-\alpha)^{1/M} \right\} \right) \\
&\leq 1 - \prod_{m \in \mathcal{H}_0} \Pr_{Q_0}\left(P_{0n}(m) > 1 - (1-\alpha)^{1/M} \right) \\
&\leq 1 - \prod_{m \in \mathcal{H}_0} \left(1 - \left(1 - (1-\alpha)^{1/M} \right) \right) \\
&= 1 - (1-\alpha)^{h_0/M} \leq \alpha,
\end{aligned}$$

where the first inequality follows from the p-value version of Šidák's Inequality, in Equation (3.4), and the second from Proposition 1.2, whereby $\Pr_{Q_0}(P_{0n}(m) > z) \geq 1 - z$, $\forall\, z \in [0,1]$ and $m \in \mathcal{H}_0$.

□

From a first-order Taylor series expansion of $(1-\alpha)^{1/M}$ about $\alpha = 0$, and when M is large, one has $1 - (1-\alpha)^{1/M} = \alpha/M + O(\alpha^2/M)$. Thus, for small α and large M, the Šidák cut-offs $a_m^{SSSidak}(\alpha)$ can be approximated by the Bonferroni cut-offs $a_m^{SSBonf}(\alpha)$, that is,

$$a_m^{SSSidak}(\alpha) = 1 - (1-\alpha)^{1/M} \approxeq \alpha/M = a_m^{SSBonf}(\alpha).$$

In general, the smaller α^2/M, the smaller the p-value cut-offs $a_m^{SSSidak}(\alpha)$, and, hence, the more conservative Šidák Procedure 3.3.

FWER-controlling single-step common-cut-off maxT and common-quantile minP procedures

The above single-step procedures are based solely on the *marginal distributions* of the test statistics. Bonferroni Procedure 3.1 and Šidák Procedure 3.3 can be very conservative for a large number of tested hypotheses M and a small nominal Type I error level α. In addition, Type I error control for Šidák Procedure 3.3 is established under the assumption that the test statistics null distribution Q_0 satisfies the so-called Šidák Inequality (Equations (3.3) and (3.4)).

In many situations, the test statistics, and hence the corresponding unadjusted p-values, have complex and unknown dependence structures. This is the case, for example, in microarray data analysis (Chapter 9), where groups of genes tend to have highly correlated expression measures due to co-regulation.

Gains in power may be achieved by taking into account the *joint distribution* of the test statistics.

Chapter 4 discusses two less conservative multiple testing procedures that account for the dependence structure of the test statistics and that control the FWER for arbitrary test statistics joint null distributions. These procedures are special cases of Procedures 4.1 and 4.2, for the FWER-specific mapping $\Theta(F) = 1 - F(0)$, and are based, respectively, on minima of unadjusted p-values (single-step common-quantile minP MTP) and maxima of test statistics (single-step common-cut-off maxT MTP). Here, for the purpose of comparing marginal and joint MTPs, we simply state the single-step maxT and minP procedures in terms of their adjusted p-values; details are given in Chapter 4.

Procedure 3.5. [FWER-controlling single-step common-cut-off maxT procedure]
The *single-step common-cut-off maxT procedure* is based on the *maximum test statistic*, $Z^\circ(1) \equiv \max_m Z(m)$, for the M-vector $Z = (Z(m) : m = 1, \ldots, M) \sim Q_0$. Adjusted p-values are given by

$$\widetilde{p}_{0n}(m) = \Pr_{Q_0}\left(\max_{m=1,\ldots,M} Z(m) \geq t_n(m)\right), \qquad m = 1, \ldots, M. \tag{3.7}$$

[Details in Chapter 4: Procedure 4.2, Section 4.2.2; Corollary 4.9, Section 4.3.3.]

Procedure 3.6. [FWER-controlling single-step common-quantile minP procedure]
The *single-step common-quantile minP procedure* is based on the *minimum unadjusted p-value*, $P_0^\circ(1) \equiv \min_m P_0(m)$, where $P_0(m) \equiv \bar{Q}_{0,m}(Z(m))$ denote unadjusted p-values under the test statistics null distribution Q_0, i.e., for $Z = (Z(m) : m = 1, \ldots, M) \sim Q_0$. Adjusted p-values are given by

$$\widetilde{p}_{0n}(m) = \Pr_{Q_0}\left(\min_{m=1,\ldots,M} P_0(m) \leq p_{0n}(m)\right), \qquad m = 1, \ldots, M. \tag{3.8}$$

[Details in Chapter 4: Procedure 4.1, Section 4.2.1; Corollary 4.8, Section 4.3.3.]

Note that, in Equations (3.7) and (3.8), the lowercase notation $t_n(m)$, $p_{0n}(m)$, and $\widetilde{p}_{0n}(m)$, is adopted to avoid confusion in the interpretation of the probabilities. Specifically, the probabilities for the single-step maxT and minP adjusted p-values refer, respectively, to the distributions of the maximum test statistic $Z^{\circ}(1)$ and minimum unadjusted p-value $P_0^{\circ}(1)$, for $Z \sim Q_0$. The lowercase notation also emphasizes that the test statistics $t_n(m)$, unadjusted p-values $p_{0n}(m)$, and adjusted p-values $\widetilde{p}_{0n}(m)$, are computed for a particular realization of the random sample $\mathcal{X}_n = \{X_i : i = 1, \ldots, n\}$.

Comparison of FWER-controlling single-step procedures

The following points should be noted regarding connections among the single-step procedures introduced above.

Single-step minP Procedure 3.6 vs. single-step Bonferroni Procedure 3.1

Applying Boole's Inequality (Equation (B.1)) and Proposition 1.2 to the single-step minP adjusted p-values of Equation (3.8) yields as upper bounds the single-step Bonferroni adjusted p-values of Equation (3.2).

$$\begin{aligned}
\widetilde{p}_{0n}^{SSminP}(m) &= \Pr_{Q_0}\left(\min_{h=1,\ldots,M} P_0(h) \leq p_{0n}(m)\right) \\
&= \Pr_{Q_0}\left(\bigcup_{h=1}^{M}\{P_0(h) \leq p_{0n}(m)\}\right) \\
&\leq \min\left\{\sum_{h=1}^{M} \Pr_{Q_0}\left(P_0(h) \leq p_{0n}(m)\right), 1\right\} \\
&\leq \min\{Mp_{0n}(m), 1\} = \widetilde{p}_{0n}^{SSBonf}(m).
\end{aligned}$$

In other words, *joint* single-step minP Procedure 3.6 is less conservative than *marginal* single-step Bonferroni Procedure 3.1.

Single-step minP Procedure 3.6 vs. single-step Šidák Procedure 3.3

If the unadjusted p-values $P_0(m) = \bar{Q}_{0,m}(Z(m))$ are IID U(0,1) for $Z \sim Q_0$, the single-step minP adjusted p-values of Equation (3.8) coincide with the single-step Šidák adjusted p-values of Equation (3.6).

$$
\begin{aligned}
\widetilde{p}_{0n}^{SSminP}(m) &= \Pr_{Q_0}\left(\min_{h=1,\ldots,M} P_0(h) \le p_{0n}(m)\right) \\
&= 1 - \Pr_{Q_0}\left(\min_{h=1,\ldots,M} P_0(h) > p_{0n}(m)\right) \\
&= 1 - \Pr_{Q_0}\left(\bigcap_{h=1}^{M} \{P_0(h) > p_{0n}(m)\}\right) \\
&= 1 - \prod_{h=1}^{M} \Pr_{Q_0}\left(P_0(h) > p_{0n}(m)\right) \\
&= 1 - \prod_{h=1}^{M} (1 - p_{0n}(m)) \\
&= 1 - (1 - p_{0n}(m))^M = \widetilde{p}_{0n}^{SSSidak}(m).
\end{aligned}
$$

In other words, *joint* single-step minP Procedure 3.6 reduces to *marginal* single-step Šidák Procedure 3.3 when the unadjusted p-values $P_0(m)$ are IID $\mathrm{U}(0,1)$ under Q_0.

Single-step minP Procedure 3.6 vs. single-step Šidák Procedure 3.3

Applying Šidák's Inequality (Equations (3.3) and (3.4)) and Proposition 1.2 to the single-step minP adjusted p-values of Equation (3.8) yields as upper bounds the single-step Šidák adjusted p-values of Equation (3.6).

$$
\begin{aligned}
\widetilde{p}_{0n}^{SSminP}(m) &= 1 - \Pr_{Q_0}\left(\bigcap_{h=1}^{M} \{P_0(h) > p_{0n}(m)\}\right) \\
&\le 1 - \prod_{h=1}^{M} \Pr_{Q_0}\left(P_0(h) > p_{0n}(m)\right) \\
&\le 1 - \prod_{h=1}^{M} (1 - p_{0n}(m)) \\
&= 1 - (1 - p_{0n}(m))^M = \widetilde{p}_{0n}^{SSSidak}(m).
\end{aligned}
$$

In other words, for test statistics null distributions Q_0 that satisfy Šidák's Inequality, *joint* single-step minP Procedure 3.6 is less conservative than *marginal* single-step Šidák Procedure 3.3. In the special case of IID $\mathrm{U}(0,1)$ unadjusted p-values, the Šidák and minP MTPs coincide, as discussed in item 2, above.

Single-step minP Procedure 3.6 vs. single-step maxT Procedure 3.5

In general, cut-offs and adjusted p-values for common-cut-off (maxT) procedures are easier to compute than those for common-quantile (minP) procedures. Common-cut-off and common-quantile MTPs coincide when the test statistics $T_n(m)$ are identically distributed. The two types of procedures are contrasted in Section 4.2.4.

The reader is referred to Chapter 4 for further discussion and comparison of single-step MTPs.

3.2.3 FWER-controlling step-down procedures

While the above single-step procedures are simple to implement, they tend to be conservative for control of the FWER. Improvement in power, while preserving Type I error control, may be achieved by step-down procedures. Recall from Section 1.2.13 that *stepwise* MTPs apply the testing procedure to a *sequence of successively smaller nested random (i.e., data-dependent) subsets of null hypotheses*, defined by the *ordering* of the test statistics (common-cut-off MTPs) or unadjusted p-values (common-quantile MTPs). *Step-down* MTPs start with the *most significant* null hypothesis; as soon as one fails to reject a null hypothesis, no further hypotheses are rejected (Procedure 1.3). In contrast, *step-up* MTPs start with the *least significant* null hypothesis; as soon as one rejects a null hypothesis, all remaining more significant hypotheses are rejected (Procedure 1.4).

A detailed treatment of FWER-controlling step-down procedures is given in Chapter 5. Below are the step-down analogues of the single-step procedures introduced in Section 3.2.2. These step-down MTPs are similar in spirit to their above single-step counterparts, with the important distinction that null hypotheses are considered successively, from most significant to least significant, with further tests depending on the outcome of earlier ones.

FWER-controlling step-down Holm procedure

Procedure 3.7. [FWER-controlling step-down Holm (1979) procedure]
For controlling the FWER at level α, the unadjusted p-value cut-offs for the *step-down Holm (1979) procedure* are as follows,

$$a_m(\alpha) \equiv \frac{1}{M-m+1}\alpha, \qquad m = 1, \ldots, M, \tag{3.9}$$

and the set of rejected null hypotheses is

$$\mathcal{R}_n(\alpha) \equiv \left\{ O_n(m) : P_{0n}(O_n(h)) \leq \frac{1}{M-h+1}\alpha,\ \forall\, h \leq m \right\}. \tag{3.10}$$

The corresponding adjusted p-values are thus given by

$$\widetilde{P}_{0n}(O_n(m)) = \max_{h=1,\ldots,m} \{\min\{(M-h+1)\,P_{0n}(O_n(h)), 1\}\},\ m = 1,\ldots,M. \tag{3.11}$$

Adjusted p-values may be derived straightforwardly from the general definition in Equation (1.58) or from Equation (1.65) in generic step-down Procedure 1.3.

Holm Procedure 3.7 is the step-down analogue of classical single-step Bonferroni Procedure 3.1 and also controls the FWER for arbitrary test statistics joint null distributions. The step-down Holm p-value cut-offs, $a_m(\alpha) = \alpha/(M-m+1)$, are greater (i.e., less conservative) than the single-step Bonferroni cut-offs, $a_m(\alpha) = \alpha/M$.

Proposition 3.8. [FWER control for step-down Holm (1979) Procedure 3.7] *Consider an M-vector of test statistics $T_n = (T_n(m) : m = 1,\ldots,M)$ and an associated M-vector of unadjusted p-values $P_{0n} = (P_{0n}(m) : m = 1,\ldots,M)$, defined as in Equation (1.45), based on a test statistics null distribution Q_0. Then, step-down Holm (1979) Procedure 3.7 controls the FWER under arbitrary null distributions Q_0. That is,*

$$\begin{aligned}&\Pr(V_0 > 0)\\ &= \Pr_{Q_0}\left(\sum_{m=1}^{M} \mathrm{I}\left(O_n(m) \in \mathcal{H}_0,\ P_{0n}(O_n(h)) \le a_h(\alpha),\ \forall\, h \le m\right) > 0\right) \le \alpha,\end{aligned}$$

where $a_m(\alpha) = \alpha/(M-m+1)$.

Proof of Proposition 3.8. As in Proposition 1.5, in the special case where $\mathcal{J} = \mathcal{H}_0$, let $m_n^0(1) \equiv \min\{m : O_n(m) \in \mathcal{H}_0\}$ denote the rank, among all M hypotheses, of the true null hypothesis with the smallest unadjusted p-value. Then, $1 \le m_n^0(1) \le M - h_0 + 1$, $P_{0n}^{\circ}(m_n^0(1)) = P_{0n}(O_n(m_n^0(1))) = \min_{m\in\mathcal{H}_0} P_{0n}(m)$, and $\{O_n(1),\ldots,O_n(m_n^0(1)-1)\} \subseteq \mathcal{H}_1$.

From Proposition 1.5, if marginal step-down Procedure 3.7 rejects at least one true null hypothesis, then

$$\min_{m\in\mathcal{H}_0} P_{0n}(m) = P_{0n}^{\circ}(m_n^0(1)) \le a_{M-h_0+1}(\alpha) = \frac{1}{h_0}\,\alpha.$$

Thus, one has

$$\begin{aligned}
\Pr(V_0 > 0) &\leq \Pr_{Q_0}\left(\min_{m \in \mathcal{H}_0} P_{0n}(m) \leq a_{M-h_0+1}(\alpha)\right) \\
&= \Pr_{Q_0}\left(\bigcup_{m \in \mathcal{H}_0} \{P_{0n}(m) \leq a_{M-h_0+1}(\alpha)\}\right) \\
&\leq \sum_{m \in \mathcal{H}_0} \Pr_{Q_0}\left(P_{0n}(m) \leq a_{M-h_0+1}(\alpha)\right) \\
&\leq \sum_{m \in \mathcal{H}_0} a_{M-h_0+1}(\alpha) \\
&= h_0 \frac{1}{h_0} \alpha = \alpha.
\end{aligned}$$

The inequality in line 3 follows from Boole's Inequality (Equation (B.1)); the inequality in line 4 from Proposition 1.2, whereby $\Pr_{Q_0}(P_{0n}(m) \leq z) \leq z$, $\forall\, z \in [0,1]$ and $m \in \mathcal{H}_0$.

□

FWER-controlling step-down Šidák-like procedure

One can derive a step-down analogue of single-step Šidák Procedure 3.3 (Holland and Copenhaver, 1987).

Procedure 3.9. [FWER-controlling step-down Šidák-like procedure]
For controlling the FWER at level α, the unadjusted p-value cut-offs for the *step-down Šidák-like procedure* are as follows,

$$a_m(\alpha) \equiv 1 - (1-\alpha)^{1/(M-m+1)}, \qquad m = 1, \ldots, M, \tag{3.12}$$

and the set of rejected null hypotheses is

$$\mathcal{R}_n(\alpha) \equiv \left\{O_n(m) : P_{0n}(O_n(h)) \leq 1 - (1-\alpha)^{1/(M-h+1)},\ \forall\, h \leq m\right\}. \tag{3.13}$$

The corresponding adjusted p-values are thus given by

$$\widetilde{P}_{0n}(O_n(m)) = \max_{h=1,\ldots,m} \left\{1 - (1 - P_{0n}(O_n(h)))^{(M-h+1)}\right\}, \qquad m = 1, \ldots, M. \tag{3.14}$$

Adjusted p-values may be derived straightforwardly from the general definition in Equation (1.58) or from Equation (1.65) in generic step-down Procedure 1.3.

As for single-step Šidák Procedure 3.3, step-down Procedure 3.9 does not guarantee control of the FWER for arbitrary test statistics distributions. However, it can be shown to control the FWER for null distributions Q_0 that satisfy Šidák's Inequality (Equations (3.3) and (3.4)). In particular, under the complete null hypothesis ($h_0 = M$) and for independent $\mathrm{U}(0,1)$ unadjusted p-values $P_{0n}(m)$, Procedure 3.9 provides exact FWER control, in the sense that $\Theta(F_{V_0}) = \Theta(F_{R_0}) = \alpha$.

Proposition 3.10. [FWER control for step-down Šidák-like Procedure 3.9] *Consider an M-vector of test statistics $T_n = (T_n(m) : m = 1, \ldots, M)$ and an associated M-vector of unadjusted p-values $P_{0n} = (P_{0n}(m) : m = 1, \ldots, M)$, defined as in Equation* (1.45)*, based on a test statistics null distribution Q_0. Then, step-down Šidák-like Procedure 3.9 controls the FWER under null distributions Q_0 that satisfy Šidák's Inequality (Equations* (3.3) *and* (3.4)*) for the true null hypotheses $\mathcal{H}_0$. That is,*

$$\Pr(V_0 > 0)$$
$$= \Pr_{Q_0}\left(\sum_{m=1}^{M} \mathrm{I}\left(O_n(m) \in \mathcal{H}_0,\ P_{0n}(O_n(h)) \leq a_h(\alpha),\ \forall\ h \leq m\right) > 0\right) \leq \alpha,$$

where $a_m(\alpha) = 1 - (1-\alpha)^{1/(M-m+1)}$.

Proof of Proposition 3.10. As for step-down Holm Procedure 3.7, the proof of FWER control appeals to Proposition 1.5 and involves the rank $m_n^0(1) \equiv \min\{m : O_n(m) \in \mathcal{H}_0\}$, among all M hypotheses, of the true null hypothesis with the smallest unadjusted p-value.

From Proposition 1.5, if marginal step-down Procedure 3.9 rejects at least one true null hypothesis, then

$$\min_{m \in \mathcal{H}_0} P_{0n}(m) = P_{0n}^{\circ}(m_n^0(1)) \leq a_{M-h_0+1}(\alpha) = 1 - (1-\alpha)^{1/h_0}.$$

Thus, one has

$$
\begin{aligned}
\Pr(V_0 > 0) &\leq \Pr_{Q_0}\left(\min_{m \in \mathcal{H}_0} P_{0n}(m) \leq a_{M-h_0+1}(\alpha)\right) \\
&= 1 - \Pr_{Q_0}\left(\min_{m \in \mathcal{H}_0} P_{0n}(m) > a_{M-h_0+1}(\alpha)\right) \\
&= 1 - \Pr_{Q_0}\left(\bigcap_{m \in \mathcal{H}_0} \{P_{0n}(m) > a_{M-h_0+1}(\alpha)\}\right) \\
&\leq 1 - \prod_{m \in \mathcal{H}_0} \Pr_{Q_0}\left(P_{0n}(m) > a_{M-h_0+1}(\alpha)\right) \\
&\leq 1 - \prod_{m \in \mathcal{H}_0} \left(1 - a_{M-h_0+1}(\alpha)\right) \\
&= 1 - \prod_{m \in \mathcal{H}_0} \left(1 - \left(1 - (1-\alpha)^{1/h_0}\right)\right) \\
&= 1 - (1-\alpha)^{h_0/h_0} = \alpha.
\end{aligned}
$$

The inequality in line 4 follows from the p-value version of Šidák's Inequality, in Equation (3.4), and the inequality in line 5 from Proposition 1.2, whereby $\Pr_{Q_0}(P_{0n}(m) > z) \geq 1 - z$, $\forall\, z \in [0, 1]$ and $m \in \mathcal{H}_0$.

In the special case of the complete null hypothesis ($h_0 = M$) and when unadjusted p-values $(P_{0n}(m) : m = 1, \ldots, M)$ are IID U(0, 1) under Q_0, the MTP provides exact control of the FWER at level α. Indeed,

$$
\begin{aligned}
\Pr(V_0 > 0) &= 1 - \Pr(R_0 = 0) \\
&= 1 - \Pr_{Q_0}\left(\min_{m=1,\ldots,M} P_{0n}(m) > a_1(\alpha)\right) \\
&= 1 - \Pr_{Q_0}\left(\bigcap_{m=1}^{M} \{P_{0n}(m) > a_1(\alpha)\}\right) \\
&= 1 - \prod_{m=1}^{M} \Pr_{Q_0}\left(P_{0n}(m) > a_1(\alpha)\right) \\
&= 1 - \prod_{m=1}^{M} \left(1 - \left(1 - (1-\alpha)^{1/M}\right)\right) \\
&= \alpha.
\end{aligned}
$$

□

FWER-controlling step-down common-cut-off maxT and common-quantile minP procedures

The above step-down procedures are based solely on the *marginal distributions* of the test statistics. Holm Procedure 3.7 and Šidák-like Procedure 3.9 can be very conservative for a large number of tested hypotheses M and a small nominal Type I error level α. In addition, Type I error control for Šidák-like Procedure 3.9 is established under the assumption that the test statistics null distribution Q_0 satisfies the so-called Šidák Inequality (Equations (3.3) and (3.4)). Gains in power may be achieved by taking into account the *joint distribution* of the test statistics.

Chapter 5 discusses two less conservative multiple testing procedures that account for the dependence structure of the test statistics and that control the FWER for arbitrary test statistics joint null distributions. Here, for the purpose of comparing marginal and joint MTPs, we simply state step-down maxT Procedure 5.1 and minP Procedure 5.6 in terms of their adjusted p-values; details are given in Chapter 5.

Procedure 3.11. [FWER-controlling step-down common-cut-off maxT procedure]
Let $O_n(m)$ denote the indices for the ordered test statistics, $T_n^\circ(m) \equiv T_n(O_n(m))$, so that $T_n(O_n(1)) \geq \cdots \geq T_n(O_n(M))$. The *step-down common-cut-off maxT procedure* is based on the distributions of *maxima of test statistics* over the nested subsets of ordered null hypotheses $\overline{\mathcal{O}}_n(h) \equiv \{O_n(h), \ldots, O_n(M)\}$. The adjusted p-values are given by

$$\widetilde{p}_{0n}(o_n(m)) = \max_{h=1,\ldots,m} \left\{ \Pr_{Q_0}\left(\max_{l \in \overline{o}_n(h)} Z(l) \geq t_n(o_n(h)) \right) \right\}, \tag{3.15}$$

where $Z = (Z(m) : m = 1, \ldots, M) \sim Q_0$.
[Details in Chapter 5: Procedure 5.1 and Proposition 5.5, Section 5.2.]

Procedure 3.12. [FWER-controlling step-down common-quantile minP procedure]
Let $O_n(m)$ denote the indices for the ordered unadjusted p-values, $P_{0n}^\circ(m) \equiv P_{0n}(O_n(m))$, so that $P_{0n}(O_n(1)) \leq \cdots \leq P_{0n}(O_n(M))$. The *step-down common-quantile minP procedure* is based on the distributions of *minima of unadjusted p-values* over the nested subsets of ordered null hypotheses $\overline{\mathcal{O}}_n(h) \equiv \{O_n(h), \ldots, O_n(M)\}$. The adjusted p-values are given by

$$\widetilde{p}_{0n}(o_n(m)) = \max_{h=1,\ldots,m} \left\{ \Pr_{Q_0}\left(\min_{l \in \overline{\mathcal{O}}_n(h)} P_0(l) \leq p_{0n}(o_n(h)) \right) \right\}, \qquad (3.16)$$

where $P_0(m) \equiv \bar{Q}_{0,m}(Z(m))$ denote unadjusted p-values under the test statistics null distribution Q_0, i.e., for $Z = (Z(m) : m = 1, \ldots, M) \sim Q_0$. [Details in Chapter 5: Procedure 5.6 and Proposition 5.11, Section 5.3.]

As for single-step maxT Procedure 3.5 and minP Procedure 3.6 (Equations (3.7) and (3.8)), the lowercase notation $t_n(m)$, $p_{0n}(m)$, $o_n(m)$, and $\widetilde{p}_{0n}(m)$, is adopted in Equations (3.15) and (3.16) to avoid confusion in the interpretation of the probabilities. Specifically, the probabilities for the step-down maxT and minP adjusted p-values refer, respectively, to the joint distributions of the M-vectors of test statistics $Z = (Z(m) : m = 1, \ldots, M) \sim Q_0$ and unadjusted p-values $P_0 = (P_0(m) = \bar{Q}_{0,m}(Z(m)) : m = 1, \ldots, M)$. The lowercase notation also emphasizes that the test statistics $t_n(m)$, unadjusted p-values $p_{0n}(m)$, indices $o_n(m)$, and adjusted p-values $\widetilde{p}_{0n}(m)$, are computed for a particular realization of the random sample $\mathcal{X}_n$.

Note that *single-step* common-cut-off maxT Procedure 3.5 and common-quantile minP Procedure 3.6 are based, respectively, on the distributions of the maximum test statistic and minimum unadjusted p-value *over all M null hypotheses.* In contrast, *step-down* common-cut-off maxT Procedure 3.11 and common-quantile minP Procedure 3.12 are based, respectively, on the distributions of maxima of test statistics and minima of unadjusted p-values *over successively smaller nested random subsets of ordered null hypotheses,* $\overline{\mathcal{O}}_n(h)$.

Also note that applying Boole's Inequality to the step-down minP adjusted p-values of Equation (3.16) yields as upper bounds the step-down Holm adjusted p-values of Equation (3.11) (see Section 5.3.5 for details). In other words, *joint* step-down minP Procedure 3.12 is less conservative than *marginal* step-down Holm Procedure 3.7. This property is the step-down analogue of that observed for single-step Bonferroni Procedure 3.1 and minP Procedure 3.6.

Step-down procedures, such as Holm Procedure 3.7, may be further improved by accounting for logically related hypotheses, as described in Shaffer (1986).

3.2.4 FWER-controlling step-up procedures

Recall from Section 1.2.13 that, in contrast to step-down procedures, step-up procedures first consider the least significant null hypotheses and, as soon as one hypothesis is rejected, reject all remaining more significant hypotheses. Thus, by giving each null hypothesis "several chances at rejection", step-up procedures generally lead to a greater number of rejected hypotheses than their step-down counterparts.

However, it is important to keep in mind that proofs of Type I error control for step-up MTPs typically involve a number of assumptions concerning the joint distribution of the test statistics (e.g., unadjusted p-values that satisfy Simes' Inequality). More research is needed to determine circumstances in which such methods are applicable.

This section focuses on FWER-controlling marginal step-up procedures; step-up procedures are further discussed in Section 5.4.

Simes' Inequality

Type I error control for commonly-used step-up procedures is typically established under the assumption that the test statistics satisfy the following inequality, known as Simes' Inequality (Simes, 1986).

Simes' Inequality. (Simes, 1986)
Consider a random M-vector $Z = (Z(m) : m = 1, \ldots, M)$ with joint distribution Q_0, unadjusted p-values $P_0 = (P_0(m) : m = 1, \ldots, M)$ defined as in Equation (1.45) based on Q_0, and ordered unadjusted p-values $P_0^\circ(m)$ such that $P_0^\circ(1) \leq \cdots \leq P_0^\circ(M)$. Then, *Simes' Inequality* states that

$$\Pr_{Q_0}\left(\bigcap_{m=1}^{M}\left\{P_0^\circ(m) > \frac{m}{M}\alpha\right\}\right) \geq 1-\alpha \tag{3.17}$$

or, equivalently,

$$\Pr_{Q_0}\left(\bigcup_{m=1}^{M}\left\{P_0^\circ(m) \leq \frac{m}{M}\alpha\right\}\right) \leq \alpha.$$

Simes (1986) proved the above inequality for *independent* test statistics, with equality in the continuous case. Although Equation (3.17) does not hold for all joint distributions Q_0, the simulation studies in Simes (1986) suggest that the inequality is conservative for a variety of multivariate Gaussian and Gamma test statistics distributions.

Until recently, Simes' Inequality had been mostly validated empirically by simulation studies. Sarkar and Chang (1997) show that the inequality holds for test statistics with *exchangeable positively dependent* multivariate distributions, such has, equi-correlated multivariate Gaussian distributions, absolute equi-correlated multivariate Gaussian distributions, and certain multivariate F-, Gamma, and t-distributions (including those considered by Simes (1986) in his simulation studies). Sarkar (1998) further establishes that Simes' Inequality holds for *multivariate totally positive of order two* (MTP_2) distributions and scale mixtures of MTP_2 distributions, thereby covering central multivariate t-distributions with common non-negative correlation coefficient and absolute multivariate t-distributions with common correlation coefficient. Sarkar (2005) notes that Equation (3.17) also holds for test statistics that satisfy a *positive regression dependence on subset* (PRDS) condition, as considered by Benjamini and Yekutieli (2001) in the context of FDR-controlling step-up procedures. In addition, Sarkar (2005) extends Simes' Inequality to allow so-called *tests of order* k, where the complete null hypothesis is rejected once at least k of the null hypotheses are rejected.

FWER-controlling step-up Hochberg procedure

Hochberg (1988) proposes the following FWER-controlling step-up procedure based on Simes' Inequality.

Procedure 3.13. [FWER-controlling step-up Hochberg (1988) procedure]
For controlling the FWER at level α, the unadjusted p-value cut-offs for the *step-up Hochberg (1988) procedure* are as follows,

$$a_m(\alpha) \equiv \frac{1}{M-m+1}\,\alpha, \qquad m = 1, \ldots, M, \tag{3.18}$$

and the set of rejected null hypotheses is

$$\mathcal{R}_n(\alpha) \equiv \left\{ O_n(m) : \exists\; h \geq m \text{ such that } P_{0n}(O_n(h)) \leq \frac{1}{M-h+1}\,\alpha \right\}. \tag{3.19}$$

The corresponding adjusted p-values are thus given by

$$\widetilde{P}_{0n}(O_n(m)) = \min_{h=m,\ldots,M} \left\{ \min\left\{ (M-h+1)\, P_{0n}(O_n(h)), 1 \right\} \right\}, \; m = 1, \ldots, M. \tag{3.20}$$

Adjusted p-values may be derived straightforwardly from the general definition in Equation (1.58) or from Equation (1.68) in generic step-up Procedure 1.4.

Hochberg Procedure 3.13 can be viewed as the step-up analogue of step-down Holm Procedure 3.7, because the ordered unadjusted p-values are compared to the same cut-offs in both procedures, namely, $a_m(\alpha) = \alpha/(M-m+1)$.

Proposition 3.14. [FWER control for step-up Hochberg (1988) Procedure 3.13] *Consider an M-vector of test statistics $T_n = (T_n(m) : m = 1, \ldots, M)$ and an associated M-vector of unadjusted p-values $P_{0n} = (P_{0n}(m) :$*

$m = 1, \ldots, M$), defined as in Equation (1.45), based on a test statistics null distribution Q_0. Then, step-up Hochberg (1988) Procedure 3.13 controls the FWER under null distributions Q_0 that satisfy Simes' Inequality (Equation (3.17)) for the true null hypotheses $\mathcal{H}_0$. That is,

$$\begin{aligned}
&\Pr(V_0 > 0) \\
&= \Pr_{Q_0}\left(\sum_{m=1}^{M} \mathrm{I}\left(O_n(m) \in \mathcal{H}_0,\ \exists\ h \geq m\ P_{0n}(O_n(h)) \leq a_h(\alpha)\right) > 0\right) \leq \alpha,
\end{aligned}$$

where $a_m(\alpha) = \alpha/(M - m + 1)$.

Proof of Proposition 3.14. As in Proposition 1.6, in the special case where $\mathcal{J} = \mathcal{H}_0$, let $m_n^0(h) \equiv \min\{m : |\{O_n(1), \ldots, O_n(m)\} \cap \mathcal{H}_0| = h\}$, $h = 1, \ldots, h_0$, denote the rank, among all M hypotheses, of the true null hypothesis with the hth smallest unadjusted p-value among all h_0 true null hypotheses $\mathcal{H}_0$. That is, $O_n(m_n^0(h)) \in \mathcal{H}_0$ and the unadjusted p-values $P_{0n}^{\circ}(m_n^0(h)) = P_{0n}(O_n(m_n^0(h)))$ are non-decreasing in h, $h = 1, \ldots, h_0$. By definition of the ranks $m_n^0(h)$, one must have $h \leq m_n^0(h) \leq M - h_0 + h$.

From Proposition 1.6, if marginal step-up Procedure 3.13 rejects at least one true null hypothesis, then

$$P_{0n}^{\circ}(m_n^0(h)) \leq a_{M-h_0+h}(\alpha) = \frac{1}{h_0 - h + 1}\alpha, \qquad \text{for some } h = 1, \ldots, h_0.$$

Thus, one has

$$\begin{aligned}
\Pr(V_0 > 0) &\leq \Pr_{Q_0}\left(\exists\ h = 1, \ldots, h_0 \text{ such that } P_{0n}^{\circ}(m_n^0(h)) \leq a_{M-h_0+h}(\alpha)\right) \\
&= \Pr_{Q_0}\left(\bigcup_{h=1}^{h_0}\left\{P_{0n}^{\circ}(m_n^0(h)) \leq \frac{1}{h_0 - h + 1}\alpha\right\}\right) \\
&\leq \Pr_{Q_0}\left(\bigcup_{h=1}^{h_0}\left\{P_{0n}^{\circ}(m_n^0(h)) \leq \frac{h}{h_0}\alpha\right\}\right) \\
&\leq \alpha.
\end{aligned}$$

The inequality in line 3 follows by noting that $1/(h_0 - h + 1) \leq h/h_0$ for each $h = 1, \ldots, h_0$ (i.e., as detailed below, the Simes cut-offs, $a_m^{Simes}(\alpha) = \alpha m/M$, are uniformly greater in m than the Holm/Hochberg cut-offs, $a_m^{HH}(\alpha) = \alpha/(M - m + 1)$). The inequality in line 4 follows for test statistics null distributions Q_0 that satisfy Simes' Inequality for the true null hypotheses, $\mathcal{H}_0 = \{O_n(m_n^0(h)) : h = 1, \ldots, h_0\}$.

□

Related step-up procedures include those of Hommel (1988) and Rom (1990). Troendle (1996) proposes a permutation-based step-up multiple testing procedure that takes into account the dependence structure of the test statistics and is related to the permutation-based step-down maxT procedure presented in Westfall and Young (1993).

Figures 1.3–1.5 compare the following three related FWER-controlling marginal procedures, in terms of sets of rejected hypotheses and adjusted p-values: single-step Bonferroni Procedure 3.1, step-down Holm Procedure 3.7, and step-up Hochberg Procedure 3.13. Interpretation of the figures is provided in Section 1.2.13.

Candidate FWER-controlling step-up procedure based on Simes cut-offs

Comparison of Holm/Hochberg and Simes unadjusted p-value cut-offs

In general, the Simes unadjusted p-value cut-offs $a_m^{Simes}(\alpha) \equiv \alpha m/M$ are greater than the Holm/Hochberg cut-offs $a_m^{HH}(\alpha) \equiv \alpha/(M-m+1)$. Specifically,

$$\frac{a_m^{Simes}(\alpha)}{a_m^{HH}(\alpha)} = m\left(1 - \frac{m-1}{M}\right) \geq 1 \tag{3.21}$$

and

$$a_m^{Simes}(\alpha) - a_m^{HH}(\alpha) = \frac{(m-1)(M-m)}{M(M-m+1)}\alpha \geq 0.$$

The two types of cut-offs are equal in the extreme cases of the most ($m=1$) and least ($m=M$) significant null hypotheses. The greatest differences between the two types of cut-offs are observed for intermediate values of m, i.e., moderately significant null hypotheses, and large numbers M of tested hypotheses. Specifically, the ratio $a_m^{Simes}(\alpha)/a_m^{HH}(\alpha)$ achieves a maximum value of $(M+1)^2/4M$ at $m = (M+1)/2$.

Figure 3.1 displays plots of the difference and ratio between the Holm/Hochberg cut-offs $a_m^{HH}(\alpha)$ and the Simes cut-offs $a_m^{Simes}(\alpha)$, for $\alpha = 1$ and total number of hypotheses $M = 10, 25, 50, 100$.

A natural question is whether a step-up procedure based on the Simes cut-offs provides control of the FWER in general or, more specifically, for test statistics distributions that satisfy Simes' Inequality. Because the Simes cut-offs are greater than the Holm/Hochberg cut-offs, such a procedure would be less conservative than step-up Hochberg Procedure 3.13.

Step-up procedure based on Simes cut-offs: Complete null hypothesis, $h_0 = M$

In the special case of the complete null hypothesis ($h_0 = M$), Simes (1986) argues that a step-up procedure based on the cut-offs $a_m^{Simes}(\alpha) = \alpha m/M$ provides control of the FWER for distributions that satisfy Simes' Inequality. Indeed, for a test statistics null distribution Q_0 that satisfies Simes' Inequality, one has

$$\Pr(V_0 > 0) = \Pr(R_0 > 0)$$
$$= \Pr_{Q_0}\left(\bigcup_{m=1}^{M}\left\{P_{0n}(O_n(m)) \le \frac{m}{M}\alpha\right\}\right)$$
$$\le \alpha.$$

Step-up procedure based on Simes cut-offs: General null hypotheses, $h_0 \le M$

In general, when $h_0 \le M$, a step-up procedure based on the Simes cut-offs does not control the FWER, even for test statistics distributions that satisfy Simes' Inequality.

As in Hommel (1988, p. 384), consider independent test statistics $T_n \sim Q_n$, with IID U(0,1) unadjusted p-values $(P_{0n}(m) : m \in \mathcal{H}_0)$ for the true null hypotheses. Assume *asymptotic separation* of the true and false null hypotheses, $\mathcal{H}_0$ and $\mathcal{H}_1 = \mathcal{H}_0^c$, so that, with probability tending to one, the smallest $h_1 = |\mathcal{H}_1|$ unadjusted p-values correspond to the false null hypotheses, i.e.,

$$\lim_{n\to\infty} \Pr\left(\{O_n(1), \ldots, O_n(h_1)\} = \mathcal{H}_1\right) = 1. \tag{3.22}$$

As in Proposition 1.6, in the special case where $\mathcal{J} = \mathcal{H}_0$, let $m_n^0(1) \equiv \min\{m : O_n(m) \in \mathcal{H}_0\}$ denote the rank, among all M hypotheses, of the true null hypothesis with the smallest unadjusted p-value. Then, $1 \le m_n^0(1) \le M - h_0 + 1$, $P_{0n}^{\circ}(m_n^0(1)) = P_{0n}(O_n(m_n^0(1))) = \min_{m\in\mathcal{H}_0} P_{0n}(m)$, and $\{O_n(1), \ldots, O_n(m_n^0(1) - 1)\} \subseteq \mathcal{H}_1$.

The asymptotic separation assumption implies that

$$\lim_{n\to\infty} \Pr\left(m_n^0(1) = M - h_0 + 1\right) = 1. \tag{3.23}$$

Then, under the actual test statistics joint distribution Q_n, one has

$$
\begin{aligned}
\lim_{n\to\infty} \Pr(V_n > 0) &= 1 - \lim_{n\to\infty} \Pr(V_n = 0) \\
&= 1 - \lim_{n\to\infty} \sum_{m=1}^{M-h_0+1} \Pr\left(V_n = 0 | m_n^0(1) = m\right) \Pr\left(m_n^0(1) = m\right) \\
&= 1 - \lim_{n\to\infty} \Pr\left(V_n = 0,\ m_n^0(1) = M - h_0 + 1\right) \\
&= 1 - \lim_{n\to\infty} \Pr\left(\bigcap_{m=M-h_0+1}^{M} \{P_{0n}(O_n(m)) > a_m^{Simes}(\alpha)\},\right. \\
&\qquad\qquad \left. m_n^0(1) = M - h_0 + 1 \right) \\
&\geq 1 - \lim_{n\to\infty} \Pr\left(\bigcap_{m=M-h_0+1}^{M} \{P_{0n}(O_n(m)) > a_{M-h_0+1}^{Simes}(\alpha)\},\right. \\
&\qquad\qquad \left. m_n^0(1) = M - h_0 + 1 \right) \\
&= 1 - \lim_{n\to\infty} \Pr\left(\bigcap_{m\in\mathcal{H}_0} \{P_{0n}(m) > a_{M-h_0+1}^{Simes}(\alpha)\} \right) \\
&= 1 - \lim_{n\to\infty} \prod_{m\in\mathcal{H}_0} \Pr\left(P_{0n}(m) > a_{M-h_0+1}^{Simes}(\alpha)\right) \\
&= 1 - (1 - a_{M-h_0+1}^{Simes}(\alpha))^{h_0} \\
&= 1 - \left(1 - \frac{M - h_0 + 1}{M}\alpha\right)^{h_0}.
\end{aligned}
\tag{3.24}
$$

The equalities in lines 3, 4, and 6 follow by definition of $m_n^0(1)$ and by asymptotic separation of the true and false null hypotheses; the inequality in line 5 follows by monotonicity of the Simes cut-offs, so that $a_m^{Simes}(\alpha) = \alpha m/M \geq \alpha(M - h_0 + 1)/M = a_{M-h_0+1}^{Simes}(\alpha)$, for $m \geq M - h_0 + 1$; the equalities in lines 7 and 8 follow from the assumption that the unadjusted p-values $(P_{0n}(m) : m \in \mathcal{H}_0)$ for the true null hypotheses are IID $U(0,1)$.

To show lack of FWER control, Hommel (1988) then sets $\alpha = 0.05$, $M = 100$, and $h_0 = 50$, so that $\lim_n \Pr(V_n > 0) \geq 0.725 > 0.05 = \alpha$.

Note that the above argument does not apply for the smaller, more conservative Holm/Hochberg unadjusted p-value cut-offs, $a_m^{HH}(\alpha) = \alpha/(M - m + 1)$, which indeed lead to proper FWER-controlling step-down/step-up MTPs. For these cut-offs, one has

$$
\lim_{n\to\infty} \Pr(V_n > 0) \geq 1 - \left(1 - \frac{1}{h_0}\alpha\right)^{h_0}, \tag{3.25}
$$

where the lower bound on the right-hand side is always less than or equal to α.

3.3 Multiple testing procedures for controlling the number of Type I errors: gFWER

This section presents four main types of gFWER-controlling procedures: (i) the marginal single-step and step-down procedures of Lehmann and Romano (2005); (ii) the joint single-step common-cut-off and common-quantile procedures of Dudoit et al. (2004b) and Pollard and van der Laan (2004) (Chapter 4); (iii) the (marginal/joint single-step/stepwise) augmentation multiple testing procedures of van der Laan et al. (2004b) (Chapter 6); (iv) the joint resampling-based empirical Bayes approach of van der Laan et al. (2005) (Chapter 7).

3.3.1 gFWER-controlling single-step and step-down Lehmann and Romano procedures

For controlling the generalized family-wise error rate, $gFWER(k) = \Theta(F_{V_n}) = 1 - F_{V_n}(k) = \Pr(V_n > k)$, Section 2 of Lehmann and Romano (2005) provides the following extensions of FWER-controlling single-step Bonferroni Procedure 3.1 and step-down Holm Procedure 3.7. As noted by Lehmann and Romano (2005), similar results are stated in the earlier publication of Hommel and Hoffman (1988).

Procedure 3.15. [gFWER-controlling single-step Bonferroni-like Lehmann and Romano (2005) procedure]
For controlling $gFWER(k)$ at level α, the *single-step Bonferroni-like Lehmann and Romano (2005) procedure* rejects any null hypothesis $H_0(m)$ with unadjusted p-value $P_{0n}(m)$ less than or equal to the common single-step cut-off $a_m(\alpha) \equiv \alpha(k+1)/M$. That is, the set of rejected null hypotheses is

$$\mathcal{R}_n(\alpha) \equiv \left\{ m : P_{0n}(m) \leq \frac{k+1}{M}\,\alpha \right\}. \tag{3.26}$$

The corresponding adjusted p-values are thus given by

$$\widetilde{P}_{0n}(m) = \min\left\{ \frac{M}{k+1}\,P_{0n}(m), 1 \right\}, \qquad m = 1, \ldots, M. \tag{3.27}$$

Adjusted p-values are derived straightforwardly from the general definition in Equation (1.58). Indeed,

$$\begin{aligned}\widetilde{P}_{0n}(m) &= \inf\{\alpha \in [0,1] : m \in \mathcal{R}_n(\alpha)\}\\ &= \inf\left\{\alpha \in [0,1] : P_{0n}(m) \leq \frac{k+1}{M}\alpha\right\}\\ &= \min\left\{\frac{M}{k+1} P_{0n}(m), 1\right\}.\end{aligned}$$

Note that in the special case of FWER control ($k = 0$), Procedure 3.15 reduces to single-step Bonferroni Procedure 3.1, i.e., the p-value cut-offs are the well-known Bonferroni cut-offs $a_m(\alpha) = \alpha/M$.

Proposition 3.16. [gFWER control for single-step Bonferroni-like Lehmann and Romano (2005) Procedure 3.15] *Consider an M-vector of test statistics $T_n = (T_n(m) : m = 1, \ldots, M)$ and an associated M-vector of unadjusted p-values $P_{0n} = (P_{0n}(m) : m = 1, \ldots, M)$, defined as in Equation* (1.45), *based on a test statistics null distribution Q_0. Then, single-step Bonferroni-like Lehmann and Romano (2005) Procedure 3.15 controls the gFWER under arbitrary null distributions Q_0. That is,*

$$\Pr(V_0 > k) = \Pr_{Q_0}\left(\sum_{m\in\mathcal{H}_0} \mathrm{I}\left(P_{0n}(m) \leq \frac{k+1}{M}\alpha\right) > k\right) \leq \alpha.$$

Proof of Proposition 3.16. As shown in Theorem 2.1 of Lehmann and Romano (2005), control of the gFWER, $\Theta(F_{V_0}) = 1 - F_{V_0}(k) = \Pr(V_0 > k)$, under an arbitrary null distribution $Q_{0,\mathcal{H}_0}$ for the test statistics $(T_n(m) : m \in \mathcal{H}_0)$ corresponding to the true null hypotheses, is established as follows.

$$\begin{aligned}\Pr(V_0 \geq (k+1)) &\leq \frac{1}{k+1}\mathrm{E}[V_0]\\ &= \frac{1}{k+1}\mathrm{E}_{Q_0}\left[\sum_{m\in\mathcal{H}_0} \mathrm{I}\left(P_{0n}(m) \leq \frac{k+1}{M}\alpha\right)\right]\\ &= \frac{1}{k+1}\sum_{m\in\mathcal{H}_0} \Pr_{Q_0}\left(P_{0n}(m) \leq \frac{k+1}{M}\alpha\right)\\ &\leq \frac{1}{k+1}\sum_{m\in\mathcal{H}_0} \frac{k+1}{M}\alpha\\ &= \frac{h_0}{M}\alpha \leq \alpha,\end{aligned}$$

where the first inequality results from Markov's Inequality (Equation (B.2)) and the second from Proposition 1.2, whereby $\Pr_{Q_0}(P_{0n}(m) \leq z) \leq z$, $\forall\, z \in [0,1]$ and $m \in \mathcal{H}_0$.

Note that, the smaller the proportion h_0/M of true null hypotheses, the more conservative the procedure.

□

Procedure 3.17. [gFWER-controlling step-down Holm-like Lehmann and Romano (2005) procedure]
For controlling $gFWER(k)$ at level α, the unadjusted p-value cut-offs for the *step-down Holm-like Lehmann and Romano (2005) procedure* are as follows,

$$a_m(\alpha) \equiv \begin{cases} \frac{k+1}{M}\,\alpha, & \text{if } m \leq k \\ \frac{k+1}{M+k+1-m}\,\alpha, & \text{if } m > k \end{cases}, \qquad m = 1, \ldots, M, \tag{3.28}$$

and the set of rejected null hypotheses is

$$\mathcal{R}_n(\alpha) \equiv \{O_n(m) : P_{0n}(O_n(h)) \leq a_h(\alpha),\ \forall\ h \leq m\}. \tag{3.29}$$

The corresponding adjusted p-values are thus given by

$$\widetilde{P}_{0n}(O_n(m)) \tag{3.30}$$
$$= \begin{cases} \min\left\{\frac{M}{k+1} P_{0n}(O_n(m)),\, 1\right\}, & \text{if } m \leq k \\ \max_{h=1,\ldots,m-k}\left\{\min\left\{\frac{M-h+1}{k+1} P_{0n}(O_n(h+k)),\, 1\right\}\right\}, & \text{if } m > k \end{cases}.$$

Note that in the special case of FWER control ($k = 0$), Procedure 3.17 reduces to step-down Holm Procedure 3.7, i.e., the p-value cut-offs are the Holm cut-offs $a_m(\alpha) = \alpha/(M - m + 1)$.

Proof of gFWER control is given in Theorem 2.2 of Lehmann and Romano (2005). Their Theorem 2.3 establishes an optimality property for this step-down procedure, in the sense that the cut-offs $a_m(\alpha)$, $m > k$, for the $(M - k)$ largest unadjusted p-values, cannot be increased without violating gFWER control. However, as shown in Section 3.3.6, this does not mean that Procedure 3.17 is most powerful (i.e., yields the greatest number of rejected hypotheses) among all gFWER-controlling procedures.

The adjusted p-values for Procedure 3.17 are derived next, according to Equation (1.65) in generic step-down Procedure 1.3.

$$
\begin{aligned}
&\widetilde{P}_{0n}(O_n(m)) \\
&= \max_{h=1,\ldots,m} \left\{\min\left\{a_h^{-1}(P_{0n}(O_n(h))),\, 1\right\}\right\} \\
&= \begin{cases}
\max_{h=1,\ldots,m}\left\{\min\left\{\frac{M}{k+1}P_{0n}(O_n(h)),\, 1\right\}\right\}, & \text{if } m \leq (k+1) \\
\max\left\{\max_{h=1,\ldots,k+1}\left\{\min\left\{\frac{M}{k+1}P_{0n}(O_n(h)),\, 1\right\}\right\}\right. & \text{,if } m > (k+1) \\
\qquad \left.\max_{h=k+2,\ldots,m}\left\{\min\left\{\frac{M+k+1-h}{k+1}P_{0n}(O_n(h)),\, 1\right\}\right\}\right\}, &
\end{cases} \\
&= \begin{cases}
\min\left\{\frac{M}{k+1}P_{0n}(O_n(m)),\, 1\right\}, & \text{if } m \leq (k+1) \\
\max\left\{\min\left\{\frac{M}{k+1}P_{0n}(O_n(k+1)),\, 1\right\},\right. & \text{if } m > (k+1) \\
\qquad \left.\max_{h=k+2,\ldots,m}\left\{\min\left\{\frac{M+k+1-h}{k+1}P_{0n}(O_n(h)),\, 1\right\}\right\}\right\}, &
\end{cases} \\
&= \begin{cases}
\min\left\{\frac{M}{k+1}P_{0n}(O_n(m)),\, 1\right\}, & \text{if } m \leq (k+1) \\
\max\left\{\min\left\{\frac{M}{k+1}P_{0n}(O_n(k+1)),\, 1\right\},\right. & \text{if } m > (k+1) \\
\qquad \left.\max_{h=2,\ldots,m-k}\left\{\min\left\{\frac{M-h+1}{k+1}P_{0n}(O_n(h+k)),\, 1\right\}\right\}\right\}, &
\end{cases} \\
&= \begin{cases}
\min\left\{\frac{M}{k+1}P_{0n}(O_n(m)),\, 1\right\}, & \text{if } m \leq k \\
\max_{h=1,\ldots,m-k}\left\{\min\left\{\frac{M-h+1}{k+1}P_{0n}(O_n(h+k)),\, 1\right\}\right\}, & \text{if } m > k
\end{cases}.
\end{aligned}
$$

3.3.2 gFWER-controlling single-step common-cut-off and common-quantile procedures

The above gFWER-controlling procedures are based solely on the *marginal distributions* of the test statistics and, as in the special case of FWER control, can be very conservative for a large number of tested hypotheses M and a small nominal Type I error level α. Gains in power may be achieved by taking into account the *joint distribution* of the test statistics.

Chapter 4 discusses two less conservative multiple testing procedures that account for the dependence structure of the test statistics and that control the gFWER for arbitrary test statistics joint null distributions. These procedures are special cases of Procedures 4.1 and 4.2, for the gFWER-specific mapping $\Theta(F) = 1 - F(k)$, and are based, respectively, on the $(k+1)$st smallest unadjusted p-value (single-step common-quantile $P(k+1)$ MTP) and $(k+1)$st largest test statistic (single-step common-cut-off $T(k+1)$ MTP). Here, for the purpose of comparing marginal and joint MTPs, we simply state the single-step $T(k+1)$ and $P(k+1)$ procedures in terms of their adjusted p-values; details are given in Chapter 4.

Procedure 3.18. [gFWER-controlling single-step common-cut-off $T(k+1)$ procedure, Dudoit et al. (2004b) and Pollard and van der Laan (2004)]
For controlling the Type I error rate $\Theta(F_{V_0})$ at level α, the set of rejected null hypotheses for the general *Θ-controlling single-step common-cut-off procedure* is of the form $\mathcal{R}_n(\alpha) \equiv \{m : T_n(m) > c_0\}$, where the common cut-off c_0 is the *smallest* (i.e., least conservative) value for which $\Theta(F_{R_0}) \leq \alpha$.

For *$gFWER(k)$* control (special case $\Theta(F_{V_0}) = 1 - F_{V_0}(k)$), the procedure is based on the *$(k+1)$st largest test statistic.* The adjusted p-values for the *single-step $T(k+1)$ procedure* are given by

$$\widetilde{p}_{0n}(m) = \Pr_{Q_0}\left(Z^\circ(k+1) \geq t_n(m)\right), \qquad m = 1, \ldots, M, \tag{3.31}$$

where $Z^\circ(m)$ denotes the mth largest element of the M-vector $Z = (Z(m) : m = 1, \ldots, M) \sim Q_0$, so that $Z^\circ(1) \geq \cdots \geq Z^\circ(M)$.

For FWER control ($k = 0$), one recovers *single-step $maxT$ Procedure 3.5*, based on the *maximum test statistic*, $Z^\circ(1) = \max_m Z(m)$.
[Details in Chapter 4: Procedure 4.2, Section 4.2.2; Corollary 4.9, Section 4.3.3.]

Procedure 3.19. [gFWER-controlling single-step common-quantile $P(k+1)$ procedure, Dudoit et al. (2004b) and Pollard and van der Laan (2004)]
For controlling the Type I error rate $\Theta(F_{V_0})$ at level α, the set of rejected null hypotheses for the general *Θ-controlling single-step common-quantile procedure* is of the form $\mathcal{R}_n(\alpha) \equiv \{m : T_n(m) > c_0(m)\}$, where $c_0(m) \equiv Q_{0,m}^{-1}(\delta_0) = \inf\{z \in I\!R : Q_{0,m}(z) \geq \delta_0\}$ is the δ_0-quantile of the marginal null distribution $Q_{0,m}$. The common quantile probability δ_0 is chosen as the *smallest* (i.e., least conservative) value for which $\Theta(F_{R_0}) \leq \alpha$.

For *$gFWER(k)$* control (special case $\Theta(F_{V_0}) = 1 - F_{V_0}(k)$), the procedure is based on the *$(k+1)$st smallest unadjusted p-value.* Specifically, let $\bar{Q}_{0,m} \equiv 1 - Q_{0,m}$ denote the survivor functions for the marginal null distributions $Q_{0,m}$ and define unadjusted p-values $P_0(m) \equiv \bar{Q}_{0,m}(Z(m))$ and $P_{0n}(m) \equiv \bar{Q}_{0,m}(T_n(m))$, for $Z \sim Q_0$ and $T_n \sim Q_n$, respectively. The adjusted p-values for the *single-step $P(k+1)$ procedure* are given by

$$\widetilde{p}_{0n}(m) = \Pr_{Q_0}\left(P_0^\circ(k+1) \leq p_{0n}(m)\right), \qquad m = 1, \ldots, M, \tag{3.32}$$

where $P_0^\circ(m)$ denotes the mth smallest element of the M-vector of unadjusted p-values $P_0 = (P_0(m) : m = 1, \ldots, M)$, so that $P_0^\circ(1) \leq \cdots \leq P_0^\circ(M)$.

For FWER control ($k = 0$), one recovers *single-step minP Procedure 3.6*, based on the *minimum unadjusted p-value*, $P_0^\circ(1) = \min_m P_0(m)$. [Details in Chapter 4: Procedure 4.1, Section 4.2.1; Corollary 4.8, Section 4.3.3.]

Here again the lowercase notation $t_n(m)$, $p_{0n}(m)$, and $\widetilde{p}_{0n}(m)$, is adopted to avoid confusion in the interpretation of the probabilities in Equations (3.31) and (3.32). Specifically, the probabilities for the single-step $T(k+1)$ and P(k+1) adjusted p-values refer, respectively, to the distributions of the $(k+1)$st largest test statistic $Z^\circ(k+1)$ and $(k+1)$st smallest unadjusted p-value $P_0^\circ(k+1)$, for $Z \sim Q_0$.

3.3.3 gFWER-controlling augmentation multiple testing procedures

As in van der Laan et al. (2004b), Chapter 6 shows that *any* FWER-controlling procedure can be straightforwardly augmented to control the gFWER, for general data generating distributions and, hence, arbitrary dependence structures for the test statistics. By choosing a suitable initial FWER-controlling procedure (e.g., single-step or step-down maxT or minP procedures), such gFWER-controlling *augmentation multiple testing procedures* (AMTP) can take into account the *joint distribution* of the test statistics and are therefore expected to be less conservative than gFWER-controlling marginal procedures, such as Procedures 3.15 and 3.17 of Lehmann and Romano (2005).

Procedure 3.20. [gFWER-controlling augmentation multiple testing procedure, van der Laan et al. (2004b)]
Consider any FWER-controlling procedure $\mathcal{R}_n(\alpha)$, with adjusted p-values $\widetilde{P}_{0n}(m)$ and indices $O_n(m)$ so that $\widetilde{P}_{0n}(O_n(1)) \leq \cdots \leq \widetilde{P}_{0n}(O_n(M))$. This initial FWER-controlling procedure rejects the following $R_n(\alpha) \equiv |\mathcal{R}_n(\alpha)|$ null hypotheses,

$$\mathcal{R}_n(\alpha) \equiv \left\{m : \widetilde{P}_{0n}(m) \leq \alpha\right\} = \{O_n(m) : m = 1, \ldots, R_n(\alpha)\}.$$

For controlling *gFWER*(k) at level α, the *augmentation multiple testing procedure* rejects the $R_n(\alpha)$ null hypotheses specified by the initial FWER-controlling MTP, as well as the next $A_n(\alpha)$ most significant null hypotheses, where

$$A_n(\alpha) \equiv \min\{k, M - R_n(\alpha)\}. \tag{3.33}$$

The set of rejected null hypotheses for the gFWER-controlling AMTP is

$$\mathcal{R}_n^+(\alpha) \equiv \{O_n(m) : m = 1, \ldots, R_n(\alpha) + A_n(\alpha)\} \tag{3.34}$$

and the adjusted p-values are

$$\widetilde{P}_{0n}^+(O_n(m)) = \begin{cases} 0, & \text{if } m \leq k \\ \widetilde{P}_{0n}(O_n(m-k)), & \text{if } m > k \end{cases}. \tag{3.35}$$

That is, the adjusted p-values for the AMTP are simply k-*shifted* versions of the adjusted p-values of the initial FWER-controlling MTP, with the first k adjusted p-values set to zero. The AMTP thus guarantees at least k rejected hypotheses.
[Details in Chapter 6: Procedure 6.2, Section 6.2.]

Figure 3.2 provides a graphical summary of gFWER-controlling augmentation Procedure 3.20.

3.3.4 gFWER-controlling resampling-based empirical Bayes procedures

Chapter 7 extends the TPPFP-controlling *resampling-based empirical Bayes* approach of van der Laan et al. (2005), to control generalized tail probability error rates, $gTP(q,g) = \Pr(g(V_n, R_n) > q)$. The gFWER corresponds to the special case $g(v,r) = v$. The TPPFP-controlling procedure, with $g(v,r) = v/r$, is summarized in Section 3.5.4, below.

3.3.5 Other gFWER-controlling procedures

Korn et al. (2004) provide a permutation-based step-down procedure for controlling the gFWER, in the special case where the null hypotheses concern equality of the marginal distributions of two data generating distributions.

The unpublished manuscript of Romano and Wolf (2005) considers gFWER-controlling joint single-step and step-down procedures.

In a very recent technical report, Sarkar (2005) proposes gFWER-controlling joint step-down and step-up procedures, where the step-up MTP relies on an extended version of Simes' Inequality (Equation (3.17)).

3.3.6 Comparison of gFWER-controlling procedures

It is of interest to compare the three main classes of gFWER-controlling MTPs introduced in Sections 3.3.1–3.3.3:

- the marginal single-step Bonferroni-like and step-down Holm-like procedures of Lehmann and Romano (2005) (Procedures 3.15 and 3.17, Section 3.3.1);

- the joint single-step common-cut-off $T(k+1)$ and common-quantile P(k+1) procedures of Dudoit et al. (2004b) and Pollard and van der Laan (2004) (Procedures 3.18 and 3.19, Section 3.3.2);
- the (marginal/joint single-step/stepwise) augmentation multiple testing procedures of van der Laan et al. (2004b) (Procedure 3.20, Section 3.3.3).

The different MTPs are most conveniently compared in terms of their adjusted p-values, whereby smaller adjusted p-values indicate a less conservative procedure. It is assumed, for simplicity, that all procedures yield the same significance ranking of the null hypotheses, that is, the same indices $O_n(m)$ may be used for ordering the adjusted p-values of all MTPs.

Comparison of gFWER-controlling single-step procedures

Let us first compare three simple single-step procedures for controlling the gFWER: (i) single-step Bonferroni-like Lehmann and Romano (2005) Procedure 3.15; (ii) single-step common-quantile $P(k+1)$ Procedure 3.19, based on the $(k+1)$st smallest unadjusted p-value $P_0^\circ(k+1)$; (iii) a conservative marginal version of van der Laan et al. (2004b) augmentation multiple testing Procedure 3.20, based on single-step Bonferroni Procedure 3.1.

Single-step $P(k+1)$ Procedure 3.19 vs. single-step Bonferroni-like Procedure 3.15

Proposition 3.21. [Single-step $P(k+1)$ Procedure 3.19 vs. single-step Bonferroni-like Procedure 3.15] *For gFWER(k) control, the adjusted p-values $\widetilde{P}_{0n}^{P(k+1)}(m)$ for single-step common-quantile $P(k+1)$ Procedure 3.19 are smaller than the adjusted p-values $\widetilde{P}_{0n}^{SSLR}(m)$ for single-step Bonferroni-like Lehmann and Romano (2005) Procedure 3.15. That is, for each $m = 1, \ldots, M$,*

$$\Pr_{Q_0}\left(P_0^\circ(k+1) \leq p_{0n}(m)\right) \leq \min\left\{\frac{M}{k+1} p_{0n}(m),\ 1\right\}. \tag{3.36}$$

Hence, Procedure 3.19 is more powerful than Procedure 3.15. As a corollary, for FWER control ($k=0$), single-step common-quantile minP Procedure 3.6 is more powerful than single-step Bonferroni Procedure 3.1.

Proof of Proposition 3.21. Applying Markov's Inequality (Equation (B.2)) and Proposition 1.2 to the adjusted p-values $\widetilde{p}_{0n}^{P(k+1)}(m)$ for single-step P(k+1) Procedure 3.19 yields as conservative upper bounds the adjusted p-values $\widetilde{p}_{0n}^{SSLR}(m)$ for single-step Bonferroni-like Procedure 3.15. Specifically,

$$\begin{aligned}
\widetilde{p}_{0n}^{P(k+1)}(m) &= \Pr_{Q_0}\left(P_0^\circ(k+1) \leq p_{0n}(m)\right) \\
&= \Pr_{Q_0}\left(\sum_{h=1}^{M} \mathrm{I}\left(P_0(h) \leq p_{0n}(m)\right) \geq (k+1)\right) \\
&\leq \min\left\{\frac{1}{k+1}\,\mathrm{E}_{Q_0}\left[\sum_{h=1}^{M} \mathrm{I}\left(P_0(h) \leq p_{0n}(m)\right)\right],\ 1\right\} \\
&= \min\left\{\frac{1}{k+1}\sum_{h=1}^{M} \Pr_{Q_0}\left(P_0(h) \leq p_{0n}(m)\right),\ 1\right\} \\
&\leq \min\left\{\frac{M}{k+1} p_{0n}(m),\ 1\right\} = \widetilde{p}_{0n}^{SSLR}(m), \qquad m = 1, \ldots, M.
\end{aligned}$$

That is, $\widetilde{P}_{0n}^{P(k+1)}(m) \leq \widetilde{P}_{0n}^{SSLR}(m)$, for each $m = 1, \ldots, M$.

□

Hence, single-step common-quantile $P(k+1)$ Procedure 3.19 is more powerful than single-step Bonferroni-like Procedure 3.15 of Lehmann and Romano (2005), in the sense that it always leads to at least as many rejected hypotheses as Procedure 3.15. As previously noted in Section 3.2.2, this example illustrates that joint procedures, which take into account the dependence structure of the test statistics, can lead to gains in power over marginal MTPs. Direct comparisons with single-step common-cut-off $T(k+1)$ Procedure 3.18 are not as straightforward, although, in practice, one also expects gains in power from exploiting the joint distribution of the test statistics.

Single-step Bonferroni-based augmentation Procedure 3.20 vs. single-step Bonferroni-like Procedure 3.15

Our next comparison focuses on single-step Bonferroni-like Lehmann and Romano (2005) Procedure 3.15 and a conservative marginal version of van der Laan et al. (2004b) augmentation multiple testing Procedure 3.20, based on single-step Bonferroni Procedure 3.1.

From Equations (3.2) and (3.35), the adjusted p-values for gFWER-controlling augmentation multiple testing Procedure 3.20, based on single-step Bonferroni Procedure 3.1, are given by

$$\widetilde{P}_{0n}^{+,Bonf}(O_n(m)) = \begin{cases} 0, & \text{if } m \leq k \\ \min\left\{M\, P_{0n}(O_n(m-k)), 1\right\}, & \text{if } m > k \end{cases}. \tag{3.37}$$

From Equation (3.27), the adjusted p-values for single-step Bonferroni-like Lehmann and Romano (2005) Procedure 3.15 are

$$\widetilde{P}_{0n}^{SSLR}(O_n(m)) = \min\left\{\frac{M}{k+1}\, P_{0n}(O_n(m)), 1\right\}, \qquad m = 1, \ldots, M.$$

Although $0 = \widetilde{P}_{0n}^{+,Bonf}(O_n(m)) \leq \widetilde{P}_{0n}^{SSLR}(O_n(m))$, for $m \leq k$, $M\, P_{0n}(O_n(m))/(k+1)$ may or may not exceed $M\, P_{0n}(O_n(m-k))$, for $m > k$.

While this comparison of Bonferroni-like Procedure 3.15 and augmentation Procedure 3.20 is not fully conclusive, note that the latter guarantees at least k rejected hypotheses. In addition, we considered a worst-case scenario for the augmentation procedure, i.e., a conservative marginal Bonferroni version of this procedure. A version of Procedure 3.20, which takes into account the joint distribution of the test statistics, based on either single-step maxT Procedure 3.5 or minP Procedure 3.6, is expected to be more powerful. Indeed, Section 3.2.2 shows that single-step minP Procedure 3.6 is less conservative than single-step Bonferroni Procedure 3.1.

Summary

One has the following relationships among the adjusted p-values of the different gFWER-controlling single-step MTPs.

$$\widetilde{P}_{0n}^{P(k+1)}(m) \leq \widetilde{P}_{0n}^{SSLR}(m), \qquad m = 1, \ldots, M,$$

$$\begin{cases} \widetilde{P}_{0n}^{+,Bonf}(O_n(m)) \leq \widetilde{P}_{0n}^{SSLR}(O_n(m)), & \text{if } m \leq k \\ \widetilde{P}_{0n}^{+,Bonf}(O_n(m)) \, ? \, \widetilde{P}_{0n}^{SSLR}(O_n(m)), & \text{if } m > k \end{cases},$$

$$\begin{cases} \widetilde{P}_{0n}^{+,Bonf}(O_n(m)) \leq \widetilde{P}_{0n}^{P(k+1)}(O_n(m)), & \text{if } m \leq k \\ \widetilde{P}_{0n}^{+,Bonf}(O_n(m)) \, ? \, \widetilde{P}_{0n}^{P(k+1)}(O_n(m)), & \text{if } m > k \end{cases},$$

$$\begin{cases} \widetilde{P}_{0n}^{+,minP}(O_n(m)) \leq \widetilde{P}_{0n}^{P(k+1)}(O_n(m)), & \text{if } m \leq k \\ \widetilde{P}_{0n}^{+,minP}(O_n(m)) \, ? \, \widetilde{P}_{0n}^{P(k+1)}(O_n(m)), & \text{if } m > k \end{cases},$$

$$\begin{cases} \widetilde{P}_{0n}^{+,minP}(O_n(m)) \leq \widetilde{P}_{0n}^{SSLR}(O_n(m)), & \text{if } m \leq k \\ \widetilde{P}_{0n}^{+,minP}(O_n(m)) \, ? \, \widetilde{P}_{0n}^{SSLR}(O_n(m)), & \text{if } m > k \end{cases},$$

$$\widetilde{P}_{0n}^{+,minP}(O_n(m)) \leq \widetilde{P}_{0n}^{+,Bonf}(O_n(m)), \qquad m = 1, \ldots, M.$$

Here, $\widetilde{P}_{0n}^{+,minP}(O_n(m))$ denote adjusted p-values for augmentation Procedure 3.20 based on single-step minP Procedure 3.6.

Comparison of gFWER-controlling step-down procedures

Here, we compare two simple step-down procedures for controlling the gFWER: (i) step-down Holm-like Lehmann and Romano (2005) Procedure 3.17; (ii) a conservative marginal version of van der Laan et al. (2004b) augmentation multiple testing Procedure 3.20, based on step-down Holm Procedure 3.7.

From Equations (3.11) and (3.35), the adjusted p-values for gFWER-controlling augmentation multiple testing Procedure 3.20, based on step-down Holm Procedure 3.7, are given by

$$\widetilde{P}_{0n}^{+,Holm}(O_n(m)) = \begin{cases} 0, & \text{if } m \leq k \\ \max_{h=1,\ldots,m-k} \{\min \{(M-h+1)\, P_{0n}(O_n(h)), 1\}\}, & \text{if } m > k \end{cases}. \tag{3.38}$$

From Equation (3.30), the adjusted p-values for step-down Holm-like Lehmann and Romano (2005) Procedure 3.17 are

$$\widetilde{P}_{0n}^{SDLR}(O_n(m)) = \begin{cases} \min\left\{\frac{M}{k+1} P_{0n}(O_n(m)),\, 1\right\}, & \text{if } m \leq k \\ \max_{h=1,\ldots,m-k}\left\{\min\left\{\frac{M-h+1}{k+1} P_{0n}(O_n(h+k)),\, 1\right\}\right\}, & \text{if } m > k \end{cases}.$$

Although $0 = \widetilde{P}_{0n}^{+,Holm}(O_n(m)) \leq \widetilde{P}_{0n}^{SDLR}(O_n(m))$, for $m \leq k$, $(M-h+1)P_{0n}(O_n(h+k))/(k+1)$ may or may not exceed $(M-h+1)\, P_{0n}(O_n(h))$, for $h = 1, \ldots, M-k$.

While this comparison of Holm-like Procedure 3.17 and augmentation Procedure 3.20 is not fully conclusive, note that the latter guarantees at least k rejected hypotheses. In addition, we considered a worst-case scenario for the augmentation procedure, i.e., a conservative marginal Holm version of this procedure. A version of Procedure 3.20, which takes into account the joint distribution of the test statistics, based on either step-down maxT Procedure 3.11 or minP Procedure 3.12, is expected to be more powerful. Indeed, Section 3.2.3 shows that step-down minP Procedure 3.12 is less conservative than step-down Holm Procedure 3.7.

In general, analytical comparisons between gFWER-controlling procedures are not fully conclusive, in the sense that the adjusted p-values of one procedure cannot be shown to be uniformly smaller than those of another procedure. However, one expects gains in power by taking into account the joint distribution of the test statistics, as in single-step common-cut-off $T(k+1)$ Procedure 3.18, single-step common-quantile $P(k+1)$ Procedure 3.19, augmentation multiple testing Procedure 3.20 based on a FWER-controlling joint MTP, and resampling-based empirical Bayes Procedure 7.1. Furthermore, the automatic rejection of the k most significant null hypotheses may, in some settings, confer an advantage to gFWER-controlling AMTPs, even when based on a FWER-controlling marginal MTP.

The results of a simulation study comparing different gFWER-controlling MTPs are reported in Dudoit et al. (2004a). It would be of interest to also compare the novel gFWER-controlling joint procedures of van der Laan et al. (2005), Romano and Wolf (2005), and Sarkar (2005).

3.4 Multiple testing procedures for controlling the proportion of Type I errors among the rejected hypotheses: FDR

3.4.1 Controlling the number vs. the proportion of Type I errors

A common criticism of multiple testing procedures designed to control parameters $\Theta(F_{V_n})$ of the distribution of the number of Type I errors V_n (e.g., FWER, with $\Theta(F_{V_n}) = 1 - F_{V_n}(0) = \Pr(V_n > 0)$) is their lack of power, especially for large-scale testing problems such as those encountered in biomedical and genomic research.

To see this, consider the simple case of independent test statistics $T_n = (T_n(m) : m = 1, \ldots, M) \sim Q_n$. Suppose that each hypothesis is tested individually at level α, that is, $\Pr(T_n(m) > c_n(m)) = \alpha$, for an M-vector of cut-offs $c_n = (c_n(m) : m = 1, \ldots, M)$, defining one-sided rejection regions $\mathcal{C}_n(m) = (c_n(m), +\infty)$. Then, under the complete null hypothesis ($h_0 = M$), the FWER increases exponentially with the number of tested hypotheses M (Section 1.2.11). Specifically,

$$
\begin{aligned}
\Pr(V_n > 0) &= 1 - \Pr(V_n = 0) \\
&= 1 - \Pr\left(\bigcap_{m=1}^{M} \{T_n(m) \leq c_n(m)\} \right) \\
&= 1 - \prod_{m=1}^{M} \Pr\left(T_n(m) \leq c_n(m)\right) \\
&= 1 - (1-\alpha)^M .
\end{aligned}
\tag{3.39}
$$

Thus, in many situations, control of the FWER can lead to unduly conservative procedures. In current areas of application of multiple testing procedures, such as gene expression studies based on microarray experiments (Chapter 9), thousands of tests are performed simultaneously and a fairly large proportion of null hypotheses are expected to be false. In this context, one may be prepared to tolerate some Type I errors, provided their number is small in comparison to the number of rejected hypotheses. Error rates based on the *proportion* of false positives among the rejected hypotheses (TPPFP, FDR) are especially appealing for large-scale testing problems, compared to error rates based on the *number* of false positives (gFWER, PFER), as they remain stable with an increasing number of tested hypotheses M.

These considerations have led Benjamini and Hochberg (1995), Genovese and Wasserman (2004a,b), Korn et al. (2004), van der Laan et al. (2004b, 2005), Lehmann and Romano (2005), and Romano and Wolf (2005), to consider controlling parameters of the distribution F_{V_n/R_n} of the *proportion of false positives* (PFP) among the rejected hypotheses. The remainder of the present section focuses on the false discovery rate (FDR), i.e., the expected value $E[V_n/R_n]$ of the proportion of false positives among the rejected hypotheses, while Section 3.5 considers tail probabilities for the proportion of false positives (TPPFP) among the rejected hypotheses, $\Pr(V_n/R_n > q)$. For current high-dimensional applications, these two less stringent Type I error rates may be more appropriate than error rates based on the absolute number of Type I errors and present promising alternatives to FWER-controlling approaches.

3.4.2 FDR-controlling step-up Benjamini and Hochberg procedure

In a now classical article, Benjamini and Hochberg (1995) call for controlling the *false discovery rate* (FDR), i.e., the expected proportion of Type I errors among the rejected hypotheses. Specifically, recall from Section 1.2.9 that the FDR is defined as $FDR = \Theta(F_{V_n/R_n}) = \int q dF_{V_n/R_n}(q) = E[V_n/R_n]$, with the convention that $0/0 = 0$, if $R_n = 0$, i.e., $FDR = E[V_n/R_n | R_n > 0] \Pr(R_n > 0) = E[V_n/R_n | V_n > 0] \Pr(V_n > 0)$.

Early references to the FDR can be found in Seeger (1968) and Sorić (1989). Since their 1995 article, Benjamini and co-authors have proposed a variety of multiple testing procedures for controlling the FDR (Benjamini and Braun, 2002; Benjamini and Hochberg, 1995; Benjamini and Yekutieli, 2001; Yekutieli and Benjamini, 1999). Among the many applications of FDR-controlling methods, we mention Benjamini and Yekutieli (2005), Genovese et al. (2002), Reiner et al. (2003), and Yekutieli and Benjamini (1999).

Benjamini and Hochberg (1995) propose the following FDR-controlling marginal step-up procedure, based on the Simes unadjusted p-value cut-offs of Equation (3.17).

Procedure 3.22. [FDR-controlling step-up Benjamini and Hochberg (1995) procedure]
For controlling the FDR at level α, the unadjusted p-value cut-offs for the *step-up Benjamini and Hochberg (1995) procedure* are as follows,

$$a_m(\alpha) \equiv \frac{m}{M}\alpha, \qquad m = 1, \ldots, M, \tag{3.40}$$

and the set of rejected null hypotheses is

$$\mathcal{R}_n(\alpha) \equiv \left\{ O_n(m) : \exists\, h \geq m \text{ such that } P_{0n}(O_n(h)) \leq \frac{h}{M}\alpha \right\}. \tag{3.41}$$

The corresponding adjusted p-values are thus given by

$$\widetilde{P}_{0n}(O_n(m)) = \min_{h=m,\ldots,M} \left\{ \min\left\{ \frac{M}{h} P_{0n}(O_n(h)), 1 \right\} \right\}, \qquad m = 1, \ldots, M. \tag{3.42}$$

Adjusted p-values may be derived straightforwardly from the general definition in Equation (1.58) or from Equation (1.68) in generic step-up Procedure 1.4.

Benjamini and Hochberg (1995) prove that Procedure 3.22 controls the FDR for *independent* test statistics. The subsequent article of Benjamini and Yekutieli (2001) establishes FDR control for test statistics with more general dependence structures, such as *positive regression dependence.* Note that Benjamini and Hochberg (1995) Procedure 3.22 can be conservative, even in the independence case, as it satisfies $\mathrm{E}[V_n/R_n] \leq h_0\alpha/M \leq \alpha$. The reader is referred to these two articles for details and proofs.

Section 7.7 revisits frequentist Procedure 3.22 within the context of empirical Bayes q-value-based approaches.

3.4.3 FDR-controlling step-up Benjamini and Yekutieli procedure

Benjamini and Yekutieli (2001) propose a simple conservative modification of Procedure 3.22, which controls the false discovery rate for test statistics with arbitrary joint distributions.

Procedure 3.23. [FDR-controlling step-up Benjamini and Yekutieli (2001) procedure]
For controlling the FDR at level α, the unadjusted p-value cut-offs for the *step-up Benjamini and Yekutieli (2001) procedure* are as follows,

$$a_m(\alpha) \equiv \frac{m}{M\,C(M)}\,\alpha, \qquad m = 1, \ldots, M, \tag{3.43}$$

where $C(M) \equiv \sum_{m=1}^{M} 1/m$. The set of rejected null hypotheses is

$$\mathcal{R}_n(\alpha) \equiv \left\{ O_n(m) : \exists\ h \geq m \text{ such that } P_{0n}(O_n(h)) \leq \frac{h}{M\,C(M)}\,\alpha \right\}. \tag{3.44}$$

The corresponding adjusted p-values are thus given by

$$\widetilde{P}_{0n}(O_n(m)) = \min_{h=m,\ldots,M} \left\{ \min\left\{ C(M)\frac{M}{h}\,P_{0n}(O_n(h)), 1 \right\} \right\}, \quad m = 1, \ldots, M. \tag{3.45}$$

The unadjusted p-value cut-offs $a_m(\alpha)$ for the above two step-up procedures differ only by a constant factor $C(M)$. For classical restricted Benjamini and Hochberg (1995) Procedure 3.22, the "penalty for multiplicity" is M/m (Equation (3.42)), whereas for conservative Benjamini and Yekutieli (2001) Procedure 3.23 the penalty is the greater value of $MC(M)/m$

(Equation (3.45)). For a large number M of hypotheses, the penalties differ by a factor $C(M) \approxeq \log M$.

3.4.4 FDR-controlling resampling-based empirical Bayes procedures

A number of authors have recently considered empirical Bayes approaches for controlling the false discovery rate (Efron, 2005; Efron et al., 2001a,b; Goss Tusher et al., 2001; Storey, 2002, 2003; Storey et al., 2004; Storey and Tibshirani, 2001, 2003). Most proposals assume that the test statistics are independently and identically distributed according to a non-parametric mixture model, although Storey (2003) and Storey et al. (2004) provide asymptotic control results under so-called weak or loose dependence conditions (e.g., dependence in finite blocks, ergodic dependence).

Section 7.7 discusses empirical Bayes q-value-based approaches to FDR control and connections to frequentist step-up Benjamini and Hochberg (1995) Procedure 3.22.

Section 7.8 proposes to extend the gTP-controlling resampling-based empirical Bayes methodology of Chapter 7 (Procedure 7.1) to control general Type I error rates, defined as parameters $\Theta(F_{g(V_n,R_n)})$ of the distribution of functions $g(V_n, R_n)$ of the numbers of false positives V_n and rejected hypotheses R_n. Such error rates include *generalized expected value* (gEV) error rates, $gEV(g) = \mathrm{E}[g(V_n, R_n)]$, and, in particular, the false discovery rate, with $g(v,r) = v/r$. The proposed joint procedures would account for the dependence structure of the test statistics and would control the FDR for general data generating distributions, with arbitrary dependence structures among variables.

3.4.5 Other FDR-controlling procedures

Most FDR-controlling procedures proposed thus far do not exploit the dependence structure of the test statistics, i.e., are based solely on the (marginal) unadjusted p-values $P_{0n}(m)$ (e.g., step-up Procedures 3.22 and 3.23). In addition, FDR control results are generally derived under the assumption that the test statistics are either independently distributed or have certain forms of dependence, such as positive regression dependence (Benjamini and Yekutieli, 2001).

Yekutieli and Benjamini (1999) propose FDR-controlling MTPs that use resampling-based adjusted p-values to account for certain types of dependence structures among test statistics (the procedures assume that the unadjusted p-values for the true null hypotheses $\mathcal{H}_0$ are independent of the unadjusted p-values for the false null hypotheses $\mathcal{H}_0^c$).

Abramovich et al. (2000) show that FDR-controlling procedures can be used to adapt to unknown sparsity in regression.

Sarkar (2002) derives finite sample FDR control results for both step-down and step-up procedures, under a positive regression dependence condition for the test statistics.

FDR-controlling procedures, based on an arbitrary TPPFP-controlling procedure, are proposed in van der Laan et al. (2004b) and discussed in Section 6.4.

3.5 Multiple testing procedures for controlling the proportion of Type I errors among the rejected hypotheses: TPPFP

3.5.1 Controlling the expected value vs. tail probabilities for the proportion of Type I errors

In contrast to FDR-controlling approaches, that focus on the *expected value* of the proportion of false positives (PFP) among the rejected hypotheses, Genovese and Wasserman (2004a,b), Korn et al. (2004), van der Laan et al. (2004b, 2005), Lehmann and Romano (2005), and Romano and Wolf (2005), propose procedures that control *tail probabilities* for this proportion. These authors argue that although FDR-controlling approaches control the PFP *on average*, they do not preclude large variations in the PFP. When one wishes to have high confidence (i.e., chance at least $(1-\alpha)$) that the set of rejected null hypotheses contains at most a specified proportion q of false positives, control of the *tail probability for the proportion of false positives* (TPPFP) *among the rejected hypotheses*, $TPPFP(q) = \Theta(F_{V_n/R_n}) = 1 - F_{V_n/R_n}(q) = \Pr(V_n/R_n > q)$, is the appropriate form of Type I error control. The parameter q confers flexibility to TPPFP-controlling MTPs and can be tuned to achieve a desired level of false positives. The relationship between TPPFP and FDR control is discussed in Section 6.4.

The recent manuscript of Genovese and Wasserman (2004b) considers two main approaches for *exceedance control of the false discovery proportion*, i.e., TPPFP control in our terminology: inversion and augmentation methods. Section 3, p. 4, presents a TPPFP-controlling *inversion multiple testing procedure*, obtained by inverting tests of uniformity for independently distributed unadjusted p-values. A conservative version of the inversion method is provided for general dependence structures in Section 8. Theorem 3.3 proposes a TPPFP-controlling augmentation multiple testing procedure, based on an initial gFWER-controlling procedure. Such AMTPs are discussed in depth in Chapter 6 of the present book. In particular, Section 6.5.1 notes that the AMTP in Theorem 3.3 of Genovese and Wasserman (2004b) is problematic, in that it does not enforce control of the TPPFP by the initial gFWER-controlling procedure. Genovese and Wasserman (2004b) relate their inversion method to the augmentation method of van der Laan et al. (2004b) in Theorems 3.4 and 3.5. Furthermore, these authors argue in Section 4 that, for control of the gFWER, gains in power can be achieved with so-called $P(k)$ *tests*, i.e., MTPs based on the kth smallest unadjusted p-value. The $P(k)$ test proposed in Theorem 4.1 of Genovese and Wasserman (2004b) is a special case of single-step common-quantile Procedure 4.1, discussed in detail in Chapter 4 of this book. For control of $gFWER(k)$ (i.e., for $\Theta(F) = 1 - F(k)$), Procedure 4.1 does indeed reduce to single-step $P(k+1)$ Procedure 3.19, i.e., is based on the $(k+1)$st smallest unadjusted p-value. Finally, Section 6 of Genovese and Wasserman (2004b) proposes combining $P(k+1)$ procedures for a range of values of the allowed number k of false positives.

Genovese and Wasserman (2004a) use the phrase *confidence thresholds* for the *false discovery proportion* (FDP) to refer to TPPFP-controlling approaches and provide procedures for controlling the TPPFP under the assumption that the test statistics are independently distributed. These authors also extend the theory of FDR control, for independent test statistics, by viewing the FDP as a stochastic process.

Korn et al. (2004) provide a permutation-based step-down procedure for controlling the TPPFP, in the special case where the null hypotheses concern equality of the marginal distributions of two data generating distributions.

The unpublished manuscript of Romano and Wolf (2005) considers TPPFP-controlling joint procedures.

The remainder of this section presents three main types of TPPFP-controlling procedures: (i) the marginal step-down procedures of Lehmann and Romano (2005); (ii) the (marginal/joint single-step/stepwise) augmentation multiple testing procedures of van der Laan et al. (2004b) (Chapter 6); (iii) the joint resampling-based empirical Bayes approach of van der Laan et al. (2005) (Chapter 7).

3.5.2 TPPFP-controlling step-down Lehmann and Romano procedures

Lehmann and Romano (2005) propose the following two marginal step-down procedures for controlling the TPPFP. These procedures are based solely on the marginal distributions of the test statistics and involve simple cut-off rules for unadjusted p-values.

The first procedure is shown to control the TPPFP under either one of two assumptions on the dependence structure of the unadjusted p-values (Assumptions LR.TPPFP1 and LR.TPPFP2, below), whereas the second and more conservative procedure controls the TPPFP under arbitrary test statistics joint null distributions. The reader is referred to the article by Lehmann and Romano (2005) for details and proofs of TPPFP control.

Adjusted p-values may be derived straightforwardly from Equation (1.65) in generic step-down Procedure 1.3.

Procedure 3.24. [TPPFP-controlling restricted step-down Lehmann and Romano (2005) procedure]
For controlling $TPPFP(q)$ at level α, the unadjusted p-value cut-offs for the *restricted step-down Lehmann and Romano (2005) procedure* are as follows,

$$a_m(\alpha) \equiv \frac{\lfloor qm \rfloor + 1}{M + \lfloor qm \rfloor + 1 - m}\alpha, \qquad m = 1, \ldots, M, \tag{3.46}$$

where the *floor* $\lfloor x \rfloor$ denotes the greatest integer less than or equal to x, i.e., $\lfloor x \rfloor \in \mathbb{Z}$ and $\lfloor x \rfloor \leq x < \lfloor x \rfloor + 1$ (Appendix B.3). The set of rejected null hypotheses is

$$\mathcal{R}_n(\alpha) \equiv \{O_n(m) : P_{0n}(O_n(h)) \leq a_h(\alpha),\ \forall\ h \leq m\}. \tag{3.47}$$

The corresponding adjusted p-values are thus given by

$$\widetilde{P}_{0n}(O_n(m)) = \max_{h=1,\ldots,m} \left\{ \min \left\{ \frac{M + \lfloor qh \rfloor + 1 - h}{\lfloor qh \rfloor + 1} P_{0n}(O_n(h)),\, 1 \right\} \right\}. \tag{3.48}$$

Theorem 3.1 in Lehmann and Romano (2005) shows that the above procedure controls the TPPFP under the following assumption on the joint distribution of the unadjusted p-values $P_{0n}(m)$.

Assumption LR.TPPFP1. For any true null hypothesis $H_0(m)$, $m \in \mathcal{H}_0$, the conditional distribution of the unadjusted p-value $P_{0n}(m)$, given the unadjusted p-values $(P_{0n}(m) : m \in \mathcal{H}_1)$ for the false null hypotheses $\mathcal{H}_1 = \mathcal{H}_0^c$, dominates the U$(0,1)$ distribution. That is, for each $m \in \mathcal{H}_0$,

$$\Pr_{Q_0}\left(P_{0n}(m) \leq u | (P_{0n}(m) : m \in \mathcal{H}_1)\right) \leq u, \qquad \forall\ u \in [0,1]. \tag{3.49}$$

Lehmann and Romano (2005) further show in their Theorem 3.2 that Procedure 3.24 controls the TPPFP under the following assumption on the joint distribution of the unadjusted p-values $(P_{0n}(m) : m \in \mathcal{H}_0)$ for the true null hypotheses.

Assumption LR.TPPFP2. The joint distribution of the unadjusted p-values for the true null hypotheses $\mathcal{H}_0$ satisfies Simes' Inequality (Equation (3.17)). That is,

$$\Pr_{Q_0}\left(\bigcup_{h=1}^{h_0} \left\{ P_{0n}^{\circ}(m_n^0(h)) \leq \frac{h}{h_0}\alpha \right\} \right) \leq \alpha, \tag{3.50}$$

where, as in Section 1.2.13, in the special case where $\mathcal{J} = \mathcal{H}_0$, $m_n^0(h) \equiv \min\{m : |\{O_n(1), \ldots, O_n(m)\} \cap \mathcal{H}_0| = h\}$, $h = 1, \ldots, h_0$, denotes the rank, among all M hypotheses, of the true null hypothesis with the hth smallest unadjusted p-value among all h_0 true null hypotheses $\mathcal{H}_0$. That is, $O_n(m_n^0(h)) \in \mathcal{H}_0$ and the unadjusted p-values $P_{0n}^{\circ}(m_n^0(h)) = P_{0n}(O_n(m_n^0(h)))$ are non-decreasing in h, $h = 1, \ldots, h_0$.

After deriving a generalization of Hommel's Inequality (Hommel, 1983) in Lemma 3.1, Lehmann and Romano (2005) prove in their Theorem 3.3 that the following conservative version of Procedure 3.24 controls the TPPFP for arbitrary test statistics joint null distributions.

Procedure 3.25. [TPPFP-controlling general step-down Lehmann and Romano (2005) procedure]
For controlling $TPPFP(q)$ at level α, the unadjusted p-value cut-offs for the *general step-down Lehmann and Romano (2005) procedure* are as follows,

$$a_m(\alpha) \equiv \frac{1}{C(\lfloor qM \rfloor + 1)} \frac{\lfloor qm \rfloor + 1}{M + \lfloor qm \rfloor + 1 - m} \alpha, \qquad m = 1, \ldots, M, \tag{3.51}$$

where $C(M) \equiv \sum_{m=1}^{M} 1/m$. The set of rejected null hypotheses is

$$\mathcal{R}_n(\alpha) \equiv \{O_n(m) : P_{0n}(O_n(h)) \leq a_h(\alpha), \ \forall \ h \leq m\}. \tag{3.52}$$

The corresponding adjusted p-values are thus given by

$$\widetilde{P}_{0n}(O_n(m)) = \tag{3.53}$$
$$\max_{h=1,\ldots,m} \left\{ \min \left\{ C(\lfloor qM \rfloor + 1) \frac{M + \lfloor qh \rfloor + 1 - h}{\lfloor qh \rfloor + 1} P_{0n}(O_n(h)), 1 \right\} \right\}.$$

The conservative unadjusted p-value cut-offs in Procedure 3.25 are simply obtained by dividing the cut-offs of Procedure 3.24 by $C(\lfloor qM \rfloor + 1)$, where $C(\lfloor qM \rfloor + 1) \approxeq \log(qM)$ for large M. It is interesting to note the parallels between the above two TPPFP-controlling step-down procedures and the FDR-controlling step-up procedures of Benjamini and Hochberg (1995) and Benjamini and Yekutieli (2001) (i.e., Procedures 3.22 and 3.23, respectively). The penalty to guarantee Type I error control for general dependence structures tends to be more severe for the FDR-controlling procedures than for the TPPFP-controlling procedures, i.e., one usually has $C(M) \geq C(\lfloor qM \rfloor + 1)$. Note that in the trivial $q = 0$ case, both Procedures 3.24 and 3.25 intuitively reduce to FWER-controlling step-down Holm Procedure 3.7.

3.5.3 TPPFP-controlling augmentation multiple testing procedures

Most multiple testing procedures proposed thus far for controlling a parameter (e.g., mean for FDR, survivor function for TPPFP) of the distribution of the proportion V_n/R_n of false positives among the rejected hypotheses suffer from one or both of the following limitations: (i) they are based solely on the marginal distributions of the test statistics; (ii) they rely on a number of assumptions concerning the joint distribution of the test statistics, such as, independence, positive regression dependence, ergodic dependence, or normality.

As in van der Laan et al. (2004b), Chapter 6 shows that *any* FWER-controlling procedure can be straightforwardly augmented to control the TPPFP, for general data generating distributions and, hence, arbitrary dependence structures for the test statistics. By choosing a suitable initial FWER-controlling procedure (e.g., single-step or step-down maxT or minP procedures), such TPPFP-controlling *augmentation multiple testing procedures* (AMTP) can take into account the *joint distribution* of the test statistics and are therefore expected to be less conservative than TPPFP-controlling marginal procedures, such as Procedures 3.24 and 3.25 of Lehmann and Romano (2005).

Procedure 3.26. [TPPFP-controlling augmentation multiple testing procedure, van der Laan et al. (2004b)]
Consider any FWER-controlling procedure $\mathcal{R}_n(\alpha)$, with adjusted p-values $\widetilde{P}_{0n}(m)$ and indices $O_n(m)$ so that $\widetilde{P}_{0n}(O_n(1)) \leq \cdots \leq \widetilde{P}_{0n}(O_n(M))$. This initial FWER-controlling procedure rejects the following $R_n(\alpha) \equiv |\mathcal{R}_n(\alpha)|$ null hypotheses,

$$\mathcal{R}_n(\alpha) \equiv \left\{m : \widetilde{P}_{0n}(m) \leq \alpha\right\} = \{O_n(m) : m = 1, \ldots, R_n(\alpha)\}.$$

For controlling $TPPFP(q)$ at level α, the *augmentation multiple testing procedure* rejects the $R_n(\alpha)$ null hypotheses specified by the initial FWER-controlling MTP, as well as the next $A_n(\alpha)$ most significant null hypotheses, where

$$\begin{aligned} A_n(\alpha) &\equiv \max\left\{m \in \{0, \ldots, M - R_n(\alpha)\} : \frac{m}{R_n(\alpha) + m} \leq q\right\} \qquad (3.54) \\ &= \min\left\{\left\lfloor \frac{qR_n(\alpha)}{1-q} \right\rfloor, M - R_n(\alpha)\right\}. \end{aligned}$$

That is, one keeps rejecting null hypotheses until the ratio of additional rejections to the total number of rejections reaches the allowed proportion q of false positives. The set of rejected null hypotheses for the TPPFP-controlling AMTP is

$$\mathcal{R}_n^+(\alpha) \equiv \{O_n(m) : m = 1, \ldots, R_n(\alpha) + A_n(\alpha)\} \tag{3.55}$$

and the adjusted p-values are

$$\widetilde{P}_{0n}^+(O_n(m)) = \widetilde{P}_{0n}(O_n(\lceil (1-q)m \rceil)), \qquad m = 1, \ldots, M, \tag{3.56}$$

where the *ceiling* $\lceil x \rceil$ denotes the least integer greater than or equal to x, i.e., $\lceil x \rceil \in \mathbb{Z}$ and $\lceil x \rceil - 1 < x \leq \lceil x \rceil$ (Appendix B.3). That is, the adjusted p-values for the AMTP are simply *mq-shifted* versions (up to a ceiling integer transformation) of the adjusted p-values of the initial FWER-controlling MTP.
[Details in Chapter 6: Procedure 6.4, Section 6.4.]

Section 6.5.1 extends the results of van der Laan et al. (2004b) by proposing augmentation Procedure 6.9 for controlling generalized tail probability error rates, $gTP(q, g) = \Pr(g(V_n, R_n) > q)$, based on any initial gFWER-controlling procedure.

3.5.4 TPPFP-controlling resampling-based empirical Bayes procedures

Augmentation multiple testing procedures provide a simple and general approach for controlling TPPFP, that can account for the joint distribution of the test statistics (Section 3.5.3; Chapter 6; Dudoit et al. (2004a); van der Laan et al. (2004b, 2005)). However, even joint AMTPs tend to be conservative in finite sample situations, as they count every additional rejected hypothesis as a false positive. The simulation studies in Dudoit et al. (2004a) and van der Laan et al. (2005) suggest that, although AMTPs compare favorably to TPPFP-controlling marginal procedures, they become more conservative as the number of tested hypotheses increases. The latter feature is problematic for the large-scale testing problems commonly-encountered in genomics.

Motivated by these observations, van der Laan et al. (2005) propose a new multiple testing approach for controlling TPPFP, which, as the augmentation method, provides asymptotic Type I error control for general data generating distributions, but is less conservative for finite samples. The van der Laan et al. (2005) *TPPFP-controlling resampling-based empirical Bayes procedure* involves specifying:

- a null distribution Q_0 (or estimator thereof, Q_{0n}) for M-vectors of *null test statistics* T_{0n};
- a distribution $Q_{0n}^{\mathcal{H}}$ for *random guessed sets of true null hypotheses* $\mathcal{H}_{0n}$.

The proposed working model for generating pairs of random variables $(T_{0n}, \mathcal{H}_{0n})$ is a *common marginal non-parametric mixture distribution* for the test statistics T_n. By randomly sampling null test statistics T_{0n} and guessed sets of true null hypotheses $\mathcal{H}_{0n}$, one obtains a distribution for a random variable $G(c; \mathcal{H}_{0n}, T_{0n}, T_n)$ representing the *guessed proportion of false positives*, for any given cut-off vector c. Cut-offs can then be chosen to control tail probabilities for this distribution at a user-supplied level α.

Specifically, given an M-vector of cut-offs $c = (c(m) : m = 1, \ldots, M)$, the guessed proportion of false positives is defined as

$$G(c; \mathcal{H}_{0n}, T_{0n}, T_n) \tag{3.57}$$
$$\equiv \frac{\sum_m \mathrm{I}\left(T_{0n}(m) > c(m),\ m \in \mathcal{H}_{0n}\right)}{\sum_m \mathrm{I}\left(T_{0n}(m) > c(m),\ m \in \mathcal{H}_{0n}\right) + \sum_m \mathrm{I}\left(T_n(m) > c(m),\ m \notin \mathcal{H}_{0n}\right)},$$

where $T_n \sim Q_n$ is the M-vector of observed test statistics, $T_{0n} \sim Q_{0n}$ is an M-vector of null test statistics, and $\mathcal{H}_{0n} \sim Q_{0n}^{\mathcal{H}}$ is a guessed set of true null hypotheses. The null test statistics T_{0n} and the guessed sets $\mathcal{H}_{0n}$ are sampled independently, given the empirical distribution P_n, from distributions Q_{0n} and $Q_{0n}^{\mathcal{H}}$, chosen (conservatively) so that the *guessed* proportion $G(c; \mathcal{H}_{0n}, T_{0n}, T_n)$ of false positives is asymptotically stochastically greater than the corresponding *true* proportion $G(c; \mathcal{H}_0, T_n, T_n) = \sum_m \mathrm{I}\left(T_n(m) > c(m),\ m \in \mathcal{H}_0\right) / \sum_m \mathrm{I}\left(T_n(m) > c(m)\right)$.

The reader is referred to Chapter 7 and van der Laan et al. (2005) for greater detail on the TPPFP-controlling resampling-based empirical Bayes method. Chapter 7 further extends the empirical Bayes approach to control generalized tail probability error rates, $gTP(q, g) = \Pr(g(V_n, R_n) > q)$. Adjusted p-values are derived in Section 7.3.

The simulation studies of van der Laan et al. (2005) reveal that the new TPPFP-controlling resampling-based empirical Bayes Procedure 7.1 tends to be more powerful than existing TPPFP-controlling MTPs, such as marginal step-down Lehmann and Romano (2005) Procedures 3.24 and 3.25 and augmentation Procedure 3.26. A TPPFP-controlling empirical Bayes MTP is applied to the HIV-1 dataset of Segal et al. (2004) in Chapter 11.

3.5.5 Comparison of TPPFP-controlling procedures

It is of interest to compare the following two classes of TPPFP-controlling MTPs:

- the marginal step-down procedures of Lehmann and Romano (2005) (Procedures 3.24 and 3.25, Section 3.5.2);
- the (marginal/joint single-step/stepwise) augmentation multiple testing procedures of van der Laan et al. (2004b) (Procedure 3.26, Section 3.5.3).

As for gFWER control in Section 3.3.6, it is assumed, for simplicity, that all procedures yield the same significance ranking of the null hypotheses, that is, the same indices $O_n(m)$ may be used for ordering the adjusted p-values of all MTPs.

Our comparison focuses specifically on a conservative marginal version of van der Laan et al. (2004b) augmentation multiple testing Procedure 3.26, based on step-down Holm Procedure 3.7. From Equations (3.11) and (3.56), the adjusted p-values for TPPFP-controlling Holm-based augmentation Procedure 3.26 are given by

$$\widetilde{P}_{0n}^{+,Holm}(O_n(m)) = \max_{h=1,\ldots,\lceil(1-q)m\rceil} \{\min\{(M-h+1)\, P_{0n}(O_n(h)), 1\}\}. \tag{3.58}$$

From Equations (3.48) and (3.53), the adjusted p-values for step-down Lehmann and Romano (2005) Procedures 3.24 and 3.25 are given by

$$\widetilde{P}_{0n}^{LR}(O_n(m)) = \max_{h=1,\ldots,m} \{\min\{b^{LR}(h)\, P_{0n}(O_n(h)), 1\}\},$$

where the p-value penalties $b^{LR}(m)$ are obtained from the unadjusted p-value cut-offs $a_m^{LR}(\alpha)$ as follows,

$$b^{LR}(m) \equiv \frac{\alpha}{a_m^{LR}(\alpha)} = \begin{cases} \frac{M+\lfloor qm \rfloor+1-m}{\lfloor qm \rfloor+1}, & \text{[Restricted]} \\ C(\lfloor qM \rfloor+1)\frac{M+\lfloor qm \rfloor+1-m}{\lfloor qm \rfloor+1}, & \text{[General]} \end{cases}.$$

Analytical comparisons of the above two classes of MTPs are not fully conclusive, in the sense that the adjusted p-values $\widetilde{P}_{0n}^{LR}(O_n(m))$ for Procedures 3.24 and 3.25 do not appear to be uniformly greater or smaller than the adjusted p-values $\widetilde{P}_{0n}^{+,Holm}(O_n(m))$ for Holm-based augmentation Procedure 3.26. Nonetheless, in order to get a sense of the relationship between these p-values, consider the following three special cases for the allowed proportion q of false positives: $q \gtrapprox 0$, $q \lessapprox 1$, and $q = 1/2$.

- **Case $q \gtrapprox 0$.** When $q \gtrapprox 0$, $\lfloor qm \rfloor = 0$ and $\lceil (1-q)m \rceil = m$, so that all three TPPFP-controlling procedures reduce intuitively to FWER-controlling step-down Holm Procedure 3.7.
- **Case $q \lessapprox 1$.** In the other extreme case of $q = (M-1)/M$, for a large number of hypotheses M, $\lfloor qm \rfloor = (m-1)$ and $\lceil (1-q)m \rceil = 1$, so that

$$\widetilde{P}_{0n}^{LR,Rest}(O_n(m)) = \max_{h=1,\ldots,m} \left\{\min\left\{\frac{M}{h}\, P_{0n}(O_n(h)), 1\right\}\right\},$$

$$\widetilde{P}_{0n}^{LR,Gen}(O_n(m)) = \max_{h=1,\ldots,m} \left\{\min\left\{C(M)\frac{M}{h}\, P_{0n}(O_n(h)), 1\right\}\right\},$$

$$\widetilde{P}_{0n}^{+,Holm}(O_n(m)) = \min\{M\, P_{0n}(O_n(1)), 1\}.$$

Thus, for augmentation Procedure 3.26, all hypotheses are assigned the same (Bonferroni or Holm) adjusted p-value, corresponding to the most significant hypothesis $H_0(O_n(1))$. In contrast, the unadjusted p-value cut-offs for TPPFP-controlling step-down Lehmann and Romano (2005) Procedures 3.24 and 3.25 are equal to those of FDR-controlling step-up

Benjamini and Hochberg (1995) Procedure 3.22 and Benjamini and Yekutieli (2001) Procedure 3.23, respectively. However, because of the step-down nature of Procedures 3.24 and 3.25 (enforced by taking maxima over $h \in \{1, \ldots, m\}$), one has

$$\widetilde{P}_{0n}^{+,Holm}(O_n(m)) \leq \widetilde{P}_{0n}^{LR,Rest}(O_n(m)) \leq \widetilde{P}_{0n}^{LR,Gen}(O_n(m)).$$

- **Case $q = 1/2$.** Another interesting special case is that of an allowed proportion $q = 1/2$ of false positives. Then, the unadjusted p-value penalties for restricted Lehmann and Romano (2005) Procedure 3.24 are

$$b^{LR,Rest}(m) = \begin{cases} \frac{2M-m+2}{m+2}, & \text{if } m \text{ is even} \\ \frac{2M-m+1}{m+1}, & \text{if } m \text{ is odd} \end{cases}.$$

In particular, omitting the upper bound of one constraint on adjusted p-values to shorten notation, one has

$$\begin{aligned}
\widetilde{P}_{0n}^{LR,Rest}(O_n(1)) &= M\, P_{0n}(O_n(1)), \\
\widetilde{P}_{0n}^{LR,Rest}(O_n(2)) &= \max\left\{ M\, P_{0n}(O_n(1)), \frac{M}{2}\, P_{0n}(O_n(2)) \right\}, \\
\widetilde{P}_{0n}^{LR,Rest}(O_n(3)) &= \max\left\{ M\, P_{0n}(O_n(1)), \frac{M}{2}\, P_{0n}(O_n(2)), \right. \\
&\qquad \left. \frac{M-1}{2}\, P_{0n}(O_n(3)) \right\}.
\end{aligned}$$

For Holm-based augmentation Procedure 3.26, adjusted p-values are given by

$$\begin{aligned}
\widetilde{P}_{0n}^{+,Holm}(O_n(2m-1)) &= \widetilde{P}_{0n}^{+,Holm}(O_n(2m)) \\
&= \max_{h=1,\ldots,m} \left\{ \min\left\{ (M-h+1)\, P_{0n}(O_n(h)), 1 \right\} \right\},
\end{aligned}$$

and, in particular,

$$\begin{aligned}
\widetilde{P}_{0n}^{+,Holm}(O_n(1)) &= \widetilde{P}_{0n}^{+,Holm}(O_n(2)) = M\, P_{0n}(O_n(1)), \\
\widetilde{P}_{0n}^{+,Holm}(O_n(3)) &= \widetilde{P}_{0n}^{+,Holm}(O_n(4)) \\
&= \max\left\{ M\, P_{0n}(O_n(1)), (M-1)\, P_{0n}(O_n(2)) \right\}.
\end{aligned}$$

Thus,

$$\begin{aligned}
\widetilde{P}_{0n}^{+,Holm}(O_n(1)) &= \widetilde{P}_{0n}^{LR,Rest}(O_n(1)), \\
\widetilde{P}_{0n}^{+,Holm}(O_n(2)) &\leq \widetilde{P}_{0n}^{LR,Rest}(O_n(2)), \\
\widetilde{P}_{0n}^{+,Holm}(O_n(3)) &\ ?\ \widetilde{P}_{0n}^{LR,Rest}(O_n(3)).
\end{aligned}$$

While this comparison of marginal Lehmann and Romano (2005) Procedures 3.24 and 3.25 with augmentation Procedure 3.26 is not fully conclusive, note that we considered a worst-case scenario for the augmentation procedure, i.e., a conservative marginal Holm version of this procedure. A version of Procedure 3.26, which takes into account the joint distribution of the test statistics, based on either step-down maxT Procedure 3.11 or minP Procedure 3.12, is expected to be more powerful. Indeed, Section 3.2.3 shows that step-down minP Procedure 3.12 is less conservative than step-down Holm Procedure 3.7.

The results of simulation studies comparing different TPPFP-controlling MTPs are reported in Dudoit et al. (2004a) and van der Laan et al. (2005). Note that Procedure 6.9 provides a TPPFP-controlling AMTP based on an initial $gFWER(k_0)$-controlling procedure. It would be worth assessing the performances of TPPFP-controlling AMTPs for different values of the parameter k_0 for the allowed number of false positives. The case $k_0 = 0$ corresponds to an initial FWER-controlling MTP, as considered in the present section. It would also be of interest to compare the novel TPPFP-controlling joint procedures of van der Laan et al. (2005) and Romano and Wolf (2005).

When considering different TPPFP-controlling MTPs, one should keep in mind the following two facts. Firstly, some (marginal/joint) procedures require assumptions on the dependence structure of the test statistics (e.g., Assumptions LR.TPPFP1 and LR.TPPFP2 for Lehmann and Romano (2005) Procedure 3.24). Secondly, as shown in Theorem 6.5, augmentation Procedure 3.26, based on an asymptotically exact FWER-controlling MTP (e.g., joint step-down maxT Procedure 3.11 or minP Procedure 3.12), provides exact asymptotic control of the TPPFP. Thus, although AMTPs may lack power in finite sample situations and at local alternative hypotheses (e.g., compared to the empirical Bayes procedures introduced in Section 3.5.4), they can be very powerful asymptotically at fixed alternatives.

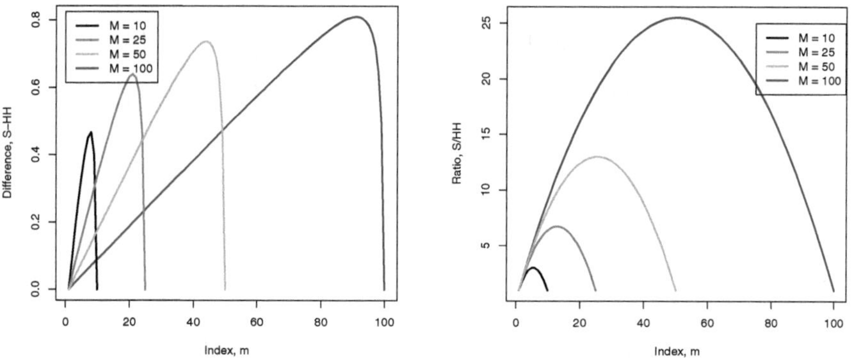

Panel (a): Difference, $a_m^{Simes}(\alpha) - a_m^{HH}(\alpha)$ Panel (b): Ratio, $a_m^{Simes}(\alpha)/a_m^{HH}(\alpha)$

Figure 3.1. *Comparison of stepwise Holm/Hochberg cut-offs and Simes cut-offs.* The figure displays plots of the difference (Panel (a)) and ratio (Panel (b)) between the Holm/Hochberg unadjusted p-value cut-offs $a_m^{HH}(\alpha) = \alpha/(M - m + 1)$ and the Simes unadjusted p-value cut-offs $a_m^{Simes}(\alpha) = \alpha m/M$, for $\alpha = 1$ and total number of hypotheses $M = 10$, 25, 50, 100 (Equation (3.21)). (Color plate p. 324)

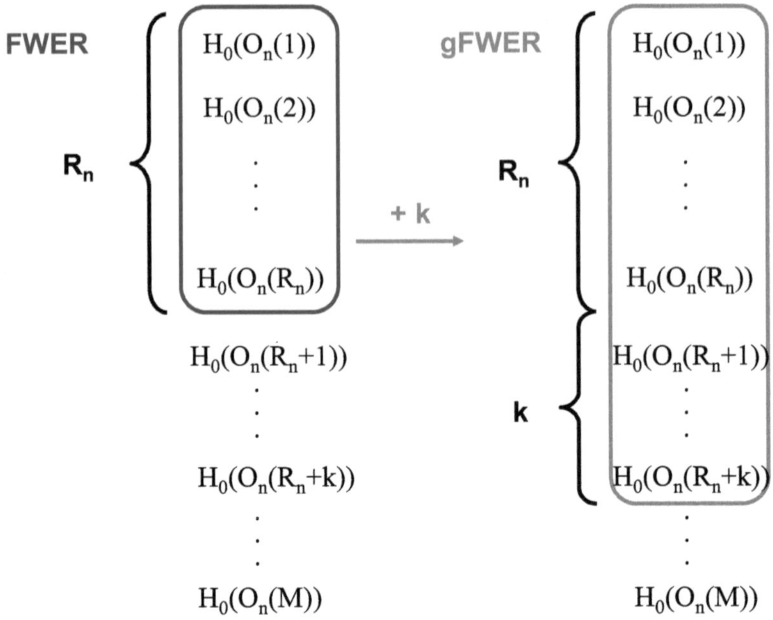

Figure 3.2. *gFWER-controlling augmentation multiple testing procedure.* This figure provides a graphical summary of $gFWER(k)$-controlling augmentation multiple testing Procedure 3.20, based on an initial FWER-controlling procedure. For an allowed number k of false positives, one simply rejects the R_n null hypotheses specified by the initial FWER-controlling MTP and the next k most significant null hypotheses.

4

Single-Step Multiple Testing Procedures for Controlling General Type I Error Rates, $\Theta(F_{V_n})$

4.1 Introduction

4.1.1 Motivation

Overview Chapter 3 focuses primarily on *marginal* multiple testing procedures (MTP), that is, procedures based solely on the marginal distributions of the test statistics. While simple, such marginal MTPs can suffer from the following limitations: (i) they can be very conservative, in order to provide Type I error control for general test statistics joint distributions, including the independence case (e.g., FWER-controlling single-step Bonferroni Procedure 3.1); (ii) some provide Type I error control only under certain assumptions concerning the dependence structure of the test statistics (e.g., FWER-controlling step-up Hochberg Procedure 3.13). In contrast, *joint* multiple testing procedures, such as single-step maxT Procedure 3.5 and minP Procedure 3.6, take into account the dependence structure of the test statistics and are generally more powerful than marginal procedures.

This chapter proposes *joint single-step* procedures for controlling Type I error rates defined as arbitrary parameters $\Theta(F_{V_n})$ of the distribution of the number of Type I errors V_n. Such error rates include the generalized family-wise error rate (gFWER), $gFWER(k) = 1 - F_{V_n}(k) = \Pr(V_n > k)$, i.e., the chance of at least $(k+1)$ Type I errors, and, in particular, the usual family-wise error rate (FWER), $FWER = gFWER(0) = 1 - F_{V_n}(0) = \Pr(V_n > 0)$. They also include the per-comparison error rate (PCER), i.e., the expected proportion of Type I errors, $PCER = \mathrm{E}[V_n]/M$.

Two main classes of multiple testing procedures are derived, *single-step common-quantile Procedure 4.1* and *single-step common-cut-off Procedure 4.2*. These procedures, inspired by the three-step road map of Procedure 2.1, are generalizations of methods proposed in Pollard and van der Laan (2004) for the test of single-parameter null hypotheses using difference or t-statistics. Given a test statistics null distribution Q_0 and nominal Type I error level α, the main idea is to substitute control of the *unknown parameter* $\Theta(F_{V_n})$, for the *true distribution* of the number of Type I errors V_n, by control of the corresponding *known parameter* $\Theta(F_{R_0})$, for the *null distribution* of the number of rejected hypotheses R_0. Among the family of MTPs that satisfy the Type I error constraint $\Theta(F_{R_0}) \leq \alpha$, two types of procedures are considered: Procedure 4.1, with common-quantile cut-offs for the test statistics, and Procedure 4.2, based on a common cut-off for all test statistics.

Specifically, single-step common-quantile Procedure 4.1 rejects null hypothesis $H_0(m)$ if the corresponding test statistic $T_n(m)$ is greater than the δ_0-quantile $Q_{0,m}^{-1}(\delta_0)$ of the marginal null distribution $Q_{0,m}$. For control of the Type I error rate $\Theta(F_{V_n})$ at level α, δ_0 is chosen as the smallest (i.e., least conservative) value such that $\Theta(F_{R_0}) \leq \alpha$. The simpler common-cut-off Procedure 4.2 rejects null hypothesis $H_0(m)$ if the corresponding test statistic $T_n(m)$ is greater than a common cut-off γ_0, chosen as the smallest value such that $\Theta(F_{R_0}) \leq \alpha$.

In the special case of family-wise error rate control ($\Theta(F_{V_n}) = 1 - F_{V_n}(0)$), one recovers *single-step maxT Procedure 3.5* and *single-step minP Procedure 3.6*, based on maxima of test statistics and minima of unadjusted p-values, respectively. Unlike classical FWER-controlling Bonferroni Procedure 3.1 and other marginal procedures discussed in Chapter 3, rejection regions for Procedures 4.1 and 4.2 are based on the joint distribution of the test statistics.

As detailed in Chapter 2, a key feature of our approach to Type I error control is the *test statistics null distribution* (rather than a data generating null distribution) used to obtain rejection regions for the test statistics, confidence regions for the parameters of interest, and adjusted p-values. For general null hypotheses (corresponding to submodels for the data generating distribution) and test statistics (e.g., t-statistics, χ^2-statistics, F-statistics), Theorem 4.3 proves that single-step common-quantile Procedure 4.1 and common-cut-off Procedure 4.2 provide asymptotic control of the Type I error rate $\Theta(F_{V_n})$, under *asymptotic null domination* Assumption NDV. Specifically, recall that asymptotic null domination Assumption NDV states that the number of Type I errors V_n, under the true distribution $Q_n = Q_n(P)$ for the test statistics T_n, is asymptotically stochastically smaller than the corresponding number of Type I errors V_0, under the assumed null distribution Q_0, i.e., $\liminf_n F_{V_n}(x) \geq F_{V_0}(x)$, $\forall\ x \in \{0, \ldots, M\}$. For a Type I error rate mapping Θ that satisfies monotonicity Assumption MΘ and continuity Assumption CΘ at F_{V_0}, one can then prove asymptotic null domination Assumption ND$\boldsymbol{\Theta}$ for the Type I error rate, i.e., $\limsup_n \Theta(F_{V_n}) \leq \Theta(F_{V_0})$. Asymptotic control of the Type I error rate $\Theta(F_{V_n})$ then follows as sketched in the three-step road map of Procedure 2.1, for arbitrary data generating distributions, without the need for assumptions such as subset pivotality. Note that analogous finite sample Type I error control results can be proved for test statistics null distributions that satisfy the finite sample version of null domination Assumption NDV.

This general characterization of a null distribution, in terms of null domination conditions for the joint distribution of the test statistics $(T_n(m) : m \in \mathcal{H}_0)$ for the true null hypotheses (Assumptions jtNDT, NDV, and ND$\boldsymbol{\Theta}$), is introduced in Section 2.2 and leads to the explicit proposal of two test statistics null distributions: the asymptotic distribution of a vector of *null shift and scale-transformed test statistics* (Section 2.3) and the asymptotic distribution of a vector of *null quantile-transformed test statistics* (Section 2.4).

As illustrated in Section 4.3, Procedures 4.1 and 4.2 can be applied to control any error rate that is a parameter of the distribution of the number of Type I errors V_n, including the gFWER and PCER. The key issue is the choice of a suitable test statistics null distribution Q_0.

4.1.2 Outline

This chapter on joint single-step procedures is organized as follows. Section 4.2 proposes general single-step common-quantile Procedure 4.1 and single-step common-cut-off Procedure 4.2, for controlling Type I error rates defined as arbitrary parameters $\Theta(F_{V_n})$ of the distribution of the number of Type I errors. Asymptotic control of the Type I error rate $\Theta(F_{V_n})$ by Procedures 4.1 and 4.2 is established in Theorem 4.3, under a general asymptotic null domination assumption concerning the joint distribution of the test statistics for the true null hypotheses. Adjusted p-values are derived in Section 4.3 for general Type I error rates $\Theta(F_{V_n})$ and in the special cases of the PCER and gFWER. Section 4.4 proves that single-step Procedures 4.1 and 4.2, based on a consistent estimator of the test statistics null distribution, also provide asymptotic control of the Type I error rate $\Theta(F_{V_n})$ (Theorems 4.10 and 4.17, Corollaries 4.18 and 4.19). Bootstrap Procedures 4.20 and 4.21 are supplied to conveniently obtain consistent estimators of the test statistics null distribution and of the corresponding test statistic cut-offs and adjusted p-values for Procedures 4.1 and 4.2, respectively. Section 4.5 proposes symmetric two-sided versions of Procedures 4.1 and 4.2, i.e., procedures for two-sided tests based on absolute values of the test statistics. Section 4.6 establishes equivalence results between Θ-specific single-step multiple testing procedures and parameter confidence regions. Finally, Section 4.7 addresses the issue of test optimality, i.e., the maximization of power subject to a Type I error constraint.

4.2 $\boldsymbol{\Theta(F_{V_n})}$-controlling single-step procedures

Consider the simultaneous test of M null hypotheses $H_0(m)$, $m = 1, \ldots, M$, based on test statistics $T_n = (T_n(m) : m = 1, \ldots, M)$, with finite sample joint distribution $Q_n = Q_n(P)$, under the true, unknown data generating distribution P. Assume that large values of the test statistic $T_n(m)$ provide evidence against the corresponding null hypothesis $H_0(m)$, that is, consider one-sided rejection regions of the form $\mathcal{C}_n(m;\alpha) = (c_n(m;\alpha), +\infty)$, where $c_n(\alpha) = (c_n(m;\alpha) : m = 1, \ldots, M) \in I\!R^M$ is an M-vector of single-step cut-offs $c_n(m;\alpha) = c(m; Q_0, \alpha) = c(m; T_n, Q_0, \alpha) \in I\!R$, computed under a null distribution Q_0 (or estimator thereof, Q_{0n}) for the test statistics T_n.

As in Equation (2.2), given a random M-vector $Z = (Z(m) : m = 1,\ldots,M)$, with joint distribution Q, and an M-vector of cut-offs $c = (c(m) : m = 1,\ldots,M) \in I\!R^M$, denote the numbers of rejected hypotheses and Type I errors by

$$R(c|Q) \equiv \sum_{m=1}^{M} \mathrm{I}\left(Z(m) > c(m)\right) \text{ and } V(c|Q) \equiv \sum_{m\in\mathcal{H}_0} \mathrm{I}\left(Z(m) > c(m)\right), \quad (4.1)$$

respectively. For a given cut-off vector c, adopt the shorthand notation of Equation (2.3), for the special cases where Q corresponds to the test statistics true distribution Q_n and null distribution Q_0,

$$\begin{aligned} R_n &\equiv R(c|Q_n), & R_0 &\equiv R(c|Q_0), \\ V_n &\equiv V(c|Q_n), & V_0 &\equiv V(c|Q_0). \end{aligned} \quad (4.2)$$

According to single-step Procedures 4.1 and 4.2, hypothesis $H_0(m)$ is rejected at nominal Type I error level α if $T_n(m) > c_n(m;\alpha)$, where the cut-offs $c_n(m;\alpha)$ satisfy the Type I error constraint $\Theta(F_{R_0}) \leq \alpha$ and are defined as either common quantiles of the marginal null distributions $Q_{0,m}$ (Procedure 4.1) or common cut-offs (Procedure 4.2).

4.2.1 Single-step common-quantile procedure

Procedure 4.1. [$\boldsymbol{\Theta(F_{V_n})}$-controlling single-step common-quantile procedure]
Given an M-variate test statistics null distribution Q_0 and a constant $\delta \in [0,1]$, define an M-vector $q_0^{-1}(\delta) \equiv (Q_{0,m}^{-1}(\delta) : m = 1,\ldots,M)$ of δ-quantiles for the marginal null distributions $Q_{0,m}$ by

$$Q_{0,m}^{-1}(\delta) \equiv \inf\left\{z \in I\!R : Q_{0,m}(z) \geq \delta\right\}, \qquad m = 1,\ldots,M. \quad (4.3)$$

For a test at nominal Type I error level $\alpha \in (0,1)$, define $\delta_0(\alpha)$ by

$$\delta_0(\alpha) \equiv \inf\left\{\delta \in [0,1] : \Theta(F_{R(q_0^{-1}(\delta)|Q_0)}) \leq \alpha\right\}, \quad (4.4)$$

where $R(q_0^{-1}(\delta)|Q_0)$ denotes the number of rejected hypotheses for an M-vector of common-quantile cut-offs $q_0^{-1}(\delta)$, under the null distribution Q_0 for the test statistics T_n. The *single-step common-quantile* multiple testing procedure, for controlling the Type I error rate $\Theta(F_{V_n})$ at nominal level α, is defined in terms of the common-quantile cut-offs $c(Q_0,\alpha) \equiv q_0^{-1}(\delta_0(\alpha))$ by the following rule. Reject $H_0(m)$ if $T_n(m) > Q_{0,m}^{-1}(\delta_0(\alpha))$, $m = 1,\ldots,M$, that is,

$$\mathcal{R}(T_n, Q_0, \alpha) \equiv \left\{m : T_n(m) > Q_{0,m}^{-1}(\delta_0(\alpha))\right\}. \quad (4.5)$$

4.2.2 Single-step common-cut-off procedure

Procedure 4.2. [$\boldsymbol{\Theta(F_{V_n})}$-controlling single-step common-cut-off procedure]
Given an M-variate test statistics null distribution Q_0 and for a test at nominal Type I error level $\alpha \in (0,1)$, define a common cut-off $\gamma_0(\alpha)$ by

$$\gamma_0(\alpha) \equiv \inf\left\{\gamma \in I\!R : \Theta(F_{R(\gamma^{(M)}|Q_0)}) \leq \alpha\right\}, \tag{4.6}$$

where $R(\gamma^{(M)}|Q_0)$ denotes the number of rejected hypotheses for an M-vector $\gamma^{(M)}$ of common cut-offs γ, under the null distribution Q_0 for the test statistics T_n. The *single-step common-cut-off* multiple testing procedure, for controlling the Type I error rate $\Theta(F_{V_n})$ at nominal level α, is defined in terms of the common cut-offs $c(Q_0,\alpha) \equiv \gamma_0(\alpha)^{(M)}$ by the following rule. Reject $H_0(m)$ if $T_n(m) > \gamma_0(\alpha)$, $m = 1,\ldots,M$, that is,

$$\mathcal{R}(T_n, Q_0, \alpha) \equiv \{m : T_n(m) > \gamma_0(\alpha)\}. \tag{4.7}$$

4.2.3 Asymptotic control of Type I error rate and test statistics null distribution

Theorem 4.3. [Asymptotic control of the Type I error rate $\boldsymbol{\Theta(F_{V_n})}$ by single-step common-quantile Procedure 4.1 and common-cut-off Procedure 4.2]
Asymptotic null domination assumption for the number of Type I errors. *Suppose there exists an M-variate null distribution $Q_0 = Q_0(P)$ such that the null test statistics $Z \sim Q_0$ and the original test statistics $T_n \sim Q_n = Q_n(P)$ satisfy the following* asymptotic null domination *condition for the number of Type I errors. For each $c = (c(m) : m = 1,\ldots,M) \in I\!R^M$ and $x \in \{0,\ldots,M\}$,*

$$\liminf_{n\to\infty} \Pr_{Q_n}\left(\sum_{m\in\mathcal{H}_0} \mathrm{I}\left(T_n(m) > c(m)\right) \leq x\right)$$
$$\geq \Pr_{Q_0}\left(\sum_{m\in\mathcal{H}_0} \mathrm{I}\left(Z(m) > c(m)\right) \leq x\right).$$

That is, for one-sided rejection regions of the form $\mathcal{C}(m) = (c(m), +\infty)$, the number of Type I errors V_n, under the true distribution Q_n for the test statistics T_n, is asymptotically stochastically smaller than the corresponding number of Type I errors V_0, under the null distribution Q_0,

$$\liminf_{n\to\infty} F_{V_n}(x) \geq F_{V_0}(x), \qquad \forall\, x \in \{0, \ldots, M\}. \tag{ANDV}$$

Monotonicity and continuity assumptions for the Type I error rate mapping. *In addition, assume that the Type I error rate mapping Θ satisfies monotonicity Assumption MΘ and continuity Assumption CΘ at F_{V_0}.*
Asymptotic control of the Type I error rate $\boldsymbol{\Theta(F_{V_n})}$. *Then, single-step Procedure 4.1, with common-quantile cut-offs $c(m; Q_0, \alpha) = Q_{0,m}^{-1}(\delta_0(\alpha))$, defined according to Equation (4.4), and single-step Procedure 4.2, with common cut-offs $c(m; Q_0, \alpha) = \gamma_0(\alpha)$, defined according to Equation (4.6), provide asymptotic control of the Type I error rate $\Theta(F_{V_n})$ at level α. That is,*

$$\limsup_{n\to\infty} \Theta(F_{V_n}) \leq \alpha,$$

where $V_n = V(c(Q_0, \alpha)|Q_n) \equiv \sum_{m\in\mathcal{H}_0} \mathrm{I}\,(T_n(m) > c(m; Q_0, \alpha))$ denotes the number of Type I errors for single-step cut-offs $c(Q_0, \alpha)$ and $T_n \sim Q_n$.

Proof of Theorem 4.3. Since $V_0 \leq R_0$ Q_0-a.s., then $F_{V_0}(x) \geq F_{R_0}(x)$, $\forall\, x$. Hence, by monotonicity Assumption MΘ and by definition of the cut-offs $c(Q_0, \alpha)$, so that $\Theta(F_{R_0}) \leq \alpha$, one has

$$\Theta(F_{V_0}) \leq \Theta(F_{R_0}) \leq \alpha. \tag{4.8}$$

Note that $F_{V_n} \geq \min\{F_{V_0}, F_{V_n}\}$ and again apply monotonicity Assumption MΘ to show that

$$\Theta(F_{V_n}) \leq \Theta(\min\{F_{V_0}, F_{V_n}\}).$$

Now, by asymptotic null domination Assumption ANDV, $\liminf_n F_{V_n}(x) \geq F_{V_0}(x)$, $\forall\, x$, so that

$$\lim_{n\to\infty} \min\{F_{V_0}(x), F_{V_n}(x)\} = F_{V_0}(x), \qquad \forall\, x.$$

By continuity Assumption CΘ at F_{V_0},

$$\limsup_{n\to\infty} \Theta(F_{V_n}) \leq \lim_{n\to\infty} \Theta(\min\{F_{V_0}, F_{V_n}\}) = \Theta(F_{V_0}). \tag{4.9}$$

Finally, combining Equations (4.8) and (4.9), one obtains the desired asymptotic control of the Type I error rate,

$$\limsup_{n\to\infty} \Theta(F_{V_n}) \leq \alpha.$$

□

Note that asymptotic null domination Assumption ANDV only concerns the joint distribution of the $\mathcal{H}_0$-specific subvector $(T_n(m) : m \in \mathcal{H}_0)$ of test statistics for the true null hypotheses. Null domination conditions for the Type I error rate, the number of Type I errors, and the $\mathcal{H}_0$-specific test statistics are discussed in greater detail in Section 2.2 (Assumptions $\mathsf{ND}\boldsymbol{\Theta}$, NDV, and jtNDT, respectively). Sections 2.3 and 2.4 provide two explicit proposals of test statistics null distributions that satisfy asymptotic null domination Assumption ANDV of Theorem 4.3.

The first null distribution of Section 2.3 is defined as the asymptotic distribution of the M-vector Z_n of *null shift and scale-transformed test statistics*,

$$Z_n(m) \equiv \sqrt{\min\left\{1, \frac{\tau_0(m)}{\text{Var}[T_n(m)]}\right\}} (T_n(m) - \text{E}[T_n(m)]) + \lambda_0(m), \qquad (4.10)$$

where $\lambda_0(m)$ and $\tau_0(m)$ are, respectively, user-supplied upper bounds for the means and variances of the $\mathcal{H}_0$-specific test statistics (Dudoit et al., 2004b). In this construction, the location null values $\lambda_0(m)$ are chosen such that the joint distribution of $(Z_n(m) : m \in \mathcal{H}_0)$ is asymptotically stochastically greater than that of $(T_n(m) : m \in \mathcal{H}_0)$. Hence, for tests based on one-sided rejection regions, asymptotic null domination Assumption ANDV is satisfied for the number of Type I errors. The scale null values $\tau_0(m)$ are chosen to prevent a degenerate limit for the false null hypotheses ($m \in \mathcal{H}_1$); an important issue for power considerations.

The second and most recent proposal of Section 2.4 is defined as the asymptotic distribution of the M-vector $\breve{Z}_n$ of *null quantile-transformed test statistics*,

$$\breve{Z}_n(m) \equiv q_{0,m}^{-1} Q_{n,m}^{\Delta}(T_n(m)), \qquad (4.11)$$

where $q_{0,m}$ are user-supplied marginal test statistics null distributions that satisfy marginal null domination Assumption mgNDT (van der Laan and Hubbard, 2006).

We stress the generality of these two test statistics null distributions: Type I error control does not rely on restrictive assumptions such as subset pivotality and holds for general data generating distributions (with arbitrary dependence structures among variables), null hypotheses, and test statistics. The latest proposal of Section 2.4 has the additional advantage that the marginal test statistics null distributions may be set to the optimal (i.e., most powerful) null distributions one would use in single hypothesis testing (e.g., permutation marginal null distributions, Gaussian or other parametric marginal null distributions).

In practice, since the data generating distribution P is unknown, then so are the aforementioned null distributions $Q_0 = Q_0(P)$. Section 4.4.1 shows that single-step Procedures 4.1 and 4.2, based on a consistent estimator Q_{0n} of the null distribution Q_0, provide asymptotic control of the Type I error rate $\Theta(F_{V_n})$ (Theorems 4.10 and 4.17, Corollaries 4.18 and 4.19). Section 4.4.2 provides bootstrap Procedures 4.20 and 4.21 for obtaining consistent estimators of the test statistic cut-offs and adjusted p-values for Procedures 4.1 and 4.2, respectively.

For greater detail on Type I error control and the definition and estimation of a test statistics null distribution, the reader is referred to Chapter 2. In particular, Sections 2.6 and 2.7 provide null values $\lambda_0(m)$ and $\tau_0(m)$ for a broad range of testing problems and also discuss test statistic-specific null distributions (e.g., for t-statistics, F-statistics). In many testing problems of interest, the null distribution Q_0 is continuous. For instance, for the test of single-parameter null hypotheses using t-statistics, the null values are $\lambda_0(m) = 0$ and $\tau_0(m) = 1$ and the null distribution Q_0 is an M-variate Gaussian distribution with mean vector zero (Section 2.6). For testing the equality of K population mean vectors using F-statistics, the null values are $\lambda_0(m) = 1$ and $\tau_0(m) = 2/(K-1)$, under the assumption of equal variances in the different populations. An F-specific null distribution Q_0^F is proposed as the joint distribution of an M-vector of quadratic forms of Gaussian random variables (Section 2.7). Analogous results are provided in van der Laan and Hubbard (2006) for the new null quantile-transformed null distribution.

Finally, note that similar finite sample Type I error control results can be proved for test statistics null distributions that satisfy a finite sample version of null domination Assumption ANDV.

4.2.4 Common-cut-off vs. common-quantile procedures

As discussed in Section 4.3.3, for control of the FWER (i.e., for $\Theta(F_{V_n}) = 1 - F_{V_n}(0)$), single-step common-quantile Procedure 4.1 and common-cut-off Procedure 4.2 reduce to single-step minP Procedure 3.6 and maxT Procedure 3.5, based on minima of unadjusted p-values and maxima of test statistics, respectively.

Procedures based on common cut-offs and common quantiles are equivalent when the test statistics $T_n(m)$, $m = 1, \ldots, M$, are identically distributed under Q_0, i.e., when the marginal null distributions $Q_{0,m}$ do not depend on m. Indeed, for common marginal null distributions, the significance rankings based on test statistics $T_n(m)$ and unadjusted p-values $P_{0n}(m) = \bar{Q}_{0,m}(T_n(m))$ coincide. In general, however, the two types of methods produce different results and considerations of balance, power, and computational feasibility should dictate the choice between the two approaches (Dudoit et al., 2003; Ge et al., 2003).

For non-identically distributed test statistics $T_n(m)$ (e.g., some null hypotheses tested with F-statistics and others with t-statistics; null hypotheses tested with t-statistics having different degrees of freedom), common-cut-off procedures do not weight all hypotheses equally and can lead to unbalanced adjustments (Beran, 1988; Westfall, 2003; Westfall and Young, 1993). In contrast, procedures based on p-values place the null hypotheses on an "equal footing", i.e., are more balanced than their common-cut-off counterparts, and may therefore be preferable.

When the null distribution Q_0 is estimated by resampling (e.g., bootstrap, permutation), quantile-based procedures tend to be more sensitive than common-cut-off procedures to the discreteness of the estimated null distribution and the number of resampling steps. This can result in more conservative

MTPs than those based directly on the test statistics. Also, quantile-based approaches are more computationally intensive, as the unadjusted p-values $P_{0n}(m)$ must be estimated before one can consider their joint distribution.

The reader is referred to Dudoit et al. (2003), Ge et al. (2003), and Pollard and van der Laan (2004), for further discussion of the relative merits of common-cut-off vs. common-quantile procedures.

4.3 Adjusted p-values for $\Theta(F_{V_n})$-controlling single-step procedures

Rather than simply reporting the rejection of a subset of null hypotheses at a prespecified nominal Type I error level α, one can report *adjusted p-values* for single-step Procedures 4.1 and 4.2 (Section 1.2.12). Although the definition of adjusted p-values in Equation (1.58) holds for arbitrary test statistics null distributions, in this section, we consider for simplicity a null distribution Q_0 with continuous and strictly monotone marginal CDFs, $Q_{0,m}$, and survivor functions, $\bar{Q}_{0,m} = 1 - Q_{0,m}$, $m = 1, \ldots, M$.

Section 4.3.1 provides adjusted p-values for general Type I error rates $\Theta(F_{V_n})$. Adjusted p-values for commonly-used MTPs are shown to correspond to particular choices for the Type I error rate mapping Θ. Explicit formulae for the adjusted p-values of PCER- and gFWER-controlling MTPs are given in Sections 4.3.2 and 4.3.3, respectively.

As usual, denote realizations of the test statistic, unadjusted p-value, and adjusted p-value for null hypothesis $H_0(m)$ by the lowercase letters $t_n(m)$, $p_{0n}(m)$, and $\widetilde{p}_{0n}(m)$, respectively. This lowercase notation is used in some of the equations below to avoid confusion in the interpretation of the probabilities. The probabilities for adjusted p-values are computed with respect to the joint distribution Q_0 of a random M-vector Z, for fixed values of the unadjusted p-values $p_{0n}(m)$ and test statistics $t_n(m)$ corresponding to a particular realization of the random sample $\mathcal{X}_n = \{X_i : i = 1, \ldots, n\}$.

4.3.1 General Type I error rates, $\Theta(F_{V_n})$

Proposition 4.4. [Adjusted p-values for $\Theta(F_{V_n})$-controlling single-step common-quantile Procedure 4.1] *The adjusted p-values for single-step common-quantile Procedure 4.1, based on a test statistics null distribution Q_0 with continuous and strictly monotone marginal distributions, are given by*

$$\widetilde{P}_{0n}(m) = \Theta(F_{R(q_0^{-1}(1-P_{0n}(m))|Q_0)}), \qquad m = 1, \ldots, M, \tag{4.12}$$

where $P_{0n}(m)$ is the unadjusted p-value for null hypothesis $H_0(m)$,

$$P_{0n}(m) = \bar{Q}_{0,m}(T_n(m)) = 1 - Q_{0,m}(T_n(m)), \tag{4.13}$$

and $q_0^{-1}(\delta) = (Q_{0,m}^{-1}(\delta) : m = 1, \ldots, M)$ *denotes an* M*-vector of* δ*-quantiles for the marginal null distributions* $Q_{0,m}$. *The common-quantile cut-offs* $q_0^{-1}(1 - P_{0n}(m))$ *are*

$$Q_{0,l}^{-1}(1 - P_{0n}(m)) = \bar{Q}_{0,l}^{-1}(P_{0n}(m)) = \bar{Q}_{0,l}^{-1}(\bar{Q}_{0,m}(T_n(m)))$$

and, in particular, $Q_{0,m}^{-1}(1 - P_{0n}(m)) = T_n(m)$, *for* $l = m$. *For controlling the Type I error rate* $\Theta(F_{V_n})$ *at nominal level* α, *Procedure 4.1 can then be stated equivalently as*

$$\mathcal{R}(T_n, Q_0, \alpha) = \left\{ m : \widetilde{P}_{0n}(m) \leq \alpha \right\}.$$

Proof of Proposition 4.4. The common-quantile cut-offs in Procedure 4.1 can be represented as

$$c(m; Q_0, \alpha) = Q_{0,m}^{-1}(\delta_0(\alpha)) = Q_{0,m}^{-1}(\theta_0^{-1}(\alpha)),$$

where θ_0^{-1} is the inverse of the non-increasing function $\delta \to \theta_0(\delta) \equiv \Theta(F_{R(q_0^{-1}(\delta)|Q_0)})$, that is, $\theta_0^{-1}(\alpha) \equiv \inf\{\delta \in [0,1] : \theta_0(\delta) \leq \alpha\}$. Then,

$$\begin{aligned}
\widetilde{P}_{0n}(m) &= \inf\{\alpha \in [0,1] : c(m; Q_0, \alpha) < T_n(m)\} \\
&= \inf\left\{\alpha \in [0,1] : Q_{0,m}^{-1}(\theta_0^{-1}(\alpha)) < T_n(m)\right\} \\
&= \inf\left\{\alpha \in [0,1] : \theta_0^{-1}(\alpha) \leq Q_{0,m}(T_n(m))\right\} \\
&= \inf\left\{\alpha \in [0,1] : \alpha \geq \theta_0(Q_{0,m}(T_n(m)))\right\} \\
&= \theta_0(Q_{0,m}(T_n(m))) \\
&= \theta_0(1 - P_{0n}(m)) \\
&= \Theta(F_{R(q_0^{-1}(1-P_{0n}(m))|Q_0)}).
\end{aligned}$$

□

Proposition 4.5. [Adjusted p-values for $\Theta(F_{V_n})$-controlling single-step common-cut-off Procedure 4.2] *The adjusted p-values for single-step common-cut-off Procedure 4.2, based on a test statistics null distribution* Q_0 *with continuous and strictly monotone marginal distributions, are given by*

$$\widetilde{P}_{0n}(m) = \Theta(F_{R(T_n(m)^{(M)}|Q_0)}), \qquad m = 1, \ldots, M, \tag{4.14}$$

where $T_n(m)^{(M)}$ *denotes an* M*-vector of common cut-offs equal to* $T_n(m)$.

The proof of this result is similar to that of Proposition 4.4 for common-quantile adjusted p-values and is therefore omitted.

4.3.2 Per-comparison error rate, PCER

Corollary 4.6. [Adjusted p-values for PCER-controlling single-step common-quantile Procedure 4.1] *For controlling the PCER, the adjusted p-values for single-step common-quantile Procedure 4.1, based on a test statistics null distribution Q_0 with continuous and strictly monotone marginal distributions, reduce to the unadjusted p-values*

$$\widetilde{P}_{0n}(m) = P_{0n}(m) = \bar{Q}_{0,m}(T_n(m)), \qquad m = 1, \ldots, M. \tag{4.15}$$

Proof of Corollary 4.6. Let $Z \sim Q_0$ and consider the Type I error rate mapping $\Theta(F) = \int v dF(v)/M$. Then,

$$\begin{aligned}
\widetilde{p}_{0n}(m) &= \Theta(F_{R(q_0^{-1}(1-p_{0n}(m))|Q_0)}) \\
&= \frac{1}{M} \mathrm{E}\left[R(q_0^{-1}(1 - p_{0n}(m))|Q_0)\right] \\
&= \frac{1}{M} \sum_{l=1}^{M} \Pr_{Q_0}\left(Z(l) > \bar{Q}_{0,l}^{-1}(p_{0n}(m))\right) \\
&= \frac{1}{M} \sum_{l=1}^{M} \bar{Q}_{0,l}\left(\bar{Q}_{0,l}^{-1}(p_{0n}(m))\right) \\
&= p_{0n}(m).
\end{aligned}$$

□

The above result, whereby unadjusted and adjusted p-values coincide, is consistent with the fact that, as noted in Section 1.2.11, PCER-controlling procedures generally do not account for the test of multiple hypotheses.

Corollary 4.7. [Adjusted p-values for PCER-controlling single-step common-cut-off Procedure 4.2] *For controlling the PCER, the adjusted p-values for single-step common-cut-off Procedure 4.2, based on a test statistics null distribution Q_0 with continuous and strictly monotone marginal distributions, are given by*

$$\widetilde{P}_{0n}(m) = \frac{1}{M} \sum_{l=1}^{M} \bar{Q}_{0,l}\left(T_n(m)\right), \qquad m = 1, \ldots, M. \tag{4.16}$$

Proof of Corollary 4.7. Again, let $Z \sim Q_0$ and consider the Type I error rate mapping $\Theta(F) = \int v dF(v)/M$. Then,

$$\begin{aligned}\widetilde{p}_{0n}(m) &= \Theta(F_{R(t_n(m)^{(M)}|Q_0)}) \\ &= \frac{1}{M}\,\mathrm{E}\left[R(t_n(m)^{(M)}|Q_0)\right] \\ &= \frac{1}{M}\sum_{l=1}^{M}\mathrm{Pr}_{Q_0}\left(Z(l) > t_n(m)\right) \\ &= \frac{1}{M}\sum_{l=1}^{M}\bar{Q}_{0,l}\left(t_n(m)\right).\end{aligned}$$

□

4.3.3 Generalized family-wise error rate, gFWER

Corollary 4.8. [Adjusted p-values for gFWER-controlling single-step common-quantile Procedure 4.1] *For controlling the gFWER, the adjusted p-values for single-step common-quantile Procedure 4.1, based on a test statistics null distribution Q_0 with continuous and strictly monotone marginal distributions, are given by*

$$\widetilde{p}_{0n}(m) = \mathrm{Pr}_{Q_0}\left(P_0^\circ(k+1) \leq p_{0n}(m)\right), \qquad m = 1, \ldots, M. \tag{4.17}$$

Here, $P_0(m) = \bar{Q}_{0,m}(Z(m))$ denote unadjusted p-values under the test statistics null distribution Q_0 (i.e., for $Z(m) \sim Q_{0,m}$) and $P_0^\circ(m)$ denotes the mth smallest unadjusted p-value, so that $P_0^\circ(1) \leq \cdots \leq P_0^\circ(M)$.

Proof of Corollary 4.8. Let $Z \sim Q_0$ and consider the Type I error rate mapping $\Theta(F) = 1 - F(k)$. Then,

$$\begin{aligned}\widetilde{p}_{0n}(m) &= \Theta(F_{R(q_0^{-1}(1-p_{0n}(m))|Q_0)}) \\ &= \mathrm{Pr}\left(R(q_0^{-1}(1-p_{0n}(m))|Q_0) > k\right) \\ &= \mathrm{Pr}_{Q_0}\left(\sum_{l=1}^{M}\mathrm{I}\left(Z(l) > Q_{0,l}^{-1}(1-p_{0n}(m))\right) > k\right) \\ &= \mathrm{Pr}_{Q_0}\left(\sum_{l=1}^{M}\mathrm{I}\left(\bar{Q}_{0,l}(Z(l)) \leq \bar{Q}_{0,l}(\bar{Q}_{0,l}^{-1}(p_{0n}(m)))\right) > k\right) \\ &= \mathrm{Pr}_{Q_0}\left(\sum_{l=1}^{M}\mathrm{I}\left(P_0(l) \leq p_{0n}(m)\right) > k\right) \\ &= \mathrm{Pr}_{Q_0}\left(P_0^\circ(k+1) \leq p_{0n}(m)\right).\end{aligned}$$

□

For control of the gFWER, Procedure 4.1 is thus based on the distribution of the $(k+1)$*st smallest unadjusted p-value* $P_0^\circ(k+1)$ and $(1-\delta_0(\alpha))$ is chosen as the α-quantile of the distribution of $P_0^\circ(k+1)$. We refer to this procedure as the *single-step* $P(k+1)$ *procedure* (Procedure 3.19). In the special case of FWER control ($k=0$), the procedure is based on the distribution of the *minimum* of the M unadjusted p-values $P_0(m)$, i.e., on $P_0^\circ(1) = \min_m P_0(m)$. Thus, for FWER control, the *single-step minP procedure* corresponds to Procedure 4.1 with $(1-\delta_0(\alpha))$ chosen as the α-quantile of the distribution of the minimum unadjusted p-value $P_0^\circ(1)$ (Procedure 3.6).

Consider the special case where the random vector $Z \sim Q_0$ has independent elements $Z(m)$, with continuous marginal distributions $Q_{0,m}$, $m = 1, \ldots, M$. Then, $P_0(m) = \bar{Q}_{0,m}(Z(m))$ are independent $\mathrm{U}(0,1)$ random variables and the $(k+1)$st smallest unadjusted p-value $P_0^\circ(k+1)$ has a $\mathrm{Beta}(k+1, M-k)$ distribution. Thus, in this independence situation, gFWER-controlling Procedure 4.1 is very simple and is based only on the marginal null distributions $Q_{0,m}$. Adjusted p-values are given by

$$\widetilde{P}_{0n}(m) = \frac{\Gamma(M+1)}{\Gamma(k+1)\Gamma(M-k)} \int_0^{P_{0n}(m)} z^k (1-z)^{M-k-1} dz, \tag{4.18}$$

where $\Gamma(j) \equiv (j-1)!$ for a positive integer j. In particular, for FWER control ($k=0$), the adjusted p-values for Procedure 4.1 reduce to the adjusted p-values for single-step Šidák Procedure 3.3,

$$\widetilde{P}_{0n}(m) = 1 - (1 - P_{0n}(m))^M . \tag{4.19}$$

Corollary 4.9. [Adjusted p-values for gFWER-controlling single-step common-cut-off Procedure 4.2] *For controlling the gFWER, the adjusted p-values for single-step common-cut-off Procedure 4.2, based on a test statistics null distribution Q_0 with continuous and strictly monotone marginal distributions, are given by*

$$\widetilde{p}_{0n}(m) = \Pr_{Q_0}\left(Z^\circ(k+1) \geq t_n(m)\right), \qquad m = 1, \ldots, M, \tag{4.20}$$

where $Z^\circ(m)$ denotes the mth largest element of $Z = (Z(m) : m = 1, \ldots, M) \sim Q_0$, so that $Z^\circ(1) \geq \cdots \geq Z^\circ(M)$.

Proof of Corollary 4.9. Again, let $Z \sim Q_0$ and consider the Type I error rate mapping $\Theta(F) = 1 - F(k)$. Then,

$$\begin{aligned}
\widetilde{p}_{0n}(m) &= \Theta(F_{R(t_n(m)^{(M)}|Q_0)}) \\
&= \Pr\left(R(t_n(m)^{(M)}|Q_0) > k\right) \\
&= \Pr_{Q_0}\left(\sum_{l=1}^{M} \mathrm{I}(Z(l) > t_n(m)) > k\right) \\
&= \Pr_{Q_0}\left(Z^\circ(k+1) > t_n(m)\right).
\end{aligned}$$

□

For control of the gFWER, Procedure 4.2 is thus based on the distribution of the $(k+1)st$ *largest element* $Z^\circ(k+1)$ of $Z \sim Q_0$ and the common cut-off $\gamma_0(\alpha)$ is chosen as the $(1-\alpha)$-quantile of the distribution of $Z^\circ(k+1)$. We refer to this procedure as the *single-step* $T(k+1)$ *procedure* (Procedure 3.18). In the special case of FWER control ($k = 0$), the procedure is based on the distribution of the *maximum* of the M random variables $Z(m)$, i.e., on $Z^\circ(1) = \max_m Z(m)$. Thus, for FWER control, the *single-step maxT procedure* corresponds to Procedure 4.2 with common cut-off chosen as the $(1-\alpha)$-quantile of the distribution of the maximum $Z^\circ(1)$ (Procedure 3.5).

Other gFWER-controlling procedures

Section 3.3 provides an overview of gFWER-controlling marginal/joint single-step/stepwise procedures.

In particular, as indicated in Procedure 3.20, one can control the gFWER using simple modifications of FWER-controlling Procedures 4.1 and 4.2. The main idea is to follow the FWER-controlling procedure exactly until the first non-rejection and then reject the null hypotheses specified by this procedure as well as the next k most significant null hypotheses (i.e., the k hypotheses with the next k smallest FWER-controlling adjusted p-values). Such an approach to gFWER control is appealing, because it only requires working with FWER-controlling MTPs and it guarantees at least k rejected hypotheses. Greater detail and formal results on augmentation multiple testing procedures are provided in Chapter 6 and in articles by Dudoit et al. (2004a) and van der Laan et al. (2004b).

4.4 $\Theta(F_{V_n})$-controlling bootstrap-based single-step procedures

Single-step common-quantile Procedure 4.1 and common-cut-off Procedure 4.2 provide asymptotic control of general Type I error rates $\Theta(F_{V_n})$, when based on either of the two main types of test statistics null distributions proposed in Chapter 2: the null shift and scale-transformed null distribution (Section 2.3) and the null quantile-transformed null distribution (Section 2.4). In practice, however, both test statistics null distributions $Q_0 = Q_0(P)$ are unknown, as they depend on the unknown data generating distribution P. Estimation of Q_0 is then needed, especially to deal with the unknown dependence structure of the test statistics. Sections 2.3.2, 2.4.2, 2.6, and 2.7 provide a variety of bootstrap procedures for obtaining consistent estimators Q_{0n} of the null distribution Q_0.

Section 4.4.1, below, establishes asymptotic Type I error control results for Procedures 4.1 and 4.2 based on a consistent estimator Q_{0n} of the null distribution Q_0. Given a bootstrap estimator of the null distribution Q_0,

Section 4.4.2 provides Procedures 4.20 and 4.21 for estimating cut-offs and adjusted p-values for Procedures 4.1 and 4.2, respectively.

4.4.1 Asymptotic control of Type I error rate for single-step procedures based on consistent estimator of test statistics null distribution

In this section, we consider analogues of single-step Procedures 4.1 and 4.2, based on a consistent estimator Q_{0n} of the null distribution Q_0. In such multiple testing procedures, the estimator Q_{0n} is used in place of Q_0, to estimate the cut-offs for the test statistics and the corresponding adjusted p-values. Theorems 4.10 and 4.17 establish consistency of the single-step common-quantile cut-offs and common cut-offs for Procedures 4.1 and 4.2, respectively. Corollaries 4.18 and 4.19 prove consistency of the resulting Type I error rates.

Theorem 4.10. [Consistency of single-step common-quantile cut-offs for Procedure 4.1]
Set-up and assumptions. *Given an M-variate distribution Q and a constant $\delta \in [0,1]$, define an M-vector $q^{-1}(\delta) \equiv (Q_m^{-1}(\delta) : m = 1,\ldots,M)$ of δ-quantiles for the marginal distributions Q_m by*

$$Q_m^{-1}(\delta) \equiv \inf\left\{z \in I\!R : Q_m(z) \geq \delta\right\}, \qquad m = 1,\ldots,M. \tag{4.21}$$

Consider a Type I error rate mapping Θ that satisfies monotonicity Assumption MΘ and define a non-increasing function

$$\delta \to \theta_Q(\delta) \equiv \Theta\left(F_{R(q^{-1}(\delta)|Q)}\right), \tag{4.22}$$

where

$$R(q^{-1}(\delta)|Q) \equiv \sum_{m=1}^{M} \mathrm{I}\left(Z(m) > Q_m^{-1}(\delta)\right) \tag{4.23}$$

denotes the number of rejected hypotheses for an M-vector $q^{-1}(\delta)$ of δ-quantiles and for $Z \sim Q$. For a fixed Type I error level $\alpha \in (0,1)$, define α-quantiles of θ_Q by

$$\delta_Q(\alpha) \equiv \theta_Q^{-1}(\alpha) = \inf\left\{\delta \in [0,1] : \Theta\left(F_{R(q^{-1}(\delta)|Q)}\right) \leq \alpha\right\}. \tag{4.24}$$

Let Q_0 be a specified M-variate null distribution and let Q_{0n} converge weakly to Q_0. Assume that: (i) Q_0 is absolutely continuous with respect to the Lebesgue measure on $I\!R^M$, with uniformly bounded density; (ii) each marginal distribution $Q_{0,m}$ has continuous Lebesgue density $f_{0,m}$, with interval support, that is, $\{z : f_{0,m}(z) > 0\} = (a(m), b(m))$, where $a(m)$ and $b(m)$ are allowed to equal $-\infty$ and $+\infty$, respectively; (iii) $\delta_{Q_0}(\alpha) \in (0,1)$; (iv) the function $\theta_{Q_0}(\delta)$ is continuous and has a positive derivative at $\delta_{Q_0}(\alpha)$; (v) the Type I error rate mapping Θ satisfies continuity Assumption CΘ at $F_{R(q_0^{-1}(\delta)|Q_0)}$ for $\delta \in (0,1)$.

Result. *Then, one has the following consistency results for the quantiles of* Q_0 *and* Q_{0n}. *Given* $(P_n : n \geq 1)$,

$$\lim_{n\to\infty} (\delta_{Q_{0n}}(\alpha) - \delta_{Q_0}(\alpha)) = 0 \tag{4.25}$$

and

$$\lim_{n\to\infty} (Q_{0n,m}^{-1}(\delta_{Q_{0n}}(\alpha)) - Q_{0,m}^{-1}(\delta_{Q_0}(\alpha))) = 0, \qquad \forall\ m = 1,\ldots,M.$$

In order to prove Theorem 4.10, we begin by establishing the following lemmas. Adopt the shorthand notation $\theta_0(\delta) \equiv \theta_{Q_0}(\delta)$, $\theta_{0n}(\delta) \equiv \theta_{Q_{0n}}(\delta)$, $\delta_0 \equiv \delta_{Q_0}(\alpha) = \theta_{Q_0}^{-1}(\alpha)$, and $\delta_{0n} \equiv \delta_{Q_{0n}}(\alpha) = \theta_{Q_{0n}}^{-1}(\alpha)$.

Lemma 4.11. *For each* $m = 1,\ldots,M$, $Q_{0,m}^{-1}$ *is uniformly continuous on any interval* $[a,b] \subset (0,1)$. *That is, if* $x_n - y_n \to 0$, *for sequences* $\{x_n\}$ *and* $\{y_n\} \in [a,b]$, *then* $Q_{0,m}^{-1}(x_n) - Q_{0,m}^{-1}(y_n) \to 0$.

Proof of Lemma 4.11. This lemma follows from the assumption that each marginal distribution $Q_{0,m}$ has continuous Lebesgue density $f_{0,m}$ with interval support.

□

Lemma 4.12. *For each* $m = 1,\ldots,M$, $Q_{0n,m} - Q_{0,m}$ *converges uniformly to zero over the support* $(a(m), b(m))$ *of* $Q_{0,m}$. *That is,* $\sup_{z\in(a(m),b(m))} |Q_{0n,m}(z) - Q_{0,m}(z)| \to 0$, *as* $n \to \infty$.

Proof of Lemma 4.12. By the weak convergence of Q_{0n} to Q_0, $Q_{0n,m}$ converges pointwise to $Q_{0,m}$. Because pointwise convergence of monotone functions $Q_{0n,m}$ to a continuous monotone function $Q_{0,m}$ implies uniform convergence, it follows that $Q_{0n,m} - Q_{0,m}$ converges uniformly to zero.

□

Lemma 4.13. *For each* $m = 1,\ldots,M$, $Q_{0n,m}^{-1} - Q_{0,m}^{-1}$ *converges uniformly to zero over any interval contained in* $(0,1)$. *That is,* $\forall\ \epsilon > 0$, $\sup_{\delta\in[\epsilon,1-\epsilon]} \left|Q_{0n,m}^{-1}(\delta) - Q_{0,m}^{-1}(\delta)\right| \to 0$, *as* $n \to \infty$.

Proof of Lemma 4.13. This statement follows from the facts that: (i) for an M-variate distribution Q, the quantile mapping $Q_m \to Q_m^{-1}(\delta)$, for each margin Q_m, is continuous with respect to the supremum norm convergence at a $Q_{0,m}$ at which $f_{0,m}(Q_{0,m}^{-1}(\delta)) > 0$; (ii) pointwise convergence of monotone functions $Q_{0n,m}^{-1}$ to a continuous monotone function $Q_{0,m}^{-1}$ at each $\delta \in (0,1)$ implies uniform convergence; (iii) Lemma 4.12.

□

Lemma 4.14. *Consider CDFs* F_n *and* F *such that* $F_n - F$ *converges uniformly to zero and* F^{-1} *is uniformly continuous on an interval* $[a,b] \subset (0,1)$. *For a sequence* $\{x_n\} \in (0,1)$, *suppose there is an integer* $N > 0$, *such that* $x_n \in [a,b]$ *and* $FF_n^{-1}(x_n) \in [a,b]$, $\forall\ n > N$. *Then,* $F_n^{-1}(x_n) - F^{-1}(x) = F^{-1}(x_n) - F^{-1}(x) + o(1)$.

Proof of Lemma 4.14. This fourth lemma follows from the expansion

$$\begin{aligned} F_n^{-1}(x_n) - F^{-1}(x) &= (F_n^{-1}(x_n) - F^{-1}(x_n)) \\ &\quad + (F^{-1}(x_n) - F^{-1}(x)) \\ &= (F^{-1}FF_n^{-1}(x_n) - F^{-1}F_nF_n^{-1}(x_n)) \\ &\quad + (F^{-1}(x_n) - F^{-1}(x)). \end{aligned}$$

The first term converges to zero, by uniform convergence of $F_n - F$ to zero, uniform continuity of F^{-1} on $[a,b]$, and the fact that x_n and $FF_n^{-1}(x_n) \in [a,b]$, $\forall\, n > N$.

□

In order to apply Lemma 4.14 with $F = Q_{0,m}$, $F_n = Q_{0n,m}$, $x = \delta_0$, and $x_n = \delta_{0n}$, one needs to prove the following Lemma 4.15.

Lemma 4.15. *Suppose that $\delta_{0n} \to \delta_0$. Then, there is an interval $[a,b] \subset (0,1)$ and an integer $N_0 > 0$, such that, for all $n > N_0$, $\delta_{0n} \in [a,b]$ and $Q_{0,m}(Q_{0n,m}^{-1}(\delta_{0n})) \in [a,b]$.*

Proof of Lemma 4.15. The first statement follows from the convergence of δ_{0n} to δ_0 and from the assumption that $\delta_0 \in (0,1)$. Thus, there exist $\epsilon > 0$ and an integer $N(\epsilon) > 0$, such that $\epsilon < \delta_{0n} < 1-\epsilon$, $\forall\, n > N(\epsilon)$. By monotonicity of $Q_{0n,m}^{-1}$, $Q_{0n,m}^{-1}(\epsilon) \leq Q_{0n,m}^{-1}(\delta_{0n}) \leq Q_{0n,m}^{-1}(1-\epsilon)$, $\forall\, n > N(\epsilon)$. Next, by weak convergence of $Q_{0n,m}$ to $Q_{0,m}$, for each $\epsilon' > 0$, there is an integer $N(\epsilon') > 0$, such that

$$Q_{0,m}^{-1}(\epsilon) - \epsilon' \leq Q_{0n,m}^{-1}(\delta_{0n}) \leq Q_{0,m}^{-1}(1-\epsilon) + \epsilon', \qquad \forall\, n > \max\{N(\epsilon), N(\epsilon')\}.$$

Hence, by monotonicity of $Q_{0,m}$,

$$Q_{0,m}(Q_{0,m}^{-1}(\epsilon) - \epsilon') \leq Q_{0,m}(Q_{0n,m}^{-1}(\delta_{0n})) \leq Q_{0,m}(Q_{0,m}^{-1}(1-\epsilon) + \epsilon'),$$

$\forall\, n > \max\{N(\epsilon), N(\epsilon')\}$. Finally, by continuity of $Q_{0,m}$ and letting $\epsilon' \to 0$, there is an interval $[a,b] \subset (0,1)$ and an integer $N_0 > 0$, such that, as required, $Q_{0,m}(Q_{0n,m}^{-1}(\delta_{0n})) \in [a,b]$, for $n > N_0$.

□

Lemma 4.16. *Consider a sequence of M-vectors $\{c_{0n}\} \in I\!R^M$, with limit $c_0 \in I\!R^M$, i.e., $c_{0n}(m) \to c_0(m)$, as $n \to \infty$, $\forall\, m = 1, \ldots, M$. Define subsets $\mathcal{A}_n \subseteq I\!R^M$ by*

$$\mathcal{A}_n \equiv \left\{ z \in I\!R^M : \left| \sum_{m=1}^{M} \mathrm{I}(z(m) > c_{0n}(m)) - \mathrm{I}(z(m) > c_0(m)) \right| > 0 \right\} \tag{4.26}$$

and, for $\epsilon > 0$, define subsets $\mathcal{A}(\epsilon) \subseteq I\!R^M$ by

$$\mathcal{A}(\epsilon) \equiv \left\{ z \in I\!R^M : \sup_{\{c:||c-c_0||<\epsilon\}} \left| \sum_{m=1}^{M} \mathrm{I}\left(z(m) > c(m)\right) - \mathrm{I}\left(z(m) > c_0(m)\right) \right| > 0 \right\}. \tag{4.27}$$

Let Q_0 be a specified M-variate distribution and let Q_{0n} converge weakly to Q_0. Further assume that there exists a sequence $\{\epsilon_k\} \to 0$, such that $\{\mathcal{A}(\epsilon_k)\}$ are continuity sets of Q_0 (i.e., the boundary sets $\partial\mathcal{A}(\epsilon_k)$ have mass zero under Q_0, $\Pr_{Q_0}(Z \in \partial\mathcal{A}(\epsilon_k)) = 0$) and $\lim_k \Pr_{Q_0}(Z \in \mathcal{A}(\epsilon_k)) = 0$. Then, $\lim_n \Pr_{Q_{0n}}(Z_n \in \mathcal{A}_n) = 0$.

Proof of Lemma 4.16. By the definition of weak convergence (van der Vaart and Wellner, 1996), for each continuity set $\mathcal{A}(\epsilon_k) \subseteq I\!R^M$ of Q_0, $\Pr_{Q_{0n}}(Z_n \in \mathcal{A}(\epsilon_k)) - \Pr_{Q_0}(Z \in \mathcal{A}(\epsilon_k)) \to 0$, as $n \to \infty$. Thus, for each k and each $\epsilon > 0$, there exists an integer $N(k,\epsilon) > 0$, such that $\Pr_{Q_{0n}}(Z_n \in \mathcal{A}(\epsilon_k)) \le \Pr_{Q_0}(Z \in \mathcal{A}(\epsilon_k)) + \epsilon$, for each $n > N(k,\epsilon)$. Next, by convergence of c_{0n} to c_0, for each k, there exists an integer $N(k) > 0$, such that $\mathcal{A}_n \subseteq \mathcal{A}(\epsilon_k)$ and hence $\Pr_{Q_{0n}}(Z_n \in \mathcal{A}_n) \le \Pr_{Q_{0n}}(Z_n \in \mathcal{A}(\epsilon_k))$, for each $n > N(k)$. Thus, for each k and each $\epsilon > 0$,

$$\Pr_{Q_{0n}}\left(Z_n \in \mathcal{A}_n\right) \le \Pr_{Q_{0n}}\left(Z_n \in \mathcal{A}(\epsilon_k)\right) \le \Pr_{Q_0}\left(Z \in \mathcal{A}(\epsilon_k)\right) + \epsilon,$$

$\forall\, n > \max\{N(k), N(k,\epsilon)\}$. But $\lim_k \Pr_{Q_0}(Z \in \mathcal{A}(\epsilon_k)) = 0$, hence, as required, $\lim_n \Pr_{Q_{0n}}(Z_n \in \mathcal{A}_n) = 0$.

□

Proof of Theorem 4.10.
Convergence of ($\delta_{0n} - \delta_0$) to zero. In order to apply Lemmas 4.11–4.15 and establish consistency of the single-step common-quantile cut-offs, the main task is to prove that $\delta_{0n} - \delta_0$ converges to zero. Recall that $\delta_{0n} - \delta_0 = \theta_{0n}^{-1}(\alpha) - \theta_0^{-1}(\alpha)$ and consider the functions $\theta_0(\delta)$ and $\theta_{0n}(\delta)$.

By monotonicity Assumption MΘ on the mapping Θ, θ_{0n} and θ_0 are non-increasing functions of δ. Also, for any M-variate distribution Q, $R(q^{-1}(\delta)|Q) \le M$, thus θ_{0n} and θ_0 are bounded above by $\Theta(F_{\{M\}})$, where $F_{\{M\}}$ denotes the CDF for the distribution with unit mass at the singleton $\{M\}$.

Under the assumption that θ_0 has a positive derivative at $\delta_0 = \theta_0^{-1}(\alpha)$, it follows that $\theta_{0n}^{-1}(\alpha) - \theta_0^{-1}(\alpha)$ converges to zero provided $\theta_{0n} - \theta_0$ converges uniformly to zero. Because pointwise convergence of monotone functions to a continuous monotone function implies uniform convergence, it suffices to show that $\theta_{0n}(\delta) - \theta_0(\delta)$ converges to zero for $\delta \in (0,1)$. By continuity Assumption CΘ on the mapping Θ at $F_{R(q_0^{-1}(\delta)|Q_0)}$ for $\delta \in (0,1)$, the latter holds if the number of rejected hypotheses $R(q_{0n}^{-1}(\delta)|Q_{0n})$, under Q_{0n}, converges weakly to the corresponding quantity $R(q_0^{-1}(\delta)|Q_0)$, under Q_0.

One has

$$R(q_{0n}^{-1}(\delta)|Q_{0n}) = R(q_{0n}^{-1}(\delta)|Q_{0n}) - R(q_0^{-1}(\delta)|Q_{0n}) + R(q_0^{-1}(\delta)|Q_{0n}).$$

Let $Z_n \sim Q_{0n}$ and $Z \sim Q_0$. By the Continuous Mapping Theorem (Theorem B.3), $R(q_0^{-1}(\delta)|Q_{0n}) = \sum_{m=1}^M \mathrm{I}\left(Z_n(m) > Q_{0,m}^{-1}(\delta)\right)$ converges weakly to $R(q_0^{-1}(\delta)|Q_0) = \sum_{m=1}^M \mathrm{I}\left(Z(m) > Q_{0,m}^{-1}(\delta)\right)$.

Thus, it remains to prove that the difference $R(q_{0n}^{-1}(\delta)|Q_{0n}) - R(q_0^{-1}(\delta)|Q_{0n})$ converges to zero Q_{0n}-a.s., that is, $\lim_n \Pr(R(q_{0n}^{-1}(\delta)|Q_{0n}) \neq R(q_0^{-1}(\delta)|Q_{0n})) = 0$. For a fixed $\delta \in (0,1)$, use the following shorthand notation for the common quantiles: $c_0(m) = Q_{0,m}^{-1}(\delta)$ and $c_{0n}(m) = Q_{0n,m}^{-1}(\delta)$. Note that, from Lemma 4.13, $c_{0n}(m) - c_0(m) \to 0$, as $n \to \infty$, $\forall\, m = 1, \ldots, M$. As in Lemma 4.16, below, define subsets $\mathcal{A}_n \subseteq I\!R^M$ by

$$\mathcal{A}_n = \left\{ z \in I\!R^M : \left| \sum_{m=1}^M \mathrm{I}\left(z(m) > c_{0n}(m)\right) - \mathrm{I}\left(z(m) > c_0(m)\right) \right| > 0 \right\}.$$

Then, $R(q_{0n}^{-1}(\delta)|Q_{0n}) \neq R(q_0^{-1}(\delta)|Q_{0n})$ if and only if $Z_n \in \mathcal{A}_n$. By absolute continuity of Q_0 with respect to the Lebesgue measure on $I\!R^M$, with uniformly bounded density, one has $\Pr_{Q_0}(Z \in \mathcal{A}(\epsilon)) \to 0$, as $\epsilon \to 0$, for subsets $\mathcal{A}(\epsilon) \subseteq I\!R^M$ defined as in Lemma 4.16. Thus, it follows from this lemma that $\lim_n \Pr_{Q_{0n}}(Z_n \in \mathcal{A}_n) = 0$. That is, as required,

$$\lim_{n\to\infty} \Pr_{Q_{0n}}\left(Z_n \in \mathcal{A}_n\right) = \lim_{n\to\infty} \Pr\left(R(q_{0n}^{-1}(\delta)|Q_{0n}) \neq R(q_0^{-1}(\delta)|Q_{0n})\right) = 0.$$

Convergence of $(Q_{0n,m}^{-1}(\delta_{0n}) - Q_{0,m}^{-1}(\delta_0))$ to zero. We are now in a position to apply Lemmas 4.11–4.15 to establish consistency of the single-step common-quantile cut-offs, i.e., convergence of $Q_{0n,m}^{-1}(\delta_{0n}) - Q_{0,m}^{-1}(\delta_0)$ to zero. The assumptions of Lemma 4.14, with $F = Q_{0,m}$, $F_n = Q_{0n,m}$, $x = \delta_0$, and $x_n = \delta_{0n}$, are satisfied from Lemmas 4.12, 4.11, and 4.15, respectively. Thus,

$$Q_{0n,m}^{-1}(\delta_{0n}) - Q_{0,m}^{-1}(\delta_0) = Q_{0,m}^{-1}(\delta_{0n}) - Q_{0,m}^{-1}(\delta_0) + o(1).$$

By continuity of $Q_{0,m}^{-1}$ (Lemma 4.11) and convergence of $\delta_{0n} - \delta_0$ to zero, it follows that $Q_{0,m}^{-1}(\delta_{0n}) - Q_{0,m}^{-1}(\delta_0)$ converges to zero.

□

An (simpler) analogue of Theorem 4.10 can be obtained for consistency of the single-step common cut-offs in Procedure 4.2.

Theorem 4.17. [Consistency of single-step common cut-offs for Procedure 4.2]
Set-up and assumptions. *Given an M-variate distribution Q and a Type I error rate mapping Θ that satisfies monotonicity Assumption MΘ, define a non-increasing function*

$$\gamma \to \theta_Q(\gamma) \equiv \Theta\left(F_{R(\gamma^{(M)}|Q)}\right), \tag{4.28}$$

where

$$R(\gamma^{(M)}|Q) \equiv \sum_{m=1}^{M} \mathrm{I}\left(Z(m) > \gamma\right) \tag{4.29}$$

denotes the number of rejected hypotheses for an M-vector $\gamma^{(M)}$ of common cut-offs γ and for $Z \sim Q$. For a fixed Type I error level $\alpha \in (0,1)$, define common cut-offs as the α-quantiles of θ_Q

$$\gamma_Q(\alpha) \equiv \theta_Q^{-1}(\alpha) = \inf\left\{\gamma \in I\!R : \Theta\left(F_{R(\gamma^{(M)}|Q)}\right) \leq \alpha\right\}. \tag{4.30}$$

Let Q_0 be a specified M-variate null distribution and let Q_{0n} converge weakly to Q_0. Assume that: (i) the function $\theta_{Q_0}(\gamma)$ is continuous and has a positive derivative at $\gamma_{Q_0}(\alpha)$; (ii) the Type I error rate mapping Θ satisfies continuity Assumption $C\Theta$ at $F_{R(\gamma^{(M)}|Q_0)}$ for $\gamma \in I\!R$.
Result. *Then, one has the following consistency result for the common cut-offs. Given $(P_n : n \geq 1)$,*

$$\lim_{n\to\infty} (\gamma_{Q_{0n}}(\alpha) - \gamma_{Q_0}(\alpha)) = 0. \tag{4.31}$$

Proof of Theorem 4.17. Adopt the shorthand notation $\theta_0(\gamma) \equiv \theta_{Q_0}(\gamma)$, $\theta_{0n}(\gamma) \equiv \theta_{Q_{0n}}(\gamma)$, $\gamma_0 \equiv \gamma_{Q_0}(\alpha) = \theta_{Q_0}^{-1}(\alpha)$, and $\gamma_{0n} \equiv \gamma_{Q_{0n}}(\alpha) = \theta_{Q_{0n}}^{-1}(\alpha)$.

Convergence of $\gamma_{0n} - \gamma_0 = \theta_{0n}^{-1}(\alpha) - \theta_0^{-1}(\alpha)$ to zero follows from the first part of the proof of Theorem 4.10, whereby $\delta_{0n} - \delta_0 \to 0$. One simply needs to show that $\theta_{0n}(\gamma) - \theta_0(\gamma)$ converges to zero for $\gamma \in I\!R$. By the Continuous Mapping Theorem (Theorem B.3), the number of rejected hypotheses $R(\gamma^{(M)}|Q_{0n})$, under Q_{0n}, converges weakly to the corresponding quantity $R(\gamma^{(M)}|Q_0)$, under Q_0. Hence, from continuity Assumption $C\Theta$ on the mapping Θ, $\theta_{0n}(\gamma) - \theta_0(\gamma) = \Theta(F_{R(\gamma^{(M)}|Q_{0n})}) - \Theta(F_{R(\gamma^{(M)}|Q_0)})$ converges to zero. □

Having established *consistency of the cut-offs* for common-quantile Procedure 4.1 and common-cut-off Procedure 4.2, based on a consistent estimator Q_{0n} of the null distribution Q_0, the following corollaries prove *consistency of the resulting Type I error rates.*

Corollary 4.18. [Consistency of Type I error rate for single-step common-quantile Procedure 4.1 and common-cut-off Procedure 4.2] *Let Q_0 be a specified M-variate null distribution and let Q_{0n} converge weakly to Q_0. Denote the number of Type I errors for Procedure 4.1/4.2, based on the null distribution Q_0 and its estimator Q_{0n}, by*

$$V(c_{0n}|Q_n) \equiv \sum_{m\in\mathcal{H}_0} \mathrm{I}\left(T_n(m) > c_{0n}(m)\right) \tag{4.32}$$

and

$$V(c_0|Q_n) \equiv \sum_{m\in\mathcal{H}_0} \mathrm{I}\left(T_n(m) > c_0(m)\right),$$

respectively, where $T_n \sim Q_n = Q_n(P)$. *For Procedure 4.1, the common-quantile cut-offs* $c_0(m) = c(m;Q_0,\alpha) = Q_{0,m}^{-1}(\delta_{Q_0}(\alpha))$ *and* $c_{0n}(m) = c(m;Q_{0n},\alpha) = Q_{0n,m}^{-1}(\delta_{Q_{0n}}(\alpha))$ *are defined as in Theorem 4.10. For Procedure 4.2, the common cut-offs* $c_0(m) = c(m;Q_0,\alpha) = \gamma_{Q_0}(\alpha)$ *and* $c_{0n}(m) = c(m;Q_{0n},\alpha) = \gamma_{Q_{0n}}(\alpha)$ *are defined as in Theorem 4.17. Assume that the conditions of Theorem 4.10/4.17 hold, so that, given* $(P_n : n \geq 1)$, $c_{0n}(m) - c_0(m) \to 0$, *as* $n \to \infty$, *for each* $m = 1, \ldots, M$. *Further assume that the joint distribution* Q_n *of the test statistics* T_n *is such that*

$$\Pr_{Q_n}(T_n \in \mathcal{A}_{\mathcal{H}_0}(\epsilon)) \to 0, \qquad \text{as } \epsilon \to 0, \tag{4.33}$$

where, for $\epsilon > 0$, *the subsets* $\mathcal{A}_{\mathcal{H}_0}(\epsilon) \subseteq I\!R^M$ *are defined as*

$$\mathcal{A}_{\mathcal{H}_0}(\epsilon) \equiv \left\{ z \in I\!R^M : \sup_{\{c : ||c - c_0|| < \epsilon\}} \left| \sum_{m \in \mathcal{H}_0} \mathrm{I}\,(z(m) > c(m)) - \mathrm{I}\,(z(m) > c_0(m)) \right| > 0 \right\}. \tag{4.34}$$

Finally, assume that the Type I error rate mapping Θ *satisfies uniform continuity Assumption C*Θ. *Then,*

$$\lim_{n \to \infty} \Pr\left(V(c_{0n}|Q_n) \neq V(c_0|Q_n)\right) = 0,$$

so that asymptotic control of the Type I error rate $\Theta(F_{V(c_0|Q_n)})$, *as in Theorem 4.3, for Procedure 4.1/4.2 based on cut-offs* $c(Q_0,\alpha)$, *implies asymptotic control of the corresponding Type I error rate* $\Theta(F_{V(c_{0n}|Q_n)})$, *for Procedure 4.1/4.2 based on estimated cut-offs* $c(Q_{0n},\alpha)$. *That is,* $\limsup_n \Theta(F_{V(c_0|Q_n)}) \leq \alpha$ *implies*

$$\limsup_{n \to \infty} \Theta(F_{V(c_{0n}|Q_n)}) \leq \alpha.$$

Proof of Corollary 4.18. Define subsets $\mathcal{A}_{n,\mathcal{H}_0} \subseteq I\!R^M$ by

$$\mathcal{A}_{n,\mathcal{H}_0} \equiv \left\{ z \in I\!R^M : \left| \sum_{m \in \mathcal{H}_0} \mathrm{I}\,(z(m) > c_{0n}(m)) - \mathrm{I}\,(z(m) > c_0(m)) \right| > 0 \right\} \tag{4.35}$$

and note that

$$\Pr\left(V(c_{0n}|Q_n) \neq V(c_0|Q_n)\right) = \Pr_{Q_n}\left(T_n \in \mathcal{A}_{n,\mathcal{H}_0}\right).$$

From Theorem 4.10/4.17, the cut-offs for Procedure 4.1/4.2 are such that $c_{0n}(m) - c_0(m) \to 0$, as $n \to \infty$, $\forall\ m = 1, \ldots, M$. Thus, for each $\epsilon > 0$, there is an integer $N(\epsilon) > 0$, such that $\mathcal{A}_{n,\mathcal{H}_0} \subseteq \mathcal{A}_{\mathcal{H}_0}(\epsilon)$ and hence $\Pr_{Q_n}(T_n \in \mathcal{A}_{n,\mathcal{H}_0}) \leq \Pr_{Q_n}(T_n \in \mathcal{A}_{\mathcal{H}_0}(\epsilon))$, for each $n > N(\epsilon)$. By assumption, $\Pr_{Q_n}(T_n \in \mathcal{A}_{\mathcal{H}_0}(\epsilon)) \to 0$, as $\epsilon \to 0$, thus $\Pr(V(c_{0n}|Q_n) \neq V(c_0|Q_n)) = \Pr_{Q_n}(T_n \in \mathcal{A}_{n,\mathcal{H}_0}) \to 0$, as $n \to \infty$, i.e., $V(c_{0n}|Q_n) - V(c_0|Q_n)$ converges to zero Q_n-a.s. Hence, by uniform continuity Assumption CΘ for the mapping Θ, $\Theta(F_{V(c_{0n}|Q_n)}) - \Theta(F_{V(c_0|Q_n)}) \to 0$. Thus, asymptotic control of the Type I error rate $\Theta(F_{V(c_{0n}|Q_n)})$ follows from asymptotic control of the Type I error rate $\Theta(F_{V(c_0|Q_n)})$, as established in Theorem 4.3.

□

The next corollary can be applied to prove asymptotic Type I error control for Procedures 4.1 and 4.2, based on a consistent estimator Q_{0n} of the null distribution Q_0, without requiring the assumption in Equation (4.33) of Corollary 4.18. Specifically, the corollary is based on an intermediate distribution $Q_{0,n}$, that converges weakly to the test statistics null distribution Q_0 and that dominates the test statistics true distribution Q_n on the set of true null hypotheses (Section 2.2.3). Such a distribution is used, for example, in the construction of the null shift and scale-transformed and null quantile-transformed null distributions (Sections 2.3 and 2.4). For estimated single-step cut-offs $c_{0n} = c(Q_{0n}, \alpha)$, that converge to $c_0 = c(Q_0, \alpha)$ (as in Theorems 4.10 and 4.17), the corollary shows that $\limsup_n \Theta(F_{V(c_{0n}|Q_n)}) \leq \Theta(F_{V(c_0|Q_0)})$.

Corollary 4.19. [Asymptotic Type I error control for single-step common-quantile Procedure 4.1 and common-cut-off Procedure 4.2 based on consistent estimator of the test statistics null distribution] *Consider a sequence of M-vectors $\{c_{0n}\} \in I\!R^M$, with limit $c_0 \in I\!R^M$, i.e., $c_{0n}(m) \to c_0(m)$, as $n \to \infty$, $\forall\, m = 1, \ldots, M$. Define subsets $\mathcal{A}_{n,\mathcal{H}_0} \subseteq I\!R^M$ by*

$$\mathcal{A}_{n,\mathcal{H}_0} \equiv \left\{ z \in I\!R^M : \left| \sum_{m \in \mathcal{H}_0} \mathrm{I}\left(z(m) > c_{0n}(m)\right) - \mathrm{I}\left(z(m) > c_0(m)\right) \right| > 0 \right\} \tag{4.36}$$

and, for $\epsilon > 0$, define subsets $\mathcal{A}_{\mathcal{H}_0}(\epsilon) \subseteq I\!R^M$ by

$$\mathcal{A}_{\mathcal{H}_0}(\epsilon) \equiv \left\{ z \in I\!R^M : \sup_{\{c: ||c - c_0|| < \epsilon\}} \left| \sum_{m \in \mathcal{H}_0} \mathrm{I}\left(z(m) > c(m)\right) - \mathrm{I}\left(z(m) > c_0(m)\right) \right| > 0 \right\}. \tag{4.37}$$

Let Q_0 be an M-variate distribution such that, as in Lemma 4.16, there exist a sequence $\{\epsilon_k\} \to 0$ and continuity sets $\{\mathcal{A}_{\mathcal{H}_0}(\epsilon_k)\}$ of Q_0 with $\lim_k \Pr_{Q_0}(Z \in \mathcal{A}_{\mathcal{H}_0}(\epsilon_k)) = 0$. Suppose that $Q_{0,n}$ is an M-variate distribution that converges weakly to Q_0 and that dominates a third M-variate distribution Q_n on the set $\mathcal{H}_0$ of true null hypotheses, in the sense that the $\mathcal{H}_0$-specific joint distributions $Q_{n,\mathcal{H}_0}$ and $Q_{0,n,\mathcal{H}_0}$ satisfy the asymptotic version of either one of the three domination Assumptions jtNDT, NDV, or NDΘ. Further assume that the Type I error rate mapping Θ satisfies monotonicity Assumption MΘ and uniform continuity Assumption CΘ. Then,

$$\limsup_{n \to \infty} \Theta(F_{V(c_{0n}|Q_n)}) \leq \Theta(F_{V(c_0|Q_0)}).$$

Proof of Corollary 4.19. By the null domination of Q_n by $Q_{0,n}$, it follows that $\limsup_n \Theta(F_{V(c_{0n}|Q_n)}) \leq \limsup_n \Theta(F_{V(c_{0n}|Q_{0,n})})$. From Lemma 4.16, $\Pr(V(c_{0n}|Q_{0,n}) \neq V(c_0|Q_{0,n})) = \Pr_{Q_{0,n}}(Z_n \in \mathcal{A}_{n,\mathcal{H}_0}) \to 0$. Thus, by uniform continuity Assumption CΘ for the mapping Θ, one has $\Theta(F_{V(c_{0n}|Q_{0,n})}) - \Theta(F_{V(c_0|Q_{0,n})}) \to 0$. Also, by weak convergence of $Q_{0,n}$ to Q_0, $\Theta(F_{V(c_0|Q_{0,n})}) - \Theta(F_{V(c_0|Q_0)}) \to 0$. Hence,

$$\limsup_{n\to\infty} \Theta(F_{V(c_{0n}|Q_n)}) \leq \limsup_{n\to\infty} \Theta(F_{V(c_{0n}|Q_{0,n})}) = \Theta(F_{V(c_0|Q_0)}).$$

□

Recall that for one-sided rejection regions of the form $\mathcal{C}_n(m;\alpha) = (c_n(m;\alpha), +\infty)$, domination Assumption jtNDT, for the $\mathcal{H}_0$-specific test statistics, implies domination Assumption NDV, for the number of Type I errors. For Type I error rate mappings Θ that satisfy monotonicity Assumption MΘ and continuity Assumption CΘ, Assumption NDV in turn implies domination Assumption ND$\boldsymbol{\Theta}$, for the Type I error rate. In addition, Corollary 4.19 obviously holds under the stronger finite sample versions of domination Assumptions jtNDT, NDV, and ND$\boldsymbol{\Theta}$.

Note that, for an estimator Q_{0n} of the null distribution Q_0, the consistency results in Theorems 4.10 and 4.17 are *conditional* on the empirical distribution $P^\infty = (P_n : n \geq 1)$ for an infinite sequence $X^\infty = (X_1, X_2, \ldots)$ of IID random variables $X_i \sim P$. That is, the results apply for every P^∞ for which Q_{0n} converges weakly to Q_0. Consequently, if $Q_{0n} \overset{\mathcal{L}}{\Rightarrow} Q_0$ P^∞-a.s., then the above consistency results hold P^∞-a.s. Under regularity conditions, bootstrap estimators Q_{0n} of the null distribution Q_0 are consistent, in the sense that Q_{0n} converges weakly to Q_0, conditional on the empirical distribution P^∞ (van der Vaart and Wellner, 1996). Thus, under such regularity conditions, the consistency results in Theorems 4.10 and 4.17 hold P^∞-a.s. for bootstrap-based analogues of Procedures 4.1 and 4.2.

4.4.2 Bootstrap-based single-step procedures

As detailed in Sections 2.3.2, 2.4.2, 2.6, and 2.7, one may use a variety of bootstrap procedures to obtain consistent estimators Q_{0n} of the proposed null shift and scale-transformed and null quantile-transformed test statistics null distributions $Q_0 = Q_0(P)$.

For instance, for the null distribution of Section 2.3, based on user-supplied null shift and scale values $\lambda_0(m)$ and $\tau_0(m)$, one may apply Procedure 2.3 to generate an $M \times B$ matrix $\mathbf{Z}_n^B = \left(Z_n^B(m,b)\right)$ of null shift and scale-transformed bootstrap test statistics $Z_n^B(m,b)$. For the null distribution of Section 2.4, based on user-supplied marginal null distributions $q_{0,m}$, one may apply Procedure 2.4 to generate an $M \times B$ matrix $\mathbf{Z}_n^B = \left(Z_n^B(m,b)\right)$ of null quantile-transformed bootstrap test statistics $Z_n^B(m,b)$. In either case, the bootstrap estimator Q_{0n} of the null distribution Q_0 is the empirical distribution of the B columns $\{Z_n^B(\cdot,b) : b = 1,\ldots,B\}$ of matrix $\mathbf{Z}_n^B$.

Given such a bootstrap estimator Q_{0n} of the null distribution Q_0, Procedures 4.20 and 4.21 may be applied to estimate cut-offs and adjusted p-values for $\Theta(F_{V_n})$-controlling single-step common-quantile Procedure 4.1 and common-cut-off Procedure 4.2, respectively.

Procedure 4.20. [Bootstrap estimation of cut-offs and adjusted p-values for single-step common-quantile Procedure 4.1]

0. Apply Procedure 2.3 or 2.4 to generate an $M \times B$ matrix $\mathbf{Z}_n^B = \left(Z_n^B(m,b) : m = 1,\ldots,M;\ b = 1,\ldots,B\right)$, of null-transformed bootstrap test statistics $Z_n^B(m,b)$. The bootstrap estimator Q_{0n} of the null distribution Q_0 is the empirical distribution of the B columns $\{Z_n^B(\cdot,b) : b = 1,\ldots,B\}$ of matrix $\mathbf{Z}_n^B$.
1. For $\delta \in [0,1]$, the bootstrap estimators $q_{0n}^{-1}(\delta)$, of the common-quantile cut-offs $q_0^{-1}(\delta)$, are simply the row quantiles of the matrix $\mathbf{Z}_n^B$. That is, $Q_{0n,m}^{-1}(\delta)$ is the δ-quantile of the B null-transformed bootstrap test statistics $\{Z_n^B(m,b) : b = 1,\ldots,B\}$ for null hypothesis $H_0(m)$,
$$Q_{0n,m}^{-1}(\delta) \equiv \inf\left\{z \in I\!R : \frac{1}{B}\sum_{b=1}^{B} \mathrm{I}\left(Z_n^B(m,b) \le z\right) \ge \delta\right\}, \quad m = 1,\ldots,M. \tag{4.38}$$
2. For a test at nominal Type I error level $\alpha \in (0,1)$, the bootstrap estimator of $\delta_0(\alpha)$ is chosen as
$$\delta_{0n}(\alpha) \equiv \inf\left\{\delta \in [0,1] : \Theta(F_{R(q_{0n}^{-1}(\delta)|Q_{0n})}) \le \alpha\right\}. \tag{4.39}$$
That is, the bootstrap estimator $\delta_{0n}(\alpha)$ corresponds to the smallest common-quantile cut-offs $q_{0n}^{-1}(\delta)$, such that the value of the Type I error rate mapping Θ, applied to the distribution of the number of rejected hypotheses $R(q_{0n}^{-1}(\delta)|Q_{0n})$ under the bootstrap distribution Q_{0n}, is at most α.
In the special case of gFWER control ($\Theta(F_{V_n}) = 1 - F_{V_n}(k)$), and for a null distribution Q_0 with continuous and strictly monotone marginal distributions, $(1 - \delta_{0n}(\alpha))$ is the α-quantile of the bootstrap estimator of the distribution of the $(k+1)$st smallest unadjusted p-value (Corollary 4.8). Specifically, $\delta_{0n}(\alpha)$ is obtained as follows.
   a) Compute an $M \times B$ matrix $\mathbf{P}_n^B = (P_n^B(m,b) : m = 1,\ldots,M;\ b = 1,\ldots,B)$, of bootstrap unadjusted p-values, by row-ranking the matrix $\mathbf{Z}_n^B$. That is, for each row m (i.e., null hypothesis $H_0(m)$),

replace $Z_n^B(m,b)$ by its (standardized) rank over the B bootstrap samples, where $1/B$ corresponds to the largest value of $Z_n^B(m,b)$ and B/B to the smallest.

b) For each column b of the matrix $\mathbf{P}_n^B$ (i.e., bootstrap sample b), compute the $(k+1)$st smallest unadjusted p-value, $P_n^{B\circ}(k+1,b)$, $b=1,\ldots,B$. For FWER control ($k=0$), simply compute column minima, $P_n^{B\circ}(1,b) = \min_m P_n^B(m,b)$.

c) The bootstrap estimator $(1-\delta_{0n}(\alpha))$ is the α-quantile of the B $(k+1)$st smallest unadjusted p-values $\{P_n^{B\circ}(k+1,b) : b=1,\ldots,B\}$.

3. Following Proposition 4.4, the bootstrap estimated adjusted p-values are given by

$$\widetilde{p}_{0n}(m) = \Theta(F_{R(q_{0n}^{-1}(1-p_{0n}(m))|Q_{0n})}), \qquad m=1,\ldots,M, \tag{4.40}$$

where $p_{0n}(m)$ are the bootstrap estimated unadjusted p-values,

$$p_{0n}(m) = \frac{1}{B}\sum_{b=1}^{B} \mathrm{I}\left(Z_n^B(m,b) \geq t_n(m)\right), \qquad m=1,\ldots,M. \tag{4.41}$$

In the special case of gFWER control (Corollary 4.8), one has

$$\widetilde{p}_{0n}(m) = \frac{1}{B}\sum_{b=1}^{B} \mathrm{I}\left(P_n^{B\circ}(k+1,b) \leq p_{0n}(m)\right), \quad m=1,\ldots,M. \tag{4.42}$$

Procedure 4.21. [Bootstrap estimation of cut-offs and adjusted p-values for single-step common-cut-off Procedure 4.2]

0. Apply Procedure 2.3 or 2.4 to generate an $M \times B$ matrix $\mathbf{Z}_n^B = \big(Z_n^B(m,b) : m=1,\ldots,M;\, b=1,\ldots,B\big)$, of null-transformed bootstrap test statistics $Z_n^B(m,b)$. The bootstrap estimator Q_{0n} of the null distribution Q_0 is the empirical distribution of the B columns $\{Z_n^B(\cdot,b) : b = 1,\ldots,B\}$ of matrix $\mathbf{Z}_n^B$.
1. For a test at nominal Type I error level $\alpha \in (0,1)$, the bootstrap estimator of the common cut-off $\gamma_0(\alpha)$ is chosen as

$$\gamma_{0n}(\alpha) \equiv \inf\left\{\gamma \in I\!R : \Theta(F_{R(\gamma^{(M)}|Q_{0n})}) \leq \alpha\right\}. \tag{4.43}$$

That is, the bootstrap estimator $\gamma_{0n}(\alpha)$ corresponds to the smallest common cut-off, such that the value of the Type I error rate mapping Θ, applied to the distribution of the number of rejected hypotheses

$R(\gamma^{(M)}|Q_{0n})$ under the bootstrap distribution Q_{0n}, is at most α.
In the special case of gFWER control ($\Theta(F_{V_n}) = 1 - F_{V_n}(k)$), and for a null distribution Q_0 with continuous and strictly monotone marginal distributions, $\gamma_{0n}(\alpha)$ is the $(1-\alpha)$-quantile of the bootstrap estimator of the distribution of the $(k+1)$st largest test statistic (Corollary 4.9). Specifically, $\gamma_{0n}(\alpha)$ is obtained as follows.
   a) For each column b of the matrix $\mathbf{Z}_n^B$ (i.e., bootstrap sample b), compute the $(k+1)$st largest statistic, $Z_n^{B\circ}(k+1, b)$, $b = 1, \ldots, B$. For FWER control ($k = 0$), simply compute column maxima, $Z_n^{B\circ}(1, b) = \max_m Z_n^B(m, b)$.
   b) The bootstrap estimator $\gamma_{0n}(\alpha)$ is the $(1-\alpha)$-quantile of the B $(k+1)$st largest statistics $\{Z_n^{B\circ}(k+1, b) : b = 1, \ldots, B\}$.
2. Following Proposition 4.5, the bootstrap estimated adjusted p-values are given by

$$\widetilde{p}_{0n}(m) = \Theta(F_{R(t_n(m)^{(M)}|Q_{0n})}), \qquad m = 1, \ldots, M. \tag{4.44}$$

In the special case of gFWER control (Corollary 4.9), one has

$$\widetilde{p}_{0n}(m) = \frac{1}{B}\sum_{b=1}^{B} \mathrm{I}\left(Z_n^{B\circ}(k+1, b) \geq t_n(m)\right), \qquad m = 1, \ldots, M. \tag{4.45}$$

Figures 2.1, 2.2, and 4.1, illustrate, respectively, the bootstrap estimation of the null shift and scale-transformed null distribution Q_0, unadjusted p-values $P_{0n}(m)$, and single-step maxT adjusted p-values $\widetilde{P}_{0n}(m)$ (FWER-controlling version of Procedure 4.21, with $\Theta(F_{V_n}) = 1 - F_{V_n}(0)$).

As indicated in Section 2.3.2, bootstrap estimation of the null distribution Q_0 and corresponding unadjusted p-values raises a number of computational issues. Additional complexities arise when estimating cut-offs and adjusted p-values for joint procedures, such as common-quantile Procedure 4.1 and common-cut-off Procedure 4.2. Indeed, even in the case of a continuous estimated joint null distribution Q_{0n}, such as the M-variate Gaussian distribution $\mathrm{N}(0, \sigma_n^*)$ of Section 2.6, one has the following numerical analysis problem when estimating single-step cut-offs. The problem involves identifying the least conservative cut-offs $c = (c(m) : m = 1, \ldots, M)$, such that $\Theta(F_{R(c|Q_{0n})}) \leq \alpha$, where $R(c|Q_{0n}) = \sum_{m=1}^{M} \mathrm{I}(Z(m) > c(m))$ denotes the number of rejected hypotheses under the estimated null distribution Q_{0n} (i.e., for $Z \sim Q_{0n}$). Specifically, for control of $gFWER(k)$ at level α, based on single-step common-cut-off Procedure 4.2 and the corresponding bootstrap Procedure 4.21, one needs to solve

$$\begin{aligned}
&\inf\{\gamma \in I\!R : \Theta(F_{R(\gamma^{(M)}|Q_{0n})}) \leq \alpha\} \\
&= \inf\left\{\gamma \in I\!R : \Pr\left(R(\gamma^{(M)}|Q_{0n}) > k\right) \leq \alpha\right\} \\
&= \inf\left\{\gamma \in I\!R : \Pr_{Q_{0n}}\left(\sum_{m=1}^{M} \mathrm{I}\left(Z(m) > \gamma\right) > k\right) \leq \alpha\right\}.
\end{aligned}$$

4.5 $\Theta(F_{V_n})$-controlling two-sided single-step procedures

As usual, consider the simultaneous test of M null hypotheses $H_0(m)$, $m = 1, \ldots, M$, based on test statistics $T_n = (T_n(m) : m = 1, \ldots, M)$, with finite sample joint distribution $Q_n = Q_n(P)$, under the true, unknown data generating distribution P. However, now assume that large and small values of the test statistic $T_n(m)$ provide evidence against the corresponding null hypothesis $H_0(m)$, that is, consider *symmetric two-sided rejection regions* of the form $\mathcal{C}_n(m;\alpha) = (-\infty, -c_n(m;\alpha)) \cup (c_n(m;\alpha), +\infty)$, where $c_n(\alpha) = (c_n(m;\alpha) : m = 1, \ldots, M) \in I\!R^{+M}$ is an M-vector of non-negative single-step cut-offs $c_n(m;\alpha) = c(m; Q_0, \alpha) = c(m; T_n, Q_0, \alpha) \in I\!R^+$, computed under a null distribution Q_0 (or estimator thereof, Q_{0n}) for the test statistics T_n. In other words, reject null hypothesis $H_0(m)$ for large values of the *absolute test statistic* $|T_n(m)|$ and let

$$\begin{aligned}
\mathcal{R}^{||}(T_n, Q_0, \alpha) &\equiv \{m : T_n(m) < -c_n(m;\alpha) \text{ or } T_n(m) > c_n(m;\alpha)\} \quad (4.46)\\
&= \{m : |T_n(m)| > c_n(m;\alpha)\}.
\end{aligned}$$

In this section, we propose symmetric two-sided versions of single-step common-quantile Procedure 4.1 and common-cut-off Procedure 4.2, based on a symmetric test statistics null distribution Q_0, so that $\bar{Q}_{0,m}(z) = 1 - Q_{0,m}(z) = Q_{0,m}(-z)$, for each $z \in I\!R$ and $m = 1, \ldots, M$. The two-sided versions of Procedures 4.1 and 4.2 are very similar to their one-sided counterparts, with absolute test statistics $|T_n(m)|$ replacing signed test statistics $T_n(m)$. Type I error control is again established according to the three-step road map of Procedure 2.1. For a null distribution Q_0, with continuous and strictly increasing marginal CDFs $Q_{0,m}$, adjusted p-values may be derived as in Propositions 4.4 and 4.5, based on the general definition of Equation (1.58). That is,

$$\begin{aligned}
\widetilde{P}_{0n}(m) &= \inf\{\alpha \in [0,1] : c_n(m;\alpha) < |T_n(m)|\} \\
&= c_n^{-1}(m; |T_n(m)|),
\end{aligned}$$

where $c_n^{-1}(m;\cdot)$ are the inverses of the non-increasing functions of α, $c_n(m;\cdot) : \alpha \rightarrow c_n(m;\alpha)$.

As discussed in Section 2.5, the above symmetric two-sided testing problem corresponds to tests based on transformations of the test statistics with the absolute value function, $\ell(z) = |z|$. Proposition 2.5 specifies conditions under which MTPs based on the transformed test statistics and a null distribution for the original test statistics provide proper Type I error control.

As in Equation (4.1), given a random M-vector $Z = (Z(m) : m = 1, \ldots, M)$, with joint distribution Q, and an M-vector of non-negative cut-offs $c = (c(m) : m = 1, \ldots, M) \in I\!R^{+M}$, let $R^{||}(c|Q)$ and $V^{||}(c|Q)$ denote, respectively, symmetric two-sided numbers of rejected hypotheses and Type I errors, i.e., numbers of rejected hypotheses and Type I errors for the absolute test statistics $|Z|$,

$$R^{||}(c|Q) \equiv \sum_{m=1}^{M} \mathrm{I}\left(|Z(m)| > c(m)\right) \text{ and } V^{||}(c|Q) \equiv \sum_{m \in \mathcal{H}_0} \mathrm{I}\left(|Z(m)| > c(m)\right). \tag{4.47}$$

For a given cut-off vector c, also adopt the shorthand notation of Equation (4.2), for the special cases where Q corresponds to the test statistics true distribution Q_n and null distribution Q_0.

4.5.1 Symmetric two-sided single-step common-quantile procedure

A symmetric two-sided version of Procedure 4.1, based on a symmetric null distribution Q_0 and common-quantile cut-offs $Q_{0,m}^{-1}(\delta) = \inf\{z \in I\!R : Q_{0,m}(z) \geq \delta\}$, rejects the following null hypotheses,

$$\begin{aligned} \mathcal{R}^{||}(T_n, Q_0, \alpha) &\equiv \left\{m : |T_n(m)| > Q_{0,m}^{-1}(\delta_0(\alpha))\right\} \qquad (4.48)\\ &= \left\{m : T_n(m) < -Q_{0,m}^{-1}(\delta_0(\alpha)) \text{ or } T_n(m) > Q_{0,m}^{-1}(\delta_0(\alpha))\right\} \\ &= \left\{m : T_n(m) < Q_{0,m}^{-1}(1-\delta_0(\alpha)) \text{ or } T_n(m) > Q_{0,m}^{-1}(\delta_0(\alpha))\right\}, \end{aligned}$$

where $\delta_0(\alpha)$ is defined as the least conservative value of δ such that the Type I error constraint $\Theta(F_{R^{||}(q_0^{-1}(\delta)|Q_0)}) \leq \alpha$ is satisfied. That is,

$$\delta_0(\alpha) \equiv \inf\left\{\delta \in [0,1] : \Theta(F_{R^{||}(q_0^{-1}(\delta)|Q_0)}) \leq \alpha\right\}. \tag{4.49}$$

A similar argument as in Proposition 4.4 shows that the adjusted p-values for the symmetric two-sided version of Procedure 4.1 are given by

$$\widetilde{P}_{0n}(m) = \Theta(F_{R^{||}(q_0^{-1}(1-P_{0n}(m)/2)|Q_0)}), \tag{4.50}$$

where $P_{0n}(m)$ are symmetric two-sided unadjusted p-values defined as

$$P_{0n}(m) = 2\bar{Q}_{0,m}(|T_n(m)|). \tag{4.51}$$

In particular, for gFWER control, the single-step common-quantile adjusted p-values are given by

$$\widetilde{p}_{0n}(m) = \Pr_{Q_0}\left(P_0^\circ(k+1) \leq p_{0n}(m)\right), \tag{4.52}$$

where $P_0(m) = 2\bar{Q}_{0,m}(|Z(m)|)$ denote symmetric two-sided unadjusted p-values under the test statistics null distribution Q_0 (i.e., for $Z \sim Q_0$) and $P_0^\circ(m)$ denotes the mth smallest unadjusted p-value, so that $P_0^\circ(1) \leq \cdots \leq P_0^\circ(M)$.

4.5.2 Symmetric two-sided single-step common-cut-off procedure

A symmetric two-sided version of Procedure 4.2, based on a symmetric null distribution Q_0 and a common cut-off $\gamma_0(\alpha)$, rejects the following null hypotheses,

$$\mathcal{R}^{||}(T_n, Q_0, \alpha) \equiv \{m : |T_n(m)| > \gamma_0(\alpha)\}, \tag{4.53}$$

where the common cut-off $\gamma_0(\alpha)$ is defined as the least conservative value of γ such that the Type I error constraint $\Theta(F_{R^{||}(\gamma^{(M)}|Q_0)})$ is satisfied. That is,

$$\gamma_0(\alpha) \equiv \inf\left\{\gamma \in I\!R : \Theta(F_{R^{||}(\gamma^{(M)}|Q_0)}) \leq \alpha\right\}. \tag{4.54}$$

A similar argument as in Proposition 4.5 shows that the adjusted p-values for the symmetric two-sided version of Procedure 4.2 are given by

$$\widetilde{P}_{0n}(m) = \Theta(F_{R^{||}(|T_n(m)|^{(M)}|Q_0)}). \tag{4.55}$$

In particular, for gFWER control, the single-step common-cut-off adjusted p-values are given by

$$\widetilde{p}_{0n}(m) = \Pr_{Q_0}\left(|Z|^\circ(k+1) \geq |t_n(m)|\right), \tag{4.56}$$

where $Z \sim Q_0$ and $|Z|^\circ(m)$ denotes the mth largest element of the M-vector $|Z| = (|Z(m)| : m = 1, \ldots, M)$, so that $|Z|^\circ(1) \geq \cdots \geq |Z|^\circ(M)$.

4.5.3 Asymptotic control of Type I error rate and test statistics null distribution

The above symmetric two-sided versions of single-step common-quantile Procedure 4.1 and common-cut-off Procedure 4.2 correspond to MTPs based on transformations of the test statistics with the absolute value function, $\ell(z) = |z|$. Procedures based on transformed test statistics are discussed in greater detail in Section 2.5.

In particular, Proposition 2.5 provides conditions under which a null distribution Q_0 for the original test statistics T_n leads to procedures that provide Type I error control for the transformed test statistics T_n^ℓ. Because the absolute value function $\ell(z) = |z|$ is not monotone (it is decreasing for $z < 0$ and increasing for $z \geq 0$), one needs to impose additional conditions on the null distribution Q_0 of the original test statistics T_n. Specifically, as stated in Proposition 2.5, Part 2, a multiple testing procedure based on the transformed test statistics T_n^ℓ and the null distribution Q_0 for the original test statistics T_n provides control of the Type I error rate $\Theta(F_{V_n^\ell})$, if joint null domination Assumption jtNDT holds with equality for the original test statistics T_n. In other words, the Type I error rate $\Theta(F_{V_n^\ell})$ is controlled asymptotically, if the true distribution $Q_{n,\mathcal{H}_0} = Q_{n,\mathcal{H}_0}(P)$ of the $\mathcal{H}_0$-specific subvector of test statistics $(T_n(m) : m \in \mathcal{H}_0)$ converges to the $\mathcal{H}_0$-specific null distribution $Q_{0,\mathcal{H}_0}$.

As detailed in Section 2.6, such a condition is satisfied for the two-sided test of single-parameter null hypotheses $H_0(m) = \mathrm{I}(\psi(m) = \psi_0(m))$, using the t-statistics T_n of Equation (2.34) and the null distribution $Q_0 = \mathrm{N}(0, \sigma^*)$ of Theorem 2.6. Indeed, a similar argument as in the proof of Theorem 2.6 shows that

$$(T_n(m) : m \in \mathcal{H}_0) \stackrel{\mathcal{L}}{\Rightarrow} Q_{0,\mathcal{H}_0} = \mathrm{N}(0, \sigma^*_{\mathcal{H}_0}).$$

Hence, for all $c = (c(m) : m = 1, \ldots, M) \in I\!R^{+M}$ and $x \in \{0, \ldots, M\}$,

$$\begin{aligned}
\lim_{n\to\infty} F_{V_n^{||}}(x) &= \lim_{n\to\infty} \Pr_{Q_n}\left(\sum_{m\in\mathcal{H}_0} \mathrm{I}(|T_n(m)| > c(m)) \leq x\right) \\
&= \Pr_{Q_0}\left(\sum_{m\in\mathcal{H}_0} \mathrm{I}(|Z(m)| > c(m)) \leq x\right) \\
&= F_{V_0^{||}}(x),
\end{aligned}$$

and the main asymptotic null domination Assumption ANDV of Theorem 4.3 is satisfied with equality for absolute t-statistics $|T_n|$. Under monotonicity Assumption MΘ and continuity Assumption CΘ at $F_{V_0^{||}}$ for the Type I error rate mapping Θ, one has

$$\lim_{n\to\infty} \Theta(F_{V_n^{||}}) = \Theta(F_{V_0^{||}}) \leq \Theta(F_{R_0^{||}}) \leq \alpha.$$

According to the three-step road map of Procedure 2.1, it follows that Procedures 4.1 and 4.2, based on absolute t-statistics $|T_n|$ and cut-offs $c_n(m;\alpha) = c(m; Q_0, \alpha)$ derived under the symmetric Gaussian null distribution $Q_0 = \mathrm{N}(0, \sigma^*)$ of Theorem 2.6, do indeed provide the desired Type I error control.

Note that, for the absolute value function and two-sided rejection regions, the stronger requirement of asymptotic *equality* of the test statistics true distribution $Q_{n,\mathcal{H}_0}$ and null distribution $Q_{0,\mathcal{H}_0}$ is essential, as the weaker domination property would only guarantee Type I error control for one of the tails.

An alternative and more general approach would be to derive a null distribution $Q_0^{||}$ directly for the absolute test statistics $|T_n|$, using the general constructions of Sections 2.3 and 2.4. However, this method could in principle lead to very different null distributions. For instance, for the null shift and scale-transformed null distribution of Section 2.3, the null values $\lambda_0(m)$ and $\tau_0(m)$ would no longer be 0 and 1 and the null distribution $Q_0^{||}$ would no longer be Gaussian.

4.5.4 Bootstrap-based symmetric two-sided single-step procedures

Regarding the bootstrap estimation of cut-offs and adjusted p-values for the symmetric two-sided versions of Procedures 4.1 and 4.2, one could first use general Procedure 2.3 or 2.4 (or a related procedure from Section 2.6 or 2.7) to derive a matrix $\mathbf{Z}_n^B = \big(Z_n^B(m,b)\big)$, of null-transformed bootstrap test statistics $Z_n^B(m,b)$, based on the original test statistics T_n. The null distribution Q_0, for the original test statistics T_n, is estimated by the empirical distribution Q_{0n} of the B columns of matrix $\mathbf{Z}_n^B$. An estimated null distribution $Q_{0n}^{||}$, for the absolute test statistics $|T_n|$, is given by the empirical distribution of the B columns of the matrix $|\mathbf{Z}_n^B| = \big(|Z_n^B(m,b)|\big)$.

Bootstrap versions of single-step common-quantile Procedure 4.1 and common-cut-off Procedure 4.2 may then be implemented as in Procedures 4.20 and 4.21, respectively, using absolute test statistics $|T_n|$ and the estimated null distribution $Q_{0n}^{||}$.

4.6 Multiple hypothesis testing and confidence regions

This section builds on Pollard and van der Laan (2004) and presents a generalization of the correspondence between multiple hypothesis testing procedures and parameter confidence regions. Specifically, for Type I error rates $\Theta(F_{V_n})$, defined as arbitrary parameters of the distribution of the number of Type I errors, equivalence results are derived between Θ-specific confidence regions and single-step procedures, such as common-quantile Procedure 4.1 and common-cut-off Procedure 4.2.

Open questions concern equivalence results for step-down procedures and for more general Type I error rates $\Theta(F_{V_n,R_n})$, defined in terms of the joint distribution of the numbers of Type I errors V_n and rejected hypotheses R_n.

Note that, in the special case of FWER control, Hsu (1996) and Westfall and Young (1993) present various equivalence results between multiple testing procedures and simultaneous confidence regions.

4.6.1 Confidence regions for general Type I error rates, $\Theta(F_{V_n})$

Consider the simultaneous test of M null hypotheses, each concerning one of the elements $\psi(m)$ of an M-dimensional parameter vector $\Psi(P) = \psi = (\psi(m) : m = 1,\ldots,M)$. Given an M-vector of null values $\psi_0 = (\psi_0(m) : m = 1,\ldots,M)$, two-sided tests evaluate the null hypotheses $H_0(m) = \mathrm{I}(\psi(m) = \psi_0(m))$ against the alternative hypotheses $H_1(m) = \mathrm{I}(\psi(m) \neq \psi_0(m))$, whereas one-sided tests evaluate the null hypotheses $H_0(m) = \mathrm{I}(\psi(m) \leq \psi_0(m))$ against the alternative hypotheses $H_1(m) = \mathrm{I}(\psi(m) > \psi_0(m))$.

As detailed in Sections 1.2.5 and 2.6, suppose one has an asymptotically linear estimator $\hat{\Psi}(P_n) = \psi_n = (\psi_n(m) : m = 1,\ldots,M)$ of the parameter vector ψ, based on a random sample $\mathcal{X}_n = \{X_i : i = 1,\ldots,n\}$ from the data generating distribution P. Define an M-vector $T_n = (T_n(m) : m = 1,\ldots,M)$ of (observable) t-statistics,

$$T_n(m) \equiv \frac{\sqrt{n}(\psi_n(m) - \psi_0(m))}{\sigma_n(m)}, \qquad m = 1,\ldots,M, \tag{4.57}$$

and an M-vector $Z_n^t = (Z_n^t(m) : m = 1, \ldots, M)$ of (unobservable) standardized statistics,

$$Z_n^t(m) \equiv \frac{\sqrt{n}(\psi_n(m) - \psi(m))}{\sigma_n(m)}, \qquad m = 1, \ldots, M, \tag{4.58}$$

where $\sigma_n^2(m)$ are consistent estimators of the variances $\sigma^2(m) = \mathrm{E}[IC^2(X|P)(m)]$ of the M-dimensional vector influence curve $IC(X|P) = (IC(X|P)(m) : m = 1, \ldots, M)$. Note that Z_n^t depend on the *unknown parameter* ψ, whereas the test statistics T_n are defined in terms of the corresponding *known null values* ψ_0. Denote the joint distributions of T_n and Z_n^t, under the true data generating distribution P, by $Q_n = Q_n(P)$ and $Q_n^t = Q_n^t(P)$, respectively.

As in Theorem 2.6, define a test statistics null distribution as the asymptotic distribution $Q_0^t = Q_0^t(P) = \mathrm{N}(0, \sigma^*)$ of the standardized statistics Z_n^t, where $\sigma^* = \Sigma^*(P)$ is the correlation matrix of the influence curve for the asymptotically linear estimator ψ_n.

Following Equation (2.2), given a random M-vector $Z = (Z(m) : m = 1, \ldots, M)$, with joint distribution Q, and M-vectors of lower and upper cut-offs $l = (l(m) : m = 1, \ldots, M) \in I\!R^M$ and $u = (u(m) : m = 1, \ldots, M) \in I\!R^M$, denote the numbers of rejected hypotheses and Type I errors by

$$R(l, u|Q) \equiv \sum_{m=1}^{M} \mathrm{I}\left(Z(m) \notin [l(m), u(m)]\right) \tag{4.59}$$

and

$$V(l, u|Q) \equiv \sum_{m \in \mathcal{H}_0} \mathrm{I}\left(Z(m) \notin [l(m), u(m)]\right),$$

respectively. For given cut-off vectors l and u, adopt the shorthand notation of Equation (2.3), for the special cases where Q corresponds to the true distribution Q_n of the test statistics T_n, the true distribution Q_n^t of the statistics Z_n^t, and the null distribution Q_0^t,

$$\begin{aligned} R_n &\equiv R(l, u|Q_n), & V_n &\equiv V(l, u|Q_n), \\ R_n^t &\equiv R(l, u|Q_n^t), & V_n^t &\equiv V(l, u|Q_n^t), \\ R_0 &\equiv R(l, u|Q_0^t), & V_0 &\equiv V(l, u|Q_0^t). \end{aligned} \tag{4.60}$$

The above definitions, based on possibly asymmetric cut-off vectors l and u, accommodate both one-sided and two-sided tests. For two-sided tests, the cut-offs are typically symmetric, $l(m) = -u(m)$, whereas for one-sided tests, one of the cut-offs is usually infinite, $l(m) = -\infty$ or $u(m) = +\infty$.

Definition 4.22. [Θ-specific confidence regions] *Consider standardized statistics $Z_n^t \sim Q_n^t$, defined as in Equation (4.58), based on an asymptotically linear estimator ψ_n of the parameter ψ. Given an assumed distribution Q for the standardized statistics Z_n^t, select M-vectors of lower and upper cut-offs*

$l(Q, \alpha) = (l(m; Q, \alpha) : m = 1, \ldots, M)$ *and* $u(Q, \alpha) = (u(m; Q, \alpha) : m = 1, \ldots, M)$, *such that*

$$\begin{aligned} \Theta(F_{R_n^t}) &\leq \alpha \qquad \textit{[finite sample confidence region]} \\ \limsup_{n \to \infty} \Theta(F_{R_n^t}) &\leq \alpha \qquad \textit{[asymptotic confidence region]}, \end{aligned} \tag{4.61}$$

where R_n^t *denotes the number of rejected hypotheses under the true distribution* Q_n^t *for the standardized statistics* Z_n^t,

$$R_n^t = R(l(Q, \alpha), u(Q, \alpha)|Q_n^t) = \sum_{m=1}^{M} \mathrm{I}\left(Z_n^t(m) \notin [l(m; Q, \alpha), u(m; Q, \alpha)]\right). \tag{4.62}$$

Then, a (finite sample or asymptotic) Θ-specific $(1-\alpha)100\%$ confidence region *for* ψ *is given by the following random subset of* $\mathbb{R}^M$

$$\begin{aligned} \mathcal{CR}_n &= \mathcal{CR}(\mathcal{X}_n, Q, \alpha) \\ &\equiv \left\{\psi \in \mathbb{R}^M : Z_n^t(m) \in [l(m; Q, \alpha), u(m; Q, \alpha)], \ \forall\, m = 1, \ldots, M\right\} \\ &= \Bigg\{\psi \in \mathbb{R}^M : \psi(m) \in \Bigg[\psi_n(m) - u(m; Q, \alpha)\frac{\sigma_n(m)}{\sqrt{n}}, \\ &\qquad \psi_n(m) - l(m; Q, \alpha)\frac{\sigma_n(m)}{\sqrt{n}}\Bigg], \ \forall\, m = 1, \ldots, M\Bigg\}. \end{aligned} \tag{4.63}$$

In the case of (unstandardized) difference statistics, $T_n(m) \equiv \sqrt{n}(\psi_n(m) - \psi_0(m))$, multiple testing procedures and confidence regions may be obtained as above by setting $\sigma_n(m) = 1$.

Definition 4.22 generalizes the notion of a simultaneous confidence region for the family-wise error rate to arbitrary Type I error rates $\Theta(F_{V_n})$. In the special case of the FWER, $\Theta(F_{V_n}) = 1 - F_{V_n}(0) = \Pr(V_n > 0)$ and Equation (4.61) becomes

$$\begin{aligned} &\Pr\Bigg(\psi(m) \in \Bigg[\psi_n(m) - u(m; Q, \alpha)\frac{\sigma_n(m)}{\sqrt{n}}, \\ &\qquad \psi_n(m) - l(m; Q, \alpha)\frac{\sigma_n(m)}{\sqrt{n}}\Bigg], \ \forall\, m = 1, \ldots, M\Bigg) \geq 1 - \alpha, \\ &\liminf_{n \to \infty} \Pr\Bigg(\psi(m) \in \Bigg[\psi_n(m) - u(m; Q, \alpha)\frac{\sigma_n(m)}{\sqrt{n}}, \\ &\qquad \psi_n(m) - l(m; Q, \alpha)\frac{\sigma_n(m)}{\sqrt{n}}\Bigg], \ \forall\, m = 1, \ldots, M\Bigg) \geq 1 - \alpha. \end{aligned} \tag{4.64}$$

4.6.2 Equivalence between Θ-specific single-step multiple testing procedures and confidence regions

We first show in Theorem 4.23 that the cut-off vectors for a Θ-specific $(1-\alpha)100\%$ confidence region yield a single-step multiple testing procedure that provides control of the Type I error rate $\Theta(F_{V_n})$ at level α.

Theorem 4.23. [Θ-specific single-step MTPs implied by Θ-specific confidence regions] *Suppose that, as in Definition 4.22, $l = l(Q, \alpha)$ and $u = u(Q, \alpha)$ are M-vectors of lower and upper cut-offs defining a (finite sample or asymptotic) Θ-specific $(1-\alpha)100\%$ confidence region $\mathcal{CR}_n = \mathcal{CR}(\mathcal{X}_n, Q, \alpha)$ for the parameter ψ. Define a single-step multiple testing procedure as follows,*

$$\begin{aligned} \mathcal{R}_n = \mathcal{R}(T_n, Q, \alpha) &\equiv \{m : T_n(m) \notin [l(m), u(m)]\} \qquad (4.65) \\ &= \left\{ m : \psi_0(m) \notin \left[\psi_n(m) - u(m)\frac{\sigma_n(m)}{\sqrt{n}}, \psi_n(m) - l(m)\frac{\sigma_n(m)}{\sqrt{n}} \right] \right\}. \end{aligned}$$

In particular, for two-sided tests of the null hypotheses $H_0(m) = \mathrm{I}(\psi(m) = \psi_0(m))$, set $l(m) = -u(m)$, and for one-sided tests of the null hypotheses $H_0(m) = \mathrm{I}(\psi(m) \leq \psi_0(m))$, set $l(m) = -\infty$. Then, for Type I error rate mappings Θ that satisfy monotonicity Assumption $\mathsf{M}\Theta$, $\mathcal{R}_n$ provides (finite sample or asymptotic) control of the Type I error rate $\Theta(F_{V_n})$ at level α. That is,

$$\begin{aligned} &\Theta(F_{R_n^t}) \leq \alpha \ \Rightarrow\ \Theta(F_{V_n}) \leq \alpha \ \textit{[finite sample control]} \\ &\limsup_{n\to\infty} \Theta(F_{R_n^t}) \leq \alpha \ \Rightarrow\ \limsup_{n\to\infty} \Theta(F_{V_n}) \leq \alpha \ \textit{[asymptotic control]}. \end{aligned} \qquad (4.66)$$

Proof of Theorem 4.23. For two-sided tests of the null hypotheses $H_0(m) = \mathrm{I}(\psi(m) = \psi_0(m))$, $T_n(m) = Z_n^t(m)$ for $m \in \mathcal{H}_0$. For one-sided tests of the null hypotheses $H_0(m) = \mathrm{I}(\psi(m) \leq \psi_0(m))$, $T_n(m) \leq Z_n^t(m)$ for $m \in \mathcal{H}_0$. Thus,

$$\begin{aligned} \sum_{m\in\mathcal{H}_0} \mathrm{I}(T_n(m) \notin [l(m), u(m)]) &\leq \sum_{m\in\mathcal{H}_0} \mathrm{I}\left(Z_n^t(m) \notin [l(m), u(m)]\right) \\ &\leq \sum_{m=1}^{M} \mathrm{I}\left(Z_n^t(m) \notin [l(m), u(m)]\right). \end{aligned}$$

That is, in the previously introduced shorthand notation, $V_n \leq V_n^t \leq R_n^t$. From monotonicity Assumption $\mathsf{M}\Theta$ for the mapping Θ, one has $\Theta(F_{V_n}) \leq \Theta(F_{R_n^t})$. Thus, by definition of the confidence region in Equations (4.61) and (4.63), it follows that

$$\begin{aligned} \Theta(F_{V_n}) \leq \Theta(F_{R_n^t}) &\leq \alpha \qquad \text{[finite sample control]} \\ \limsup_{n\to\infty} \Theta(F_{V_n}) \leq \limsup_{n\to\infty} \Theta(F_{R_n^t}) &\leq \alpha \qquad \text{[asymptotic control]}. \end{aligned}$$

□

The above result shows that any Θ-specific $(1-\alpha)100\%$ confidence region implies a single-step multiple testing procedure that provides control of the Θ-specific Type I error rate at level α. The converse (i.e., a level–α multiple testing procedure implies a $(1-\alpha)100\%$ confidence region), however, is not true in general, except under the complete null hypothesis H_0^C for which $\mathcal{H}_0 = \{1, \ldots, M\}$. Indeed, confidence regions require the more stringent control of the parameter $\Theta(F_{R_n})$, for the number of rejected hypotheses, which, by monotonicity Assumption MΘ for the mapping Θ, is greater than the corresponding parameter $\Theta(F_{V_n})$, for the number of Type I errors (i.e., $R_n \geq V_n$ with equality for the complete null hypothesis H_0^C).

The next theorem relates asymptotic Θ-specific confidence regions and single-step procedures, such as common-quantile Procedure 4.1 and common-cut-off Procedure 4.2, based on the null distribution of Theorem 2.6.

Theorem 4.24. [Equivalence between Θ-specific single-step MTPs and confidence regions] *Suppose $l_0 = l(Q_0^t, \alpha)$ and $u_0 = u(Q_0^t, \alpha)$ are M-vectors of lower and upper single-step cut-offs, selected as in common-quantile Procedure 4.1 or common-cut-off Procedure 4.2, so that*

$$\Theta(F_{R_0}) \leq \alpha, \tag{4.67}$$

where $R_0 = R(l_0, u_0|Q_0^t)$ denotes the number of rejected hypotheses under the asymptotic distribution $Q_0^t = Q_0^t(P) = N(0, \sigma^*)$ of the standardized statistics Z_n^t of Theorem 2.6. In particular, for two-sided tests of the null hypotheses $H_0(m) = \mathrm{I}(\psi(m) = \psi_0(m))$, let $l_0(m) = l(m; Q_0^t, \alpha) = -u(m; Q_0^t, \alpha) = -u_0(m)$, and for one-sided tests of the null hypotheses $H_0(m) = \mathrm{I}(\psi(m) \leq \psi_0(m))$, let $l_0(m) = l(m; Q_0^t, \alpha) = -\infty$. Then, for Type I error rate mappings Θ that satisfy monotonicity Assumption MΘ and continuity Assumption CΘ at F_{R_0},

$$\limsup_{n\to\infty} \Theta(F_{R_n^t}) \leq \alpha \qquad \textit{and} \qquad \limsup_{n\to\infty} \Theta(F_{V_n}) \leq \alpha, \tag{4.68}$$

where $R_n^t = R(l_0, u_0|Q_n^t)$ is the number of rejected hypotheses under the distribution $Q_n^t = Q_n^t(P)$ of the standardized statistics Z_n^t and $V_n = V(l_0, u_0|Q_n)$ is the number of Type I errors under the distribution $Q_n = Q_n(P)$ of the t-statistics T_n. It follows that

$$\begin{aligned} \mathcal{CR}_n &= \mathcal{CR}(\mathcal{X}_n, Q_0^t, \alpha) \\ &\equiv \left\{\psi \in I\!R^M : Z_n^t(m) \in [l_0(m), u_0(m)]\,,\ \forall\ m = 1, \ldots, M\right\} \end{aligned} \tag{4.69}$$

is an asymptotic Θ-specific $(1-\alpha)100\%$ confidence region for the parameter ψ and

$$\mathcal{R}_n = \mathcal{R}(T_n, Q_0^t, \alpha) \equiv \{m : T_n(m) \notin [l_0(m), u_0(m)]\} \tag{4.70}$$

is a single-step multiple testing procedure that provides asymptotic control of the Type I error rate $\Theta(F_{V_n})$ at level α. That is, cut-offs based on the *asymptotic distribution Q_0^t of the standardized statistics Z_n^t and such that $\Theta(F_{R_0}) \leq \alpha$ provide both asymptotic Θ-specific confidence regions and single-step multiple testing procedures.*

Proof of Theorem 4.24. From Theorem 2.6, the asymptotic distribution of the standardized statistics $Z_n^t \sim Q_n^t$ is $Q_0^t = \mathrm{N}(0, \sigma^*)$. It follows from the Continuous Mapping Theorem (Theorem B.3) and continuity Assumption CΘ for Θ that $\lim_n \Theta(F_{R_n^t}) = \Theta(F_{R_0})$. Furthermore, as in Theorem 4.23, note that $\Theta(F_{V_n}) \leq \Theta(F_{R_n^t})$. Hence, choosing cut-off vectors l_0 and u_0 so that $\Theta(F_{R_0}) \leq \alpha$ implies

$$\limsup_{n\to\infty} \Theta(F_{V_n}) \leq \limsup_{n\to\infty} \Theta(F_{R_n^t}) = \Theta(F_{R_0}) \leq \alpha.$$

□

Thus, the null distribution Q_0^t of Theorem 2.6 yields multiple testing procedures that asymptotically control the Type I error rate $\Theta(F_{V_n})$ at level α as well as asymptotic Θ-specific $(1-\alpha)100\%$ confidence regions for the parameter ψ. The set of rejected null hypotheses corresponds to hypotheses $H_0(m)$ for which the null values $\psi_0(m)$ fall outside the confidence region.

For two-sided tests, with $l_0(m) = -u_0(m)$, one obtains symmetric confidence regions

$$\mathcal{CR}_n \equiv \left\{\psi \in I\!R^M : \psi(m) \in \left[\psi_n(m) \pm u_0(m)\frac{\sigma_n(m)}{\sqrt{n}}\right], \ \forall\, m = 1, \ldots, M\right\} \tag{4.71}$$

and rejection regions

$$\mathcal{R}_n \equiv \left\{m : \psi_0(m) \notin \left[\psi_n(m) \pm u_0(m)\frac{\sigma_n(m)}{\sqrt{n}}\right]\right\}. \tag{4.72}$$

For one-sided tests, with $l_0(m) = -\infty$, one obtains one-sided confidence regions

$$\mathcal{CR}_n \equiv \left\{\psi \in I\!R^M : \psi(m) \geq \psi_n(m) - u_0(m)\frac{\sigma_n(m)}{\sqrt{n}}, \ \forall\, m = 1, \ldots, M\right\} \tag{4.73}$$

and rejection regions

$$\mathcal{R}_n \equiv \left\{m : \psi_0(m) < \psi_n(m) - u_0(m)\frac{\sigma_n(m)}{\sqrt{n}}\right\}. \tag{4.74}$$

4.6.3 Bootstrap-based confidence regions for general Type I error rates, $\Theta(F_{V_n})$

In practice, the null distribution Q_0^t of Theorem 2.6 is unknown and can be estimated consistently by the bootstrap, as detailed in Sections 2.3.2, 2.4.2, and 2.6. Let $l_{0n} = l(Q_{0n}^t, \alpha)$ and $u_{0n} = u(Q_{0n}^t, \alpha)$ denote M-vectors of lower and upper cut-offs corresponding to an estimated null distribution Q_{0n}^t and defined so that

$$\Theta(F_{R(l_{0n}, u_{0n}|Q_{0n}^t)}) \leq \alpha. \tag{4.75}$$

Then,

$$\begin{aligned} \mathcal{CR}_n &= \mathcal{CR}(\mathcal{X}_n, Q_{0n}^t, \alpha) \qquad (4.76) \\ &\equiv \left\{\psi \in I\!\!R^M : Z_n^t(m) \in [l_{0n}(m), u_{0n}(m)], \ \forall\, m = 1, \ldots, M\right\} \end{aligned}$$

is an asymptotic Θ-specific $(1-\alpha)100\%$ confidence region for the parameter ψ, in the sense that

$$\limsup_{n\to\infty} \Theta(F_{R(l_{0n}, u_{0n}|Q_n^t)}) \leq \alpha. \tag{4.77}$$

Single-step Procedures 4.1 and 4.2, based on the estimated null distribution Q_{0n}^t, can be stated equivalently as

$$\mathcal{R}_n = \mathcal{R}(T_n, Q_{0n}^t, \alpha) \equiv \{m : T_n(m) \notin [l_{0n}(m), u_{0n}(m)]\}. \tag{4.78}$$

4.7 Optimal multiple testing procedures

The recent manuscript by Rubin et al. (2006) concerns *optimal multiple testing procedures*, i.e., MTPs that seek to maximize power subject to a Type I error constraint. Specifically, optimal single-step cut-offs are derived to maximize the average power (i.e., the expected number of true positives, Equation (1.35)) subject to controlling the per-family error rate (i.e., the expected number of false positives, Equation (1.20)). Closed form solutions are obtained for shift alternative hypotheses, in the special cases of Gaussian, logistic, t-, and χ^2-distributions for the test statistics. Sample-splitting procedures are proposed for estimating optimal cut-offs. In addition, when prior information is available on the parameters of interest, pooling procedures are suggested for estimating optimal cut-offs. Simulation studies, comparing the estimated optimal cut-offs to common cut-offs, illustrate that the proposed MTPs perform well when one has accurate information about the parameters of interest.

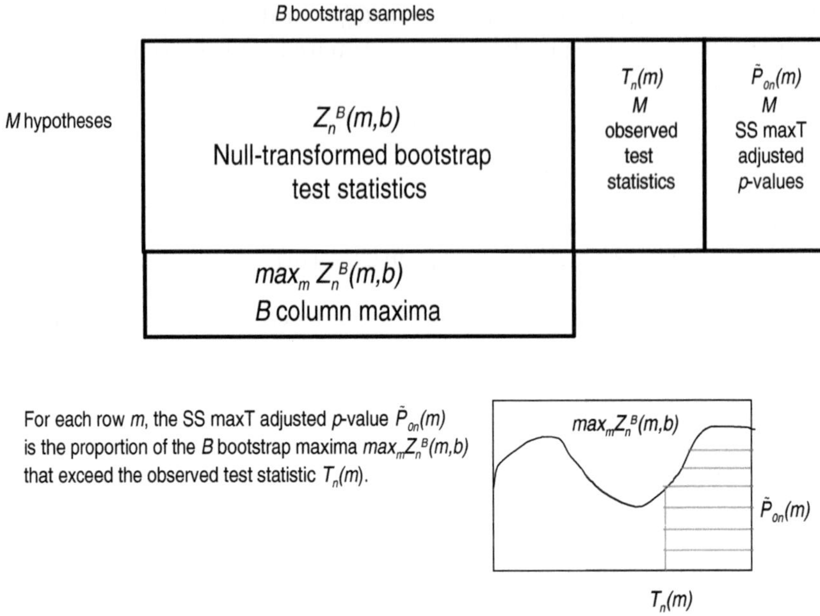

Figure 4.1. *Bootstrap estimation of the single-step maxT adjusted p-values $\widetilde{P}_{0n}(m)$ (Procedure 4.21).* Given a matrix $\mathbf{Z}_n^B = \left(Z_n^B(m,b)\right)$, of null-transformed bootstrap test statistics, bootstrap estimators of the single-step maxT adjusted p-values $\widetilde{P}_{0n}(m)$ are obtained as follows: (i) for each column b, compute the maximum statistic $Z_n^{B\circ}(1,b) = \max_m Z_n^B(m,b)$; (ii) the adjusted p-value for null hypothesis $H_0(m)$ is the proportion of the B bootstrap maxima $Z_n^{B\circ}(1,b)$ that are greater than or equal to the observed test statistic $T_n(m)$. Figure 2.1 illustrates the bootstrap estimation of the null shift and scale-transformed null distribution Q_0 of Section 2.3.

5

Step-Down Multiple Testing Procedures for Controlling the Family-Wise Error Rate

5.1 Introduction

5.1.1 Motivation

The present chapter is concerned with *joint step-down* multiple testing procedures (MTP) for controlling the *family-wise error rate* (FWER), i.e., the probability of at least one Type I error. We follow the general multiple testing framework described in Chapters 1 and 2 and refer the reader to Section 1.2 for basic definitions and notation.

Recall from Section 1.2.13 that one usually distinguishes between two main classes of multiple testing procedures, single-step and stepwise procedures, depending on whether the rejection regions for the test statistics $T_n = (T_n(m) : m = 1, \ldots, M)$ are constant or random (given a test statistics null distribution Q_0 or an estimator thereof, Q_{0n}), i.e., are independent or not of the data $\mathcal{X}_n = \{X_i : i = 1, \ldots, n\}$. In *single-step* procedures, the subject of Chapter 4, each null hypothesis $H_0(m)$ is tested using a rejection region that is independent of the results of the tests of other hypotheses and is not a function of the data $\mathcal{X}_n$ (unless these data are used to estimate the null distribution Q_0). Improvement in power, while preserving Type I error control, may be achieved by *stepwise* procedures, in which the decision to reject a particular null hypothesis depends on the outcome of the tests of other hypotheses. That is, the (single-step) test procedure is applied to a *sequence of successively smaller nested random (i.e., data-dependent) subsets of ordered null hypotheses*, defined by the *ordering* of the test statistics (common-cut-off MTPs) or unadjusted p-values (common-quantile MTPs). The rejection regions are therefore allowed to depend on the data $\mathcal{X}_n$ via the test statistics T_n. In *step-down* procedures, the *most significant* null hypotheses (i.e., the null hypotheses with the largest test statistics for common-cut-off MTPs or smallest unadjusted p-values for common-quantile MTPs) are considered successively, with further tests depending on the outcome of earlier ones. As soon as one fails to reject a null hypothesis, no further hypotheses are rejected.

As described in Section 3.2.3, a simple step-down procedure for FWER control is Holm Procedure 3.7, the step-down analogue of classical single-step Bonferroni Procedure 3.1. In this chapter, we consider more general FWER-controlling step-down procedures, that are step-down analogues of single-step common-quantile (minP) Procedure 4.1 and common-cut-off (maxT) Procedure 4.2. Like their single-step counterparts of Chapter 4, *step-down maxT Procedure 5.1* and *step-down minP Procedure 5.6* take into account the *joint* distribution of the test statistics and, hence, are generally more powerful than *marginal* procedures such as Holm Procedure 3.7. Single-step and step-down maxT and minP procedures are introduced in overview Chapter 3: single-step maxT Procedure 3.5 and minP Procedure 3.6, Section 3.2.2; step-down maxT Procedure 3.11 and minP Procedure 3.12, Section 3.2.3.

As for the single-step procedures of Chapter 4, a crucial ingredient in our step-down procedures is the *test statistics null distribution* used to derive rejection regions for the test statistics and adjusted p-values. Section 2.2 outlines the main features of our approach to Type I error control and the key choice of a test statistics null distribution based on the notion of *null domination*. This general characterization leads to two explicit constructions for a proper test statistics null distribution: the asymptotic distribution of a vector of *null shift and scale-transformed test statistics* (Section 2.3) and the asymptotic distribution of a vector of *null quantile-transformed test statistics* (Section 2.4). Both null distributions yield procedures that provide control of the Type I error rate, for general data generating distributions (with arbitrary dependence structures among variables), null hypotheses (defined in terms of submodels for the data generating distribution), and test statistics (e.g., t-statistics, χ^2-statistics, F-statistics), without the need for restrictive assumptions such as subset pivotality (Westfall and Young, 1993, p. 42–43).

The step-down procedures proposed in this chapter are similar in spirit to their single-step counterparts of Chapter 4, with the important step-down distinction that null hypotheses are considered successively, from most significant to least significant, with further tests depending on the outcome of earlier ones. Thus, rather than being based solely on the distribution of the maximum test statistic/minimum unadjusted p-value *over all M null hypotheses*, the step-down common cut-offs/common-quantile cut-offs and corresponding adjusted p-values are based on the distributions of maxima of test statistics/minima of unadjusted p-values *over successively smaller nested random subsets of ordered null hypotheses*.

The reader is referred to Korn et al. (2004), Lehmann and Romano (2005), Romano and Wolf (2005), Sarkar (2005), Troendle (1995, 1996), and Westfall and Young (1993), for recent work on stepwise methods. In particular, Korn et al. (2004) provide permutation-based step-down procedures for controlling the generalized family-wise error rate (gFWER) and tail probabilities for the proportion of false positives (TPPFP) among the rejected hypotheses, in the special case where the null hypotheses concern equality of the marginal distributions of two data generating distributions. The unpublished manuscript of Romano and Wolf (2005) considers gFWER-controlling joint single-step and step-down procedures. In a very recent technical report, Sarkar (2005) proposes gFWER-controlling joint step-down and step-up procedures, where the step-up MTP relies on an extended version of Simes' Inequality (Equation (B.5)). Troendle (1996) proposes a permutation-based step-up multiple testing procedure that takes into account the dependence structure of the test statistics and is related to the step-down maxT procedure presented in Westfall and Young (1993).

5.1.2 Outline

Sections 5.2 and 5.3 propose, respectively, step-down common-cut-off and common-quantile multiple testing procedures for controlling the family-wise error rate (FWER). The common-cut-off approach, in step-down maxT Procedure 5.1, relies on the distributions of successive maxima of test statistics, while the common-quantile approach, in step-down minP Procedure 5.6, involves the distributions of successive minima of unadjusted p-values. Two main types of results are derived concerning asymptotic control of the FWER by Procedures 5.1 and 5.6. The more general Theorems 5.2 and 5.7 prove that step-down maxT Procedure 5.1 and minP Procedure 5.6 provide asymptotic control of the FWER, under general asymptotic null domination assumptions for the test statistics null distribution (Assumptions ANDmaxT and ANDminP). By making additional asymptotic separation assumptions for the test statistics for the true and false null hypotheses (Assumptions AST and ASP), Theorems 5.3 and 5.8 provide sharper Type I error control results. These four theorems imply that gains in power from step-down procedures, relative to their single-step counterparts, do not come at the expense of Type I error control. Theorem 5.4 proposes the null shift and scale-transformed test statistics null distribution of Section 2.3, as an explicit null distribution for use in step-down Procedures 5.1 and 5.6. It is argued in Section 3 of van der Laan and Hubbard (2006) that the new null quantile-transformed test statistics null distribution, introduced in Section 2.4, also leads to proper asymptotic control of the FWER by the proposed step-down procedures. Note that analogous finite sample Type I error control results can be proved for test statistics null distributions that satisfy finite sample versions of null domination Assumptions ANDmaxT and ANDminP, respectively. Step-up procedures are discussed in Section 5.4. Section 5.5 shows that step-down Procedures 5.1 and 5.6, based on a consistent estimator of the test statistics null distribution, provide asymptotic control of the FWER (Theorems 5.12–5.14). General bootstrap procedures are supplied to conveniently obtain consistent estimators of the test statistics null distribution and of the resulting step-down maxT and minP cut-offs and adjusted p-values (Procedure 5.15).

5.2 FWER-controlling step-down common-cut-off procedure based on maxima of test statistics

As for the single-step procedures of Chapter 4, consider the simultaneous test of M null hypotheses $H_0(m)$, $m = 1, \ldots, M$, based on test statistics $T_n = (T_n(m) : m = 1, \ldots, M)$, with finite sample joint distribution $Q_n = Q_n(P)$, under the true, unknown data generating distribution P. Assume that large values of the test statistic $T_n(m)$ provide evidence against the corresponding null hypothesis $H_0(m)$, that is, consider one-sided rejection regions of the form $\mathcal{C}_n(m; \alpha) = (c_n(m; \alpha), +\infty)$, where $c_n(\alpha) = (c_n(m; \alpha) : m = 1, \ldots, M) \in \mathbb{R}^M$ is an M-vector of step-down cut-offs $c_n(m; \alpha) = c(m; T_n, Q_0, \alpha) \in \mathbb{R}$, computed under a null distribution Q_0 (or estimator thereof, Q_{0n}) for the test statistics T_n.

5.2.1 Step-down maxT procedure

Procedure 5.1. [FWER-controlling step-down maxT procedure] Let $O_n(m)$ denote the index of the mth largest test statistic $T_n^{\circ}(m) \equiv T_n(O_n(m))$, so that $T_n^{\circ}(1) \geq \cdots \geq T_n^{\circ}(M)$. For a test at nominal FWER level $\alpha \in (0, 1)$, given an M-variate test statistics null distribution Q_0 and a random M-vector $Z = (Z(m) : m = 1, \ldots, M) \sim Q_0$, define $(1 - \alpha)$-quantiles, $c(\mathcal{A}) = c(\mathcal{A}; Q_0, \alpha) \in \mathbb{R}$, for the distributions of maxima, $\max_{m \in \mathcal{A}} Z(m)$, of random variables $Z(m)$ over subsets $\mathcal{A} \subseteq \{1, \ldots, M\}$. That is, let

$$c(\mathcal{A}; Q_0, \alpha) \equiv F_{\mathcal{A}, Q_0}^{-1}(1 - \alpha) = \inf \{z \in \mathbb{R} : F_{\mathcal{A}, Q_0}(z) \geq 1 - \alpha\}, \tag{5.1}$$

where $F_{\mathcal{A}, Q_0}(z) \equiv \Pr_{Q_0}(\max_{m \in \mathcal{A}} Z(m) \leq z)$ denotes the CDF of $\max_{m \in \mathcal{A}} Z(m)$ for $Z \sim Q_0$. Next, given the indices $O_n(m)$ for the ordered test statistics $T_n^{\circ}(m)$, define $(1-\alpha)$-quantiles, $C_n(m)$, for subsets of the form $\overline{\mathcal{O}}_n(m) \equiv \{O_n(m), \ldots, O_n(M)\}$,

$$C_n(m) \equiv c(\overline{\mathcal{O}}_n(m); Q_0, \alpha) = F_{\overline{\mathcal{O}}_n(m), Q_0}^{-1}(1 - \alpha), \tag{5.2}$$

and data-dependent step-down cut-offs

$$C_n^{\star}(1) \equiv C_n(1), \tag{5.3}$$

$$C_n^{\star}(m) \equiv \begin{cases} C_n(m), & \text{if } T_n^{\circ}(m-1) > C_n^{\star}(m-1) \\ +\infty, & \text{otherwise} \end{cases}, \qquad m = 2, \ldots, M.$$

The *step-down maxT* multiple testing procedure, for controlling the FWER at nominal level α, is defined by the following rule. Reject null hypothesis $H_0(O_n(m))$, corresponding to the mth most significant test statistic

$T_n^\circ(m) = T_n(O_n(m))$, if $T_n^\circ(m) > C_n^\star(m)$, $m = 1, \ldots, M$. That is, let

$$\mathcal{R}_n = \mathcal{R}(T_n, Q_0, \alpha) \equiv \{O_n(m) : T_n^\circ(m) > C_n^\star(m),\ m = 1, \ldots, M\}. \quad (5.4)$$

Step-down maxT Procedure 5.1 is introduced in overview Chapter 3 (Procedure 3.11) and can be stated in a more compact manner as

$$\mathcal{R}_n \equiv \{O_n(1), \ldots, O_n(R_n)\}, \quad (5.5)$$

where R_n, the number of rejected null hypotheses, is defined as

$$R_n \equiv \max\left\{ m : \left(\sum_{h=1}^{m} \mathrm{I}\left(T_n^\circ(h) > C_n(h)\right) \right) = m \right\}, \quad (5.6)$$

with the understanding that $R_n = 0$ if $T_n^\circ(1) \leq C_n(1)$. Adjusted p-values are derived in Section 5.2.4, below.

Note that the definition $C_n^\star(m) = +\infty$, if $T_n^\circ(m-1) \leq C_n^\star(m-1)$, ensures that the procedure is indeed step-down, that is, one can only reject a particular null hypothesis provided all hypotheses with more significant (i.e., greater) test statistics were rejected beforehand. In addition, the cut-offs $C_n(m)$ are random variables that depend on the data $\mathcal{X}_n$ via the ranks of the test statistics T_n (i.e., via the random subsets $\overline{\mathcal{O}}_n(m)$), again reflecting the stepwise nature of the method. This is in contrast to the constant cut-offs used in single-step common-cut-off Procedure 4.2.

Procedure 5.1 is a step-down analogue of single-step maxT Procedure 3.5, that arises as a special case of single-step common-cut-off Procedure 4.2. However, rather than being based solely on the distribution of the maximum test statistic over all M null hypotheses, the step-down common cut-offs and corresponding adjusted p-values are based on the distributions of maxima of test statistics over successively smaller nested random subsets of ordered null hypotheses.

Similar step-down maxT procedures are discussed in Dudoit et al. (2003) and Westfall and Young (1993, Algorithm 4.1, p. 116–117), with an important distinction in the choice of the null distribution Q_0 used to derive the quantiles $C_n(m)$ and the resulting adjusted p-values (Sections 5.2.2 and 5.2.3).

5.2.2 Asymptotic control of the FWER

In order to establish asymptotic control of the FWER by Procedure 5.1, we rely on one or both of the following two assumptions concerning the true joint distribution $Q_n = Q_n(P)$ of the test statistics T_n and the null distribution Q_0.

Assumption ANDmaxT. [Asymptotic null domination for the maximum of the $\mathcal{H}_0$-specific test statistics] There exists an M-variate test statistics null distribution $Q_0 = Q_0(P)$ that satisfies the following *asymptotic null domination* condition for the *maximum of the $\mathcal{H}_0$-specific test statistics.* For each $x \in I\!R$,

$$\limsup_{n\to\infty} \Pr_{Q_n}\left(\max_{m\in\mathcal{H}_0} T_n(m) > x\right) \le \Pr_{Q_0}\left(\max_{m\in\mathcal{H}_0} Z(m) > x\right), \tag{5.7}$$

where T_n and Z are random M-vectors with $T_n \sim Q_n = Q_n(P)$ and $Z \sim Q_0$. That is, the maximum $\max_{m\in\mathcal{H}_0} T_n(m)$ of the $\mathcal{H}_0$-specific test statistics is asymptotically stochastically greater under the null distribution Q_0 than under the true distribution Q_n.

Note that Assumption ANDmaxT follows from asymptotic joint null domination Assumption jtNDT, for the $\mathcal{H}_0$-specific subvector $(T_n(m) : m \in \mathcal{H}_0)$ of test statistics for the true null hypotheses (Section 2.2.3). In Chapter 4, in accordance with the three-step road map of Procedure 2.1, such an assumption is applied to asymptotically control general Type I error rates of the form $\Theta(F_{V_n})$, using single-step common-quantile Procedure 4.1 and common-cut-off Procedure 4.2. General Assumption jtNDT is sufficient, but not necessary, for control of the FWER by Procedure 5.1. Assumption ANDmaxT can be viewed as a weaker, FWER-specific asymptotic null domination condition for the test statistics, that leads to asymptotic null domination of the Type I error rate (Assumption ND$\boldsymbol{\Theta}$, with $\Theta(F_{V_n}) = 1 - F_{V_n}(0)$). Explicit guidelines for constructing a test statistics null distribution Q_0 that satisfies Assumption ANDmaxT are given in Section 5.2.3, below.

Assumption AST. [Asymptotic separation of the test statistics for the true and false null hypotheses] The $\mathbf{H}_0$-specific test statistics $(T_n(m) : m \in \mathcal{H}_0)$ and $\mathcal{H}_1$-specific test statistics $(T_n(m) : m \in \mathcal{H}_1)$ satisfy the following *asymptotic separation* conditions. Let K_0 be a possibly degenerate (e.g., $+\infty$) maximal value, so that $\Pr_{Q_n}(\max_m T_n(m) < K_0) = 1$, for all n. For each $K < K_0$, assume that

$$\lim_{n\to\infty} \Pr_{Q_n}\left(\min_{m\in\mathcal{H}_1} T_n(m) \ge K\right) = 1 \tag{5.8}$$

and

$$\lim_{K\to K_0} \lim_{n\to\infty} \Pr_{Q_n}\left(\max_{m\in\mathcal{H}_0} T_n(m) \ge K\right) = 0. \tag{5.9}$$

In addition, for $\alpha \in (0,1)$ and $Z = (Z(m) : m = 1,\ldots,M) \sim Q_0$, the distributions of maxima $\max_{m\in\mathcal{A}} Z(m)$ of random variables $Z(m)$ over subsets $\mathcal{A} \subseteq \{1,\ldots,M\}$ are assumed to have $(1-\alpha)$-quantiles bounded above by K_0. That is,

$$\max_{\mathcal{A}\subseteq\{1,\ldots,M\}} c(\mathcal{A}; Q_0, \alpha) < K_0, \tag{5.10}$$

where the quantiles $c(\mathcal{A}) = c(\mathcal{A}; Q_0, \alpha)$ are defined as in Equation (5.1) of Procedure 5.1.

Also note that Equations (5.8) and (5.9) in Assumption AST only require that $T_n = (T_n(m) : m = 1, \ldots, M)$ be a sensible choice of test statistics, that asymptotically separate into two groups, $(T_n(m) : m \in \mathcal{H}_0)$ and $(T_n(m) : m \in \mathcal{H}_1)$, depending on the truth or falsity of the null hypotheses. That is, asymptotically, the $h_1 = |\mathcal{H}_1|$ largest test statistics correspond to the h_1 false null hypotheses.

We derive two main results concerning asymptotic control of the FWER by Procedure 5.1. The more general Theorem 5.2 establishes that Procedure 5.1 provides asymptotic control of the FWER under asymptotic null domination Assumption ANDmaxT only. By making the additional asymptotic separation Assumption AST, Theorem 5.3 provides a sharper Type I error control result. In particular, consistent identification of the set $\mathcal{H}_1$ of false null hypotheses, as in Assumption AST, leads to exact asymptotic control of the FWER, when Assumption ANDmaxT holds with equality. However, asymptotic separation of the test statistics for the true and false null hypotheses does not hold for local alternative hypotheses.

Because step-down maxT cut-offs are always less conservative than (i.e., less than or equal to) the corresponding single-step maxT cut-offs, Theorems 5.2 and 5.3 show that gains in power from step-down maxT Procedure 5.1, relative to single-step maxT Procedure 3.5, do not come at the expense of Type I error control.

Theorem 5.2. [Asymptotic control of the FWER by step-down maxT Procedure 5.1, under Assumption ANDmaxT] *Suppose that asymptotic null domination Assumption ANDmaxT is satisfied by the distribution $Q_n = Q_n(P)$ of the test statistics $T_n = (T_n(m) : m = 1, \ldots, M)$ and by the null distribution Q_0. Denote the number of Type I errors for Procedure 5.1 by*

$$V_n \equiv \sum_{m=1}^{M} \mathrm{I}\left(T_n^{\circ}(m) > C_n^{\star}(m),\ O_n(m) \in \mathcal{H}_0\right).$$

Then, step-down maxT Procedure 5.1 provides asymptotic control of the family-wise error rate at level α, that is,

$$\limsup_{n \to \infty} \Pr(V_n > 0) \leq \alpha.$$

Proof of Theorem 5.2. Let $m_n \equiv \min\{m : O_n(m) \in \mathcal{H}_0\}$, that is, $O_n(m_n)$ is the index of the true null hypothesis with the largest test statistic. Thus, by definition of m_n, $T_n^{\circ}(m_n) = T_n(O_n(m_n)) = \max_{m \in \mathcal{H}_0} T_n(m)$ and $\{O_n(1), \ldots, O_n(m_n - 1)\} = \overline{\mathcal{O}}_n^c(m_n) \subseteq \mathcal{H}_1$. It then follows that

$$\begin{aligned}
\Pr(V_n > 0) &= \Pr\left(O_n(m_n) \in \mathcal{R}_n\right) \\
&\le \Pr\left(T_n^\circ(m_n) > c(\overline{\mathcal{O}}_n(m_n); Q_0, \alpha)\right) \\
&= \Pr\left(\max_{m \in \mathcal{H}_0} T_n(m) > c(\overline{\mathcal{O}}_n(m_n); Q_0, \alpha)\right) \\
&\le \Pr\left(\max_{m \in \mathcal{H}_0} T_n(m) > c(\mathcal{H}_0; Q_0, \alpha)\right),
\end{aligned}$$

where the first inequality results from the step-down property and the last inequality from the fact that $\mathcal{H}_0 \subseteq \overline{\mathcal{O}}_n(m_n)$ implies $c(\mathcal{H}_0; Q_0, \alpha) \le c(\overline{\mathcal{O}}_n(m_n); Q_0, \alpha)$.

Finally, under Assumption ANDmaxT and for $Z \sim Q_0$,

$$\begin{aligned}
\limsup_{n\to\infty} \Pr(V_n > 0) &\le \limsup_{n\to\infty} \Pr\left(\max_{m \in \mathcal{H}_0} T_n(m) > c(\mathcal{H}_0; Q_0, \alpha)\right) \\
&\le \Pr\left(\max_{m \in \mathcal{H}_0} Z(m) > c(\mathcal{H}_0; Q_0, \alpha)\right) \\
&\le \alpha,
\end{aligned}$$

which completes the proof.

□

Theorem 5.3. [Asymptotic control of the FWER by step-down maxT Procedure 5.1, under Assumptions ANDmaxT and AST] *Suppose that Assumptions ANDmaxT and AST are satisfied by the distribution $Q_n = Q_n(P)$ of the test statistics $T_n = (T_n(m) : m = 1, \ldots, M)$ and by the null distribution Q_0. Then, as under the weaker assumptions of Theorem 5.2, step-down maxT Procedure 5.1 provides asymptotic control of the family-wise error rate at level α. In addition, if Assumption ANDmaxT holds with equality and Q_0 is continuous (so that $\max_{m\in\mathcal{H}_0} Z(m)$ is a continuous random variable for $Z \sim Q_0$), then asymptotic control is exact, i.e.,*

$$\lim_{n\to\infty} \Pr(V_n > 0) = \alpha.$$

Proof of Theorem 5.3. Procedure 5.1 can be stated equivalently as follows. Let

$$T_n^\star(1) \equiv T_n^\circ(1), \tag{5.11}$$

$$T_n^\star(m) \equiv \begin{cases} T_n^\circ(m), & \text{if } T_n^\star(m-1) > C_n(m-1) \\ -\infty, & \text{otherwise} \end{cases}, \qquad m = 2, \ldots, M,$$

and reject the following set of null hypotheses,

$$\mathcal{R}_n \equiv \{O_n(m) : T_n^\star(m) > C_n(m),\ m = 1, \ldots, M\}. \tag{5.12}$$

We begin by stating the main ideas of the proof. Firstly, from asymptotic separation Assumption AST, with probability one in the limit, the $h_1 = |\mathcal{H}_1|$

false null hypotheses have the largest h_1 test statistics and are rejected by Procedure 5.1 (see argument with indicator B_n, below). Thus, no Type I errors are committed for these first h_1 rejected hypotheses and one can focus on the remaining h_0 least significant test statistics $(T_n^\circ(m) : m = h_1 + 1, \ldots, M)$, which correspond to the test statistics $(T_n(m) : m \in \mathcal{H}_0)$ for the true null hypotheses. By definition of the step-down procedure, a Type I error is then committed if and only if $\max_{m \in \mathcal{H}_0} T_n(m) = T_n^\star(h_1+1) > C_n(h_1+1) = c(\mathcal{H}_0)$. Thus, one needs to control $\Pr\left(\max_{m \in \mathcal{H}_0} T_n(m) > c(\mathcal{H}_0)\right)$, which the procedure indeed asymptotically controls at level α, under asymptotic null domination Assumption ANDmaxT. Details of the proof are given next.

Define Bernoulli random variables

$$B_n \equiv \mathrm{I}\left(\{O_n(1), \ldots, O_n(h_1)\} = \mathcal{H}_1, \quad T_n^\star(1) > C_n(1), \ldots, T_n^\star(h_1) > C_n(h_1)\right). \tag{5.13}$$

Under Assumption AST, $\Pr(B_n = 1) \to 1$, as $n \to \infty$. Then, the FWER for Procedure 5.1 is given by

$$\begin{aligned}
\Pr(V_n > 0) &= \Pr\left(\bigcup_{m=1}^{M} \{T_n^\star(m) > C_n(m),\ O_n(m) \in \mathcal{H}_0\}\right) \\
&= \Pr\left(\bigcup_{m=1}^{M} \{T_n^\star(m) > C_n(m), O_n(m) \in \mathcal{H}_0\} \,\middle|\, B_n = 1\right) + o(1) \\
&= \Pr\left(\bigcup_{m=h_1+1}^{M} \{T_n^\star(m) > C_n(m)\} \,\middle|\, B_n = 1\right) + o(1) \\
&= \Pr\left(T_n^\star(h_1+1) > C_n(h_1+1) \,|\, B_n = 1\right) + o(1),
\end{aligned}$$

where the last equality follows by noting that if $T_n^\star(h_1+1) \leq C_n(h_1+1)$, then $T_n^\star(m) = -\infty$ for $m = h_1 + 2, \ldots, M$, so that $\bigcup_{m=h_1+1}^{M}\{T_n^\star(m) > C_n(m)\} = \{T_n^\star(h_1+1) > C_n(h_1+1)\}$. Now, given $B_n = 1$, then $\overline{\mathcal{O}}_n(h_1+1) = \mathcal{H}_0$, $T_n^\star(h_1+1) = T_n^\circ(h_1+1) = T_n(O_n(h_1+1)) = \max_{m \in \mathcal{H}_0} T_n(m)$, and $C_n(h_1+1) = c(\mathcal{H}_0)$. Hence,

$$\begin{aligned}
\Pr(V_n > 0) &= \Pr\left(\max_{m \in \mathcal{H}_0} T_n(m) > c(\mathcal{H}_0) \,\middle|\, B_n = 1\right) + o(1) \\
&= \Pr\left(\max_{m \in \mathcal{H}_0} T_n(m) > c(\mathcal{H}_0)\right) + o(1),
\end{aligned}$$

where we again use the fact that $\Pr(B_n = 1) \to 1$.

Finally, under Assumption ANDmaxT and for $Z \sim Q_0$,

$$\limsup_{n \to \infty} \Pr\left(\max_{m \in \mathcal{H}_0} T_n(m) > c(\mathcal{H}_0)\right) \leq \Pr\left(\max_{m \in \mathcal{H}_0} Z(m) > c(\mathcal{H}_0)\right) \leq \alpha.$$

In addition, if Assumption ANDmaxT holds with equality and the null distribution Q_0 is continuous, so that quantiles $c(\mathcal{A}) = c(\mathcal{A}; Q_0, \alpha)$ yield exact

α survival probabilities, then Procedure 5.1 provides exact asymptotic FWER control at level α,

$$\lim_{n\to\infty} \Pr\left(\max_{m\in\mathcal{H}_0} T_n(m) > c(\mathcal{H}_0)\right) = \Pr\left(\max_{m\in\mathcal{H}_0} Z(m) > c(\mathcal{H}_0)\right) = \alpha.$$

□

5.2.3 Test statistics null distribution

Sections 2.3 and 2.4 provide two explicit proposals of test statistics null distributions that satisfy asymptotic null domination Assumption ANDmaxT.

The first null distribution of Section 2.3 is defined as the asymptotic distribution of the M-vector Z_n of *null shift and scale-transformed test statistics*,

$$Z_n(m) \equiv \sqrt{\min\left\{1, \frac{\tau_0(m)}{\mathrm{Var}[T_n(m)]}\right\}}\,(T_n(m) - \mathrm{E}[T_n(m)]) + \lambda_0(m), \qquad (5.14)$$

where $\lambda_0(m)$ and $\tau_0(m)$ are, respectively, user-supplied upper bounds for the means and variances of the $\mathcal{H}_0$-specific test statistics (Dudoit et al., 2004b; van der Laan et al., 2004a). In this construction, the location null values $\lambda_0(m)$ are chosen such that the joint distribution of $(Z_n(m) : m \in \mathcal{H}_0)$ is asymptotically stochastically greater than that of $(T_n(m) : m \in \mathcal{H}_0)$. The scale null values $\tau_0(m)$ are chosen to prevent a degenerate limit for the false null hypotheses ($m \in \mathcal{H}_1$); an important issue for power considerations.

The second and most recent proposal of Section 2.4 is defined as the asymptotic distribution of the M-vector $\breve{Z}_n$ of *null quantile-transformed test statistics*,

$$\breve{Z}_n(m) \equiv q_{0,m}^{-1} Q_{n,m}^{\Delta}(T_n(m)), \qquad (5.15)$$

where $q_{0,m}$ are user-supplied marginal test statistics null distributions that satisfy marginal null domination Assumption mgNDT (van der Laan and Hubbard, 2006).

Theorem 5.4 (and related results in Section 5.3.3, below) proves that the null shift and scale-transformed null distribution Q_0 of Section 2.3 does indeed provide asymptotic Type I error control for step-down maxT Procedure 5.1 (and minP Procedure 5.6). Similar results are discussed in Section 3 of van der Laan and Hubbard (2006) for the new null quantile-transformed test statistics null distribution introduced in Section 2.4.

We stress the generality of these two test statistics null distributions: Type I error control does not rely on restrictive assumptions such as subset pivotality and holds for general data generating distributions (with arbitrary dependence structures among variables), null hypotheses, and test statistics. The latest proposal of Section 2.4 has the additional advantage that the marginal test statistics null distributions may be set to the optimal (i.e., most powerful) null distributions one would use in single hypothesis testing (e.g., permutation marginal null distributions, Gaussian or other parametric marginal null distributions).

In practice, since the data generating distribution P is unknown, then so are the aforementioned null distributions $Q_0 = Q_0(P)$. Section 5.5.1 shows that step-down Procedures 5.1 and 5.6, based on a consistent estimator Q_{0n} of the null distribution Q_0, provide asymptotic control of the family-wise error rate (Theorems 5.12–5.14). Section 5.5.2 provides bootstrap Procedure 5.15 for obtaining consistent estimators of the test statistic cut-offs and adjusted p-values for Procedures 5.1 and 5.6.

For greater detail on Type I error control and the definition and estimation of a test statistics null distribution, the reader is referred to Chapter 2. In particular, Sections 2.6 and 2.7 provide null values $\lambda_0(m)$ and $\tau_0(m)$ for a broad range of testing problems and also discuss test statistic-specific null distributions (e.g., for t-statistics, F-statistics). In many testing problems of interest, the null distribution Q_0 is continuous. For instance, for the test of single-parameter null hypotheses using t-statistics, the null values are $\lambda_0(m) = 0$ and $\tau_0(m) = 1$ and the null distribution Q_0 is an M-variate Gaussian distribution with mean vector zero (Section 2.6). For testing the equality of K population mean vectors using F-statistics, the null values are $\lambda_0(m) = 1$ and $\tau_0(m) = 2/(K-1)$, under the assumption of equal variances in the different populations. An F-specific null distribution Q_0^F is proposed as the joint distribution of an M-vector of quadratic forms of Gaussian random variables (Section 2.7). Analogous results are provided in van der Laan and Hubbard (2006) for the new null quantile-transformed null distribution.

Theorem 5.4. [Null shift and scale-transformed test statistics null distribution: Asymptotic null domination for the maximum of the $\mathcal{H}_0$-specific test statistics] *Suppose there exist known M-vectors $\lambda_0 \in I\!R^M$ and $\tau_0 \in I\!R^{+M}$ of null values, so that, for each $m \in \mathcal{H}_0$,*

$$\limsup_{n \to \infty} \mathrm{E}[T_n(m)] \leq \lambda_0(m) \tag{5.16}$$

and

$$\limsup_{n \to \infty} \mathrm{Var}[T_n(m)] \leq \tau_0(m).$$

Let

$$\nu_{0,n}(m) \equiv \sqrt{\min\left\{1, \frac{\tau_0(m)}{\mathrm{Var}[T_n(m)]}\right\}} \tag{5.17}$$

and define an M-vector of null shift and scale-transformed test statistics $Z_n = (Z_n(m) : m = 1, \ldots, M)$ by

$$Z_n(m) \equiv \nu_{0,n}(m)\,(T_n(m) - \mathrm{E}[T_n(m)]) + \lambda_0(m), \qquad m = 1, \ldots, M. \tag{5.18}$$

Suppose that the random M-vector Z_n converges weakly to a random M-vector Z, with continuous joint distribution $Q_0 = Q_0(P)$,

$$Z_n \stackrel{\mathcal{L}}{\Rightarrow} Z \sim Q_0(P). \tag{5.19}$$

Then, the null distribution Q_0 satisfies asymptotic null domination Assumption ANDmaxT for the maximum $\max_{m \in \mathcal{H}_0} T_n(m)$ of the $\mathcal{H}_0$-specific subvector of test statistics $(T_n(m) : m \in \mathcal{H}_0)$. That is, for all $x \in I\!R$,

$$\limsup_{n \to \infty} \Pr_{Q_n}\left(\max_{m \in \mathcal{H}_0} T_n(m) > x\right) \leq \Pr_{Q_0}\left(\max_{m \in \mathcal{H}_0} Z(m) > x\right). \tag{5.20}$$

In particular, if $\lim_n \mathrm{E}[T_n(m)] = \lambda_0(m)$, for all $m \in \mathcal{H}_0$, then Equation (5.20) *holds with equality.*

One can then appeal to Theorems 5.2 and 5.3 to show that step-down maxT Procedure 5.1 provides asymptotic control of the family-wise error rate at level α, when based on the null shift and scale-transformed test statistics null distribution of Section 2.3, that is, $\limsup_n \Pr(V_n > 0) \leq \alpha$.

Proof of Theorem 5.4. The proof is straightforward and is based on an intermediate random vector $\widetilde{Z}_n = (\widetilde{Z}_n(m) : m = 1, \ldots, M)$, defined as

$$\widetilde{Z}_n(m) = T_n(m) + \max\left\{0, \lambda_0(m) - \mathrm{E}[T_n(m)]\right\}, \qquad m = 1, \ldots, M. \tag{5.21}$$

First, note that $T_n(m) \leq \widetilde{Z}_n(m)$ for each $m = 1, \ldots, M$. Next, for $m \in \mathcal{H}_0$, since $\limsup_n \mathrm{E}[T_n(m)] \leq \lambda_0(m)$ and $\limsup_n \mathrm{Var}[T_n(m)] \leq \tau_0(m)$, then $\lim_n \nu_{0,n}(m) = 1$ and the $\mathcal{H}_0$-specific subvectors $(\widetilde{Z}_n(m) : m \in \mathcal{H}_0)$ and $(Z_n(m) : m \in \mathcal{H}_0)$ have the same asymptotic joint distribution. That is,

$$(\widetilde{Z}_n(m) : m \in \mathcal{H}_0) \stackrel{\mathcal{L}}{\Rightarrow} (Z(m) : m \in \mathcal{H}_0) \sim Q_{0,\mathcal{H}_0}.$$

Thus, by the Continuous Mapping Theorem (Theorem B.3), for all $x \in I\!R$,

$$\begin{aligned}\limsup_{n \to \infty} \Pr\left(\max_{m \in \mathcal{H}_0} T_n(m) > x\right) &\leq \limsup_{n \to \infty} \Pr\left(\max_{m \in \mathcal{H}_0} \widetilde{Z}_n(m) > x\right) \\ &= \Pr\left(\max_{m \in \mathcal{H}_0} Z(m) > x\right).\end{aligned}$$

In addition, if Equation (5.16) holds with equality for the location null values $\lambda_0(m)$, then Equation (5.20) also holds with equality.

□

5.2.4 Adjusted p-values

Rather than simply reporting the rejection of a subset of null hypotheses at a prespecified nominal Type I error level α, one can report *adjusted p-values* for step-down maxT Procedure 5.1 (Section 1.2.12). Although the definition of adjusted p-values in Equation (1.58) holds for arbitrary test statistics null distributions, in this section, we consider for simplicity a null distribution Q_0 with continuous and strictly monotone marginal CDFs, $Q_{0,m}$, and survivor functions, $\bar{Q}_{0,m} = 1 - Q_{0,m}$, $m = 1, \ldots, M$.

Proposition 5.5. [**Adjusted p-values for step-down maxT Procedure 5.1**] *The adjusted p-values for step-down maxT Procedure 5.1, based on a test statistics null distribution Q_0 with continuous and strictly monotone marginal distributions, are given by*

$$\widetilde{p}_{0n}(o_n(m)) = \max_{h=1,\ldots,m} \left\{ \Pr_{Q_0}\left(\max_{l \in \{o_n(h),\ldots,o_n(M)\}} Z(l) \geq t_n(o_n(h)) \right) \right\}, \tag{5.22}$$

where $Z = (Z(m) : m = 1, \ldots, M) \sim Q_0$ and $O_n(m)$ is the index of the mth largest test statistic, that is, $T_n^{\circ}(m) = T_n(O_n(m))$ and $T_n^{\circ}(1) \geq \cdots \geq T_n^{\circ}(M)$. For controlling the FWER at level α, step-down maxT Procedure 5.1 can then be stated equivalently as

$$\mathcal{R}_n = \left\{ O_n(m) : \widetilde{P}_{0n}(O_n(m)) \leq \alpha, \; m = 1, \ldots, M \right\}. \tag{5.23}$$

Note that the adjusted p-values $\widetilde{P}_{0n}(O_n(m))$ are computed *conditionally* on the test statistics $T_n(m)$ and their ranks (i.e., the indices $O_n(m)$). The lowercase notation $t_n(m)$, $o_n(m)$, and $\widetilde{p}_{0n}(m)$, is used in Equation (5.22) to avoid confusion in the interpretation of the probabilities. These probabilities refer to the joint distribution Q_0 of the M-vector Z. The lowercase notation also emphasizes that the test statistics $t_n(m)$, indices $o_n(m)$, and adjusted p-values $\widetilde{p}_{0n}(m)$, are computed for a particular realization of the random sample $\mathcal{X}_n$.

Note also that taking successive maxima of the probabilities in Equation (5.22) enforces the step-down property and monotonicity of the adjusted p-values, $\widetilde{P}_{0n}(O_n(1)) \leq \cdots \leq \widetilde{P}_{0n}(O_n(M))$.

The adjusted p-values in Equation (5.22) correspond to those given in Westfall and Young (1993, Algorithm 4.1, p. 116–117), with an important distinction in the choice of the null distribution Q_0.

Proof of Proposition 5.5. As in Procedure 5.1, let $F_{\mathcal{A},Q_0}(z) = \Pr_{Q_0}(\max_{m \in \mathcal{A}} Z(m) \leq z)$ denote the CDF of $\max_{m \in \mathcal{A}} Z(m)$ for $Z \sim Q_0$. Also, let $\overline{\mathcal{O}}_n(m) = \{O_n(m), \ldots, O_n(M)\}$ and $C_n(m) = F^{-1}_{\overline{\mathcal{O}}_n(m),Q_0}(1-\alpha)$. Then, from the definition of adjusted p-values in Equation (1.58) and the expression for the number of rejected hypotheses R_n in Equation (5.6), one has

$$
\begin{aligned}
\widetilde{p}_{0n}(o_n(m)) &= \inf\left\{\alpha \in [0,1] : \sum_{h=1}^{m} \mathrm{I}\left(t_n(o_n(h)) > c_n(h)\right) = m\right\} \\
&= \inf\left\{\alpha \in [0,1] : t_n(o_n(h)) > c_n(h), \ \forall\, h = 1,\ldots,m\right\} \\
&= \max_{h=1,\ldots,m} \left\{\inf\left\{\alpha \in [0,1] : t_n(o_n(h)) > c_n(h)\right\}\right\} \\
&= \max_{h=1,\ldots,m} \left\{\inf\left\{\alpha \in [0,1] : t_n(o_n(h)) > F^{-1}_{\overline{o}_n(h),Q_0}(1-\alpha)\right\}\right\} \\
&= \max_{h=1,\ldots,m} \left\{\inf\left\{\alpha \in [0,1] : \bar{F}_{\overline{o}_n(h),Q_0}(t_n(o_n(h))) \leq \alpha\right\}\right\} \qquad (*) \\
&= \max_{h=1,\ldots,m} \left\{\bar{F}_{\overline{o}_n(h),Q_0}(t_n(o_n(h)))\right\} \\
&= \max_{h=1,\ldots,m} \left\{\Pr_{Q_0}\left(\max_{l \in \{o_n(h),\ldots,o_n(M)\}} Z(l) > t_n(o_n(h))\right)\right\},
\end{aligned}
$$

where $(*)$ follows from the fact that, for a CDF F and corresponding survivor function $\bar{F}$, $\bar{F}^{-1}(\alpha) = F^{-1}(1-\alpha)$.

□

5.3 FWER-controlling step-down common-quantile procedure based on minima of unadjusted p-values

One can also prove asymptotic control of the FWER for an analogue of Procedure 5.1, where maxima of the test statistics $T_n(m)$ are replaced by minima of the unadjusted p-values $P_{0n}(m)$, computed under the test statistics null distribution Q_0.

As noted in Section 4.2.4, procedures based on maxima of test statistics (maxT) and minima of unadjusted p-values (minP) are equivalent when the test statistics $T_n(m)$, $m = 1,\ldots,M$, are identically distributed under Q_0, i.e., when the marginal null distributions $Q_{0,m}$ do not depend on m. Indeed, for common marginal null distributions, the significance rankings based on test statistics $T_n(m)$ and unadjusted p-values $P_{0n}(m) = \bar{Q}_{0,m}(T_n(m))$ coincide. In general, however, the two types of methods produce different results and considerations of balance, power, and computational feasibility should dictate the choice between the two approaches (Dudoit et al., 2003; Ge et al., 2003). For non-identically distributed test statistics $T_n(m)$ (e.g., some null hypotheses tested with F-statistics and others with t-statistics; null hypotheses tested with t-statistics having different degrees of freedom), common-cut-off procedures do not weight all hypotheses equally and can lead to unbalanced adjustments (Beran, 1988; Westfall, 2003; Westfall and Young, 1993). In contrast, procedures based on p-values place the null hypotheses on an "equal footing", i.e., are more balanced than their common-cut-off counterparts, and may therefore be preferable.

Also note that although nominal p-values computed from a standard normal or some other distribution may not be correct, a step-down procedure based on minima of such transformed test statistics nonetheless provides asymptotic control of the FWER. That is, these p-values can be viewed as just another type of test statistic, for example, $-\bar{\Phi}(T_n(m))$, where $\bar{\Phi} = 1 - \Phi$ is the standard normal survivor function. One can then apply Procedure 5.1 and appeal to Theorems 5.2–5.4 for FWER control.

Here, however, we propose a step-down minP multiple testing procedure where unadjusted p-values are also defined in terms of the test statistics null distribution Q_0. Specifically, the unadjusted p-values are defined as $P_{0n}(m) = \bar{Q}_{0,m}(T_n(m))$, where $\bar{Q}_{0,m} = 1 - Q_{0,m}$ are the marginal survivor functions corresponding to Q_0, $m = 1, \ldots, M$. We therefore have more specific methods and assumptions than in Section 5.2, when it comes to proving asymptotic control of the family-wise error rate.

Asymptotic Type I error control by step-down minP Procedure 5.6 relies on Assumptions ANDminP and ASP, below; guidelines for constructing the null distribution Q_0 are given in Section 5.3.3 (Lemmas 5.9 and 5.10) and are as in Section 5.2.3, with a few additional requirements. Thus, while similar, step-down maxT Procedure 5.1 and minP Procedure 5.6 are not equivalent and require a distinct treatment if one is to use the same test statistics null distribution Q_0 for both approaches (as opposed to a different null distribution for defining the unadjusted p-values in Procedure 5.6).

5.3.1 Step-down minP procedure

Procedure 5.6. [FWER-controlling step-down minP procedure]
Given an M-variate test statistics null distribution Q_0, with marginal CDFs $Q_{0,m}$ and survivor functions $\bar{Q}_{0,m} = 1 - Q_{0,m}$, $m = 1, \ldots, M$, define unadjusted p-values

$$P_{0n}(m) \equiv \bar{Q}_{0,m}(T_n(m)) \tag{5.24}$$

and

$$P_0(m) \equiv \bar{Q}_{0,m}(Z(m)), \tag{5.25}$$

for random M-vectors $T_n = (T_n(m) : m = 1, \ldots, M) \sim Q_n = Q_n(P)$ and $Z = (Z(m) : m = 1, \ldots, M) \sim Q_0$. Let $O_n(m)$ denote the index of the mth smallest unadjusted p-value $P^{\circ}_{0n}(m) \equiv P_{0n}(O_n(m))$, so that $P^{\circ}_{0n}(1) \leq \cdots \leq P^{\circ}_{0n}(M)$. For a test at nominal FWER level $\alpha \in (0, 1)$, define α-quantiles, $c(\mathcal{A}) = c(\mathcal{A}; Q_0, \alpha) \in [0, 1]$, for the distributions of minima, $\min_{m \in \mathcal{A}} P_0(m)$, of unadjusted p-values $P_0(m)$ over subsets $\mathcal{A} \subseteq \{1, \ldots, M\}$. That is, let

$$c(\mathcal{A}; Q_0, \alpha) \equiv F^{-1}_{\mathcal{A},Q_0}(\alpha) = \inf\left\{z \in [0,1] : F_{\mathcal{A},Q_0}(z) \geq \alpha\right\}, \tag{5.26}$$

where $F_{\mathcal{A},Q_0}(z) \equiv \Pr_{Q_0}(\min_{m\in\mathcal{A}} P_0(m) \leq z)$ denotes the CDF of $\min_{m\in\mathcal{A}} P_0(m)$ for $Z \sim Q_0$. Next, given the indices $O_n(m)$ for the ordered unadjusted p-values $P^\circ_{0n}(m)$, define α-quantiles, $C_n(m)$, for subsets of the form $\overline{\mathcal{O}}_n(m) \equiv \{O_n(m), \ldots, O_n(M)\}$,

$$C_n(m) \equiv c(\overline{\mathcal{O}}_n(m); Q_0, \alpha) = F^{-1}_{\overline{\mathcal{O}}_n(m),Q_0}(\alpha), \tag{5.27}$$

and data-dependent step-down cut-offs

$$C^\star_n(1) \equiv C_n(1), \tag{5.28}$$

$$C^\star_n(m) \equiv \begin{cases} C_n(m), & \text{if } P^\circ_{0n}(m-1) < C^\star_n(m-1) \\ 0, & \text{otherwise} \end{cases}, \qquad m = 2, \ldots, M.$$

The *step-down minP* multiple testing procedure, for controlling the FWER at nominal level α, is defined by the following rule. Reject null hypothesis $H_0(O_n(m))$, corresponding to the mth most significant unadjusted p-value $P^\circ_{0n}(m) = P_{0n}(O_n(m)) = \bar{Q}_{0,O_n(m)}(T_n(O_n(m)))$, if $P^\circ_{0n}(m) < C^\star_n(m)$, $m = 1, \ldots, M$. That is, let

$$\mathcal{R}_n = \mathcal{R}(T_n, Q_0, \alpha) \equiv \{O_n(m) : P^\circ_{0n}(m) < C^\star_n(m),\ m = 1, \ldots, M\}. \tag{5.29}$$

Step-down minP Procedure 5.6 is introduced in overview Chapter 3 Procedure 3.12) and, as for Procedure 5.1, can be stated more succinctly as

$$\mathcal{R}_n \equiv \{O_n(1), \ldots, O_n(R_n)\}, \tag{5.30}$$

where R_n, the number of rejected null hypotheses, is defined as

$$R_n \equiv \max\left\{m : \left(\sum_{h=1}^{m} \mathrm{I}\left(P^\circ_{0n}(h) < C_n(h)\right)\right) = m\right\}, \tag{5.31}$$

with the understanding that $R_n = 0$ if $P^\circ_{0n}(1) \geq C_n(1)$. Adjusted p-values are derived in Section 5.3.4, below.

The definition $C^\star_n(m) = 0$, if $P^\circ_{0n}(m-1) \geq C^\star_n(m-1)$, ensures that the procedure is indeed step-down, that is, one can only reject a particular null hypothesis provided all hypotheses with more significant (i.e., smaller) unadjusted p-values were rejected beforehand.

Procedure 5.6 is a step-down analogue of single-step minP Procedure 3.6, that arises as a special case of single-step common-quantile Procedure 4.1. However, rather than being based solely on the distribution of the minimum unadjusted p-value over all M null hypotheses, the step-down common-quantile cut-offs and corresponding adjusted p-values are based on the distributions of minima of unadjusted p-values over successively smaller nested random subsets of ordered null hypotheses.

Similar step-down minP procedures are discussed in Dudoit et al. (2003) and Westfall and Young (1993, Algorithm 2.8, p. 66–67), with an important distinction in the choice of the null distribution Q_0 used to derive the quantiles $C_n(m)$ and the resulting adjusted p-values (Sections 5.3.2 and 5.3.3).

Note that for a null distribution Q_0 with continuous marginal distributions $Q_{0,m}$, the unadjusted p-values $P_0(m) = \bar{Q}_{0,m}(Z(m))$ have $\mathrm{U}(0,1)$ marginal distributions. However, the p-values $P_0(m)$ are not independent, therefore, the quantiles $C_n(m)$ cannot be obtained trivially from the $\mathrm{Beta}(1, M-m+1)$ distribution for the minimum of $(M-m+1)$ independent $\mathrm{U}(0,1)$ random variables (as in marginal step-down Šidák-like Procedure 3.9, discussed in Section 5.3.5). A key feature of Sections 5.2.3 and 5.3.3 is the general characterization and explicit proposal of joint null distributions Q_0 for multiple testing procedures that take into account the dependence structure of the test statistics.

5.3.2 Asymptotic control of the FWER

As for step-down maxT Procedure 5.1, we prove two main theorems concerning asymptotic control of the FWER by Procedure 5.6, under the following p-value analogues of Assumptions ANDmaxT and AST. The more general result in Theorem 5.7 is proved under only asymptotic null domination Assumption ANDminP for the minimum of the $\mathcal{H}_0$-specific unadjusted p-values. The sharper result in Theorem 5.8 is proved under both asymptotic null domination Assumption ANDminP and asymptotic separation Assumption ASP.

Assumption ANDminP. [Asymptotic null domination for the minimum of the $\mathcal{H}_0$-specific unadjusted p-values] There exists an M-variate test statistics null distribution $Q_0 = Q_0(P)$ that satisfies the following *asymptotic null domination* condition for the *minimum of the $\mathcal{H}_0$-specific unadjusted p-values*. For each $x \in [0,1]$,

$$\limsup_{n\to\infty} \Pr_{Q_n}\left(\min_{m\in\mathcal{H}_0} P_{0n}(m) < x\right) \leq \Pr_{Q_0}\left(\min_{m\in\mathcal{H}_0} P_0(m) < x\right), \tag{5.32}$$

where $P_{0n}(m) = \bar{Q}_{0,m}(T_n(m))$ and $P_0(m) = \bar{Q}_{0,m}(Z(m))$ are unadjusted p-values defined for random M-vectors $T_n \sim Q_n = Q_n(P)$ and $Z \sim Q_0$, respectively, and $\bar{Q}_{0,m}$ denote the marginal survivor functions corresponding to the null distribution Q_0, $m = 1, \ldots, M$. That is, the minimum $\min_{m\in\mathcal{H}_0} P_{0n}(m)$ of the $\mathcal{H}_0$-specific unadjusted p-values is asymptotically stochastically smaller under the null distribution Q_0 than under the true distribution Q_n.

Note that, like Assumption ANDmaxT for step-down maxT Procedure 5.1, Assumption ANDminP follows from asymptotic joint null domination Assumption jtNDT, for the $\mathcal{H}_0$-specific subvector $(T_n(m) : m \in \mathcal{H}_0)$ of test statistics for the true null hypotheses (Section 2.2.3). Assumption ANDminP can be viewed as a weaker form of null domination, that is specific to FWER-controlling procedures based on unadjusted p-values and that leads to asymptotic null domination of the Type I error rate (Assumption NDΘ, with $\Theta(F_{V_n}) = 1 - F_{V_n}(0)$).

Assumption ASP. [Asymptotic separation of the unadjusted p-values for the true and false null hypotheses] Let $P_{0n}(m) = \bar{Q}_{0,m}(T_n(m))$ and $P_0(m) = \bar{Q}_{0,m}(Z(m))$ denote unadjusted p-values, defined for random M-vectors $T_n \sim Q_n = Q_n(P)$ and $Z \sim Q_0$, respectively, where $\bar{Q}_{0,m}$ are the marginal survivor functions corresponding to the null distribution Q_0, $m = 1, \ldots, M$. The $\mathcal{H}_0$-specific unadjusted p-values $(P_{0n}(m) : m \in \mathcal{H}_0)$ and $\mathcal{H}_1$-specific unadjusted p-values $(P_{0n}(m) : m \in \mathcal{H}_1)$ satisfy the following *asymptotic separation* conditions. For each $\epsilon > 0$, assume that

$$\lim_{n\to\infty} \Pr_{Q_n}\left(\max_{m\in\mathcal{H}_1} P_{0n}(m) \leq \epsilon\right) = 1 \tag{5.33}$$

and

$$\lim_{\epsilon\to 0}\lim_{n\to\infty} \Pr_{Q_n}\left(\min_{m\in\mathcal{H}_0} P_{0n}(m) \leq \epsilon\right) = 0. \tag{5.34}$$

In addition, for $\alpha \in (0,1)$ and $Z = (Z(m) : m = 1, \ldots, M) \sim Q_0$, the distributions of minima $\min_{m\in\mathcal{A}} P_0(m)$ of unadjusted p-values $P_0(m) = \bar{Q}_{0,m}(Z(m))$ over subsets $\mathcal{A} \subseteq \{1, \ldots, M\}$ are assumed to have positive α-quantiles. That is,

$$\min_{\mathcal{A}\subseteq\{1,\ldots,M\}} c(\mathcal{A}; Q_0, \alpha) > 0, \tag{5.35}$$

where the quantiles $c(\mathcal{A}) = c(\mathcal{A}; Q_0, \alpha)$ are defined as in Equation (5.26) of Procedure 5.6.

Guidelines for constructing a null distribution Q_0 that satisfies Assumptions ANDminP and ASP are as for step-down maxT Procedure 5.1, in Section 5.2.3, with a few additional conditions stated in Section 5.3.3, below (Lemmas 5.9 and 5.10). Again, we refer to Section 3 of van der Laan and Hubbard (2006) for results concerning the new null quantile-transformed test statistics null distribution introduced in Section 2.4.

Theorem 5.7. [Asymptotic control of the FWER by step-down minP Procedure 5.6, under Assumption ANDminP] *Suppose that asymptotic null domination Assumption ANDminP is satisfied by the unadjusted p-values $P_{0n} = (P_{0n}(m) = \bar{Q}_{0,m}(T_n(m)) : m = 1, \ldots, M)$, i.e., by the distribution $Q_n = Q_n(P)$ of the test statistics $T_n = (T_n(m) : m = 1, \ldots, M)$ and by the null distribution Q_0. Denote the number of Type I errors for Procedure 5.6 by*

$$V_n \equiv \sum_{m=1}^{M} \mathrm{I}\left(P_{0n}^{\circ}(m) < C_n^{\star}(m),\ O_n(m) \in \mathcal{H}_0\right).$$

Then, step-down minP Procedure 5.6 provides asymptotic control of the family-wise error rate at level α, that is,

$$\limsup_{n\to\infty} \Pr(V_n > 0) \leq \alpha.$$

Proof of Theorem 5.7. The proof follows that of Theorem 5.2, with unadjusted p-values $P_{0n}(m)$ replacing test statistics $T_n(m)$.

□

Theorem 5.8. [Asymptotic control of the FWER by step-down minP Procedure 5.6, under Assumptions ANDminP and ASP] *Suppose that Assumptions ANDminP and ASP are satisfied by the unadjusted p-values $P_{0n} = (P_{0n}(m) = \bar{Q}_{0,m}(T_n(m)) : m = 1,\ldots,M)$, i.e., by the distribution $Q_n = Q_n(P)$ of the test statistics $T_n = (T_n(m) : m = 1,\ldots,M)$ and by the null distribution Q_0. Then, as under the weaker assumptions of Theorem 5.7, step-down minP Procedure 5.6 provides asymptotic control of the family-wise error rate at level α. In addition, if Assumption ANDminP holds with equality and Q_0 is continuous (so that $\min_{m\in\mathcal{H}_0} \bar{Q}_{0,m}(Z(m))$ is a continuous random variable for $Z \sim Q_0$), then asymptotic control is exact, i.e.,*

$$\lim_{n\to\infty} \Pr(V_n > 0) = \alpha.$$

Proof of Theorem 5.8. The proof is analogous to that of Theorem 5.3 for step-down maxT Procedure 5.1. Therefore, only the main steps where minP-specific Assumptions ANDminP and ASP come into play are highlighted. Compared to this previous proof, maxima of test statistics are replaced by minima of unadjusted p-values and the direction of the cut-off rules is reversed.

Procedure 5.6 can be stated equivalently as follows. Let

$$P^{\star}_{0n}(1) \equiv P^{\circ}_{0n}(1), \tag{5.36}$$

$$P^{\star}_{0n}(m) \equiv \begin{cases} P^{\circ}_{0n}(m), & \text{if } P^{\star}_{0n}(m-1) < C_n(m-1) \\ 1, & \text{otherwise} \end{cases}, \qquad m = 2,\ldots,M,$$

and reject the following set of null hypotheses

$$\mathcal{R}_n \equiv \{O_n(m) : P^{\star}_{0n}(m) < C_n(m),\ m = 1,\ldots,M\}. \tag{5.37}$$

As before, define Bernoulli random variables

$$B_n \equiv \mathrm{I}\left(\{O_n(1),\ldots,O_n(h_1)\} = \mathcal{H}_1,\right. \tag{5.38}$$
$$\left. P^{\star}_{0n}(1) < C_n(1),\ldots,P^{\star}_{0n}(h_1) < C_n(h_1)\right)$$

and note that, under asymptotic separation Assumption ASP, $\Pr(B_n = 1) \to 1$, as $n \to \infty$.

Asymptotic null domination Assumption ANDminP is used at the very last step of the proof to show that

$$\limsup_{n\to\infty} \Pr\left(\min_{m\in\mathcal{H}_0} P_{0n}(m) < c(\mathcal{H}_0)\right) \leq \alpha.$$

□

5.3.3 Test statistics null distribution

A proper test statistics null distribution Q_0 for step-down minP Procedure 5.6 can be constructed as for step-down maxT Procedure 5.1 (Section 5.2.3), with a few additional requirements in order to meet minP-specific Assumptions ANDminP and ASP of Theorems 5.7 and 5.8. Lemmas 5.9 and 5.10, stated below, are concerned with providing sufficient conditions (in terms of continuity and monotonicity of the marginal null distributions $Q_{0,m}$) for Assumptions ANDminP and ASP. These conditions could be imposed on either of our two main test statistics null distributions: the null shift and scale-transformed null distribution (Section 2.3) and the null quantile-transformed null distribution (Section 2.4).

Lemma 5.9. [Asymptotic null domination] *Consider a test statistics null distribution $Q_0 = Q_0(P)$ that satisfies asymptotic joint null domination Assumption jtNDT for the $\mathcal{H}_0$-specific subvector $(T_n(m) : m \in \mathcal{H}_0)$ of test statistics for the true null hypotheses. Further suppose that Q_0 has continuous and strictly monotone marginal distributions $Q_{0,m}$, $m = 1, \ldots, M$. Then, Q_0 satisfies asymptotic null domination Assumption ANDminP for the minimum $\min_{m \in \mathcal{H}_0} P_{0n}(m)$ of the $\mathcal{H}_0$-specific unadjusted p-values. In particular, if Assumption jtNDT holds with equality, then so does Assumption ANDminP.*

Proof of Lemma 5.9. For all $x \in [0,1]$ and $Z \sim Q_0$,

$$
\begin{aligned}
&\liminf_{n\to\infty} \Pr_{Q_n}\left(\min_{m\in\mathcal{H}_0} P_{0n}(m) \geq x\right) \\
&\quad = \liminf_{n\to\infty} \Pr_{Q_n}\left(\bar{Q}_{0,m}(T_n(m)) \geq x, \ \forall\, m \in \mathcal{H}_0\right) \\
&\quad = \liminf_{n\to\infty} \Pr_{Q_n}\left(T_n(m) \leq \bar{Q}_{0,m}^{-1}(x), \ \forall\, m \in \mathcal{H}_0\right) \\
&\quad \geq \Pr_{Q_0}\left(Z(m) \leq \bar{Q}_{0,m}^{-1}(x), \ \forall\, m \in \mathcal{H}_0\right) \\
&\quad = \Pr_{Q_0}\left(\min_{m\in\mathcal{H}_0} P_0(m) \geq x\right),
\end{aligned}
$$

where the inequality follows from asymptotic joint null domination Assumption jtNDT.

□

This lemma may be applied to either the null shift and scale-transformed null distribution (Section 2.3) or the null quantile-transformed null distribution (Section 2.4), as they both satisfy asymptotic joint null domination Assumption jtNDT.

Lemma 5.10. [Asymptotic separation] *Consider a test statistics null distribution Q_0, with marginal survivor functions $\bar{Q}_{0,m}$, $m = 1, \ldots, M$, that satisfy the following: (i) $\bar{Q}_{0,m}$ is continuous; (ii) $\bar{Q}_{0,m}$ is strictly decreasing; (iii) there exists a constant K_0 (possibly degenerate), such that $\lim_{\epsilon \to 0} \bar{Q}_{0,m}^{-1}(\epsilon) = K_0$ for each m. Then, asymptotic separation Assumption AST for the test statistics $T_n(m)$ implies asymptotic separation Assumption ASP for the unadjusted p-values $P_{0n}(m) = \bar{Q}_{0,m}(T_n(m))$, $m = 1, \ldots, M$.*

Proof of Lemma 5.10. Equation (5.33) follows from Equation (5.8) by noting that, for each $\epsilon > 0$, $K_1(\epsilon) \equiv \max_{m\in\mathcal{H}_1} \bar{Q}_{0,m}^{-1}(\epsilon) < K_0$ and

$$\begin{aligned}
\Pr\left(\max_{m\in\mathcal{H}_1} P_{0n}(m) \le \epsilon\right) &= \Pr\left(\bar{Q}_{0,m}(T_n(m)) \le \epsilon,\ \forall\, m \in \mathcal{H}_1\right) \\
&= \Pr\left(T_n(m) \ge \bar{Q}_{0,m}^{-1}(\epsilon),\ \forall\, m \in \mathcal{H}_1\right) \\
&\ge \Pr\left(T_n(m) \ge K_1(\epsilon),\ \forall\, m \in \mathcal{H}_1\right) \\
&= \Pr\left(\min_{m\in\mathcal{H}_1} T_n(m) \ge K_1(\epsilon)\right).
\end{aligned}$$

Similarly, Equation (5.34) follows from Equation (5.9) by noting that, for each $\epsilon > 0$ and $k_0(\epsilon) \equiv \min_{m\in\mathcal{H}_0} \bar{Q}_{0,m}^{-1}(\epsilon)$, then

$$\begin{aligned}
\Pr\left(\min_{m\in\mathcal{H}_0} P_{0n}(m) \le \epsilon\right) &= \Pr\left(\bar{Q}_{0,m}(T_n(m)) \le \epsilon,\ \text{for some } m \in \mathcal{H}_0\right) \\
&= \Pr\left(T_n(m) \ge \bar{Q}_{0,m}^{-1}(\epsilon),\ \text{for some } m \in \mathcal{H}_0\right) \\
&\le \Pr\left(T_n(m) \ge k_0(\epsilon),\ \text{for some } m \in \mathcal{H}_0\right) \\
&= \Pr\left(\max_{m\in\mathcal{H}_0} T_n(m) \ge k_0(\epsilon)\right).
\end{aligned}$$

Also, $k_0(\epsilon) \to K_0$, as $\epsilon \to 0$, thus, by Equation (5.9),

$$\lim_{\epsilon\to 0}\lim_{n\to\infty} \Pr\left(\min_{m\in\mathcal{H}_0} P_{0n}(m) \le \epsilon\right) \le \lim_{\epsilon\to 0}\lim_{n\to\infty} \Pr\left(\max_{m\in\mathcal{H}_0} T_n(m) \ge k_0(\epsilon)\right) = 0.$$

□

5.3.4 Adjusted p-values

Adjusted p-values are derived similarly as for Procedure 5.1, in Section 5.2.4. Again, for simplicity (and as in Lemmas 5.9 and 5.10), consider a null distribution Q_0 with continuous and strictly monotone marginal CDFs, $Q_{0,m}$, and survivor functions, $\bar{Q}_{0,m} = 1 - Q_{0,m}$, $m = 1, \ldots, M$.

Proposition 5.11. [Adjusted p-values for step-down minP Procedure 5.6] *The adjusted p-values for step-down minP Procedure 5.6, based on a test statistics null distribution Q_0 with continuous and strictly monotone marginal distributions, are given by*

$$\widetilde{p}_{0n}(o_n(m)) = \max_{h=1,\ldots,m} \left\{ \Pr_{Q_0}\left(\min_{l\in\{o_n(h),\ldots,o_n(M)\}} P_0(l) \le p_{0n}(o_n(h)) \right) \right\}, \tag{5.39}$$

where $P_{0n}(m) = \bar{Q}_{0,m}(T_n(m))$ and $P_0(m) = \bar{Q}_{0,m}(Z(m))$ denote unadjusted p-values for random M-vectors $T_n \sim Q_n = Q_n(P)$ and $Z \sim Q_0$, respectively, *and $O_n(m)$ denotes the index of the mth smallest unadjusted p-value $P^\circ_{0n}(m) = P_{0n}(O_n(m))$, so that, $P^\circ_{0n}(1) \leq \cdots \leq P^\circ_{0n}(M)$. For controlling the FWER at level α, step-down minP Procedure 5.6 can then be stated equivalently as*

$$\mathcal{R}_n = \left\{ O_n(m) : \widetilde{P}_{0n}(O_n(m)) \leq \alpha, \ m = 1, \ldots, M \right\}. \qquad (5.40)$$

Here again the lowercase notation $p_{0n}(m)$, $o_n(m)$, and $\widetilde{p}_{0n}(m)$, is used to avoid confusion in the interpretation of the probabilities in Equation (5.39). These probabilities refer to the joint distribution Q_0 of the M-vector Z.

The adjusted p-values in Equation (5.39) correspond to those given in Westfall and Young (1993, Equation (2.10), p. 66), again with an important distinction in the choice of the null distribution Q_0.

Proof of Proposition 5.11. As in Procedure 5.6, let $F_{\mathcal{A},Q_0}(z) = \Pr_{Q_0}(\min_{m\in\mathcal{A}} P_0(m) \leq z)$ denote the CDF of $\min_{m\in\mathcal{A}} P_0(m)$ for $P_0(m) = \bar{Q}_{0,m}(Z(m))$ and $Z \sim Q_0$. Also, let $\overline{\mathcal{O}}_n(m) = \{O_n(m), \ldots, O_n(M)\}$ and $C_n(m) = F^{-1}_{\overline{\mathcal{O}}_n(m),Q_0}(\alpha)$. Then, from the definition of adjusted p-values in Equation (1.58) and the expression for the number of rejected hypotheses R_n in Equation (5.31), one has

$$\begin{aligned}
\widetilde{p}_{0n}(o_n(m)) &= \inf\left\{ \alpha \in [0,1] : \sum_{h=1}^{m} \mathrm{I}\left(p_{0n}(o_n(h)) < c_n(h)\right) = m \right\} \\
&= \inf\left\{ \alpha \in [0,1] : p_{0n}(o_n(h)) < c_n(h), \ \forall\, h = 1, \ldots, m \right\} \\
&= \max_{h=1,\ldots,m} \left\{ \inf\left\{ \alpha \in [0,1] : p_{0n}(o_n(h)) < c_n(h) \right\}\right\} \\
&= \max_{h=1,\ldots,m} \left\{ \inf\left\{ \alpha \in [0,1] : p_{0n}(o_n(h)) < F^{-1}_{\overline{o}_n(h),Q_0}(\alpha) \right\}\right\} \\
&= \max_{h=1,\ldots,m} \left\{ \inf\left\{ \alpha \in [0,1] : F_{\overline{o}_n(h),Q_0}(p_{0n}(o_n(h))) \leq \alpha \right\}\right\} \\
&= \max_{h=1,\ldots,m} \left\{ F_{\overline{o}_n(h),Q_0}(p_{0n}(o_n(h))) \right\} \\
&= \max_{h=1,\ldots,m} \left\{ \Pr_{Q_0}\left(\min_{l\in\{o_n(h),\ldots,o_n(M)\}} P_0(l) \leq p_{0n}(o_n(h)) \right)\right\}.
\end{aligned}$$

□

5.3.5 Comparison of joint step-down minP procedure to marginal step-down procedures

Recall from Procedure 5.6 that the step-down minP cut-offs are based on the α-quantiles $c(\mathcal{A}) = c(\mathcal{A}; Q_0, \alpha)$, for the distributions of minima $\min_{m\in\mathcal{A}} P_0(m)$ of unadjusted p-values $P_0(m)$ over subsets $\mathcal{A} \subseteq \{1, \ldots, M\}$. Specifically,

$$c(\mathcal{A}; Q_0, \alpha) = F^{-1}_{\mathcal{A},Q_0}(\alpha) = \inf\{z \in [0,1] : F_{\mathcal{A},Q_0}(z) \geq \alpha\},$$

where $F_{\mathcal{A},Q_0}(z) = \Pr_{Q_0}(\min_{m\in\mathcal{A}} P_0(m) \leq z)$ denotes the CDF of $\min_{m\in\mathcal{A}} P_0(m)$, for unadjusted p-values $P_0(m) = \bar{Q}_{0,m}(Z(m))$ based on a random M-vector $Z \sim Q_0$.

Comparison of joint step-down minP Procedure 5.6 to marginal step-down Holm Procedure 3.7

Applying Boole's Inequality (Equation (B.1)) and Proposition 1.2 to the CDF $F_{\mathcal{A},Q_0}(z)$ yields

$$F_{\mathcal{A},Q_0}(z) = \Pr_{Q_0}\left(\bigcup_{m\in\mathcal{A}} \{P_0(m) \leq z\}\right) \leq \sum_{m\in\mathcal{A}} \Pr_{Q_0}\left(P_0(m) \leq z\right) \leq |\mathcal{A}|z,$$

so that,

$$c(\mathcal{A}; Q_0, \alpha) = F^{-1}_{\mathcal{A},Q_0}(\alpha) \geq \frac{1}{|\mathcal{A}|}\alpha.$$

Thus, the joint step-down minP cut-offs $C_n^{SDminP}(m)$, defined in terms of subsets $\overline{\mathcal{O}}_n(m) = \{O_n(m), \ldots, O_n(M)\}$ of cardinality $(M - m + 1)$, are bounded below by, i.e., are less conservative than, the cut-offs $C_n^{SDHolm}(m)$ for marginal step-down Holm Procedure 3.7. That is,

$$C_n^{SDminP}(m) = F^{-1}_{\overline{\mathcal{O}}_n(m),Q_0}(\alpha) \geq \frac{1}{M-m+1}\alpha = C_n^{SDHolm}(m). \tag{5.41}$$

Equivalently, the step-down minP adjusted p-values of Equation (5.39) are bounded above by the adjusted p-values of Equation (3.11) for step-down Holm Procedure 3.7. That is,

$$\begin{aligned}
&\widetilde{p}_{0n}^{SDminP}(o_n(m)) \qquad\qquad (5.42)\\
&= \max_{h=1,\ldots,m}\left\{\Pr_{Q_0}\left(\min_{l\in\{o_n(h),\ldots,o_n(M)\}} P_0(l) \leq p_{0n}(o_n(h))\right)\right\}\\
&= \max_{h=1,\ldots,m}\left\{\Pr_{Q_0}\left(\bigcup_{l\in\{o_n(h),\ldots,o_n(M)\}} \{P_0(l) \leq p_{0n}(o_n(h))\}\right)\right\}\\
&\leq \max_{h=1,\ldots,m}\left\{\min\left\{\sum_{l\in\{o_n(h),\ldots,o_n(M)\}} \Pr_{Q_0}\left(P_0(l) \leq p_{0n}(o_n(h))\right), 1\right\}\right\}\\
&\leq \max_{h=1,\ldots,m}\left\{\min\left\{(M-h+1)\, p_{0n}(o_n(h)), 1\right\}\right\}\\
&= \widetilde{p}_{0n}^{SDHolm}(o_n(m)),
\end{aligned}$$

where the inequality in line 3 follows from Boole's Inequality and that in line 4 from Proposition 1.2, whereby $\Pr_{Q_0}(P_0(m) \leq z) \leq z$, $\forall\ z \in [0,1]$.

Joint step-down minP Procedure 5.6 is thus less conservative than *marginal* step-down Holm Procedure 3.7.

Comparison of joint step-down minP Procedure 5.6 to marginal step-down Šidák-like Procedure 3.9

Test statistics null distributions that satisfy Šidák's Inequality

Assume that the random M-vector $Z = (Z(m) : m = 1, \ldots, M) \sim Q_0$ and corresponding unadjusted p-values $P_0 = (P_0(m) = \bar{Q}_{0,m}(Z(m)) : m = 1, \ldots, M)$ satisfy Šidák's Inequality (Equations (B.3) and (B.4)).

Then, applying the p-value version of Šidák's Inequality and Proposition 1.2 to the CDF $F_{\mathcal{A},Q_0}(z)$ yields

$$\begin{aligned} F_{\mathcal{A},Q_0}(z) &= 1 - \Pr_{Q_0}\left(\min_{m\in\mathcal{A}} P_0(m) > z\right) \\ &= 1 - \Pr_{Q_0}\left(\bigcap_{m\in\mathcal{A}} \{P_0(m) > z\}\right) \\ &\le 1 - \prod_{m\in\mathcal{A}} \Pr_{Q_0}(P_0(m) > z) \\ &\le 1 - (1-z)^{|\mathcal{A}|}, \end{aligned}$$

so that,

$$c(\mathcal{A}; Q_0, \alpha) = F^{-1}_{\mathcal{A},Q_0}(\alpha) \ge 1 - (1-\alpha)^{1/|\mathcal{A}|}.$$

Thus, the joint step-down minP cut-offs $C_n^{SDminP}(m)$, defined in terms of subsets $\overline{\mathcal{O}}_n(m) = \{O_n(m), \ldots, O_n(M)\}$ of cardinality $(M - m + 1)$, are bounded below by, i.e., are less conservative than, the cut-offs $C_n^{SDSidak}(m)$ for marginal step-down Šidák-like Procedure 3.9. That is,

$$C_n^{SDminP}(m) = F^{-1}_{\overline{\mathcal{O}}_n(m),Q_0}(\alpha) \ge 1 - (1-\alpha)^{1/(M-m+1)} = C_n^{SDSidak}(m). \quad (5.43)$$

Equivalently, the step-down minP adjusted p-values of Equation (5.39) are bounded above by the adjusted p-values of Equation (3.14) for step-down Šidák-like Procedure 3.9. That is,

$$\begin{aligned}
&\widetilde{p}_{0n}^{SDminP}(o_n(m)) && (5.44)\\
&= \max_{h=1,\ldots,m} \left\{ \Pr_{Q_0}\left(\min_{l\in\{o_n(h),\ldots,o_n(M)\}} P_0(l) \leq p_{0n}(o_n(h)) \right) \right\}\\
&= \max_{h=1,\ldots,m} \left\{ 1 - \Pr_{Q_0}\left(\min_{l\in\{o_n(h),\ldots,o_n(M)\}} P_0(l) > p_{0n}(o_n(h)) \right) \right\}\\
&= \max_{h=1,\ldots,m} \left\{ 1 - \Pr_{Q_0}\left(\bigcap_{l\in\{o_n(h),\ldots,o_n(M)\}} \{P_0(l) > p_{0n}(o_n(h))\} \right) \right\}\\
&\leq \max_{h=1,\ldots,m} \left\{ 1 - \prod_{l\in\{o_n(h),\ldots,o_n(M)\}} \Pr_{Q_0}\left(P_0(l) > p_{0n}(o_n(h))\right) \right\}\\
&\leq \max_{h=1,\ldots,m} \left\{ 1 - (1 - p_{0n}(o_n(h)))^{(M-h+1)} \right\}\\
&= \widetilde{p}_{0n}^{SDSidak}(o_n(m)),
\end{aligned}$$

where the inequality in line 4 follows from the p-value version of Šidák's Inequality, in Equation (B.4), and that in line 5 from Proposition 1.2, whereby $\Pr_{Q_0}(P_0(m) > z) \geq 1 - z$, $\forall\ z \in [0,1]$.

Joint step-down minP Procedure 5.6 is thus less conservative than *marginal* step-down Šidák-like Procedure 3.9.

Test statistics null distributions with independent and continuous marginal distributions

Consider now the special case where the random M-vector $Z \sim Q_0$ has independent and continuous marginal distributions $Q_{0,m}$, $m = 1, \ldots, M$.

Then, the unadjusted p-values $P_0(m) = \bar{Q}_{0,m}(Z(m))$ are independent $\mathrm{U}(0,1)$ random variables and the minima $\min_{m\in\mathcal{A}} P_0(m)$ have $\mathrm{Beta}(1, |\mathcal{A}|)$ distributions. The adjusted p-values for step-down minP Procedure 5.6 reduce to the adjusted p-values for step-down Šidák-like Procedure 3.9. That is,

$$\begin{aligned}
&\widetilde{p}_{0n}^{SDminP}(o_n(m)) && (5.45)\\
&= \max_{h=1,\ldots,m} \left\{ 1 - \Pr_{Q_0}\left(\bigcap_{l\in\{o_n(h),\ldots,o_n(M)\}} \{P_0(l) > p_{0n}(o_n(h))\} \right) \right\}\\
&= \max_{h=1,\ldots,m} \left\{ 1 - \prod_{l\in\{o_n(h),\ldots,o_n(M)\}} \Pr_{Q_0}\left(P_0(l) > p_{0n}(o_n(h))\right) \right\}\\
&= \max_{h=1,\ldots,m} \left\{ 1 - (1 - p_{0n}(o_n(h)))^{(M-h+1)} \right\}\\
&= \widetilde{p}_{0n}^{SDSidak}(o_n(m)).
\end{aligned}$$

Thus, for a test statistics null distribution Q_0 with independent and continuous marginal distributions, step-down minP Procedure 5.6 is very simple and is based solely on the marginal null distributions $Q_{0,m}$. In general, however, the test statistics are not independent and Procedure 5.6, based on a joint null distribution Q_0 constructed as in Section 5.3.3, takes into account the dependence structure of the test statistics when computing cut-offs $C_n(m)$ and adjusted p-values $\widetilde{P}_{0n}(O_n(m))$.

5.4 FWER-controlling step-up common-cut-off and common-quantile procedures

5.4.1 Candidate step-up maxT and minP procedures

It is interesting to consider step-up versions of step-down maxT Procedure 5.1 and minP Procedure 5.6 (see Sections 1.2.13 and 3.2.4, for an introduction to and examples of step-up procedures). One can view the step-down/step-up maxT and step-down/step-up minP procedure pairs as *joint* analogues of the *marginal* step-down Holm/step-up Hochberg procedure pair (Procedures 3.7 and 3.13). Joint step-up maxT and minP procedures could potentially be more powerful than corresponding step-down procedures and existing marginal step-up procedures, such as Hochberg Procedure 3.13. Indeed, as shown in Section 1.2.13, a step-up procedure always leads to more rejected hypotheses, i.e., is less conservative, than its step-down counterpart based on the same cut-offs. Unfortunately, as argued below, the step-up maxT and minP procedures do not in general control the FWER.

As discussed previously in Section 3.2.4, establishing Type I error control for step-up procedures is generally more difficult than for step-down procedures and often requires a number of assumptions concerning the joint distribution of the test statistics. For instance, for step-up Bonferroni-like Hochberg Procedure 3.13, FWER control is proved by appealing to Simes' Inequality (Equation (B.5)). Similar difficulties arise with FDR-controlling step-up procedures (Section 3.4).

In a very recent technical report, Sarkar (2005) proposes gFWER-controlling joint step-down and step-up procedures, where the step-up MTP relies on an extended version of Simes' Inequality.

Candidate step-up maxT procedure

The candidate *step-up maxT* procedure is based on the same ordered test statistics $T_n^\circ(m)$ and $(1-\alpha)$-quantiles $C_n(m)$ as step-down maxT Procedure 5.1, but the stepwise cut-offs $C_n^\star(m)$ are redefined as follows.

$$C_n^\star(M) \equiv C_n(M), \tag{5.46}$$

$$C_n^\star(m) \equiv \begin{cases} C_n(m), & \text{if } T_n^\circ(m+1) \leq C_n^\star(m+1) \\ -\infty, & \text{otherwise} \end{cases}, \qquad m = 1, \ldots, M-1.$$

This step-up maxT procedure then rejects the following null hypotheses

$$\mathcal{R}(T_n, \mathcal{Q}_0, \alpha) \equiv \{O_n(m) : T_n^\circ(m) > C_n^\star(m),\ m = 1, \ldots, M\}. \tag{5.47}$$

Note that the definition $C_n^\star(m) = -\infty$, if $T_n^\circ(m+1) > C_n^\star(m+1)$, ensures that the procedure is indeed step-up, that is, as soon as one hypothesis is rejected, all remaining more significant hypotheses are rejected.

The adjusted p-values for the step-up maxT procedure are as those for step-down maxT Procedure 5.1, in Equation (5.22), but with $\min_{h=m,\ldots,M}$ replacing $\max_{h=1,\ldots,m}$. That is,

$$\widetilde{p}_{0n}(o_n(m)) = \min_{h=m,\ldots,M} \left\{ \Pr_{Q_0}\left(\max_{l \in \{o_n(h),\ldots,o_n(M)\}} Z(l) \geq t_n(o_n(h)) \right) \right\}. \tag{5.48}$$

Candidate step-up minP procedure

Likewise, the candidate *step-up minP* procedure is based on the same ordered unadjusted p-values $P_{0n}^\circ(m)$ and α-quantiles $C_n(m)$ as step-down minP Procedure 5.6, but the stepwise cut-offs $C_n^\star(m)$ are redefined as follows.

$$C_n^\star(M) \equiv C_n(M), \tag{5.49}$$

$$C_n^\star(m) \equiv \begin{cases} C_n(m), & \text{if } P_{0n}^\circ(m+1) \geq C_n^\star(m+1) \\ 1, & \text{otherwise} \end{cases}, \qquad m = 1, \ldots, M-1.$$

This step-up minP procedure then rejects the following null hypotheses

$$\mathcal{R}(T_n, \mathcal{Q}_0, \alpha) \equiv \{O_n(m) : P_{0n}^\circ(m) < C_n^\star(m),\ m = 1, \ldots, M\}. \tag{5.50}$$

Note that the definition $C_n^\star(m) = 1$, if $P_{0n}^\circ(m+1) < C_n^\star(m+1)$, ensures that the procedure is indeed step-up, that is, as soon as one hypothesis is rejected, all remaining more significant hypotheses are rejected.

The adjusted p-values for the step-up minP procedure are as those for step-down minP Procedure 5.6, in Equation (5.39), but with $\min_{h=m,\ldots,M}$ replacing $\max_{h=1,\ldots,m}$. That is,

$$\widetilde{p}_{0n}(o_n(m)) = \min_{h=m,\ldots,M} \left\{ \Pr_{Q_0}\left(\min_{l \in \{o_n(h),\ldots,o_n(M)\}} P_0(l) \leq p_{0n}(o_n(h)) \right) \right\}. \tag{5.51}$$

Failure of FWER control for candidate step-up maxT and minP procedures

Next, we illustrate how FWER control can fail for the candidate step-up minP procedure. A similar argument could be applied to the candidate step-up maxT procedure.

Consider the complete null hypothesis $H_0^C = \prod_{m=1}^M H_0(m) = \mathrm{I}(P \in \cap_{m=1}^M \mathcal{M}(m))$, i.e., the case where all M null hypotheses are true and $\mathcal{H}_0 = \{1, \ldots, M\}$. For the sake of simplicity, further assume that the test statistics null distribution Q_0 is equal to the true distribution $Q_n = Q_n(P)$ and that $\Pr(P_{0n}^\circ(m) < C_n(m)) = \alpha$ for each $m = 1, \ldots, M$. Then, $V_n = R_n$ and, for each $m = 1, \ldots, M$,

$$\begin{aligned}
\Pr(V_n > 0) &= 1 - \Pr(R_n = 0) \qquad (5.52)\\
&= 1 - \Pr\left(\bigcap_{h=1}^M \{P_{0n}^\circ(h) \geq C_n^\star(h)\}\right)\\
&= 1 - \Pr\left(\bigcap_{h=1}^M \{P_{0n}^\circ(h) \geq C_n(h)\}\right)\\
&\geq 1 - \Pr\left(P_{0n}^\circ(m) \geq C_n(m)\right) \qquad (*)\\
&= \alpha.
\end{aligned}$$

If $(*)$ holds with equality, then the step-up minP procedure controls the FWER exactly at level α; otherwise, if the inequality in $(*)$ is strict, $\Pr(V_n > 0) > \alpha$ and FWER control fails.

Now let A_m denote the event $\{P_{0n}^\circ(m) \geq C_n(m)\}$, where, by assumption, $\Pr(A_m) = 1 - \alpha$ for each $m = 1, \ldots, M$. For simplicity, further consider the case of $M = 2$ null hypotheses. For monotone stepwise cut-offs, $C_n(1) \leq C_n(2)$, and ordered unadjusted p-values, $P_{0n}^\circ(1) \leq P_{0n}^\circ(2)$, the event $\{P_{0n}^\circ(1) \geq C_n(1)\}$ does not imply $\{P_{0n}^\circ(2) \geq C_n(2)\}$ and vice versa. In other words, there is no reason to expect either $A_1 \subseteq A_2$ or $A_2 \subseteq A_1$, that is, either $A_1 = A_1 \cap A_2$ or $A_2 = A_1 \cap A_2$. Thus, one expects $\Pr(A_1 \cap A_2) < \Pr(A_1)$ and $\Pr(A_1 \cap A_2) < \Pr(A_2)$. This argument extends to an arbitrary number of null hypotheses M and suggests that, in general, $\Pr(V_n > 0) > \alpha$, that is, the candidate step-up minP procedure fails to control the FWER.

In contrast, under the complete null hypothesis, proof of FWER control for step-down minP Procedure 5.6 only requires consideration of the minimum unadjusted p-value $P_{0n}^\circ(1)$ ($m = 1$ case), i.e., only requires that

$$\Pr(V_n > 0) = \Pr\left(P_{0n}^\circ(1) < C_n(1)\right) \leq \alpha.$$

In particular, FWER control is exact for test statistics with continuous distributions.

The previous argument shows that, except under pathological circumstances, the candidate step-up minP procedure cannot improve upon step-down minP Procedure 5.6 based on sharp joint common-quantile cut-offs $C_n(m)$ defined such that $\Pr(P_{0n}^\circ(m) < C_n(m)) = \alpha$.

5.4.2 Comparison of joint stepwise minP procedures to marginal stepwise Holm and Hochberg procedures

As detailed in Section 5.3.5, above, one can show by Boole's Inequality (Equation (B.1)) that the joint step-down/step-up minP cut-offs $C_n^{minP}(m)$ are bounded below by, i.e., are less conservative than, the cut-offs $C_n^{HH}(m)$ for marginal step-down Holm Procedure 3.7 and step-up Hochberg Procedure 3.13. That is,

$$C_n^{minP}(m) = F^{-1}_{\overline{\mathcal{O}}_n(m),Q_0}(\alpha) \geq \frac{1}{M-m+1}\alpha = C_n^{HH}(m). \tag{5.53}$$

Accordingly, the step-down and step-up minP adjusted p-values are bounded above by the step-down Holm and step-up Hochberg adjusted p-values, respectively. That is,

$$\begin{aligned}
\widetilde{p}_{0n}^{SDminP}(o_n(m)) &= \max_{h=1,\dots,m} \left\{ \Pr_{Q_0}\left(\min_{l\in\{o_n(h),\dots,o_n(M)\}} P_0(l) \leq p_{0n}(o_n(h)) \right) \right\} \\
&\leq \max_{h=1,\dots,m} \left\{ \min\left\{ (M-h+1)\, p_{0n}(o_n(h)), 1 \right\} \right\} \\
&= \widetilde{p}_{0n}^{SDHolm}(o_n(m))
\end{aligned} \tag{5.54}$$

and

$$\begin{aligned}
\widetilde{p}_{0n}^{SUminP}(o_n(m)) &= \min_{h=m,\dots,M} \left\{ \Pr_{Q_0}\left(\min_{l\in\{o_n(h),\dots,o_n(M)\}} P_0(l) \leq p_{0n}(o_n(h)) \right) \right\} \\
&\leq \min_{h=m,\dots,M} \left\{ \min\left\{ (M-h+1)\, p_{0n}(o_n(h)), 1 \right\} \right\} \\
&= \widetilde{p}_{0n}^{SUHoch}(o_n(m)).
\end{aligned} \tag{5.55}$$

5.5 FWER-controlling bootstrap-based step-down procedures

Step-down maxT Procedure 5.1 and minP Procedure 5.6 provide asymptotic control of the FWER, when based on either of the two main types of test statistics null distributions proposed in Chapter 2: the null shift and scale-transformed null distribution (Section 2.3) and the null quantile-transformed null distribution (Section 2.4). In practice, however, both test statistics null distributions $Q_0 = Q_0(P)$ are unknown, as they depend on the unknown data generating distribution P. Estimation of Q_0 is then needed, especially to deal with the unknown dependence structure of the test statistics. Sections 2.3.2, 2.4.2, 2.6, and 2.7 provide a variety of bootstrap procedures for obtaining consistent estimators Q_{0n} of the null distribution Q_0.

Section 5.5.1, below, establishes asymptotic FWER control results for Procedures 5.1 and 5.6 based on a consistent estimator Q_{0n} of the null distribution Q_0. Given a bootstrap estimator of the null distribution Q_0, Section 5.5.2 provides Procedure 5.15 for estimating cut-offs and adjusted p-values for Procedures 5.1 and 5.6.

5.5.1 Asymptotic control of FWER for step-down procedures based on consistent estimator of test statistics null distribution

In this section, we consider analogues of step-down Procedures 5.1 and 5.6, based on a consistent estimator Q_{0n} of the null distribution Q_0. In such multiple testing procedures, the estimator Q_{0n} is used in place of Q_0, to estimate the cut-offs for the test statistics and the corresponding adjusted p-values. Theorems 5.12 and 5.14 establish consistency of the step-down maxT and minP cut-offs for Procedures 5.1 and 5.6, respectively.

Theorem 5.12. [Consistency of step-down maxT cut-offs for Procedure 5.1]
Set-up and assumptions. *Consider an M-variate distribution Q, a random M-vector $Z = (Z(m) : m = 1, \ldots, M) \sim Q$, and a level $\alpha \in (0,1)$. Define $(1-\alpha)$-quantiles, $c(\mathcal{A}; Q, \alpha) \in I\!R$, for the distributions of maxima, $\max_{m\in\mathcal{A}} Z(m)$, of random variables $Z(m)$ over subsets $\mathcal{A} \subseteq \{1, \ldots, M\}$. That is, let*

$$c(\mathcal{A}; Q, \alpha) \equiv F_{\mathcal{A},Q}^{-1}(1-\alpha) = \inf\{z \in I\!R : F_{\mathcal{A},Q}(z) \geq 1-\alpha\}, \tag{5.56}$$

where $F_{\mathcal{A},Q}(z) \equiv \Pr_Q(\max_{m\in\mathcal{A}} Z(m) \leq z)$ denotes the CDF of $\max_{m\in\mathcal{A}} Z(m)$ for $Z \sim Q$.
Let Q_0 be a specified M-variate null distribution and let Q_{0n} converge weakly to Q_0. Assume that, for each subset $\mathcal{A} \subseteq \{1, \ldots, M\}$, $F_{\mathcal{A},Q_0}$ is continuous and has Lebesgue density $f_{\mathcal{A},Q_0}$ with interval support, that is, $\{z : f_{\mathcal{A},Q_0}(z) > 0\} = (a(\mathcal{A}), b(\mathcal{A}))$, where $a(\mathcal{A})$ and $b(\mathcal{A})$ are allowed to equal $-\infty$ and $+\infty$, respectively.
Result. *Then, one has the following consistency result for the step-down maxT cut-offs. Given $(P_n : n \geq 1)$,*

$$\lim_{n\to\infty} c(\mathcal{A}; Q_{0n}, \alpha) = c(\mathcal{A}; Q_0, \alpha), \qquad \forall\ \mathcal{A} \subseteq \{1, \ldots, M\}. \tag{5.57}$$

Proof of Theorem 5.12. Consider random M-vectors $Z_n \sim Q_{0n}$ and $Z \sim Q_0$. By assumption, $(Z_n(m) : m \in \mathcal{A})$ converges weakly to $(Z(m) : m \in \mathcal{A})$, $\forall\ \mathcal{A} \subseteq \{1, \ldots, M\}$. Hence, by the Continuous Mapping Theorem (Theorem B.3), $\max_{m\in\mathcal{A}} Z_n(m)$ converges weakly to $\max_{m\in\mathcal{A}} Z(m)$, so that $F_{\mathcal{A},Q_{0n}}$ converges pointwise to $F_{\mathcal{A},Q_0}$ at each continuity point of $F_{\mathcal{A},Q_0}$. Because pointwise convergence of monotone functions to a continuous monotone function implies uniform convergence, $F_{\mathcal{A},Q_{0n}}$ converges uniformly to $F_{\mathcal{A},Q_0}$. By continuity of the quantile mapping $F \to F^{-1}(1-\alpha)$, with respect to the supremum norm convergence at $F_{\mathcal{A},Q_0}$, with $f_{\mathcal{A},Q_0}(F_{\mathcal{A},Q_0}^{-1}(1-\alpha)) > 0$ (by assumption, $F_{\mathcal{A},Q_0}^{-1}(1-\alpha) \in (a(\mathcal{A}), b(\mathcal{A}))$), then $F_{\mathcal{A},Q_{0n}}^{-1}(1-\alpha) = c(\mathcal{A}; Q_{0n}, \alpha)$ converges to $F_{\mathcal{A},Q_0}^{-1}(1-\alpha) = c(\mathcal{A}; Q_0, \alpha)$, as $n \to \infty$, $\forall\ \mathcal{A} \subseteq \{1, \ldots, M\}$.

□

Note that the above proof corresponds to the proof of Theorem 4.17, for consistency of the single-step common cut-offs for Procedure 4.2, with the following modifications: (i) Θ is the FWER-specific mapping, $\Theta(F) = 1 - F(0)$; (ii) the number of rejected hypotheses R is computed over subsets $\mathcal{A} \subseteq \{1, \ldots, M\}$, rather than over the entire set $\{1, \ldots, M\}$ of null hypotheses, that is, one considers $\mathcal{A}$-specific numbers of rejected hypotheses, $R_{\mathcal{A}}(\gamma^{(M)}|Q) \equiv \sum_{m \in \mathcal{A}} \mathrm{I}(Z(m) > \gamma)$, for $Z \sim Q$. One can then define $\mathcal{A}$-specific functions, $\gamma \to \theta_{\mathcal{A},Q}(\gamma) \equiv \Theta\left(F_{R_{\mathcal{A}}(\gamma^{(M)}|Q)}\right)$, and note that $\theta_{\mathcal{A},Q}(\gamma) = 1 - F_{R_{\mathcal{A}}(\gamma^{(M)}|Q)}(0) = 1 - F_{\mathcal{A},Q}(\gamma)$. Thus, assumptions regarding θ_{Q_0}, in single-step Theorem 4.17, translate into assumptions on the CDF $F_{\mathcal{A},Q_0}$ of $\max_{m \in \mathcal{A}} Z(m)$, in step-down Theorem 5.12, above.

Consistency of the step-down minP cut-offs for Procedure 5.6 follows from Theorem 4.10, on consistency of the single-step common-quantile cut-offs for Procedure 4.1, with modifications (i) and (ii), above. A general consistency result for $\mathcal{A}$-specific common quantiles is stated below, for an arbitrary Type I error rate mapping Θ. The proof is identical to that of Theorem 4.10, with $\mathcal{A}$-specific numbers of rejected hypotheses $R_{\mathcal{A}}$, and is therefore omitted.

Theorem 5.13. [Consistency of $\mathcal{A}$-specific common quantiles]
Set-up and assumptions. *Given an M-variate distribution Q, a subset $\mathcal{A} \subseteq \{1, \ldots, M\}$, and a constant $\delta \in [0,1]$, define an $\mathcal{A}$-specific vector $q^{-1}(\mathcal{A}; \delta) \equiv (Q_m^{-1}(\delta) : m \in \mathcal{A})$ of δ-quantiles for the marginal distributions Q_m by*

$$Q_m^{-1}(\delta) \equiv \inf \{z \in I\!R : Q_m(z) \geq \delta\}, \qquad m = 1, \ldots, M. \tag{5.58}$$

Consider a Type I error rate mapping Θ that satisfies monotonicity Assumption M$\Theta$and define non-increasing functions

$$\delta \to \theta_{\mathcal{A},Q}(\delta) \equiv \Theta\left(F_{R_{\mathcal{A}}(q^{-1}(\mathcal{A};\delta)|Q)}\right), \tag{5.59}$$

where

$$R_{\mathcal{A}}(q^{-1}(\mathcal{A}; \delta)|Q) \equiv \sum_{m \in \mathcal{A}} \mathrm{I}\left(Z(m) > Q_m^{-1}(\delta)\right) \tag{5.60}$$

denotes the $\mathcal{A}$-specific number of rejected hypotheses for an $\mathcal{A}$-specific vector $q^{-1}(\mathcal{A}; \delta)$ of δ-quantiles and for $Z \sim Q$. For a fixed Type I error level $\alpha \in (0,1)$ and any subset $\mathcal{A} \subseteq \{1, \ldots, M\}$, define α-quantiles of $\theta_{\mathcal{A},Q}$ by

$$\delta_{\mathcal{A},Q}(\alpha) \equiv \theta_{\mathcal{A},Q}^{-1}(\alpha) = \inf\left\{\delta \in [0,1] : \Theta\left(F_{R_{\mathcal{A}}(q^{-1}(\mathcal{A};\delta)|Q)}\right) \leq \alpha\right\}. \tag{5.61}$$

Let Q_0 be a specified M-variate null distribution and let Q_{0n} converge weakly to Q_0. Assume that: (i) Q_0 is absolutely continuous with respect to the Lebesgue measure on $I\!R^M$, with uniformly bounded density; (ii) each marginal distribution $Q_{0,m}$ has continuous Lebesgue density $f_{0,m}$, with interval support, that is, $\{z : f_{0,m}(z) > 0\} = (a(m), b(m))$, where $a(m)$ and $b(m)$ are allowed to equal $-\infty$ and $+\infty$, respectively; (iii) $\delta_{\mathcal{A},Q_0}(\alpha) \in (0,1)$; (iv) the function $\theta_{\mathcal{A},Q_0}(\delta)$ is continuous and has a positive derivative at $\delta_{\mathcal{A},Q_0}(\alpha)$; (v) the Type I error rate mapping Θ satisfies continuity Assumption CΘat $F_{R_{\mathcal{A}}(q_0^{-1}(\mathcal{A};\delta)|Q_0)}$ for $\delta \in (0,1)$.

Result. *Then, one has the following consistency results for the $\mathcal{A}$-specific quantiles of Q_0 and Q_{0n}. Given $(P_n : n \geq 1)$, for each $\mathcal{A} \subseteq \{1,\ldots,M\}$,*

$$\lim_{n\to\infty} (\delta_{\mathcal{A},Q_{0n}}(\alpha) - \delta_{\mathcal{A},Q_0}(\alpha)) = 0 \tag{5.62}$$

and

$$\lim_{n\to\infty} (Q_{0n,m}^{-1}(\delta_{\mathcal{A},Q_{0n}}(\alpha)) - Q_{0,m}^{-1}(\delta_{\mathcal{A},Q_0}(\alpha))) = 0, \qquad \forall\, m = 1,\ldots,M.$$

Theorem 5.14, below, establishes consistency of the step-down minP cut-offs, by noting that these cut-offs are equal to the quantiles of the functions $\theta_{\mathcal{A},Q}$ in Theorem 5.13, that is, $c(\mathcal{A}; Q, \alpha) = \delta_{\mathcal{A},Q}(\alpha)$.

Theorem 5.14. [Consistency of step-down minP cut-offs for Procedure 5.6]
Set-up and assumptions. *Consider an M-variate distribution Q, a random M-vector $Z = (Z(m) : m = 1,\ldots,M) \sim Q$, and a level $\alpha \in (0,1)$. Define α-quantiles, $c(\mathcal{A}; Q, \alpha) \in [0,1]$, for the distributions of minima, $\min_{m\in\mathcal{A}} P(m)$, of unadjusted p-values $P(m) \equiv \bar{Q}_m(Z(m))$ over subsets $\mathcal{A} \subseteq \{1,\ldots,M\}$. That is, let*

$$c(\mathcal{A}; Q, \alpha) \equiv F_{\mathcal{A},Q}^{-1}(\alpha) = \inf\{z \in [0,1] : F_{\mathcal{A},Q}(z) \geq \alpha\}, \tag{5.63}$$

where $F_{\mathcal{A},Q}(z) \equiv \Pr_Q(\min_{m\in\mathcal{A}} P(m) \leq z)$ denotes the CDF of $\min_{m\in\mathcal{A}} P(m)$ for $Z \sim Q$.
As in Theorem 5.13, define: $\mathcal{A}$-specific vectors $q^{-1}(\mathcal{A};\delta) \equiv (Q_m^{-1}(\delta) : m \in \mathcal{A})$ of δ-quantiles $Q_m^{-1}(\delta)$ for the marginal distributions Q_m; $\mathcal{A}$-specific numbers of rejected hypotheses $R_{\mathcal{A}}(q^{-1}(\mathcal{A};\delta)|Q) \equiv \sum_{m\in\mathcal{A}} \mathrm{I}\left(Z(m) > Q_m^{-1}(\delta)\right)$ for $Z \sim Q$; non-increasing functions $\delta \to \theta_{\mathcal{A},Q}(\delta) \equiv \Theta\left(F_{R_{\mathcal{A}}(q^{-1}(\mathcal{A};\delta)|Q)}\right)$; α-quantiles $\delta_{\mathcal{A},Q}(\alpha) \equiv \theta_{\mathcal{A},Q}^{-1}(\alpha)$.
Let Q_0 be an M-variate null distribution, satisfying Assumptions (i)–(v) in Theorem 5.13, and let Q_{0n} converge weakly to Q_0.
Result. *For an M-variate distribution Q, with continuous and strictly increasing marginal CDFs Q_m, and for the FWER-specific mapping $\Theta(F) = 1 - F(0)$, one has*

$$\theta_{\mathcal{A},Q}(\delta) = \Theta\left(F_{R_{\mathcal{A}}(q^{-1}(\mathcal{A};\delta)|Q)}\right) = F_{\mathcal{A},Q}(1-\delta). \tag{5.64}$$

Hence, the step-down minP cut-offs are equal to the α-quantiles $\delta_{\mathcal{A},Q}(\alpha)$ of Theorem 5.13, that is,

$$c(\mathcal{A}; Q, \alpha) = \delta_{\mathcal{A},Q}(\alpha), \qquad \forall\, \mathcal{A} \subseteq \{1,\ldots,M\}. \tag{5.65}$$

Consequently, by Theorem 5.13, one has the following consistency result for the step-down minP cut-offs. Given $(P_n : n \geq 1)$,

$$\lim_{n\to\infty} c(\mathcal{A}; Q_{0n}, \alpha) = c(\mathcal{A}; Q_0, \alpha), \qquad \forall\, \mathcal{A} \subseteq \{1,\ldots,M\}. \tag{5.66}$$

Proof of Theorem 5.14. For an M-variate distribution Q, with continuous and strictly increasing marginal CDFs Q_m,

$$\begin{aligned}
\theta_{\mathcal{A},Q}(\delta) &= 1 - F_{R_{\mathcal{A}}(q^{-1}(\mathcal{A};\delta)|Q)}(0) \\
&= \Pr\left(R_{\mathcal{A}}(q^{-1}(\mathcal{A};\delta)|Q) > 0\right) \\
&= \Pr_Q\left(\sum_{m\in\mathcal{A}} \mathrm{I}\left(Z(m) > Q_m^{-1}(\delta)\right) > 0\right) \\
&= \Pr_Q\left(\exists\ m \in \mathcal{A} \text{ such that } Z(m) > Q_m^{-1}(\delta)\right) \\
&= \Pr_Q\left(\exists\ m \in \mathcal{A} \text{ such that } \bar{Q}_m(Z(m)) \leq \bar{Q}_m(Q_m^{-1}(\delta))\right) \\
&= \Pr_Q\left(\exists\ m \in \mathcal{A} \text{ such that } \bar{Q}_m(Z(m)) \leq 1-\delta\right) \\
&= \Pr_Q\left(\min_{m\in\mathcal{A}} P(m) \leq 1-\delta\right) \\
&= F_{\mathcal{A},Q}(1-\delta),
\end{aligned}$$

where $F_{\mathcal{A},Q}$ denotes the CDF of $\min_{m\in\mathcal{A}} P(m)$ for $P(m) = \bar{Q}_m(Z(m))$ and $Z \sim Q$. Hence, for each $\mathcal{A} \subseteq \{1,\ldots,M\}$,

$$\begin{aligned}
c(\mathcal{A};Q,\alpha) &= \inf\left\{z \in [0,1] : F_{\mathcal{A},Q}(z) \geq \alpha\right\} \\
&= \sup\left\{z \in [0,1] : F_{\mathcal{A},Q}(z) \leq \alpha\right\} \\
&= \sup\left\{z \in [0,1] : \theta_{\mathcal{A},Q}(1-z) \leq \alpha\right\} \\
&= \inf\left\{\delta \in [0,1] : \theta_{\mathcal{A},Q}(\delta) \leq \alpha\right\} \\
&= \delta_{\mathcal{A},Q}(\alpha).
\end{aligned}$$

Convergence of the estimated step-down minP cut-offs $c(\mathcal{A};Q_{0n},\alpha)$ to $c(\mathcal{A};Q_0,\alpha)$ is then a direct consequence of the convergence of $\delta_{\mathcal{A},Q_{0n}}(\alpha)$ to $\delta_{\mathcal{A},Q_0}(\alpha)$, as established in Theorem 5.13.

□

Having derived consistency of the cut-offs for step-down Procedures 5.1 and 5.6, based on a consistent estimator Q_{0n} of the null distribution Q_0, Corollary 4.18 can be applied to prove consistency of the resulting Type I error rates.

Note that, for a broad class of testing problems, the null distribution $Q_0 = Q_0(P)$ has continuous and strictly monotone marginal distributions. For example, for the test of single-parameter null hypotheses using t-statistics, the null shift and scale-transformed null distribution of Section 2.3 is an M-variate Gaussian distribution with mean vector zero (Section 2.6). In such cases, consistent estimators Q_{0n} can also be defined in terms of Gaussian distributions, with a suitable estimator of the covariance matrix. The assumptions of Theorem 5.14 are therefore satisfied by both Q_0 and Q_{0n}. In the case when Q_{0n} is not continuous (e.g., obtained from general bootstrap Procedure 2.3), but converges weakly to a continuous Q_0, Theorem 5.14 strongly suggests asymptotic equality of $c(\mathcal{A};Q_{0n},\alpha)$ and $\delta_{\mathcal{A},Q_{0n}}(\alpha)$.

5.5.2 Bootstrap-based step-down procedures

As detailed in Sections 2.3.2, 2.4.2, 2.6, and 2.7, one may use a variety of bootstrap procedures to obtain consistent estimators Q_{0n} of the proposed null shift and scale-transformed and null quantile-transformed test statistics null distributions $Q_0 = Q_0(P)$.

For instance, for the null distribution of Section 2.3, based on user-supplied null shift and scale values $\lambda_0(m)$ and $\tau_0(m)$, one may apply Procedure 2.3 to generate an $M \times B$ matrix $\mathbf{Z}_n^B = \left(Z_n^B(m,b)\right)$ of null shift and scale-transformed bootstrap test statistics $Z_n^B(m,b)$. For the null distribution of Section 2.4, based on user-supplied marginal null distributions $q_{0,m}$, one may apply Procedure 2.4 to generate an $M \times B$ matrix $\mathbf{Z}_n^B = \left(Z_n^B(m,b)\right)$ of null quantile-transformed bootstrap test statistics $Z_n^B(m,b)$. In either case, the bootstrap estimator Q_{0n} of the null distribution Q_0 is the empirical distribution of the B columns $\{Z_n^B(\cdot,b) : b = 1,\ldots,B\}$ of matrix $\mathbf{Z}_n^B$.

Given such a bootstrap estimator Q_{0n} of the null distribution Q_0, Procedure 5.15 may be applied to estimate cut-offs and adjusted p-values for FWER-controlling step-down maxT Procedure 5.1 and minP Procedure 5.6.

Procedure 5.15. [Bootstrap estimation of cut-offs and adjusted p-values for step-down maxT Procedure 5.1 and minP Procedure 5.6]

- Apply Procedure 2.3 or 2.4 to generate an $M \times B$ matrix, $\mathbf{Z}_n^B = \left(Z_n^B(m,b) : m = 1,\ldots,M;\ b = 1,\ldots,B\right)$, of null-transformed bootstrap test statistics $Z_n^B(m,b)$. The bootstrap estimator Q_{0n} of the null distribution Q_0 is the empirical distribution of the B columns $\{Z_n^B(\cdot,b) : b = 1,\ldots,B\}$ of matrix $\mathbf{Z}_n^B$.
- **Step-down maxT Procedure 5.1.** Compute maxima $\max_{m\in\mathcal{A}} Z_n^B(m,b)$ of bootstrap test statistics $Z_n^B(m,b)$ over rows corresponding to subsets of null hypotheses $\mathcal{A}$. The bootstrap estimators of the $(1-\alpha)$-quantiles $c(\mathcal{A}; Q_0, \alpha)$ are the empirical $(1-\alpha)$-quantiles of the B maxima $\{\max_{m\in\mathcal{A}} Z_n^B(m,b) : b = 1,\ldots,B\}$. That is,

$$c(\mathcal{A}; Q_{0n}, \alpha) \equiv \inf\left\{z \in I\!R : \frac{1}{B}\sum_{b=1}^{B} \mathrm{I}\left(\max_{m\in\mathcal{A}} Z_n^B(m,b) \leq z\right) \geq 1-\alpha\right\}. \tag{5.67}$$

The bootstrap estimated step-down maxT adjusted p-values are given by

$$\begin{aligned}\widetilde{p}_{0n}(o_n(m)) &= \max_{h=1,\ldots,m} \left\{ \Pr_{Q_{0n}} \left(\max_{l \in \{o_n(h),\ldots,o_n(M)\}} Z(l) \geq t_n(o_n(h)) \right) \right\} \\ &= \max_{h=1,\ldots,m} \left\{ \frac{1}{B} \sum_{b=1}^{B} \mathrm{I} \left(\max_{l \in \{o_n(h),\ldots,o_n(M)\}} Z_n^B(l,b) \geq t_n(o_n(h)) \right) \right\}.\end{aligned} \tag{5.68}$$

- **Step-down minP Procedure 5.6.** One must first estimate unadjusted p-values using Q_{0n}, before considering the distribution of their successive minima. Estimated unadjusted p-values are obtained by row-ranking the matrix $\mathbf{Z}_n^B$ and are given by

$$p_{0n}(m) = \Pr_{Q_{0n}} \left(Z(m) \geq t_n(m) \right) = \frac{1}{B} \sum_{b=1}^{B} \mathrm{I} \left(Z_n^B(m,b) \geq t_n(m) \right). \tag{5.69}$$

The reader is referred to Ge et al. (2003) for a fast algorithm for computing resampling-based (bootstrap or permutation) adjusted p-values for step-down minP procedures.

6

Augmentation Multiple Testing Procedures for Controlling Generalized Tail Probability Error Rates

6.1 Introduction

6.1.1 Motivation

This chapter is concerned with *augmentation multiple testing procedures* (AMTP), that is, multiple testing procedures (MTP) obtained by adding suitably chosen null hypotheses to the set of hypotheses already rejected by an initial MTP. Specifically, suppose one has available a multiple testing procedure $\mathcal{R}_n(\alpha)$, that controls a Type I error rate $\Theta(F_{V_n,R_n})$ at level α. Our goal is to derive a random (i.e., data-dependent) subset $\mathcal{A}_n(\alpha) \subseteq \mathcal{R}_n^c(\alpha)$, so that the new MTP defined by the set of rejected hypotheses $\mathcal{R}_n^+(\alpha) = \mathcal{R}_n(\alpha) \cup \mathcal{A}_n(\alpha)$ controls a new Type I error rate $\Theta^+(F_{V_n^+,R_n^+})$ at level α.

After a general introduction to AMTPs, we focus on procedures for controlling two main classes of Type I error rates: tail probabilities for the number of false positives and for the proportion of false positives among the rejected hypotheses. The generalized family-wise error rate (gFWER) is a relaxed version of the family-wise error rate (FWER), that allows $k \geq 0$ Type I errors, that is, $gFWER(k)$ is defined as the chance of at least $(k+1)$ Type I errors, $FWER(k) = \Pr(V_n > k)$. The usual FWER corresponds to the special case $k = 0$. In contrast, the tail probability for the proportion of false positives (TPPFP) allows a user-supplied proportion $q \in (0,1)$ of Type I errors among the rejected hypotheses, i.e., $TPPFP(q) = \Pr(V_n/R_n > q)$. The popular false discovery rate (FDR) is the expected value of the proportion of Type I errors, i.e., $FDR = \mathrm{E}[V_n/R_n]$. As discussed in Sections 1.2.9, 1.2.11, and 3.4.1, error rates based on the proportion of false positives (e.g., TPPFP, FDR) are especially appealing for large-scale testing problems, compared to error rates based on the number of false positives (e.g., gFWER), as they remain stable with an increasing number of tested hypotheses M.

The few currently available gFWER-controlling procedures are summarized in Section 3.3 and include the following main approaches: (i) marginal single-step Bonferroni-like Procedure 3.15 and step-down Holm-like Procedure 3.17 (Lehmann and Romano, 2005); (ii) joint single-step common-cut-off $T(k+1)$ Procedure 3.18 and common-quantile $P(k+1)$ Procedure 3.19, discussed in detail in Chapter 4 (Dudoit et al., 2004b; Pollard and van der Laan, 2004); (iii) general (marginal/joint single-step/stepwise) augmentation multiple testing Procedure 3.20, discussed in detail in the present chapter (van der Laan et al., 2004b); (iv) joint resampling-based empirical Bayes Procedure 7.1, discussed in detail in Chapter 7 (van der Laan et al., 2005).

Existing TPPFP-controlling procedures are introduced in Section 3.5 and include: (i) marginal step-down Procedures 3.24 and 3.25 (Lehmann and Romano, 2005); (ii) the marginal inversion method for independent test statistics and its conservative version for test statistics with general dependence structures (Genovese and Wasserman, 2004a,b); (iii) general (marginal/joint single-step/stepwise) augmentation multiple testing Procedure 3.26, discussed in detail in the present chapter (van der Laan et al., 2004b); (iv) joint resampling-based empirical Bayes Procedure 7.1, discussed in detail in Chapter 7 (van der Laan et al., 2005).

Most multiple testing procedures proposed thus far for controlling a parameter (e.g., mean for FDR, survivor function for TPPFP) of the distribution of the proportion V_n/R_n of false positives among the rejected hypotheses suffer from one or both of the following limitations: they are based solely on the marginal distributions of the test statistics and they rely on a number of assumptions concerning the joint distribution of the test statistics (e.g., independence, positive regression dependence, ergodic dependence, or normality). The first limitation typically translates into low power, whereas the second can result in Type I error control failure.

The present chapter, following van der Laan et al. (2004b), shows that any procedure controlling the FWER can be straightforwardly augmented to control the gFWER and TPPFP, for general data generating distributions and, hence, arbitrary dependence structures among the test statistics (Sections 6.2 and 6.3). In addition, by choosing a suitable initial FWER-controlling MTP (e.g., single-step maxT Procedure 3.5 and minP Procedure 3.6; step-down maxT Procedure 3.11 and minP Procedure 3.12), such gFWER- and TPPFP-controlling procedures can take into account the joint distribution of the test statistics and are therefore expected to be less conservative than marginal procedures such as those of Lehmann and Romano (2005). We also demonstrate how two (conservative) FDR-controlling procedures can be obtained from a TPPFP-controlling MTP (Section 6.4).

Furthermore, as detailed in Section 6.5, our previous results on gFWER- and TPPFP-controlling augmentation procedures can be extended to a broad class of Type I error rates, defined as generalized tail probability (gTP) error rates, $gTP(q,g) = \Pr(g(V_n, R_n) > q)$, for arbitrary functions $g(V_n, R_n)$ of the numbers of false positives V_n and rejected hypotheses R_n. Adjusted p-values for the AMTP are shown to be simply shifted versions of the adjusted p-values of the initial MTP. The gFWER and TPPFP correspond, respectively, to the special cases $g(v, r) = v$ and $g(v, r) = v/r$.

The important practical implication of the results presented in this chapter is that *any* FWER-controlling (marginal/joint single-step/stepwise) MTP provides immediately and trivially multiple testing procedures controlling a broad class of Type I error rates, such as the gFWER and TPPFP. One can therefore build on the large pool of available FWER-controlling procedures, such as the single-step and step-down maxT and minP procedures introduced in Chapter 3 and discussed in detail in Chapters 4 and 5. In addition, as shown in Theorem 6.10, augmentation procedures based on an asymptotically exact gFWER-controlling MTP (e.g., joint step-down maxT Procedure 3.11 and minP Procedure 3.12), provide exact asymptotic control of generalized tail probability error rates. Thus, although AMTPs may lack power in finite sample situations and at local alternative hypotheses (e.g., compared to the empirical Bayes procedures discussed in Chapter 7), they can be very powerful asymptotically at fixed alternatives. Finally, we wish to stress the generality of our proposed augmentation approach to multiple testing: the AMTPs apply to general data generating distributions (with arbitrary dependence structures among variables), null hypotheses (defined as submodels for the data generating distribution), and test statistics (e.g., t-statistics, χ^2-statistics, F-statistics).

6.1.2 Outline

The present chapter focuses on augmentation multiple testing procedures, that is, procedures obtained by adding suitably chosen null hypotheses to the set of hypotheses already rejected by an initial MTP. The augmentation approach to multiple testing is introduced in Section 6.1.4, after a brief review of Type I error rates in Section 6.1.3. Sections 6.2 and 6.3 show, respectively, how one can straightforwardly augment any FWER-controlling (marginal/joint single-step/stepwise) procedure to control the gFWER (Procedure 6.2) and the TPPFP (Procedure 6.4). Theorems 6.3 and 6.5 establish finite sample and exact asymptotic control results for these two types of augmentation procedures, under general data generating distributions, with arbitrary dependence structures among variables. Control of the false discovery rate, based on a TPPFP-controlling procedure, is discussed in Section 6.4. Given an initial gFWER-controlling procedure, Section 6.5 proposes augmentation procedures for controlling a broad class of Type I error rates, defined as generalized tail probability (gTP) error rates, $gTP(q, g) = \Pr(g(V_n, R_n) > q)$, for arbitrary functions $g(V_n, R_n)$ of the numbers of false positives V_n and rejected hypotheses R_n (Procedure 6.9). Theorem 6.10 shows that augmentation procedures inherit the Type I error control properties of the initial MTP, i.e., finite sample/asymptotic control of the gFWER translates into finite sample/asymptotic control of the gTP. The asymptotic control results are relevant, because in many situations one can only count on asymptotic control of the gFWER for the initial MTP. Theorem 6.11 proves that the adjusted *p*-values for an AMTP are simply shifted versions of the adjusted *p*-values for the initial MTP. Control of generalized expected value (gEV) error rates, $gEV(g) = \mathrm{E}[g(V_n, R_n)]$, based on a gTP-controlling procedure, is discussed in Section 6.6. Section 6.7 addresses the choice of an initial gFWER-controlling MTP. Finally, Section 6.8 summarizes our findings and discusses ongoing work.

We follow the general multiple testing framework described in Section 1.2 and refer the reader to this section for basic definitions and notation. The key choice of a test statistics null distribution is treated in Chapter 2. Sections 3.2, 3.3, 3.4, and 3.5 provide summaries of available MTPs for controlling the FWER, gFWER, FDR, and TPPFP, respectively. Detailed discussions of FWER- and gFWER-controlling joint single-step and step-down procedures are given in Chapters 4 and 5.

The multiple testing procedures proposed in the present chapter are summarized in the flowchart of Figure 6.1 and in Appendix A. gFWER- and TPPFP-controlling augmentation procedures are compared to other recently proposed MTPs in the simulation studies of Dudoit et al. (2004a) and van der Laan et al. (2005). The augmentation approach is applied in Chapters 9 and 12 to address various multiple testing problems in biomedical and genomic research.

6.1.3 Type I error rates

Following the general framework of Chapter 1, define Type I error rates as parameters $\theta_n = \Theta(F_{V_n,R_n})$ of the joint distribution F_{V_n,R_n} of the numbers of Type I errors V_n and rejected hypotheses R_n.

The present chapter considers a broad class of Type I error rates, defined as tail probabilities (i.e., survivor functions) and expected values for arbitrary functions $g(V_n, R_n)$ of the numbers of Type I errors V_n and rejected hypotheses R_n. Specifically, for a given function g and constant q, the *generalized tail probability* (gTP) error rate is defined as

$$gTP(q, g) \equiv \Pr(g(V_n, R_n) > q) \tag{6.1}$$

and the *generalized expected value* (gEV) error rate as

$$gEV(g) \equiv \mathrm{E}[g(V_n, R_n)]. \tag{6.2}$$

Sections 6.2 and 6.3 focus, respectively, on procedures that control tail probabilities for the *number* V_n of false positives and the *proportion* V_n/R_n of false positives among the rejected hypotheses, i.e., on the special cases $g(v, r) = v$ and $g(v, r) = v/r$, respectively.

Specifically, the *generalized family-wise error rate* (gFWER), for a user-supplied integer k, $k \in \{0, \ldots, M\}$, is the probability of at least $(k + 1)$ Type I errors. That is,

$$gFWER(k) \equiv \Pr(V_n > k) = 1 - F_{V_n}(k), \tag{6.3}$$

where F_{V_n} is the discrete cumulative distribution function on $\{0, \ldots, M\}$ for the number of Type I errors V_n. When $k = 0$, the gFWER is the usual *family-wise error rate* (FWER), or probability of at least one Type I error,

$$FWER \equiv \Pr(V_n > 0) = 1 - F_{V_n}(0). \tag{6.4}$$

The *tail probability for the proportion of false positives* (TPPFP) among the rejected hypotheses, for a user-supplied constant $q \in (0, 1)$, is defined as

$$TPPFP(q) \equiv \Pr\left(\frac{V_n}{R_n} > q\right) = 1 - F_{V_n/R_n}(q), \tag{6.5}$$

where F_{V_n/R_n} is the CDF for the proportion V_n/R_n of Type I errors among the rejected hypotheses, with the convention that $V_n/R_n \equiv 0$ if $R_n = 0$. The *false discovery rate* (FDR) is defined as the expected value of the proportion of Type I errors among the rejected hypotheses,

$$FDR \equiv \mathrm{E}\left[\frac{V_n}{R_n}\right] = \int q dF_{V_n/R_n}(q), \tag{6.6}$$

again with the convention that $V_n/R_n \equiv 0$ if $R_n = 0$ (Benjamini and Hochberg, 1995).

Section 6.5, which provides augmentation procedures for controlling generalized tail probability error rates $gTP(q, g)$, considers a modified version of the TPPFP, corresponding to the function $g(v, r) = \mathrm{I}(k_0/r \leq q)\, v/r$. Specifically, given an integer $k_0 \geq 0$, the *generalized tail probability for the proportion of false positives* (gTPPFP) is defined as

$$gTPPFP(k_0, q) \equiv \Pr\left(\mathrm{I}\left(\frac{k_0}{R_n} \leq q\right) \frac{V_n}{R_n} > q\right). \tag{6.7}$$

Note that while the gFWER is a parameter of only the *marginal* distribution of the number of Type I errors V_n, the TPPFP, gTPPFP, and FDR are parameters of the *joint* distribution of (V_n, R_n). We use the shorter phrase *proportion of false positives* (PFP) to refer to the proportion V_n/R_n of false positives *among the R_n rejected hypotheses*, rather than among the M null hypotheses. Controlling the latter proportion would reduce to controlling the number of false positives, i.e., error rates of the form $\Theta(F_{V_n})$, such as the gFWER.

6.1.4 Augmentation multiple testing procedures

Suppose one has available a (marginal/joint single-step/stepwise) multiple testing procedure, such as Procedure 6.1, below, that controls a Type I error rate $\Theta(F_{V_n,R_n})$ at level α. The reader is referred to other chapters and articles for details on the choice of test statistics T_n and corresponding null distribution Q_0 and for specific proposals of gTP-controlling procedures (Chapters 1–5 and 7; Dudoit et al. (2004a,b); van der Laan et al. (2004a,b, 2005); van der Laan and Hubbard (2006); Pollard et al. (2005a); Pollard and van der Laan (2004)). Accordingly, the test statistics T_n and their null distribution Q_0 (or estimator thereof, Q_{0n}) are viewed as given, and the set of rejected hypotheses $\mathcal{R}(T_n, Q_0, \alpha)$ for the initial Θ-controlling MTP is denoted simply by $\mathcal{R}_n(\alpha)$ or $\mathcal{R}_n$, to only emphasize the dependence on the nominal Type I error level α.

Procedure 6.1. [Initial Θ-controlling multiple testing procedure] Suppose one has available a (marginal/joint single-step/stepwise) multiple testing procedure $\mathcal{R}_n = \mathcal{R}_n(\alpha)$, that provides finite sample control of a Type I error rate $\Theta(F_{V_n,R_n})$ at level $\alpha_n \in (0,1)$ and asymptotic control of this error rate at level $\alpha \in (0,1)$. That is, the *initial Θ-controlling multiple testing procedure* is such that

$$\Theta(F_{V_n,R_n}) = \alpha_n, \qquad \forall\ n, \tag{6.8}$$

and

$$\limsup_{n\to\infty} \Theta(F_{V_n,R_n}) = \limsup_{n\to\infty} \alpha_n = \alpha^\star \leq \alpha, \tag{6.9}$$

where $R_n = R_n(\alpha) \equiv |\mathcal{R}_n(\alpha)|$ and $V_n = V_n(\alpha) \equiv |\mathcal{R}_n(\alpha) \cap \mathcal{H}_0|$ denote, respectively, the number of rejected hypotheses and the number of Type I errors for procedure $\mathcal{R}_n(\alpha)$.

The subject of this chapter is the construction of a MTP that controls a new target Type I error rate $\Theta^+(F_{V_n^+,R_n^+})$, by augmenting the set of rejected hypotheses of the initial Θ-controlling procedure $\mathcal{R}_n$. Specifically, a Θ^+-*controlling augmentation multiple testing procedure* (AMTP), $\mathcal{R}_n^+ = \mathcal{R}_n^+(\alpha)$, is defined by forming the union of the set of rejected hypotheses $\mathcal{R}_n(\alpha)$ for the initial Θ-controlling procedure with an *augmentation set* $\mathcal{A}_n(\alpha) \subseteq \{1,\ldots,M\}$,

$$\mathcal{R}_n^+(\alpha) \equiv \mathcal{R}_n(\alpha) \cup \mathcal{A}_n(\alpha). \tag{6.10}$$

The augmentation set $\mathcal{A}_n(\alpha)$ is a random subset, i.e., a deterministic function of the data $\mathcal{X}_n = \{X_i : i = 1,\ldots,n\}$, and satisfies $\Pr(\mathcal{A}_n(\alpha) \subseteq \mathcal{R}_n^c(\alpha)) = 1$, where $\mathcal{R}_n^c(\alpha) = \{1,\ldots,M\}/\mathcal{R}_n(\alpha)$ is the set of non-rejected null hypotheses. Thus, the new MTP $\mathcal{R}_n^+(\alpha)$ identifies $A_n(\alpha) = |\mathcal{A}_n(\alpha)|$ additional rejections among the null hypotheses that are not rejected by $\mathcal{R}_n(\alpha)$. Define

$$R_n^+ = R_n^+(\alpha) \equiv |\mathcal{R}_n^+(\alpha)| = R_n(\alpha) + A_n(\alpha), \tag{6.11}$$

as the number of rejected hypotheses, and

$$V_n^+ = V_n^+(\alpha) \equiv |\mathcal{R}_n^+(\alpha) \cap \mathcal{H}_0|, \tag{6.12}$$

as the number of Type I errors for the augmentation multiple testing procedure $\mathcal{R}_n^+(\alpha)$. One seeks an augmentation set $\mathcal{A}_n(\alpha)$ so that the new MTP $\mathcal{R}_n^+(\alpha)$ controls the Type I error rate $\Theta^+(F_{V_n^+,R_n^+})$ at level α.

Sections 6.2 and 6.3 present simple augmentation procedures for controlling the gFWER and TPPFP, respectively, based on any initial FWER-controlling procedure. Given any initial gFWER-controlling MTP, Section 6.5 proposes augmentation procedures for controlling gTP error rates, i.e., tail probabilities for arbitrary functions $g(V_n, R_n)$ of the numbers of false positives V_n and rejected hypotheses R_n. Adjusted p-values for the AMTP are shown to be simply shifted versions of the adjusted p-values of the initial MTP. Conditions on the Type I error rate mappings Θ and Θ^+, that ensure Type I error control by the augmentation procedure, are also stated in Section 6.5.

We focus on augmentation procedures that reject null hypotheses in order of their *increasing Θ-specific adjusted p-values*, i.e., starting with the null hypothesis with the smallest adjusted p-value for the initial Θ-controlling MTP (see Section 1.2.12 for an introduction to adjusted p-values). Specifically, denote the adjusted p-values for the initial Θ-controlling procedure by $\widetilde{P}_{0n}(m)$. Order the M null hypotheses according to these p-values, from smallest to largest, that is, define indices $O_n(m)$, so that $\widetilde{P}_{0n}(O_n(1)) \leq \cdots \leq \widetilde{P}_{0n}(O_n(M))$. Then, for a test at nominal level α, the initial Θ-controlling procedure rejects the following $R_n(\alpha)$ null hypotheses,

$$\mathcal{R}_n(\alpha) \equiv \left\{m : \widetilde{P}_{0n}(m) \leq \alpha\right\} = \{O_n(m) : m = 1, \ldots, R_n(\alpha)\} . \qquad (6.13)$$

The augmentation procedure rejects the next $A_n(\alpha)$ most significant null hypotheses, i.e., the augmentation set has the form

$$\mathcal{A}_n(\alpha) \equiv \{O_n(m) : m = R_n(\alpha) + 1, \ldots, R_n(\alpha) + A_n(\alpha)\} . \qquad (6.14)$$

Notation

To emphasize the dependence of a MTP on parameters defining the Type I error rate mappings Θ or Θ^+, we may adopt a longer notation, whereby sets of null hypotheses and their cardinality are indexed by these parameters as well as by the nominal Type I error level α. For instance, for a gFWER-controlling AMTP with k allowed false positives, we may use: $\mathcal{A}_n(k;\alpha)$ and $A_n(k;\alpha) = |\mathcal{A}_n(k;\alpha)|$, for the augmentation set and its cardinality, respectively; $\mathcal{R}_n^+(k;\alpha)$ and $R_n^+(k;\alpha) = |\mathcal{R}_n^+(k;\alpha)|$, for the augmented set of rejected hypotheses and its cardinality, respectively; and $V_n^+(k;\alpha) = |\mathcal{R}_n^+(k;\alpha) \cap \mathcal{H}_0|$, for the number of Type I errors.

When clear from the context, we may omit references to the nominal Type I error level α and/or parameters of the Type I error rate mappings, e.g., use $\mathcal{R}_n(\alpha)$ or $\mathcal{R}_n$.

In what follows, the augmentation set $\mathcal{A}_n$ always denotes a subset of $\mathcal{R}_n^c$, the set of non-rejected null hypotheses for the initial Θ-controlling MTP.

6.2 Augmentation multiple testing procedures for controlling the generalized family-wise error rate, $gFWER(k) = \Pr(V_n > k)$

6.2.1 gFWER-controlling augmentation multiple testing procedures

In this section, we derive augmentation procedures for which the initial Type I error rate is the FWER and the target error rate is the gFWER (Procedure 6.2). Specifically, consider an initial multiple testing procedure $\mathcal{R}_n(\alpha)$, that provides finite sample control of the FWER at level α_n and asymptotic control of this error rate at level α (Procedure 6.1, with $\Theta(F_{V_n,R_n}) = 1 - F_{V_n}(0)$).

Theorem 6.3, below, states that, for any non-negative integer k and random set $\mathcal{A}_n(k;\alpha)$ with $\Pr(A_n(k;\alpha) = \min\{k, M - R_n(\alpha)\}) = 1$, the augmentation multiple testing procedure $\mathcal{R}_n^+(k;\alpha) = \mathcal{R}_n(\alpha) \cup \mathcal{A}_n(k;\alpha)$ provides finite sample control of $gFWER(k)$ at level α_n, that is, $\Pr(V_n^+(k;\alpha) > k) \leq \alpha_n$, $\forall\ n$. In addition, if $\mathcal{R}_n(\alpha)$ controls the FWER asymptotically exactly at level $\alpha^\star$ and the false null hypotheses $\mathcal{H}_1 = \mathcal{H}_0^c$ are asymptotically always rejected by $\mathcal{R}_n(\alpha)$ (Equation (6.20)), then the augmentation procedure $\mathcal{R}_n^+(k;\alpha)$ also provides exact asymptotic control of $gFWER(k)$ at level $\alpha^\star$.

Note that Theorem 6.3 applies to *any* random set $\mathcal{A}_n(k;\alpha)$ satisfying the specified size constraints. However, for power considerations, we propose the following specific construction for the augmentation set $\mathcal{A}_n(k;\alpha)$, based on the *ordered adjusted p-values* for the initial FWER-controlling procedure $\mathcal{R}_n(\alpha)$. It would be of interest to provide a formal result concerning the optimality (for a suitable definition of power) of the augmentation set $\mathcal{A}_n(k;\alpha)$ of Procedure 6.2.

Procedure 6.2. [Augmentation procedure for controlling the gFWER based on a FWER-controlling procedure]
Consider a multiple testing procedure $\mathcal{R}_n = \mathcal{R}_n(\alpha)$, that provides finite sample control of the FWER at level α_n and asymptotic control of the FWER at level α (Procedure 6.1, with $\Theta(F_{V_n,R_n}) = 1 - F_{V_n}(0)$).

1. First, order the M null hypotheses according to their FWER adjusted p-values $\widetilde{P}_{0n}(m)$, from smallest to largest, that is, define indices $O_n(m)$, so that $\widetilde{P}_{0n}(O_n(1)) \leq \cdots \leq \widetilde{P}_{0n}(O_n(M))$. The initial FWER-controlling procedure rejects the following $R_n(\alpha) \equiv |\mathcal{R}_n(\alpha)|$ null hypotheses,
$$\mathcal{R}_n(\alpha) \equiv \left\{m : \widetilde{P}_{0n}(m) \leq \alpha\right\} = \{O_n(m) : m = 1, \ldots, R_n(\alpha)\}. \quad (6.15)$$
2. For a given bound $k \in \{0, \ldots, M\}$ on the number of Type I errors, define a *gFWER-controlling augmentation multiple testing procedure*

$\mathcal{R}_n^+ = \mathcal{R}_n^+(k;\alpha)$ by

$$\mathcal{R}_n^+(k;\alpha) \equiv \mathcal{R}_n(\alpha) \cup \mathcal{A}_n(k;\alpha), \tag{6.16}$$

where $\mathcal{A}_n(k;\alpha)$ is an augmentation set of cardinality

$$A_n(k;\alpha) \equiv \min\{k, M - R_n(\alpha)\}, \tag{6.17}$$

defined by

$$\mathcal{A}_n(k;\alpha) \equiv \{O_n(m) : m = R_n(\alpha)+1, \ldots, R_n(\alpha) + A_n(k;\alpha)\}. \tag{6.18}$$

That is, the set $\mathcal{A}_n(k;\alpha)$ corresponds to the $\min\{k, M - R_n(\alpha)\}$ most significant null hypotheses that are not rejected by the initial FWER-controlling procedure $\mathcal{R}_n(\alpha)$.

Figure 3.2 provides a graphical summary of a gFWER-controlling augmentation procedure.

6.2.2 Finite sample and asymptotic control of the gFWER

Theorem 6.3. [Control of the gFWER] *Consider a multiple testing procedure $\mathcal{R}_n = \mathcal{R}_n(\alpha)$, that provides finite sample control of the FWER at level α_n and asymptotic control of the FWER at level α (Procedure 6.1, with $\Theta(F_{V_n,R_n}) = 1 - F_{V_n}(0)$). That is, $\Pr(V_n(\alpha) > 0) = \alpha_n$, $\forall\ n$, and $\limsup_n \Pr(V_n(\alpha) > 0) = \limsup_n \alpha_n = \alpha^\star \leq \alpha$. For a user-supplied constant $k \in \{0, \ldots, M\}$, suppose $\mathcal{A}_n(k;\alpha)$ is a random subset such that the events $\mathcal{A}_n(k;\alpha) \subseteq \mathcal{R}_n^c(\alpha)$ and $|\mathcal{A}_n(k;\alpha)| = \min\{k, M - R_n(\alpha)\}$ have joint probability one.*

(a) Finite sample control. *The augmentation multiple testing procedure $\mathcal{R}_n^+ = \mathcal{R}_n^+(k;\alpha) = \mathcal{R}_n(\alpha) \cup \mathcal{A}_n(k;\alpha)$ provides finite sample control of gFWER(k) at level α_n, that is,*

$$\Pr(V_n^+(k;\alpha) > k) \leq \Pr(V_n(\alpha) > 0) = \alpha_n, \qquad \forall\ n. \tag{6.19}$$

(b) Exact asymptotic control. *Asymptotic control at level α follows immediately from finite sample control above. Further suppose that the initial FWER-controlling MTP $\mathcal{R}_n(\alpha)$ satisfies*

$$\lim_{n\to\infty} \Pr(\mathcal{H}_1 \subseteq \mathcal{R}_n(\alpha)) = 1, \tag{6.20}$$

that is, the false null hypotheses $\mathcal{H}_1 = \mathcal{H}_0^c$ are asymptotically always rejected by $\mathcal{R}_n(\alpha)$, and

$$\lim_{n\to\infty} \Pr(R_n(\alpha) \leq M - k) = 1. \tag{6.21}$$

Then, the augmentation multiple testing procedure $\mathcal{R}_n^+(k;\alpha) = \mathcal{R}_n(\alpha) \cup \mathcal{A}_n(k;\alpha)$ *is such that*

$$\liminf_{n\to\infty} \Pr(V_n^+(k;\alpha) = k) = 1 - \alpha^\star. \tag{6.22}$$

In particular, $\mathcal{R}_n^+(k;\alpha)$ *provides exact asymptotic control of gFWER(k) at level* $\alpha^\star \leq \alpha$*, that is,*

$$\limsup_{n\to\infty} \Pr(V_n^+(k;\alpha) > k) = \alpha^\star. \tag{6.23}$$

Note that the augmentation set $\mathcal{A}_n(k;\alpha)$ of Procedure 6.2 trivially satisfies the conditions of Theorem 6.3.

Proof of Theorem 6.3. The shorter notation $\mathcal{R}_n = \mathcal{R}_n(\alpha)$ and $\mathcal{R}_n^+ = \mathcal{R}_n^+(k;\alpha)$ is adopted for this proof.

(a) Finite sample control. Since $\Pr(|\mathcal{A}_n| \leq k) = 1$, one has $\Pr(V_n^+ \leq V_n + k) = 1$. Thus, $\Pr(V_n^+ > k) \leq \Pr(V_n + k > k) = \Pr(V_n > 0)$, which equals α_n by assumption.

(b) Exact asymptotic control. Define $B_n \equiv \mathrm{I}(\mathcal{H}_1 \subseteq \mathcal{R}_n)$. By assumption, $\lim_n \Pr(B_n = 1) = 1$. Thus,

$$\Pr(V_n^+ = k) = \Pr(V_n^+ = k | B_n = 1)\Pr(B_n = 1) + o(1).$$

Note that Equation (6.21) implies that $A_n = |\mathcal{A}_n| = k$ with probability one in the limit and that $B_n = 1$ implies $\mathcal{A}_n \subseteq \mathcal{R}_n^c \subseteq \mathcal{H}_0$. Thus, $V_n^+ = V_n + A_n = V_n + k$ with probability one in the limit. Hence,

$$\begin{aligned}\Pr(V_n^+ = k) &= \Pr(V_n + k = k | B_n = 1)\Pr(B_n = 1) + o(1)\\ &= \Pr(V_n = 0 | B_n = 1)\Pr(B_n = 1) + o(1)\\ &= \Pr(V_n = 0) - \Pr(V_n = 0, B_n = 0) + o(1)\\ &= \Pr(V_n = 0) + o(1),\end{aligned}$$

where the last equality follows by noting that $\Pr(V_n = 0, B_n = 0) \leq \Pr(B_n = 0) \to 0$, as $n \to \infty$. By assumption, one has $\liminf_n \Pr(V_n = 0) = 1 - \alpha^\star$, thus, as required, $\liminf_n \Pr(V_n^+ = k) = 1 - \alpha^\star$.

□

6.2.3 Adjusted *p*-values for gFWER-controlling augmentation multiple testing procedures

Consider augmentation multiple testing Procedure 6.2, for controlling $gFWER(k)$ at level α, where $\mathcal{R}_n^+(k;\alpha) = \mathcal{R}_n(\alpha) \cup \mathcal{A}_n(k;\alpha)$ and the augmentation set $\mathcal{A}_n(k;\alpha)$ is specified in Equation (6.18). By definition, the adjusted

p-values $\widetilde{P}^+_{0n}(m)$, for gFWER-controlling procedure $\mathcal{R}^+_n(k;\alpha)$, are given by

$$\widetilde{P}^+_{0n}(m) = \inf\left\{\alpha \in [0,1] : m \in \mathcal{R}^+_n(k;\alpha)\right\}, \qquad m = 1,\ldots,M.$$

As one might expect, these adjusted p-values are trivial functions of the adjusted p-values $\widetilde{P}_{0n}(m)$ for the initial FWER-controlling procedure $\mathcal{R}_n(\alpha)$. Because the AMTP always rejects at least k null hypotheses, the ordered adjusted p-values $\widetilde{P}^+_{0n}(O_n(m))$ for the new gFWER-controlling AMTP are simply *k-shifted* versions of the ordered adjusted p-values $P_{0n}(O_n(m))$ for the initial FWER-controlling MTP, with the first k p-values set to zero. That is,

$$\widetilde{P}^+_{0n}(O_n(m)) = \begin{cases} 0, & \text{if } m \leq k \\ \widetilde{P}_{0n}(O_n(m-k)), & \text{if } m > k \end{cases}. \tag{6.24}$$

Note that the same expression is obtained in Equation (6.68), using the general result in Equation (6.64) (Section 6.5.3).

6.3 Augmentation multiple testing procedures for controlling the tail probability for the proportion of false positives, $TPPFP(q) = \Pr(V_n/R_n > q)$

6.3.1 TPPFP-controlling augmentation multiple testing procedures

In this section, we derive augmentation procedures for which the initial Type I error rate is the FWER and the target error rate is the TPPFP (Procedure 6.4). Specifically, consider an initial multiple testing procedure $\mathcal{R}_n(\alpha)$, that provides finite sample control of the FWER at level α_n and asymptotic control of this error rate at level α (Procedure 6.1, with $\Theta(F_{V_n,R_n}) = 1 - F_{V_n}(0)$).

Given $q \in (0,1)$, consider augmentation sets of cardinality

$$\begin{aligned} A_n(q;\alpha) &\equiv \max\left\{m \in \{0,\ldots,M - R_n(\alpha)\} : \frac{m}{R_n(\alpha)+m} \leq q\right\} \\ &= \min\left\{\left\lfloor \frac{qR_n(\alpha)}{1-q} \right\rfloor, M - R_n(\alpha)\right\}, \end{aligned}$$

corresponding to a proportion of additional rejected hypotheses

$$q^\star = q^\star_n(q;\alpha) \equiv \frac{A_n(q;\alpha)}{R_n(\alpha) + A_n(q;\alpha)} \leq q,$$

where $R_n(\alpha) = |\mathcal{R}_n(\alpha)|$ denotes the number of rejected hypotheses for FWER-controlling procedure $\mathcal{R}_n(\alpha)$. The *floor* $\lfloor x \rfloor$ denotes the greatest integer less than or equal to x, i.e., $\lfloor x \rfloor \in \mathbb{Z}$ and $\lfloor x \rfloor \leq x < \lfloor x \rfloor + 1$. Basic properties of the floor function are given in Appendix B.3. In other words, the augmentation set is obtained by rejecting null hypotheses until the ratio of additional rejections to the total number of rejections reaches the allowed proportion q of false positives.

Theorem 6.5, below, states that, for a random set $\mathcal{A}_n(q;\alpha)$ with $\Pr(|\mathcal{A}_n(q;\alpha)| = A_n(q;\alpha)) = 1$, the augmentation multiple testing procedure $\mathcal{R}_n^+(q;\alpha) = \mathcal{R}_n(\alpha) \cup \mathcal{A}_n(q;\alpha)$ provides finite sample control of $TPPFP(q^\star)$ at level α_n, that is, $\Pr\left(V_n^+(q;\alpha)/R_n^+(q;\alpha) > q^\star\right) \leq \alpha_n, \forall\, n$. In addition, if $\mathcal{R}_n(\alpha)$ controls the FWER asymptotically exactly at level $\alpha^\star$ and the false null hypotheses $\mathcal{H}_1$ are asymptotically always rejected by $\mathcal{R}_n(\alpha)$ (Equation (6.31)), then the augmentation procedure $\mathcal{R}_n^+(q;\alpha)$ also provides exact asymptotic control of $TPPFP(q^\star)$ at level $\alpha^\star$.

As Theorem 6.3 for gFWER control, Theorem 6.5 applies to *any* random set $\mathcal{A}_n(q;\alpha)$ satisfying the specified size constraints. However, for power considerations, we propose the following specific construction for the augmentation set $\mathcal{A}_n(q;\alpha)$, based on the *ordered adjusted p-values* for the initial FWER-controlling procedure $\mathcal{R}_n(\alpha)$. Again, it would be of interest to provide a formal result concerning the optimality of the augmentation set $\mathcal{A}_n(q;\alpha)$ of Procedure 6.4.

Procedure 6.4. [Augmentation procedure for controlling the TPPFP based on a FWER-controlling procedure]
Consider a multiple testing procedure $\mathcal{R}_n = \mathcal{R}_n(\alpha)$, that provides finite sample control of the FWER at level α_n and asymptotic control of the FWER at level α (Procedure 6.1, with $\Theta(F_{V_n,R_n}) = 1 - F_{V_n}(0)$).

1. First, order the M null hypotheses according to their FWER adjusted p-values $\widetilde{P}_{0n}(m)$, from smallest to largest, that is, define indices $O_n(m)$, so that $\widetilde{P}_{0n}(O_n(1)) \leq \cdots \leq \widetilde{P}_{0n}(O_n(M))$. The initial FWER-controlling procedure rejects the following $R_n(\alpha) \equiv |\mathcal{R}_n(\alpha)|$ null hypotheses,

$$\mathcal{R}_n(\alpha) \equiv \left\{m : \widetilde{P}_{0n}(m) \leq \alpha\right\} = \{O_n(m) : m = 1, \ldots, R_n(\alpha)\}. \quad (6.25)$$

2. For a given bound $q \in (0,1)$ on the proportion of Type I errors, define a *TPPFP-controlling augmentation multiple testing procedure* $\mathcal{R}_n^+ = \mathcal{R}_n^+(q;\alpha)$ by

$$\mathcal{R}_n^+(q;\alpha) \equiv \mathcal{R}_n(\alpha) \cup \mathcal{A}_n(q;\alpha), \quad (6.26)$$

where $\mathcal{A}_n(q;\alpha)$ is an augmentation set of cardinality

$$A_n(q;\alpha) \equiv \max\left\{m \in \{0,\ldots,M-R_n(\alpha)\} : \frac{m}{R_n(\alpha)+m} \leq q\right\}$$
$$= \min\left\{\left\lfloor \frac{qR_n(\alpha)}{1-q}\right\rfloor, M - R_n(\alpha)\right\}, \tag{6.27}$$

defined by

$$\mathcal{A}_n(q;\alpha) \equiv \{O_n(m) : m = R_n(\alpha)+1,\ldots,R_n(\alpha)+A_n(q;\alpha)\}. \tag{6.28}$$

That is, the set $\mathcal{A}_n(q;\alpha)$ corresponds to the $A_n(q;\alpha)$ most significant null hypotheses that are not rejected by the initial FWER-controlling procedure $\mathcal{R}_n(\alpha)$. The proportion of additional rejected hypotheses is

$$q^\star = q_n^\star(q;\alpha) \equiv \frac{A_n(q;\alpha)}{R_n(\alpha)+A_n(q;\alpha)} \leq q. \tag{6.29}$$

Note that in their Theorem 3.3, Genovese and Wasserman (2004b) claim, without providing a proof, that one can readily generalize Procedure 6.4 to produce a $TPPFP(q)$-controlling procedure from an initial $gFWER(k_0)$-controlling MTP. However, one immediately faces the problem that $TPPFP(q)$ may exceed $gFWER(k_0)$, e.g., having already rejected k_0 true null hypotheses with the initial gFWER-controlling procedure may lead to an excessive proportion of false positives. Section 6.5 addresses this issue, by considering a modified version of the TPPFP, the generalized tail probability for the proportion of false positives (gTPPFP), defined in Equation (6.7). Theorem 6.10 establishes finite sample and exact asymptotic control results for a general class of augmentation procedures, aimed at controlling generalized tail probability error rates of the form $gTP(q,g) = \Pr(g(V_n,R_n) > q)$, based on an initial $gFWER(k_0)$-controlling MTP. Such error rates include the gTPPFP, with $g(v,r) = \mathrm{I}(k_0/r \leq q)\, v/r$.

6.3.2 Finite sample and asymptotic control of the TPPFP

Theorem 6.5. [Control of the TPPFP] *Consider a multiple testing procedure $\mathcal{R}_n = \mathcal{R}_n(\alpha)$, that provides finite sample control of the FWER at level α_n and asymptotic control of the FWER at level α (Procedure 6.1, with $\Theta(F_{V_n,R_n}) = 1 - F_{V_n}(0)$). That is, $\Pr(V_n(\alpha) > 0) = \alpha_n$, $\forall\, n$, and $\limsup_n \Pr(V_n(\alpha) > 0) = \limsup_n \alpha_n = \alpha^\star \leq \alpha$. For a user-supplied constant $q \in (0,1)$, suppose $\mathcal{A}_n(q;\alpha)$ is a random subset such that the events $\mathcal{A}_n(q;\alpha) \subseteq \mathcal{R}_n^c(\alpha)$ and $|\mathcal{A}_n(q;\alpha)| = A_n(q;\alpha)$ have joint probability one, where $A_n(q;\alpha)$ and $q^\star = q_n^\star(q;\alpha)$ are defined as in Equations (6.27) and (6.29), respectively.*

(a) Finite sample control. *The augmentation multiple testing procedure $\mathcal{R}_n^+ = \mathcal{R}_n^+(q;\alpha) = \mathcal{R}_n(\alpha) \cup \mathcal{A}_n(q;\alpha)$ provides finite sample control of*

$TPPFP(q^\star)$ at level α_n, that is,

$$\Pr\left(\frac{V_n^+(q;\alpha)}{R_n^+(q;\alpha)} > q^\star\right) \leq \Pr(V_n(\alpha) > 0) = \alpha_n, \qquad \forall\ n. \tag{6.30}$$

(b) Exact asymptotic control. *Asymptotic control at level α follows immediately from finite sample control above. Further suppose that the initial FWER-controlling MTP $\mathcal{R}_n(\alpha)$ satisfies*

$$\lim_{n\to\infty} \Pr(\mathcal{H}_1 \subseteq \mathcal{R}_n(\alpha)) = 1, \tag{6.31}$$

that is, the false null hypotheses $\mathcal{H}_1 = \mathcal{H}_0^c$ are asymptotically always rejected by $\mathcal{R}_n(\alpha)$. Then, the augmentation multiple testing procedure $\mathcal{R}_n^+(q;\alpha) = \mathcal{R}_n(\alpha) \cup \mathcal{A}_n(q;\alpha)$ is such that

$$\liminf_{n\to\infty} \Pr\left(\frac{V_n^+(q;\alpha)}{R_n^+(q;\alpha)} = q^\star\right) = 1 - \alpha^\star. \tag{6.32}$$

In particular, $\mathcal{R}_n^+(q;\alpha)$ provides exact asymptotic control of $TPPFP(q^\star)$ at level $\alpha^\star \leq \alpha$, that is,

$$\limsup_{n\to\infty} \Pr\left(\frac{V_n^+(q;\alpha)}{R_n^+(q;\alpha)} > q^\star\right) = \alpha^\star. \tag{6.33}$$

Proof of Theorem 6.5. The shorter notation $\mathcal{R}_n = \mathcal{R}_n(\alpha)$ and $\mathcal{R}_n^+ = \mathcal{R}_n^+(q;\alpha)$ is adopted for this proof.

(a) Finite sample control. Since $\Pr(|\mathcal{A}_n| = A_n) = 1$, for A_n defined as in Equation (6.27), one has $V_n^+ \leq V_n + A_n$ and $R_n^+ = R_n + A_n$. Thus,

$$\frac{V_n^+}{R_n^+} \leq \frac{V_n + A_n}{R_n + A_n}.$$

Define the indicator $C_n \equiv \mathrm{I}(V_n = 0)$. Given $C_n = 1$, and by definition of $q^\star$ in Equation (6.29), one has

$$\frac{V_n + A_n}{R_n + A_n} = \frac{A_n}{R_n + A_n} = q^\star \leq q.$$

Thus,

$$\begin{aligned}
\Pr\left(\frac{V_n^+}{R_n^+} > q^\star\right) &\leq \Pr\left(\frac{V_n + A_n}{R_n + A_n} > q^\star\right) \\
&= \Pr\left(\frac{V_n + A_n}{R_n + A_n} > q^\star \middle| C_n = 1\right) \Pr(C_n = 1) \\
&\quad + \Pr\left(\frac{V_n + A_n}{R_n + A_n} > q^\star \middle| C_n = 0\right) \Pr(C_n = 0) \\
&= 0 \times \Pr(C_n = 1) + \Pr\left(\frac{V_n + A_n}{R_n + A_n} > q^\star \middle| C_n = 0\right) \times \alpha_n \\
&\leq \alpha_n.
\end{aligned}$$

(b) Exact asymptotic control. As in the proof of Theorem 6.3, let $B_n \equiv \mathrm{I}(\mathcal{H}_1 \subseteq \mathcal{R}_n)$. Given $B_n = 1$, one has $\mathcal{A}_n \subseteq \mathcal{R}_n^c \subseteq \mathcal{H}_0$ and, hence,

$$\frac{V_n^+}{R_n^+} = \frac{V_n + A_n}{R_n + A_n}. \tag{6.34}$$

Now define $C_n \equiv \mathrm{I}(V_n = 0)$ and $D_n \equiv \mathrm{I}(B_n = C_n = 1)$. Since, by assumption, $\lim_n \Pr(B_n = 1) = 1$ and $\liminf_n \Pr(C_n = 1) = 1 - \alpha^\star$, then $\liminf_n \Pr(D_n = 1) = 1 - \alpha^\star$. Given $D_n = 1$ (i.e., $V_n = 0$ and $\mathcal{A}_n \subseteq \mathcal{R}_n^c \subseteq \mathcal{H}_0$), and by definition of $q^\star$, one has

$$\frac{V_n^+}{R_n^+} = \frac{V_n + A_n}{R_n + A_n} = \frac{A_n}{R_n + A_n} = q^\star.$$

Thus,

$$\begin{aligned}
\Pr\left(\frac{V_n^+}{R_n^+} = q^\star\right) &= \Pr\left(\frac{V_n^+}{R_n^+} = q^\star \middle| D_n = 1\right) \Pr(D_n = 1) \\
&\quad + \Pr\left(\frac{V_n^+}{R_n^+} = q^\star \middle| D_n = 0\right) \Pr(D_n = 0) \\
&= 1 \times \Pr(D_n = 1) \\
&\quad + \Pr\left(\frac{V_n^+}{R_n^+} = q^\star \middle| D_n = 0\right) \times \Pr(D_n = 0)
\end{aligned}$$

and, hence,

$$\begin{aligned}
&\liminf_{n \to \infty} \Pr\left(\frac{V_n^+}{R_n^+} = q^\star\right) \\
&\quad = (1 - \alpha^\star) + \liminf_{n \to \infty} \Pr\left(\frac{V_n^+}{R_n^+} = q^\star \middle| D_n = 0\right) \Pr(D_n = 0).
\end{aligned}$$

In the remainder of the proof, we show that the second term equals zero. The event $\{D_n = 0\}$ is the union of three events, namely: $\{B_n = 1, C_n = 0\}$,

$\{B_n = 0, C_n = 1\}$, and $\{B_n = 0, C_n = 0\}$. Because $\lim_n \Pr(B_n = 0) = 0$, it follows that the probabilities of the last two events tend to zero. Consequently,

$$\Pr\left(\frac{V_n^+}{R_n^+} = q^\star \middle| D_n = 0\right) \Pr(D_n = 0)$$
$$= \Pr\left(\frac{V_n^+}{R_n^+} = q^\star \middle| B_n = 1, C_n = 0\right) \Pr(B_n = 1, C_n = 0) + o(1).$$

Finally, by Equation (6.34), one has

$$\begin{aligned}
\Pr&\left(\frac{V_n^+}{R_n^+} = q^\star \middle| B_n = 1, C_n = 0\right) \\
&= \Pr\left(\frac{V_n + A_n}{R_n + A_n} = q^\star \middle| B_n = 1, C_n = 0\right) \\
&= \Pr(V_n = 0 | B_n = 1, C_n = 0) \\
&= 0.
\end{aligned}$$

This completes the proof that $\liminf_n \Pr(V_n^+/R_n^+ = q^\star) = 1 - \alpha^\star$.

□

6.3.3 Adjusted p-values for TPPFP-controlling augmentation multiple testing procedures

Consider augmentation multiple testing Procedure 6.4, for controlling $TPPFP(q)$ at level α, where $\mathcal{R}_n^+(q;\alpha) = \mathcal{R}_n(\alpha) \cup \mathcal{A}_n(q;\alpha)$ and the augmentation set $\mathcal{A}_n(q;\alpha)$ is specified in Equation (6.28). By definition, the adjusted p-values $\widetilde{P}_{0n}^+(O_n(m))$, for TPPFP-controlling procedure $\mathcal{R}_n^+(q;\alpha)$, are given by

$$\begin{aligned}
\widetilde{P}_{0n}^+(O_n(m)) &= \inf\left\{\alpha \in [0,1] : O_n(m) \in \mathcal{R}_n^+(q;\alpha)\right\} \qquad (6.35)\\
&= \inf\left\{\alpha \in [0,1] : R_n(\alpha) + A_n(q;\alpha) \geq m\right\} \\
&= \inf\left\{\alpha \in [0,1] : \frac{m - R_n(\alpha)}{m} \leq q\right\} \\
&= \inf\left\{\alpha \in [0,1] : R_n(\alpha) \geq (1-q)m\right\} \\
&= \inf\left\{\alpha \in [0,1] : R_n(\alpha) \geq \lceil (1-q)m \rceil\right\} \\
&= \widetilde{P}_{0n}(O_n(\lceil (1-q)m \rceil)), \qquad m = 1, \ldots, M,
\end{aligned}$$

where the *ceiling* $\lceil x \rceil$ denotes the least integer greater than or equal to x, i.e., $\lceil x \rceil \in \mathbb{Z}$ and $\lceil x \rceil - 1 < x \leq \lceil x \rceil$. Basic properties of the ceiling function are given in Appendix B.3. The fifth equality follows by noting that $R_n(\alpha) \in \mathbb{N}$ and the ceiling function property $\lceil x \rceil \leq n$ i.f.f. $x \leq n$, for any $x \in \mathbb{R}$ and $n \in \mathbb{Z}$. The last equality follows by definition of an adjusted p-value.

Thus, the ordered adjusted p-values $\widetilde{P}^{+}_{0n}(O_n(m))$ for the new TPPFP-controlling AMTP are simply *mq-shifted* versions (up to a ceiling integer transformation) of the ordered adjusted p-values $P_{0n}(O_n(m))$ for the initial FWER-controlling MTP.

Note that the same expression is obtained in Equation (6.72), using the general result in Equation (6.64) (Section 6.5.4).

6.4 TPPFP-based multiple testing procedures for controlling the false discovery rate, $FDR = \mathrm{E}[V_n/R_n]$

This section shows how one can straightforwardly obtain (conservative) FDR-controlling procedures from *any* TPPFP-controlling procedure (e.g., augmentation Procedure 6.4). Specifically, two FDR-controlling procedures are derived: the first procedure is based on a general, but conservative upper bound for the FDR of the TPPFP-controlling MTP (Bound 1), while the second procedure is based on a less conservative upper bound, obtained by assuming exact asymptotic TPPFP control (Bound 2).

6.4.1 FDR-controlling TPPFP-based multiple testing procedures

Suppose one has a multiple testing procedure $\mathcal{R}_n = \mathcal{R}_n(q;\alpha)$, that provides finite sample control of $TPPFP(q)$ at level α_n and asymptotic control of $TPPFP(q)$ at level α, for $q \in (0,1)$. That is,

$$\Pr\left(\frac{V_n}{R_n} > q\right) = \alpha_n, \qquad \forall\ n, \tag{6.36}$$

and

$$\limsup_{n\to\infty} \Pr\left(\frac{V_n}{R_n} > q\right) = \limsup_{n\to\infty} \alpha_n = \alpha^\star \leq \alpha. \tag{6.37}$$

General conservative bound for FDR: Bound 1

One can bound the FDR of TPPFP-controlling procedure $\mathcal{R}_n$ as follows.

$$\begin{aligned}
\mathrm{E}\left[\frac{V_n}{R_n}\right] &= \mathrm{E}\left[\frac{V_n}{R_n}\middle| \frac{V_n}{R_n} \leq q\right] \Pr\left(\frac{V_n}{R_n} \leq q\right) \\
&\quad + \mathrm{E}\left[\frac{V_n}{R_n}\middle| \frac{V_n}{R_n} > q\right] \Pr\left(\frac{V_n}{R_n} > q\right) \\
&\leq q \times (1-\alpha_n) + 1 \times \alpha_n \\
&\leq q \times 1 + 1 \times \alpha_n,
\end{aligned} \tag{6.38}$$

where the first inequality follows by replacing the conditional expected values $\mathrm{E}[V_n/R_n|V_n/R_n \leq q]$ and $\mathrm{E}[V_n/R_n|V_n/R_n > q]$ by the upper bounds of q and one, respectively. Thus, in the limit,

$$\limsup_{n\to\infty} \mathrm{E}\left[\frac{V_n}{R_n}\right] \leq q + \alpha. \tag{6.39}$$

Hence, setting, for example, $\alpha = q = \alpha'/2$, results in asymptotic control of the FDR at level α'.

A TPPFP-controlling procedure $\mathcal{R}_n^+ = \mathcal{R}_n^+(q;\alpha)$, satisfying Equations (6.36) and (6.37), can be obtained as in Section 6.3, by augmenting a FWER-controlling procedure $\mathcal{R}_n = \mathcal{R}_n(\alpha)$ (e.g., according to Procedure 6.4) and appealing to Theorem 6.5.

Restricted sharper bound for FDR: Bound 2

A less conservative FDR-controlling procedure, based on a sharper upper bound for the FDR of the TPPFP-controlling procedure $\mathcal{R}_n = \mathcal{R}_n(q;\alpha)$, can be obtained by making the following slightly stronger assumption of exact asymptotic control of $TPPFP(q)$ at level α,

$$\lim_{n\to\infty} \Pr\left(\frac{V_n}{R_n} > q\right) = \lim_{n\to\infty} \alpha_n = \alpha. \tag{6.40}$$

Then, in the limit, the FDR of procedure $\mathcal{R}_n$ can be bounded above as follows.

$$\begin{aligned} \lim_{n\to\infty} \mathrm{E}\left[\frac{V_n}{R_n}\right] &\leq \lim_{n\to\infty} \left(q(1-\alpha_n) + \alpha_n\right) \\ &= q(1-\alpha) + \alpha. \end{aligned} \tag{6.41}$$

Hence, setting, for example, $\alpha = q = 1 - \sqrt{1-\alpha'}$, results in asymptotic control of the FDR at level α', i.e., in $\lim_n \mathrm{E}[V_n/R_n] \leq \alpha'$.

A TPPFP-controlling procedure $\mathcal{R}_n^+ = \mathcal{R}_n^+(q;\alpha)$, satisfying Equation (6.40), can be obtained as in Section 6.3, by augmenting a FWER-controlling procedure $\mathcal{R}_n = \mathcal{R}_n(\alpha)$ (e.g., according to Procedure 6.4) and appealing to Theorem 6.5, with the following two slightly stronger assumptions: (i) the limit of the FWER exists and, in particular, $\lim_n \Pr(V_n = 0) = 1 - \alpha$; (ii) the false null hypotheses $\mathcal{H}_1$ are asymptotically always rejected by $\mathcal{R}_n$, that is, $\lim_n \Pr(\mathcal{H}_1 \subseteq \mathcal{R}_n) = 1$. The first assumption implies that the asymptotic statements of Theorem 6.5 hold with $\limsup_n$ and $\liminf_n$ replaced by $\lim_n$. Then, by Theorem 6.5, augmentation procedure $\mathcal{R}_n^+$ satisfies

$$\lim_{n\to\infty} \Pr\left(\frac{V_n^+}{R_n^+} = q^\star\right) = 1 - \alpha \qquad \text{and} \qquad \lim_{n\to\infty} \Pr\left(\frac{V_n^+}{R_n^+} > q^\star\right) = \alpha,$$

where $q^\star$ is defined in Equation (6.29) and is such that $q^\star \leq q$.

FDR-controlling TPPFP-based multiple testing procedures

The preceding arguments provide the following procedures for controlling the FDR at a user-supplied level α'. The procedures are first stated in terms of a general TPPFP-controlling procedure (Theorem 6.6), and then in terms of a TPPFP-controlling augmentation procedure obtained from a FWER-controlling procedure (Theorem 6.7).

Theorem 6.6. [Control of the FDR based on a general TPPFP-controlling MTP] *Consider a multiple testing procedure* $\mathcal{R}_n = \mathcal{R}_n(q;\alpha)$*, that provides finite sample control of* $TPPFP(q)$ *at level* α_n *and asymptotic control of* $TPPFP(q)$ *at level* α*. That is,* $\Pr(V_n/R_n > q) = \alpha_n$*,* $\forall\ n$*, and* $\limsup_n \Pr(V_n/R_n > q) = \limsup_n \alpha_n = \alpha^\star \leq \alpha$.

(a) Finite sample control. *Selecting* q *and* α_n *such that* $q(1-\alpha_n)+\alpha_n = \alpha'$ *(e.g.,* $\alpha_n = q = 1-\sqrt{1-\alpha'}$*) leads to finite sample control of the FDR at level* α'*, that is,*

$$\mathrm{E}\left[\frac{V_n}{R_n}\right] \leq \alpha'. \tag{6.42}$$

(b) Asymptotic control.
Bound 1. *Selecting* q *and* α *such that* $q+\alpha = \alpha'$ *(e.g.,* $\alpha = q = \alpha'/2$*) leads to asymptotic control of the FDR at level* α'*, that is,*

$$\limsup_{n\to\infty} \mathrm{E}\left[\frac{V_n}{R_n}\right] \leq \alpha'. \tag{6.43}$$

Bound 2. *If asymptotic control of* $TPPFP(q)$ *is exact, i.e.,* $\lim_n \Pr(V_n/R_n > q) = \alpha$*, then selecting* q *and* α *such that* $q(1-\alpha)+\alpha = \alpha'$ *(e.g.,* $\alpha = q = 1-\sqrt{1-\alpha'}$*) leads to asymptotic control of the FDR at level* α'.

Theorem 6.7. [Control of the FDR based on a TPPFP-controlling augmentation MTP] *Consider a multiple testing procedure* $\mathcal{R}_n = \mathcal{R}_n(\alpha)$*, that provides finite sample control of the FWER at level* α_n *and asymptotic control of the FWER at level* α*. That is,* $\Pr(V_n > 0) = \alpha_n$*,* $\forall\ n$*, and* $\limsup_n \Pr(V_n > 0) = \limsup_n \alpha_n = \alpha^\star \leq \alpha$.

(a) Finite sample control. *The corresponding* $TPPFP(q)$*-controlling augmentation multiple testing procedure* $\mathcal{R}_n^+ = \mathcal{R}_n^+(q;\alpha)$ *(e.g., Procedure 6.4), with* q *and* α_n *such that* $q(1-\alpha_n)+\alpha_n = \alpha'$ *(e.g.,* $\alpha_n = q = 1-\sqrt{1-\alpha'}$*), provides finite sample control of the FDR at level* α'*, that is,*

$$\mathrm{E}\left[\frac{V_n^+}{R_n^+}\right] \leq \alpha'. \tag{6.44}$$

(b) Asymptotic control.
Bound 1. *The corresponding* $TPPFP(q)$*-controlling augmentation multiple testing procedure* $\mathcal{R}_n^+ = \mathcal{R}_n^+(q;\alpha)$ *(e.g., Procedure 6.4), with* q *and* α *such that* $q+\alpha = \alpha'$ *(e.g.,* $\alpha = q = \alpha'/2$*), asymptotically controls the FDR at level* α'*, that is,*

$$\limsup_{n\to\infty} \mathrm{E}\left[\frac{V_n^+}{R_n^+}\right] \leq \alpha'. \tag{6.45}$$

Bound 2. *In addition, suppose that the FWER-controlling MTP $\mathcal{R}_n$ satisfies: (i)* $\lim_n \Pr(V_n = 0) = 1 - \alpha^\star$*; (ii)* $\alpha^\star = \alpha$*; (iii)* $\lim_n \Pr(\mathcal{H}_1 \subseteq \mathcal{R}_n) = 1$. *Then, augmentation procedure* $\mathcal{R}_n^+$ *provides exact asymptotic control of* $TPPFP(q)$ *at level* α *(Theorem 6.5). Hence, selecting* q *and* α *such that* $q(1-\alpha)+\alpha = \alpha'$ *(e.g.,* $\alpha = q = 1 - \sqrt{1-\alpha'}$*) leads to asymptotic control of the FDR at level* α'.

Note that the FDR bounds in Equations (6.38), (6.39), and (6.41), are *non-parametric*. In particular, they do not rely on independence assumptions concerning the joint distribution of the test statistics.

In addition, for both Bounds 1 and 2, the $q = 0$ and $\alpha = \alpha'$ case corresponds to control of the FDR based on a FWER-controlling MTP ($TPPFP(0) = FWER$). As shown in Section 1.2.9, for any given MTP, $FDR \leq FWER$. Thus, one expects FWER-controlling MTPs to be conservative for FDR control. However, the degree of conservativeness depends on the sharpness of the inequality $FDR \leq FWER$ and, in some settings, FWER-controlling *joint* MTPs (e.g., as in Chapters 4, 5, and 7) could in fact be more powerful than FDR-controlling *marginal* MTPs (e.g., step-up Benjamini and Hochberg (1995) Procedure 3.22 and Benjamini and Yekutieli (2001) Procedure 3.23).

Another issue worthy of investigation is the optimal selection of the pair (α, q), i.e., the choice of (α, q) leading to the greatest number of rejected hypotheses, subject to either constraint $q+\alpha = \alpha'$ (Bound 1) or $q(1-\alpha)+\alpha = \alpha'$ (Bound 2). This would entail constrained minimization of the adjusted p-values of Proposition 6.8, below, with respect to (α, q). If one could identify an optimal pair (α, q) and a powerful $TPPFP(q)$-controlling MTP (such as the resampling-based empirical Bayes MTPs of Chapter 7), then the resulting FDR-controlling procedure might improve upon existing MTPs, in the case of test statistics with general dependence structures.

It would indeed be interesting to compare the new FDR-controlling procedures of Theorems 6.6 and 6.7 to the conservative step-up procedure of Benjamini and Yekutieli (2001), which is, to our knowledge, the only fully non-parametric FDR-controlling procedure in the literature (Procedure 3.23). The adjusted p-values for this conservative procedure are given in Equation (3.45) and have a $\approxeq \log M$ penalty compared to the adjusted p-values for the original step-up procedure of Benjamini and Hochberg (1995) (Equation (3.42), Procedure 3.22). For further discussion of FDR-controlling MTPs, the reader is referred to Sections 3.4, 7.7, and 7.8.

Finally, note that similar bounds on the FDR of a TPPFP-controlling procedure are provided in Equation (34), p. 1153, of Lehmann and Romano (2005).

6.4.2 Adjusted p-values for FDR-controlling TPPFP-based multiple testing procedures

Adjusted p-values for the FDR-controlling TPPFP-based procedures of Theorems 6.6 and 6.7 are derived next.

Proposition 6.8. [Adjusted p-values for FDR-controlling TPPFP-based procedures] *Consider a multiple testing procedure $\mathcal{R}_n^{TPPFP}(q;\alpha)$, that controls $TPPFP(q) = \Pr(V_n/R_n > q)$ at nominal level α (e.g., Procedure 6.4). Let $\widetilde{P}_{0n}^{TPPFP(q)}(m)$ denote the adjusted p-values for this TPPFP-controlling MTP. Then, the adjusted p-values $\widetilde{P}_{0n}^{FDR}(m)$, for the FDR-controlling TPPFP-based procedures $\mathcal{R}_n^{FDR}(\alpha)$ of Theorem 6.6, are given by*

$$\widetilde{P}_{0n}^{FDR}(m) = \inf\left\{\alpha \in [0,1] : \widetilde{P}_{0n}^{TPPFP(q(\alpha))}(m) \leq \alpha^{TPPFP}(\alpha)\right\}, \tag{6.46}$$

where, for control of the FDR at nominal level α, the allowed proportion q of false positives and the nominal level α^{TPPFP} of the TPPFP-controlling procedure satisfy either of the following two conditions,

$$q(\alpha) + \alpha^{TPPFP}(\alpha) = \alpha \qquad \textit{[Bound 1]} \tag{6.47}$$

or

$$q(\alpha)(1 - \alpha^{TPPFP}(\alpha)) + \alpha^{TPPFP}(\alpha) = \alpha \qquad \textit{[Bound 2]}.$$

In particular, the adjusted p-values for the FDR-controlling procedures $\mathcal{R}_n^{FDR}(\alpha)$ of Theorem 6.7, based on TPPFP-controlling augmentation multiple testing Procedure 6.4, are given by

$$\widetilde{P}_{0n}^{FDR}(O_n(m)) = \inf\left\{\alpha \in [0,1] : \widetilde{P}_{0n}^{FWER}(O_n(\lceil(1-q(\alpha))m\rceil)) \leq \alpha^{TPPFP}(\alpha)\right\}, \tag{6.48}$$

where $\widetilde{P}_{0n}^{FWER}(m)$ denote the adjusted p-values for the initial FWER-controlling MTP and $O_n(m)$ denote corresponding indices so that $\widetilde{P}_{0n}^{FWER}(O_n(1)) \leq \cdots \leq \widetilde{P}_{0n}^{FWER}(O_n(M))$.

For example, for Bound 1, with $\alpha^{TPPFP}(\alpha) = q(\alpha) = \alpha/2$, one has

$$\widetilde{P}_{0n}^{FDR}(m) = \inf\left\{\alpha \in [0,1] : \widetilde{P}_{0n}^{TPPFP(\alpha/2)}(m) \leq \alpha/2\right\},$$

and for Bound 2, with $\alpha^{TPPFP}(\alpha) = q(\alpha) = 1 - \sqrt{1-\alpha}$, one has

$$\widetilde{P}_{0n}^{FDR}(m) = \inf\left\{\alpha \in [0,1] : \widetilde{P}_{0n}^{TPPFP(1-\sqrt{1-\alpha})}(m) \leq 1 - \sqrt{1-\alpha}\right\}.$$

In general, one cannot obtain closed form expressions for the adjusted p-values $\widetilde{P}_{0n}^{FDR}(m)$ of the FDR-controlling TPPFP-based procedures, as one cannot solve explicitly for α in Equations (6.46) and (6.48).

Proof of Proposition 6.8. The proof follows straightforwardly from the general definition of an adjusted p-value in Equation (1.58).

$$\begin{aligned}
\widetilde{P}_{0n}^{FDR}(m) &= \inf\{\alpha \in [0,1] : m \in \mathcal{R}_n^{FDR}(\alpha)\} \\
&= \inf\{\alpha \in [0,1] : m \in \mathcal{R}_n^{TPPFP}(q(\alpha); \alpha^{TPPFP}(\alpha))\} \\
&= \inf\left\{\alpha \in [0,1] : \widetilde{P}_{0n}^{TPPFP(q(\alpha))}(m) \leq \alpha^{TPPFP}(\alpha)\right\}.
\end{aligned}$$

□

6.5 General results on augmentation multiple testing procedures

Consider as target Type I error rate Θ^+, the *generalized tail probability* (gTP) error rate, introduced in Equation (6.1),

$$\Theta^+(F_{V_n^+, R_n^+}) = gTP(q, g) = \Pr(g(V_n^+, R_n^+) > q),$$

that is, tail probabilities for arbitrary functions $g(V_n^+, R_n^+)$ of the numbers of Type I errors V_n^+ and rejected hypotheses R_n^+. Error rates covered by this representation include: the generalized family-wise error rate (gFWER), where $g(v,r) = v$; the tail probability for the proportion of false positives (TPPFP), where $g(v,r) = v/r$; the generalized tail probability for the proportion of false positives (gTPPFP), where $g(v,r) = \mathrm{I}(k_0/r \leq q)\, v/r$, for a user-supplied non-negative integer k_0. One may also consider error rates based on the function $g(v,r) = \mathrm{I}(r > r_0)\, v/r$, for a user-supplied non-negative integer r_0. Controlling tail probabilities $\Pr(\mathrm{I}(R_n > r_0)\, V_n/R_n > q)$ amounts to considering multiple testing procedures for which the proportion of false positives V_n/R_n does not exceed q when more than r_0 hypotheses are rejected (i.e., when $R_n > r_0$).

Given an initial gFWER-controlling procedure, this section proposes general augmentation multiple testing Procedure 6.9, for controlling generalized tail probability error rates $\Theta^+(F_{V_n^+, R_n^+}) = gTP(q,g) = \Pr(g(V_n^+, R_n^+) > q)$. Finite sample and exact asymptotic control results are established in Theorem 6.10. Adjusted p-values for general AMTPs are derived in Section 6.5.2 and are shown to be simply shifted versions of the adjusted p-values of the initial MTP. The adjusted p-values and corresponding shift functions for gFWER-, TPPFP-, and gTPPFP-controlling augmentation procedures are examined in detail in Sections 6.5.3, 6.5.4, and 6.5.5, respectively. Control of the generalized expected value (gEV) error rate, $gEV(g) = \mathrm{E}[g(V_n, R_n)]$, based on a gTP-controlling procedure, is discussed in Section 6.6.

6.5.1 Augmentation multiple testing procedures for controlling the generalized tail probability error rate, $gTP(q, g) = \Pr(g(V_n, R_n) > q)$

Set-up and assumptions

Consider a multiple testing procedure $\mathcal{R}_n = \mathcal{R}_n(k_0; \alpha)$, that controls $gFWER(k_0)$ at level α, i.e., such that $\Pr(V_n(k_0; \alpha) > k_0) \leq \alpha$. Given $\mathcal{R}_n(k_0; \alpha)$, we wish to derive an augmentation multiple testing procedure $\mathcal{R}_n^+ = \mathcal{R}_n^+(k_0, q, g; \alpha)$, that controls the generalized tail probability error rate $\Theta^+(F_{V_n^+, R_n^+}) = gTP(q, g) = \Pr(g(V_n^+, R_n^+) > q)$. Suppose that the function g and the initial MTP $\mathcal{R}_n(k_0; \alpha)$ satisfy the following three assumptions.

Assumption AMTP.MgV. [Monotonicity of g] The function $v \to g(v, r)$ is non-decreasing for any given r.

Assumption AMTP.MgA. [Monotonicity of g] The function $a \to g(v + a, r + a)$ is non-decreasing for any given (v, r).

Assumption AMTP.gTP0. [Initial gTP control] One has

$$\Pr(g(k_0, R_n(k_0; \alpha)) \leq q) = 1, \tag{6.49}$$

so that the initial gFWER-controlling MTP also controls the target gTP error rate.

Monotonicity Assumption AMTP.MgV is used in Theorem 6.10, to prove gTP control by an AMTP such as Procedure 6.9, below. Monotonicity Assumption AMTP.MgA guarantees that the cardinality of the augmentation set for Procedure 6.9 increases with the bound q for false positives (Equation (6.54)). Finally, Assumption AMTP.gTP0 ensures that the initial gFWER-controlling procedure also controls the target gTP error rate at level α. Indeed, one can show that $\Pr(g(V_n, R_n) > q) \leq \alpha$ as follows.

$$\begin{aligned}
\Pr(g(V_n, R_n) > q) &= \Pr(g(V_n, R_n) > q | V_n > k_0) \Pr(V_n > k_0) \\
&\quad + \Pr(g(V_n, R_n) > q | V_n \leq k_0) \Pr(V_n \leq k_0) \\
&\leq 1 \times \Pr(V_n > k_0) + \Pr(g(V_n, R_n) > q | V_n \leq k_0) \times 1 \\
&\leq 1 \times \alpha + \Pr(g(k_0, R_n) > q | V_n \leq k_0) \times 1 \\
&= \alpha.
\end{aligned}$$

For the last inequality, note that, by monotonicity Assumption AMTP.MgV, $g(V_n, R_n) \leq g(k_0, R_n)$ if $V_n \leq k_0$. Hence, $\Pr(g(V_n, R_n) > q | V_n \leq k_0) \leq \Pr(g(k_0, R_n) > q | V_n \leq k_0)$. By Assumption AMTP.gTP0, $\Pr(g(k_0, R_n) > q) = 0$ and, hence, $\Pr(g(k_0, R_n) > q | V_n \leq k_0) = 0$. Thus, one can always obtain a gTP-controlling AMTP, even in the worst-case scenario of an empty augmentation set.

For instance, for control of the proportion of false positives V_n/R_n, one can let

$$g(v,r) \equiv \mathrm{I}\left(\frac{k_0}{r} \leq q\right) \frac{v}{r} \tag{6.50}$$

and consider as target Type I error rate the *generalized tail probability for the proportion of false positives* (gTPPFP)

$$gTPPFP(k_0, q) \equiv \Pr\left(\mathrm{I}\left(\frac{k_0}{R_n} \leq q\right) \frac{V_n}{R_n} > q\right). \tag{6.51}$$

Assumption AMTP.gTP0 allows us to derive an augmentation procedure that controls $gTPPFP(k_0, q)$, even when the PFP exceeds q for the initial $gFWER(k_0)$-controlling procedure, i.e., even when $k_0/R_n > q$. As detailed in Procedure 6.9 and Section 6.5.5, below, one simply keeps rejecting null hypotheses until $g(k_0 + a, R_n + a)$ exceeds the bound q for false positives, i.e., the following two conditions are both met: $(k_0 + a)/(R_n + a) > q$ and $k_0/(R_n + a) \leq q$, where a denotes the number of additional rejections.

Note that for their TPPFP-controlling AMTP of Theorem 3.3, Genovese and Wasserman (2004b) do not enforce control of the TPPFP by the initial gFWER-controlling procedure.

gTP-controlling augmentation multiple testing procedures

Procedure 6.9 provides an explicit construction for an augmentation multiple testing procedure $\mathcal{R}_n^+ = \mathcal{R}_n^+(k_0, q, g; \alpha)$, that controls the generalized tail probability error rate $\Theta^+(F_{V_n^+, R_n^+}) = gTP(q, g) = \Pr(g(V_n^+, R_n^+) > q)$, based on an initial $gFWER(k_0)$-controlling procedure $\mathcal{R}_n = \mathcal{R}_n(k_0; \alpha)$.

Procedure 6.9. [Augmentation procedure for controlling the gTP based on a gFWER-controlling procedure]
Consider a multiple testing procedure $\mathcal{R}_n = \mathcal{R}_n(k_0; \alpha)$, that provides finite sample control of $gFWER(k_0)$ at level α_n and asymptotic control of $gFWER(k_0)$ at level α (Procedure 6.1, with $\Theta(F_{V_n, R_n}) = 1 - F_{V_n}(k_0)$).

1. First, order the M null hypotheses according to their gFWER adjusted p-values $\widetilde{P}_{0n}(m)$, from smallest to largest, that is, define indices $O_n(m)$, so that $\widetilde{P}_{0n}(O_n(1)) \leq \cdots \leq \widetilde{P}_{0n}(O_n(M))$. The initial gFWER-controlling procedure rejects the following $R_n(k_0; \alpha) \equiv |\mathcal{R}_n(k_0; \alpha)|$ null hypotheses,

$$\mathcal{R}_n(k_0; \alpha) \equiv \left\{m : \widetilde{P}_{0n}(m) \leq \alpha\right\} = \{O_n(m) : m = 1, \ldots, R_n(k_0; \alpha)\}. \tag{6.52}$$

2. For a given Type I error bound q, define a gTP-controlling augmentation multiple testing procedure $\mathcal{R}_n^+ = \mathcal{R}_n^+(k_0, q, g; \alpha)$ by

$$\mathcal{R}_n^+(k_0, q, g; \alpha) \equiv \mathcal{R}_n(k_0; \alpha) \cup \mathcal{A}_n(k_0, q, g; \alpha), \tag{6.53}$$

where $\mathcal{A}_n(k_0, q, g; \alpha)$ is an augmentation set of cardinality

$$A_n(k_0, q, g; \alpha) \equiv \tag{6.54}$$
$$\max \{m \in \{0, \ldots, M - R_n(k_0; \alpha)\} : g(k_0 + m, R_n(k_0; \alpha) + m) \leq q\},$$

defined by

$$\mathcal{A}_n(k_0, q, g; \alpha) \equiv \tag{6.55}$$
$$\{O_n(m) : m = R_n(k_0; \alpha) + 1, \ldots, R_n(k_0; \alpha) + A_n(k_0, q, g; \alpha)\}.$$

That is, the set $\mathcal{A}_n(k_0, q, g; \alpha)$ corresponds to the $A_n(k_0, q, g; \alpha)$ most significant null hypotheses that are not rejected by the initial gFWER-controlling procedure $\mathcal{R}_n(k_0; \alpha)$. The g-specific function of the number of additional rejected hypotheses is

$$q^\star = q_n^\star(k_0, q, g; \alpha) \tag{6.56}$$
$$\equiv g(k_0 + A_n(k_0, q, g; \alpha), R_n(k_0; \alpha) + A_n(k_0, q, g; \alpha)) \leq q.$$

Note that gFWER- and TPPFP-controlling augmentation Procedures 6.2 and 6.4, based on an initial FWER-controlling MTP, are special cases of Procedure 6.9, corresponding to $k_0 = 0$ and to $g(v, r) = v$ and $g(v, r) = v/r$, respectively.

The augmentation set for gTP-controlling augmentation Procedure 6.9 is obtained by rejecting additional null hypotheses until $g(k_0 + a, R_n + a)$ exceeds the bound q for false positives, where a denotes the number of additional rejections. As Procedures 6.2 and 6.4, general Procedure 6.9 augments the set of rejected null hypotheses conservatively, in the sense that every additional rejected hypothesis is counted as a false positive.

Finite sample and asymptotic control of the gTP

Theorem 6.10. [Control of the gTP] *Consider a multiple testing procedure $\mathcal{R}_n = \mathcal{R}_n(k_0; \alpha)$, that provides finite sample control of $gFWER(k_0)$ at level α_n and asymptotic control of $gFWER(k_0)$ at level α (Procedure 6.1, with $\Theta(F_{V_n, R_n}) = 1 - F_{V_n}(k_0)$). That is,* $\Pr(V_n(k_0; \alpha) > k_0) = \alpha_n, \ \forall \ n$, *and* $\limsup_n \Pr(V_n(k_0; \alpha) > k_0) = \limsup_n \alpha_n = \alpha^\star \leq \alpha$. *For a user-supplied function g, that satisfies Assumptions AMTP.MgV, AMTP.MgA, and AMTP.gTP0, and a user-supplied constant q, consider as target Type I error rate the generalized tail probability $\Theta^+(F_{V_n^+, R_n^+}) = gTP(q, g) =$* $\Pr(g(V_n^+, R_n^+) > q)$. *Suppose $\mathcal{A}_n(k_0, q, g; \alpha)$ is a random subset such that the events $\mathcal{A}_n(k_0, q, g; \alpha) \subseteq \mathcal{R}_n^c(k_0; \alpha)$ and $|\mathcal{A}_n(k_0, q, g; \alpha)| = A_n(k_0, q, g; \alpha)$ have* *joint probability one, where* $A_n(k_0, q, g; \alpha)$ *and* $q^\star = q_n^\star(k_0, q, g; \alpha)$ *are defined as in Equations (6.54) and (6.56), respectively.*

(a) Finite sample control. *The augmentation multiple testing procedure* $\mathcal{R}_n^+ = \mathcal{R}_n^+(k_0, q, g; \alpha) = \mathcal{R}_n(k_0; \alpha) \cup \mathcal{A}_n(k_0, q, g; \alpha)$ *provides finite sample control of* $gTP(q^\star, g)$ *at level* α_n. *That is,* $\forall\, n$,

$$\Pr\left(g(V_n^+(k_0, q, g; \alpha), R_n^+(k_0, q, g; \alpha)) > q^\star\right) \leq \Pr(V_n(k_0; \alpha) > k_0) = \alpha_n. \tag{6.57}$$

(b) Exact asymptotic control. *Asymptotic control at level* α *follows immediately from finite sample control above. Further suppose that the initial gFWER-controlling MTP* $\mathcal{R}_n(k_0; \alpha)$ *satisfies*

$$\lim_{n\to\infty} \Pr(\mathcal{H}_1 \subseteq \mathcal{R}_n(k_0; \alpha)) = 1, \tag{6.58}$$

that is, the false null hypotheses $\mathcal{H}_1 = \mathcal{H}_0^c$ *are asymptotically always rejected by* $\mathcal{R}_n(k_0; \alpha)$, *and is such that*

$$\liminf_{n\to\infty} \Pr(V_n(k_0; \alpha) = k_0) = 1 - \alpha^\star, \tag{6.59}$$

that is, $\mathcal{R}_n(k_0; \alpha)$ *controls* $gFWER(k_0)$ *asymptotically exactly at level* $\alpha^\star$. *In addition, assume that the function* $v \to g(v + a, r + a)$ *is continuous for each* (v, r, a) *and strictly increasing at* $v = k_0$ *for any* (a, r) *such that* $g(k_0 + a, r + a) = q^\star$. *Then, the augmentation multiple testing procedure* $\mathcal{R}_n^+(k_0, q, g; \alpha) = \mathcal{R}_n(k_0; \alpha) \cup \mathcal{A}_n(k_0, q, g; \alpha)$ *is such that*

$$\liminf_{n\to\infty} \Pr\left(g(V_n^+(k_0, q, g; \alpha), R_n^+(k_0, q, g; \alpha)) = q^\star\right) = 1 - \alpha^\star. \tag{6.60}$$

In particular, $\mathcal{R}_n^+(k_0, q, g; \alpha)$ *provides exact asymptotic control of* $gTP(q^\star, g)$ *at level* $\alpha^\star \leq \alpha$, *that is,*

$$\limsup_{n\to\infty} \Pr\left(g(V_n^+(k_0, q, g; \alpha), R_n^+(k_0, q, g; \alpha)) > q^\star\right) = \alpha^\star. \tag{6.61}$$

Proof of Theorem 6.10. The shorter notation $\mathcal{R}_n = \mathcal{R}_n(k_0; \alpha)$ and $\mathcal{R}_n^+ = \mathcal{R}_n^+(k_0, q, g; \alpha)$ is adopted for this proof.

By Assumption AMTP.gTP0, the initial $gFWER(k_0)$-controlling procedure also controls the target error rate $gTP(q, g)$. Monotonicity Assumption AMTP.MgA guarantees that the cardinality of the augmentation set for Procedure 6.9, in Equation (6.54), increases with q. Thus, one can always obtain a gTP-controlling AMTP, even in the worst-case scenario of an empty augmentation set.

(a) Finite sample control. Since $\Pr(|\mathcal{A}_n| = A_n) = 1$, for A_n defined as in Equation (6.54), one has $V_n^+ \leq V_n + A_n$ and $R_n^+ = R_n + A_n$. Thus, by monotonicity Assumption AMTP.MgV,

$$g(V_n^+, R_n^+) \leq g(V_n + A_n, R_n + A_n).$$

Furthermore, if $V_n \leq k_0$, then, by monotonicity Assumption AMTP.MgV and by definition of $q^\star$ in Equation (6.56), one has

$$g(V_n + A_n, R_n + A_n) \leq g(k_0 + A_n, R_n + A_n) = q^\star \leq q.$$

Hence,

$$\begin{aligned} gTP(q^\star, g) &= \Pr(g(V_n^+, R_n^+) > q^\star) \\ &\leq \Pr(g(V_n + A_n, R_n + A_n) > q^\star) \\ &= \Pr(g(V_n + A_n, R_n + A_n) > q^\star | V_n > k_0) \Pr(V_n > k_0) \\ &\quad + \Pr(g(V_n + A_n, R_n + A_n) > q^\star | V_n \leq k_0) \Pr(V_n \leq k_0) \\ &\leq \Pr(g(V_n + A_n, R_n + A_n) > q^\star | V_n > k_0) \Pr(V_n > k_0) \\ &\quad + \Pr(g(k_0 + A_n, R_n + A_n) > q^\star | V_n \leq k_0) \Pr(V_n \leq k_0) \\ &\leq 1 \times \Pr(V_n > k_0) + 0 \times \Pr(V_n \leq k_0) = \alpha_n. \end{aligned}$$

Therefore, the augmentation procedure $\mathcal{R}_n^+$ provides finite sample control of $gTP(q^\star, g)$ at level α_n.

(b) Exact asymptotic control. As in the proofs of Theorems 6.3 and 6.5, let $B_n \equiv \mathrm{I}(\mathcal{H}_1 \subseteq \mathcal{R}_n)$. Given $B_n = 1$, one has $V_n^+ = V_n + A_n$ and, hence, $g(V_n^+, R_n^+) = g(V_n + A_n, R_n + A_n)$. Now define $C_n \equiv \mathrm{I}(V_n = k_0)$ and $D_n \equiv \mathrm{I}(B_n = C_n = 1)$. Since, by assumption, $\lim_n \Pr(B_n = 1) = 1$ and $\liminf_n \Pr(C_n = 1) = 1 - \alpha^\star$, then $\liminf_n \Pr(D_n = 1) = 1 - \alpha^\star$. Given $D_n = 1$ (i.e., $V_n = k_0$ and $\mathcal{A}_n \subseteq \mathcal{R}_n^c \subseteq \mathcal{H}_0$), and by definition of $q^\star$, one has

$$g(V_n^+, R_n^+) = g(V_n + A_n, R_n + A_n) = g(k_0 + A_n, R_n + A_n) = q^\star.$$

Thus,

$$\begin{aligned} \Pr\left(g(V_n^+, R_n^+) = q^\star\right) &= \Pr\left(g(V_n^+, R_n^+) = q^\star | D_n = 1\right) \Pr(D_n = 1) \\ &\quad + \Pr\left(g(V_n^+, R_n^+) = q^\star | D_n = 0\right) \Pr(D_n = 0) \\ &= 1 \times \Pr(D_n = 1) \\ &\quad + \Pr\left(g(V_n^+, R_n^+) = q^\star | D_n = 0\right) \times \Pr(D_n = 0) \end{aligned}$$

and, hence,

$$\begin{aligned} &\liminf_{n \to \infty} \Pr\left(g(V_n^+, R_n^+) = q^\star\right) \\ &\quad = (1 - \alpha^\star) + \liminf_{n \to \infty} \Pr\left(g(V_n^+, R_n^+) = q^\star | D_n = 0\right) \Pr(D_n = 0). \end{aligned}$$

In the remainder of the proof, we show that the second term equals zero. The event $\{D_n = 0\}$ is the union of three events, namely: $\{B_n = 1, C_n = 0\}$, $\{B_n = 0, C_n = 1\}$, and $\{B_n = 0, C_n = 0\}$. Because $\lim_n \Pr(B_n = 0) = 0$, it follows that the probabilities of the last two events tend to zero. Consequently,

$$\begin{aligned}
&\Pr\left(g(V_n^+, R_n^+) = q^\star | D_n = 0\right) \Pr(D_n = 0) \\
&\quad = \Pr\left(g(V_n^+, R_n^+) = q^\star | B_n = 1, C_n = 0\right) \Pr(B_n = 1, C_n = 0) + o(1).
\end{aligned}$$

Finally,

$$\begin{aligned}
&\Pr\left(g(V_n^+, R_n^+) = q^\star | B_n = 1, C_n = 0\right) \\
&\quad = \Pr\left(g(V_n + A_n, R_n + A_n) = g(k_0 + A_n, R_n + A_n) | B_n = 1, C_n = 0\right) \\
&\quad = \Pr(V_n = k_0 | B_n = 1, C_n = 0) \\
&\quad = 0,
\end{aligned}$$

where, by continuity and strict monotonicity of $v \to g(v + a, r + a)$ at $v = k_0$, $g(V_n + A_n, R_n + A_n) = g(k_0 + A_n, R_n + A_n)$ if and only if $V_n = k_0$. This completes the proof that $\liminf_n \Pr(g(V_n^+, R_n^+) = q^\star) = 1 - \alpha^\star$.

□

As Theorems 6.3 and 6.5, for gFWER- and TPPFP-controlling AMTPs, respectively, note that Theorem 6.10 applies to *any* random set $\mathcal{A}_n(k_0, q, g; \alpha)$ satisfying the specified size constraints. However, for power considerations, the explicit construction of Procedure 6.9 is based on the *ordered adjusted p-values* for the initial gFWER-controlling procedure $\mathcal{R}_n(k_0; \alpha)$.

6.5.2 Adjusted *p*-values for general augmentation multiple testing procedures

This section presents general results concerning the adjusted *p*-values of an augmentation multiple testing procedure. We stress the level of generality of these results: they apply to any procedure controlling an *arbitrary initial Type I error rate* $\Theta(F_{V_n,R_n})$ (i.e., not only the gFWER) and to any augmentation procedure controlling an *arbitrary target Type I error rate* $\Theta^+(F_{V_n^+,R_n^+})$ (i.e., not only generalized tail probability error rates such as the gFWER or TPPFP).

Let $\widetilde{P}_{0n}(m)$ denote the adjusted *p*-values for the initial Θ-controlling procedure $\mathcal{R}_n(\alpha)$ and let $O_n(m)$ denote the indices for the ordered adjusted *p*-values, so that $\widetilde{P}_{0n}(O_n(1)) \leq \cdots \leq \widetilde{P}_{0n}(O_n(M))$. As before, focus on augmentation multiple testing procedures that reject null hypotheses in order of their *increasing Θ-specific adjusted p-values*, i.e., starting with the null hypothesis $H_0(O_n(1))$ with the smallest adjusted *p*-value for the initial Θ-controlling MTP.

If the nominal Type I error level α is set at the mth ordered Θ-specific adjusted *p*-value, i.e., $\alpha = \widetilde{P}_{0n}(O_n(m))$, the initial Θ-controlling procedure rejects the following $R_n(\widetilde{P}_{0n}(O_n(m))) = m$ null hypotheses,

$$\mathcal{R}_n(\widetilde{P}_{0n}(O_n(m))) = \left\{h : \widetilde{P}_{0n}(h) \leq \widetilde{P}_{0n}(O_n(m))\right\} = \{O_n(h) : h = 1, \ldots, m\}$$

and the Θ^+-controlling augmentation multiple testing procedure $\mathcal{R}_n^+(\alpha)$ rejects the following $R_n^+(\widetilde{P}_{0n}(O_n(m))) = R_n(\widetilde{P}_{0n}(O_n(m))) + A_n(\widetilde{P}_{0n}(O_n(m))) = m + A_n(\widetilde{P}_{0n}(O_n(m)))$ null hypotheses,

$$\begin{aligned}\mathcal{R}_n^+(\widetilde{P}_{0n}(O_n(m))) &= \mathcal{R}_n(\widetilde{P}_{0n}(O_n(m))) \cup \mathcal{A}_n(\widetilde{P}_{0n}(O_n(m))) \\ &= \left\{O_n(h) : h = 1, \ldots, m + A_n(\widetilde{P}_{0n}(O_n(m)))\right\}.\end{aligned}$$

Hence, as stated in Theorem 6.11, below, the mth ordered Θ-specific adjusted p-value $\widetilde{P}_{0n}(O_n(m))$ is the $(m + A_n(\widetilde{P}_{0n}(O_n(m))))$th ordered Θ^+-specific adjusted p-value.

Theorem 6.11. [Adjusted p-values for a general augmentation multiple testing procedure] *Consider a multiple testing procedure $\mathcal{R}_n(\alpha)$, that controls a Type I error rate $\Theta(F_{V_n,R_n})$ at nominal level α (e.g., Procedure 6.1). Let $\widetilde{P}_{0n}(m)$ denote the adjusted p-values for the initial Θ-controlling procedure $\mathcal{R}_n(\alpha)$ and let $O_n(m)$ denote the indices for the ordered adjusted p-values, so that $\widetilde{P}_{0n}(O_n(1)) \leq \cdots \leq \widetilde{P}_{0n}(O_n(M))$. Then, the adjusted p-values $\widetilde{P}_{0n}^+(O_n(m))$, for the Θ^+-controlling augmentation multiple testing procedure $\mathcal{R}_n^+(\alpha) = \mathcal{R}_n(\alpha) \cup \mathcal{A}_n(\alpha)$, satisfy*

$$\widetilde{P}_{0n}(O_n(m)) = \widetilde{P}_{0n}^+(O_n(S_n(m))), \tag{6.62}$$

where $S_n : \{1, \ldots, M\} \rightarrow \{1, \ldots, M\}$ is an integer shift function *defined by*

$$S_n(m) \equiv R_n^+(\widetilde{P}_{0n}(O_n(m))) = m + A_n(\widetilde{P}_{0n}(O_n(m))) \tag{6.63}$$

and $A_n(\widetilde{P}_{0n}(O_n(m)))$ denotes the cardinality of the augmentation set for the AMTP $\mathcal{R}_n^+(\alpha)$ with nominal Type I error level $\alpha = \widetilde{P}_{0n}(O_n(m))$. Thus, the adjusted p-values $\widetilde{P}_{0n}^+(O_n(m))$ for the Θ^+-controlling AMTP are simply shifted *versions of the adjusted p-values $\widetilde{P}_{0n}(O_n(m))$ for the initial Θ-controlling MTP.*

Note that getting a closed form expression for the Θ^+-specific adjusted p-values $\widetilde{P}_{0n}^+(O_n(m))$ may or may not be straightforward, depending on the complexity of the function $A_n(\alpha)$ for the cardinality of the augmentation set, i.e., on how easily one can invert the shift function $S_n : m \rightarrow m + A_n(\widetilde{P}_{0n}(O_n(m)))$. In general, one cannot simply shift the Θ-specific adjusted p-values by the cardinality $A_n(\widetilde{P}_{0n}(O_n(m)))$ of the augmentation set, as this quantity is a function of m. Furthermore, the shift function S_n is not necessarily one-to-one or onto (e.g., Figures 6.2–6.5, for gFWER-controlling AMTPs; Figures 6.6–6.9, for TPPFP-controlling AMTPs; Figures 6.10–6.13, for gTPPFP-controlling AMTPs).

As illustrated in Sections 6.5.3–6.5.5, below, one can rely on the general definition of an adjusted p-value in Equation (1.58) to derive Θ^+-specific adjusted p-values $\widetilde{P}_{0n}^+(O_n(m))$.

$$\widetilde{P}^{+}_{0n}(O_n(m)) = \inf\left\{\alpha \in [0,1] : O_n(m) \in \mathcal{R}^{+}_n(\alpha)\right\} \qquad (6.64)$$
$$= \inf\left\{\alpha \in [0,1] : R^{+}_n(\alpha) \geq m\right\}$$
$$= \inf\left\{\alpha \in [0,1] : R_n(\alpha) + A_n(\alpha) \geq m\right\}.$$

The general results in Equations (6.62) and (6.64) cover the special cases of gFWER- and TPPFP-controlling AMTPs, treated in detail in Sections 6.2 and 6.3, respectively. These results can also be applied to general augmentation Procedure 6.9 of Section 6.5.1, for control of generalized tail probability error rates, $\Theta^{+}(F_{V_n^+,R_n^+}) = gTP(q,g) = \Pr(g(V_n^+, R_n^+) > q)$, and, in particular, for control of the gTPPFP defined in Equation (6.7).

6.5.3 gFWER-controlling augmentation multiple testing procedures

For gFWER-controlling augmentation multiple testing Procedure 6.2 of Section 6.2, with nominal Type I error level $\alpha = \widetilde{P}_{0n}(O_n(m))$, the initial FWER-controlling procedure rejects $R_n(\widetilde{P}_{0n}(O_n(m))) = m$ null hypotheses and the cardinality of the augmentation set is given by

$$A_n(\widetilde{P}_{0n}(O_n(m))) = \min\left\{k, M - R_n(\widetilde{P}_{0n}(O_n(m)))\right\} = \min\left\{k, M - m\right\}. \qquad (6.65)$$

Thus, the shift function is

$$S_n(m) = R^{+}_n(\widetilde{P}_{0n}(O_n(m))) = m + \min\left\{k, M - m\right\} = \min\left\{m + k, M\right\} \qquad (6.66)$$

and, from Equation (6.62), the adjusted p-values satisfy

$$\widetilde{P}_{0n}(O_n(m)) = \widetilde{P}^{+}_{0n}(O_n(\min\left\{m + k, M\right\})). \qquad (6.67)$$

From Equation (6.64), and as in Equation (6.24), one has

$$\widetilde{P}^{+}_{0n}(O_n(m)) = \inf\left\{\alpha \in [0,1] : \min\left\{M, k + R_n(\alpha)\right\} \geq m\right\} \qquad (6.68)$$
$$= \inf\left\{\alpha \in [0,1] : k + R_n(\alpha) \geq m\right\}$$
$$= \begin{cases} 0, & \text{if } m \leq k \\ \inf\left\{\alpha \in [0,1] : R_n(\alpha) \geq m - k\right\}, & \text{if } m > k \end{cases}$$
$$= \begin{cases} 0, & \text{if } m \leq k \\ \widetilde{P}_{0n}(O_n(m-k)), & \text{if } m > k \end{cases}.$$

Thus, one can define an inverse shift function as $S_n^{-1}(m) \equiv \max\left\{0, m - k\right\}$ and let $\widetilde{P}^{+}_{0n}(O_n(m)) = \widetilde{P}_{0n}(O_n(S_n^{-1}(m)))$, with the convention that $\widetilde{P}_{0n}(O_n(0)) = 0$.

Figures 6.2–6.4 display plots of the adjusted p-value shift function S_n and inverse shift function S_n^{-1}, for $gFWER(k)$-controlling augmentation procedures based on an initial FWER-controlling MTP, with $M = 25$ null hypotheses and an allowed number $k \in \{1, 5, 10, 15\}$ of Type I errors.

Figure 6.5 displays plots of the cardinality of the augmentation and rejection sets and of the adjusted p-value shift and inverse shift functions, for $gFWER(k)$-controlling augmentation procedures based on an initial FWER-controlling MTP, with $M = 25$ null hypotheses and an allowed number $k \in \{0, 2, 4, 6, 8, 10\}$ of Type I errors. Specifically, Panel (a) plots the adjusted p-value shift function $S_n(m)$ vs. m. Panel (b) plots the adjusted p-value inverse shift function $S_n^{-1}(m)$ vs. m. Panel (c) plots the cardinality $A_n(\widetilde{P}_{0n}(O_n(m)))$ of the augmentation set for the AMTP vs. the number of rejected hypotheses $R_n(\widetilde{P}_{0n}(O_n(m))) = m$ for the initial MTP, for nominal Θ and Θ^+ Type I error level $\alpha = \widetilde{P}_{0n}(O_n(m))$. Note that the plot in Panel (a) is equivalent to a plot of the total number of rejected hypotheses $R_n^+(\widetilde{P}_{0n}(O_n(m)))$ for the AMTP vs. the number of rejected hypotheses $R_n(\widetilde{P}_{0n}(O_n(m))) = m$ for the initial MTP, for nominal Θ and Θ^+ Type I error level $\alpha = \widetilde{P}_{0n}(O_n(m))$.

Intuitively, the greater the allowed number k of false positives, the larger the augmentation set and the smaller the adjusted p-values for the AMTP. Indeed, the number of rejected hypotheses and the adjusted p-values for the AMTP are simply shifted by the allowed number k of false positives, provided this does not lead to more than M rejected hypotheses. The shift function $S_n(m)$ is piecewise linear and is neither one-to-one nor onto.

6.5.4 TPPFP-controlling augmentation multiple testing procedures

For TPPFP-controlling augmentation multiple testing Procedure 6.4 of Section 6.3, with nominal Type I error level $\alpha = \widetilde{P}_{0n}(O_n(m))$, the initial FWER-controlling procedure rejects $R_n(\widetilde{P}_{0n}(O_n(m))) = m$ null hypotheses and the cardinality of the augmentation set is given by

$$A_n(\widetilde{P}_{0n}(O_n(m))) = \min\left\{\left\lfloor \frac{qm}{1-q} \right\rfloor, M - m\right\}, \tag{6.69}$$

where the floor $\lfloor x \rfloor$ denotes the greatest integer less than or equal to x (Appendix B.3). Thus, the shift function is

$$\begin{aligned} S_n(m) &= R_n^+(\widetilde{P}_{0n}(O_n(m))) = m + A_n(\widetilde{P}_{0n}(O_n(m))) \qquad (6.70) \\ &= m + \min\left\{\left\lfloor \frac{qm}{1-q} \right\rfloor, M - m\right\} \\ &= \min\left\{\left\lfloor \frac{m}{1-q} \right\rfloor, M\right\}, \end{aligned}$$

where we use the following property of the floor function: for any $x \in I\!R$ and $n \in Z\!\!Z$, $\lfloor x \rfloor + n = \lfloor x + n \rfloor$. From Equation (6.62), the adjusted p-values satisfy

$$\widetilde{P}_{0n}(O_n(m)) = \widetilde{P}^+_{0n}\left(O_n\left(\min\left\{\left\lfloor \frac{m}{1-q} \right\rfloor, M\right\}\right)\right). \tag{6.71}$$

From Equation (6.64), and as in Equation (6.35), one has

$$\begin{aligned}
\widetilde{P}^+_{0n}(O_n(m)) &= \inf\left\{\alpha \in [0,1] : \min\left\{M, \left\lfloor \frac{R_n(\alpha)}{1-q} \right\rfloor\right\} \geq m\right\} \qquad (6.72)\\
&= \inf\left\{\alpha \in [0,1] : \left\lfloor \frac{R_n(\alpha)}{1-q} \right\rfloor \geq m\right\}\\
&= \inf\left\{\alpha \in [0,1] : \frac{R_n(\alpha)}{1-q} \geq m\right\}\\
&= \inf\left\{\alpha \in [0,1] : R_n(\alpha) \geq \lceil (1-q)m \rceil\right\}\\
&= \widetilde{P}_{0n}(O_n(\lceil (1-q)m \rceil)),
\end{aligned}$$

where the ceiling $\lceil x \rceil$ denotes the least integer greater than or equal to x. The third equality results from the following property of the floor function: for any $x \in I\!R$ and $n \in Z\!\!Z$, $\lfloor x \rfloor \geq n$ i.f.f. $x \geq n$. The fourth equality follows by noting that $R_n(\alpha) \in I\!N$ and the ceiling function property $\lceil x \rceil \leq n$ i.f.f. $x \leq n$. The last equality follows by definition of an adjusted p-value.

Figures 6.6–6.8 display plots of the adjusted p-value shift function S_n and inverse shift function S_n^{-1}, for $TPPFP(q)$-controlling augmentation procedures based on an initial FWER-controlling MTP, with $M = 25$ null hypotheses and an allowed proportion $q \in \{0.05, 0.10, 0.50, 0.75\}$ of Type I errors.

Figure 6.9 displays plots of the cardinality of the augmentation and rejection sets and of the adjusted p-value shift and inverse shift functions, for $TPPFP(q)$-controlling augmentation procedures based on an initial FWER-controlling MTP, with $M = 25$ null hypotheses and an allowed proportion $q \in \{0, 0.05, 0.10, 0.25, 0.50, 0.75\}$ of Type I errors. Specifically, Panel (a) plots the adjusted p-value shift function $S_n(m)$ vs. m. Panel (b) plots the adjusted p-value inverse shift function $S_n^{-1}(m)$ vs. m. Panel (c) plots the cardinality $A_n(\widetilde{P}_{0n}(O_n(m)))$ of the augmentation set for the AMTP vs. the number of rejected hypotheses $R_n(\widetilde{P}_{0n}(O_n(m))) = m$ for the initial MTP, for nominal Θ and Θ^+ Type I error level $\alpha = \widetilde{P}_{0n}(O_n(m))$. Note that the plot in Panel (a) is equivalent to a plot of the total number of rejected hypotheses $R^+_n(\widetilde{P}_{0n}(O_n(m)))$ for the AMTP vs. the number of rejected hypotheses $R_n(\widetilde{P}_{0n}(O_n(m))) = m$ for the initial MTP, for nominal Θ and Θ^+ Type I error level $\alpha = \widetilde{P}_{0n}(O_n(m))$.

Intuitively, the greater the allowed proportion q of false positives, the larger the augmentation set and the smaller the adjusted p-values for the AMTP.

As evidenced in the figures and in the above equations, TPPFP-controlling AMTPs are more complex than gFWER-controlling AMTPs. Indeed, while the adjusted p-values for gFWER-controlling augmentation Procedure 6.2 are shifted by a constant k, the shift mq for the adjusted p-values of TPPFP-controlling augmentation Procedure 6.4 increases with m, i.e., as the hypotheses become less significant. As in the gFWER case, the TPPFP shift functions are neither one-to-one nor onto. Furthermore, the shift and inverse shift functions are step functions, as they are defined in terms of floor and ceiling functions, respectively.

6.5.5 gTPPFP-controlling augmentation multiple testing procedures

For gTP-controlling augmentation multiple testing Procedure 6.9 of Section 6.5.1, with nominal Type I error level $\alpha = \widetilde{P}_{0n}(O_n(m))$, the initial $gFWER(k_0)$-controlling procedure rejects $R_n(\widetilde{P}_{0n}(O_n(m))) = m$ null hypotheses and the cardinality of the augmentation set is given by

$$A_n(\widetilde{P}_{0n}(O_n(m))) = \max\left\{a \in \{0, \ldots, M-m\} : g(k_0 + a, m + a) \leq q\right\}. \tag{6.73}$$

In particular, for control of $gTPPFP(k_0, q)$, with $g(v, r) = \mathrm{I}(k_0/r \leq q)\, v/r$, then

$$\begin{aligned} g(k_0 + a, m + a) \leq q &\iff \frac{k_0}{m+a} > q \quad \text{or} \quad \frac{k_0 + a}{m+a} \leq q \\ &\iff a < \frac{k_0 - qm}{q} \quad \text{or} \quad a \leq \frac{qm - k_0}{1-q}. \end{aligned}$$

Hence,

$$\begin{aligned} &A_n(\widetilde{P}_{0n}(O_n(m))) \qquad (6.74) \\ &= \max\left\{a \in \{0, \ldots, M-m\} : \mathrm{I}\left(\frac{k_0}{m+a} \leq q\right) \frac{k_0 + a}{m+a} \leq q\right\} \\ &= \min\left\{M - m,\ \max\left\{\left\lceil \frac{k_0}{q} - (m+1)\right\rceil, \left\lfloor \frac{qm - k_0}{1-q} \right\rfloor\right\}\right\}, \end{aligned}$$

where we use the following properties of the floor and ceiling functions: for any $x \in \mathbb{R}$ and $n \in \mathbb{Z}$, $n \leq \lfloor x \rfloor$ i.f.f. $n \leq x$ and $n \leq \lceil x - 1 \rceil$ i.f.f. $n < x$ (Appendix B.3). Thus, the shift function is

$$S_n(m) = R_n^+(\widetilde{P}_{0n}(O_n(m))) = m + A_n(\widetilde{P}_{0n}(O_n(m))) \tag{6.75}$$
$$= m + \min\left\{M - m,\ \max\left\{\left\lceil \frac{k_0}{q} - (m+1)\right\rceil, \left\lfloor \frac{qm - k_0}{1-q} \right\rfloor\right\}\right\}$$
$$= \min\left\{M,\ \max\left\{\left\lceil \frac{k_0}{q} - 1\right\rceil, \left\lfloor \frac{m - k_0}{1-q} \right\rfloor\right\}\right\}$$
$$= \begin{cases} \min\left\{M,\ \left\lceil \frac{k_0}{q} - 1\right\rceil\right\}, & \text{if } m < k_0 + (1-q)\left\lceil \frac{k_0}{q} - 1\right\rceil \\ \min\left\{M,\ \left\lfloor \frac{m-k_0}{1-q}\right\rfloor\right\}, & \text{if } m \geq k_0 + (1-q)\left\lceil \frac{k_0}{q} - 1\right\rceil \end{cases},$$

where we rely on the following properties of the floor and ceiling functions: for any $x \in \mathbb{R}$ and $n \in \mathbb{Z}$, $\lfloor x \rfloor + n = \lfloor x + n \rfloor$, $\lceil x \rceil + n = \lceil x + n \rceil$, and $\lfloor x \rfloor \geq n$ i.f.f. $x \geq n$. From Equation (6.62), the adjusted p-values satisfy

$$\widetilde{P}_{0n}(O_n(m)) = \tag{6.76}$$
$$\begin{cases} \widetilde{P}_{0n}^+\left(O_n\left(\min\left\{M,\ \left\lceil \frac{k_0}{q} - 1\right\rceil\right\}\right)\right), & \text{if } m < k_0 + (1-q)\left\lceil \frac{k_0}{q} - 1\right\rceil \\ \widetilde{P}_{0n}^+\left(O_n\left(\min\left\{M,\ \left\lfloor \frac{m-k_0}{1-q}\right\rfloor\right\}\right)\right), & \text{if } m \geq k_0 + (1-q)\left\lceil \frac{k_0}{q} - 1\right\rceil \end{cases}.$$

From Equation (6.64), one has

$$\widetilde{P}_{0n}^+(O_n(m)) = \inf\left\{\alpha \in [0,1] : R_n^+(\alpha) \geq m\right\} \tag{6.77}$$
$$= \inf\left\{\alpha \in [0,1] : \max\left\{\left\lceil \frac{k_0}{q} - 1\right\rceil, \left\lfloor \frac{R_n(\alpha) - k_0}{1-q}\right\rfloor\right\} \geq m\right\}$$
$$= \min\left\{\inf\left\{\alpha \in [0,1] : \left\lceil \frac{k_0}{q} - 1\right\rceil \geq m\right\},\right.$$
$$\left.\inf\left\{\alpha \in [0,1] : \left\lfloor \frac{R_n(\alpha) - k_0}{1-q}\right\rfloor \geq m\right\}\right\}$$
$$= \min\left\{\inf\left\{\alpha \in [0,1] : \left\lceil \frac{k_0}{q}\right\rceil > m\right\},\right.$$
$$\left.\widetilde{P}_{0n}(O_n(\lceil (1-q)m + k_0\rceil))\right\}$$
$$= \begin{cases} 0, & \text{if } m < \left\lceil \frac{k_0}{q}\right\rceil \\ \widetilde{P}_{0n}(O_n(\lceil (1-q)m + k_0\rceil)), & \text{if } m \geq \left\lceil \frac{k_0}{q}\right\rceil \end{cases}.$$

The fourth equality follows from Equation (6.72).

The adjusted p-values of Equation (6.77), for the gTPPFP-controlling version of augmentation Procedure 6.9, are hybrids of the adjusted p-values of Equations (6.68) and (6.72), for, respectively, the gFWER- and TPPFP-controlling versions of augmentation Procedure 6.9. In particular, as a result of starting from an initial gFWER-controlling MTP, that allows k_0 false positives, the first $\lceil k_0/q \rceil - 1$ adjusted p-values are set to zero, i.e., one automatically rejects $\lceil k_0/q \rceil - 1$ null hypotheses, compared to k for the $gFWER(k)$-controlling AMTP. The remaining adjusted p-values are similar in form to those of the $TPPFP(q)$-controlling AMTP, but with a shift of $qm - k_0$ instead of qm.

Note that for $k_0 = 0$, i.e., for an initial FWER-controlling procedure, one recovers the shift function and adjusted p-values for TPPFP-controlling Procedure 6.4, given in Equations (6.70) and (6.72), respectively.

Figures 6.10–6.12 display plots of the adjusted p-value shift function S_n and inverse shift function S_n^{-1}, for $gTPPFP(k_0, q)$-controlling augmentation procedures based on an initial $gFWER(k_0)$-controlling MTP, with $M = 25$ null hypotheses, an allowed number $k_0 = 2$ of Type I errors, and an allowed proportion $q \in \{0.05, 0.10, 0.50, 0.75\}$ of Type I errors.

Figure 6.13 displays plots of the cardinality of the augmentation and rejection sets and of the adjusted p-value shift and inverse shift functions, for $gTPPFP(k_0, q)$-controlling augmentation procedures based on an initial $gFWER(k_0)$-controlling MTP, with $M = 25$ null hypotheses, an allowed number $k_0 = 2$ of Type I errors, and an allowed proportion $q \in \{0, 0.05, 0.10, 0.25, 0.50, 0.75\}$ of Type I errors. Specifically, Panel (a) plots the adjusted p-value shift function $S_n(m)$ vs. m. Panel (b) plots the adjusted p-value inverse shift function $S_n^{-1}(m)$ vs. m. Panel (c) plots the cardinality $A_n(\widetilde{P}_{0n}(O_n(m)))$ of the augmentation set for the AMTP vs. the number of rejected hypotheses $R_n(\widetilde{P}_{0n}(O_n(m))) = m$ for the initial MTP, for nominal Θ and Θ^+ Type I error level $\alpha = \widetilde{P}_{0n}(O_n(m))$. Note that the plot in Panel (a) is equivalent to a plot of the total number of rejected hypotheses $R_n^+(\widetilde{P}_{0n}(O_n(m)))$ for the AMTP vs. the number of rejected hypotheses $R_n(\widetilde{P}_{0n}(O_n(m))) = m$ for the initial MTP, for nominal Θ and Θ^+ Type I error level $\alpha = \widetilde{P}_{0n}(O_n(m))$.

6.6 gTP-based multiple testing procedures for controlling the generalized expected value, $gEV(g) = \mathrm{E}[g(V_n, R_n)]$

Section 6.4 shows how one can derive (conservative) procedures controlling the expected value of the proportion V_n/R_n of false positives among the rejected hypotheses, based on a procedure controlling tail probabilities for this proportion. That is, Theorems 6.6 and 6.7 provide FDR-controlling MTPs based on TPPFP-controlling MTPs.

Here, we generalize the results of Section 6.4 to tail probabilities and expected values of arbitrary functions $g(V_n, R_n)$ of the numbers of Type I errors V_n and rejected hypotheses R_n. Specifically, we show that given *any* procedure controlling the generalized tail probability (gTP) error rate, $gTP(q, g) =$

$\Pr(g(V_n, R_n) > q)$, one can derive (conservative) procedures controlling the generalized expected value (gEV) error rate, $gEV(g) = \mathrm{E}[g(V_n, R_n)]$.

6.6.1 gEV-controlling gTP-based multiple testing procedures

Theorem 6.12. [Control of the gEV based on a gTP-controlling MTP] *Consider a multiple testing procedure $\mathcal{R}_n = \mathcal{R}_n(q, g; \alpha)$, that provides finite sample control of the generalized tail probability error rate $gTP(q, g)$ at level α_n and asymptotic control of $gTP(q, g)$ at level α. That is, $\Pr(g(V_n, R_n) > q) = \alpha_n$, $\forall\ n$, and $\limsup_n \Pr(g(V_n, R_n) > q) = \limsup_n \alpha_n = \alpha^\star \leq \alpha$. Suppose the function g is bounded above by G_0, i.e., let $G_0 \equiv \max_{(v,r) \in \{0,\ldots,M\}^2} g(v, r)$.*

(a) Finite sample control. *Selecting q and α_n such that $q(1-\alpha_n)+G_0\alpha_n = \alpha'$ (e.g., $\alpha_n = 1 - \sqrt{1 - \alpha'/G_0}$, $q = \alpha_n G_0$) leads to finite sample control of $gEV(g)$ at level α', that is,*

$$\mathrm{E}[g(V_n, R_n)] \leq \alpha'. \tag{6.78}$$

(b) Asymptotic control.
Bound 1. *Selecting q and α such that $q + G_0\alpha = \alpha'$ (e.g., $\alpha = \alpha'/(2G_0)$, $q = \alpha G_0 = \alpha'/2$) leads to asymptotic control of $gEV(g)$ at level α', that is,*

$$\limsup_{n \to \infty} \mathrm{E}[g(V_n, R_n)] \leq \alpha'. \tag{6.79}$$

Bound 2. *If asymptotic control of $gTP(q, g)$ is exact, i.e., $\lim_n \Pr(g(V_n, R_n) > q) = \alpha$, then selecting q and α such that $q(1 - \alpha) + G_0\alpha = \alpha'$ (e.g., $\alpha = 1 - \sqrt{1 - \alpha'/G_0}$, $q = \alpha G_0$) leads to asymptotic control of $gEV(g)$ at level α'.*

In many applications, the function g is bounded above by one, i.e., $G_0 = 1$. Such functions cover, for example, error rates based on the proportion $V_n/R_n \in [0, 1]$ of false positives, with the convention that $0/0 = 0$ (Equations (6.5) and (6.7)). In the special case $g(v, r) = v/r$, one recovers Theorem 6.6, for control of the FDR based on a TPPFP-controlling MTP.

Proof of Theorem 6.12. Note that

$$\begin{aligned} \mathrm{E}[g(V_n, R_n)] &= \mathrm{E}\left[g(V_n, R_n) | g(V_n, R_n) \leq q\right] \Pr(g(V_n, R_n) \leq q) \\ &\quad + \mathrm{E}\left[g(V_n, R_n) | g(V_n, R_n) > q\right] \Pr(g(V_n, R_n) > q) \\ &\leq q \times (1 - \alpha_n) + G_0 \times \alpha_n \\ &\leq q \times 1 + G_0 \times \alpha_n. \end{aligned} \tag{6.80}$$

The first inequality follows by replacing the conditional expected values $\mathrm{E}\left[g(V_n, R_n) | g(V_n, R_n) \leq q\right]$ and $\mathrm{E}\left[g(V_n, R_n) | g(V_n, R_n) > q\right]$ by the upper bounds of q and G_0, respectively. The last inequality results from conservatively bounding the probability $\Pr(g(V_n, R_n) \leq q) = 1 - \alpha_n$ by one.

Bound 1. One has the following conservative upper bound on the asymptotic gEV,

$$\limsup_{n\to\infty} \mathrm{E}\left[g(V_n, R_n)\right] \leq q + G_0\alpha. \tag{6.81}$$

Bound 2. If, in addition, $\lim_n \Pr(g(V_n, R_n) > q) = \lim_n \alpha_n = \alpha$, then one has a less conservative upper bound on the asymptotic gEV,

$$\begin{aligned}\lim_{n\to\infty} \mathrm{E}\left[g(V_n, R_n)\right] &\leq \lim_{n\to\infty} (q(1-\alpha_n) + G_0\alpha_n) \\ &= q(1-\alpha) + G_0\alpha.\end{aligned} \tag{6.82}$$

□

6.6.2 Adjusted p-values for gEV-controlling gTP-based multiple testing procedures

Adjusted p-values for the gEV-controlling gTP-based procedures of Theorem 6.12 are derived next.

Proposition 6.13. [Adjusted p-values for gEV-controlling gTP-based procedures] *Consider a multiple testing procedure $\mathcal{R}_n^{gTP}(q, g; \alpha)$, that controls $gTP(q, g) = \Pr(g(V_n, R_n) > q)$ at nominal level α (e.g., Procedure 6.9). Let $\widetilde{P}_{0n}^{gTP(q)}(m)$ denote the adjusted p-values for this gTP-controlling MTP. Then, the adjusted p-values $\widetilde{P}_{0n}^{gEV}(m)$, for the gEV-controlling gTP-based procedures $\mathcal{R}_n^{gEV}(g; \alpha)$ of Theorem 6.12, are given by*

$$\widetilde{P}_{0n}^{gEV}(m) = \inf\left\{\alpha \in [0,1] : \widetilde{P}_{0n}^{gTP(q(\alpha))}(m) \leq \alpha^{gTP}(\alpha)\right\}, \tag{6.83}$$

where, for control of $gEV(g)$ at nominal level α, the Type I error bound q and the nominal level α^{gTP} of the gTP-controlling procedure satisfy either of the following two conditions,

$$q(\alpha) + G_0\alpha^{gTP}(\alpha) = \alpha \qquad \textit{[Bound 1]} \tag{6.84}$$

or

$$q(\alpha)(1 - \alpha^{gTP}(\alpha)) + G_0\alpha^{gTP}(\alpha) = \alpha \qquad \textit{[Bound 2]}.$$

In general, one cannot obtain closed form expressions for the adjusted p-values $\widetilde{P}_{0n}^{gEV}(m)$ of the gEV-controlling gTP-based procedures, as one cannot solve explicitly for α in Equation (6.83).

Proof of Proposition 6.13. As for Proposition 6.8, the proof follows straightforwardly from the general definition of an adjusted p-value in Equation (1.58).

$$\begin{aligned}
\widetilde{P}_{0n}^{gEV}(m) &= \inf \{\alpha \in [0,1] : m \in \mathcal{R}_n^{gEV}(g;\alpha)\} \\
&= \inf \{\alpha \in [0,1] : m \in \mathcal{R}_n^{gTP}(q(\alpha), g; \alpha^{gTP}(\alpha))\} \\
&= \inf \left\{\alpha \in [0,1] : \widetilde{P}_{0n}^{gTP(q(\alpha))}(m) \leq \alpha^{gTP}(\alpha)\right\}.
\end{aligned}$$

□

6.7 Initial FWER- and gFWER-controlling multiple testing procedures

The performance of an augmentation multiple testing procedure, in terms of Type I error control, power, and computational efficiency, is clearly related to that of the initial MTP (Dudoit et al., 2004a; van der Laan et al., 2005). Hence, great care should be exercised in choosing a suitable initial procedure.

For control of the generalized tail probability error rate, $gTP(q,g) = \Pr(g(V_n, R_n) > q)$, using augmentation Procedure 6.9, a number of decisions must be made regarding the initial $gFWER(k_0)$-controlling MTP. Choices include the allowed number k_0 of false positives and, given k_0, the selection of a suitable $gFWER(k_0)$-controlling MTP. For example, one could select FWER-controlling step-down maxT Procedure 3.11 ($k_0 = 0$) or $gFWER(k_0)$-controlling single-step common-quantile $P(k+1)$ Procedure 3.19 ($k_0 \geq 0$). The empirical Bayes framework of Chapter 7 offers promising new approaches for obtaining a powerful initial gFWER-controlling procedure. Furthermore, $gFWER(k)$ control could be achieved in two (or more) steps, by augmenting an initial $gFWER(k_0)$-controlling MTP ($k_0 < k$).

For initial FWER control (i.e., for $k_0 = 0$), single-step and stepwise procedures are discussed in depth in other chapters and articles (Chapters 3–5 and 7; Dudoit et al. (2004a,b); van der Laan et al. (2004a,b, 2005); Pollard and van der Laan (2004)). Specifically, basic FWER-controlling procedures are summarized in Section 3.2, with emphasis on marginal MTPs, that rely solely on the marginal distributions of the test statistics (e.g., Bonferroni Procedure 3.1). Section 3.2 also introduces four procedures that take into account the joint distribution of the test statistics: single-step maxT Procedure 3.5, single-step minP Procedure 3.6, step-down maxT Procedure 3.11, and step-down minP Procedure 3.12. As the names suggest, the maxT (or common-cut-off) procedures are based on the distributions of maxima of test statistics over sets of null hypotheses, while the minP (or common-quantile) procedures are based on the distributions of minima of unadjusted p-values. Detailed accounts of the single-step maxT and minP and step-down maxT and minP procedures are given in Chapters 4 and 5, respectively. One could also obtain an initial FWER-controlling MTP using the resampling-based empirical Bayes approach of Chapter 7.

Currently available gFWER-controlling procedures are summarized in Section 3.3 and include the following main approaches: (i) marginal single-step Bonferroni-like Procedure 3.15 and step-down Holm-like Procedure 3.17

(Lehmann and Romano, 2005); (ii) joint single-step common-cut-off $T(k+1)$ Procedure 3.18 and common-quantile $P(k+1)$ Procedure 3.19, discussed in detail in Chapter 4 (Dudoit et al., 2004b; Pollard and van der Laan, 2004); (iii) general (marginal/joint single-step/stepwise) augmentation multiple testing Procedure 3.20, discussed in detail in the present chapter (van der Laan et al., 2004b); (iv) joint resampling-based empirical Bayes Procedure 7.1, discussed in detail in Chapter 7 (van der Laan et al., 2005).

In addition to proper Type I error control, one of the main issues in choosing an initial MTP is power. Joint procedures, such as the FWER-controlling single-step and step-down maxT and minP MTPs, can be substantially more powerful than marginal procedures and are thus recommended as initial MTPs (Dudoit et al., 2003, 2004a).

6.8 Discussion

Building on van der Laan et al. (2004b), this chapter proposes multiple testing procedures for controlling generalized tail probability (gTP) error rates, $gTP(q,g) = \Pr(g(V_n, R_n) > q)$, for arbitrary functions $g(V_n, R_n)$ of the numbers of false positives V_n and rejected hypotheses R_n. Special cases include tail probabilities for the number of false positives and the proportion of false positives among the rejected hypotheses, i.e., the generalized family-wise error rate (gFWER), with $g(v,r) = v$, and tail probabilities for the proportion of false positives (TPPFP), with $g(v,r) = v/r$.

Specifically, Section 6.5 shows that one can map any multiple testing procedure $\mathcal{R}_n$, controlling $gFWER(k_0)$ at level α_n (for a fixed sample size n), directly and computationally trivially, into a multiple testing procedure $\mathcal{R}_n^+$, controlling $gTP(q,g)$ at level α_n, for any user-supplied function g and Type I error bound q. Moreover, if the initial multiple testing procedure $\mathcal{R}_n$ (i) provides exact asymptotic control of the gFWER at level α and (ii) rejects all false null hypotheses $\mathcal{H}_1$ with probability tending to one, then the corresponding augmentation multiple testing procedure $\mathcal{R}_n^+$ provides exact asymptotic control of the gTP at level α. Assumptions (i) and (ii) are satisfied, under general conditions, by the FWER-controlling step-down maxT and minP procedures proposed in van der Laan et al. (2004a) and described in Chapter 5. The asymptotic control results are relevant, because in many situations one can only establish that an initial MTP provides asymptotic (vs. finite sample) control of the gFWER. As shown in Section 6.5.2, the adjusted p-values for an augmentation multiple testing procedure are simply shifted versions of the ordered adjusted p-values for the initial MTP. Section 6.6 demonstrates that one can readily derive (conservative) procedures controlling generalized expected value (gEV) error rates, $gEV(g) = \mathrm{E}[g(V_n, R_n)]$, based on procedures controlling generalized tail probability error rates, $gTP(q,g) = \Pr(g(V_n, R_n) > q)$.

Augmentation procedures, Type I error control results, and adjusted p-values for the special cases of the gFWER and TPPFP are discussed in detail in Sections 6.2 and 6.5.3 and Sections 6.3 and 6.5.4, respectively. The false discovery rate (FDR), corresponding to the generalized expected value $gEV(g) = \mathrm{E}[g(V_n, R_n)]$ for the function $g(v, r) = v/r$, is treated in Section 6.4.

Unlike existing results on TPPFP and FDR control, which assume either independence or specific dependence structures for the joint distribution of the test statistics, the results presented in this chapter hold for general data generating distributions (i.e., arbitrary joint distributions for the test statistics), null hypotheses, and test statistics.

We stress the generality and important practical implications of the augmentation approach to multiple testing. As proved in Section 6.5, *any* multiple testing procedure $\mathcal{R}_n$ controlling the gFWER immediately provides multiple testing procedures controlling a wide variety of error rates defined as tail probabilities $\Pr(g(V_n, R_n) > q)$ for arbitrary functions $g(V_n, R_n)$ of the numbers of false positives V_n and rejected hypotheses R_n. One can therefore capitalize on the large pool of available FWER-controlling procedures, such as the single-step and step-down maxT and minP procedures.

Augmentation multiple testing procedures provide a simple and general approach for controlling generalized tail probability error rates and compare favorably to previously proposed gFWER- and TPPFP-controlling marginal procedures (Dudoit et al., 2004a; van der Laan et al., 2005). However, AMTPs tend to be conservative in finite sample situations, because every additional rejected hypothesis is counted as a false positive when deriving the augmentation set. Motivated by these observations, van der Laan et al. (2005) propose a TPPFP-controlling resampling-based empirical Bayes procedure, which is extended in Chapter 7 for controlling gTP error rates.

A number of open questions remain to be addressed. Firstly, by expanding the class of Type I error rates one can control to gTP error rates, augmentation procedures raise the issue of the selection of an appropriate function g and Type I error bound q. Although such choices are obviously subject matter-related, data-adaptive methods may also be relevant. Secondly, as discussed in Section 6.7, various options are available for the initial gFWER-controlling procedure. Preliminary simulation studies and data analyses indicate, as expected, that the performance of an augmentation multiple testing procedure can be greatly affected by that of the initial MTP (Dudoit et al., 2004a; van der Laan et al., 2005). Finally, it would be of interest to combine the augmentation and resampling-based empirical Bayes procedures of Chapters 6 and 7, which are, to our knowledge, the only currently available approaches for controlling generalized tail probability error rates.

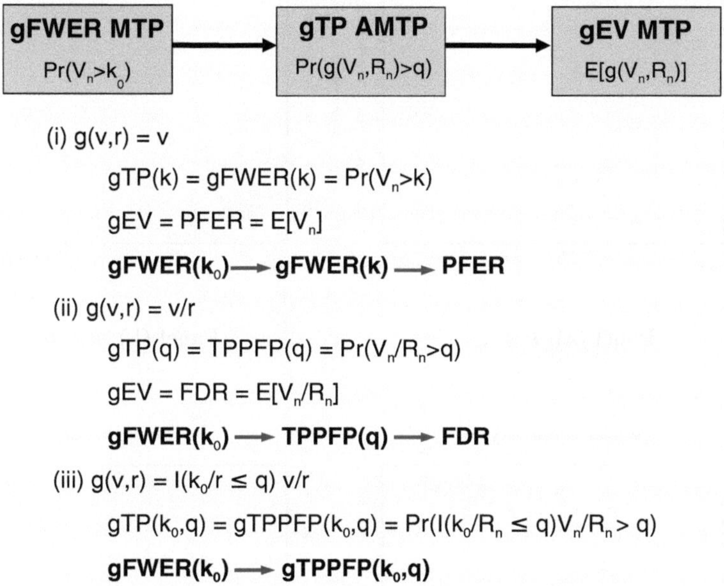

Figure 6.1. *Multiple testing procedures for controlling generalized tail probability error rates and generalized expected value error rates.* The flowchart represents augmentation multiple testing procedures for controlling generalized tail probability (gTP) error rates, $gTP(q, g) = \Pr(g(V_n, R_n) > q)$, based on an initial procedure controlling the generalized family-wise error rate, $gFWER(k_0) = \Pr(V_n > k_0)$. The gTP-controlling procedure may then be used to obtain a procedure controlling the generalized expected value (gEV) error rate, $gEV(g) = \mathrm{E}[g(V_n, R_n)]$. Special cases of gTP- and gEV-controlling MTP pairs, corresponding to different choices for the g-function, include the following. (i) $g(v, r) = v$: generalized family-wise error rate, $gTP(k) = gFWER(k) = \Pr(V_n > k)$, and per-family error rate, $gEV = PFER = \mathrm{E}[V_n]$. (ii) $g(v, r) = v/r$: tail probability for the proportion of false positives, $gTP(q) = TPPFP(q) = \Pr(V_n/R_n > q)$, and false discovery rate, $gEV = FDR = \mathrm{E}[V_n/R_n]$. (iii) $g(v, r) = \mathrm{I}(k_0/r \le q)\, v/r$: generalized tail probability for the proportion of false positives, $gTP(k_0, q) = gTPPFP(k_0, q) = \Pr(\mathrm{I}(k_0/R_n \le q)\, V_n/R_n > q)$.

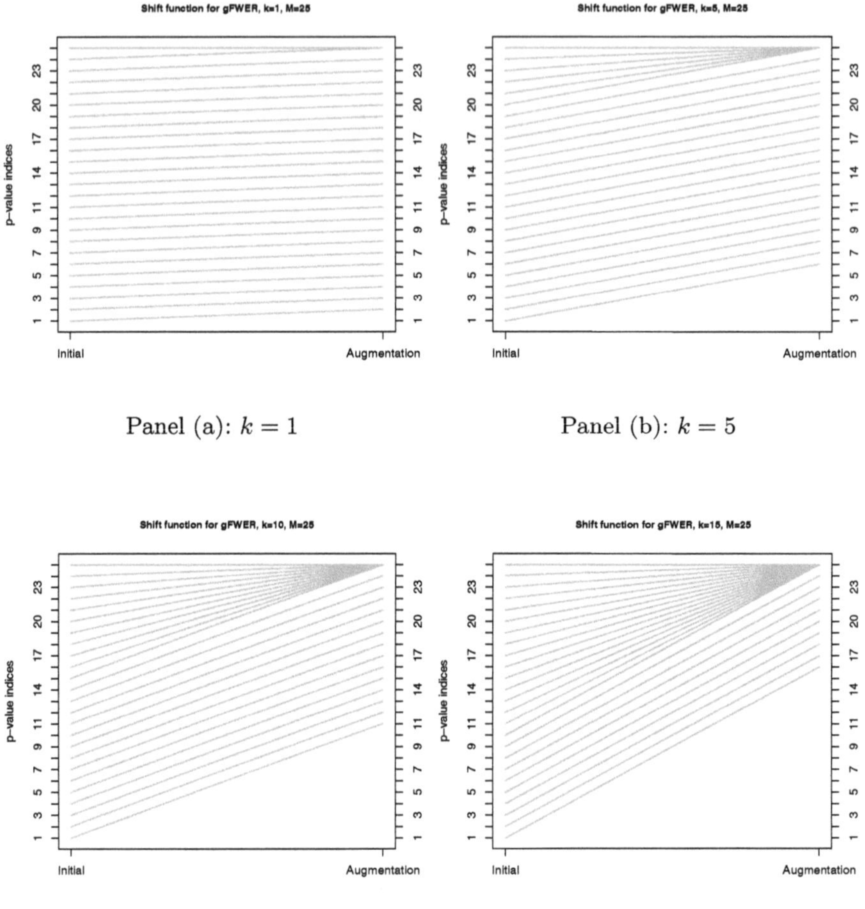

Figure 6.2. *Adjusted p-value shift function for a gFWER-controlling AMTP.* Parallel coordinate plots of shift function, $m \to S_n(m)$, for a $gFWER(k)$-controlling augmentation multiple testing procedure based on an initial FWER-controlling procedure, with $M = 25$ null hypotheses and an allowed number $k \in \{1, 5, 10, 15\}$ of Type I errors.

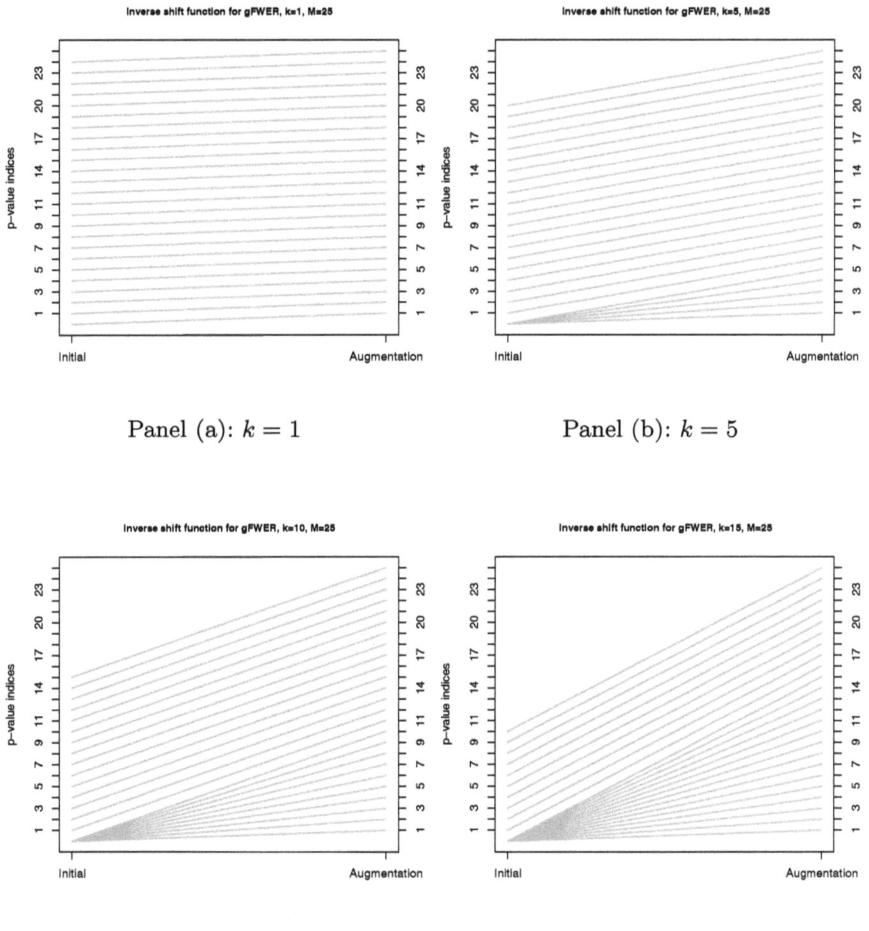

Panel (a): $k = 1$

Panel (b): $k = 5$

Panel (c): $k = 10$

Panel (d): $k = 15$

Figure 6.3. *Adjusted p-value inverse shift function for a gFWER-controlling AMTP.* Parallel coordinate plots of inverse shift function, $S_n^{-1}(m) \to m$, for a $gFWER(k)$-controlling augmentation multiple testing procedure based on an initial FWER-controlling procedure, with $M = 25$ null hypotheses and an allowed number $k \in \{1, 5, 10, 15\}$ of Type I errors.

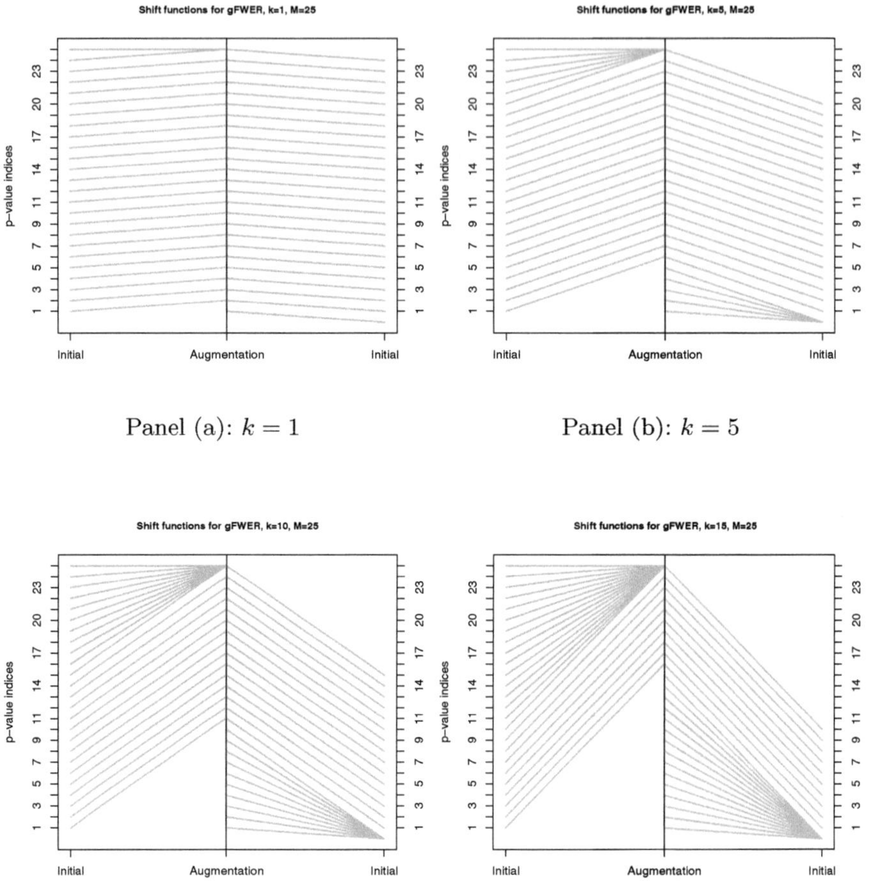

Panel (a): $k = 1$ Panel (b): $k = 5$

Panel (c): $k = 10$ Panel (d): $k = 15$

Figure 6.4. *Adjusted p-value shift and inverse shift functions for a gFWER-controlling AMTP.* Parallel coordinate plots of shift function, $m \to S_n(m)$, and inverse shift function, $m \to S_n^{-1}(m)$, for a $gFWER(k)$-controlling augmentation multiple testing procedure based on an initial FWER-controlling procedure, with $M = 25$ null hypotheses and an allowed number $k \in \{1, 5, 10, 15\}$ of Type I errors.

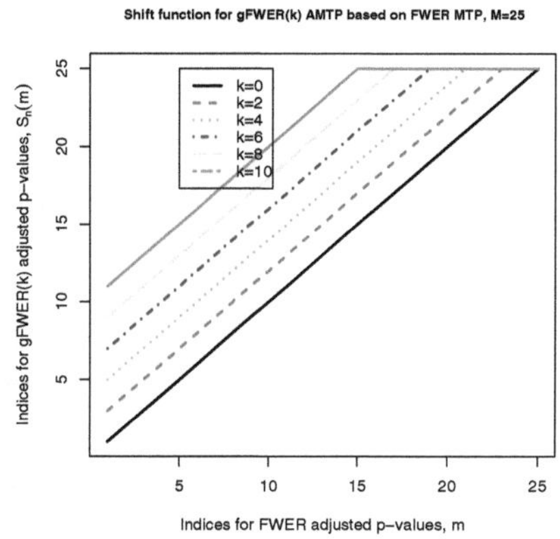

Panel (a): Shift function

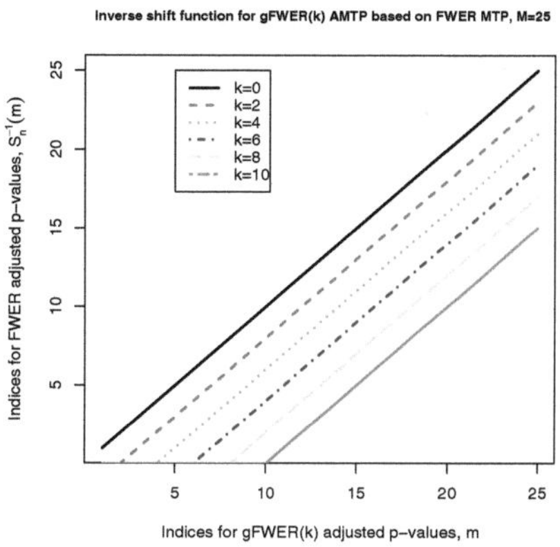

Panel (b): Inverse shift function

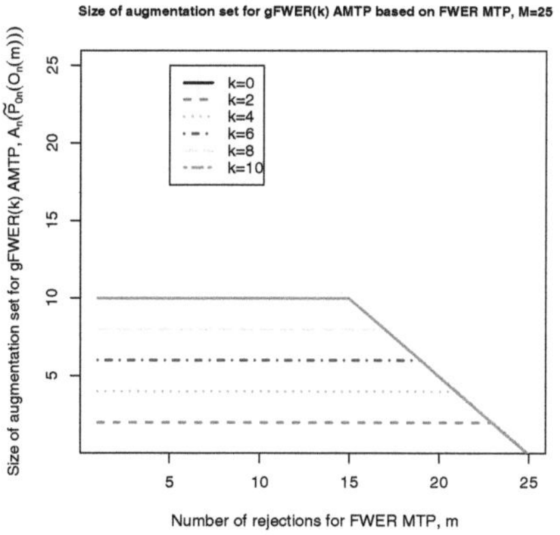

Panel (c): Augmentation set

Figure 6.5. *Sets of rejected hypotheses and adjusted p-values for a gFWER-controlling AMTP.* $gFWER(k)$-controlling augmentation multiple testing procedure based on an initial FWER-controlling procedure, with $M = 25$ null hypotheses and an allowed number $k \in \{0, 2, 4, 6, 8, 10\}$ of Type I errors. Panel (a): Adjusted p-value shift function $S_n(m)$ vs. m. Panel (b): Adjusted p-value inverse shift function $S_n^{-1}(m)$ vs. m. Panel (c): Cardinality $A_n(\widetilde{P}_{0n}(O_n(m)))$ of the augmentation set for the AMTP vs. number of rejected hypotheses $R_n(\widetilde{P}_{0n}(O_n(m))) = m$ for the initial MTP, for nominal Θ and Θ^+ Type I error level $\alpha = \widetilde{P}_{0n}(O_n(m))$. (Color plate p. 325)

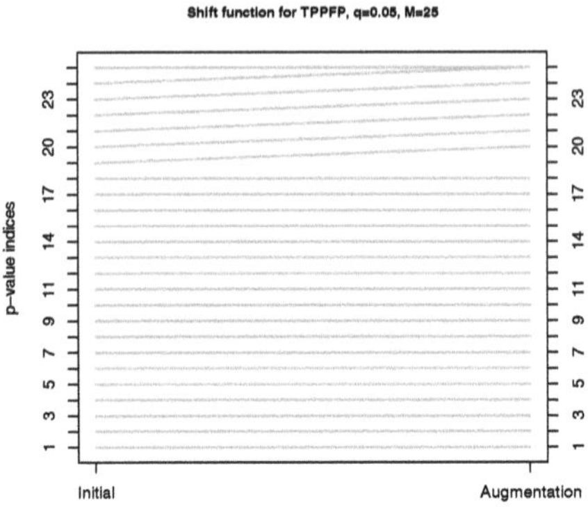

Panel (a): $q = 0.05$

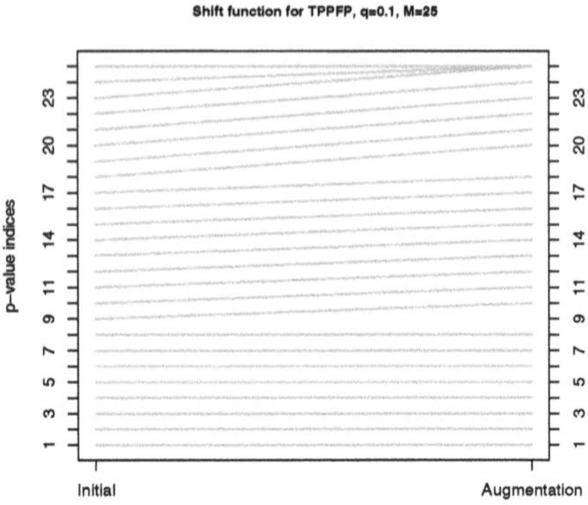

Panel (b): $q = 0.10$

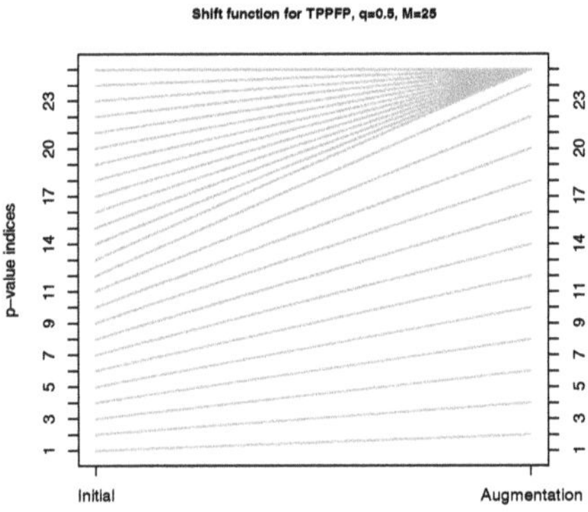

Panel (c): $q = 0.50$

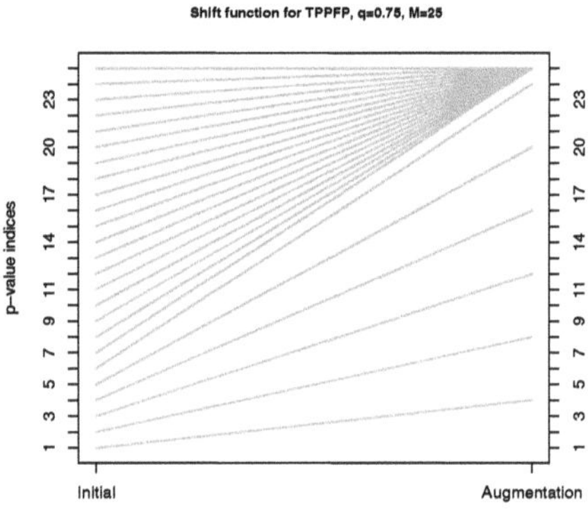

Panel (d): $q = 0.75$

Figure 6.6. *Adjusted p-value shift function for a TPPFP-controlling AMTP.* Parallel coordinate plots of shift function, $m \rightarrow S_n(m)$, for a $TPPFP(q)$-controlling augmentation multiple testing procedure based on an initial FWER-controlling procedure, with $M = 25$ null hypotheses and an allowed proportion $q \in \{0.05, 0.10, 0.50, 0.75\}$ of Type I errors.

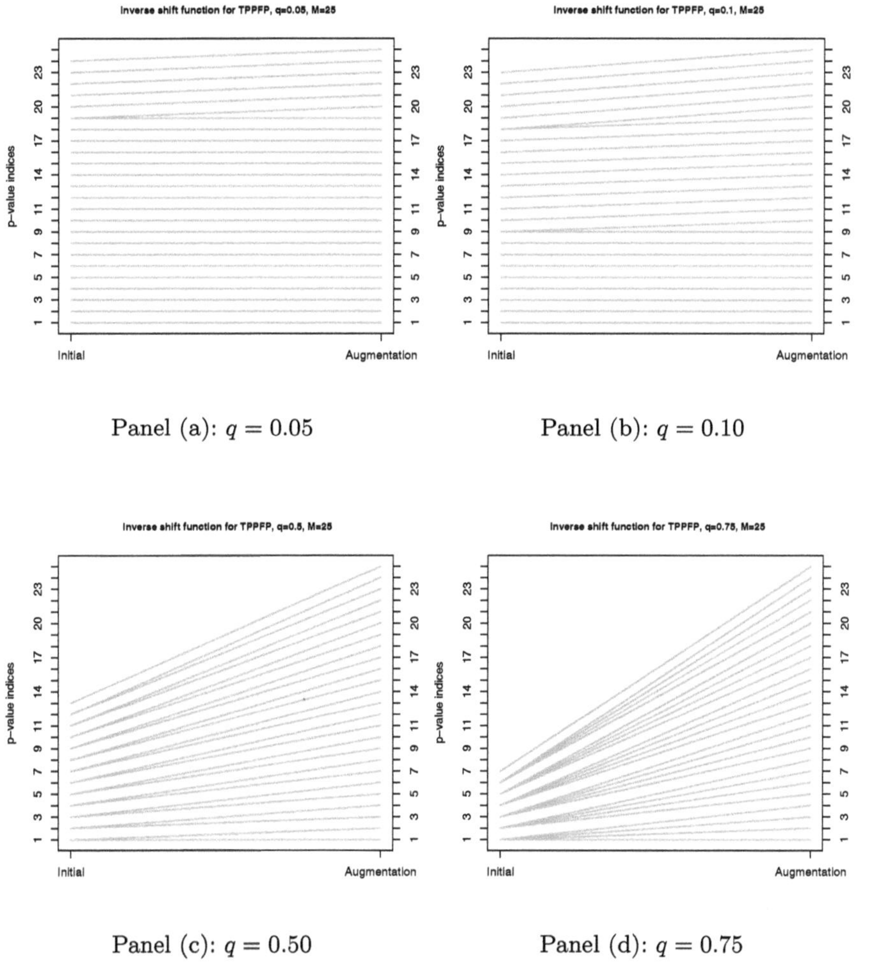

Figure 6.7. *Adjusted p-value inverse shift function for a TPPFP-controlling AMTP.* Parallel coordinate plots of inverse shift function, $S_n^{-1}(m) \to m$, for a $TPPFP(q)$-controlling augmentation multiple testing procedure based on an initial FWER-controlling procedure, with $M = 25$ null hypotheses and an allowed proportion $q \in \{0.05, 0.10, 0.50, 0.75\}$ of Type I errors.

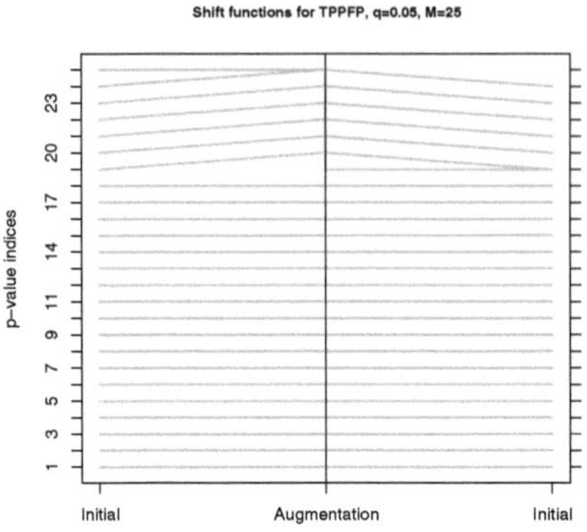

Panel (a): $q = 0.05$

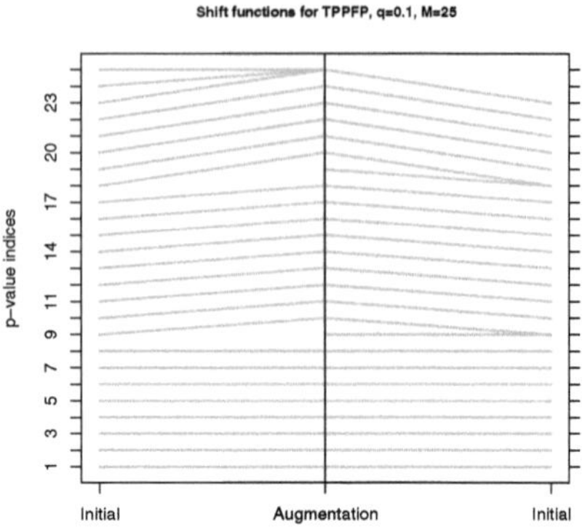

Panel (b): $q = 0.10$

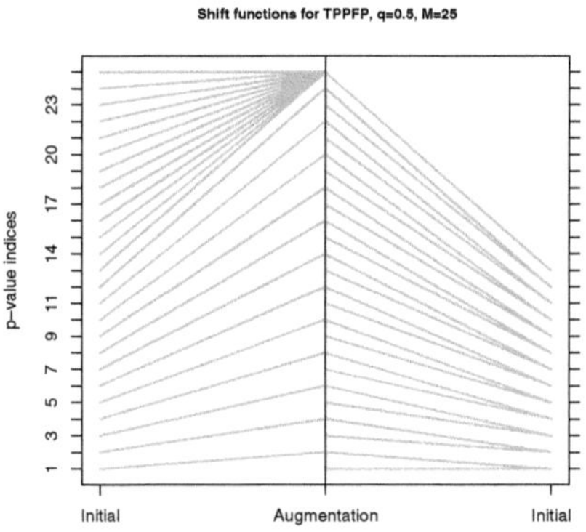

Panel (c): $q = 0.50$

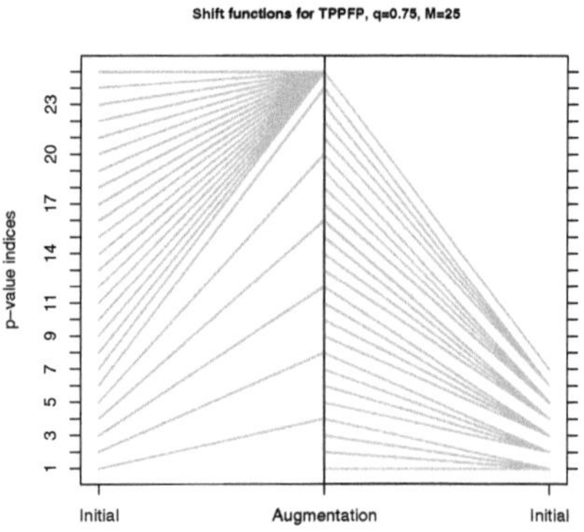

Panel (d): $q = 0.75$

Figure 6.8. *Adjusted p-value shift and inverse shift functions for a TPPFP-controlling AMTP.* Parallel coordinate plots of shift function, $m \rightarrow S_n(m)$, and inverse shift function, $m \rightarrow S_n^{-1}(m)$, for a $TPPFP(q)$-controlling augmentation multiple testing procedure based on an initial FWER-controlling procedure, with $M = 25$ null hypotheses and an allowed proportion $q \in \{0.05, 0.10, 0.50, 0.75\}$ of Type I errors.

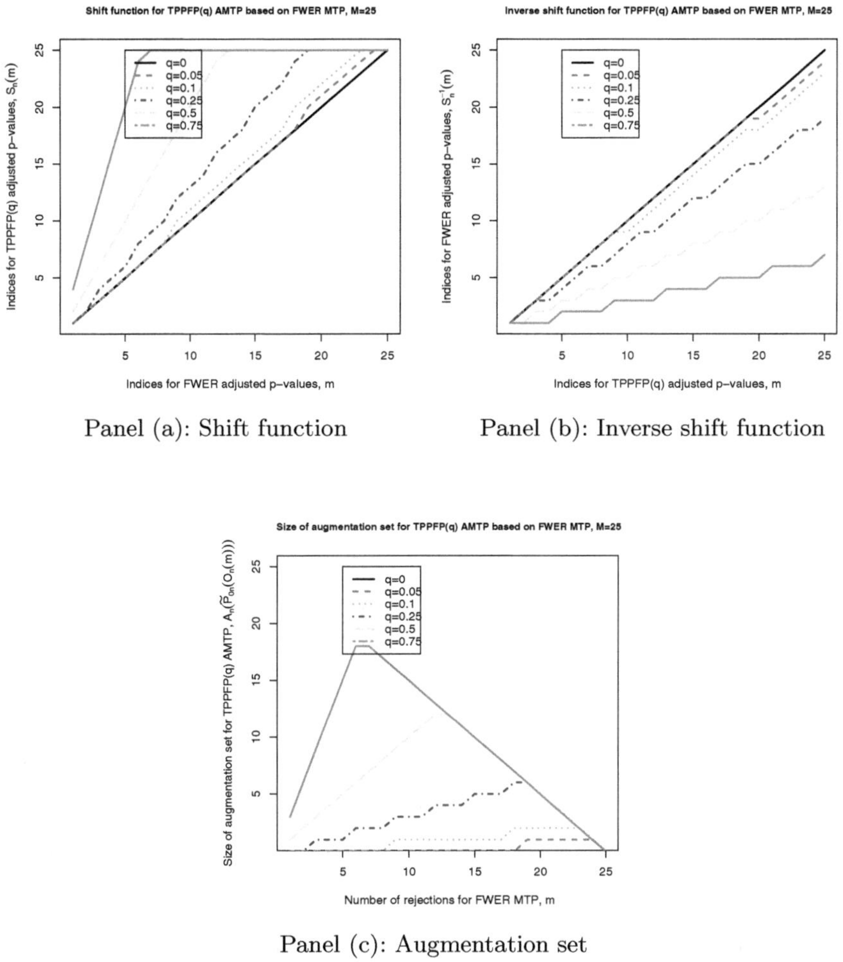

Figure 6.9. *Sets of rejected hypotheses and adjusted p-values for a TPPFP-controlling AMTP.* $TPPFP(q)$-controlling augmentation multiple testing procedure based on an initial FWER-controlling procedure, with $M = 25$ null hypotheses and an allowed proportion $q \in \{0, 0.05, 0.10, 0.25, 0.50, 0.75\}$ of Type I errors. Panel (a): Adjusted p-value shift function $S_n(m)$ vs. m. Panel (b): Adjusted p-value inverse shift function $S_n^{-1}(m)$ vs. m. Panel (c): Cardinality $A_n(\widetilde{P}_{0n}(O_n(m)))$ of the augmentation set for the AMTP vs. number of rejected hypotheses $R_n(\widetilde{P}_{0n}(O_n(m))) = m$ for the initial MTP, for nominal Θ and Θ^+ Type I error level $\alpha = \widetilde{P}_{0n}(O_n(m))$. (Color plate p. 326)

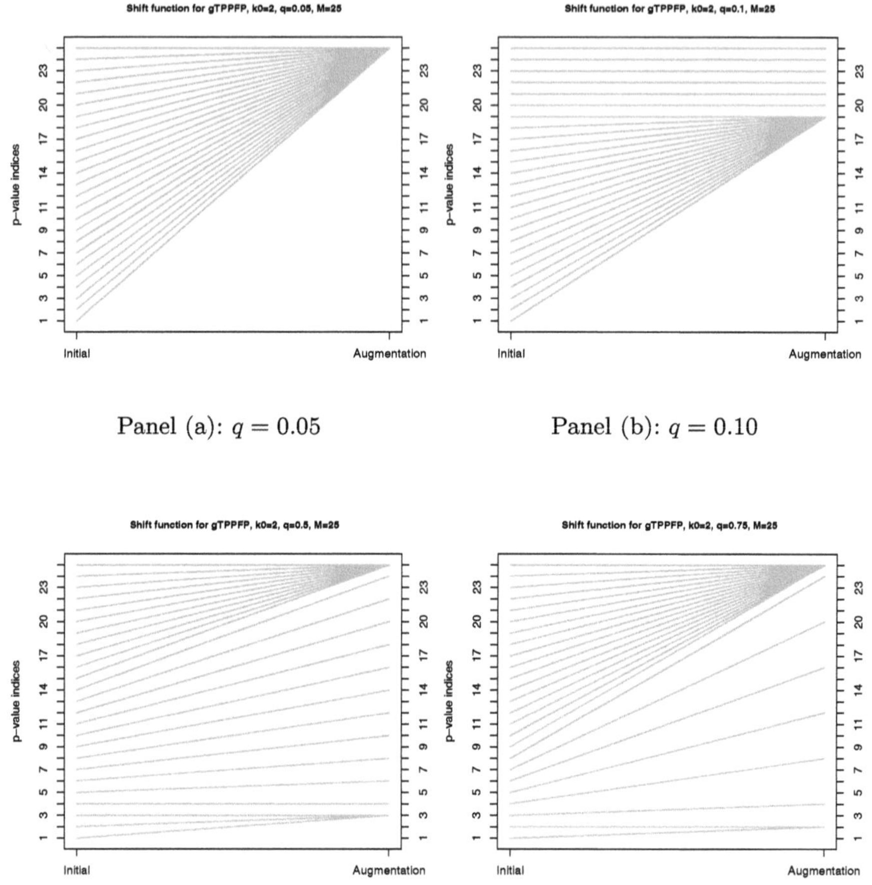

Panel (a): $q = 0.05$ Panel (b): $q = 0.10$

Panel (c): $q = 0.50$ Panel (d): $q = 0.75$

Figure 6.10. *Adjusted p-value shift function for a gTPPFP-controlling AMTP.* Parallel coordinate plots of shift function, $m \rightarrow S_n(m)$, for a $gTPPFP(k_0, q)$-controlling augmentation multiple testing procedure based on an initial $gFWER(k_0)$-controlling procedure, with $M = 25$ null hypotheses, an allowed number $k_0 = 2$ of Type I errors, and an allowed proportion $q \in \{0.05, 0.10, 0.50, 0.75\}$ of Type I errors.

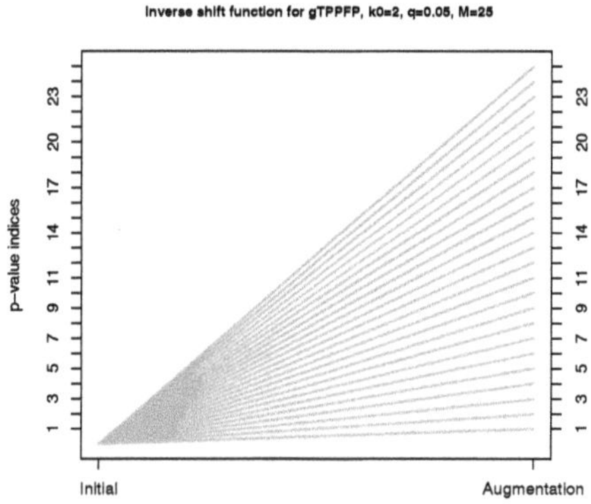

Panel (a): $q = 0.05$

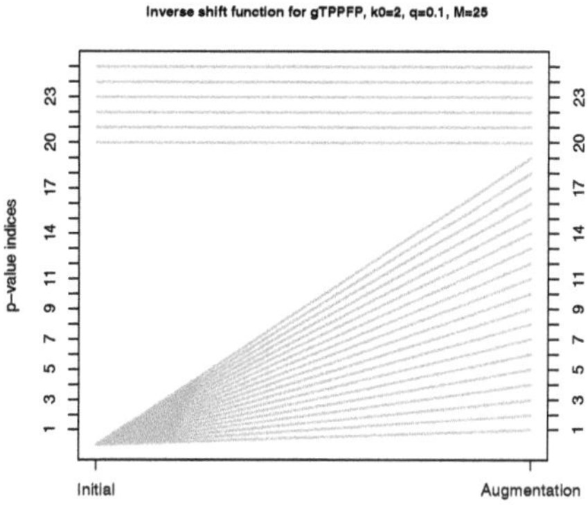

Panel (b): $q = 0.10$

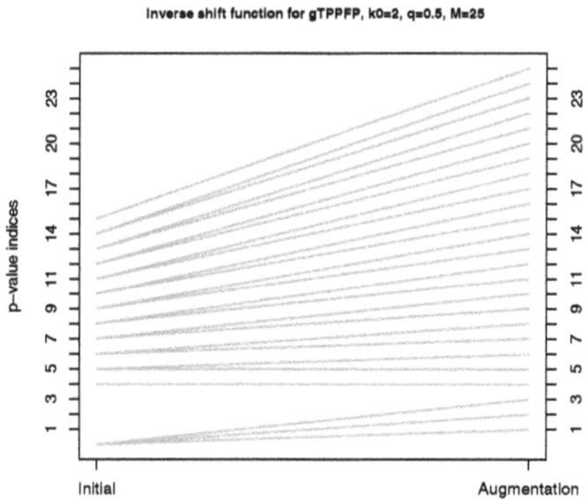

Panel (c): $q = 0.50$

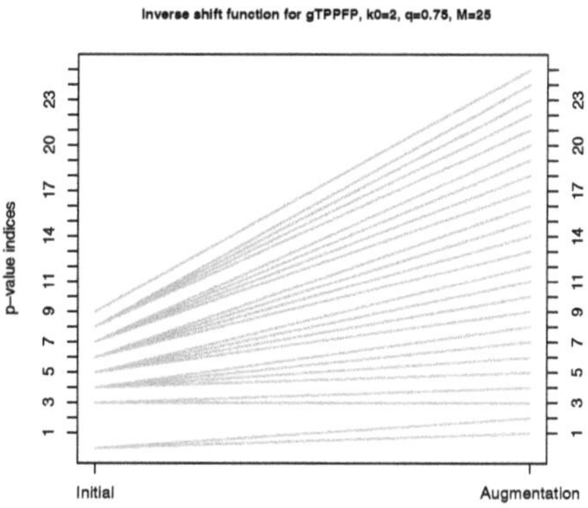

Panel (d): $q = 0.75$

Figure 6.11. *Adjusted p-value inverse shift function for a gTPPFP-controlling AMTP.* Parallel coordinate plots of inverse shift function, $S_n^{-1}(m) \to m$, for a $gTPPFP(k_0, q)$-controlling augmentation multiple testing procedure based on an initial $gFWER(k_0)$-controlling procedure, with $M = 25$ null hypotheses, an allowed number $k_0 = 2$ of Type I errors, and an allowed proportion $q \in \{0.05, 0.10, 0.50, 0.75\}$ of Type I errors.

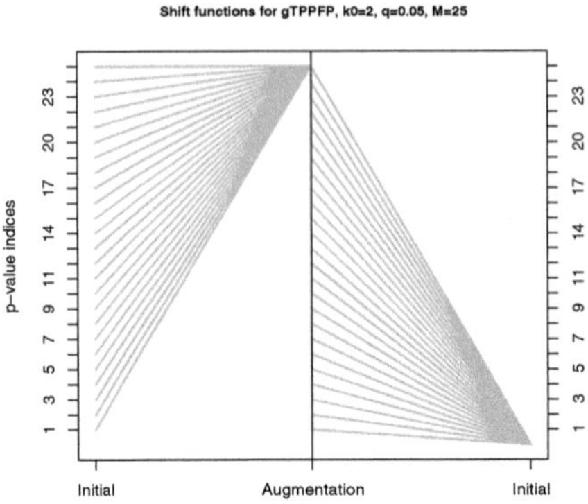

Panel (a): $q = 0.05$

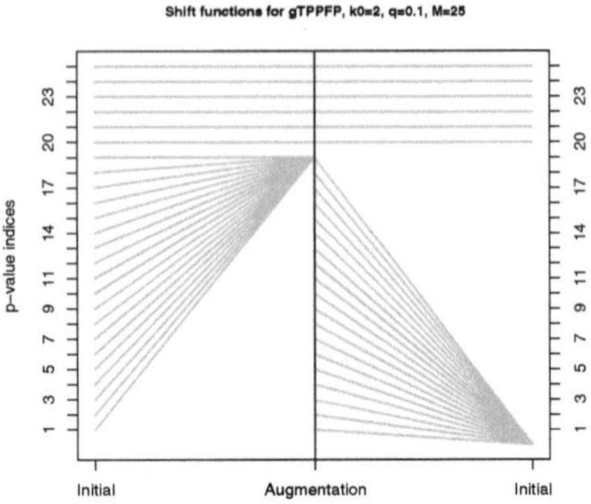

Panel (b): $q = 0.10$

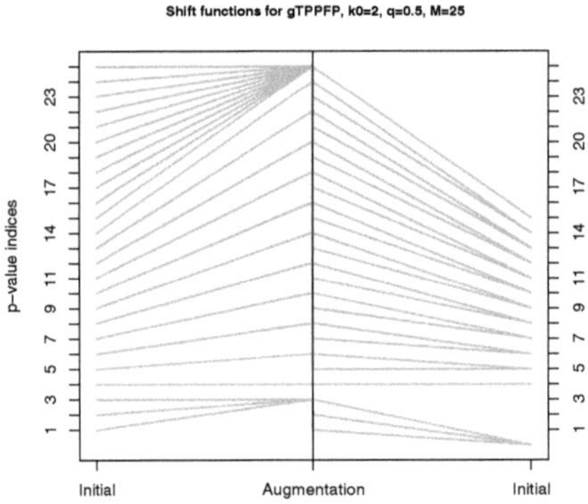

Panel (c): $q = 0.50$

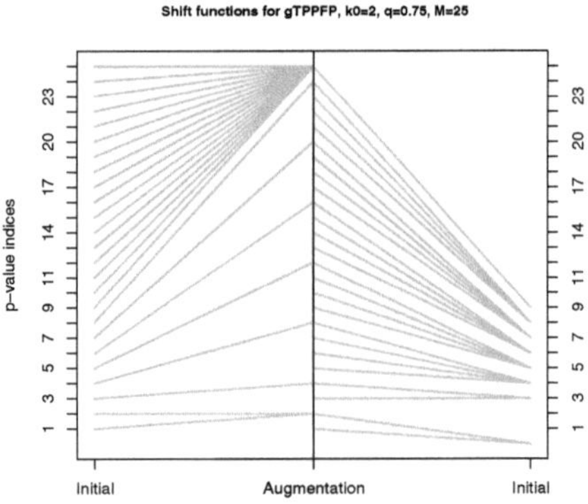

Panel (d): $q = 0.75$

Figure 6.12. *Adjusted p-value shift and inverse shift functions for a gTPPFP-controlling AMTP.* Parallel coordinate plots of shift function, $m \rightarrow S_n(m)$, and inverse shift function, $m \rightarrow S_n^{-1}(m)$, for a $gTPPFP(k_0, q)$-controlling augmentation multiple testing procedure based on an initial $gFWER(k_0)$-controlling procedure, with $M = 25$ null hypotheses, an allowed number $k_0 = 2$ of Type I errors, and an allowed proportion $q \in \{0.05, 0.10, 0.50, 0.75\}$ of Type I errors.

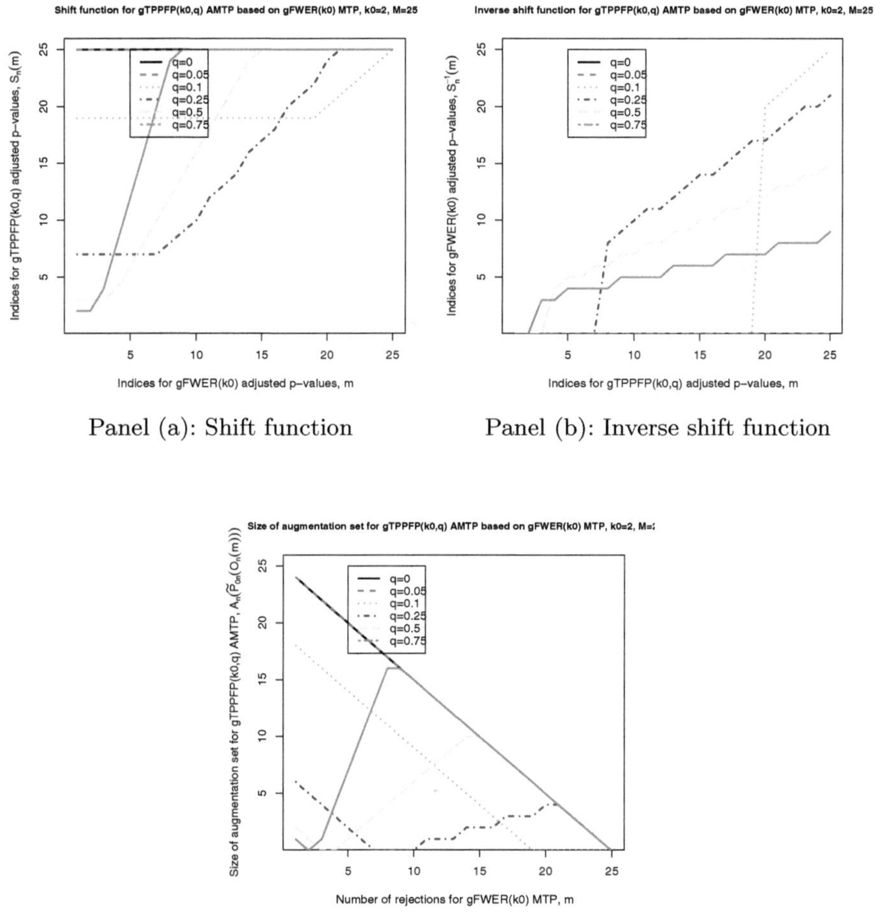

Panel (a): Shift function

Panel (b): Inverse shift function

Panel (c): Augmentation set

Figure 6.13. *Sets of rejected hypotheses and adjusted p-values for a gTPPFP-controlling AMTP.* $gTPPFP(k_0, q)$-controlling augmentation multiple testing procedure based on an initial $gFWER(k_0)$-controlling procedure, with $M = 25$ null hypotheses, an allowed number $k_0 = 2$ of Type I errors, and an allowed proportion $q \in \{0, 0.05, 0.10, 0.25, 0.50, 0.75\}$ of Type I errors. Panel (a): Adjusted p-value shift function $S_n(m)$ vs. m. Panel (b): Adjusted p-value inverse shift function $S_n^{-1}(m)$ vs. m. Panel (c): Cardinality $A_n(\widetilde{P}_{0n}(O_n(m)))$ of the augmentation set for the AMTP vs. number of rejected hypotheses $R_n(\widetilde{P}_{0n}(O_n(m))) = m$ for the initial MTP, for nominal Θ and Θ^+ Type I error level $\alpha = \widetilde{P}_{0n}(O_n(m))$. (Color plate p. 327)

7

Resampling-Based Empirical Bayes Multiple Testing Procedures for Controlling Generalized Tail Probability Error Rates

7.1 Introduction

7.1.1 Motivation

Multiple testing problems currently encountered in biomedical and genomic research are characterized by a large number of hypotheses (in the thousands, or even millions), concerning high-dimensional multivariate distributions, with complex and unknown dependence structures among variables. For instance, for the identification of differentially expressed or co-expressed genes as in Chapter 9, microarray datasets typically consist of thousands of expression measures for fewer than one hundred observational units. As argued in Section 3.4.1, multiple testing procedures (MTP) controlling the *proportion of false positives among the rejected hypotheses* are particularly appealing for large-scale testing problems. However, only a handful of approaches are currently available for controlling such Type I error rates.

Procedures controlling the *expected value of the proportion of false positives*, i.e., the *false discovery rate* (FDR), are summarized in Section 3.4. Existing MTPs controlling the *tail probability for the proportion of false positives* (TPPFP) are reviewed in Section 3.5 and include: (i) marginal step-down Procedures 3.24 and 3.25 (Lehmann and Romano, 2005); (ii) the marginal inversion method for independent test statistics and its conservative version for test statistics with general dependence structures (Genovese and Wasserman, 2004a,b); (iii) general (marginal/joint single-step/stepwise) augmentation multiple testing Procedure 3.26, discussed in detail in Chapter 6 (van der Laan et al., 2004b); (iv) joint resampling-based empirical Bayes Procedure 7.1, discussed in detail in the present chapter (van der Laan et al., 2005).

The first two types of TPPFP-controlling procedures, based solely on the *marginal* distributions of the test statistics (i.e., on unadjusted p-values), either rely on assumptions concerning the joint distribution of these test statistics (e.g., independence, Simes' Inequality) or err on the conservative side by using Bonferroni-like adjustments. *Augmentation multiple testing procedures*

(AMTP) provide a simple and general approach for controlling generalized tail probability error rates, that can account for the *joint* distribution of the test statistics. However, even joint AMTPs tend to be conservative in finite sample situations, as they count every additional rejected hypothesis as a false positive. The simulation studies in Dudoit et al. (2004a) and van der Laan et al. (2005) suggest that, although AMTPs compare favorably to TPPFP-controlling marginal procedures, they become more conservative as the number of tested hypotheses increases. The latter feature is problematic for the large-scale testing problems commonly-encountered in genomics.

Motivated by these observations, van der Laan et al. (2005) propose a new multiple testing approach for controlling TPPFP, which, as does the augmentation method, provides asymptotic Type I error control for general data generating distributions, but is less conservative for finite samples. The van der Laan et al. (2005) *TPPFP-controlling resampling-based empirical Bayes procedure* involves specifying:

- a null distribution Q_0 (or estimator thereof, Q_{0n}) for M-vectors of *null test statistics* T_{0n};
- a distribution $Q_{0n}^{\mathcal{H}}$ for *random guessed sets of true null hypotheses* $\mathcal{H}_{0n}$.

A proposed working model for generating pairs of random variables $(T_{0n}, \mathcal{H}_{0n})$ is a *common marginal non-parametric mixture distribution* for the test statistics T_n. By randomly sampling null test statistics T_{0n} and guessed sets of true null hypotheses $\mathcal{H}_{0n}$, one obtains a distribution for a random variable $G_n(c)$ representing the *guessed proportion of false positives* (given the empirical distribution P_n), for any given cut-off vector c. Cut-offs can then be chosen to control tail probabilities for this distribution at a user-supplied level α.

The present chapter extends the TPPFP-controlling empirical Bayes approach of van der Laan et al. (2005) to a broad class of Type I error rates, defined as *generalized tail probability* (gTP) error rates, $gTP(q,g) = \Pr(g(V_n, S_n) > q)$, for arbitrary functions $g(V_n, S_n)$ of the numbers of false positives V_n and true positives $S_n = R_n - V_n$. Such error rates, expressed as $gTP(q,g) = \Pr(g(V_n, R_n) > q)$, in terms of the number of rejected hypotheses $R_n = V_n + S_n$, are discussed in detail in Chapter 6 in the context of augmentation multiple testing procedures. The *generalized family-wise error rate* (gFWER) and the *tail probability for the proportion of false positives* (TPPFP) correspond, respectively, to the special cases of the *number* $g(V_n, S_n) = V_n$ of false positives and *proportion* $g(V_n, S_n) = V_n/(V_n + S_n)$ of false positives among the rejected hypotheses.

7.1.2 Outline

Section 7.2 motivates and states the proposed gTP-controlling resampling-based empirical Bayes method (Procedure 7.1). Adjusted p-values are derived in Section 7.3. Section 7.4 gives a finite sample rationale for this new class of MTPs and Section 7.5 provides formal asymptotic Type I error control results. Section 7.6 proposes a weighted version of the gTP-controlling resampling-based empirical Bayes method. Section 7.7 concerns empirical Bayes q-value-based approaches to FDR control and connections to frequentist step-up Benjamini and Hochberg (1995) Procedure 3.22. Finally, Section 7.8 summarizes our findings and discusses ongoing work.

7.2 gTP-controlling resampling-based empirical Bayes procedures

7.2.1 Notation

Before describing the proposed gTP-controlling resampling-based empirical Bayes Procedure 7.1, we introduce the following notation for the number of false positives (i.e., Type I errors), the number of true positives, the number of rejected hypotheses, and a function g of the numbers of false positives and true positives,

$$\begin{aligned} V(c;\mathcal{H},Z) &\equiv \sum_{m\in\mathcal{H}} \mathrm{I}\left(Z(m) > c(m)\right), \\ S(c;\mathcal{H},Z) &\equiv \sum_{m\notin\mathcal{H}} \mathrm{I}\left(Z(m) > c(m)\right), \\ R(c;\mathcal{H},Z_0,Z) &\equiv V(c;\mathcal{H},Z_0) + S(c;\mathcal{H},Z), \end{aligned} \tag{7.1}$$

and

$$\begin{aligned} G(c;\mathcal{H},Z_0,Z) &\equiv g(V(c;\mathcal{H},Z_0), S(c;\mathcal{H},Z)) \\ &= g(V(c;\mathcal{H},Z_0), R(c;\mathcal{H},Z_0,Z) - V(c;\mathcal{H},Z_0)), \end{aligned}$$

respectively, where $c = (c(m) : m = 1,\dots,M) \in I\!R^M$ denotes an M-dimensional cut-off vector that defines one-sided rejection regions of the form $\mathcal{C}(m) = (c(m),+\infty)$, $\mathcal{H} \subseteq \{1,\dots,M\}$ denotes a set of null hypotheses, and $Z_0 = (Z_0(m) : m = 1,\dots,M)$ and $Z = (Z(m) : m = 1,\dots,M)$ denote random M-vectors of test statistics.

For $c \in I\!R$, let $c^{(M)}$ denote the M-vector with all elements equal to c, i.e., $c^{(M)}(m) \equiv c$, $\forall\ m = 1,\dots,M$.

Given an M-variate distribution Q, with marginal cumulative distribution functions Q_m, and $\delta \in [0,1]$, let

$$q^{-1}(\delta) \equiv (Q_m^{-1}(\delta) : m = 1,\dots,M) \tag{7.2}$$

denote the M-vector of δ-quantiles of Q, defined as

$$Q_m^{-1}(\delta) \equiv \inf\left\{z \in I\!R : Q_m(z) \geq \delta\right\}. \tag{7.3}$$

7.2.2 gTP control and optimal test statistic cut-offs

Let $\mathcal{X}_n = \{X_i : i = 1, \ldots, n\}$ be a random sample of n independent and identically distributed (IID) random variables from a data generating distribution $P \in \mathcal{M}$. Let P_n denote the corresponding empirical distribution, which places probability $1/n$ on each realization of X. Consider the simultaneous test of M null hypotheses, $H_0(m)$, $m = 1, \ldots, M$, based on an M-vector of test statistics $T_n = (T_n(m) : m = 1, \ldots, M)$, with (unknown) true distribution $Q_n = Q_n(P)$. Let $\mathcal{H}_0 = \mathcal{H}_0(P) = \{m : H_0(m) = 1\}$ be the set of $h_0 = |\mathcal{H}_0|$ true null hypotheses, where we note that $\mathcal{H}_0$ depends on the data generating distribution P. Assume that large values of the test statistic $T_n(m)$ provide evidence against the corresponding null hypothesis $H_0(m)$, that is, consider one-sided rejection regions of the form $\mathcal{C}_n(m) = (c_n(m), +\infty)$, corresponding to sets of rejected null hypotheses $\mathcal{R}_n = \{m : T_n(m) > c_n(m)\}$.

For a user-supplied function $g(V_n, S_n)$, of the numbers of false positives $V_n \equiv V(c_n; \mathcal{H}_0, T_n)$ and true positives $S_n \equiv S(c_n; \mathcal{H}_0, T_n)$, and a Type I error bound q, the *generalized tail probability* (gTP) error rate is defined as

$$
\begin{aligned}
gTP(q, g) &\equiv \Pr\left(g(V_n, S_n) > q\right) \qquad (7.4)\\
&= \Pr\left(g(V(c_n; \mathcal{H}_0, T_n), S(c_n; \mathcal{H}_0, T_n)) > q\right)\\
&= \Pr\left(G(c_n; \mathcal{H}_0, T_n, T_n) > q\right)\\
&= \bar{F}_{G(c_n; \mathcal{H}_0, T_n, T_n)}(q),
\end{aligned}
$$

where $\bar{F}_{G(c_n; \mathcal{H}_0, T_n, T_n)}$ denotes the survivor function of the random variable $G(c_n; \mathcal{H}_0, T_n, T_n)$. Note that in this chapter, the gTP is expressed for mere convenience in terms of a function of (V_n, S_n), rather than $(V_n, R_n) = (V_n, V_n + S_n)$, as in Chapter 6 on augmentation multiple testing procedures.

van der Laan et al. (2005) address the special case of TPPFP control, with $g(v, s) = v/(v + s) = v/r$, and propose a resampling-based empirical Bayes common-cut-off procedure. The present chapter extends the results of van der Laan et al. (2005) to generalized tail probability error rates, defined in terms of a function g that satisfies the following two monotonicity assumptions.

Assumption EB.MgV. [Monotonicity of g] The function $g_{S=s} : v \rightarrow g(v, s)$ is continuous and strictly increasing for any given s.

Assumption EB.MgS. [Monotonicity of g] The function $g_{V=v} : s \rightarrow g(v, s)$ is continuous and non-increasing for any given v.

The functions $g(v, s) = v$ and $g(v, s) = v/(v + s)$, defining, respectively, the gFWER and TPPFP, clearly satisfy these two monotonicity assumptions.

Given a user-supplied Type I error level $\alpha \in (0, 1)$, one seeks cut-offs $c_n = (c_n(m) : m = 1, \ldots, M)$, for the test statistics $T_n = (T_n(m) : m = 1, \ldots, M) \sim Q_n$, so that $gTP(q, g)$ is controlled at level α. That is,

$$\bar{F}_{G(c_n;\mathcal{H}_0,T_n,T_n)}(q) \leq \alpha \qquad \text{[finite sample control]}$$
$$\limsup_{n\to\infty} \bar{F}_{G(c_n;\mathcal{H}_0,T_n,T_n)}(q) \leq \alpha \qquad \text{[asymptotic control]}. \tag{7.5}$$

Among the collection of cut-offs that satisfy the above gTP constraints, we distinguish between common cut-offs and common-quantile cut-offs.

Common cut-offs

For simplicity, focus initially on *common-cut-off procedures*, that is, let $c_n = \gamma_n^{(M)}$. As discussed in Section 4.2.4, common-cut-off MTPs are appropriate for identically distributed test statistics $T_n(m)$, i.e., for marginal distributions $Q_{n,m}$ that do not depend on m. The *optimal common cut-off* $\gamma_n^\star$ is defined as

$$\begin{aligned}
\gamma_n^\star &= \gamma(g, q, \alpha; \mathcal{H}_0, Q_n) \qquad (7.6)\\
&\equiv \inf \left\{\gamma \in I\!R : \bar{F}_{G(\gamma^{(M)};\mathcal{H}_0,T_n,T_n)}(q) \leq \alpha\right\}\\
&= \inf \Bigg\{\gamma \in I\!R : \Pr\Bigg(g\Bigg(\sum_{m\in\mathcal{H}_0} \mathrm{I}\left(T_n(m) > \gamma\right),\\
&\qquad \sum_{m\notin\mathcal{H}_0} \mathrm{I}\left(T_n(m) > \gamma\right)\Bigg) > q\Bigg) \leq \alpha\Bigg\}.
\end{aligned}$$

This cut-off is optimal, in the sense that it maximizes the number of rejected hypotheses, while controlling $gTP(q, g)$ at level α.

Common-quantile cut-offs

Given a null distribution Q_0 (or estimator thereof, Q_{0n}) for the test statistics T_n, one can also consider *common-quantile procedures*, specified in terms of cut-off vectors of the form $c_n = q_0^{-1}(\delta_n) = (Q_{0,m}^{-1}(\delta_n) : m = 1, \ldots, M)$. The *optimal common-quantile cut-offs* are based on the *optimal common quantile probability* $\delta_n^\star$, defined as

$$\begin{aligned}
\delta_n^\star &= \delta(g, q, \alpha; \mathcal{H}_0, Q_n, Q_0) \qquad (7.7)\\
&\equiv \inf \left\{\delta \in [0,1] : \bar{F}_{G(q_0^{-1}(\delta);\mathcal{H}_0,T_n,T_n)}(q) \leq \alpha\right\}\\
&= \inf \Bigg\{\delta \in [0,1] : \Pr\Bigg(g\Bigg(\sum_{m\in\mathcal{H}_0} \mathrm{I}\left(T_n(m) > Q_{0,m}^{-1}(\delta)\right),\\
&\qquad \sum_{m\notin\mathcal{H}_0} \mathrm{I}\left(T_n(m) > Q_{0,m}^{-1}(\delta)\right)\Bigg) > q\Bigg) \leq \alpha\Bigg\}.
\end{aligned}$$

Note that the common-quantile cut-offs could in principle be defined in terms of any distribution Q, i.e., not just the null distribution Q_0 for the test statistics.

The long notation $\gamma(g, q, \alpha; \mathcal{H}_0, Q_n)$ and $\delta(g, q, \alpha; \mathcal{H}_0, Q_n, Q_0)$ emphasizes that optimal common cut-offs and common-quantile cut-offs depend on: the set of true null hypotheses $\mathcal{H}_0$; the true distribution Q_n of the test statistics; the function g defining the Type I error rate; the Type I error bound q; and the level α of the MTP. Thus, when attempting to control gTP, one is immediately faced with the problem that the set $\mathcal{H}_0 = \mathcal{H}_0(P)$ of true null hypotheses and the true distribution $Q_n = Q_n(P)$ of the test statistics T_n depend on the unknown data generating distribution P and are therefore unknown.

7.2.3 Overview of gTP-controlling resampling-based empirical Bayes procedures

The optimal gTP-controlling common cut-offs and common-quantile cut-offs introduced in Section 7.2.2 are functions of the *unknown* data generating distribution P. Building on the recent work of van der Laan et al. (2005), we propose an empirical Bayes approach for estimating these cut-offs. *gTP-controlling resampling-based empirical Bayes Procedure 7.1* involves specifying:

- a null distribution Q_0 (or estimator thereof, Q_{0n}) for M-vectors of *null test statistics* T_{0n};
- a distribution $Q_{0n}^{\mathcal{H}}$ for *random guessed sets of true null hypotheses* $\mathcal{H}_{0n}$.

A possible model for generating pairs of random variables $(T_{0n}, \mathcal{H}_{0n})$ is a *common marginal non-parametric mixture distribution* for the test statistics T_n (Section 7.2.4). By randomly sampling null test statistics T_{0n} and guessed sets of true null hypotheses $\mathcal{H}_{0n}$, one obtains a distribution for a random variable $G_n(c)$ representing the *guessed g-specific function of the numbers of false positives and true positives* (given the empirical distribution P_n), for any given cut-off vector c. Cut-offs can then be chosen to control tail probabilities for this distribution at a user-supplied level α.

Specifically, given an M-vector of cut-offs $c = (c(m) : m = 1, \ldots, M)$, the function $G_n(c)$ is defined as

$$
\begin{aligned}
G_n(c) &\equiv G(c; \mathcal{H}_{0n}, T_{0n}, T_n) = g(V(c; \mathcal{H}_{0n}, T_{0n}), S(c; \mathcal{H}_{0n}, T_n)) \qquad (7.8)\\
&= g\left(\sum_{m \in \mathcal{H}_{0n}} \mathrm{I}\left(T_{0n}(m) > c(m)\right), \sum_{m \notin \mathcal{H}_{0n}} \mathrm{I}\left(T_n(m) > c(m)\right) \right),
\end{aligned}
$$

where $T_n \sim Q_n$ is the M-vector of observed test statistics, $T_{0n} \sim Q_{0n}$ is an M-vector of null test statistics, and $\mathcal{H}_{0n} \sim Q_{0n}^{\mathcal{H}}$ is a guessed set of true null hypotheses.

The null test statistics T_{0n} and the guessed sets $\mathcal{H}_{0n}$ are sampled independently, given the empirical distribution P_n, from distributions Q_{0n}

and $Q_{0n}^{\mathcal{H}}$, chosen (conservatively) so that the *guessed* function $G_n(c) = G(c; \mathcal{H}_{0n}, T_{0n}, T_n)$ of the numbers of false positives and true positives is asymptotically stochastically greater than the corresponding *true* function $G(c; \mathcal{H}_0, T_n, T_n)$.

The main components of this new gTP-controlling approach, including the non-parametric mixture model and corresponding sampling distributions Q_{0n} and $Q_{0n}^{\mathcal{H}}$, for null test statistics and guessed sets of true null hypotheses, respectively, are discussed in Section 7.2.4, below, and summarized in Procedure 7.1.

Note that the empirical Bayes procedure outlined above is very general, in the sense that it can be applied to *any* distribution pair $(Q_{0n}, Q_{0n}^{\mathcal{H}})$. The working model of Section 7.2.4 is only one among many reasonable candidate models that does not assume independence of the test statistics.

7.2.4 Working model for distributions of null test statistics and guessed sets of true null hypotheses

Common marginal non-parametric mixture model

A proposed working model for the distribution pair $(Q_{0n}, Q_{0n}^{\mathcal{H}})$ in the gTP-controlling empirical Bayes approach posits M *identically distributed pairs of test statistics and null hypotheses* $((T_n(m), H_0(m)) : m = 1, \ldots, M)$. Test statistics are assumed to have the following *common marginal non-parametric mixture distribution*,

$$T_n(m) \sim f \equiv \pi_0 f_0 + (1 - \pi_0) f_1, \qquad m = 1, \ldots, M, \tag{7.9}$$

where $\pi_0 \equiv \Pr(H_0(m) = 1)$ denotes the *prior probability of a true null hypothesis*, f_0 the *marginal null density of the test statistics*, and f_1 the *marginal alternative density of the test statistics*, i.e., $T_n(m)|\{H_0(m) = 1\} \sim f_0$ and $T_n(m)|\{H_0(m) = 0\} \sim f_1$[1].

Recall that, as defined in Section 1.2.4, null hypotheses are indicators $H_0(m) = \mathrm{I}(P \in \mathcal{M}(m))$, for submodels $\mathcal{M}(m) \subseteq \mathcal{M}$ for the data generating distribution P. That is, $H_0(m) = 1$ corresponds to a true null hypothesis and $H_0(m) = 0$ to a false null hypothesis. Under the mixture model of Equation (7.9), the null hypotheses $H_0(m)$ are identically distributed Bernoulli random variables, with $H_0(m) \sim \text{Bernoulli}(\pi_0)$, i.e., $\Pr(H_0(m) = 1) = \pi_0$, for each $m = 1, \ldots, M$. The full data consist of pairs of random variables $((T_n(m), H_0(m)) : m = 1, \ldots, M)$, where $T_n(m)$ given $H_0(m)$ has density $f_{1-H_0(m)}$. However, one only observes the test statistics $T_n(m)$ and the goal is to estimate the set $\mathcal{H}_0 = \{m : H_0(m) = 1\}$ of true null hypotheses.

[1] N.B. The densities f, f_0, and f_1, for the mixture model of Equation (7.9), and, hence, the associated q-value functions $\Pi_0(t)$ and $\pi_0(t)$, in Equations (7.10), (7.61), and (7.62), depend on the size n of the sample $\mathcal{X}_n$ used to compute the test statistics T_n. However, for simplicity, references to the sample size n are omitted from the notation.

q-values

A parameter of interest, for generating guessed sets of true null hypotheses under the marginal non-parametric mixture model of Equation (7.9), is the *posterior probability function for a true null hypothesis* $H_0(m)$, given the corresponding test statistic $T_n(m)$,

$$\pi_0(t) \equiv \Pr(H_0(m) = 1 | T_n(m) = t) = \frac{\pi_0 f_0(t)}{f(t)}, \qquad m = 1, \ldots, M. \qquad (7.10)$$

The random variables $\pi_0(T_n(m))$ have been referred to in the FDR literature as *local q-values* (Efron, 2005; Storey, 2002, 2003; Storey et al., 2004); we therefore refer to $\pi_0(t)$ as a *local q-value function*[1].

Empirical Bayes q-values are similar in spirit to frequentist p-values: the smaller the q-value $\pi_0(T_n(m))$, the stronger the evidence against the corresponding null hypothesis $H_0(m)$.

For known local q-value function $\pi_0(t)$, one can generate guessed sets $\mathcal{H}_{0n}$ of true null hypotheses as the realizations of M independent Bernoulli$(\pi_0(T_n(m)))$ random variables, $m = 1, \ldots, M$.

However, in practice, the prior probability π_0, the densities f_0 and f, and hence the local q-value function $\pi_0(t)$ are unknown and must be estimated from the available data $\mathcal{X}_n$. Various estimation approaches are outlined below.

The reader is referred to Section 7.7 for further discussion of q-values and FDR-controlling empirical Bayes q-value-based procedures.

Distribution Q_{0n} for the null test statistics

As detailed in Chapter 2, valid test statistics null distributions Q_0 include: the asymptotic distribution of a vector of null shift and scale-transformed test statistics (Section 2.3) and the asymptotic distribution of a vector of null quantile-transformed test statistics (Section 2.4).

Resampling methods, such as bootstrap Procedures 2.3 and 2.4, may be used to conveniently obtain consistent estimators Q_{0n} of these null distributions.

For a broad class of testing problems, such as the test of single-parameter null hypotheses using t-statistics, a proper null distribution is the M-variate Gaussian distribution $\mathrm{N}(0, \sigma^*)$, with mean vector zero and covariance matrix $\sigma^* = \Sigma^*(P)$ equal to the correlation matrix of the vector influence curve for the estimator ψ_n of the parameter of interest ψ (Section 2.6).

Distribution $Q_{0n}^{\mathcal{H}}$ for the guessed sets of true null hypotheses

In order to generate guessed sets $\mathcal{H}_{0n}$ of true null hypotheses, one needs to estimate the local q-value function $\pi_0(t)$. Estimators of $\pi_0(t)$ may be obtained by the *plug-in* method, from estimators of the three main components of the mixture model of Equation (7.9): the prior probability of the true null hypotheses π_0, the test statistics marginal null density f_0, and the test statistics marginal density f.

Estimation of π_0

A simple estimator π_{0n} of the prior probability π_0 of the true null hypotheses is the conservative value of one, i.e., $\pi_{0n} = 1$.

Estimation of f_0

For the test of single-parameter null hypotheses using t-statistics, f_0 is simply a standard normal density, i.e., $T_n(m)|\{H_0(m) = 1\} \sim \mathrm{N}(0, 1)$ (Section 2.6). For other types of test statistics, one may estimate f_0 as in Procedures 2.3 and 2.4, from null-transformed bootstrap test statistics, by applying a kernel density estimator to the $M \times B$ pooled elements of the matrix $\mathbf{Z}_n^B$.

Estimation of f

Likewise, the marginal density f may be estimated as in Procedures 2.3 and 2.4, by applying a kernel density estimator to the $M \times B$ pooled elements of the matrix $\mathbf{T}_n^B$ of raw bootstrap test statistics, before the null transformation.

Guessed sets of true null hypotheses

Given estimators π_{0n}, f_{0n}, and f_n, of the prior null hypotheses probability π_0 and of the marginal densities f_0 and f, respectively, one can estimate the local q-value function $\pi_0(t)$ by

$$\pi_{0n}(t) \equiv \min\left\{1, \frac{\pi_{0n} f_{0n}(t)}{f_n(t)}\right\}. \tag{7.11}$$

Guessed sets of true null hypotheses $\mathcal{H}_{0n}$ can then be generated from a distribution $Q_{0n}^{\mathcal{H}}$ that corresponds to M independent Bernoulli random variables with parameters $\pi_{0n}(T_n(m))$. That is, generate binary random M-vectors $H_{0n} = (H_{0n}(m) : m = 1, \ldots, M)$ of null hypotheses as

$$H_{0n}(m) \stackrel{\perp}{\sim} \mathrm{Bernoulli}\left(\pi_{0n}(T_n(m))\right), \qquad m = 1, \ldots, M, \tag{7.12}$$

so that for any $h \in \{0, 1\}^M$,

$$\Pr(H_{0n} = h) = \prod_{m=1}^{M} \pi_{0n}(T_n(m))^{h(m)}(1 - \pi_{0n}(T_n(m)))^{1-h(m)},$$

and define sets

$$\mathcal{H}_{0n} \equiv \{m : H_{0n}(m) = 1\}. \tag{7.13}$$

Distribution for the g-specific function of the numbers of false positives and true positives

Following van der Laan et al. (2005), the gTP-controlling resampling-based empirical Bayes procedure generates B pairs of random variables $(T_{0n}^b, \mathcal{H}_{0n}^b)$, $b = 1, \ldots, B$, where: (i) $T_{0n}^b \sim Q_{0n}$ is an M-vector of null test statistics, with null distribution Q_{0n} (e.g., defined as in Procedure 2.3 or 2.4); (ii) $\mathcal{H}_{0n}^b$ is a guessed set of true null hypotheses, with distribution $Q_{0n}^{\mathcal{H}}$ (e.g., defined as in Equations (7.11)–(7.13)); (iii) T_{0n}^b and $\mathcal{H}_{0n}^b$ are independent, given the empirical distribution P_n.

For any given test statistic cut-off vector $c = (c(m) : m = 1, \ldots, M)$ and for each of the B pairs of null test statistics T_{0n}^b and guessed sets $\mathcal{H}_{0n}^b$, compute, as in Equation (7.8), a corresponding guessed g-specific function of the numbers of false positives and true positives, $G_n^b(c) \equiv G(c; \mathcal{H}_{0n}^b, T_{0n}^b, T_n)$. The empirical distribution of the B realizations $\{G_n^b(c) : b = 1, \ldots, B\}$ of $G_n(c)$ provides an estimator of the distribution of the unobservable $G(c; \mathcal{H}_0, T_n, T_n)$.

For user-supplied Type I error bound q and Type I error level $\alpha \in (0,1)$, select a cut-off vector c_n, such that the gTP Type I error constraint $\sum_b \mathrm{I}\left(G_n^b(c_n) > q\right)/B \leq \alpha$ is satisfied.

7.2.5 gTP-controlling resampling-based empirical Bayes procedures

The gTP-controlling resampling-based empirical Bayes procedure is summarized in the box below. Procedure 7.1 explicitly provides two types of cut-off vectors among the collection of cut-off vectors that satisfy the gTP Type I error constraint: vectors of common cut-offs and vectors of common-quantile cut-offs.

Procedure 7.1. [gTP-controlling resampling-based empirical Bayes procedure]
Consider the simultaneous test of M null hypotheses $H_0(m)$, $m = 1, \ldots, M$, based on an M-vector of test statistics $T_n = (T_n(m) : m = 1, \ldots, M)$, with true distribution $Q_n = Q_n(P)$. Given a function g, that satisfies monotonicity Assumptions EB.MgV and EB.MgS, and a Type I error bound q, the following *resampling-based empirical Bayes procedure* may be used to control the generalized tail probability error rate, $gTP(q, g) = \Pr(g(V_n, S_n) > q)$.

1. Generate B pairs, $\{(T_{0n}^b, \mathcal{H}_{0n}^b) : b = 1, \ldots, B\}$, of null test statistics T_{0n}^b and random guessed sets $\mathcal{H}_{0n}^b$ of true null hypotheses, as follows.
   a) The M-vectors of *null test statistics* T_{0n}^b have a null distribution Q_{0n}, such as the bootstrap distributions of Procedures 2.3 and 2.4, i.e., T_{0n}^b is a column of an $M \times B$ matrix $\mathbf{Z}_n^B$ of null-transformed bootstrap test statistics.

b) The *random guessed sets of true null hypotheses* $\mathcal{H}_{0n}^b$ have a distribution $Q_{0n}^{\mathcal{H}}$ that corresponds to M independent Bernoulli random variables with parameters $\pi_{0n}(T_n(m))$. That is, generate binary random M-vectors $H_{0n}^b = (H_{0n}^b(m) : m = 1, \ldots, M)$ of null hypotheses as

$$H_{0n}^b(m) \stackrel{\perp}{\sim} \text{Bernoulli}\left(\pi_{0n}(T_n(m))\right), \qquad m = 1, \ldots, M, \tag{7.14}$$

and define sets

$$\mathcal{H}_{0n}^b \equiv \left\{m : H_{0n}^b(m) = 1\right\}. \tag{7.15}$$

Here, $\pi_{0n}(t)$ is an estimated true null hypothesis posterior probability function, such as the estimated local q-value function

$$\pi_{0n}(t) = \min\left\{1, \frac{\pi_{0n} f_{0n}(t)}{f_n(t)}\right\}, \tag{7.16}$$

corresponding to the marginal non-parametric mixture model of Section 7.2.4.

c) Null test statistics T_{0n}^b and guessed sets $\mathcal{H}_{0n}^b$ are *independent*, given the empirical distribution P_n.

2. For any given test statistic cut-off vector $c = (c(m) : m = 1, \ldots, M)$, compute, for each of the B pairs $(T_{0n}^b, \mathcal{H}_{0n}^b)$, the corresponding *guessed g-specific function of the numbers of false positives and true positives*,

$$\begin{aligned} G_n^b(c) &\equiv G(c; \mathcal{H}_{0n}^b, T_{0n}^b, T_n) \\ &= g(V(c; \mathcal{H}_{0n}^b, T_{0n}^b), S(c; \mathcal{H}_{0n}^b, T_n)). \end{aligned} \tag{7.17}$$

3. For user-supplied Type I error bound q and Type I error level $\alpha \in (0, 1)$, derive a cut-off vector c_n that satisfies the empirical gTP Type I error constraint

$$\frac{1}{B}\sum_{b=1}^{B} \mathrm{I}\left(G_n^b(c_n) > q\right) \leq \alpha. \tag{7.18}$$

Common-cut-off procedure. The *common cut-off* γ_n is the *smallest* (i.e., least conservative) value γ for which the gTP constraint in Equation (7.18) is satisfied. That is,

$$\gamma_n \equiv \inf\left\{\gamma \in I\!R : \frac{1}{B}\sum_{b=1}^{B} \mathrm{I}\left(G_n^b(\gamma^{(M)}) > q\right) \leq \alpha\right\}, \tag{7.19}$$

where $\gamma^{(M)}$ denotes the M-vector with all elements equal to γ, i.e., $\gamma^{(M)}(m) = \gamma$, $\forall\, m = 1, \ldots, M$.

Common-quantile procedure. The *common quantile probability* δ_n, corresponding to the test statistics null distribution Q_{0n}, is the *smallest* (i.e., least conservative) value δ for which the gTP constraint in Equation (7.18) is satisfied. That is,

$$\delta_n \equiv \inf\left\{\delta \in [0,1] : \frac{1}{B}\sum_{b=1}^{B} \mathrm{I}\left(G_n^b(q_{0n}^{-1}(\delta)) > q\right) \leq \alpha\right\}, \tag{7.20}$$

where $q_{0n}^{-1}(\delta) = (Q_{0n,m}^{-1}(\delta) : m = 1, \ldots, M)$ denotes the M-vector of δ-quantiles for the null distribution Q_{0n}.

Again, we stress the generality of Procedure 7.1, in that it can be applied to *any* distribution pair $(Q_{0n}, Q_{0n}^{\mathcal{H}})$. The common marginal non-parametric mixture model of Section 7.2.4 is only one among many reasonable working models that does not assume independence of the test statistics.

7.3 Adjusted p-values for gTP-controlling resampling-based empirical Bayes procedures

Adjusted p-values for resampling-based empirical Bayes Procedure 7.1 may be obtained from the general definition of Section 1.2.12, whereby the adjusted p-value $\widetilde{P}_{0n}(m)$, for null hypothesis $H_0(m)$, is the smallest nominal Type I error level at which one would reject $H_0(m)$, given the test statistics T_n. Specifically, from Equation (1.58),

$$\begin{aligned}\widetilde{P}_{0n}(m) &= \inf\{\alpha \in [0,1] : \text{Reject } H_0(m) \text{ at nominal gTP level } \alpha\} \\ &= \inf\{\alpha \in [0,1] : m \in \mathcal{R}_n(\alpha)\} \\ &= \inf\{\alpha \in [0,1] : T_n(m) > c_n(m)\}, \qquad m = 1, \ldots, M.\end{aligned}$$

In what follows, we may use the lowercase notation $t_n(m)$, $p_{0n}(m)$, and $\widetilde{p}_{0n}(m)$, for realizations of the random variables corresponding, respectively, to the test statistics $T_n(m)$, unadjusted p-values $P_{0n}(m)$, and adjusted p-values $\widetilde{P}_{0n}(m)$.

7.3.1 Adjusted p-values for common-cut-off procedure

For the common-cut-off version of Procedure 7.1, the adjusted p-value for null hypothesis $H_0(m)$, with observed test statistic $t_n(m)$, is

$$\begin{aligned}\widetilde{p}_{0n}(m) &= \inf\{\alpha \in [0,1] : t_n(m) > \gamma_n\} \qquad (7.21)\\ &\approx \inf\Bigg\{\alpha \in [0,1] : t_n(m) \\ &\qquad \geq \inf\left\{\gamma \in I\!R : \frac{1}{B}\sum_{b=1}^{B} \mathrm{I}\left(G_n^b(\gamma^{(M)}) > q\right) \leq \alpha\right\}\Bigg\}.\end{aligned}$$

A naive approximation for the adjusted p-values is

$$\widetilde{p}_{0n}(m) \approxeq \frac{1}{B}\sum_{b=1}^{B} \mathrm{I}\left(G_n^b((t_n(m))^{(M)}) > q\right). \tag{7.22}$$

According to this approximation, the adjusted p-value $\widetilde{p}_{0n}(m)$ is an empirical gTP error rate for common cut-off γ_n set at the observed test statistic $t_n(m)$ for null hypothesis $H_0(m)$. That is, $\widetilde{p}_{0n}(m)$ is the proportion (out of B) of random guessed functions $G_n^b((t_n(m))^{(M)})$ that exceed the Type I error bound q, for a vector of common cut-offs $c_n = (t_n(m))^{(M)}$. This approximation is sensible when *each* $G_n^b(\gamma^{(M)})$ is non-increasing in the common cut-off γ or under the weaker condition that the *average* $\sum_b \mathrm{I}\left(G_n^b(\gamma^{(M)}) > q\right)/B$ is non-increasing in γ.

However, in practice, the random G-functions are not monotone in γ. A better approximation, which enforces the desirable property of monotonicity of the common-cut-off adjusted p-values as functions of the test statistics $T_n(m)$, is as follows. Define the function

$$\bar{G}_n^B(\gamma) \equiv \frac{1}{B}\sum_{b=1}^{B} \mathrm{I}\left(G_n^b(\gamma^{(M)}) > q\right).$$

Then, one can show that

$$\inf\left\{\alpha \in [0,1] : \inf\left\{\gamma \in I\!R : \bar{G}_n^B(\gamma) \le \alpha\right\} \le c\right\} = \inf_{\gamma \le c} \bar{G}_n^B(\gamma). \tag{7.23}$$

Indeed, firstly note that $\inf\left\{\gamma \in I\!R : \bar{G}_n^B(\gamma) \le \inf_{\gamma \le c} \bar{G}_n^B(\gamma)\right\} \le c$ follows trivially by definition of the infimum and setting $\gamma = c$. Secondly, suppose that $\alpha < \inf_{\gamma \le c} \bar{G}_n^B(\gamma)$. Then, for each $\gamma \le c$, $\bar{G}_n^B(\gamma) > \alpha$, thus $\inf\left\{\gamma \in I\!R : \bar{G}_n^B(\gamma) \le \alpha\right\} > c$.

Thus, the adjusted p-values are given by

$$\widetilde{p}_{0n}(m) \approxeq \inf_{\gamma \le t_n(m)} \bar{G}_n^B(\gamma) = \inf_{\gamma \le t_n(m)} \frac{1}{B}\sum_{b=1}^{B} \mathrm{I}\left(G_n^b(\gamma^{(M)}) > q\right). \tag{7.24}$$

The infimum over the interval $(-\infty, t_n(m)]$ may be approximated by the minimum over the finite set of observed test statistics that are less than or equal to (i.e., not more significant than) the test statistic $t_n(m)$, for the null hypothesis $H_0(m)$ under consideration. Specifically, let $O_n(m)$ denote the indices for the *ordered test statistics* $T_n(O_n(m))$, so that $T_n(O_n(1)) \ge \cdots \ge T_n(O_n(M))$, and define nested subsets of ordered null hypotheses $\overline{\mathcal{O}}_n(m) \equiv \{O_n(m), \ldots, O_n(M)\}$. Then, one can approximate the adjusted p-values by

$$\widetilde{p}_{0n}(o_n(m)) \approxeq \min_{h \in \overline{\mathcal{O}}_n(m)} \bar{G}_n^B(t_n(h)) = \min_{h \in \overline{\mathcal{O}}_n(m)} \frac{1}{B}\sum_{b=1}^{B} \mathrm{I}\left(G_n^b((t_n(h))^{(M)}) > q\right). \tag{7.25}$$

The adjusted p-values of Equations (7.24) and (7.25) satisfy the wished monotonicity property: $\widetilde{p}_{0n}(o_n(1)) \leq \cdots \leq \widetilde{p}_{0n}(o_n(M))$.

7.3.2 Adjusted p-values for common-quantile procedure

For the common-quantile version of Procedure 7.1, the adjusted p-value for null hypothesis $H_0(m)$, with observed test statistic $t_n(m)$, is

$$\begin{aligned}
\widetilde{p}_{0n}(m) &= \inf\left\{\alpha \in [0,1] : t_n(m) > Q_{0n,m}^{-1}(\delta_n)\right\} \qquad (7.26)\\
&\approxeq \inf\left\{\alpha \in [0,1] : Q_{0n,m}(t_n(m)) \geq \delta_n\right\}\\
&= \inf\Bigg\{\alpha \in [0,1] : Q_{0n,m}(t_n(m))\\
&\qquad \geq \inf\left\{\delta \in [0,1] : \frac{1}{B}\sum_{b=1}^{B} \mathrm{I}\left(G_n^b(q_{0n}^{-1}(\delta)) > q\right) \leq \alpha\right\}\Bigg\}.
\end{aligned}$$

As in the common-cut-off case, a naive approximation for the adjusted p-values is

$$\widetilde{p}_{0n}(m) \approxeq \frac{1}{B}\sum_{b=1}^{B} \mathrm{I}\left(G_n^b(q_{0n}^{-1}(1 - p_{0n}(m))) > q\right). \qquad (7.27)$$

According to this approximation, the adjusted p-value $\widetilde{p}_{0n}(m)$ is an empirical gTP error rate for common quantile probability δ_n set at $(1 - p_{0n}(m))$, where $p_{0n}(m) = 1 - Q_{0n,m}(t_n(m))$ is the unadjusted p-value for null hypothesis $H_0(m)$. That is, $\widetilde{p}_{0n}(m)$ is the proportion (out of B) of random guessed functions $G_n^b(q_{0n}^{-1}(1 - p_{0n}(m)))$ that exceed the Type I error bound q, for a vector of common-quantile cut-offs $c_n = q_{0n}^{-1}(1 - p_{0n}(m)) = (Q_{0n,l}^{-1}(1 - p_{0n}(m)) : l = 1, \ldots, M)$. This approximation is sensible when *each* $G_n^b(q_{0n}^{-1}(\delta))$ is non-increasing in the common quantile probability δ or under the weaker condition that the *average* $\sum_b \mathrm{I}\left(G_n^b(q_{0n}^{-1}(\delta)) > q\right)/B$ is non-increasing in δ.

Again, as in the common-cut-off case, a better approximation, which enforces the desirable property of monotonicity of the common-quantile adjusted p-values as functions of the unadjusted p-values $P_{0n}(m) = 1 - Q_{0n,m}(T_n(m))$, is as follows. Define the function

$$\bar{G}_n^B(\delta) \equiv \frac{1}{B}\sum_{b=1}^{B} \mathrm{I}\left(G_n^b(q_{0n}^{-1}(\delta)) > q\right).$$

Then, one can show that

$$\inf\left\{\alpha \in [0,1] : \inf\left\{\delta \in [0,1] : \bar{G}_n^B(\delta) \leq \alpha\right\} \leq d\right\} = \inf_{\delta \leq d} \bar{G}_n^B(\delta). \qquad (7.28)$$

Thus, the adjusted p-values are given by

$$\widetilde{p}_{0n}(m) \approxeq \inf_{\delta \le 1 - p_{0n}(m)} \bar{G}_n^B(\delta) = \inf_{\delta \le 1 - p_{0n}(m)} \frac{1}{B} \sum_{b=1}^{B} \mathrm{I}\left(G_n^b(q_{0n}^{-1}(\delta)) > q\right). \quad (7.29)$$

The infimum over the interval $[0, 1 - p_{0n}(m)]$ may be approximated by the minimum over the finite set of observed unadjusted p-values that are greater than or equal to (i.e., not more significant than) the p-value $p_{0n}(m)$, for the null hypothesis $H_0(m)$ under consideration. Specifically, let $O_n(m)$ denote the indices for the *ordered unadjusted p-values* $P_{0n}(O_n(m))$, so that $P_{0n}(O_n(1)) \le \cdots \le P_{0n}(O_n(M))$, and define nested subsets of ordered null hypotheses $\overline{\mathcal{O}}_n(m) \equiv \{O_n(m), \ldots, O_n(M)\}$. Then, one can approximate the adjusted p-values by

$$\begin{aligned}\widetilde{p}_{0n}(o_n(m)) &\approxeq \min_{h \in \overline{o}_n(m)} \bar{G}_n^B(1 - p_{0n}(h)) \qquad (7.30)\\ &= \min_{h \in \overline{o}_n(m)} \frac{1}{B} \sum_{b=1}^{B} \mathrm{I}\left(G_n^b(q_{0n}^{-1}(1 - p_{0n}(h))) > q\right).\end{aligned}$$

The adjusted p-values of Equations (7.29) and (7.30) satisfy the wished monotonicity property: $\widetilde{p}_{0n}(o_n(1)) \le \cdots \le \widetilde{p}_{0n}(o_n(M))$.

7.4 Finite sample rationale for gTP control by resampling-based empirical Bayes procedures

This section provides a semi-formal finite sample rationale for the resampling-based empirical Bayes procedure described above.

7.4.1 Procedures based on constant guessed set of true null hypotheses and observed test statistics

Firstly, suppose one has available a *constant conservative guessed set of true null hypotheses* $\mathcal{H}_0^+$, in the sense that the guessed set $\mathcal{H}_0^+$ contains the actual set $\mathcal{H}_0$ of true null hypotheses, i.e., $\mathcal{H}_0 \subseteq \mathcal{H}_0^+$. Then,

$$\begin{aligned}G(c; \mathcal{H}_0^+, T_n, T_n) &= g(V(c; \mathcal{H}_0^+, T_n), S(c; \mathcal{H}_0^+, T_n))\\ &\ge g(V(c; \mathcal{H}_0, T_n), S(c; \mathcal{H}_0^+, T_n))\\ &\ge g(V(c; \mathcal{H}_0, T_n), S(c; \mathcal{H}_0, T_n))\\ &= G(c; \mathcal{H}_0, T_n, T_n).\end{aligned}$$

The first inequality follows from the facts that for $\mathcal{H}_0 \subseteq \mathcal{H}_0^+$, the number of false positives for the guessed set $\mathcal{H}_0^+$ dominates the number of false positives for the actual set $\mathcal{H}_0$, that is, $V(c; \mathcal{H}_0^+, T_n) = \sum_m \mathrm{I}\left(T_n(m) > c(m),\ m \in \mathcal{H}_0^+\right) \ge \sum_m \mathrm{I}\left(T_n(m) > c(m),\ m \in \mathcal{H}_0\right) = V(c; \mathcal{H}_0, T_n)$, and the function $g_{S=s}$:

$v \to g(v, s)$ is non-decreasing for any constant s (Assumption EB.MgV). The second inequality holds by noting that the number of true positives for the guessed set $\mathcal{H}_0^+$ is dominated by the number of true positives for the actual set $\mathcal{H}_0$, that is, $S(c; \mathcal{H}_0^+, T_n) = \sum_m \mathrm{I}\left(T_n(m) > c(m),\ m \notin \mathcal{H}_0^+\right) \leq \sum_m \mathrm{I}\left(T_n(m) > c(m),\ m \notin \mathcal{H}_0\right) = S(c; \mathcal{H}_0, T_n)$, and the function $g_{V=v} : s \to g(v, s)$ is non-increasing for any constant v (Assumption EB.MgS).

Under the above conditions, it follows that the *true* function $G(c; \mathcal{H}_0, T_n, T_n)$ of the numbers of false positives and true positives is bounded above by the corresponding *guessed* function $G(c; \mathcal{H}_0^+, T_n, T_n)$. Thus, one can simply choose the cut-offs c_n so that the guessed function $G(c_n; \mathcal{H}_0^+, T_n, T_n)$ equals the allowed Type I error bound q.

Although simple, this approach is sensitive to the choice of guessed set $\mathcal{H}_0^+$, in terms of both Type I error control and power.

Regarding Type I error control, if the guessed set $\mathcal{H}_0^+$ does not contain the actual set $\mathcal{H}_0$ of true null hypotheses, then the procedure tends to be *anti-conservative*. Indeed, for $m \in \mathcal{H}_0^{+c} \cap \mathcal{H}_0$, the corresponding chosen cut-off $c_n(m)$ tends to be too small, i.e., leads to an excess of false positives.

Regarding power, if the guessed set $\mathcal{H}_0^+$ is "too large", i.e., contains the actual set $\mathcal{H}_0$ of true null hypotheses as well as some false null hypotheses, then the procedure tends to be *conservative*. Indeed, for $m \in \mathcal{H}_0^+ \cap \mathcal{H}_0^c$, the corresponding chosen cut-off $c_n(m)$ tends to be too large, i.e., leads to an excess of false negatives.

In order to reduce sensitivity to estimators of $\mathcal{H}_0$, the resampling-based empirical Bayes approach involves the following key steps.

1. Replacing the observed test statistics $(T_n(m) : m \in \mathcal{H}_0^+)$ by a random vector of *null test statistics* $(T_{0n}(m) : m \in \mathcal{H}_0^+)$, sampled from a null distribution Q_0 (or estimator thereof, Q_{0n}) that satisfies null domination Assumption NDV for the number of Type I errors, i.e., such that $V(c; \mathcal{H}_0, T_{0n})$ is stochastically greater than $V(c; \mathcal{H}_0, T_n)$.
2. Replacing the fixed guessed set $\mathcal{H}_0^+$ by a *random guessed set* $\mathcal{H}_{0n} \sim Q_{0n}^{\mathcal{H}}$, sampled (conditionally on P_n) independently of T_{0n}, from a distribution $Q_{0n}^{\mathcal{H}}$ that is asymptotically degenerate at the actual set $\mathcal{H}_0$ and conservative in finite sample situations.
3. Controlling tail probabilities (conditionally on P_n) of the random guessed function $G_n(c) = G(c; \mathcal{H}_{0n}, T_{0n}, T_n)$, defined by the pair of random variables $(T_{0n}, \mathcal{H}_{0n})$.

As detailed in Section 7.2.4, suitable sampling distributions for the null test statistics T_{0n} are provided by bootstrap estimators Q_{0n} of the null distributions Q_0 introduced in Sections 2.3 and 2.4 (e.g., Procedure 2.3 or 2.4). The distribution $Q_{0n}^{\mathcal{H}}$ for the guessed sets $\mathcal{H}_{0n}$ of true null hypotheses may be based on the common marginal non-parametric mixture model of Equation (7.9).

7.4.2 Procedures based on constant guessed set of true null hypotheses and null test statistics

Suppose the following three conditions are satisfied: $\mathcal{H}_0 \subseteq \mathcal{H}_0^+$, $V(c;\mathcal{H}_0,T_{0n})$ is stochastically greater than $V(c;\mathcal{H}_0,T_n)$ for all M-vectors c, and $(T_{0n}(m) : m \in \mathcal{H}_0)$ and $(T_n(m) : m \in \mathcal{H}_0^c)$ are independent. Then,

$$\begin{aligned}
G(c;\mathcal{H}_0^+,T_{0n},T_n) &= g(V(c;\mathcal{H}_0^+,T_{0n}),S(c;\mathcal{H}_0^+,T_n)) \\
&\geq g(V(c;\mathcal{H}_0,T_{0n}),S(c;\mathcal{H}_0^+,T_n)) \\
&\geq g(V(c;\mathcal{H}_0,T_{0n}),S(c;\mathcal{H}_0,T_n)) \\
&\geq g(V(c;\mathcal{H}_0,T_n),S(c;\mathcal{H}_0,T_n)) \\
&= G(c;\mathcal{H}_0,T_n,T_n), \qquad \text{conditional on } S(c;\mathcal{H}_0,T_n).
\end{aligned}$$

Thus, selecting a cut-off vector c_n such that the conditional (given the empirical distribution P_n, i.e., the test statistics T_n) tail probability at q of the *guessed* function $G(c_n;\mathcal{H}_0^+,T_{0n},T_n)$ equals α, yields a multiple testing procedure that controls the tail probability at q of the *actual* function $G(c_n;\mathcal{H}_0,T_n,T_n)$ at level α.

Note that the assumption that the number of false positives $V(c;\mathcal{H}_0,T_{0n})$ is independent of the number of true positives $S(c;\mathcal{H}_0,T_n)$ is sufficient, but not necessary, to obtain the wished stochastic domination of $G(c;\mathcal{H}_0,T_n,T_n)$ by $G(c;\mathcal{H}_0^+,T_{0n},T_n)$. Indeed, in the limit, the null test statistics $T_{0n} \sim Q_{0n}$ are independent of the empirical distribution P_n and, hence, of the actual test statistics T_n. In particular, $V(c;\mathcal{H}_0,T_{0n})$ is independent of $S(c;\mathcal{H}_0,T_n)$. Furthermore, at a fixed data generating distribution P, the number of true positives $S(c;\mathcal{H}_0,T_n)$ converges to the constant $M - h_0 = |\mathcal{H}_0^c|$, so that this independence condition is asymptotically void.

7.4.3 Procedures based on random guessed sets of true null hypotheses and null test statistics

The previous procedure still relies on the guessed set $\mathcal{H}_0^+$ containing the actual set $\mathcal{H}_0$ of true null hypotheses. As in van der Laan et al. (2005), we further propose to replace the fixed set $\mathcal{H}_0^+$ by a random set $\mathcal{H}_{0n}$, sampled (conditionally on P_n) independently of T_{0n}, from a distribution $Q_{0n}^{\mathcal{H}}$ that is asymptotically degenerate at the actual set $\mathcal{H}_0$ and conservative in finite sample situations. The cut-off vector c_n is then selected to control tail probabilities for the random guessed function $G(c_n;\mathcal{H}_{0n},T_{0n},T_n)$.

If the distribution $Q_{0n}^{\mathcal{H}}$ is conservative for finite samples (and asymptotically degenerate at $\mathcal{H}_0$), in the sense that the guessed sets $\mathcal{H}_{0n}$ "typically contain or are larger than" the actual set $\mathcal{H}_0$ (e.g., $|\mathcal{H}_0| \leq \mathrm{E}[|\mathcal{H}_{0n}||P_n]$ for almost every $(P_n : n \geq 1)$), then one would expect the finite sample rationale for a fixed set $\mathcal{H}_0^+$, with $\mathcal{H}_0 \subseteq \mathcal{H}_0^+$, to approximately hold for random sets $\mathcal{H}_{0n}$. Furthermore, replacing the fixed set $\mathcal{H}_0^+$ by random sets $\mathcal{H}_{0n}$ should lead to more robust procedures in finite sample situations.

7.5 Formal asymptotic gTP control results for resampling-based empirical Bayes procedures

7.5.1 Asymptotic control of gTP by resampling-based empirical Bayes Procedure 7.1

The following theorem formally establishes the asymptotic validity of the common-cut-off and common-quantile versions of gTP-controlling resampling-based empirical Bayes Procedure 7.1. We stress that Procedure 7.1 provides Type I error control for general data generating distributions (with general dependence structures among variables), null hypotheses (defined in terms of submodels for the data generating distribution), and test statistics (e.g., t-statistics, χ^2-statistics, F-statistics).

Theorem 7.2. [Asymptotic control of gTP by resampling-based empirical Bayes Procedure 7.1]
Set-up. *Let $\mathcal{X}_n = \{X_i : i = 1,\dots,n\}$ denote a random sample from a data generating distribution $P \in \mathcal{M}$ and let P_n denote the corresponding empirical distribution. Consider the simultaneous test of M null hypotheses $H_0(m)$, $m = 1,\dots,M$, based on an M-vector of test statistics $T_n = (T_n(m) : m = 1,\dots,M)$, with true distribution $Q_n = Q_n(P)$. Further consider null test statistics $T_{0n} = (T_{0n}(m) : m = 1,\dots,M)$ and random guessed sets $\mathcal{H}_{0n} \subseteq \{1,\dots,M\}$ of true null hypotheses, where, given the empirical distribution P_n, T_{0n} and $\mathcal{H}_{0n}$ are sampled independently from conditional distributions Q_{0n} and $Q_{0n}^{\mathcal{H}}$, respectively. Given an M-vector of cut-offs $c = (c(m) : m = 1,\dots,M)$, observed test statistics T_n, null test statistics T_{0n}, and a random guessed set $\mathcal{H}_{0n}$ of true null hypotheses, define a g-specific function $G_n(c)$ of the numbers of false positives and true positives by*

$$\begin{aligned} G_n(c) &\equiv G(c;\mathcal{H}_{0n},T_{0n},T_n) \\ &= g(V(c;\mathcal{H}_{0n},T_{0n}),S(c;\mathcal{H}_{0n},T_n)). \end{aligned} \tag{7.31}$$

For user-supplied function g, Type I error bound q, and Type I error level $\alpha \in (0,1)$, the corresponding (conditional on P_n) gTP Type I error constraint is

$$\bar{F}_{G_n(c)|P_n}(q) = \Pr(G_n(c) > q|P_n) \leq \alpha. \tag{7.32}$$

Among the family of cut-offs that satisfy the above gTP constraint, consider common cut-offs and common-quantile cut-offs.
Common cut-offs. *Define a common cut-off γ_n, as the smallest (i.e., least conservative) value γ that satisfies the conditional gTP constraint in Equation (7.32). That is,*

$$\gamma_n = \gamma(g,q,\alpha;Q_{0n}^{\mathcal{H}},Q_{0n},P_n) \equiv \inf\left\{\gamma \in I\!R : \bar{F}_{G_n(\gamma^{(M)})|P_n}(q) \leq \alpha\right\}, \tag{7.33}$$

where $\gamma^{(M)}$ denotes the M-vector with all elements equal to γ, i.e., $\gamma^{(M)}(m) = \gamma$, $\forall\ m = 1,\dots,M$.

Common-quantile cut-offs. *Define a common quantile probability δ_n, corresponding to the test statistics null distribution Q_{0n}, as the smallest (i.e., least conservative) value δ that satisfies the conditional gTP constraint in Equation (7.32). That is,*

$$\delta_n = \delta(g, q, \alpha; Q_{0n}^{\mathcal{H}}, Q_{0n}, P_n) \equiv \inf\left\{\delta \in [0,1] : \bar{F}_{G_n(q_{0n}^{-1}(\delta))|P_n}(q) \leq \alpha\right\}, \tag{7.34}$$

where $q_{0n}^{-1}(\delta) = (Q_{0n,m}^{-1}(\delta) : m = 1, \ldots, M)$ denotes the M-vector of δ-quantiles for the null distribution Q_{0n}.

Asymptotic Type I error control result. *Under the assumptions stated in Section 7.5.2, the common-cut-off and common-quantile versions of resampling-based empirical Bayes Procedure 7.1 provide asymptotic control of $gTP(q, g)$ at level α. That is,*

$$\limsup_{n\to\infty} \bar{F}_{G(\gamma_n^{(M)};\mathcal{H}_0,T_n,T_n)}(q) \leq \alpha \qquad \textit{[common cut-offs]} \tag{7.35}$$

$$\limsup_{n\to\infty} \bar{F}_{G(q_{0n}^{-1}(\delta_n);\mathcal{H}_0,T_n,T_n)}(q) \leq \alpha \qquad \textit{[common-quantile cut-offs]},$$

where $G(c; \mathcal{H}_0, T_n, T_n)$ is the g-specific function of the actual numbers of false positives and true positives for an M-vector of cut-offs c,

$$G(c; \mathcal{H}_0, T_n, T_n) = g\left(\sum_{m\in\mathcal{H}_0} \mathrm{I}\left(T_n(m) > c(m)\right), \sum_{m\notin\mathcal{H}_0} \mathrm{I}\left(T_n(m) > c(m)\right)\right).$$

The long notation $\gamma(g, q, \alpha; Q_{0n}^{\mathcal{H}}, Q_{0n}, P_n)$ and $\delta(g, q, \alpha; Q_{0n}^{\mathcal{H}}, Q_{0n}, P_n)$ reflects the dependence of the cut-offs on: the conditional (given P_n) distribution $Q_{0n}^{\mathcal{H}}$ of the guessed sets $\mathcal{H}_{0n}$ of true null hypotheses; the conditional (given P_n) distribution Q_{0n} of the null test statistics T_{0n}; the empirical distribution P_n (i.e., the actual test statistics T_n); the user-supplied function g defining the gTP Type I error rate; the user-supplied Type I error bound q; and the user-supplied Type I error level α.

Section 7.5.2 states and interprets the main assumptions for Theorem 7.2; the proof of the theorem is given in Section 7.5.3.

7.5.2 Assumptions for Theorem 7.2

Statement of Theorem 7.2 assumptions

Asymptotic $gTP(q, g)$ control by Procedure 7.1 is established under the following assumptions.

Assumption EB.MgV. [**Monotonicity of g**] The function $g_{S=s} : v \rightarrow g(v, s)$ is continuous and strictly increasing for any given s.

Assumption EB.MgS. [**Monotonicity of g**] The function $g_{V=v} : s \rightarrow g(v,s)$ is continuous and non-increasing for any given v.

Assumption EB.Cons$\mathcal{H}_{0n}$. [**Consistency of guessed sets $\mathcal{H}_{0n}$**] The distribution $Q_{0n}^{\mathcal{H}}$, for the random guessed sets of true null hypotheses $\mathcal{H}_{0n}$, converges to the degenerate distribution that places probability one on the actual (constant) set of true null hypotheses $\mathcal{H}_0 = \mathcal{H}_0(P)$. That is, for almost every $(P_n : n \geq 1)$,

$$\lim_{n\rightarrow\infty} \mathcal{H}_{0n} = \mathcal{H}_0. \tag{7.36}$$

Assumption EB.AS. [**Asymptotic separation**] Let T_∞ denote a random M-vector of test statistics, which is independent of P_n and for which the number of true positives is always equal to the actual number of false null hypotheses, i.e., $S(c;\mathcal{H}_0,T_\infty) = \sum_m \mathrm{I}\left(T_\infty(m) > c(m),\ m \notin \mathcal{H}_0\right) = |\mathcal{H}_0^c|$, for any M-vector of cut-offs c (e.g., one could set $T_\infty(m) = +\infty$ for $m \in \mathcal{H}_0^c$). Define a g-specific function $G_{0n}(c)$ of the numbers of false positives and true positives by

$$\begin{aligned} G_{0n}(c) &\equiv G(c;\mathcal{H}_0,T_{0n},T_\infty) \\ &= g(V(c;\mathcal{H}_0,T_{0n}),S(c;\mathcal{H}_0,T_\infty)) \\ &= g(V(c;\mathcal{H}_0,T_{0n}),|\mathcal{H}_0^c|). \end{aligned} \tag{7.37}$$

Common cut-offs. Define the optimal common cut-off corresponding to G_{0n} by

$$\gamma_{0n} \equiv \inf\left\{\gamma \in I\!R : \bar{F}_{G_{0n}(\gamma^{(M)})|P_n}(q) \leq \alpha\right\}. \tag{7.38}$$

Assume there exists a constant τ so that, for almost every $(P_n : n \geq 1)$,

$$\limsup_{n\rightarrow\infty} \gamma_{0n} \leq \tau \tag{7.39}$$

and

$$\lim_{n\rightarrow\infty} S(\tau^{(M)};\mathcal{H}_0,T_n) = \lim_{n\rightarrow\infty} \sum_{m\notin\mathcal{H}_0} \mathrm{I}\left(T_n(m) > \tau\right) = |\mathcal{H}_0^c|. \tag{7.40}$$

Common-quantile cut-offs. Define the optimal common quantile probability corresponding to G_{0n} by

$$\delta_{0n} \equiv \inf\left\{\delta \in [0,1] : \bar{F}_{G_{0n}(q_{0n}^{-1}(\delta))|P_n}(q) \leq \alpha\right\}. \tag{7.41}$$

Assume there exists a constant $\tau < 1$ so that, for almost every $(P_n : n \geq 1)$,

$$\limsup_{n\rightarrow\infty} \delta_{0n} \leq \tau \tag{7.42}$$

and

$$\lim_{n\rightarrow\infty} S(q_{0n}^{-1}(\tau);\mathcal{H}_0,T_n) = \lim_{n\rightarrow\infty} \sum_{m\notin\mathcal{H}_0} \mathrm{I}\left(T_n(m) > Q_{0n,m}^{-1}(\tau)\right) = |\mathcal{H}_0^c|. \tag{7.43}$$

Assumption EB.AND. [Asymptotic null domination]
Common cut-offs. For almost every $(P_n : n \geq 1)$ and for each $x \in \{1, \ldots, M\}$,

$$\limsup_{n\to\infty} \sup_{\gamma \leq \tau} \left(\bar{F}_{V(\gamma^{(M)};\mathcal{H}_0,T_n)}(x) - \bar{F}_{V(\gamma^{(M)};\mathcal{H}_0,T_{0n})|P_n}(x)\right) \leq 0. \tag{7.44}$$

Common-quantile cut-offs. For almost every $(P_n : n \geq 1)$ and for each $x \in \{1, \ldots, M\}$,

$$\limsup_{n\to\infty} \sup_{\delta \in [0,\tau]} \left(\bar{F}_{V(q_{0n}^{-1}(\delta);\mathcal{H}_0,T_n)}(x) - \bar{F}_{V(q_{0n}^{-1}(\delta);\mathcal{H}_0,T_{0n})|P_n}(x)\right) \leq 0. \tag{7.45}$$

Assumption EB.gTP1.
Common cut-offs. Given $(P_n : n \geq 1)$, if

$$\lim_{n\to\infty} \sup_{\gamma \leq \tau} \left|\bar{F}_{G_n(\gamma^{(M)})|P_n}(q) - \bar{F}_{G_{0n}(\gamma^{(M)})|P_n}(q)\right| = 0$$

and $\limsup_n \gamma_{0n} \leq \tau$, then

$$\liminf_{n\to\infty} (\gamma_n - \gamma_{0n}) \geq 0. \tag{7.46}$$

Common-quantile cut-offs. Given $(P_n : n \geq 1)$, if

$$\lim_{n\to\infty} \sup_{\delta \in [0,\tau]} \left|\bar{F}_{G_n(q_{0n}^{-1}(\delta))|P_n}(q) - \bar{F}_{G_{0n}(q_{0n}^{-1}(\delta))|P_n}(q)\right| = 0$$

and $\limsup_n \delta_{0n} \leq \tau$, then

$$\liminf_{n\to\infty} (\delta_n - \delta_{0n}) \geq 0. \tag{7.47}$$

Assumption EB.gTP2.
Common cut-offs. If γ_{0n} is a sequence so that, for almost every $(P_n : n \geq 1)$, $\liminf_n (\gamma_n - \gamma_{0n}) \geq 0$, then

$$\limsup_{n\to\infty} \left(\bar{F}_{G(\gamma_n^{(M)};\mathcal{H}_0,T_n,T_n)}(q) - \bar{F}_{G(\gamma_{0n}^{(M)};\mathcal{H}_0,T_n,T_n)}(q)\right) \leq 0. \tag{7.48}$$

Common-quantile cut-offs. If δ_{0n} is a sequence so that, for almost every $(P_n : n \geq 1)$, $\liminf_n (\delta_n - \delta_{0n}) \geq 0$, then

$$\limsup_{n\to\infty} \left(\bar{F}_{G(q_{0n}^{-1}(\delta_n);\mathcal{H}_0,T_n,T_n)}(q) - \bar{F}_{G(q_{0n}^{-1}(\delta_{0n});\mathcal{H}_0,T_n,T_n)}(q)\right) \leq 0. \tag{7.49}$$

Interpretation of Theorem 7.2 assumptions

Assumptions EB.MgV and EB.MgS are natural *monotonicity* assumptions for the function g defining the generalized tail probability error rate.

These assumptions are satisfied, in particular, by the following two functions, $g(v,s) = v$ and $g(v,s) = v/(v+s)$, corresponding, respectively, to the familiar gFWER and TPPFP. Note that because the random guessed sets $\mathcal{H}_{0n}$ are assumed to be asymptotically equal to the actual (constant) set of true null hypotheses $\mathcal{H}_0$ (Assumption EB.Cons$\mathcal{H}_{0n}$), monotonicity Assumption EB.MgS, for the number of true positives, is not needed for the proof of asymptotic gTP control. This natural assumption is used, however, for the finite sample rationale of Section 7.4.

Assumption EB.Cons$\mathcal{H}_{0n}$ states that the random guessed sets $\mathcal{H}_{0n}$ should be *consistent* or asymptotically "on target", that is, should converge to the actual set $\mathcal{H}_0 = \mathcal{H}_0(P)$ of true null hypotheses. As noted in Section 7.4, above, the finite sample distribution $Q_{0n}^{\mathcal{H}}$ should be chosen conservatively, so that the guessed sets $\mathcal{H}_{0n}$ "typically contain or are larger than" the actual set $\mathcal{H}_0$ (e.g., $|\mathcal{H}_0| \leq \mathrm{E}[|\mathcal{H}_{0n}||P_n]$ for almost every $(P_n : n \geq 1)$).

Assumption EB.AS naturally holds at a fixed data generating distribution P, as it states that the test statistics $(T_n(m) : m \notin \mathcal{H}_0)$, corresponding to the false null hypotheses $\mathcal{H}_0^c$, are *asymptotically separate* from the test statistics $(T_n(m) : m \in \mathcal{H}_0)$, corresponding to the true null hypotheses $\mathcal{H}_0$.

Assumption EB.AND states that the number of false positives $V(c; \mathcal{H}_0, T_{0n})$, under the test statistics null distribution Q_{0n}, *asymptotically dominates* the number of false positives $V(c; \mathcal{H}_0, T_n)$, under the true distribution $Q_n = Q_n(P)$ (Assumption NDV, Section 2.2.3). This assumption is satisfied, in particular, by (estimators of) the following two test statistics null distributions: the asymptotic distribution of the vector of null shift and scale-transformed test statistics (Section 2.3) and the asymptotic distribution of the vector of null quantile-transformed test statistics (Section 2.4).

Assumptions EB.gTP1 and EB.gTP2 are mild regularity conditions.

7.5.3 Proof of Theorem 7.2

The proof focuses on the common-cut-off version of Theorem 7.2. The proof for common-quantile cut-offs is similar and is therefore omitted.

Firstly, by Assumptions EB.Cons$\mathcal{H}_{0n}$ and EB.AS, it follows that, given almost every $(P_n : n \geq 1)$, $(G_n(\gamma^{(M)}) : \gamma \leq \tau)$ and $(G_{0n}(\gamma^{(M)}) : \gamma \leq \tau)$ are equal with probability tending to one, that is,

$$\lim_{n \to \infty} \Pr\left((G_n(\gamma^{(M)}) : \gamma \leq \tau) = (G_{0n}(\gamma^{(M)}) : \gamma \leq \tau) \middle| P_n \right) = 1.$$

As a consequence, the difference between the conditional survivor function of $G_n(\gamma^{(M)})$ at q, given P_n, and the conditional survivor function of $G_{0n}(\gamma^{(M)})$ at q, given P_n, converges to zero uniformly in $\gamma \leq \tau$. That is, for almost every $(P_n : n \geq 1)$,

$$\lim_{n\to\infty} \sup_{\gamma\leq\tau} \left| \bar{F}_{G(\gamma^{(M)};\mathcal{H}_{0n},T_{0n},T_n)|P_n}(q) - \bar{F}_{G(\gamma^{(M)};\mathcal{H}_0,T_{0n},T_\infty)|P_n}(q) \right| \quad (7.50)$$
$$= \lim_{n\to\infty} \sup_{\gamma\leq\tau} \left| \bar{F}_{G_n(\gamma^{(M)})|P_n}(q) - \bar{F}_{G_{0n}(\gamma^{(M)})|P_n}(q) \right| = 0.$$

Now, note that, given $(P_n : n \geq 1)$, γ_{0n} is a constant (i.e., non-random) sequence and, by Assumption EB.AS, $\exists\ N \in I\!N$ so that, $\forall\ n \geq N$, $\gamma_{0n} \leq \tau$. Thus, Equation (7.50) implies, in particular, that the difference between the conditional survivor function of $G_n(\gamma_{0n}^{(M)})$ at q, given P_n, and the conditional survivor function of $G_{0n}(\gamma_{0n}^{(M)})$ at q, given P_n, converges to zero. That is, for almost every $(P_n : n \geq 1)$,

$$\lim_{n\to\infty} \left| \bar{F}_{G_n(\gamma_{0n}^{(M)})|P_n}(q) - \bar{F}_{G_{0n}(\gamma_{0n}^{(M)})|P_n}(q) \right| = 0.$$

By definition of γ_{0n} in Assumption EB.AS, the conditional survivor function of $G_{0n}(\gamma_{0n}^{(M)})$ at q, given P_n, is bounded above by α, i.e., $\bar{F}_{G_{0n}(\gamma_{0n}^{(M)})|P_n}(q) \leq \alpha$. Thus, for almost every $(P_n : n \geq 1)$,

$$\limsup_{n\to\infty} \bar{F}_{G_n(\gamma_{0n}^{(M)})|P_n}(q) = \limsup_{n\to\infty} \bar{F}_{G(\gamma_{0n}^{(M)};\mathcal{H}_{0n},T_{0n},T_n)|P_n}(q) \leq \alpha. \quad (7.51)$$

Next, note that, for all $\gamma \leq \tau$,

$$\begin{aligned} \bar{F}_{G_{0n}(\gamma^{(M)})|P_n}(q) &= \Pr\left(G(\gamma^{(M)};\mathcal{H}_0,T_{0n},T_\infty) > q \,\middle|\, P_n\right) \quad (7.52)\\ &= \Pr\left(g(V(\gamma^{(M)};\mathcal{H}_0,T_{0n}),|\mathcal{H}_0^c|) > q \,\middle|\, P_n\right)\\ &= \Pr\left(V(\gamma^{(M)};\mathcal{H}_0,T_{0n}) > g^{-1}_{S=|\mathcal{H}_0^c|}(q) \,\middle|\, P_n\right), \end{aligned}$$

where, for a fixed s, the function $g^{-1}_{S=s}$ is defined as the inverse of the continuous and strictly increasing function $g_{S=s} : v \to g(v,s)$ (Assumption EB.MgV). By null domination Assumption EB.AND, the above conditional (given P_n) probability $\bar{F}_{G_{0n}(\gamma^{(M)})|P_n}(q)$ is, uniformly in $\gamma \leq \tau$, asymptotically greater than the marginal probability

$$\begin{aligned} \Pr\left(V(\gamma^{(M)};\mathcal{H}_0,T_n) > g^{-1}_{S=|\mathcal{H}_0^c|}(q)\right) &= \Pr\left(g(V(\gamma^{(M)};\mathcal{H}_0,T_n),|\mathcal{H}_0^c|) > q\right)\\ &= \bar{F}_{G(\gamma^{(M)};\mathcal{H}_0,T_n,T_\infty)}(q). \quad (7.53) \end{aligned}$$

That is, for almost every $(P_n : n \geq 1)$,

$$\limsup_{n\to\infty} \sup_{\gamma\leq\tau} \left(\bar{F}_{G(\gamma^{(M)};\mathcal{H}_0,T_n,T_\infty)}(q) - \bar{F}_{G(\gamma^{(M)};\mathcal{H}_0,T_{0n},T_\infty)|P_n}(q)\right) \leq 0. \quad (7.54)$$

In addition, by Assumption EB.AS, the probability in Equation (7.53) is, uniformly in $\gamma \leq \tau$, asymptotically equal to

$$\begin{aligned} \Pr\left(g(V(\gamma^{(M)};\mathcal{H}_0,T_n),S(\gamma^{(M)};\mathcal{H}_0,T_n)) > q\right) &= \Pr\left(G(\gamma^{(M)};\mathcal{H}_0,T_n,T_n) > q\right)\\ &= \bar{F}_{G(\gamma^{(M)};\mathcal{H}_0,T_n,T_n)}(q). \end{aligned}$$

That is,

$$\lim_{n\to\infty} \sup_{\gamma\leq\tau} \left|\bar{F}_{G(\gamma^{(M)};\mathcal{H}_0,T_n,T_\infty)}(q) - \bar{F}_{G(\gamma^{(M)};\mathcal{H}_0,T_n,T_n)}(q)\right| = 0. \tag{7.55}$$

Combining Equations (7.55), (7.54), and (7.50), proves that, for almost every $(P_n : n \geq 1)$,

$$\begin{aligned}
&\limsup_{n\to\infty} \sup_{\gamma\leq\tau} \left(\bar{F}_{G(\gamma^{(M)};\mathcal{H}_0,T_n,T_n)}(q) - \bar{F}_{G(\gamma^{(M)};\mathcal{H}_{0n},T_{0n},T_n)|P_n}(q)\right) \qquad (7.56)\\
&= \limsup_{n\to\infty} \sup_{\gamma\leq\tau} \Big(\left(\bar{F}_{G(\gamma^{(M)};\mathcal{H}_0,T_n,T_n)}(q) - \bar{F}_{G(\gamma^{(M)};\mathcal{H}_0,T_n,T_\infty)}(q)\right)\\
&\quad + \left(\bar{F}_{G(\gamma^{(M)};\mathcal{H}_0,T_n,T_\infty)}(q) - \bar{F}_{G(\gamma^{(M)};\mathcal{H}_0,T_{0n},T_\infty)|P_n}(q)\right)\\
&\quad + \left(\bar{F}_{G(\gamma^{(M)};\mathcal{H}_0,T_{0n},T_\infty)|P_n}(q) - \bar{F}_{G(\gamma^{(M)};\mathcal{H}_{0n},T_{0n},T_n)|P_n}(q)\right)\Big)\\
&\leq 0.
\end{aligned}$$

Recall that, by Assumption EB.AS, $\limsup_n \gamma_{0n} \leq \tau$, and, by Equation (7.51), $\limsup_n \bar{F}_{G(\gamma_{0n}^{(M)};\mathcal{H}_{0n},T_{0n},T_n)|P_n}(q) \leq \alpha$. Thus, Equation (7.56) implies that

$$\limsup_{n\to\infty} \bar{F}_{G(\gamma_{0n}^{(M)};\mathcal{H}_0,T_n,T_n)}(q) \leq \alpha. \tag{7.57}$$

By Equation (7.50) and $\limsup_n \gamma_{0n} \leq \tau$, Assumption EB.gTP1 implies that, for almost every $(P_n : n \geq 1)$,

$$\liminf_{n\to\infty}(\gamma_n - \gamma_{0n}) \geq 0. \tag{7.58}$$

Finally, under Equations (7.57) and (7.58), Assumption EB.gTP2 yields

$$\limsup_{n\to\infty} \bar{F}_{G(\gamma_n^{(M)};\mathcal{H}_0,T_n,T_n)}(q) \leq \alpha.$$

This completes the proof that the common-cut-off version of Procedure 7.1 asymptotically controls $gTP(q,g)$.

□

7.6 gTP-controlling resampling-based weighted empirical Bayes procedures

In the context of TPPFP control, van der Laan et al. (2005) propose a variant of Procedure 7.1, based on *weighted* numbers of false positives and true positives. In this approach, null hypotheses are weighted according to a (estimated) local q-value function, i.e., a posterior probability function for the true null hypotheses (Section 7.2.4).

Specifically, given an M-vector of weights $w = (w(m) : m = 1, \ldots, M)$, define a weighted number of false positives, a weighted number of true positives, and a g-specific function of the weighted numbers of false positives and true positives, as

$$V(c; w, Z) \equiv \sum_{m=1}^{M} w(m) \mathrm{I}\left(Z(m) > c(m)\right), \tag{7.59}$$

$$S(c; w, Z) \equiv \sum_{m=1}^{M} (1 - w(m)) \mathrm{I}\left(Z(m) > c(m)\right),$$

and

$$G(c; w, Z_0, Z) \equiv g(V(c; w, Z_0), S(c; w, Z)),$$

respectively.

In the general case of gTP control, the modified weighted procedure involves controlling tail probabilities for

$$G_n(c) \equiv G(c; \pi_{0n}(T_n), T_{0n}, T_n), \tag{7.60}$$

where $T_n \sim Q_n$ is the M-vector of observed test statistics, $T_{0n} \sim Q_{0n}$ is an M-vector of null test statistics, and $\pi_{0n}(T_n) \equiv (\pi_{0n}(T_n(m)) : m = 1, \ldots, M)$ is an M-vector of estimated *local q-values*, i.e., posterior probabilities for the true null hypotheses (Equations (7.10) and (7.11)).

Note that in this method, conditional on the empirical distribution P_n, the weights $\pi_{0n}(T_n)$ are constant and $G_n(c)$ is only random through the null test statistics T_{0n}. In contrast, for Procedure 7.1, the function $G_n(c)$ is random through both the guessed sets of true null hypotheses $\mathcal{H}_{0n}$ and the null test statistics T_{0n}. One can show, as in Theorem 7.2, that the weighted procedure asymptotically controls the gTP.

7.7 FDR-controlling empirical Bayes procedures

A number of authors have recently considered empirical Bayes approaches for controlling the *false discovery rate*, $FDR = \mathrm{E}[V_n/R_n]$ (Efron, 2005; Efron et al., 2001a,b; Goss Tusher et al., 2001; Storey, 2002, 2003; Storey et al., 2004; Storey and Tibshirani, 2001, 2003). Most proposals assume that the test statistics $T_n = (T_n(m) : m = 1, \ldots, M)$ are independently and identically distributed according to the non-parametric mixture model of Equation (7.9), although Storey (2003) and Storey et al. (2004) provide asymptotic control results under so-called weak or loose dependence conditions (e.g., dependence in finite blocks, ergodic dependence).

The present section provides a brief treatment of the empirical Bayes non-parametric mixture model approach to FDR control and its relationship to frequentist step-up Benjamini and Hochberg (1995) Procedure 3.22. Further detail are given in a forthcoming article.

7.7.1 FDR-controlling empirical Bayes q-value-based procedures

Parameters of interest for the common marginal non-parametric mixture model of Equation (7.9) are the following *posterior probability functions* for a true null hypothesis $H_0(m)$, given the corresponding test statistic $T_n(m)$,

$$\begin{aligned}\Pi_0(t) &\equiv \Pr(H_0(m) = 1 | T_n(m) > t) \\ &= \frac{\pi_0 \bar{F}_0(t)}{\bar{F}(t)} \\ &= \frac{\pi_0 \bar{F}_0(t)}{\pi_0 \bar{F}_0(t) + (1-\pi_0)\bar{F}_1(t)} \\ &= \left(1 + \frac{1-\pi_0}{\pi_0}\frac{\bar{F}_1(t)}{\bar{F}_0(t)}\right)^{-1}, \qquad m = 1, \ldots, M,\end{aligned} \tag{7.61}$$

and

$$\begin{aligned}\pi_0(t) &\equiv \Pr(H_0(m) = 1 | T_n(m) = t) \\ &= \frac{\pi_0 f_0(t)}{f(t)} \\ &= \frac{\pi_0 f_0(t)}{\pi_0 f_0(t) + (1-\pi_0) f_1(t)} \\ &= \left(1 + \frac{1-\pi_0}{\pi_0}\frac{f_1(t)}{f_0(t)}\right)^{-1}, \qquad m = 1, \ldots, M.\end{aligned} \tag{7.62}$$

The random variables $\Pi_0(T_n(m))$ and $\pi_0(T_n(m))$ have been referred to in the FDR literature as *q-values* and *local q-values*, respectively (Efron, 2005; Storey, 2002, 2003; Storey et al., 2004); we therefore refer to $\Pi_0(t)$ and $\pi_0(t)$ as a *q-value function* and a *local q-value function*, respectively[1].

Empirical Bayes q-values are similar in spirit to frequentist p-values. Indeed, the smaller the q-values $\Pi_0(T_n(m))$ and $\pi_0(T_n(m))$, the stronger the evidence against the corresponding null hypothesis $H_0(m)$. Thus, one can envisage the following two FDR-controlling q-value-based multiple testing procedures.

Procedure 7.3. [FDR-controlling empirical Bayes q-value-based procedure]
For controlling the FDR at level $\alpha \in (0,1)$, the *empirical Bayes q-value-based procedure* rejects any hypothesis $H_0(m)$ with q-value $\Pi_0(T_n(m))$ less than or equal to α. That is, the set of rejected null hypotheses is

$$\mathcal{R}_n(\alpha) \equiv \{m : \Pi_0(T_n(m)) \leq \alpha\}. \tag{7.63}$$

The adjusted p-value $\widetilde{P}_{0n}(m)$ corresponding to null hypothesis $H_0(m)$ is simply its q-value, that is,

$$\widetilde{P}_{0n}(m) = \Pi_0(T_n(m)), \qquad m = 1, \ldots, M. \tag{7.64}$$

Procedure 7.4. [FDR-controlling empirical Bayes local q-value-based procedure]
For controlling the FDR at level $\alpha \in (0, 1)$, the *empirical Bayes local q-value-based procedure* rejects any hypothesis $H_0(m)$ with local q-value $\pi_0(T_n(m))$ less than or equal to α. That is, the set of rejected null hypotheses is

$$\mathcal{R}_n^\ell(\alpha) \equiv \{m : \pi_0(T_n(m)) \leq \alpha\}. \tag{7.65}$$

The adjusted p-value $\widetilde{P}_{0n}^\ell(m)$ corresponding to null hypothesis $H_0(m)$ is simply its local q-value, that is,

$$\widetilde{P}_{0n}^\ell(m) = \pi_0(T_n(m)), \qquad m = 1, \ldots, M. \tag{7.66}$$

One can readily show that the above two procedures control the FDR, under the assumption that the test statistics $T_n = (T_n(m) : m = 1, \ldots, M)$ are independently and identically distributed according to the non-parametric mixture model of Equation (7.9).

Specifically, consider local q-value-based Procedure 7.4, with $\mathcal{R}_n^\ell = \mathcal{R}_n^\ell(\alpha)$ defined as in Equation (7.65). Firstly, note that

$$\frac{V_n^\ell}{R_n^\ell} = \mathrm{I}\left(R_n^\ell > 0\right) \frac{V_n^\ell}{R_n^\ell} = \mathrm{I}\left(R_n^\ell > 0\right) \frac{\sum_m \mathrm{I}\left(\pi_0(T_n(m)) \leq \alpha,\ H_0(m) = 1\right)}{\sum_m \mathrm{I}\left(\pi_0(T_n(m)) \leq \alpha\right)},$$

where $R_n^\ell \equiv |\mathcal{R}_n^\ell|$ and $V_n^\ell \equiv |\mathcal{R}_n^\ell \cap \mathcal{H}_0|$ denote, respectively, the numbers of rejected hypotheses and Type I errors, and one adopts the usual FDR convention that $V_n^\ell / R_n^\ell \equiv 0$ if $R_n^\ell = 0$. The conditional expected value of the proportion V_n^ℓ / R_n^ℓ of false positives, given the test statistics T_n, therefore satisfies

$$\begin{aligned}
\mathrm{E}\left[\left.\frac{V_n^\ell}{R_n^\ell}\right| T_n\right] &= \mathrm{I}\left(R_n^\ell > 0\right) \frac{\sum_m \mathrm{I}\left(\pi_0(T_n(m)) \leq \alpha\right) \Pr(H_0(m) = 1 | T_n)}{\sum_m \mathrm{I}\left(\pi_0(T_n(m)) \leq \alpha\right)} \\
&= \mathrm{I}\left(R_n^\ell > 0\right) \frac{\sum_m \mathrm{I}\left(\pi_0(T_n(m)) \leq \alpha\right) \pi_0(T_n(m))}{\sum_m \mathrm{I}\left(\pi_0(T_n(m)) \leq \alpha\right)} \\
&\leq \alpha\, \mathrm{I}\left(R_n^\ell > 0\right),
\end{aligned}$$

where the second equality follows by independence of the pairs $(T_n(m), H_0(m))$, so that $\Pr(H_0(m) = 1|T_n) = \Pr(H_0(m) = 1|T_n(m)) = \pi_0(T_n(m))$. Taking the expected value with respect to the test statistics T_n, one has

$$\mathrm{E}\left[\frac{V_n^\ell}{R_n^\ell}\right] \leq \alpha \Pr(R_n^\ell > 0) \leq \alpha.$$

A similar proof of FDR control may be given for q-value-based Procedure 7.3.

Note that one can also propose test statistic-based versions of the above two q-value-based procedures. Specifically, the set of rejected null hypotheses for the test statistic-based analogue of q-value-based Procedure 7.3 is defined as

$$\mathcal{R}_n(\alpha) \equiv \{m : T_n(m) > c_n(\alpha)\}, \tag{7.67}$$

where the common cut-off $c_n(\alpha)$ is the smallest (i.e., least conservative) value of t such that $\Pi_0(t) \leq \alpha$, that is,

$$c_n(\alpha) \equiv \inf\{t \in I\!R : \Pi_0(t) \leq \alpha\}. \tag{7.68}$$

A test statistic-based analogue of local q-value-based Procedure 7.4 can be defined likewise. The test statistic-based and original q-value-based MTPs coincide in the case of non-increasing q-value functions; otherwise, the test statistic-based MTPs are less conservative.

In practice, both q-value functions $\Pi_0(t)$ and $\pi_0(t)$ are unknown and must be estimated from the available data $\mathcal{X}_n$. Various estimation approaches are discussed in Sections 7.2.4 and 7.7.2.

In as much as the q-value function $\Pi_0(t)$ is typically easier to estimate than the local q-value function $\pi_0(t)$, Procedure 7.3 is preferred to its local analogue in Procedure 7.4.

7.7.2 Equivalence between empirical Bayes q-value-based procedure and frequentist step-up Benjamini and Hochberg procedure

Consider empirical Bayes q-value-based Procedure 7.3, defined in terms of the q-value function $\Pi_0(t)$ of Equation (7.61).

The q-value function $\Pi_0(t)$ may be estimated by the plug-in method, from estimators of π_0, F, and F_0 (Section 7.2.4). A simple estimator π_{0n}, of the prior probability π_0 of the true null hypotheses, is the conservative value of one (i.e., $\pi_{0n} = 1$). The common marginal CDF F may be estimated by the empirical CDF F_n, defined as

$$F_n(t) \equiv \frac{1}{M}\sum_{m=1}^{M} \mathrm{I}(T_n(m) \leq t). \tag{7.69}$$

The common marginal null CDF F_0 may be estimated as in bootstrap Procedures 2.3 and 2.4, by pooling the elements of the $M \times B$ matrix $\mathbf{Z}_n^B$ of null-transformed bootstrap test statistics. The estimated q-value function is then given by

$$\Pi_{0n}(t) \equiv \min\left\{1, \frac{\pi_{0n}\bar{F}_{0n}(t)}{\bar{F}_n(t)}\right\}. \tag{7.70}$$

As usual, consider one-sided rejection regions of the form $\mathcal{C}_n(m) = (c_n(m), +\infty)$ and let $O_n(m)$ denote the indices for the ordered test statistics, so that $T_n(O_n(1)) \geq \cdots \geq T_n(O_n(M))$. Then, by definition of the empirical CDF F_n and corresponding survivor function $\bar{F}_n = 1 - F_n$, one has

$$F_n(T_n(O_n(m))) = \frac{M-m+1}{M} \quad \text{and} \quad \bar{F}_n(T_n(O_n(m))) = \frac{m-1}{M}. \tag{7.71}$$

Now, note that the unadjusted p-values,

$$P_{0n}(m) = \bar{F}_{0n}(T_n(m)), \qquad m = 1, \ldots, M, \tag{7.72}$$

computed under the common marginal null survivor function $\bar{F}_{0n}$, are such that $P_{0n}(O_n(1)) \leq \cdots \leq P_{0n}(O_n(M))$.

The set of rejected hypotheses corresponding to FDR-controlling Procedure 7.3, based on estimated q-values $\Pi_{0n}(T_n(m))$, is given by

$$\begin{aligned}
\mathcal{R}_n(\alpha) &\equiv \{m : \Pi_{0n}(T_n(m)) \leq \alpha\} \qquad (7.73)\\
&= \{O_n(m) : \Pi_{0n}(T_n(O_n(m))) \leq \alpha\}\\
&= \{O_n(m) : \exists\ h \geq m \text{ such that } \Pi_{0n}(T_n(O_n(h))) \leq \alpha\}\\
&= \left\{O_n(m) : \exists\ h \geq m \text{ such that } \frac{\pi_{0n}\bar{F}_{0n}(T_n(O_n(h)))}{\bar{F}_n(T_n(O_n(h)))} \leq \alpha\right\}\\
&= \left\{O_n(m) : \exists\ h \geq m \text{ such that } \frac{\pi_{0n}P_{0n}(O_n(h))}{\frac{h-1}{M}} \leq \alpha\right\}\\
&= \left\{O_n(m) : \exists\ h \geq m \text{ such that } P_{0n}(O_n(h)) \leq \frac{1}{\pi_{0n}}\frac{h-1}{M}\alpha\right\},
\end{aligned}$$

where the third equality follows under the natural assumption that the estimated q-value function $\Pi_{0n}(t)$ is non-increasing in t, that is, the greater the test statistic $T_n(m)$, the less probable the corresponding null hypothesis $H_0(m)$. This monotonicity assumption on the q-value function can also be viewed as a form of stochastic domination, in the sense that $\bar{F}_{1n}(t)/\bar{F}_{0n}(t)$ is non-decreasing in t, that is, the test statistics are "greater" under the alternative hypotheses than under the null hypotheses.

For the conservative estimator $\pi_{0n} = 1$ and a large number M of hypotheses (so that $(m-1)/M \approxeq m/M$), then

$$\mathcal{R}_n(\alpha) = \left\{O_n(m) : \exists\ h \geq m \text{ such that } P_{0n}(O_n(h)) \leq \frac{h}{M}\alpha\right\}. \tag{7.74}$$

Thus, under the above conditions, *empirical Bayes q-value-based Procedure 7.3* coincides with *frequentist step-up Benjamini and Hochberg (1995) Procedure 3.22.*

7.8 Discussion

The present chapter has extended the TPPFP-controlling resampling-based empirical Bayes approach of van der Laan et al. (2005) to control generalized tail probability (gTP) error rates, $gTP(q,g) = \Pr(g(V_n, S_n) > q)$, for arbitrary functions $g(V_n, S_n)$ of the numbers of false positives V_n and true positives $S_n = R_n - V_n$. The generalized family-wise error rate (gFWER) and tail probabilities for the proportion of false positives (TPPFP) correspond, respectively, to the special cases of the number $g(V_n, S_n) = V_n$ of false positives and proportion $g(V_n, S_n) = V_n/(V_n + S_n)$ of false positives among the rejected hypotheses.

The simulation studies of van der Laan et al. (2005) reveal that the new TPPFP-controlling resampling-based empirical Bayes Procedure 7.1 tends to be more powerful than existing TPPFP-controlling MTPs, such as marginal step-down Lehmann and Romano (2005) Procedures 3.24 and 3.25 and augmentation Procedure 3.26. A TPPFP-controlling empirical Bayes MTP is applied to the HIV-1 dataset of Segal et al. (2004) in Chapter 11.

We wish to stress the generality of the proposed resampling-based empirical Bayes approach to gTP control: (i) it controls tail probability error rates for arbitrary functions $g(V_n, S_n)$ of the numbers of false positives and true positives; (ii) unlike most MTPs controlling the proportion of false positives, it provides Type I error control for general data generating distributions, with arbitrary dependence structures among variables; (iii) it can be applied to any distribution pair $(Q_{0n}, Q_{0n}^{\mathcal{H}})$ for the null test statistics and guessed sets of true null hypotheses, i.e., the common marginal non-parametric mixture model of Equation (7.9) is only one among many reasonable working models that does not assume independence of the test statistics.

Hence, Chapter 6 and the present Chapter 7 have developed two different general and flexible classes of MTPs for controlling generalized tail probability error rates. The augmentation approach has the advantage of simplicity, whereas the empirical Bayes approach tends to be superior in terms of power. Combining the two types of methods by, for example, augmenting a powerful initial gFWER-controlling empirical Bayes MTP, could readily lead to a variety of new powerful gTP-controlling MTPs.

Finally, note that one could envisage a similar empirical Bayes approach for controlling a broad class of Type I error rates, defined as parameters $\Theta(F_{g(V_n,S_n)})$ of the distribution of functions $g(V_n, S_n)$ of the numbers of false positives V_n and true positives S_n, where the function g satisfies monotonicity Assumptions EB.MgV and EB.MgS and the mapping Θ satisfies monotonicity

Assumption MΘ (i.e., $F_1 \geq F_2$ implies $\Theta(F_1) \leq \Theta(F_2)$, for appropriately defined CDFs F_1 and F_2). Such error rates include generalized expected value (gEV) error rates, $gEV(g) = \mathrm{E}[g(V_n, S_n)]$, and, in particular, the false discovery rate, with $g(v, s) = v/(v+s)$. A Θ-controlling MTP would essentially be the same as gTP-controlling Procedure 7.1, except that one would compute a general parameter Θ (e.g., expected value) for the distribution of the guessed function $G_n(c)$ of the numbers of false positives and true positives, rather than tail probabilities for $G_n(c)$. Specifically, given an M-vector of cut-offs c, let $F_n^B(c)$ denote the empirical CDF of the B bootstrap guessed g-specific functions $\{G_n^b(c) : b = 1, \ldots, B\}$ of the numbers of false positives and true positives (Equation (7.17)). For controlling $\Theta(F_{g(V_n,S_n)})$ at level $\alpha \in (0,1)$, a generalization of Procedure 7.1 is based on cut-off vectors c_n that satisfy the empirical Type I error constraint

$$\Theta(F_n^B(c_n)) \leq \alpha. \qquad (7.75)$$

Adjusted p-values may be derived as in Section 7.3. For instance, for common cut-offs, a generalization of Equation (7.25) is

$$\widetilde{p}_{0n}(o_n(m)) \approxeq \min_{h \in \overline{o}_n(m)} \Theta(F_n^B((t_n(h))^{(M)})). \qquad (7.76)$$

Color Plates

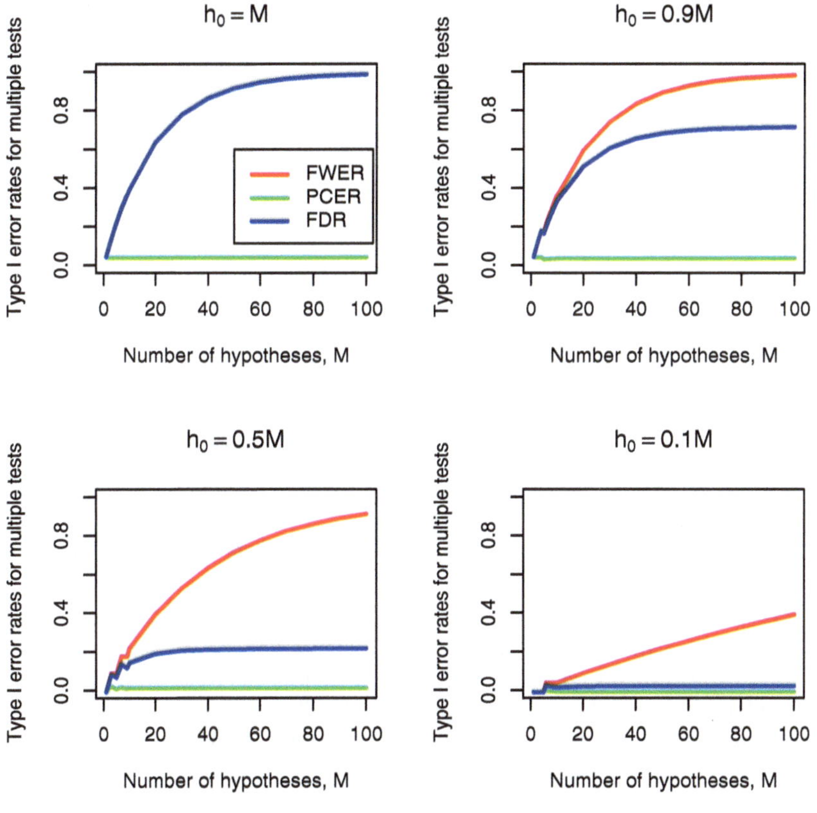

Figure 1.1. *Comparison of Type I error rates for a simple example.* Plots of Type I error rates for multiple tests (FWER: red curve; PCER: green curve; FDR: blue curve) vs. number of null hypotheses M, for different proportions h_0/M of true null hypotheses, $h_0/M = 1$, 0.9, 0.5, 0.1. The model and multiple testing procedures are described in Section 1.2.11. The single test actual Type I error rate is $\alpha = 0.05$ and the alternative shift parameter d is set to 1. The non smooth behavior for small M is due to the fact that it is not always possible to have exactly 90%, 50%, or 10% of true null hypotheses and rounding to the nearest integer may be necessary.

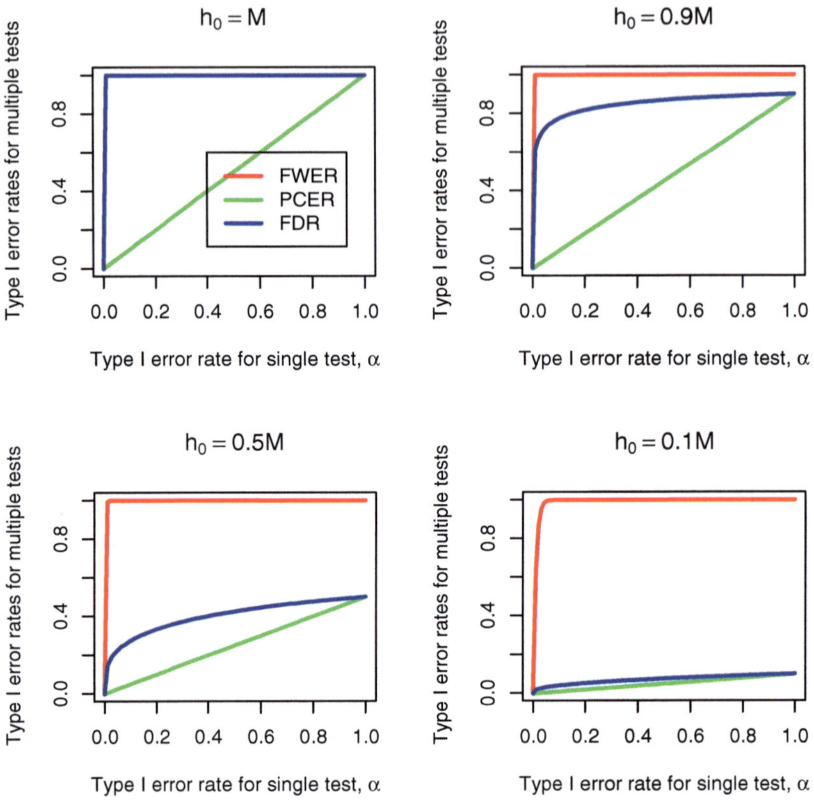

Figure 1.2. *Comparison of Type I error rates for a simple example.* Plots of Type I error rates for multiple tests (FWER: red curve; PCER: green curve; FDR: blue curve) vs. single test actual Type I error rate α, for different proportions h_0/M of true null hypotheses, $h_0/M = 1$, 0.9, 0.5, 0.1. The model and multiple testing procedures are described in Section 1.2.11. The number of hypotheses is $M = 100$ and the alternative shift parameter d is set to 1.

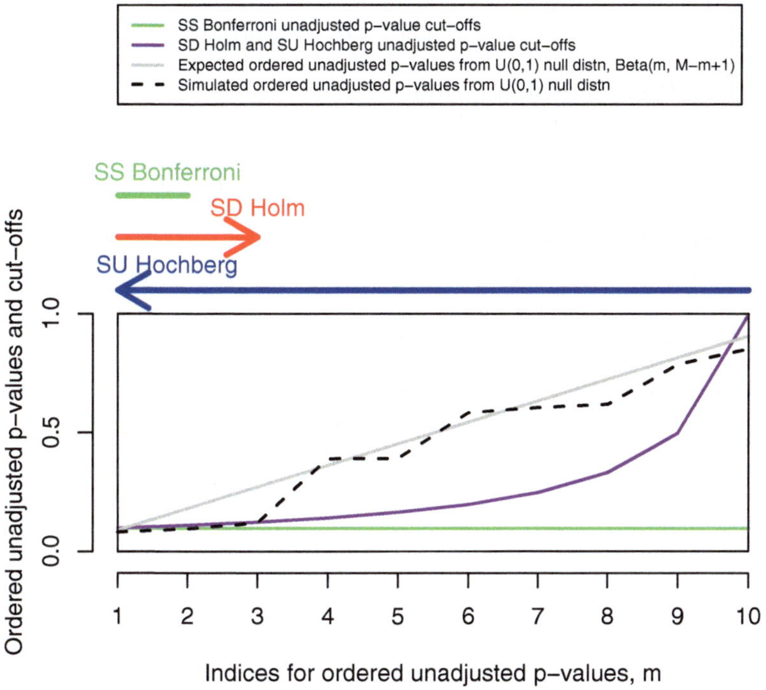

Figure 1.3. *Comparison of single-step, step-down, and step-up procedures: Cut-offs for FWER-controlling marginal Bonferroni, Holm, and Hochberg procedures.* The figure displays unadjusted p-value cut-offs for single-step Bonferroni Procedure 3.1 (SS Bonferroni; green curve), step-down Holm Procedure 3.7 (SD Holm; purple curve), and step-up Hochberg Procedure 3.13 (SU Hochberg; purple curve), for the test of $M = 10$ null hypotheses. Null unadjusted p-values are simulated as M independent realizations from the $\mathrm{U}(0,1)$ distribution. The simulated ordered unadjusted p-values and their expected values $m/(M+1)$ under the $\mathrm{Beta}(m, M-m+1)$ distribution are plotted as the dashed black curve and solid gray curve, respectively. The rejected null hypotheses are indicated by horizontal lines for each of the three procedures. An extreme nominal Type I error level $\alpha = 1$ is used for illustration purposes.

Ordered unadjusted p-values

$$P_{0n}(O_n(1)) \leq P_{0n}(O_n(2)) \leq P_{0n}(O_n(3)) \leq \cdots \leq P_{0n}(O_n(M))$$

Single-step Bonferroni adjusted p-values

$$MP_{0n}(O_n(1)) \leq MP_{0n}(O_n(2)) \leq MP_{0n}(O_n(3)) \leq \cdots \leq MP_{0n}(O_n(M))$$

$$\widetilde{P}_{0n}(O_n(m)) = \min\{MP_{0n}(O_n(m)), 1\}$$

Step-down Holm adjusted p-values

$$MP_{0n}(O_n(1)) \, ? \, (M-1)P_{0n}(O_n(2)) \, ? \, (M-2)P_{0n}(O_n(3)) \, ? \cdots ? \, 1P_{0n}(O_n(M))$$

$$\Longrightarrow$$

$$\widetilde{P}_{0n}(O_n(m)) = \max_{h=1,\ldots,m}\{\min\{(M-h+1)\,P_{0n}(O_n(h)), 1\}\}$$

Step-up Hochberg adjusted p-values

$$MP_{0n}(O_n(1)) \, ? \, (M-1)P_{0n}(O_n(2)) \, ? \, (M-2)P_{0n}(O_n(3)) \, ? \cdots ? \, 1P_{0n}(O_n(M))$$

$$\Longleftarrow$$

$$\widetilde{P}_{0n}(O_n(m)) = \min_{h=m,\ldots,M}\{\min\{(M-h+1)\,P_{0n}(O_n(h)), 1\}\}$$

Figure 1.4. *Comparison of single-step, step-down, and step-up procedures: Adjusted p-values for FWER-controlling marginal Bonferroni, Holm, and Hochberg procedures.*

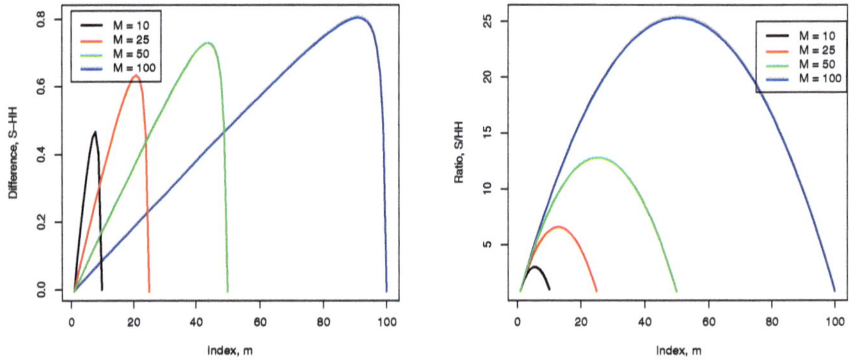

Panel (a): Difference, $a_m^{Simes}(\alpha) - a_m^{HH}(\alpha)$ Panel (b): Ratio, $a_m^{Simes}(\alpha)/a_m^{HH}(\alpha)$

Figure 3.1. *Comparison of stepwise Holm/Hochberg cut-offs and Simes cut-offs.* The figure displays plots of the difference (Panel (a)) and ratio (Panel (b)) between the Holm/Hochberg unadjusted p-value cut-offs $a_m^{HH}(\alpha) = \alpha/(M-m+1)$ and the Simes unadjusted p-value cut-offs $a_m^{Simes}(\alpha) = \alpha m/M$, for $\alpha = 1$ and total number of hypotheses $M = 10, 25, 50, 100$ (Equation (3.21)).

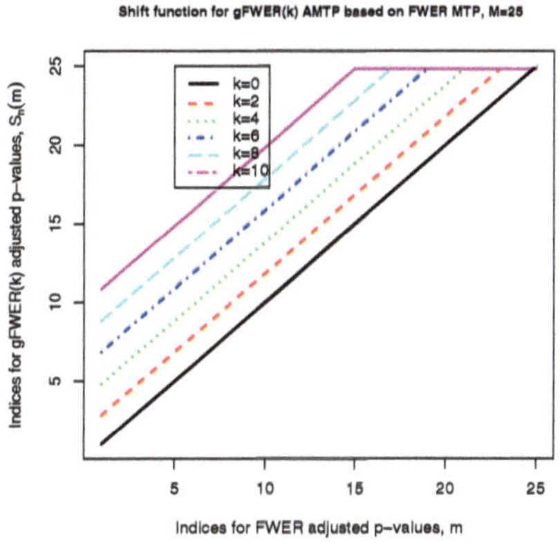

Panel (a): Shift function

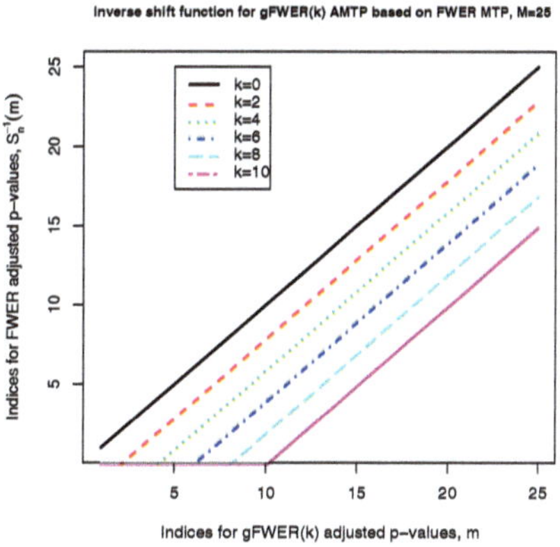

Panel (b): Inverse shift function

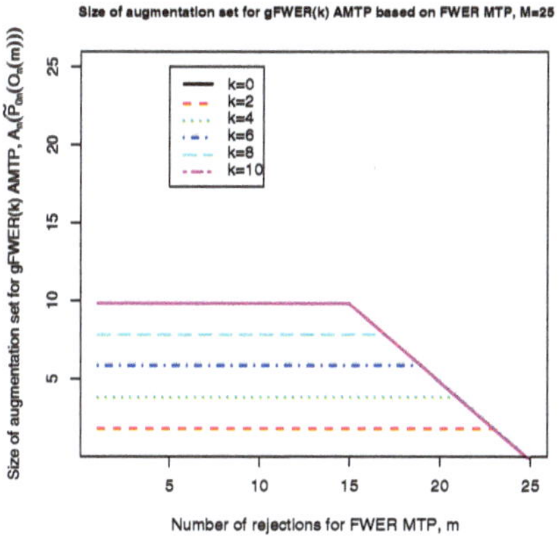

Panel (c): Augmentation set

Figure 6.5. *Sets of rejected hypotheses and adjusted p-values for a gFWER-controlling AMTP.* $gFWER(k)$-controlling augmentation multiple testing procedure based on an initial FWER-controlling procedure, with $M = 25$ null hypotheses and an allowed number $k \in \{0, 2, 4, 6, 8, 10\}$ of Type I errors. Panel (a): Adjusted p-value shift function $S_n(m)$ vs. m. Panel (b): Adjusted p-value inverse shift function $S_n^{-1}(m)$ vs. m. Panel (c): Cardinality $A_n(\widetilde{P}_{0n}(O_n(m)))$ of the augmentation set for the AMTP vs. number of rejected hypotheses $R_n(\widetilde{P}_{0n}(O_n(m))) = m$ for the initial MTP, for nominal Θ and Θ^+ Type I error level $\alpha = \widetilde{P}_{0n}(O_n(m))$.

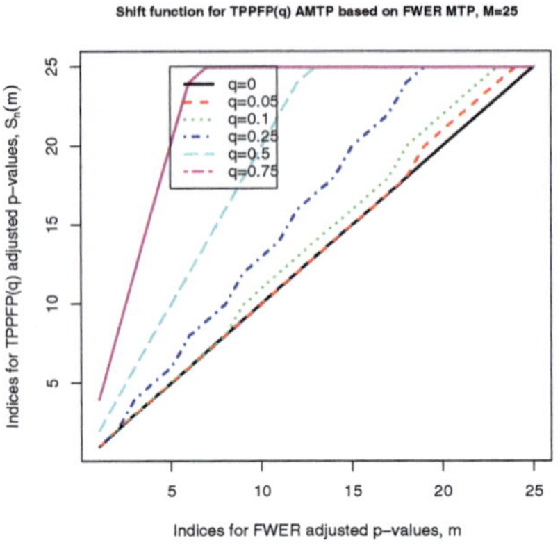

Panel (a): Shift function

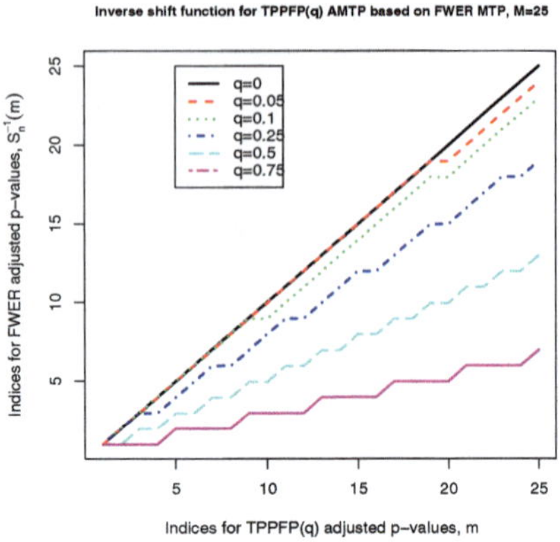

Panel (b): Inverse shift function

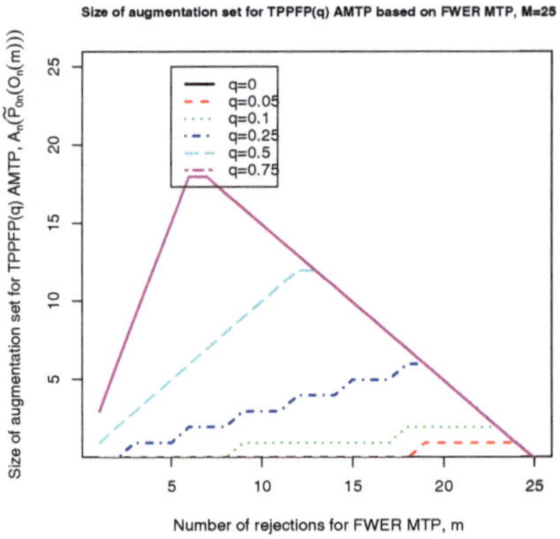

Panel (c): Augmentation set

Figure 6.9. *Sets of rejected hypotheses and adjusted p-values for a TPPFP-controlling AMTP.* $TPPFP(q)$-controlling augmentation multiple testing procedure based on an initial FWER-controlling procedure, with $M = 25$ null hypotheses and an allowed proportion $q \in \{0, 0.05, 0.10, 0.25, 0.50, 0.75\}$ of Type I errors. Panel (a): Adjusted p-value shift function $S_n(m)$ vs. m. Panel (b): Adjusted p-value inverse shift function $S_n^{-1}(m)$ vs. m. Panel (c): Cardinality $A_n(\widetilde{P}_{0n}(O_n(m)))$ of the augmentation set for the AMTP vs. number of rejected hypotheses $R_n(\widetilde{P}_{0n}(O_n(m))) = m$ for the initial MTP, for nominal Θ and Θ^+ Type I error level $\alpha = \widetilde{P}_{0n}(O_n(m))$.

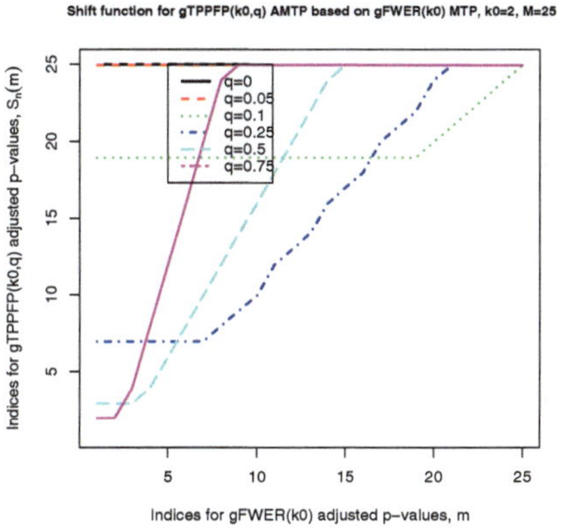

Panel (a): Shift function

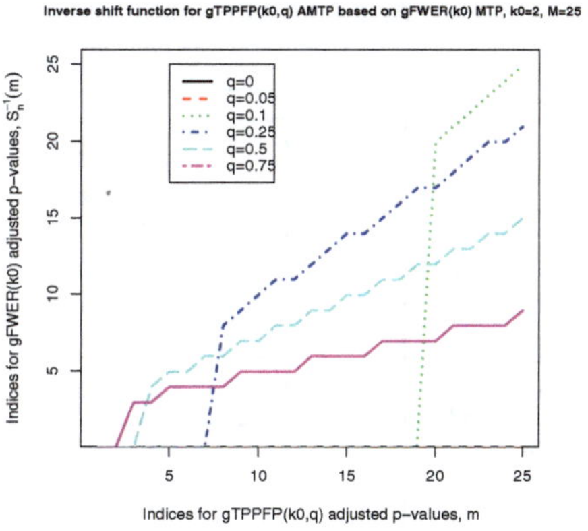

Panel (b): Inverse shift function

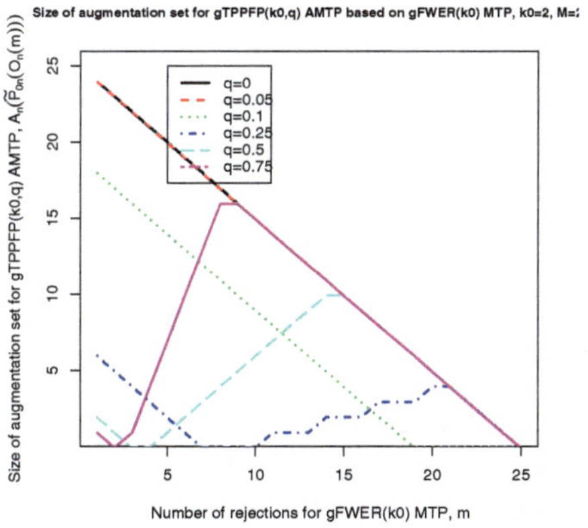

Panel (c): Augmentation set

Figure 6.13. *Sets of rejected hypotheses and adjusted p-values for a gTPPFP-controlling AMTP.* $gTPPFP(k_0, q)$-controlling augmentation multiple testing procedure based on an initial $gFWER(k_0)$-controlling procedure, with $M = 25$ null hypotheses, an allowed number $k_0 = 2$ of Type I errors, and an allowed proportion $q \in \{0, 0.05, 0.10, 0.25, 0.50, 0.75\}$ of Type I errors. Panel (a): Adjusted p-value shift function $S_n(m)$ vs. m. Panel (b): Adjusted p-value inverse shift function $S_n^{-1}(m)$ vs. m. Panel (c): Cardinality $A_n(\widetilde{P}_{0n}(O_n(m)))$ of the augmentation set for the AMTP vs. number of rejected hypotheses $R_n(\widetilde{P}_{0n}(O_n(m))) = m$ for the initial MTP, for nominal Θ and Θ^+ Type I error level $\alpha = \widetilde{P}_{0n}(O_n(m))$.

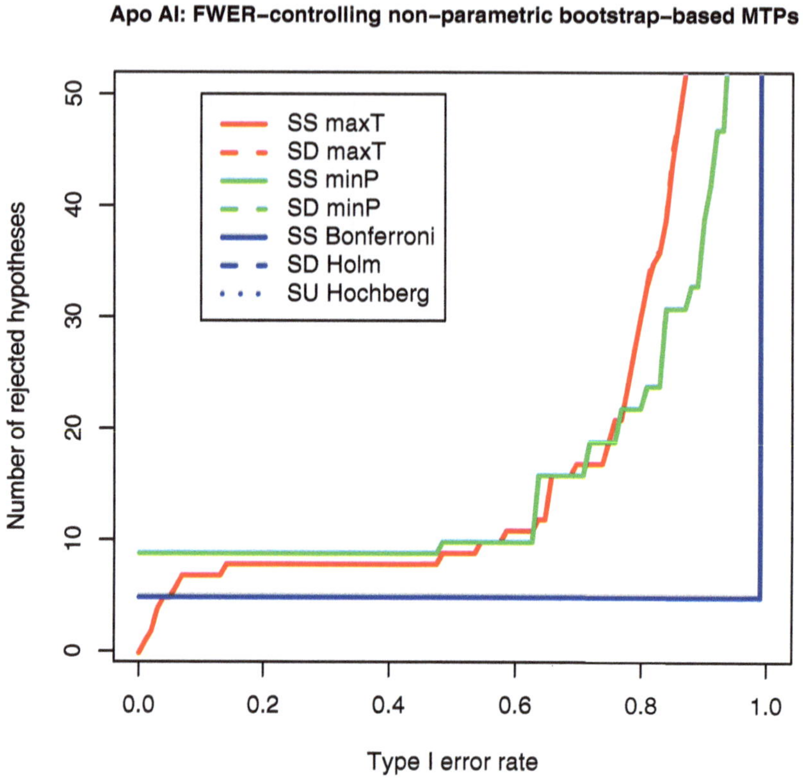

Figure 9.5. *Apo AI dataset: FWER-controlling non-parametric bootstrap-based MTPs.* Plots of number of rejected hypotheses vs. nominal Type I error level for FWER-controlling non-parametric bootstrap-based multiple testing procedures: single-step maxT Procedure 3.5 (SS maxT), single-step minP Procedure 3.6 (SS minP), step-down maxT Procedure 3.11 (SD maxT), step-down minP Procedure 3.12 (SD minP), single-step Bonferroni Procedure 3.1 (SS Bonferroni), step-down Holm Procedure 3.7 (SD Holm), and step-up Hochberg Procedure 3.13 (SU Hochberg). Adjusted p-values are computed under the non-parametric bootstrap null shift and scale-transformed test statistics null distribution of Procedure 2.3, based on either $B = 5,000$ (SS maxT, SD maxT, SS Bonferroni, SD Holm, SU Hochberg) or $B = 1,000$ (SS minP, SD minP) samples. Solid, dashed, and dotted lines represent, respectively, single-step, step-down, and step-up MTPs; red, green, and blue lines represent, respectively, joint common-cut-off, joint common-quantile, and marginal common-quantile MTPs. Results are displayed for the top 50 probes, where, for each MTP, probes are sorted in increasing order of their adjusted p-values. Note that the ordering of adjusted p-values may differ for common-cut-off (SS maxT, SD maxT) and common-quantile (SS minP, SD minP, SS Bonferroni, SD Holm, SU Hochberg) procedures.

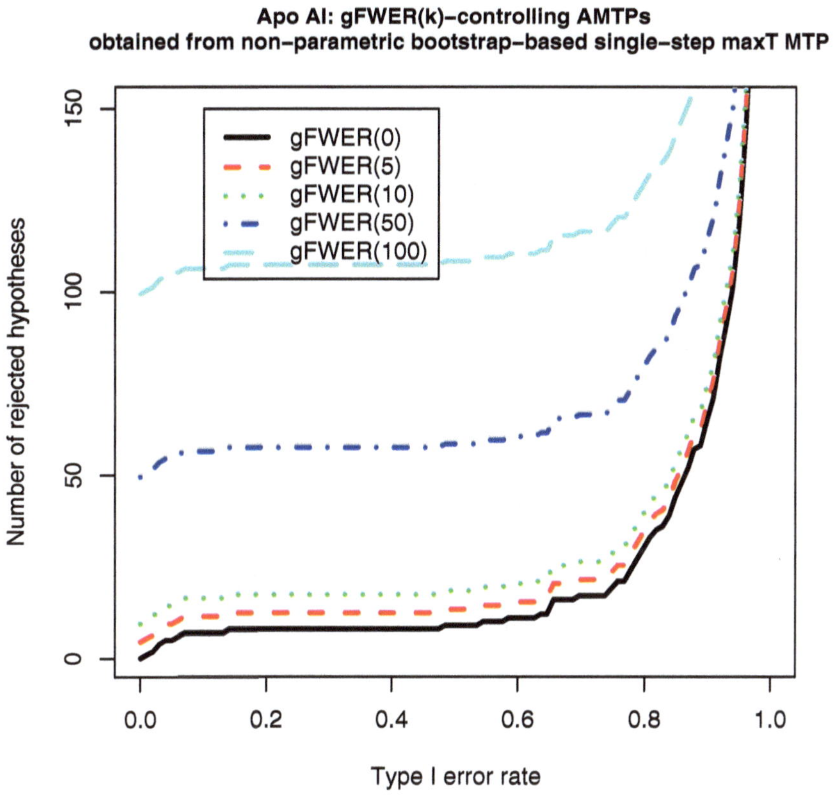

Figure 9.6. *Apo AI dataset: gFWER-controlling non-parametric bootstrap-based AMTPs.* Plots of number of rejected hypotheses vs. nominal Type I error level for gFWER-controlling augmentation multiple testing Procedure 3.20 (gFWER(k) AMTP), obtained from FWER-controlling non-parametric bootstrap-based single-step maxT Procedure 3.5 (SS maxT), with an allowed number $k \in \{0, 5, 10, 50, 100\}$ of false positives ($k = 0$ case corresponds to FWER). Adjusted p-values are computed under the non-parametric bootstrap null shift and scale-transformed test statistics null distribution of Procedure 2.3, based on $B = 5,000$ samples. Results are displayed for the top 150 probes, where, for each MTP, probes are sorted in increasing order of their adjusted p-values.

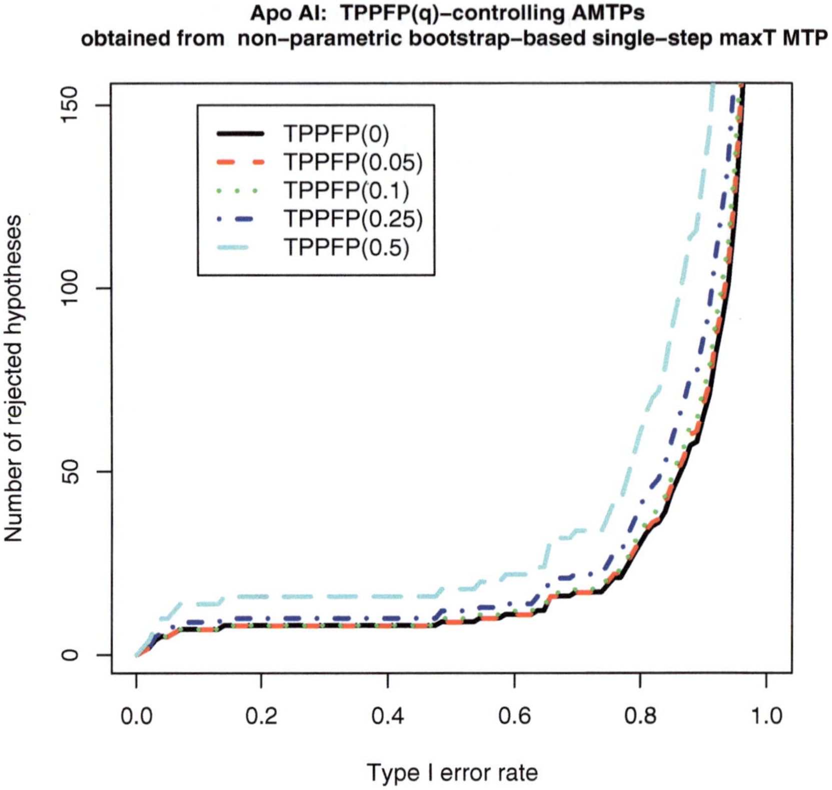

Figure 9.7. *Apo AI dataset: TPPFP-controlling non-parametric bootstrap-based AMTPs.* Plots of number of rejected hypotheses vs. nominal Type I error level for TPPFP-controlling augmentation multiple testing Procedure 3.26 (TPPFP(q) AMTP), obtained from FWER-controlling non-parametric bootstrap-based single-step maxT Procedure 3.5 (SS maxT), with an allowed proportion $q \in \{0, 0.05, 0.10, 0.25, 0.50\}$ of false positives ($q = 0$ case corresponds to FWER). Adjusted p-values are computed under the non-parametric bootstrap null shift and scale-transformed test statistics null distribution of Procedure 2.3, based on $B = 5,000$ samples. Results are displayed for the top 150 probes, where, for each MTP, probes are sorted in increasing order of their adjusted p-values.

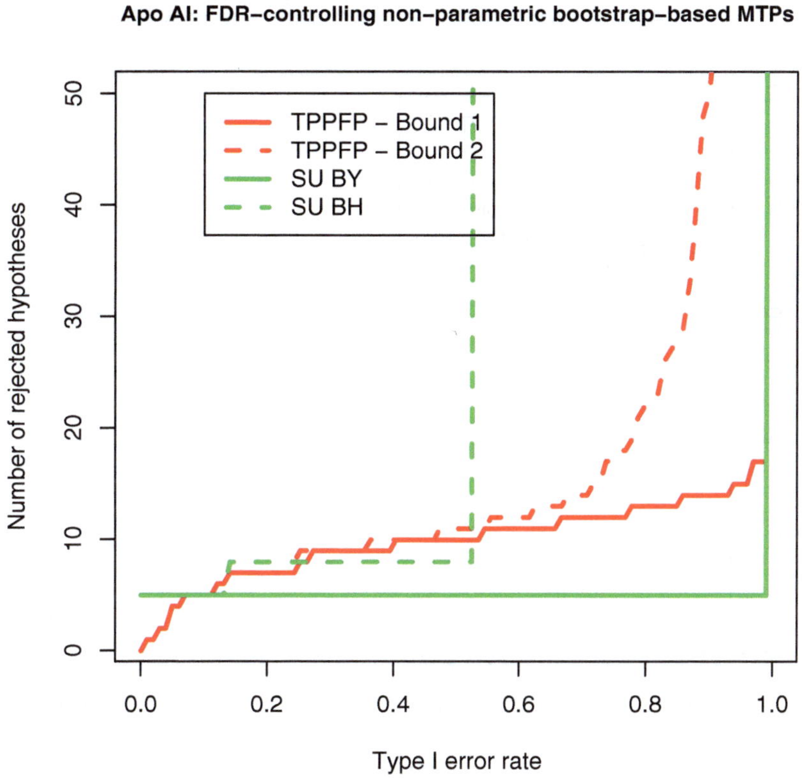

Figure 9.8. *Apo AI dataset: FDR-controlling non-parametric bootstrap-based MTPs.* Plots of number of rejected hypotheses vs. nominal Type I error level for FDR-controlling non-parametric bootstrap-based multiple testing procedures: marginal step-up Benjamini and Hochberg (1995) Procedure 3.22 (SU BH), marginal step-up Benjamini and Yekutieli (2001) Procedure 3.23 (SU BY), procedures described in Theorem 6.7, based on a TPPFP-controlling augmentation of joint single-step maxT Procedure 3.5 (TPPFP-based 1, TPPFP-based 2). Adjusted p-values are computed under the non-parametric bootstrap null shift and scale-transformed test statistics null distribution of Procedure 2.3, based on $B = 5,000$ samples. Solid and dashed lines represent, respectively, general conservative and restricted MTPs; red and green lines represent, respectively, joint common-cut-off and marginal common-quantile MTPs. Results are displayed for the top 50 probes, where, for each MTP, probes are sorted in increasing order of their adjusted p-values. Note that the ordering of adjusted p-values may differ for common-cut-off (TPPFP-based 1, TPPFP-based 2) and common-quantile (SU BH, SU BY) procedures.

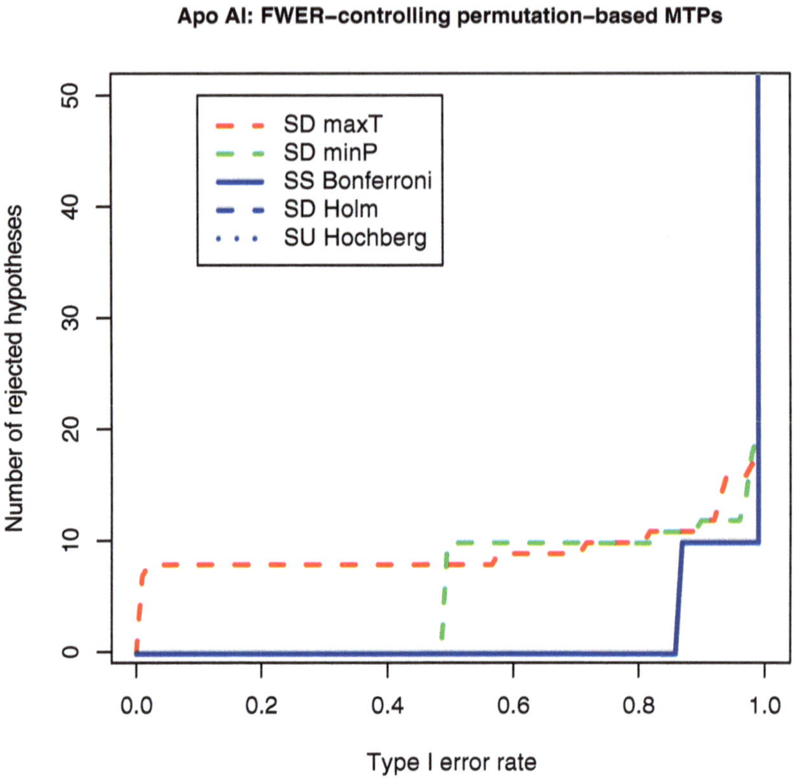

Figure 9.9. *Apo AI dataset: FWER-controlling permutation-based MTPs.* Plots of number of rejected hypotheses vs. nominal Type I error level for FWER-controlling permutation-based multiple testing procedures: step-down maxT Procedure 3.11 (SD maxT), step-down minP Procedure 3.12 (SD minP), single-step Bonferroni Procedure 3.1 (SS Bonferroni), step-down Holm Procedure 3.7 (SD Holm), and step-up Hochberg Procedure 3.13 (SU Hochberg). Adjusted p-values are computed based on all possible $\binom{16}{8} = 12,870$ permutations of the treatment and control labels. Solid, dashed, and dotted lines represent, respectively, single-step, step-down, and step-up MTPs; red, green, and blue lines represent, respectively, joint common-cut-off, joint common-quantile, and marginal common-quantile MTPs. Results are displayed for the top 50 probes, where, for each MTP, probes are sorted in increasing order of their adjusted p-values. Note that the ordering of adjusted p-values may differ for common-cut-off (SD maxT) and common-quantile (SD minP, SS Bonferroni, SD Holm, SU Hochberg) procedures.

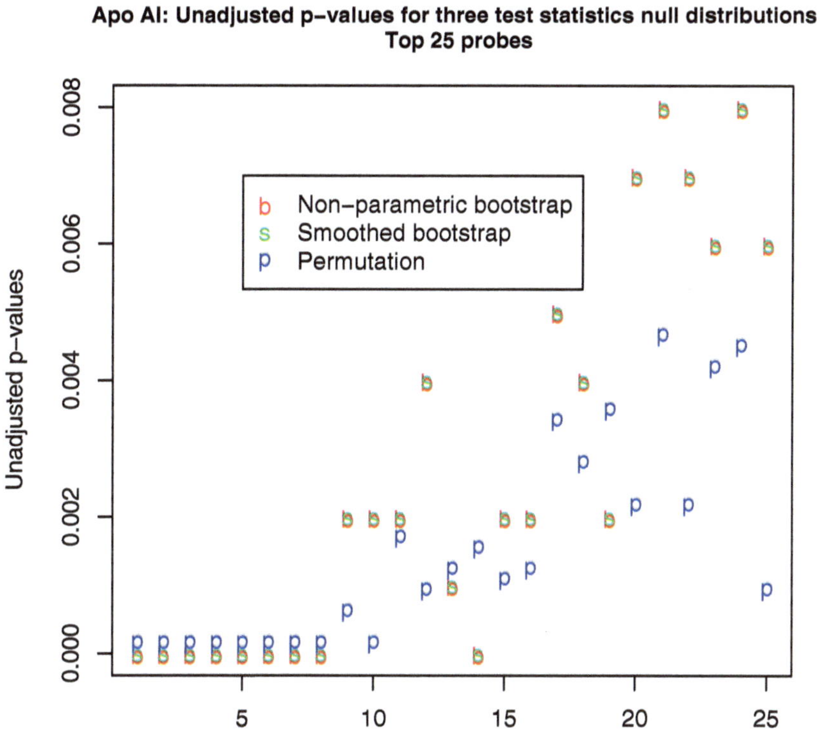

Figure 9.10. *Apo AI dataset: Unadjusted p-values for three test statistics null distributions.* Plots of unadjusted p-values for three test statistics null distributions: non-parametric bootstrap, smoothed non-parametric bootstrap, and permutation. Bootstrap p-values are computed under the non-parametric bootstrap null shift and scale-transformed test statistics null distribution of Procedure 2.3, based on $B = 1,000$ samples. Smoothed versions of the bootstrap p-values are obtained by kernel density smoothing of the marginal distributions of the null-transformed bootstrap test statistics. Permutation p-values are computed based on all possible $\binom{16}{8} = 12,870$ permutations of the treatment and control labels. Results are displayed for the top 25 probes, where probes are sorted first in increasing order of their non-parametric bootstrap SS maxT adjusted p-values, next in increasing order of their non-parametric bootstrap unadjusted p-values, and finally in decreasing order of their absolute test statistics.

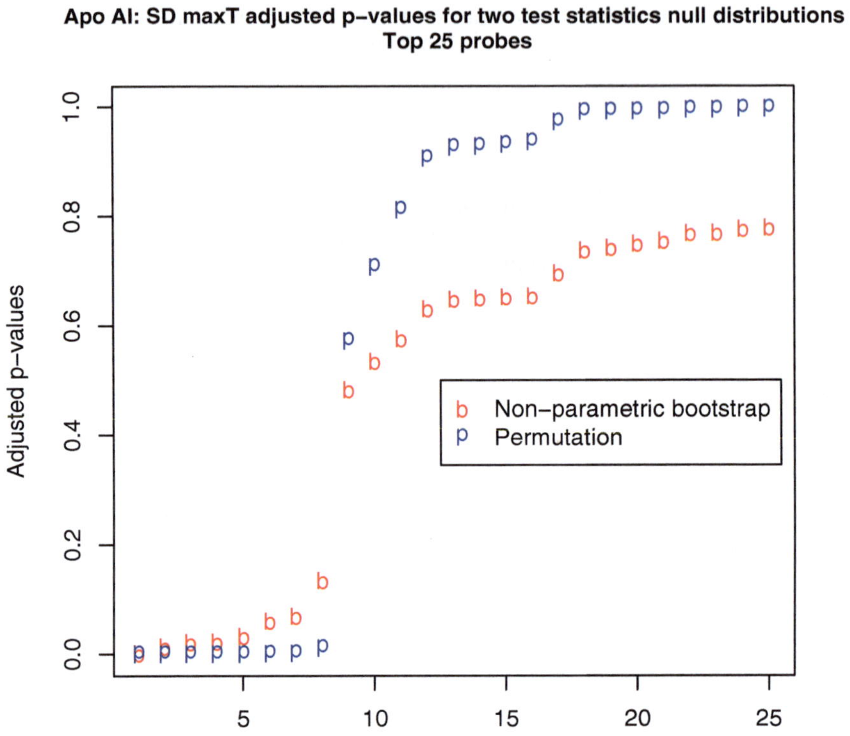

Figure 9.11. *Apo AI dataset: Step-down maxT adjusted p-values for non-parametric bootstrap and permutation test statistics null distributions.* Plots of adjusted p-values for step-down maxT Procedure 3.11 (SD maxT), for two test statistics null distributions: non-parametric bootstrap and permutation. Bootstrap p-values are computed under the non-parametric bootstrap null shift and scale-transformed test statistics null distribution of Procedure 2.3, based on $B = 5,000$ samples. Permutation p-values are computed based on all possible $\binom{16}{8} = 12,870$ permutations of the treatment and control labels. Results are displayed for the top 25 probes, where probes are sorted first in increasing order of their non-parametric bootstrap SS maxT adjusted p-values, next in increasing order of their non-parametric bootstrap unadjusted p-values, and finally in decreasing order of their absolute test statistics. Note the large jump in adjusted p-values between the 8th and 9th ordered probes, for both null distributions.

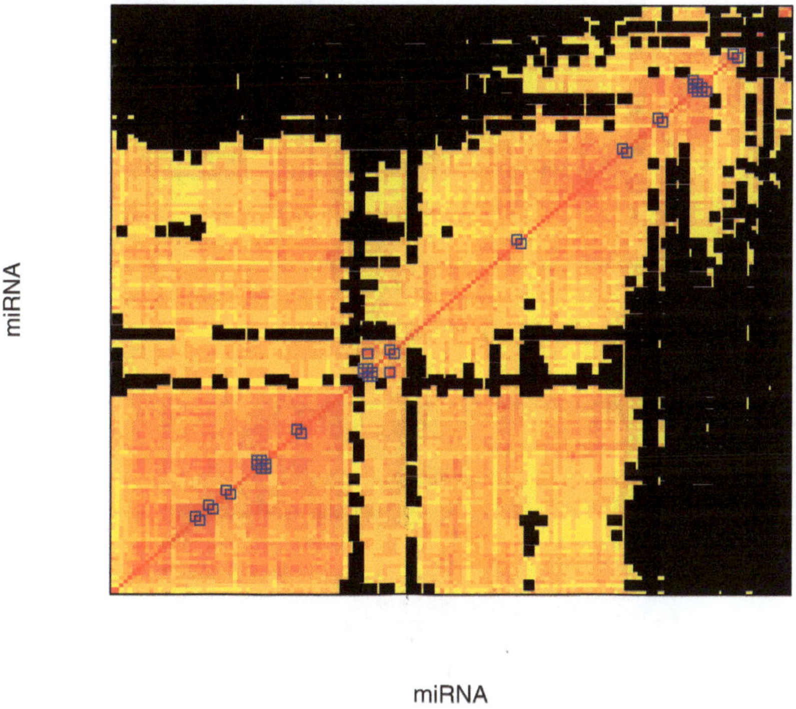

Figure 9.14. *Cancer miRNA dataset, co-expression: HOPACH clustering of miRNA expression profiles.* The figure provides a pseudo-color image of the 155×155 matrix of pairwise correlation coefficients for the expression profiles of the $J = 155$ miRNAs. Rows and columns are ordered according to the final level of the hierarchical tree of miRNA clusters produced by the HOPACH algorithm with Pearson correlation distance. Pairwise correlation coefficients not significantly different from zero are displayed in black. The remaining correlation coefficients are represented using a white (anti-correlated) to red (positively-correlated) color palette. Groups of co-expressed miRNAs appear as red blocks along the diagonal of the correlation matrix. The 20 most significantly correlated pairs of miRNAs from Table 9.11 are highlighted in blue.

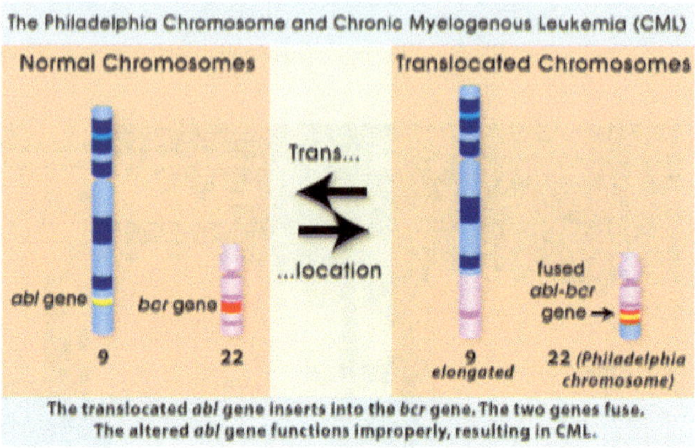

Panel (a): t(9;22) translocation

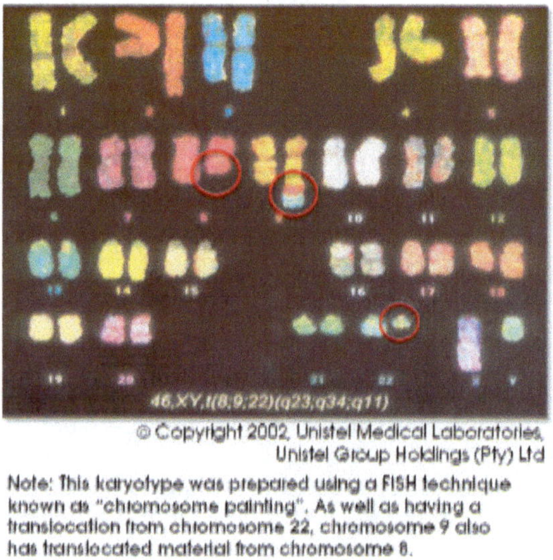

Panel (b): Karyotype

Figure 10.4. *The Philadelphia chromosome and the BCR/ABL fusion.* The BCR/ABL fusion is the molecular analogue of the Philadelphia chromosome. This t(9;22) translocation leads to a head-to-tail fusion of the **v-abl Abelson murine leukemia viral oncogene homolog 1** (**ABL1**) from chromosome 9 with the 5' half of the **breakpoint cluster region** (**BCR**) on chromosome 22. (Figure obtained from the Genetic Science Learning Center, The University of Utah, **gslc.genetics.utah.edu/units/disorders/karyotype/reciprocal.cfm**.)

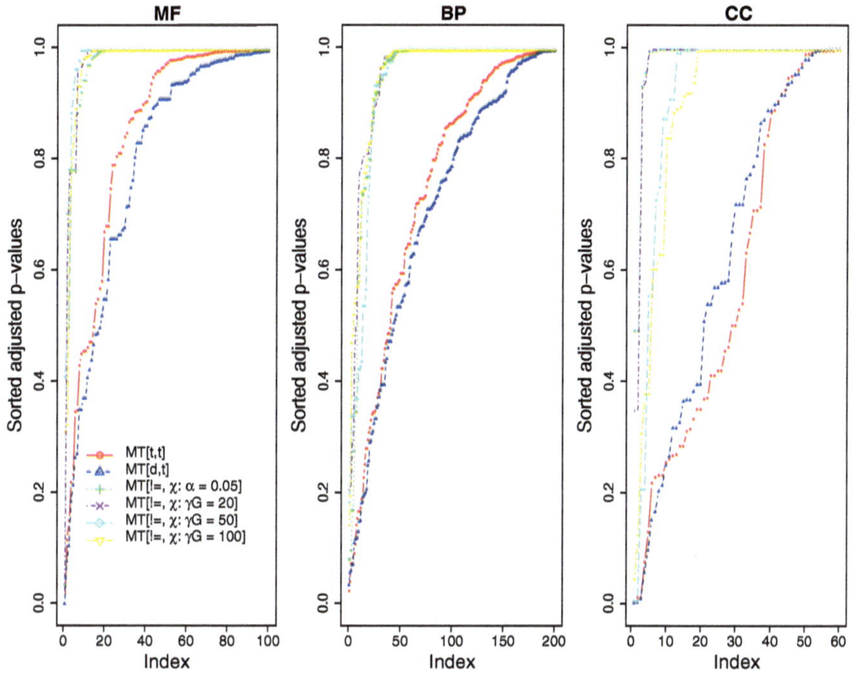

Figure 10.6. *GO terms associated with differential gene expression between BCR/ABL and NEG B-cell ALL, adjusted p-values.* Plots of sorted bootstrap-based single-step maxT adjusted p-values $\widetilde{P}_{0n}(m)$, for each of the three gene ontologies and each of the three testing scenarios.

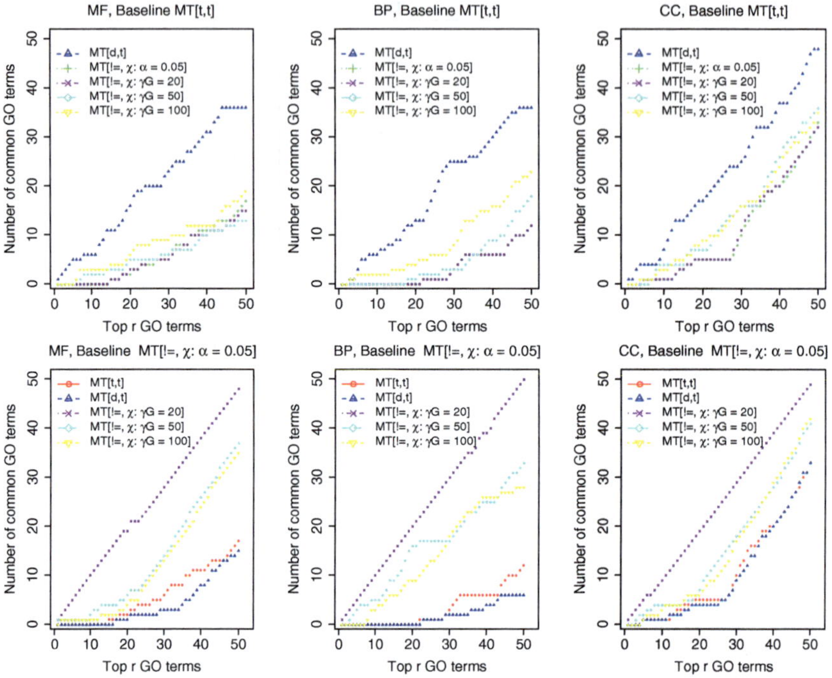

Figure 10.7. *GO terms associated with differential gene expression between BCR/ABL and NEG B-cell ALL, common terms between testing scenarios.* Plots of numbers of common GO terms among sets of ordered GO terms $\mathcal{O}_n(r)$ of various cardinality r for pairs of testing scenarios. Scenario $\mathsf{MT}[t,t]$ is used as the baseline in the top panels and Scenario $\mathsf{MT}[\neq, \chi : \alpha = 0.05]$, with adjusted p-value-based estimator $\lambda^{\neq}_{n,\alpha}$, $\alpha = 0.05$, for the binary DE gene-parameter profile $\lambda^{\neq}$, is used as the baseline in the bottom panels. For example, the blue curve in the top-left panel is a plot of $\left|\mathcal{O}^{d,t}_n(r) \cap \mathcal{O}^{t,t}_n(r)\right|$ vs. r for the MF gene ontology, i.e., of the overlap between the r most significant MF GO terms according to Scenarios $\mathsf{MT}[d,t]$ and $\mathsf{MT}[t,t]$.

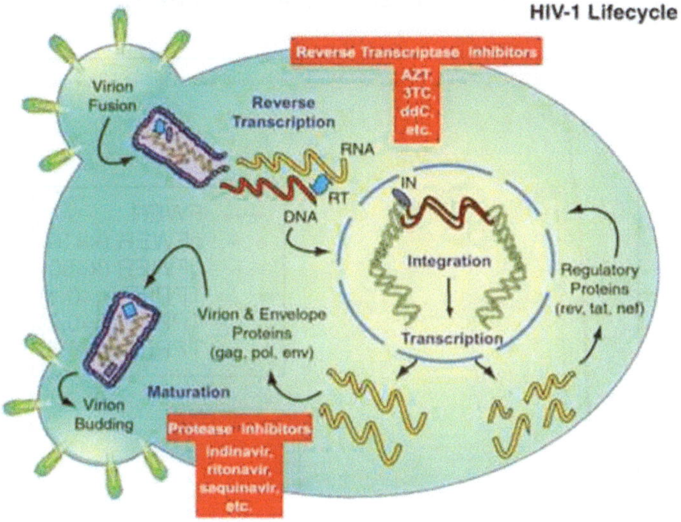

Figure 11.1. *HIV-1 lifecycle.* Diagram of the HIV-1 lifecycle and modes of action of protease and reverse transcriptase inhibitors.

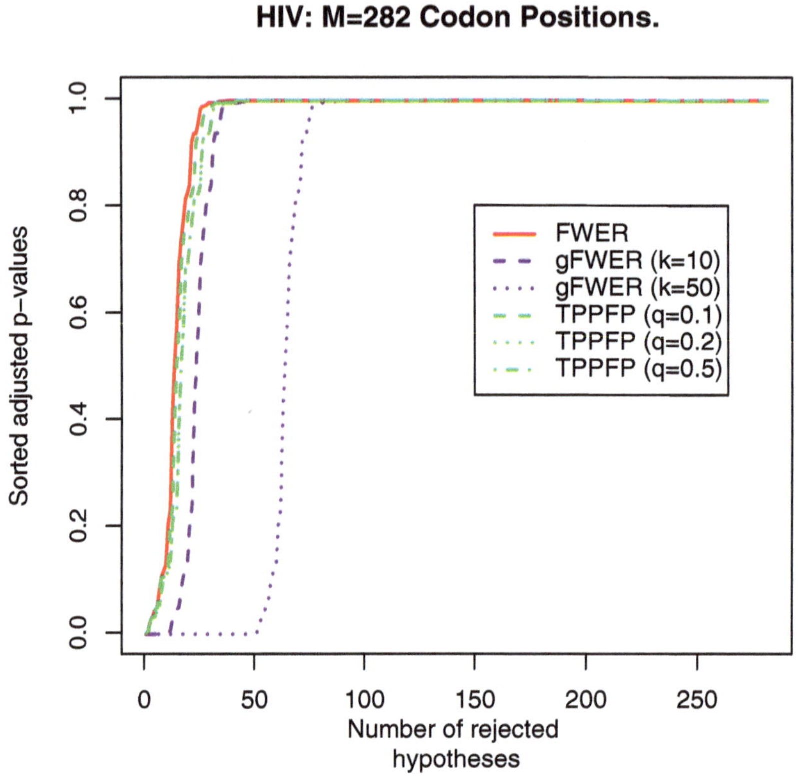

Figure 11.2. *HIV-1 dataset: Multiple testing analysis, Part I.* Sorted adjusted p-values for FWER-controlling single-step maxT Procedure 3.5, $gFWER(k)$-controlling augmentation Procedure 3.20 (allowed number of false positives, $k \in \{10, 50\}$), and $TPPFP(q)$-controlling augmentation Procedure 3.26 (allowed proportion of false positives, $q \in \{0.10, 0.20, 0.50\}$).

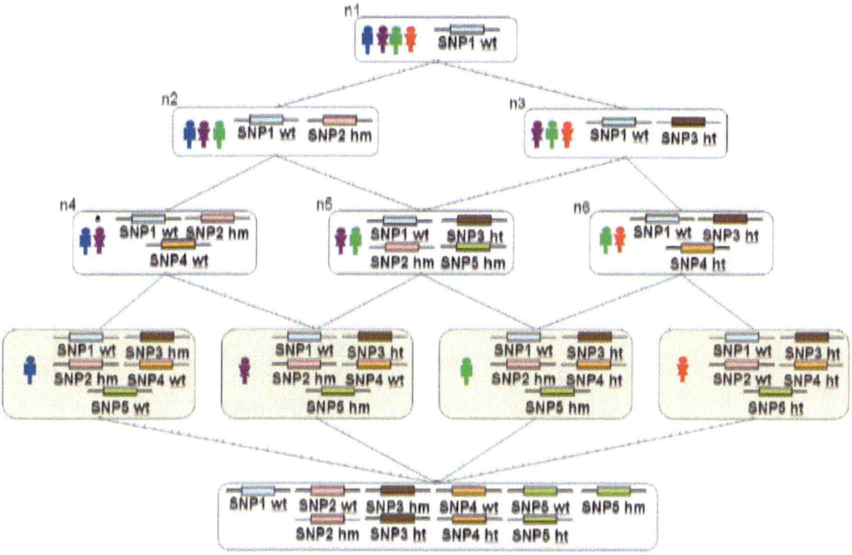

Figure 12.2. *Galois lattice for SNP genotypes.* The Hass diagram represents the Galois lattice for a simple artificial example with $n = 4$ patients genotyped at 5 SNPs. The set of objects $\mathcal{O} = \{o_1, o_2, o_3, o_4\}$ corresponds to the $n = 4$ patients and the set of descriptions $\mathcal{D} = \{$`(SNP 1 = wt)`, `(SNP 1 = ht)`, $\ldots$, `(SNP 5 = hm)`$\}$ to the 5×3 possible unphased genotypes for the 5 SNPs. Table 12.4 represents the formal context $(\mathcal{O}, \mathcal{D}, \mathcal{I})$, i.e., the binary relation $\mathcal{I}$ between the sets $\mathcal{O}$ and $\mathcal{D}$, underlying the Galois lattice. The Hass diagram has $M = 6$ nodes (in addition to the 4 leaf nodes), each corresponding to a subset of patients with a particular multilocus composite SNP genotype. For example, node $N_2 = (O_2, D_2)$ has coverage $O_2 = \{o_1, o_2, o_3\}$ of three patients and description $D_2 = \{$`(SNP 1 = wt)`, `(SNP 2 = hm)`$\}$. The leaf nodes (shaded in yellow) represent the set of all SNP genotypes (i.e., descriptions) observed in the sample of $n = 4$ individuals, in this case, 10 out of a possible 5×3 genotypes.

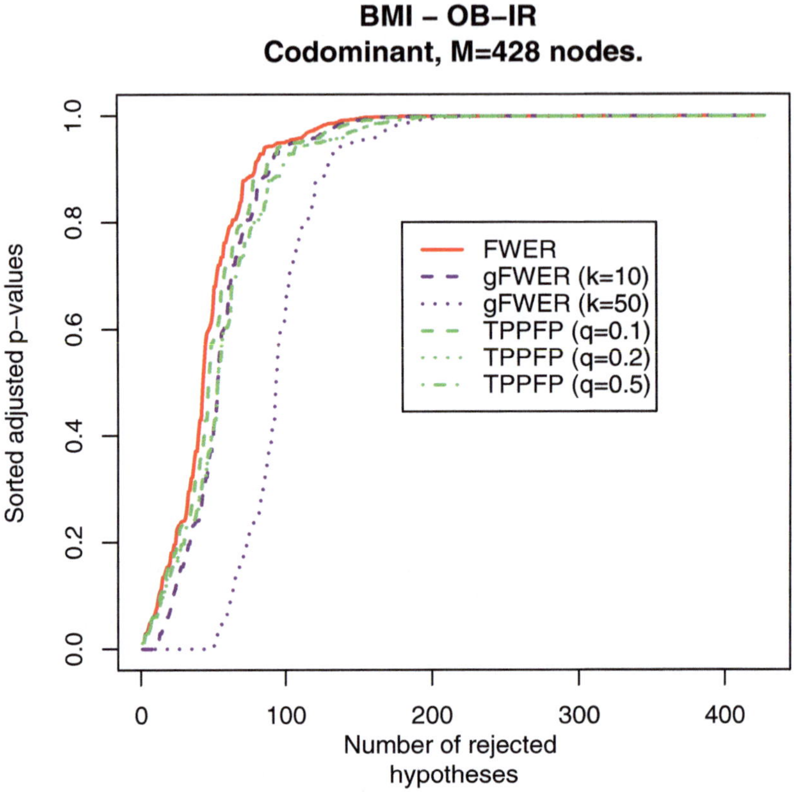

Figure 12.3. *ObeLinks dataset: BMI phenotype,* `OB-IR Codominant` *SNP genotype set.* Sorted adjusted p-values for FWER-controlling single-step maxT Procedure 3.5, $gFWER(k)$-controlling augmentation Procedure 3.20 (allowed number of false positives, $k \in \{10, 50\}$), and $TPPFP(q)$-controlling augmentation Procedure 3.26 (allowed proportion of false positives, $q \in \{0.10, 0.20, 0.50\}$).

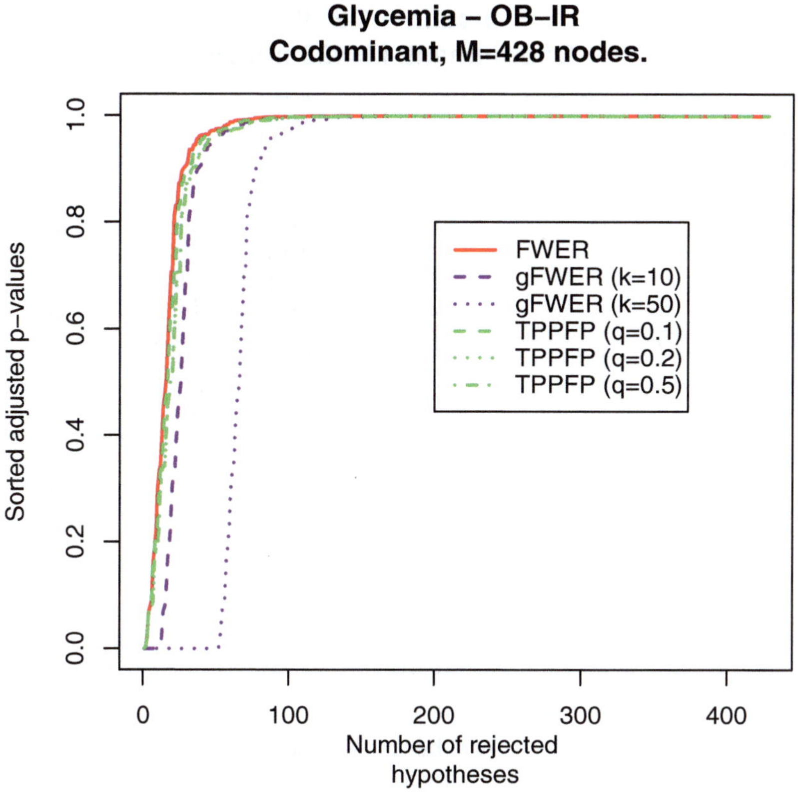

Figure 12.4. *ObeLinks dataset: Glycemia phenotype,* `OB-IR Codominant` *SNP genotype set.* Sorted adjusted p-values for FWER-controlling single-step maxT Procedure 3.5, $gFWER(k)$-controlling augmentation Procedure 3.20 (allowed number of false positives, $k \in \{10, 50\}$), and $TPPFP(q)$-controlling augmentation Procedure 3.26 (allowed proportion of false positives, $q \in \{0.10, 0.20, 0.50\}$).

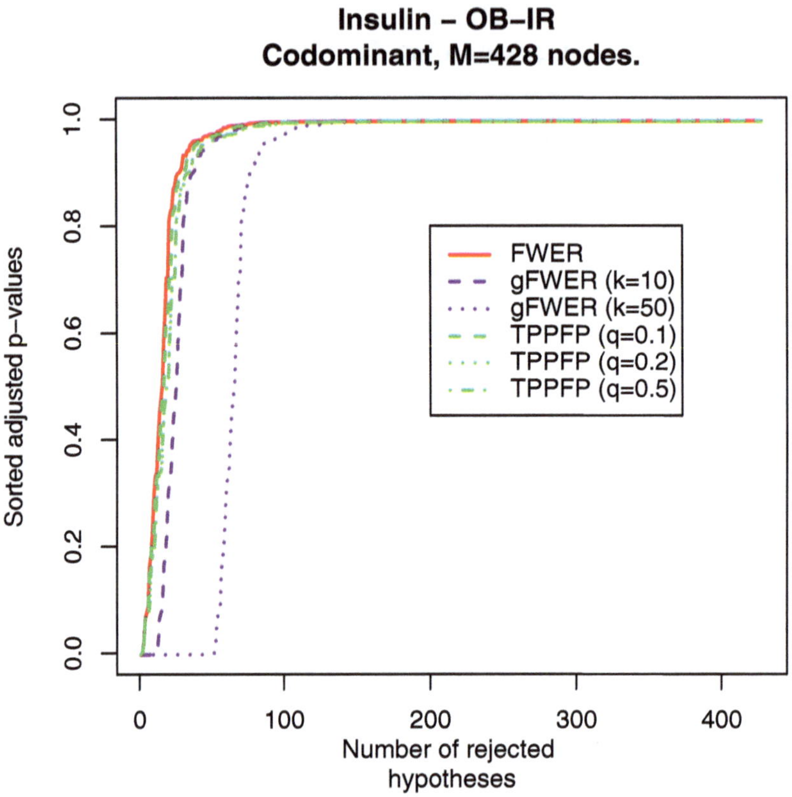

Figure 12.5. *ObeLinks dataset: Insulinemia phenotype,* `OB-IR Codominant` *SNP genotype set.* Sorted adjusted p-values for FWER-controlling single-step maxT Procedure 3.5, $gFWER(k)$-controlling augmentation Procedure 3.20 (allowed number of false positives, $k \in \{10, 50\}$), and $TPPFP(q)$-controlling augmentation Procedure 3.26 (allowed proportion of false positives, $q \in \{0.10, 0.20, 0.50\}$).

8

Simulation Studies: Assessment of Test Statistics Null Distributions

This chapter presents simulation studies assessing the performance of multiple testing procedures described in Chapters 1–7. The simulation studies focus on the choice of a test statistics null distribution in testing problems concerning correlation coefficients and regression coefficients in models where the covariates and error terms are allowed to be dependent (Pollard et al., 2005a). The reader is referred to Dudoit et al. (2004a) and van der Laan et al. (2005) for simulation studies comparing various gFWER- and TPPFP-controlling procedures.

8.1 Introduction

8.1.1 Motivation

As argued in Chapter 2, a key feature of the multiple hypothesis testing methodology proposed in Chapters 1–7 is the *test statistics null distribution* (rather than a data generating null distribution) used to obtain rejection regions (i.e., cut-offs) for the test statistics, confidence regions for the parameters of interest, and adjusted p-values. Indeed, whether testing single or multiple hypotheses, one needs the (joint) distribution of the test statistics in order to derive a procedure that probabilistically controls Type I errors. In practice, however, the true distribution of the test statistics is unknown and replaced by a null distribution. The choice of a proper null distribution is crucial in order to ensure that (finite sample or asymptotic) control of the Type I error rate under the *assumed null distribution* does indeed provide the desired control under the *true distribution.* This issue is particularly relevant for large-scale testing problems, such as those encountered in biomedical and genomic research (Chapters 9–12), which concern high-dimensional multivariate distributions, with complex and unknown dependence structures among variables.

Common approaches use a data generating distribution, such as a permutation distribution, that satisfies the *complete null hypothesis* that *all* null hypotheses are true (Section 2.8). Procedures based on such a *data generating null distribution* typically rely on the *subset pivotality* assumption, stated in Westfall and Young (1993, p. 42–43), to ensure that Type I error control under the data generating null distribution leads to the desired control under the true data generating distribution. However, subset pivotality is violated in many important testing problems, because a data generating null distribution may result in a joint distribution for the test statistics that has a different dependence structure than their true distribution. In fact, in most problems, there does not exist a data generating null distribution that correctly specifies the joint distribution of the test statistics corresponding to the true null hypotheses.

Indeed, subset pivotality fails for two types of testing problems that are highly relevant in biomedical and genomic data analysis: tests concerning *correlation coefficients* (Sections 2.6, 8.4, and 9.3) and tests concerning *regression coefficients* (Sections 2.6, 8.3, and 9.3). Tests of correlation arise, for example, when seeking to discover sets of *co-expressed* genes based on microarray expression measures. Tests concerning regression coefficients in linear and non-linear models (e.g., logistic model, Cox proportional hazards model) are commonly-used, particularly in medical applications, to investigate genotype/phenotype associations, e.g., identify genes or genomic regions associated with a possibly censored outcome (e.g., survival time, tumor class, response to treatment). The identification of *differentially expressed* genes, between different types of cell samples, based on microarray expression measures, is a canonical example for the test of multiple hypotheses concerning regression coefficients. Although subset pivotality may hold for some regression models, it fails for many models used in practice (e.g., linear regression model with dependent covariates and error terms, in Section 8.3, and logistic regression model, in Section 9.3).

As detailed in Chapter 2, one of our main contributions is the general characterization and explicit construction of proper test statistics null distributions. The first original proposal of Dudoit et al. (2004b), van der Laan et al. (2004a), and Pollard and van der Laan (2004), defines the null distribution as the asymptotic distribution of a vector of *null shift and scale-transformed test statistics*, based on user-supplied upper bounds for the means and variances of the test statistics for the true null hypotheses (Section 2.3). The second and most recent proposal of van der Laan and Hubbard (2006) defines the null distribution as the asymptotic distribution of a vector of *null quantile-transformed test statistics*, based on user-supplied marginal test statistics null distributions (Section 2.4). Resampling procedures (e.g., non-parametric or model-based bootstrap) are provided to conveniently obtain consistent estimators of the null distribution and of the corresponding test statistic cut-offs, parameter confidence regions, and adjusted p-values.

We stress the generality of the two test statistics null distributions proposed in Chapter 2: Type I error control does not rely on restrictive assumptions such as subset pivotality and holds for general data generating distributions (with arbitrary dependence structures among variables), null hypotheses (defined in terms of submodels for the data generating distribution), and test statistics (e.g., t-statistics, χ^2-statistics, F-statistics). In particular, these two null distributions allow one to address testing problems that cannot be handled by existing approaches, such as tests concerning correlation coefficients and parameters in general regression models (e.g., linear regression models where the covariates and error terms are allowed to be dependent, logistic regression models, Cox proportional hazards models).

8.1.2 Outline

The present chapter investigates the Type I error and power properties of multiple testing procedures (MTP) based on our general *non-parametric bootstrap null shift and scale-transformed test statistics null distribution* (Section 2.3) and various *parameter-specific bootstrap data generating null distributions* (Westfall and Young, 1993).

For the purpose of comparing null distributions, we focus on control of the *family-wise error rate* (FWER), using the *single-step maxT procedure*, a common-cut-off procedure exploiting the joint distribution of the test statistics (Procedure 3.5, details in Chapter 4). Note, however, that each null distribution could be employed with any other MTP, including joint augmentation and empirical Bayes procedures, for controlling generalized tail probability (gTP) error rates, $gTP(q,g) = \Pr(g(V_n, R_n) > q)$, for arbitrary functions $g(V_n, R_n)$ of the numbers of false positives V_n and rejected hypotheses R_n (Chapters 3–7).

We expect the new null quantile-transformed test statistics null distribution (Section 2.4; van der Laan and Hubbard (2006)) to yield similar or better results than the original null shift and scale-transformed test statistics null distribution. Indeed, the simulation results in van der Laan and Hubbard (2006) suggest that the null quantile-transformed distribution provides more accurate Type I error control and is more powerful than the null shift and scale-transformed distribution.

Section 8.2 summarizes, for convenience, algorithms for the bootstrap estimation of the null shift and scale-transformed test statistics null distribution (Procedure 8.1) and the corresponding resampling version of single-step maxT Procedure 3.5 (Procedure 8.2). Two separate simulation studies are performed to compare our general non-parametric bootstrap null shift and scale-transformed test statistics null distribution to parameter-specific bootstrap data generating null distributions proposed in Westfall and Young (1993). The first simulation study, in Section 8.3, considers tests for *regression coefficients*, in *linear models where the covariates and error terms are allowed to be dependent*, and compares the bootstrap null distribution of Procedure 8.1 to a bootstrap null distribution which involves *resampling residuals* (Westfall and Young, 1993, Section 3.4.1, p. 106–109). The second study, in Section 8.4, considers tests for *correlation coefficients* and compares the bootstrap null distribution of Procedure 8.1 to a bootstrap null distribution which involves *resampling individual variables independently* (Westfall and Young, 1993, Section 6.3, p. 194). As detailed in Sections 8.3.4 and 8.4.4, the simulation results demonstrate that the choice of null distribution can have a substantial impact on the Type I error properties of a given multiple testing procedure, such as the single-step maxT MTP. Procedures based on our general non-parametric bootstrap null shift and scale-transformed test statistics null distribution typically control the Type I error rate "on target" at the nominal level. In contrast, comparable procedures, based on parameter-specific bootstrap data generating null distributions, can be severely anti-conservative (bootstrapping residuals for testing regression coefficients) or conservative (independent covariates bootstrap for testing correlation coefficients).

8.2 Bootstrap-based multiple testing procedures

8.2.1 Null shift and scale-transformed test statistics null distribution

As detailed in Chapter 2, one of our main contributions is the general characterization (Section 2.2) and explicit construction (Sections 2.3 and 2.4) of proper test statistics null distributions.

Specifically, the first null distribution $Q_0 = Q_0(P)$ of Section 2.3 is defined as the asymptotic distribution of the M-vector $Z_n = (Z_n(m) : m = 1, \ldots, M)$ of *null shift and scale-transformed test statistics*,

$$Z_n(m) \equiv \sqrt{\min\left\{1, \frac{\tau_0(m)}{\mathrm{Var}[T_n(m)]}\right\}}\,(T_n(m) - \mathrm{E}[T_n(m)]) + \lambda_0(m), \qquad (8.1)$$

where $\lambda_0(m)$ and $\tau_0(m)$ are, respectively, user-supplied upper bounds for the means and variances of the $\mathcal{H}_0$-specific test statistics.

For the test of single-parameter null hypotheses using t-statistics, the null values are $\lambda_0(m) = 0$ and $\tau_0(m) = 1$ (Section 2.6). For testing the equality of K population mean vectors using F-statistics, the null values are $\lambda_0(m) = 1$ and $\tau_0(m) = 2/(K-1)$, under the assumption of equal variances in the different populations (Section 2.7). Furthermore, for a broad class of testing problems, such as the test of single-parameter null hypotheses using t-statistics, the null distribution Q_0 is an M-variate Gaussian distribution, with mean vector zero and covariance matrix $\sigma^* = \Sigma^*(P)$ equal to the correlation matrix of the vector influence curve (Section 2.6).

The test statistics null distribution $Q_0 = Q_0(P)$, defined according to Equation (8.1), depends on the true data generating distribution P and is therefore typically *unknown*. It can be estimated with the (non-parametric or model-based) *bootstrap*, as in Procedure 8.1, below (details in Sections 2.3.2, 2.6, and 2.7).

As established in earlier chapters, single-step and stepwise procedures based on the null shift and scale-transformed null distribution Q_0 (or a consistent estimator thereof, Q_{0n}) do indeed provide the desired Type I error control, for general data generating distributions (with arbitrary dependence structures among variables), null hypotheses, and test statistics.

8.2.2 Bootstrap estimation of the null shift and scale-transformed test statistics null distribution

Procedure 8.1. [Bootstrap estimation of the null shift and scale-transformed test statistics null distribution]
Let $P_n^\star$ denote an estimator of the true data generating distribution P. For the non-parametric bootstrap, $P_n^\star$ is simply the empirical distribution P_n, that is, samples of size n are drawn at random, with replacement from the observed data, $\mathcal{X}_n = \{X_i : i = 1, \ldots, n\}$. For the model-based bootstrap, $P_n^\star$ belongs to a model $\mathcal{M}$ for the data generating distribution P, such as a family of multivariate Gaussian distributions. One then proceeds as follows to generate the bootstrap test statistics null distribution.

1. Obtain the bth bootstrap sample, $\mathcal{X}_n^b \equiv \{X_i^b : i = 1, \ldots, n\}$, $b = 1, \ldots, B$, by generating n independent and identically distributed random variables X_i^b with distribution $P_n^\star$.
2. For each bootstrap sample $\mathcal{X}_n^b$, compute an M-vector of test statistics, $T_n^B(\cdot, b) = (T_n^B(m, b) : m = 1, \ldots, M)$, that can be arranged in an $M \times B$ matrix, $\mathbf{T}_n^B = \big(T_n^B(m, b) : m = 1, \ldots, M;\, b = 1, \ldots, B\big)$, with rows corresponding to the M null hypotheses and columns to the B bootstrap samples.
3. Compute row means and variances of the matrix $\mathbf{T}_n^B$, to yield estimators of the means, $\mathrm{E}[T_n(m)]$, and variances, $\mathrm{Var}[T_n(m)]$, of the test statistics under the true data generating distribution P. That is, compute

$$\mathrm{E}[T_n^B(m, \cdot)] \equiv \frac{1}{B} \sum_{b=1}^{B} T_n^B(m, b), \tag{8.2}$$

$$\mathrm{Var}[T_n^B(m, \cdot)] \equiv \frac{1}{B} \sum_{b=1}^{B} (T_n^B(m, b) - \mathrm{E}[T_n^B(m, \cdot)])^2.$$

4. Obtain an $M \times B$ matrix, $\mathbf{Z}_n^B = \big(Z_n^B(m, b) : m = 1, \ldots, M;\, b = 1, \ldots, B\big)$, of null shift and scale-transformed bootstrap test statistics $Z_n^B(m, b)$, as in Equation (8.1), by row-shifting and scaling the matrix $\mathbf{T}_n^B$ using the bootstrap estimators of $\mathrm{E}[T_n(m)]$ and $\mathrm{Var}[T_n(m)]$ and the user-supplied null values $\lambda_0(m)$ and $\tau_0(m)$. That is, define

$$Z_n^B(m,b) \equiv \sqrt{\min\left\{1, \frac{\tau_0(m)}{\mathrm{Var}[T_n^B(m,\cdot)]}\right\}}\left(T_n^B(m,b) - \mathrm{E}[T_n^B(m,\cdot)]\right) + \lambda_0(m). \tag{8.3}$$

5. The bootstrap estimator Q_{0n} of the null distribution Q_0 defined according to Equation (8.1) is the empirical distribution of the B columns $\{Z_n^B(\cdot,b) : b = 1,\ldots,B\}$ of matrix $\mathbf{Z}_n^B$.

As detailed in Sections 8.3.2 and 8.4.2, below, general bootstrap Procedure 8.1 differs in a number of key aspects from commonly-used bootstrap procedures. The latter procedures typically derive a test statistics null distribution $Q_n(P_{0n})$ by first creating a data generating distribution P_{0n} that satisfies the complete null hypothesis, $H_0^C = \mathrm{I}\left(P \in \cap_{m=1}^M \mathcal{M}(m)\right)$, that all M null hypotheses are true. For example, for tests concerning regression coefficients, in Section 8.3.2, Procedure Bootstrap e resamples residuals to generate bootstrap samples for which the outcome Y is independent of each covariate $X(j)$. Raw test statistics T_n are then computed, rather than null-transformed test statistics Z_n.

8.2.3 Bootstrap-based single-step maxT procedure

Bootstrap-based test statistic common cut-offs and adjusted p-values for the FWER-controlling single-step maxT procedure may be obtained as follows (Procedure 3.5, details in Chapter 4).

Procedure 8.2. [Bootstrap estimation of common cut-offs and adjusted p-values for single-step maxT Procedure 3.5]

0. Apply Procedure 8.1 to generate an $M \times B$ matrix, $\mathbf{Z}_n^B = \left(Z_n^B(m,b) : m = 1,\ldots,M;\, b = 1,\ldots,B\right)$, of null-transformed bootstrap test statistics $Z_n^B(m,b)$.
1. Compute the maximum statistic, $\max_m Z_n^B(m,b)$, $b = 1,\ldots,B$, for each bootstrap sample $\mathcal{X}_n^b$, i.e., each column of the matrix $\mathbf{Z}_n^B$.
2. For controlling the FWER at nominal level $\alpha \in (0,1)$, the bootstrap single-step maxT common cut-off $\gamma_{0n}(\alpha)$ is the $(1-\alpha)$-quantile of the empirical distribution of the B maxima $\{\max_m Z_n^B(m,b) : b = 1,\ldots,B\}$. That is,

$$\gamma_{0n}(\alpha) \equiv \inf\left\{\gamma \in I\!R : \frac{1}{B}\sum_{b=1}^{B} \mathrm{I}\left(\max_{m=1,\ldots,M} Z_n^B(m,b) \leq \gamma\right) \geq 1-\alpha\right\}. \tag{8.4}$$

3. The bootstrap single-step maxT adjusted p-value for null hypothesis $H_0(m)$ is the proportion of maxima $\{\max_m Z_n^B(m,b) : b = 1,\ldots,B\}$ that are greater than or equal to the corresponding observed test statistic $T_n(m)$. That is,

$$\widetilde{P}_{0n}(m) = \frac{1}{B}\sum_{b=1}^{B} \mathrm{I}\left(\max_{m=1,\ldots,M} Z_n^B(m,b) \geq T_n(m)\right), \qquad m = 1,\ldots,M. \tag{8.5}$$

Note that Procedure 8.2 could be applied, as in Sections 8.3 and 8.4, below, to any matrix $\mathbf{Z}_n^B$ of resampled statistics (e.g., from other bootstrap or permutation procedures). In particular, one could apply Procedure 8.2 to estimators of the new null quantile-transformed test statistics null distribution (Section 2.4; van der Laan and Hubbard (2006)).

8.3 Simulation Study 1: Tests for regression coefficients in linear models with dependent covariates and error terms

The first simulation study concerns tests for *regression coefficients* in *linear models where the covariates and error terms are allowed to be dependent.* This represents an important and practical testing scenario, because in many biomedical and genomic applications, covariates and error terms cannot be assumed to be independent and may have a complex and unknown joint distribution (e.g., logistic regression model relating cancer status to miRNA expression measures in Section 9.3).

8.3.1 Simulation model

Data generating distribution

Consider a data structure $(X, Y) \sim P$, where X is an M-dimensional covariate row vector and Y a univariate outcome. Assume that the pair (X, Y) has an $(M+1)$-dimensional Gaussian distribution P, that satisfies

$$\begin{aligned} \mathrm{E}[X] &= 0, & \mathrm{Cov}[X] &= \sigma_{xx}, \\ \mathrm{E}[Y|X] &= X\psi, & \mathrm{Var}[Y|X] &= \sigma_{y|X} = s(X), \end{aligned} \tag{8.6}$$

where the parameter ψ is an M-dimensional column vector of *regression coefficients*, σ_{xx} an $M \times M$ covariance matrix, and $s(X)$ a scalar function of the covariates X. That is, one can express the outcome Y in terms of the familiar *linear regression model*

$$Y = X\psi + \epsilon, \qquad \text{where} \qquad \epsilon|X \sim \mathrm{N}(0, s(X)), \tag{8.7}$$

so that,

$$Y|X \sim \mathrm{N}(X\psi, s(X)).$$

Suppose one has a random sample, $\mathcal{XY}_n \equiv \{(X_i, Y_i) : i = 1, \ldots, n\}$, of n independent and identically distributed (IID) pairs $(X_i, Y_i) \sim P$, from the above-specified Gaussian data generating distribution P. Let $\mathbf{X}_n$ and $\mathbf{Y}_n$ denote, respectively, the $n \times M$ design matrix and the $n \times 1$ outcome vector.

Null and alternative hypotheses

The hypotheses of interest concern the M elements of the regression parameter vector ψ. Specifically, consider two-sided tests of the M null hypotheses $H_0(m) = \mathrm{I}(\psi(m) = \psi_0(m))$ vs. the alternative hypotheses $H_1(m) = \mathrm{I}(\psi(m) \neq \psi_0(m))$, $m = 1, \ldots, M$. For simplicity, and without loss of generality, set the null values equal to zero (i.e., $\psi_0(m) = 0$).

8.3.2 Multiple testing procedures

Test statistics

The M null hypotheses are tested based on *t-statistics* for *ordinary least squares* (OLS) regression,

$$T_n(m) \equiv \frac{\psi_n(m)}{\sqrt{\sigma_n(m, m)}}, \qquad m = 1, \ldots, M, \tag{8.8}$$

where $\psi_n = (\psi_n(m) : m = 1, \ldots, M)$ is an M-vector of *least squares estimators* for the regression parameters, with estimated $M \times M$ covariance matrix σ_n, such that

$$\begin{aligned} \psi_n &\equiv (\mathbf{X}_n^\top \mathbf{X}_n)^{-1} \mathbf{X}_n^\top \mathbf{Y}_n, \\ \sigma_n &\equiv \frac{(\mathbf{Y}_n - \mathbf{X}_n \psi_n)^\top (\mathbf{Y}_n - \mathbf{X}_n \psi_n)}{n - M} (\mathbf{X}_n^\top \mathbf{X}_n)^{-1}. \end{aligned} \tag{8.9}$$

Define an n-vector of *residuals* $e_i \equiv Y_i - X_i \psi_n$, $i = 1, \ldots, n$, that is, let

$$\mathbf{e}_n \equiv \mathbf{Y}_n - \mathbf{X}_n \psi_n. \tag{8.10}$$

Test statistics null distributions

The simulation study compares the Type I error and power properties of FWER-controlling single-step maxT Procedure 8.2, based on the following two different bootstrap test statistics null distributions ($B = 10{,}000$ bootstrap samples).

Procedure 8.3. [Bootstrap XY null distribution: Bootstrapping covariate/outcome pairs (X, Y)]
The general non-parametric bootstrap test statistics null distribution of Procedure 8.1 involves *resampling covariate/outcome pairs* (X_i, Y_i) and computing *null shift and scale-transformed test statistics* for each bootstrap sample. Specifically, one proceeds as follows for the bth bootstrap sample, $b = 1, \ldots, B$.

1. Sample n covariate/outcome pairs (X_i^b, Y_i^b) at random, with replacement from the set of n observations $\mathcal{XY}_n = \{(X_i, Y_i) : i = 1, \ldots, n\}$. Let $\mathcal{XY}_n^b \equiv \{(X_i^b, Y_i^b) : i = 1, \ldots, n\}$ denote the resulting bootstrap sample.
2. Compute an M-vector $T_n^B(\cdot, b) = (T_n^B(m, b) : m = 1, \ldots, M)$ of bootstrap test statistics as in Equation (8.8), based on the bootstrap sample $\mathcal{XY}_n^b$.
3. Compute an M-vector $Z_n^B(\cdot, b) = (Z_n^B(m, b) : m = 1, \ldots, M)$ of null shift and scale-transformed bootstrap test statistics,
$$Z_n^B(m, b) \equiv \sqrt{\min\left\{1, \frac{1}{\mathrm{Var}[T_n^B(m, \cdot)]}\right\}} \left(T_n^B(m, b) - \mathrm{E}[T_n^B(m, \cdot)]\right),$$
where $\lambda_0(m) = 0$, $\tau_0(m) = 1$, and $\mathrm{E}[T_n^B(m, \cdot)] \equiv \sum_b T_n^B(m, b)/B$ and $\mathrm{Var}[T_n^B(m, \cdot)] \equiv \sum_b (T_n^B(m, b) - \mathrm{E}[T_n^B(m, \cdot)])^2/B$ denote, respectively, the empirical mean and variance of the B bootstrap test statistics $T_n^B(m, b)$ for null hypothesis $H_0(m)$, $m = 1, \ldots, M$ (i.e., row means and variances of the matrix $\mathbf{T}_n^B$, as in Procedure 8.1).

The test statistics null distribution is the empirical distribution Q_{0n} of the B M-vectors $\{Z_n^B(\cdot, b) : b = 1, \ldots, B\}$, i.e., of the columns of matrix $\mathbf{Z}_n^B$.

Procedure 8.4. [Bootstrap e null distribution: Bootstrapping residuals e]
In contrast, the parameter-specific bootstrap test statistics null distribution proposed in Westfall and Young (1993, Section 3.4.1, p. 106–109) involves *resampling residuals* e_i and computing *raw test statistics* (without null transformation) for each bootstrap sample. Specifically, one proceeds as follows for the bth bootstrap sample, $b = 1, \ldots, B$.

1. Sample n residuals at random, with replacement from the set of n observed residuals $\{e_i : i = 1, \ldots, n\}$ defined in Equation (8.10). Let $\mathbf{e}_n^b \equiv (e_i^b : i = 1, \ldots, n)$ denote the resulting n-vector of bootstrap residuals.

2. Generate n bootstrap covariate/outcome pairs, by randomly pairing each of the n observed covariate vectors X_i with a bootstrap residual e_i^b, that is, by defining a bootstrap outcome n-vector $\mathbf{Y}_n^b \equiv \mathbf{e}_n^b$ as the vector of bootstrap residuals. Let $\mathcal{XY}_n^b \equiv \{(X_i, Y_i^b) : i = 1, \ldots, n\}$ denote the resulting bootstrap sample.
3. Compute an M-vector $T_n^B(\cdot, b) = (T_n^B(m, b) : m = 1, \ldots, M)$ of bootstrap test statistics as in Equation (8.8), based on the bootstrap sample $\mathcal{XY}_n^b$.

The test statistics null distribution is the empirical distribution Q_{0n} of the B M-vectors $\{T_n^B(\cdot, b) : b = 1, \ldots, B\}$, i.e., of the columns of matrix $\mathbf{T}_n^B$.

Thus, bootstrap Procedures Bootstrap XY and Bootstrap e differ in the following two key aspects.

1. The (re)sampling units: Bootstrap XY resamples covariate/outcome pairs (X_i, Y_i), whereas Bootstrap e resamples residuals e_i.
2. The bootstrap test statistics: Bootstrap XY relies on null-transformed test statistics Z_n, whereas Bootstrap e relies on raw test statistics T_n.

In other words, Procedure Bootstrap e derives the test statistics null distribution by first creating a data generating null distribution that corresponds to the complete null hypothesis that the outcome Y is independent of each covariate $X(j)$. Note that bootstrapping covariate/outcome pairs (X_i, Y_i) preserves the correlation structure of the data, while bootstrapping residuals and randomly pairing bootstrap residuals and covariates destroy this correlation structure.

Single-step maxT procedure

Adjusted p-values for single-step maxT Procedure 3.5 may be obtained by applying Procedure 8.2 with bootstrap null distributions Bootstrap XY and Bootstrap e. Specifically, adjusted p-values for Bootstrap XY and Bootstrap e are computed, respectively, from the empirical distributions of the B maxima of null-transformed test statistics $\{\max_m Z_n^B(m, b) : b = 1, \ldots, B\}$ and raw test statistics $\{\max_m T_n^B(m, b) : b = 1, \ldots, B\}$. For a test at nominal FWER level α, one rejects null hypotheses with adjusted p-values less than or equal to α.

8.3.3 Simulation study design

Simulation parameters

The following model parameters are varied in the simulation study.

- *Sample size, n.* $n = 25$, 100.

- *Number of hypotheses, M.* $M = 10, 20$.
- *Covariance matrix of the covariates, σ_{xx}.* The covariance matrix σ_{xx} of the covariates X has unit diagonal elements and off-diagonal elements set to a common value ς, i.e., $\sigma_{xx}(m,m) = 1$, for $m = 1, \ldots, M$, and $\sigma_{xx}(m,m') = \varsigma$, for $m \neq m' = 1, \ldots, M$. The following values are considered for the common covariance: $\varsigma = 0.10, 0.50, 0.80$.
- *Conditional variance of outcome Y given covariates X, $s(X)$.* $\mathrm{Var}[Y|X] = \sigma_{y|X} = s(X) = \sum_{m \notin \mathcal{H}_0} X(m)$.
- *Proportion of true null hypotheses, h_0/M.* $h_0/M = 0.50, 0.75$.
- *Alternative regression parameters, $(\psi(m) : m \notin \mathcal{H}_0)$.* For each simulation model, regression parameters $(\psi(m) : m \notin \mathcal{H}_0)$, for the false null hypotheses, are generated as $|\mathcal{H}_0^c| = M - h_0$ independent uniform random variables over the interval $[0, \mu/\sqrt{n}]$. That is, $\psi(m) \overset{IID}{\sim} \mathrm{U}(0, \mu/\sqrt{n})$, $m \notin \mathcal{H}_0$. The following values are considered for the alternative shift parameter: $\mu = 0.10, 0.25$.

Estimating Type I error rate and power

For each simulation model (i.e., each combination of parameter values n, M, ς, $s(X)$, h_0/M, and μ), generate $A = 500$ random samples, $\mathcal{XY}_n^a \equiv \{(X_i^a, Y_i^a) : i = 1, \ldots, n\}$, $a = 1, \ldots, A$, of covariate/outcome pairs $(X, Y) \sim P$. For each such simulated sample, compute adjusted p-values $\widetilde{P}_{0n}^a(m)$ for single-step maxT Procedure 8.2, based on each of the two bootstrap null distributions (Bootstrap XY and Bootstrap e).

For a given nominal Type I error level α, compute, for each MTP, the numbers of rejected hypotheses $R_n^a(\alpha)$, Type I errors $V_n^a(\alpha)$, and Type II errors $U_n^a(\alpha)$,

$$R_n^a(\alpha) \equiv \sum_{m=1}^{M} \mathrm{I}\left(\widetilde{P}_{0n}^a(m) \leq \alpha\right), \tag{8.11}$$

$$V_n^a(\alpha) \equiv \sum_{m \in \mathcal{H}_0} \mathrm{I}\left(\widetilde{P}_{0n}^a(m) \leq \alpha\right),$$

$$U_n^a(\alpha) \equiv \sum_{m \notin \mathcal{H}_0} \mathrm{I}\left(\widetilde{P}_{0n}^a(m) > \alpha\right).$$

The *actual Type I error rate* is estimated as follows and then compared to the *nominal Type I error level* α,

$$FWER(\alpha) \equiv \frac{1}{A} \sum_{a=1}^{A} \mathrm{I}\left(V_n^a(\alpha) > 0\right). \tag{8.12}$$

The *average power* of a given MTP is estimated by

$$AvgPwr(\alpha) \equiv 1 - \frac{1}{h_1}\frac{1}{A}\sum_{a=1}^{A} U_n^a(\alpha). \tag{8.13}$$

The simulation error for the actual Type I error rate and power is of the order $1/\sqrt{A} = 1/\sqrt{500} \approxeq 0.045$.

Graphical summaries

Simulation results are displayed using the following two main types of graphical summaries.

Type I error control comparison

For a given data generating model, plot, for each MTP, the difference between the nominal and actual Type I error rates vs. the nominal Type I error rate, that is, plot

$$(\alpha - FWER(\alpha)) \qquad \text{vs.} \qquad \alpha,$$

for $\alpha \in \{0, 0.01, 0.02, \ldots, 0.50\}$, i.e., values of α in `seq(from = 0, to = 0.50, by=0.01)`. Positive (negative) differences correspond to (anti-) conservative MTPs; the higher the curve, the more conservative the procedure.

Power comparison

For a given data generating model, *receiver operator characteristic* (ROC) curves may be used to compare different MTPs in terms of power. ROC curves are obtained by plotting, for each MTP, power vs. actual Type I error rate, i.e., $AvgPwr(\alpha)$ vs. $FWER(\alpha)$, for a range of nominal Type I error levels α.

However, due to possibly large variations in power between simulation models, consider instead the following modified display, which facilitates comparisons across models. For a given model, plot the difference in power between two procedures vs. the actual Type I error rate, that is, plot

$$\left(AvgPwr^{Boot\ XY}(\alpha^{Boot\ XY}(a)) - AvgPwr^{Boot\ e}(\alpha^{Boot\ e}(a))\right) \qquad \text{vs.} \qquad a,$$

where α^j is defined such that $FWER^j(\alpha^j(a)) = a$, $j \in \{Boot\ XY, Boot\ e\}$, for $a \in \{FWER^{Boot\ XY}(\alpha) : \alpha \in \{0, 0.01, 0.02, \ldots, 0.50\}\} \cap \{FWER^{Boot\ e}(\alpha) : \alpha \in \{0, 0.01, 0.02, \ldots, 0.50\}\}$. That is, $\alpha^j(a)$ is the nominal FWER that corresponds to an actual FWER of a for procedure j.

8.3.4 Simulation study results

Our comparison of the test statistics null distributions Bootstrap XY and Bootstrap e focuses primarily on Type I error control.

Figure 8.1 displays differences between nominal and actual Type I error rates for four simulation models, where one parameter is varied as the others remain constant. In general, MTPs based on the residual bootstrap null distribution Bootstrap e are *anti-conservative* over the entire range of the nominal level α, whereas MTPs based on the general non-parametric bootstrap null distribution Bootstrap XY control the Type I error rate close to the target nominal level α. In some testing scenarios, the actual Type I error rate for Bootstrap e exceeds the nominal Type I error level by as much as 0.20. The following trends are observed.

- *Covariance matrix of the covariates,* σ_{xx}. [Figure 8.1, Panels (b) vs. (a)] As the correlation ς between covariates increases, the actual Type I error rate for Bootstrap XY gets closer to the nominal level α. In contrast, single-step maxT Procedure 8.2 based on Bootstrap e becomes more anti-conservative as the correlation ς increases.
- *Sample size,* n. [Figure 8.1, Panels (c) vs. (a)] As the sample size n increases, the actual Type I error rate for Bootstrap XY gets closer to the nominal level α.
- *Alternative regression parameters,* $(\psi(m) : m \notin \mathcal{H}_0)$. [Figure 8.1, Panels (d) vs. (a)] As the magnitude of the parameter μ, defining the regression parameters $(\psi(m) : m \notin \mathcal{H}_0)$ for the false null hypotheses, increases, the actual Type I error rate for Bootstrap XY gets closer to the nominal level α. In contrast, single-step maxT Procedure 8.2 based on Bootstrap e becomes more anti-conservative as the shift μ increases.
- *Proportion of true null hypotheses,* h_0/M. No clear trends are noticeable for the proportion of true null hypotheses (data not shown).

For most simulation models, the differences in power are within simulation error (i.e., of the order $1/\sqrt{A} = 1/\sqrt{500} \approxeq 0.045$) for the two versions of bootstrap-based single-step maxT Procedure 8.2 (Figure 8.2). The main noticeable trends are, as expected, that power increases with sample size n and effect size μ.

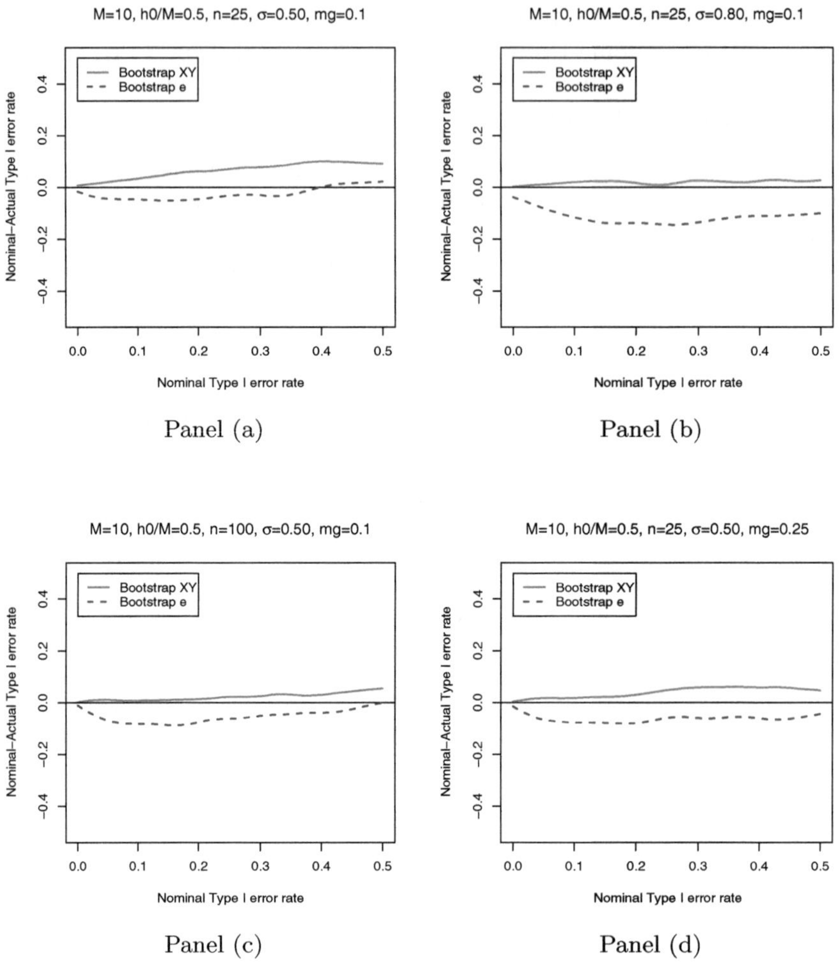

Figure 8.1. *Simulation Study 1: Tests for linear regression coefficients, Type I error control comparison.* Plots of differences between nominal and actual Type I error rates vs. nominal Type I error rate, for single-step maxT Procedure 8.2 based on general non-parametric bootstrap null distribution Bootstrap XY and residual bootstrap null distribution Bootstrap e. The null hypotheses are tested using the t-statistics of Equation (8.8). Panel (a): Model with sample size $n = 25$; $M = 10$ null hypotheses; common covariance $\varsigma = 0.50$ for the covariates; proportion $h_0/M = 0.50$ of true null hypotheses; shift parameter $\mu = 0.10$ for alternative regression coefficients. Panel (b): $n = 25$; $M = 10$; $\varsigma = 0.80$; $h_0/M = 0.50$; $\mu = 0.10$. Panel (c): $n = 100$; $M = 10$; $\varsigma = 0.50$; $h_0/M = 0.50$; $\mu = 0.10$. Panel (d): $n = 25$; $M = 10$; $\varsigma = 0.50$; $h_0/M = 0.50$; $\mu = 0.25$.

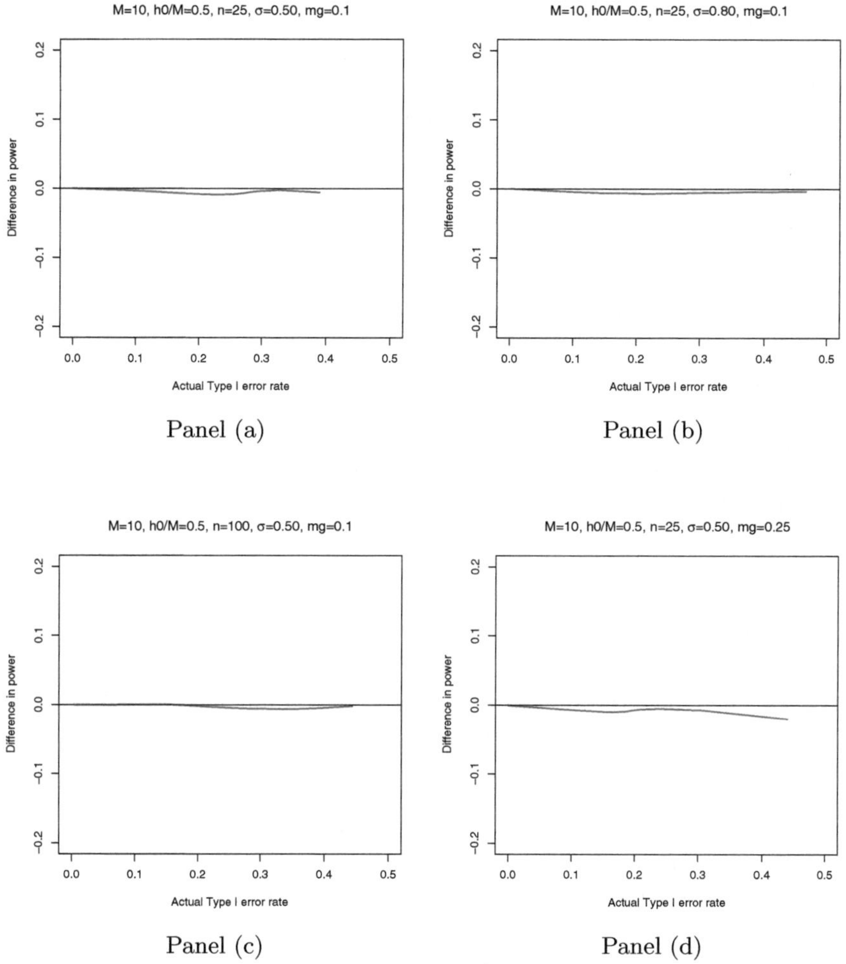

Figure 8.2. *Simulation Study 1: Tests for linear regression coefficients, power comparison.* Plots of difference in power vs. actual Type I error rate, for single-step maxT Procedure 8.2 based on general non-parametric bootstrap null distribution Bootstrap XY and residual bootstrap null distribution Bootstrap e. The null hypotheses are tested using the t-statistics of Equation (8.8). Positive differences indicate greater power for Bootstrap XY. Panel (a): Model with sample size $n = 25$; $M = 10$ null hypotheses; common covariance $\varsigma = 0.50$ for the covariates; proportion $h_0/M = 0.50$ of true null hypotheses; shift parameter $\mu = 0.10$ for alternative regression coefficients. Panel (b): $n = 25$; $M = 10$; $\varsigma = 0.80$; $h_0/M = 0.50$; $\mu = 0.10$. Panel (c): $n = 100$; $M = 10$; $\varsigma = 0.50$; $h_0/M = 0.50$; $\mu = 0.10$. Panel (d): $n = 25$; $M = 10$; $\varsigma = 0.50$; $h_0/M = 0.50$; $\mu = 0.25$.

8.4 Simulation Study 2: Tests for correlation coefficients

The second simulation study concerns tests for *correlation coefficients*, a testing scenario of great interest in genomic applications. Indeed, as illustrated in Section 9.3, a common question in microarray and other high-throughput gene expression assays, is the identification of co-expressed genes, i.e., pairs (or sets) of genes with correlated expression profiles.

8.4.1 Simulation model

Data generating distribution

Consider a J-dimensional Gaussian random row vector $X \sim P = \mathrm{N}(0, \sigma)$, with mean vector zero and *covariance matrix* $\sigma = (\sigma(j,j') : j,j' = 1,\ldots,J) \equiv \mathrm{Cov}[X]$ equal to the corresponding *correlation matrix* $\rho = (\rho(j,j') : j,j' = 1,\ldots,J) \equiv \mathrm{Cor}[X]$.

Suppose one has a random sample, $\mathcal{X}_n \equiv \{X_i : i = 1,\ldots,n\}$, of n IID random variables $X_i \sim P$, from the above-specified Gaussian data generating distribution P.

Null and alternative hypotheses

The hypotheses of interest concern the $M \equiv \binom{J}{2} = J(J-1)/2$ distinct off-diagonal elements, $\psi = (\psi(m) : m = 1,\ldots,M)$, of the $J \times J$ correlation matrix ρ. One may recode pairs of row and column indices $\{(j,j') : j = 1,\ldots,J-1,\ j' = j+1,\ldots,J\}$, for the upper-triangle of ρ, into a single index $m = 1,\ldots,M$, defined by $m \equiv (j-1)(2J-j)/2 + (j'-j)$.

Consider two-sided tests of the $M = J(J-1)/2$ null hypotheses $H_0(m) = \mathrm{I}(\psi(m) = \psi_0(m))$ vs. the alternative hypotheses $H_1(m) = \mathrm{I}(\psi(m) \neq \psi_0(m))$, $m = 1,\ldots,M$. For simplicity, and without loss of generality, set the null values $\psi_0(m)$ equal to zero, i.e., test the null hypotheses of no pairwise correlations.

8.4.2 Multiple testing procedures

Test statistics

The M null hypotheses are tested based on the following *t-statistics*,

$$T_n(m) \equiv \sqrt{n-2}\frac{\psi_n(m)}{\sqrt{1-\psi_n^2(m)}}, \qquad m = 1,\ldots,M, \tag{8.14}$$

where $\psi_n = (\psi_n(m) : m = 1,\ldots,M)$ is the M-vector of distinct *empirical correlation coefficients*. Specifically, the empirical correlation coefficient for the pair $(X(j), X(j'))$, corresponding to the mth null hypothesis, is defined as

$$\psi_n(m) = \rho_n(j, j') \equiv \frac{\sigma_n(j, j')}{\sqrt{\sigma_n(j, j)\sigma_n(j', j')}}, \tag{8.15}$$

based on empirical means $\bar{X}_n(j)$ and covariances $\sigma_n(j, j')$,

$$\bar{X}_n(j) \equiv \frac{1}{n}\sum_{i=1}^{n} X_i(j), \qquad \sigma_n(j, j') \equiv \frac{1}{n}\sum_{i=1}^{n}(X_i(j) - \bar{X}_n(j))(X_i(j') - \bar{X}_n(j')).$$

For Gaussian data generating distributions, the t-statistics of Equation (8.14) have marginal t-distributions with $(n-2)$ degrees of freedom, under the null hypotheses that the corresponding correlation coefficients are zero (i.e., $\psi(m) = 0$).

One could also use unstandardized *difference statistics*,

$$T_n(m) \equiv \sqrt{n}\psi_n(m), \qquad m = 1, \ldots, M. \tag{8.16}$$

Test statistics null distributions

The simulation study compares the Type I error and power properties of FWER-controlling single-step maxT Procedure 8.2, based on the following two different bootstrap test statistics null distributions ($B = 10,000$ bootstrap samples).

Procedure 8.5. [Bootstrap X null distribution: Bootstrapping entire J-vectors X]
The general non-parametric bootstrap test statistics null distribution of Procedure 8.1 involves *resampling entire J-vectors X_i* and computing *null shift and scale-transformed test statistics* for each bootstrap sample. Specifically, one proceeds as follows for the bth bootstrap sample, $b = 1, \ldots, B$.

1. Sample n J-vectors X_i^b at random, with replacement from the set of n observations $\mathcal{X}_n = \{X_i : i = 1, \ldots, n\}$. Let $\mathcal{X}_n^b \equiv \{X_i^b : i = 1, \ldots, n\}$ denote the resulting bootstrap sample.
2. Compute an M-vector $T_n^B(\cdot, b) = (T_n^B(m, b) : m = 1, \ldots, M)$ of bootstrap test statistics as in Equation (8.14), based on the bootstrap sample $\mathcal{X}_n^b$.
3. Compute an M-vector $Z_n^B(\cdot, b) = (Z_n^B(m, b) : m = 1, \ldots, M)$ of null shift and scale-transformed bootstrap test statistics,

$$Z_n^B(m, b) \equiv \sqrt{\min\left\{1, \frac{1}{\mathrm{Var}[T_n^B(m, \cdot)]}\right\}}\left(T_n^B(m, b) - \mathrm{E}[T_n^B(m, \cdot)]\right),$$

where $\lambda_0(m) = 0$, $\tau_0(m) = 1$, and $\mathrm{E}[T_n^B(m, \cdot)] \equiv \sum_b T_n^B(m, b)/B$ and $\mathrm{Var}[T_n^B(m, \cdot)] \equiv \sum_b (T_n^B(m, b) - \mathrm{E}[T_n^B(m, \cdot)])^2/B$ denote, respectively, the empirical mean and variance of the B bootstrap test statistics $T_n^B(m,b)$ for null hypothesis $H_0(m)$, $m = 1,\ldots,M$ (i.e., row means and variances of the matrix $\mathbf{T}_n^B$, as in Procedure 8.1).

The test statistics null distribution is the empirical distribution Q_{0n} of the B M-vectors $\{Z_n^B(\cdot,b) : b = 1,\ldots,B\}$, i.e., of the columns of matrix $\mathbf{Z}_n^B$.

Procedure 8.6. [Bootstrap X(j) null distribution: Bootstrapping independent elements $X(j)$ of the J-vectors X]
In contrast, the parameter-specific bootstrap test statistics null distribution proposed in Westfall and Young (1993, Section 6.3, p. 194) involves *resampling each element $X_i(j)$ of the J-vectors X_i independently* and computing *raw test statistics* (without null transformation) for each bootstrap sample. Specifically, one proceeds as follows for the bth bootstrap sample, $b = 1,\ldots,B$.

1. For each variable $X(j)$, $j = 1,\ldots,J$, sample n j-specific elements $X_i^b(j)$, $i = 1,\ldots,n$, at random, with replacement from the set of n j-specific observations $\{X_i(j) : i = 1,\ldots,n\}$. The ith bootstrap J-vector $X_i^b = (X_i^b(j) : j = 1,\ldots,J)$, $i = 1,\ldots,n$, is obtained by combining J such independently sampled variables. Let $\mathcal{X}_n^b \equiv \{X_i^b : i = 1,\ldots,n\}$ denote the resulting bootstrap sample.
2. Compute an M-vector $T_n^B(\cdot,b) = (T_n^B(m,b) : m = 1,\ldots,M)$ of bootstrap test statistics as in Equation (8.14), based on the bootstrap sample $\mathcal{X}_n^b$.

The test statistics null distribution is the empirical distribution Q_{0n} of the B M-vectors $\{T_n^B(\cdot,b) : b = 1,\ldots,B\}$, i.e., of the columns of matrix $\mathbf{T}_n^B$.

As in the regression example of Section 8.3, bootstrap Procedures Bootstrap X and Bootstrap X(j) differ in the following two key aspects.

1. The (re)sampling units: Bootstrap X resamples entire J-vectors X_i, whereas Bootstrap X(j) resamples independent elements $X_i(j)$.
2. The bootstrap test statistics: Bootstrap X relies on null-transformed test statistics Z_n, whereas Bootstrap X(j) relies on raw test statistics T_n.

In other words, Procedure Bootstrap X(j) derives the test statistics null distribution by first creating a data generating null distribution that corresponds to the complete null hypothesis that the J variables $X(j)$, $j = 1,\ldots,J$, are independent.

Single-step maxT procedure

Adjusted p-values for single-step maxT Procedure 3.5 may be obtained by applying Procedure 8.2 with bootstrap null distributions Bootstrap X and Bootstrap X(j). Specifically, adjusted p-values for Bootstrap X and Bootstrap X(j) are computed, respectively, from the empirical distributions of the B maxima of null-transformed test statistics $\{\max_m Z_n^B(m,b) : b = 1,\ldots,B\}$ and raw test statistics $\{\max_m T_n^B(m,b) : b = 1,\ldots,B\}$. For a test at nominal FWER level α, one rejects null hypotheses with adjusted p-values less than or equal to α.

8.4.3 Simulation study design

Simulation parameters

The following model parameters are used in the simulation study.

- *Sample size, n.* $n = 25$.
- *Number of hypotheses, M.* $M = 45$.
- *Proportion of true null hypotheses, h_0/M.* $h_0/M = 25/45 \approxeq 0.56$.
- *Correlation matrix, ρ.* The correlation matrix $\rho = (\rho(j,j') : j,j' = 1,\ldots,J)$ (here, equal to the covariance matrix σ) has the following block-diagonal form,
$$\rho = \begin{bmatrix} \varrho_{J/2\times J/2} & O_{J/2\times J/2} \\ O_{J/2\times J/2} & \varrho_{J/2\times J/2} \end{bmatrix},$$
where $O_{J/2\times J/2}$ denotes a $J/2 \times J/2$ matrix of zeros and $\varrho_{J/2\times J/2}$ a $J/2 \times J/2$ matrix with unit diagonal elements and off-diagonal elements set to a common value ϱ, i.e., $\varrho_{J/2\times J/2}(j,j) = 1$, for $j = 1,\ldots,J/2$, and $\varrho_{J/2\times J/2}(j,j') = \varrho$, for $j \neq j' = 1,\ldots,J/2$. The following values are considered for the common block correlation coefficient: $\varrho = 0.30, 0.50, 0.60$.

Note that the only parameter that is varied in the simulation study concerns the correlation matrix ρ, that is, the parameter of interest in the multiple testing problem.

Estimating Type I error rate and power

As in Section 8.3.3, above, for Simulation Study 1.

Graphical summaries

As in Section 8.3.3, above, for Simulation Study 1.

8.4.4 Simulation study results

Our comparison of the test statistics null distributions Bootstrap X and Bootstrap X(j) focuses primarily on Type I error control.

Figure 8.3 displays differences between nominal and actual Type I error rates for three simulation models, where the common block correlation coefficient ϱ is varied as the other parameters remain constant. In general, MTPs based on the independent covariates bootstrap null distribution Bootstrap X(j) are *conservative* over the entire range of the nominal level α, whereas MTPs based on the general non-parametric bootstrap null distribution Bootstrap X control the Type I error rate close to the target nominal level α. The most extreme differences are observed for large nominal Type I error levels α. In some testing scenarios, the nominal level for Bootstrap X(j) exceeds the actual Type I error rate by as much as 0.25. As the correlation parameter ϱ increases, single-step maxT Procedure 8.2 based on Bootstrap X(j) becomes more conservative.

As in the regression simulation study of Section 8.3, we find that, for most simulation models, the differences in power are within simulation error (i.e., of the order $1/\sqrt{A} = 1/\sqrt{500} \approxeq 0.045$) for the two versions of bootstrap-based single-step maxT Procedure 8.2 (Figure 8.4). The main noticeable trends are, as expected, that power increases with sample size n and effect size ϱ.

Similar trends are observed for standardized (Equation (8.14)) and unstandardized (Equation (8.16)) correlation test statistics (data not shown for unstandardized statistics).

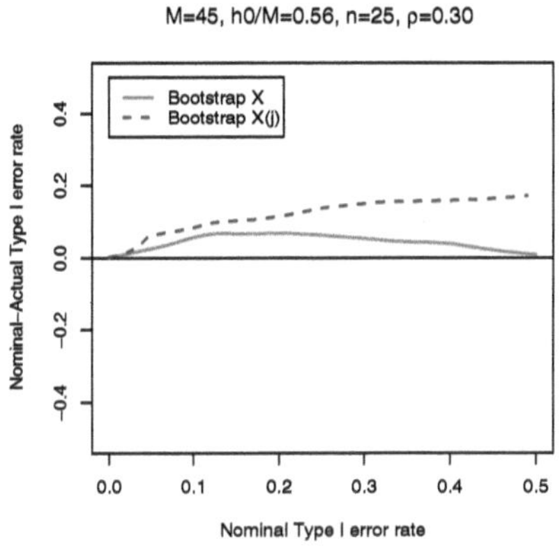

Panel (a)

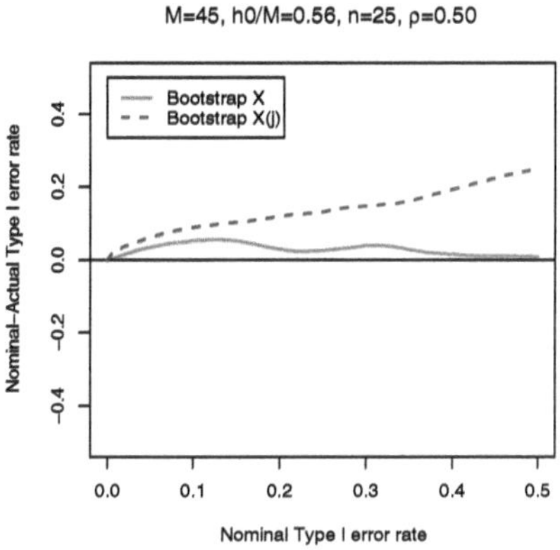

Panel (b)

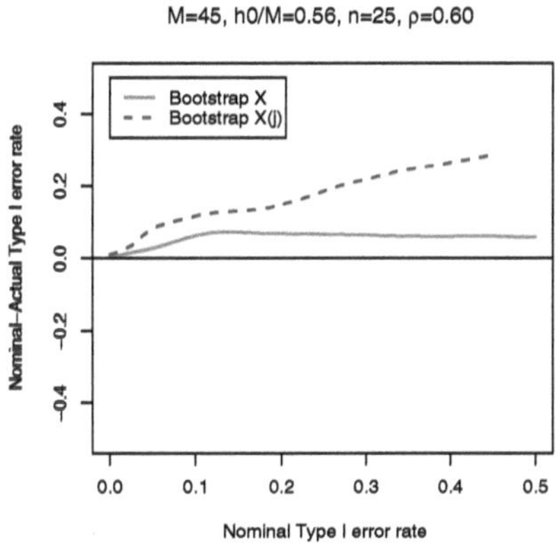

Panel (c)

Figure 8.3. *Simulation Study 2: Tests for correlation coefficients, Type I error control comparison.* Plots of differences between nominal and actual Type I error rates vs. nominal Type I error rate, for single-step maxT Procedure 8.2 based on general non-parametric bootstrap null distribution Bootstrap X and independent covariates bootstrap null distribution Bootstrap X(j). The null hypotheses are tested using the t-statistics of Equation (8.14). Model with sample size $n = 25$; $M = 45$ null hypotheses; proportion $h_0/M = 25/45$ of true null hypotheses. Panel (a): Common block correlation coefficient $\varrho = 0.30$. Panel (b): $\varrho = 0.50$. Panel (c): $\varrho = 0.60$.

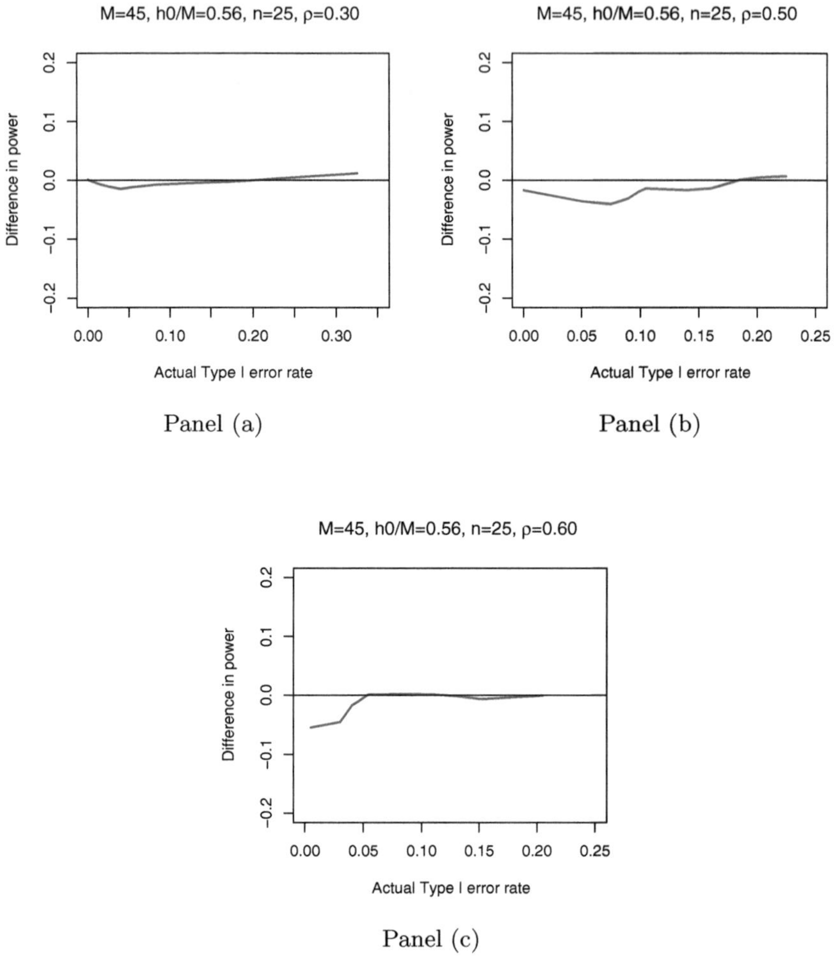

Figure 8.4. *Simulation Study 2: Tests for correlation coefficients, power comparison.* Plots of difference in power vs. actual Type I error rate, for single-step maxT Procedure 8.2 based on general non-parametric bootstrap null distribution Bootstrap X and independent covariates bootstrap null distribution Bootstrap X(j). The null hypotheses are tested using the t-statistics of Equation (8.14). Positive differences indicate greater power for Bootstrap X. Model with sample size $n = 25$; $M = 45$ null hypotheses; proportion $h_0/M = 25/45$ of true null hypotheses. Panel (a): Common block correlation coefficient $\varrho = 0.30$. Panel (b): $\varrho = 0.50$. Panel (c): $\varrho = 0.60$.

9

Identification of Differentially Expressed and Co-Expressed Genes in High-Throughput Gene Expression Experiments

9.1 Introduction

In recent years, a number of novel biotechnologies have enabled biologists to readily monitor genome-wide expression levels. For instance, *microarray experiments* provide high-throughput assays for measuring the abundance of *deoxyribonucleic acids* (DNA) and *ribonucleic acids* (RNA) in different types of cell samples for thousands of sequences simultaneously (Phimister and Cohen, 1999; Packer, 2002; Packer and Axton, 2005; Speed, 2003). The microarray technology is being applied increasingly in biological and medical research to address a wide range of problems, such as the classification of tumors or the study of host genomic responses to bacterial infections (Alizadeh et al., 2000; Barrier et al., 2005a,b,c, 2006; Birkner et al., 2005a; Blanchette et al., 2005; Boldrick et al., 2002; Callow et al., 2000; Cawley et al., 2004; Chiappini et al., 2006; Chiaretti et al., 2004; DeSantis et al., 2005; Dudoit et al., 2002, 2003, 2006; Ge et al., 2003; Golub et al., 1999; Keleş et al., 2006; Pollard et al., 2005b; Pollard and van der Laan, 2004).

A common question in microarray data analysis is the identification of *differentially expressed* genes, i.e., genes whose expression measures are associated with possibly censored biological and clinical covariates and outcomes. The biological question of differential expression can be restated as a problem in multiple hypothesis testing: the simultaneous test, for each gene, of the null hypothesis of no association between the expression measures and the covariates and outcomes.

Another common question is the identification of *co-expressed* genes, i.e., pairs (or sets) of genes with associated expression measures. This question translates into the simultaneous test, for each gene pair, of the null hypothesis of no association (e.g., zero correlation) between their expression measures.

As a typical microarray experiment reports expression levels for thousands of genes at a time, large multiplicity problems are generated.

The present chapter applies the multiple testing methodology of Chapters 1–7 to the following two gene expression datasets, with the aim of identifying differentially expressed and co-expressed genes.

- *Apo AI dataset*, from a study relating messenger RNA (mRNA) expression to lipid metabolism and artherosclerosis susceptibility in mice (Section 9.2; Callow et al. (2000); Dudoit et al. (2002, 2003); Ge et al. (2003)).
- *Cancer miRNA dataset*, from a study of microRNA (miRNA) expression in cancerous and non-cancerous tissues (Section 9.3; Lu et al. (2005); Pollard et al. (2005a)).

Our proposed multiple testing procedures have also been applied to the following microarray datasets.

- Lymphoma dataset of Alizadeh et al. (2000) (Pollard and van der Laan, 2004).
- Bacteria dataset from the Boldrick et al. (2002) study of host (human peripheral blood mononuclear cells) genomic responses to bacterial (*Bordetella pertussis*, *Staphylococcus aureus*) infection (Dudoit et al., 2003).
- Acute lymphoblastic leukemia dataset of Chiaretti et al. (2004) (Dudoit et al., 2006; Pollard et al., 2005b).
- Acute lymphoblastic (ALL) and acute myeloid (AML) leukemia dataset of Golub et al. (1999) (Dudoit et al., 2003; Ge et al., 2003).
- Airborne bacteria dataset, for monitoring spatial and temporal bacterial differential abundance in air samples from various US cities, based on measures from a 16s small-subunit ribosomal RNA (rRNA) microarray (Birkner et al., 2005a; DeSantis et al., 2005).
- p53 ChIP-Chip dataset of Cawley et al. (2004), for identifying binding sites for the p53 transcription factor, using chromatin immunoprecipitation (ChIP) of transcription factor-bound DNA followed by microarray (Chip) hybridization of the IP-enriched DNA (Keleş et al., 2006).

9.2 Apolipoprotein AI experiment of Callow et al. (2000)

9.2.1 Apo AI dataset

The Apo AI microarray experiment was carried out as part of a study of lipid metabolism and artherosclerosis susceptibility in mice (Callow et al., 2000). *Apolipoprotein AI* (Apo AI) is a gene known to play a pivotal role in high-density lipoprotein (HDL) cholesterol metabolism and mice with the Apo AI gene knocked-out have very low HDL cholesterol levels. The goal of the experiment was to identify genes with altered expression in the livers of Apo AI knock-out mice compared to inbred control mice. The treatment group consists of 8 inbred C57Bl/6 mice with the Apo AI gene knocked-out and the control group consists of 8 inbred C57Bl/6 mice.

For each of the 16 mice, target *complementary DNA* (cDNA) was obtained from *messenger RNA* (mRNA) by reverse transcription and labeled using the red-fluorescent dye Cyanine 5 (Cy5). The reference sample used in all hybridizations was prepared by pooling cDNA from the 8 control mice and was labeled with the green-fluorescent dye Cyanine 3 (Cy3). Combined red- and green-labeled target cDNA samples were hybridized to microarrays with 6,384 spots ($= 4 \times 4 \times 19 \times 21$), including 257 probe sequences related to lipid metabolism.

Each of the 16 hybridizations produced a pair of 16-bit TIFF images, which were processed by seeded region growing segmentation and morphological opening background adjustment (Yang et al. (2002); R package Spot, `experimental.act.cmis.csiro.au/Spot/index.php`). The resulting fluorescence intensity measures were normalized by within-print-tip-group loess robust local regression (Dudoit et al. (2002); Dudoit and Yang (2003); Yang et al. (2001); Yang and Paquet (2005); Bioconductor R package marray, `www.bioconductor.org`). Among the 6,384 spots on the microarray, only those 5,548 spots corresponding to actual cDNA sequences are retained for subsequent analyses. The other 836 spots are either blank or control spots.

The R package ApoAI provides microarray data objects, at various levels of processing, for the Apo AI experiment. Specifically, it includes: `rawApoAI`, an object of class *marrayRaw*, for pre-normalization intensity measures; `normApoAI`, an object of class *marrayNorm*, for normalized expression measures; and `ApoAI`, an object of class *exprSet*, combining normalized microarray expression measures (base-2 logarithm of the Cy5/Cy3 fluorescence intensity ratios) and covariates (treatment/control status).

The data for each of the $n = 16$ mice consist of a binary treatment/control *covariate/genotype* Z_i and an $M = 5,548$-dimensional *outcome/phenotype* vector $X_i = (X_i(m) : m = 1, \ldots, M)$ of microarray expression measures, $i = 1, \ldots, n$. Specifically, $X_i(m)$ denotes the base-2 logarithm of the Cy5/Cy3 fluorescence intensity ratio for probe m in microarray i, $i = 1, \ldots, n$, $m = 1, \ldots, M$. For ease of notation, recode the genotypes Z_i as 0 and 1 for the control and treatment groups, respectively, and order the data such that the first 8 mice are control mice and the last 8 mice are treatment mice, i.e., let $Z_i = 0$, $i = 1, \ldots, 8$, and $Z_i = 1$, $i = 9, \ldots, 16$.

In Section 9.2.3, functions from the Bioconductor R package multtest are applied to normalized microarray measures stored in the object `ApoAI`, with the aim of identifying differentially expressed genes between knock-out and control mice (Section 13.1; Pollard et al. (2005b); `www.bioconductor.org`).

The website supplement for the Apo AI data analysis provides the image analysis output file from Spot, a text file describing probe sequences, the experimental data R package ApoAI, R code for the multiple testing analyses of Sections 9.2.2–9.2.4, supplementary tables and figures, and references (`www.stat.berkeley.edu/~sandrine/MTBook/ApoAI/ApoAI.html`).

9.2.2 Multiple testing procedures

In order to identify *differentially expressed* genes between knock-out and control mice, we test for each of the $M = 5,548$ probes whether it differs in mean expression measures between these two groups. Thus, the parameter of interest for the mth probe (i.e., mth hypothesis) is the *difference in mean expression measures* $\psi(m)$ between treatment and control mice, that is,

$$\psi(m) \equiv \mathrm{E}[X(m)|Z = 1] - \mathrm{E}[X(m)|Z = 0], \qquad m = 1, \ldots, M. \tag{9.1}$$

We consider two-sided tests of the null hypotheses $H_0(m) = \mathrm{I}(\psi(m) = 0)$ of no differences in mean expression measures vs. the alternative hypotheses $H_1(m) = \mathrm{I}(\psi(m) \neq 0)$ of different mean expression measures, for each probe m.

The tests are based on *two-sample Welch t-statistics*,

$$T_n(m) \equiv \frac{\psi_n(m) - \psi_0(m)}{\sigma_n(m)} = \frac{\bar{X}_1(m) - \bar{X}_0(m) - 0}{\sqrt{\frac{\sigma^2_{0,n}(m)}{n_0(m)} + \frac{\sigma^2_{1,n}(m)}{n_1(m)}}}, \tag{9.2}$$

where the null values $\psi_0(m)$ are zero and $n_k(m) \equiv \sum_i \mathrm{I}(Z_i = k)$, $\bar{X}_k(m) \equiv \sum_i \mathrm{I}(Z_i = k)\, X_i(m)/n_k(m)$, and $\sigma^2_{k,n}(m) \equiv \sum_i \mathrm{I}(Z_i = k)\,(X_i(m) - \bar{X}_k(m))^2/(n_k(m) - 1)$ denote, respectively, the sample sizes, sample means, and sample variances for the expression measures of probe m in treatment ($k = 1$) and control ($k = 0$) mice (note that the sample sizes $n_k(m)$ may differ across probes m due to missing data). The estimators of the differences in mean expression measures, $\psi(m)$, are simply the corresponding differences in sample means, $\psi_n(m) \equiv \bar{X}_1(m) - \bar{X}_0(m)$. The estimated standard errors of these estimators are $\sigma_n(m) \equiv \sqrt{\sigma^2_{0,n}(m)/n_0(m) + \sigma^2_{1,n}(m)/n_1(m)}$. The null hypothesis $H_0(m)$ is rejected, i.e., the corresponding probe is declared significantly differentially expressed, for large absolute values of the test statistic $T_n(m)$.

A variety of bootstrap- and permutation-based multiple testing procedures, controlling the FWER, gFWER, TPPFP, and FDR, are applied to assess the statistical significance of the differences in expression measures between treatment and control mice. The results of each MTP are reported in terms of adjusted p-values, so that probes with adjusted p-values less than or equal to a user-supplied α are declared significantly differentially expressed at nominal Type I error level α (Section 1.2.12).

The main analyses are based on the *null shift and scale-transformed test statistics null distribution* Q_0 of Section 2.3 and its *non-parametric bootstrap* estimator Q_{0n} from Procedure 2.3 (with $B = 5,000$ samples). A *smoothed non-parametric bootstrap* version of Procedure 2.3 is also considered (with $B = 1,000$ samples). The smoothed version of Procedure 2.3 consists of applying a kernel density estimator to the marginal empirical distributions of the B null shift and scale-transformed bootstrap test statistics $\{Z_n^B(m,b) : 1,\ldots,B\}$. The smoothing procedure is applied to hypotheses with non-parametric bootstrap unadjusted p-values of zero, using the Gaussian kernel density estimator implemented in the `density` function (R package stats). As indicated in Section 2.9, in the case of equal sample sizes for the treatment and control groups, a *permutation data generating null distribution* yields a valid test statistics null distribution. A permutation test statistics null distribution, based on all possible $\binom{16}{8} = 12,870$ permutations of the treatment and control labels, is therefore also compared to the standard non-parametric bootstrap null shift and scale-transformed test statistics null distribution of Procedure 2.3.

The above estimated test statistics null distributions are used to compute unadjusted p-values and adjusted p-values for the following procedures.

- *FWER control*: Joint single-step maxT Procedure 3.5 (SS maxT) and minP Procedure 3.6 (SS minP), joint step-down maxT Procedure 3.11 (SD maxT) and minP Procedure 3.12 (SD minP), marginal single-step Bonferroni Procedure 3.1 (SS Bonferroni), marginal step-down Holm Procedure 3.7 (SD Holm), and marginal step-up Hochberg Procedure 3.13 (SU Hochberg). SS maxT and SD maxT are common-cut-off procedures, whereas the other five MTPs are common-quantile procedures.
- *gFWER control*: $gFWER(k)$-controlling augmentation multiple testing Procedure 3.20, based on FWER-controlling joint single-step maxT Procedure 3.5, for an allowed number $k \in \{5, 10, 50, 100\}$ of false positives (gFWER(k) AMTP).
- *TPPFP control*: $TPPFP(q)$-controlling augmentation multiple testing Procedure 3.26, based on FWER-controlling joint single-step maxT Procedure 3.5, for an allowed proportion $q \in \{0.05, 0.10, 0.25, 0.50\}$ of false positives (TPPFP(q) AMTP).
- *FDR control*: Marginal step-up Benjamini and Hochberg (1995) Procedure 3.22 (SU BH) and Benjamini and Yekutieli (2001) Procedure 3.23 (SU BY); procedures described in Theorem 6.7, based on a TPPFP-controlling augmentation of joint single-step maxT Procedure 3.5 (TPPFP-based 1, TPPFP-based 2).

Note that proofs of Type I error control for step-up Hochberg Procedure 3.13 and step-up Benjamini and Hochberg (1995) Procedure 3.22 rely on certain assumptions concerning the joint distribution of the test statistics (e.g., positive regression dependence). Failure to satisfy these assumptions could possibly lead to anti-conservative behavior.

The multiple testing procedures are applied as described in Section 9.2.3, below, using the Bioconductor R package multtest.

Previous analyses of the Apo AI dataset are reported in Callow et al. (2000), Dudoit et al. (2002, 2003), and Ge et al. (2003), with emphasis on permutation-based maxT and minP multiple testing procedures.

9.2.3 Software implementation using the Bioconductor R package multtest

The multiple testing procedures described in Section 9.2.2 are implemented using functions from the Bioconductor R package multtest (Pollard et al. (2005b); multtest package, Version 1.10.0, Bioconductor Release 1.8, `www.bioconductor.org`; R Release 2.3.0, `www.r-project.org`). The reader is referred to Section 13.1 for an overview of multtest functionality and to the package documentation (helpfiles, vignettes) for details on each function.

The data are stored in the object `ApoAI` of class *exprSet*, from the experimental data R package ApoAI.

The present section provides sample code and code output. The ApoAI package and the full code for the analyses reported in Section 9.2.4 may be obtained from the website supplement (`www.stat.berkeley.edu/~sandrine/MTBook/ApoAI/ApoAI.html`).

Sample R code

```
#####################################################################
## Set-up
#####################################################################
## Load packages

options(width=70)

library(multtest)
library(ApoAI)

packageDescription("multtest")$Version

##########################
## Load data

data(ApoAI)
X.ApoAI <- exprs(ApoAI)
Y.ApoAI <- pData(ApoAI)

#####################################################################
## FWER-controlling non-parametric bootstrap-based
## single-step maxT MTP
#####################################################################

B <- 5000
bootSSmaxT.ApoAI <- MTP(X=X.ApoAI, Y=Y.ApoAI,
  test = "t.twosamp.unequalvar", standardize = TRUE,
```

```
  alternative = "two.sided", psi0 = 0, typeone = "fwer",
  alpha = 0.05, smooth.null = FALSE, nulldist = "boot",
  B = B, method = "ss.maxT", get.cr = TRUE, get.cutoff = TRUE,
  get.adjp = TRUE, keep.nulldist = TRUE, seed = 9)

#########################
## Print method

bootSSmaxT.ApoAI

#########################
## Summary method

summary(bootSSmaxT.ApoAI)

#########################
## Test statistics and cut-offs

plot(bootSSmaxT.ApoAI,which=6,main="Apo AI: Test statistics
  and cut-offs for top 10 probes \n FWER-controlling
  non-parametric bootstrap-based single-step maxT MTP",
  caption="", sub.caption="", cex.main=0.9)

#########################
## Parameter estimates and confidence regions

plot(bootSSmaxT.ApoAI, which=5, main="Apo AI: Parameter
  estimates and confidence regions for top 10 probes \n
  FWER-controlling non-parametric bootstrap-based single-step
  maxT MTP", caption="", sub.caption="", cex.main=0.9)
abline(h=0, col="red", lwd=3)

###############################################################
## gFWER-controlling AMTPs, obtained from non-parametric
## bootstrap-based single-step maxT MTP
###############################################################

k <- c(5, 10, 50, 100)
gFWER.ApoAI <- fwer2gfwer(adjp=bootSSmaxT.ApoAI@adjp, k=k)
gFWER.ApoAI <- cbind(bootSSmaxT.ApoAI@adjp, gFWER.ApoAI)
gFWER <- paste("gFWER(",c(0,k),")", sep="")
dimnames(gFWER.ApoAI)[[2]] <- gFWER

#########################
## Number of rejected hypothesese vs. nominal Type I error level
```

```
mt.plot(adjp=gFWER.ApoAI, teststat=bootSSmaxT.ApoAI@statistic,
  proc=gFWER, leg=c(0.1,150),col=1:length(gFWER),
  lty=1:length(gFWER), lwd=3, ylim=c(1,150))
title("Apo AI: gFWER(k)-controlling AMTPs \n
  obtained from non-parametric bootstrap-based
  single-step maxT MTP", cex.main=0.9)

#################################################################
## TPPFP-controlling AMTPs, obtained from non-parametric
## bootstrap-based single-step maxT MTP
#################################################################

q <- c(0.05,0.1,0.25,0.5)
TPPFP.ApoAI <- fwer2tppfp(adjp=bootSSmaxT.ApoAI@adjp, q=q)
TPPFP.ApoAI <- cbind(bootSSmaxT.ApoAI@adjp, TPPFP.ApoAI)
TPPFP <- paste("TPPFP(",c(0,q),")", sep="")
dimnames(TPPFP.ApoAI)[[2]] <- TPPFP

#########################
## Number of rejected hypotheses vs. nominal Type I error level

mt.plot(adjp=TPPFP.ApoAI, teststat=bootSSmaxT.ApoAI@statistic,
  proc=TPPFP, leg=c(0.1,150),col=1:length(TPPFP),
  lty=1:length(TPPFP), lwd=3, ylim=c(1,150))
title("Apo AI:  TPPFP(q)-controlling AMTPs \n
  obtained from  non-parametric bootstrap-based
  single-step maxT MTP", cex.main=0.9)

#################################################################
```

Sample R code output

```
#################################################################
## Print method

> bootSSmaxT.ApoAI

        Multiple Testing Procedure

Object of class:  MTP
sample size = 16
number of hypotheses = 5548
test statistics = t.twosamp.unequalvar
type I error rate = fwer
```

```
nominal level alpha = 0.05
multiple testing procedure = ss.maxT

Call: MTP(X = X.ApoAI, Y = Y.ApoAI,
    test = "t.twosamp.unequalvar", standardize = TRUE,
    alternative = "two.sided", psi0 = 0, typeone = "fwer",
    alpha = 0.05, smooth.null = FALSE, nulldist = "boot", B = B,
    method = "ss.maxT", get.cr = TRUE, get.cutoff = TRUE,
    get.adjp = TRUE, keep.nulldist = TRUE, seed = 9)

Slots:
            Class    Mode   Length Dimension
statistic numeric numeric     5548
estimate  numeric numeric     5548
sampsize  integer numeric        1
rawp      numeric numeric     5548
adjp      numeric numeric     5548
conf.reg    array numeric    11096  5548,2,1
cutoff     matrix numeric     5548    5548,1
reject     matrix logical     5548    5548,1
nulldist   matrix numeric 27740000 5548,5000
call         call    call       18
seed      integer numeric        1

#########################
## Summary method

> summary(bootSSmaxT.ApoAI)
MTP:  ss.maxT
Type I error rate:  fwer

  Level Rejections
1  0.05          5

               Min.  1st Qu.    Median       Mean 3rd Qu.   Max.
adjp         0.0014  1.00000  1.000000  0.994600 1.00000 1.0000
rawp         0.0000  0.14940  0.381800  0.423400 0.67510 1.0000
statistic -22.8900 -0.84080 -0.013700 -0.061740 0.75250 4.1400
estimate   -3.1660 -0.09992 -0.001852 -0.002908 0.09295 0.7309

################################################################
```

9.2.4 Results

All multiple testing procedures single out 8 probes, out of 5,548 spotted probe sequences, as being differentially expressed between knock-out and control mice. This feature is apparent in the boxplot and normal quantile-quantile (Q-Q) plot of Figure 9.1. The 8 probes have the largest absolute test statistics and the smallest adjusted p-values for all bootstrap- and permutation-based multiple testing procedures (Tables 9.1–9.7; Figures 9.1–9.4, 9.10, and 9.11). The negative t-statistics suggest that the probes are *under-expressed* in the Apo AI knock-out mice compared to the control mice.

FWER-controlling single-step common-cut-off maxT procedure based on non-parametric bootstrap null shift and scale-transformed test statistics null distribution

The multiple testing analysis is centered around FWER-controlling single-step maxT Procedure 3.5 (SS maxT), based on the non-parametric bootstrap null shift and scale-transformed test statistics null distribution of Procedure 2.3.

Table 9.1 provides: parameter estimates, $\psi_n(m)$; two-sample Welch t-statistics, $T_n(m)$; unadjusted p-values, $P_{0n}(m)$; and SS maxT adjusted p-values, $\widetilde{P}_{0n}(m)$. Results are displayed for the top 10 probes, where probes are sorted first in increasing order of their non-parametric bootstrap SS maxT adjusted p-values, next in increasing order of their non-parametric bootstrap unadjusted p-values, and finally in decreasing order of their absolute test statistics. For a nominal Type I error level $\alpha = 0.05$, the SS maxT MTP identifies 5 probes as being differentially expressed between the knock-out and control mice. Note the large jump in adjusted p-values (from 0.13 to 0.48) between the 8th and 9th ordered probes.

Figure 9.2 provides a plot of unadjusted p-values and SS maxT adjusted p-values vs. test statistics. This figure illustrates the monotonic relationship between test statistics and adjusted p-values for the common-cut-off SS maxT MTP. The aforementioned 8 probes also stand out in this display as having small negative t-statistics and small SS maxT adjusted p-values.

Figure 9.3 displays absolute test statistics $|T_n(m)|$ and SS maxT cut-offs $c_n(m) = 9.57$ (nominal FWER level $\alpha = 0.05$) for the top 10 probes. Note that only the test statistics for the top 5 probes fall within the common-cut-off rejection regions $\mathcal{C}_n(m) = (9.57, +\infty)$.

Figure 9.4 displays parameter estimates $\psi_n(m)$ and SS maxT 95% confidence regions for the top 10 probes. Note that only the confidence regions for the top 5 probes do not include the parameter null values $\psi_0(m) = 0$ for the differences $\psi(m)$ in mean expression measures between knock-out and control mice (red line), i.e., lead to the rejection of the corresponding null hypotheses of no differential expression at nominal FWER level $\alpha = 0.05$.

Comparison of FWER-controlling procedures

Tables 9.2 and 9.6 and Figures 9.5 and 9.9 report the results of a variety of FWER-controlling bootstrap- and permutation-based MTPs in terms of their adjusted p-values and numbers of rejected hypotheses. The smaller the adjusted p-values, the less conservative the MTP, and the greater the number of probes identified as differentially expressed at any nominal FWER level α.

The relative performances of the various MTPs differ for bootstrap- and permutation-based test statistics null distributions. Differences between bootstrap- and permutation-based MTPs can be attributed to the fact that the set $\{T_n^B(\cdot, b) : b = 1, \ldots, B\}$ of B bootstrap test statistics does not necessarily include the observed test statistics T_n. This allows bootstrap unadjusted p-values to be zero for some null hypotheses. In contrast, for a null distribution based on all possible $B = \binom{n}{n_0}$ permutations of the treatment and control labels, the observed test statistics T_n are included in the set of B permutation test statistics. As a result, two-sided unadjusted p-values are at least $2/B$ (here, $2/B = 2/\binom{16}{8} \approxeq 0.000155$) (Table 9.8; Figures 9.10 and 9.12). An unadjusted p-value of $2/B$ leads to a SS Bonferroni adjusted p-value of $M \times 2/B$ (here, $M \times 2/B = 5,548 \times 2/\binom{16}{8} \approxeq 0.8622$) (Table 9.6; Figure 9.9).

For the non-parametric bootstrap null shift and scale-transformed test statistics null distribution (Procedure 2.3), common-cut-off MTPs appear to be more conservative than common-quantile MTPs for nominal FWER level $\alpha \lessapprox 0.5$ (SS maxT vs. SS minP, SD maxT vs. SD minP; Table 9.2; Figure 9.5). In contrast, for the permutation test statistics null distribution, common-cut-off MTPs appear to be less conservative than common-quantile MTPs (SD maxT vs. SD minP; Table 9.6; Figure 9.9)

The results for both bootstrap- and permutation-based test statistics null distributions clearly show the gains in power achieved by taking into account the joint distribution of the test statistics, as in the single-step and step-down maxT and minP procedures (SS minP vs. SS Bonferroni, SD minP vs. SD Holm; Tables 9.2 and 9.6; Figures 9.5 and 9.9). Among FWER-controlling permutation-based procedures, only the SD maxT MTP rejects any hypothesis at nominal level $\alpha \lessapprox 0.5$ (Table 9.6).

The results do not suggest any notable gains in power for stepwise procedures compared to their single-step analogues (SS maxT vs. SD maxT, SS minP vs. SD minP, SS Bonferroni vs. SD Holm and SU Hochberg).

Note that, for computational reasons, the bootstrap-based SS minP and SD minP MTPs are implemented with only $B = 1,000$ bootstrap samples, rather than $B = 5,000$ for all other bootstrap-based MTPs. Also note that permutation versions of the SS maxT and SS minP MTPs are currently not implemented in the functions `mt.maxT` and `mt.minP` of the multtest package.

gFWER-controlling augmentation multiple testing procedures

As expected, the number of rejected hypotheses for gFWER(k) AMTP increases linearly with the allowed number k of false positives, compared to the initial FWER-controlling SS maxT MTP ($k = 0$ case).

TPPFP-controlling augmentation multiple testing procedures

As expected, the number of rejected hypotheses for TPPFP(q) AMTP increases with the allowed proportion q of false positives, compared to the initial FWER-controlling SS maxT MTP ($q = 0$ case). However, noticeable gains in power only seem to be achieved for fairly large values of q (i.e., $q \geq 0.25$).

Comparison of FDR-controlling procedures

For the non-parametric bootstrap null shift and scale-transformed test statistics null distribution (Procedure 2.3) and small nominal FDR level ($\alpha \lessapprox 0.2$), the joint TPPFP-based 1 and TPPFP-based 2 procedures, obtained from a TPPFP-controlling augmentation of the SS maxT MTP, appear to be more conservative than the marginal SU BH procedure. However, when the bootstrap null distribution is replaced by a permutation null distribution, TPPFP-based 1 and TPPFP-based 2 become less conservative than SU BY and SU BH for small nominal FDR level α (data not shown). As discussed above in the context of FWER control, this is likely due to the fact that bootstrap unadjusted p-values may be zero, whereas permutation two-sided unadjusted p-values are at least $2/B$.

The results also show that, for small nominal FDR level α, the general conservative TPPFP-based 1 and restricted TPPFP-based 2 MTPs have similar behaviors. The classical restricted SU BH procedure seems much less conservative than its general conservative SU BY version, with $\log M = \log 5,548 \approxeq 8.6$ penalty for the adjusted p-values. As the level α increases, gains in power are achieved by the restricted procedures (especially SU BH vs. SU BY) compared to their general analogues.

Comparison of test statistics null distributions

In addition to the main non-parametric bootstrap null shift and scale-transformed test statistics null distribution of Procedure 2.3, we consider a smoothed bootstrap version of Procedure 2.3 and a permutation test statistics null distribution. For each of the three null distributions, the aforementioned 8 probes stand out as being differentially expressed, as indicated by the large jump in adjusted p-values between the 8th and 9th ordered probes (Tables 9.2, 9.6, and 9.7; Figure 9.11).

Unadjusted p-values and adjusted p-values for all MTPs are virtually identical for the standard and smoothed non-parametric bootstrap null shift and scale-transformed test statistics null distributions (Table 9.8; Figure 9.10).

For this experiment, with equal sample sizes for the treatment and control groups, one can obtain a valid test statistics null distribution from a permutation data generating null distribution (Section 2.9). As discussed above in the context of FWER control, the relative performances of the various MTPs differ for bootstrap- and permutation-based null distributions (Tables 9.2 vs. 9.6; Figures 9.5 vs. 9.9), due to the fact that bootstrap unadjusted p-values may be zero, whereas permutation two-sided unadjusted p-values are at least $2/B$ (Table 9.8; Figures 9.10 and 9.12).

Applications of permutation-based multiple testing procedures to the Apo AI dataset are further discussed in Dudoit et al. (2002, 2003) and Ge et al. (2003).

Biological interpretation

All multiple testing procedures single out 8 probes as being differentially expressed between knock-out and control mice. The negative t-statistics suggest that the probes are under-expressed in the Apo AI knock-out mice compared to the control mice.

As indicated in Tables 9.1 and 9.9, the 8 most extreme probes actually correspond to only 4 distinct genes and 1 EST: `Apo AI` (2 copies), `Apo CIII` (2 copies), `Sterol desaturase` (2 copies), `Catechol-O-methyltransferase` (1 copy), and a novel EST (1 copy). All changes were confirmed by real-time quantitative PCR (RT-PCR), as described in Callow et al. (2000).

The presence of `Apo AI` among the under-expressed genes is to be expected, as this is the gene that was knocked out in the treatment mice.

The `Apo CIII` gene, also associated with lipoprotein metabolism, is located very close to the `Apo AI` locus (Table 9.9). Callow et al. (2000) showed that the down-regulation of `Apo CIII` is actually due to genetic polymorphism rather than lack of `Apo AI`. The presence of `Apo AI` and `Apo CIII` among the under-expressed genes thus provides a validation of the statistical methodology, if not a biologically novel finding.

`Sterol desaturase` is an enzyme that catalyzes one of the terminal steps in cholesterol synthesis.

Liver membrane-bound `Catechol-O-methyltransferase` was found to be a relevant factor in blood pressure regulation in rats (Tsunoda et al., 2003).

The novel EST shares sequence similarity to a family of ATPases.

For further detail on the 4 genes found to be most significantly differentially expressed, the interested reader is invited to consult hyperlinked Supplementary Table 9.1 on the website companion and follow links to PubMed and other databases.

The Apo AI experiment is rather unusual, in the sense that 8 spotted probe sequences clearly stand out from the remaining 5,540 probes as being differentially expressed. Such a dichotomy in gene expression is seldom observed in other applications of the microarray technology. For example, in many cancer microarray studies, genes tend to exhibit a continuum of change in expression measures and it is difficult to identify distinct groups of genes (Alizadeh et al., 2000; Barrier et al., 2005a,b,c, 2006; Chiappini et al., 2006; Chiaretti et al., 2004; Dudoit et al., 2003, 2006; Ge et al., 2003; Golub et al., 1999; Pollard et al., 2005b; Pollard and van der Laan, 2004). Differences in patterns of differential expression likely reflect the nature of the target samples under investigation. The Apo AI experiment compares relatively pure cell samples (hepatocytes), from wild-type and knock-out mice with an otherwise identical genetic background. In contrast, human cancer microarray studies typically assay samples composed of a variety of cell types, from genetically diverse individuals.

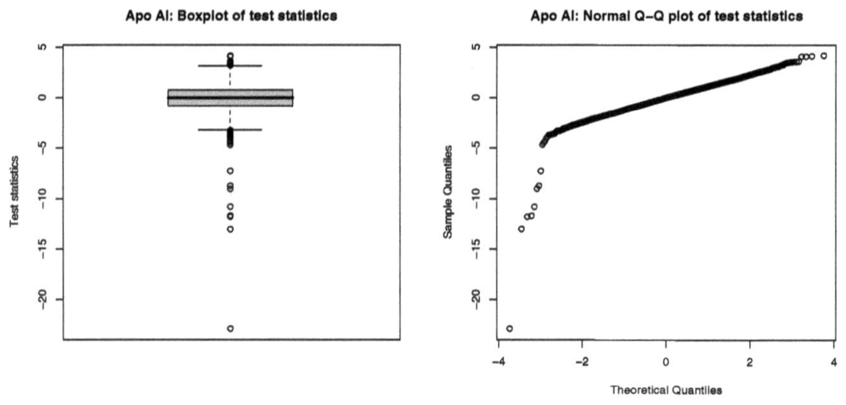

Panel (a) Panel (b)

Figure 9.1. *Apo AI dataset: Test statistics.* Graphical summaries of the distribution of the $M = 5,548$ two-sample Welch t-statistics. Panel (a): Boxplot of the test statistics. Panel (b): Normal quantile-quantile plot of the test statistics. These displays indicate that 8 probes stand out from the remaining probes in terms of their extreme negative test statistics.

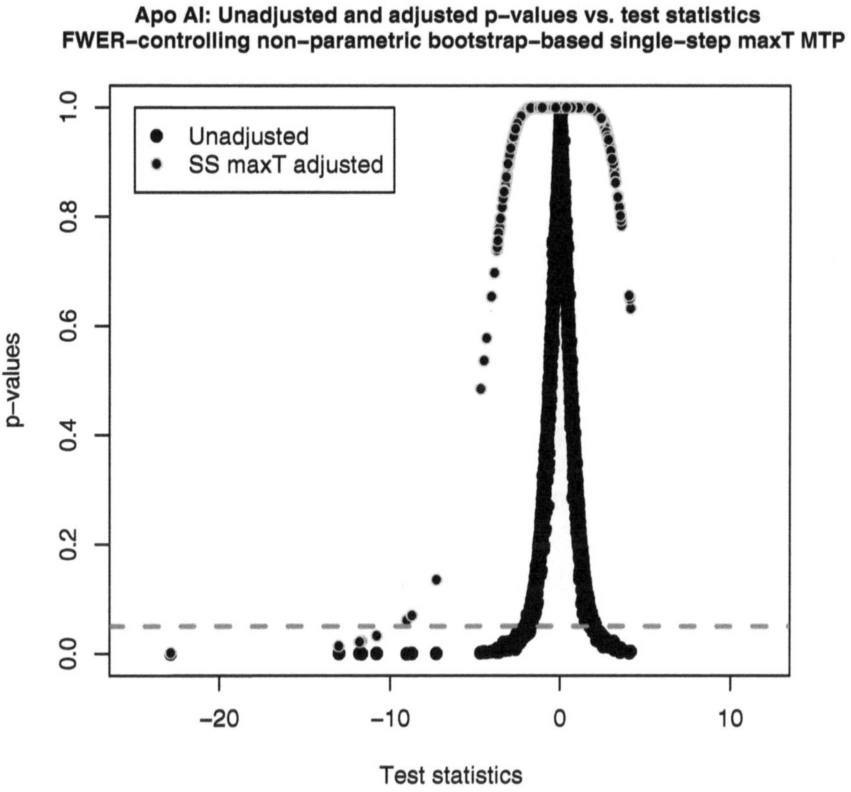

Figure 9.2. *Apo AI dataset: FWER-controlling non-parametric bootstrap-based single-step maxT MTP, test statistics and p-values.* Plots of unadjusted and adjusted p-values vs. test statistics, for FWER-controlling non-parametric bootstrap-based single-step maxT Procedure 3.5 (SS maxT). p-values are computed under the non-parametric bootstrap null shift and scale-transformed test statistics null distribution of Procedure 2.3, based on $B = 5,000$ samples. The horizontal and vertical red dashed lines indicate, respectively, the SS maxT adjusted p-value cut-off and test statistic common cut-off for a test at nominal FWER level $\alpha = 0.05$.

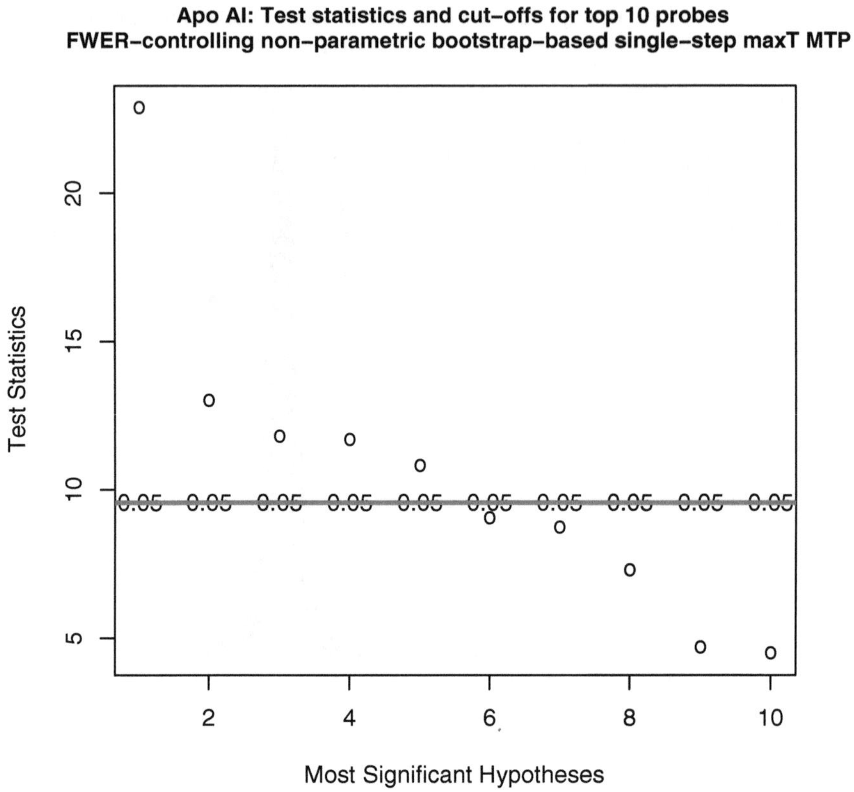

Figure 9.3. *Apo AI dataset: FWER-controlling non-parametric bootstrap-based single-step maxT MTP, test statistics and cut-offs.* Plots of absolute test statistics (open circle plotting symbols) and corresponding cut-offs for FWER-controlling non-parametric bootstrap-based single-step maxT Procedure 3.5 (SS maxT), with nominal Type I error level $\alpha = 0.05$. Cut-offs are computed under the non-parametric bootstrap null shift and scale-transformed test statistics null distribution of Procedure 2.3, based on $B = 5,000$ samples. Results are displayed for the top 10 probes, where probes are sorted first in increasing order of their non-parametric bootstrap SS maxT adjusted p-values, next in increasing order of their non-parametric bootstrap unadjusted p-values, and finally in decreasing order of their absolute test statistics. Note that only the test statistics for the top 5 probes fall within the common-cut-off rejection regions $(9.572218, +\infty)$ for nominal FWER level $\alpha = 0.05$ (above red line).

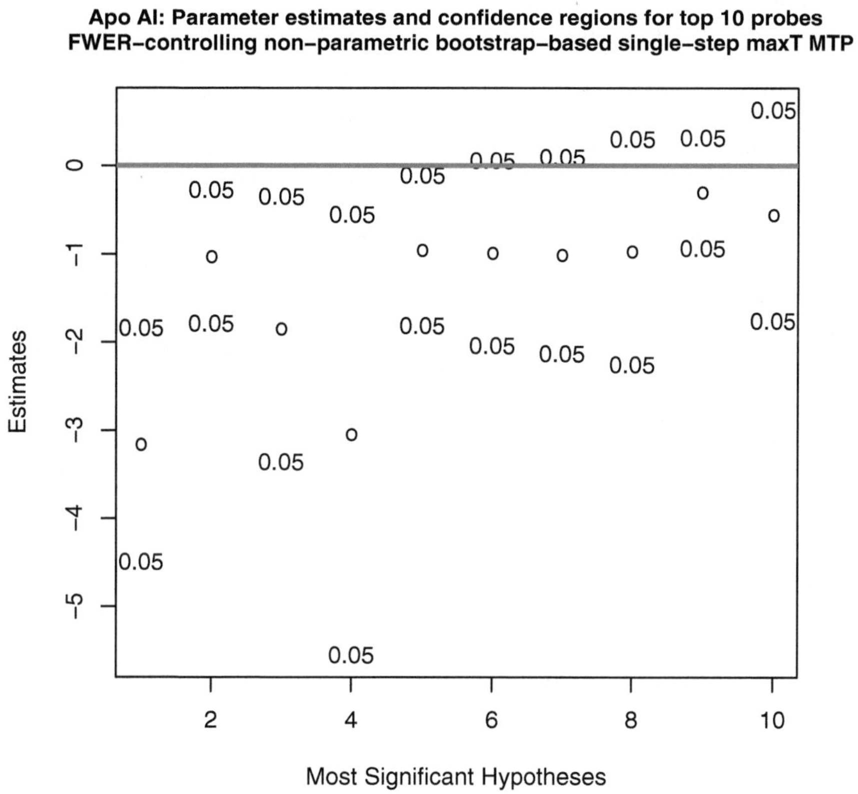

Figure 9.4. *Apo AI dataset: FWER-controlling non-parametric bootstrap-based single-step maxT MTP, parameter estimates and confidence regions.* Plots of parameter estimates (open circle plotting symbols) and 95% confidence regions for FWER-controlling non-parametric bootstrap-based single-step maxT Procedure 3.5 (SS maxT). Confidence regions are computed under the non-parametric bootstrap null shift and scale-transformed test statistics null distribution of Procedure 2.3, based on $B = 5,000$ samples. Results are displayed for the top 10 probes, where probes are sorted first in increasing order of their non-parametric bootstrap SS maxT adjusted p-values, next in increasing order of their non-parametric bootstrap unadjusted p-values, and finally in decreasing order of their absolute test statistics. Note that only the confidence regions for the top 5 probes do not include the parameter null value of zero (red line).

Table 9.1. *Apo AI dataset: FWER-controlling non-parametric bootstrap-based single-step maxT MTP.* The table provides results for FWER-controlling non-parametric bootstrap-based single-step maxT Procedure 3.5 (SS maxT): parameter estimates, $\psi_n(m)$; two-sample Welch t-statistics, $T_n(m)$; unadjusted p-values, $P_{0n}(m)$; SS maxT adjusted p-values, $\widetilde{P}_{0n}(m)$. p-values are computed under the non-parametric bootstrap null shift and scale-transformed test statistics null distribution of Procedure 2.3, based on $B = 5,000$ samples. Results are displayed for the top 10 probes, where probes are sorted first in increasing order of their non-parametric bootstrap SS maxT adjusted p-values, next in increasing order of their non-parametric bootstrap unadjusted p-values, and finally in decreasing order of their absolute test statistics. A short and a long version of the gene name, a description, and a spot ID (out of 6,384, the total number of spots on the microarray) are provided for each probe. The largest adjusted p-value not exceeding $\alpha = 0.05$ is underlined. Note that the ordering of the SS maxT adjusted p-values may differ from the ordering of the unadjusted p-values. The ranks of the probes based on their non-parametric bootstrap unadjusted p-values are indicated by the superscripts preceding the short gene names.

Gene name	Spot ID	Estimate $\psi_n(m)$	t-statistic $T_n(m)$	Unadjusted p-value $P_{0n}(m)$	Adjusted p-value $\widetilde{P}_{0n}(m)$
^{1} `Apoa1` `Apo AI` Apo AI, lipid-Img	2149	−3.17	−22.89	0.0000	0.0014
^{1} `Sc5d` `Sterol desaturase` EST, Weakly similar to C-5 STEROL DESATURASE [Saccharomyces cerevisiae], lipid-UG	4139	−1.03	−13.02	0.0000	0.0136
^{1} `Comt` `Catechol-O-methyltransferase` CATECHOL O-METHYLTRANSFERASE, MEMBRANE-BOUND FORM, Brain-Img	5356	−1.86	−11.80	0.0000	0.0214
^{1} `Apoa1` `Apo AI` EST, Highly similar to APOLIPOPROTEIN A-I PRECURSOR [Mus musculus], lipid-UG	540	−3.05	−11.69	0.0000	0.0224
^{2} `Apoc3` `Apo CIII` Apo CIII, lipid-Img	1739	−0.96	−10.81	0.0002	<u>0.0322</u>
^{1} EST est	1496	−0.99	−9.05	0.0000	0.0606

Continued on next page ...

... *continued from previous page*

Gene name	Spot ID	Estimate $\psi_n(m)$	t-statistic $T_n(m)$	Unadjusted p-value $P_{0n}(m)$	Adjusted p-value $\widetilde{P}_{0n}(m)$
[2] Apoc3 Apo CIII	2537	−1.02	−8.74	0.0002	0.0694
ESTs, Highly similar to APOLIPOPROTEIN C-III PRECURSOR [Mus musculus], lipid-UG					
[2] Sc5d Sterol desaturase	4941	−0.97	−7.29	0.0002	0.1346
similar to yeast sterol desaturase, lipid-Img					
[4] Casp7 Caspase 7	954	−0.31	−4.70	0.0018	0.4840
Caspase 7, heart-Img					
[8] EST	947	−0.56	−4.50	0.0034	0.5358
EST, Weakly similar to FATTY ACID-BINDING PROTEIN, EPIDERMAL [Mus musculus], lipid-UG					

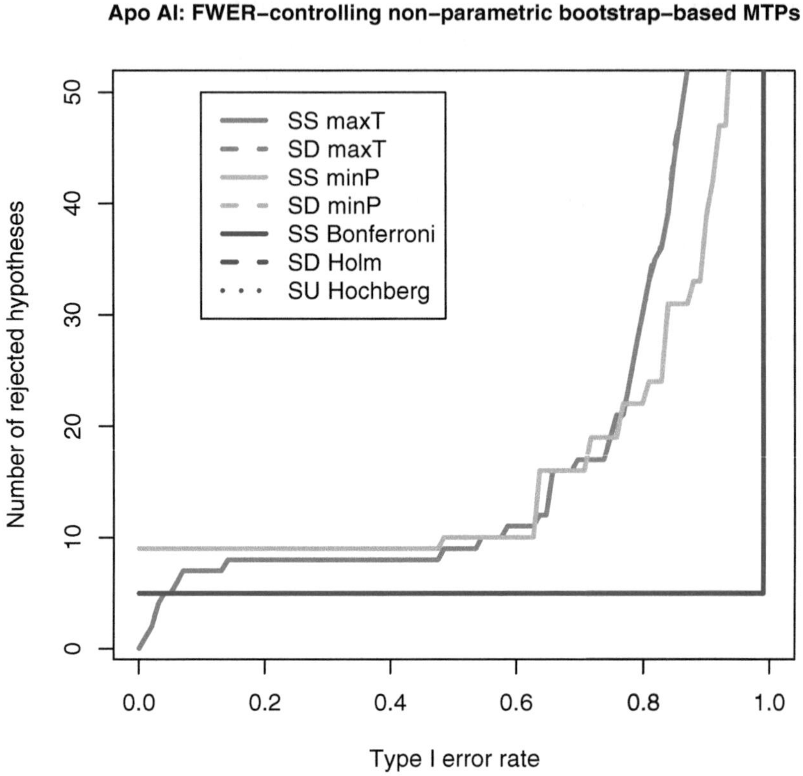

Figure 9.5. *Apo AI dataset: FWER-controlling non-parametric bootstrap-based MTPs.* Plots of number of rejected hypotheses vs. nominal Type I error level for FWER-controlling non-parametric bootstrap-based multiple testing procedures: single-step maxT Procedure 3.5 (SS maxT), single-step minP Procedure 3.6 (SS minP), step-down maxT Procedure 3.11 (SD maxT), step-down minP Procedure 3.12 (SD minP), single-step Bonferroni Procedure 3.1 (SS Bonferroni), step-down Holm Procedure 3.7 (SD Holm), and step-up Hochberg Procedure 3.13 (SU Hochberg). Adjusted p-values are computed under the non-parametric bootstrap null shift and scale-transformed test statistics null distribution of Procedure 2.3, based on either $B = 5,000$ (SS maxT, SD maxT, SS Bonferroni, SD Holm, SU Hochberg) or $B = 1,000$ (SS minP, SD minP) samples. Solid, dashed, and dotted lines represent, respectively, single-step, step-down, and step-up MTPs; red, green, and blue lines represent, respectively, joint common-cut-off, joint common-quantile, and marginal common-quantile MTPs. Results are displayed for the top 50 probes, where, for each MTP, probes are sorted in increasing order of their adjusted p-values. Note that the ordering of adjusted p-values may differ for common-cut-off (SS maxT, SD maxT) and common-quantile (SS minP, SD minP, SS Bonferroni, SD Holm, SU Hochberg) procedures. (Color plate p. 328)

Table 9.2. *Apo AI dataset: FWER-controlling non-parametric bootstrap-based MTPs.* Adjusted $\boldsymbol{p}$-values for FWER-controlling non-parametric bootstrap-based multiple testing procedures: single-step maxT Procedure 3.5 (SS maxT), single-step minP Procedure 3.6 (SS minP), step-down maxT Procedure 3.11 (SD maxT), step-down minP Procedure 3.12 (SD minP), single-step Bonferroni Procedure 3.1 (SS Bonferroni), step-down Holm Procedure 3.7 (SD Holm), and step-up Hochberg Procedure 3.13 (SU Hochberg). Adjusted p-values are computed under the non-parametric bootstrap null shift and scale-transformed test statistics null distribution of Procedure 2.3, based on either $B = 5,000$ (SS maxT, SD maxT, SS Bonferroni, SD Holm, SU Hochberg) or $B = 1,000$ (SS minP, SD minP) samples. Results are displayed for the top 10 probes, where probes are sorted first in increasing order of their non-parametric bootstrap SS maxT adjusted p-values, next in increasing order of their non-parametric bootstrap unadjusted p-values, and finally in decreasing order of their absolute test statistics. A short version of the gene name and a spot ID (out of 6,384, the total number of spots on the microarray) are provided for each probe. For each MTP, the largest adjusted p-value not exceeding $\alpha = 0.05$ is underlined. Note that the ordering of adjusted p-values may differ for common-cut-off (SS maxT, SD maxT) and common-quantile (SS minP, SD minP, SS Bonferroni, SD Holm, SU Hochberg) procedures. The ranks of the probes based on their non-parametric bootstrap unadjusted p-values are indicated by the superscripts preceding the short gene names.

		Adjusted p-values						
Gene name	**Spot ID**	SS maxT	SD maxT	SS minP	SD minP	SS Bonferroni	SD Holm	SU Hochberg
^{1} Apoa1	2149	0.0014	0.0014	0.0000	0.0000	0.0000	0.0000	0.0000
^{1} Sc5d	4139	0.0136	0.0136	0.0000	0.0000	0.0000	0.0000	0.0000
^{1} Comt	5356	0.0214	0.0214	0.0000	0.0000	0.0000	0.0000	0.0000
^{1} Apoa1	540	0.0224	0.0224	0.0000	0.0000	0.0000	0.0000	0.0000
^{2} Apoc3	1739	0.0322	0.0322	0.0000	0.0000	1.0000	1.0000	1.0000
^{1} EST	1496	0.0606	0.0606	0.0000	0.0000	0.0000	0.0000	0.0000
^{2} Apoc3	2537	0.0694	0.0694	0.0000	0.0000	1.0000	1.0000	1.0000
^{2} Sc5d	4941	0.1346	0.1346	0.0000	0.0000	1.0000	1.0000	1.0000
^{4} Casp7	954	0.4840	0.4840	0.6270	0.6270	1.0000	1.0000	1.0000
^{8} EST	947	0.5358	0.5358	0.6270	0.6270	1.0000	1.0000	1.0000

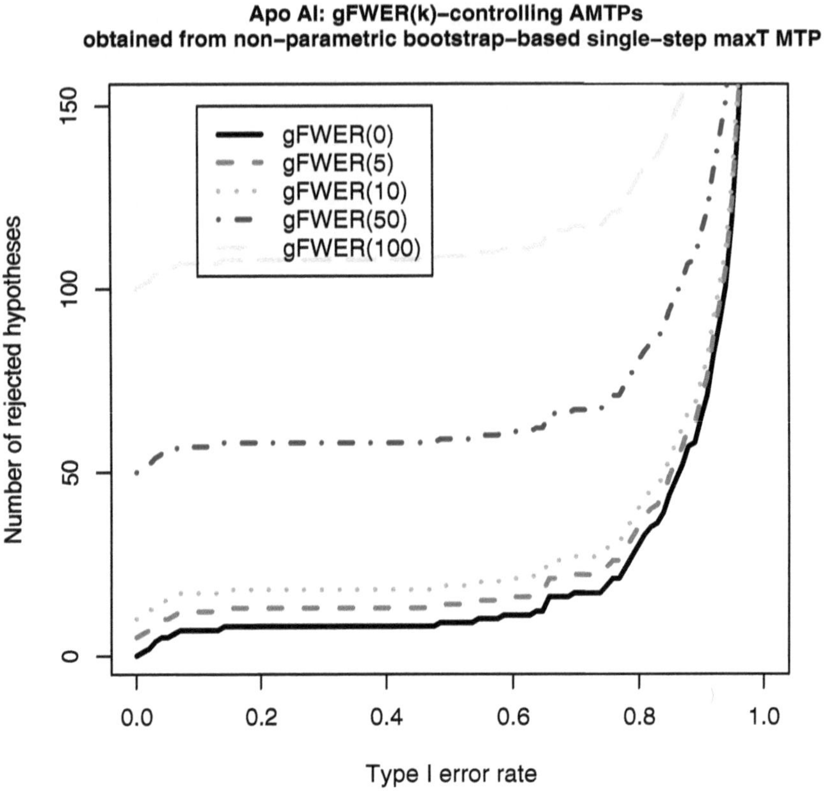

Figure 9.6. *Apo AI dataset: gFWER-controlling non-parametric bootstrap-based AMTPs.* Plots of number of rejected hypotheses vs. nominal Type I error level for gFWER-controlling augmentation multiple testing Procedure 3.20 (gFWER(k) AMTP), obtained from FWER-controlling non-parametric bootstrap-based single-step maxT Procedure 3.5 (SS maxT), with an allowed number $k \in \{0, 5, 10, 50, 100\}$ of false positives ($k = 0$ case corresponds to FWER). Adjusted p-values are computed under the non-parametric bootstrap null shift and scale-transformed test statistics null distribution of Procedure 2.3, based on $B = 5,000$ samples. Results are displayed for the top 150 probes, where, for each MTP, probes are sorted in increasing order of their adjusted p-values. (Color plate p. 329)

Table 9.3. *Apo AI dataset: gFWER-controlling non-parametric bootstrap-based AMTPs.* Adjusted p-values for gFWER-controlling augmentation multiple testing Procedure 3.20 (gFWER(k) AMTP), obtained from FWER-controlling non-parametric bootstrap-based single-step maxT Procedure 3.5 (SS maxT), with an allowed number $k \in \{0, 5, 10, 50, 100\}$ of false positives ($k = 0$ case corresponds to FWER). Adjusted p-values are computed under the non-parametric bootstrap null shift and scale-transformed test statistics null distribution of Procedure 2.3, based on $B = 5,000$ samples. Results are displayed for the top 10 probes, where probes are sorted first in increasing order of their non-parametric bootstrap SS maxT adjusted p-values, next in increasing order of their non-parametric bootstrap unadjusted p-values, and finally in decreasing order of their absolute test statistics. A short version of the gene name and a spot ID (out of 6,384, the total number of spots on the microarray) are provided for each probe. For each MTP, the largest adjusted p-value not exceeding $\alpha = 0.05$ is underlined.

		Adjusted p-values				
Gene	**Spot**	Allowed number of false positives, k				
name	**ID**	0	5	10	50	100
`Apoa1`	2149	0.0014	0.0000	0.0000	0.0000	0.0000
`Sc5d`	4139	0.0136	0.0000	0.0000	0.0000	0.0000
`Comt`	5356	0.0214	0.0000	0.0000	0.0000	0.0000
`Apoa1`	540	0.0224	0.0000	0.0000	0.0000	0.0000
`Apoc3`	1739	<u>0.0322</u>	0.0000	0.0000	0.0000	0.0000
`EST`	1496	0.0606	0.0014	0.0000	0.0000	0.0000
`Apoc3`	2537	0.0694	0.0136	0.0000	0.0000	0.0000
`Sc5d`	4941	0.1346	0.0214	0.0000	0.0000	0.0000
`Casp7`	954	0.4840	0.0224	0.0000	0.0000	0.0000
`EST`	947	0.5358	<u>0.0322</u>	0.0000	0.0000	0.0000

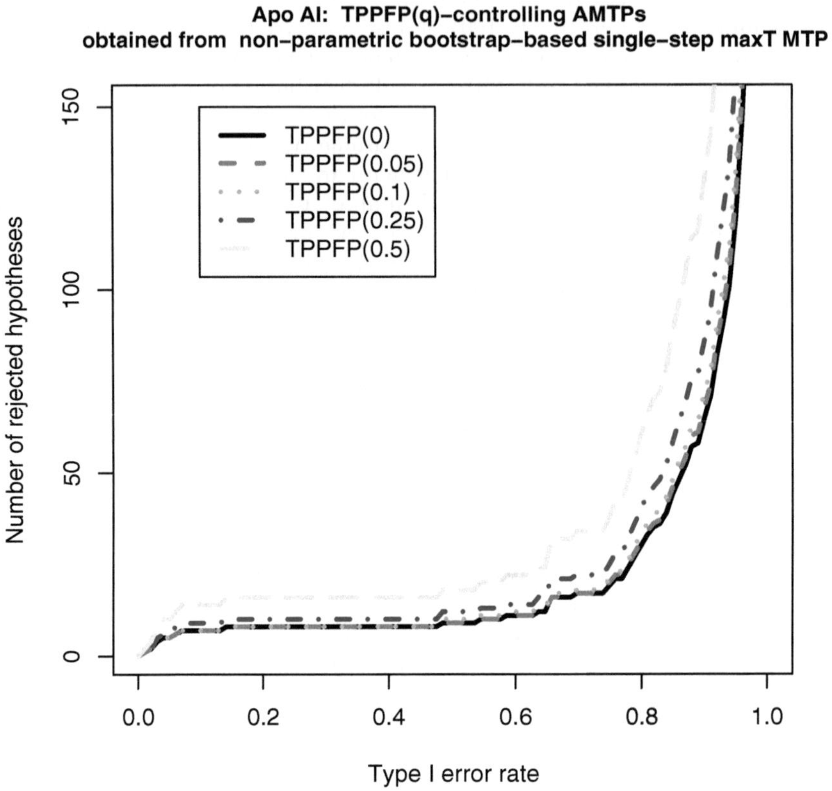

Figure 9.7. *Apo AI dataset: TPPFP-controlling non-parametric bootstrap-based AMTPs.* Plots of number of rejected hypotheses vs. nominal Type I error level for TPPFP-controlling augmentation multiple testing Procedure 3.26 (TPPFP(q) AMTP), obtained from FWER-controlling non-parametric bootstrap-based single-step maxT Procedure 3.5 (SS maxT), with an allowed proportion $q \in \{0, 0.05, 0.10, 0.25, 0.50\}$ of false positives ($q = 0$ case corresponds to FWER). Adjusted p-values are computed under the non-parametric bootstrap null shift and scale-transformed test statistics null distribution of Procedure 2.3, based on $B = 5,000$ samples. Results are displayed for the top 150 probes, where, for each MTP, probes are sorted in increasing order of their adjusted p-values. (Color plate p. 330)

Table 9.4. *Apo AI dataset: TPPFP-controlling non-parametric bootstrap-based AMTPs.* Adjusted p-values for TPPFP-controlling augmentation multiple testing Procedure 3.26 (TPPFP(q) AMTP), obtained from FWER-controlling non-parametric bootstrap-based single-step maxT Procedure 3.5 (SS maxT), with an allowed proportion $q \in \{0, 0.05, 0.10, 0.25, 0.50\}$ of false positives ($q = 0$ case corresponds to FWER). Adjusted p-values are computed under the non-parametric bootstrap null shift and scale-transformed test statistics null distribution of Procedure 2.3, based on $B = 5,000$ samples. Results are displayed for the top 10 probes, where probes are sorted first in increasing order of their non-parametric bootstrap SS maxT adjusted p-values, next in increasing order of their non-parametric bootstrap unadjusted p-values, and finally in decreasing order of their absolute test statistics. A short version of the gene name and a spot ID (out of 6,384, the total number of spots on the microarray) are provided for each probe. For each MTP, the largest adjusted p-value not exceeding $\alpha = 0.05$ is underlined.

		Adjusted p-values				
Gene	**Spot**	Allowed proportion of false positives, q				
name	**ID**	0	0.05	0.10	0.25	0.50
Apoa1	2149	0.0014	0.0014	0.0014	0.0014	0.0014
Sc5d	4139	0.0136	0.0136	0.0136	0.0136	0.0014
Comt	5356	0.0214	0.0214	0.0214	0.0214	0.0136
Apoa1	540	0.0224	0.0224	0.0224	0.0214	0.0136
Apoc3	1739	<u>0.0322</u>	<u>0.0322</u>	<u>0.0322</u>	0.0224	0.0214
EST	1496	0.0606	0.0606	0.0606	<u>0.0322</u>	0.0214
Apoc3	2537	0.0694	0.0694	0.0694	0.0606	0.0224
Sc5d	4941	0.1346	0.1346	0.1346	0.0606	0.0224
Casp7	954	0.4840	0.4840	0.4840	0.0694	0.0322
EST	947	0.5358	0.5358	0.4840	0.1346	<u>0.0322</u>

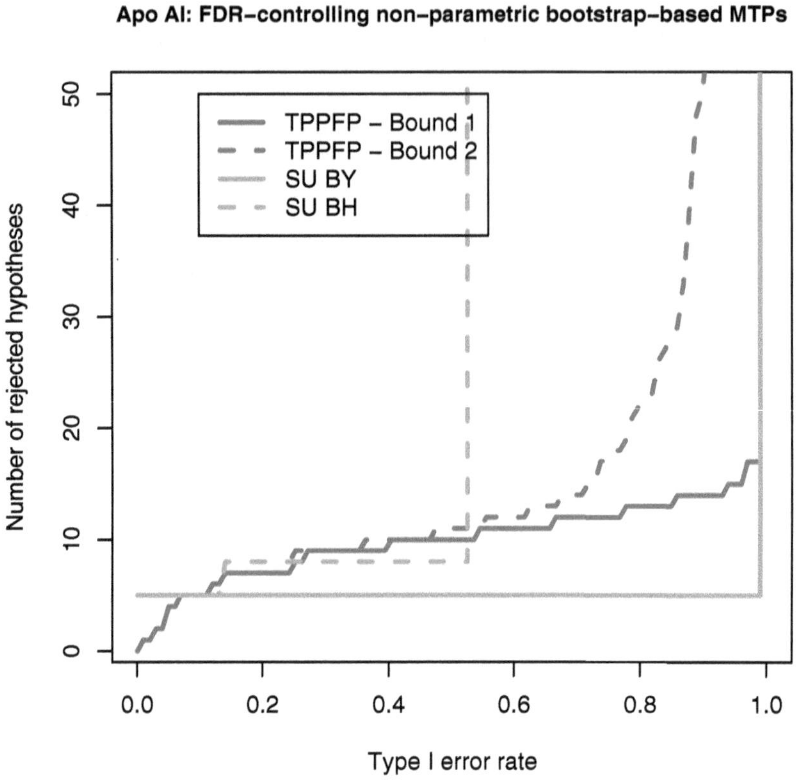

Figure 9.8. *Apo AI dataset: FDR-controlling non-parametric bootstrap-based MTPs.* Plots of number of rejected hypotheses vs. nominal Type I error level for FDR-controlling non-parametric bootstrap-based multiple testing procedures: marginal step-up Benjamini and Hochberg (1995) Procedure 3.22 (SU BH), marginal step-up Benjamini and Yekutieli (2001) Procedure 3.23 (SU BY), procedures described in Theorem 6.7, based on a TPPFP-controlling augmentation of joint single-step maxT Procedure 3.5 (TPPFP-based 1, TPPFP-based 2). Adjusted p-values are computed under the non-parametric bootstrap null shift and scale-transformed test statistics null distribution of Procedure 2.3, based on $B = 5,000$ samples. Solid and dashed lines represent, respectively, general conservative and restricted MTPs; red and green lines represent, respectively, joint common-cut-off and marginal common-quantile MTPs. Results are displayed for the top 50 probes, where, for each MTP, probes are sorted in increasing order of their adjusted p-values. Note that the ordering of adjusted p-values may differ for common-cut-off (TPPFP-based 1, TPPFP-based 2) and common-quantile (SU BH, SU BY) procedures. (Color plate p. 331)

Table 9.5. *Apo AI dataset: FDR-controlling non-parametric bootstrap-based MTPs.* Adjusted p-values for FDR-controlling non-parametric bootstrap-based multiple testing procedures: marginal step-up Benjamini and Hochberg (1995) Procedure 3.22 (SU BH), marginal step-up Benjamini and Yekutieli (2001) Procedure 3.23 (SU BY), procedures described in Theorem 6.7, based on a TPPFP-controlling augmentation of joint single-step maxT Procedure 3.5 (TPPFP-based 1, TPPFP-based 2). Adjusted p-values are computed under the non-parametric bootstrap null shift and scale-transformed test statistics null distribution of Procedure 2.3, based on $B = 5,000$ samples. Results are displayed for the top 10 probes, where probes are sorted first in increasing order of their non-parametric bootstrap SS maxT adjusted p-values, next in increasing order of their non-parametric bootstrap unadjusted p-values, and finally in decreasing order of their absolute test statistics. A short version of the gene name and a spot ID (out of 6,384, the total number of spots on the microarray) are provided for each probe. For each MTP, the largest adjusted p-value not exceeding $\alpha = 0.05$ is underlined. Note that the ordering of adjusted p-values may differ for common-cut-off (TPPFP-based 1, TPPFP-based 2) and common-quantile (SU BH, SU BY) procedures. The ranks of the probes based on their non-parametric bootstrap unadjusted p-values are indicated by the superscripts preceding the short gene names.

Gene	Spot	Adjusted p-values			
name	ID	TPPFP-based 1	TPPFP-based 2	SU BY	SU BH
^{1}Apoa1	2149	0.0028	0.0028	0.0000	0.0000
^{1}Sc5d	4139	0.0272	0.0270	0.0000	0.0000
^{1}Comt	5356	0.0428	0.0423	0.0000	0.0000
^{1}Apoa1	540	0.0448	0.0443	0.0000	0.0000
^{2}Apoc3	1739	0.0644	0.0634	1.0000	0.1387
^{1}EST	1496	0.1212	0.1175	0.0000	0.0000
^{2}Apoc3	2537	0.1388	0.1340	1.0000	0.1387
^{2}Sc5d	4941	0.2500	0.2344	1.0000	0.1387
^{4}Casp7	954	0.2692	0.2511	1.0000	0.5307
^{8}EST	947	0.4000	0.3600	1.0000	0.5307

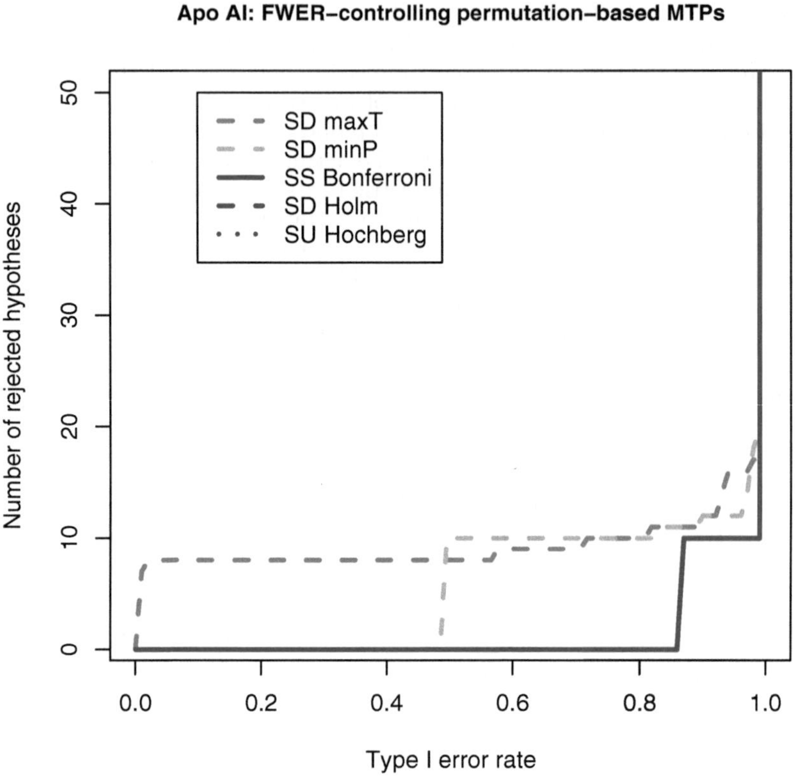

Figure 9.9. *Apo AI dataset: FWER-controlling permutation-based MTPs.* Plots of number of rejected hypotheses vs. nominal Type I error level for FWER-controlling permutation-based multiple testing procedures: step-down maxT Procedure 3.11 (SD maxT), step-down minP Procedure 3.12 (SD minP), single-step Bonferroni Procedure 3.1 (SS Bonferroni), step-down Holm Procedure 3.7 (SD Holm), and step-up Hochberg Procedure 3.13 (SU Hochberg). Adjusted p-values are computed based on all possible $\binom{16}{8} = 12,870$ permutations of the treatment and control labels. Solid, dashed, and dotted lines represent, respectively, single-step, step-down, and step-up MTPs; red, green, and blue lines represent, respectively, joint common-cut-off, joint common-quantile, and marginal common-quantile MTPs. Results are displayed for the top 50 probes, where, for each MTP, probes are sorted in increasing order of their adjusted p-values. Note that the ordering of adjusted p-values may differ for common-cut-off (SD maxT) and common-quantile (SD minP, SS Bonferroni, SD Holm, SU Hochberg) procedures. (Color plate p. 332)

Table 9.6. *Apo AI dataset: FWER-controlling permutation-based MTPs.* Adjusted p-values for FWER-controlling permutation-based multiple testing procedures: step-down maxT Procedure 3.11 (SD maxT), step-down minP Procedure 3.12 (SD minP), single-step Bonferroni Procedure 3.1 (SS Bonferroni), step-down Holm Procedure 3.7 (SD Holm), and step-up Hochberg Procedure 3.13 (SU Hochberg). Adjusted p-values are computed based on all possible $\binom{16}{8} = 12,870$ permutations of the treatment and control labels. Results are displayed for the top 10 probes, where probes are sorted first in increasing order of their non-parametric bootstrap SS maxT adjusted p-values, next in increasing order of their non-parametric bootstrap unadjusted p-values, and finally in decreasing order of their absolute test statistics. A short version of the gene name and a spot ID (out of 6,384, the total number of spots on the microarray) are provided for each probe. For each MTP, the largest adjusted p-value not exceeding $\alpha = 0.05$ is underlined. Note that the ordering of adjusted p-values may differ for common-cut-off (SD maxT) and common-quantile (SD minP, SS Bonferroni, SD Holm, SU Hochberg) procedures. The ranks of the probes based on their permutation unadjusted p-values are indicated by the superscripts preceding the short gene names.

		Adjusted p-values				
Gene name	**Spot ID**	SD maxT	SD minP	SS Bonferroni	SD Holm	SU Hochberg
^{1}Apoa1	2149	0.0002	0.4870	0.8622	0.8622	0.8608
^{1}Sc5d	4139	0.0003	0.4870	0.8622	0.8622	0.8608
^{1}Comt	5356	0.0005	0.4870	0.8622	0.8622	0.8608
^{1}Apoa1	540	0.0005	0.4870	0.8622	0.8622	0.8608
^{1}Apoc3	1739	0.0005	0.4870	0.8622	0.8622	0.8608
^{1}EST	1496	0.0014	0.4870	0.8622	0.8622	0.8608
^{1}Apoc3	2537	0.0019	0.4870	0.8622	0.8622	0.8608
^{1}Sc5d	4941	$\underline{0.0115}$	0.4870	0.8622	0.8622	0.8608
^{3}Casp7	954	0.5728	0.8985	1.0000	1.0000	1.0000
^{1}EST	947	0.7080	0.4870	0.8622	0.8622	0.8608

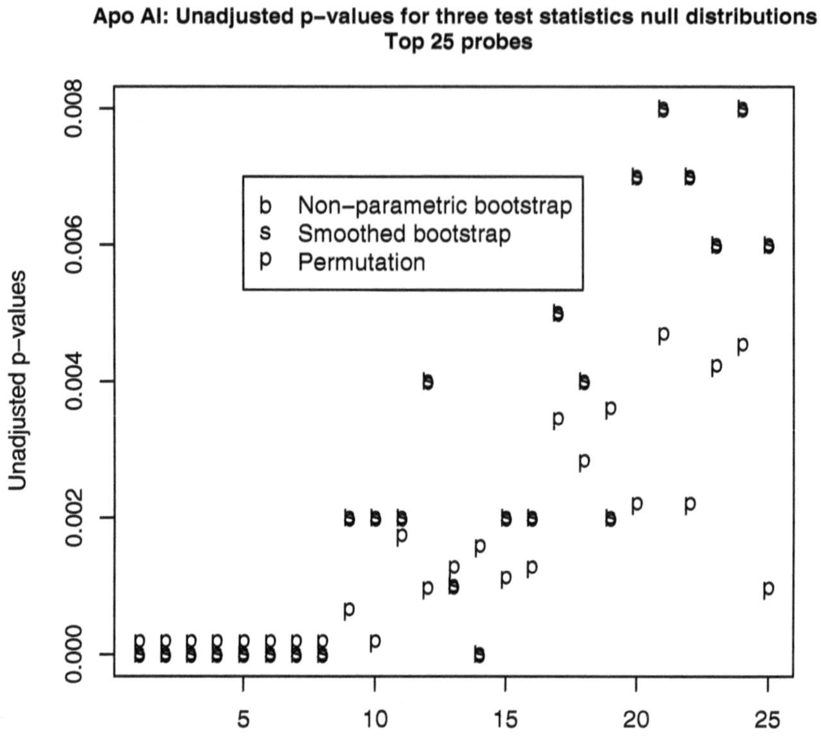

Figure 9.10. *Apo AI dataset: Unadjusted p-values for three test statistics null distributions.* Plots of unadjusted p-values for three test statistics null distributions: non-parametric bootstrap, smoothed non-parametric bootstrap, and permutation. Bootstrap p-values are computed under the non-parametric bootstrap null shift and scale-transformed test statistics null distribution of Procedure 2.3, based on $B = 1,000$ samples. Smoothed versions of the bootstrap p-values are obtained by kernel density smoothing of the marginal distributions of the null-transformed bootstrap test statistics. Permutation p-values are computed based on all possible $\binom{16}{8} = 12,870$ permutations of the treatment and control labels. Results are displayed for the top 25 probes, where probes are sorted first in increasing order of their non-parametric bootstrap SS maxT adjusted p-values, next in increasing order of their non-parametric bootstrap unadjusted p-values, and finally in decreasing order of their absolute test statistics. (Color plate p. 333)

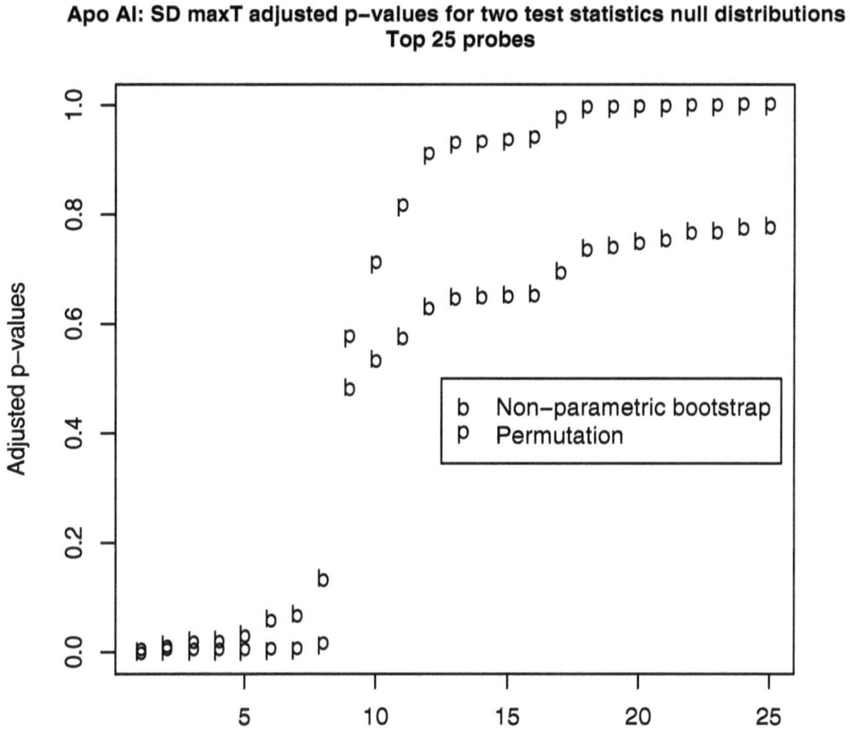

Figure 9.11. *Apo AI dataset: Step-down maxT adjusted p-values for non-parametric bootstrap and permutation test statistics null distributions.* Plots of adjusted p-values for step-down maxT Procedure 3.11 (SD maxT), for two test statistics null distributions: non-parametric bootstrap and permutation. Bootstrap p-values are computed under the non-parametric bootstrap null shift and scale-transformed test statistics null distribution of Procedure 2.3, based on $B = 5,000$ samples. Permutation p-values are computed based on all possible $\binom{16}{8} = 12,870$ permutations of the treatment and control labels. Results are displayed for the top 25 probes, where probes are sorted first in increasing order of their non-parametric bootstrap SS maxT adjusted p-values, next in increasing order of their non-parametric bootstrap unadjusted p-values, and finally in decreasing order of their absolute test statistics. Note the large jump in adjusted p-values between the 8th and 9th ordered probes, for both null distributions. (Color plate p. 334)

Table 9.7. *Apo AI dataset: FWER-controlling non-parametric bootstrap-based vs. permutation-based step-down maxT MTPs.* Adjusted p-values for FWER-controlling non-parametric bootstrap-based (Boot SD maxT) and permutation-based (Perm SD maxT) step-down maxT Procedure 3.11. Bootstrap p-values are computed under the non-parametric bootstrap null shift and scale-transformed test statistics null distribution of Procedure 2.3, based on $B = 5,000$ samples. Permutation p-values are computed based on all possible $\binom{16}{8} = 12,870$ permutations of the treatment and control labels. Results are displayed for the top 10 probes, where probes are sorted first in increasing order of their non-parametric bootstrap SS maxT adjusted p-values, next in increasing order of their non-parametric bootstrap unadjusted p-values, and finally in decreasing order of their absolute test statistics. A short version of the gene name and a spot ID (out of 6,384, the total number of spots on the microarray) are provided for each probe. For each MTP, the largest adjusted p-value not exceeding $\alpha = 0.05$ is underlined.

Gene name	Spot ID	Adjusted p-values Boot SD maxT	Perm SD maxT
Apoa1	2149	0.0014	0.0002
Sc5d	4139	0.0136	0.0003
Comt	5356	0.0214	0.0005
Apoa1	540	0.0224	0.0005
Apoc3	1739	<u>0.0322</u>	0.0005
EST	1496	0.0606	0.0014
Apoc3	2537	0.0694	0.0019
Sc5d	4941	0.1346	<u>0.0115</u>
Casp7	954	0.4840	0.5728
EST	947	0.5358	0.7080

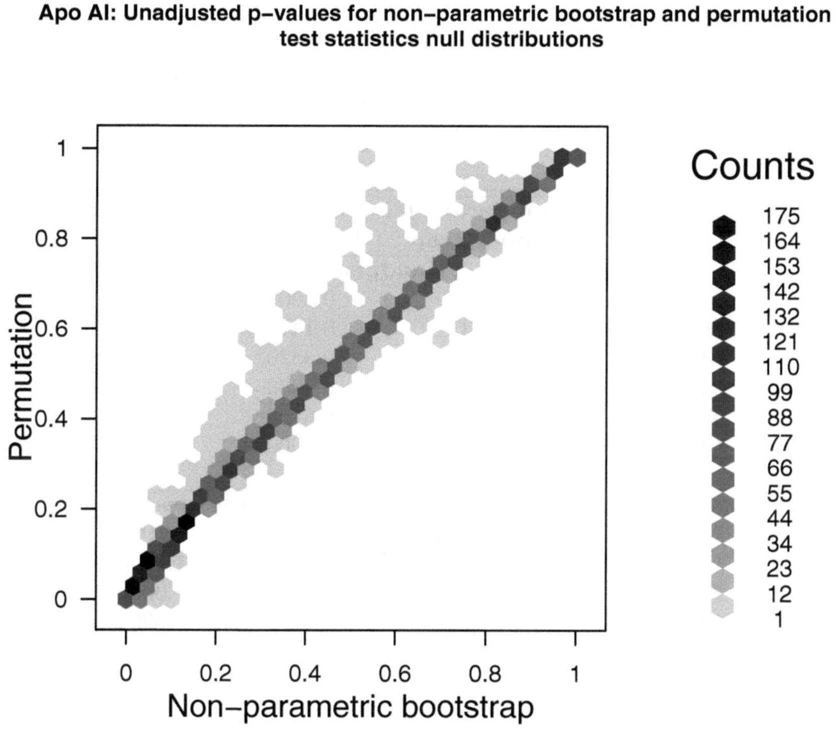

Figure 9.12. *Apo AI dataset: Unadjusted p-values for non-parametric bootstrap and permutation test statistics null distributions.* Hexagonal binning scatterplot of unadjusted p-values for two test statistics null distributions: permutation vs. non-parametric bootstrap. Bootstrap p-values are computed under the non-parametric bootstrap null shift and scale-transformed test statistics null distribution of Procedure 2.3, based on $B = 5,000$ samples. Permutation p-values are computed based on all possible $\binom{16}{8} = 12,870$ permutations of the treatment and control labels.

Table 9.8. *Apo AI dataset: Unadjusted and step-down maxT adjusted p-values for three test statistics null distributions.* Six-number summaries of the distributions of unadjusted p-values and adjusted p-values for step-down maxT Procedure 3.11 (SD maxT), for three test statistics null distributions: non-parametric bootstrap, smoothed non-parametric bootstrap, and permutation. Bootstrap p-values are computed under the non-parametric bootstrap null shift and scale-transformed test statistics null distribution of Procedure 2.3, based on either $B = 5,000$ (SS maxT adjusted p-values) or $B = 1,000$ (unadjusted p-values) samples. Smoothed versions of the bootstrap p-values are obtained by kernel density smoothing of the marginal distributions of the null-transformed bootstrap test statistics. Permutation p-values are computed based on all possible $\binom{16}{8} = 12,870$ permutations of the treatment and control labels. The SD maxT MTP was not applied for the smoothed bootstrap null distribution.

	Non-parametric bootstrap	Smoothed bootstrap	Permutation
	Unadjusted p-values		
Minimum	0.0000	0.0000	0.0002
1st quartile	0.1478	0.1478	0.1993
Median	0.3820	0.3820	0.4491
Mean	0.4236	0.4236	0.4655
3rd quartile	0.6765	0.6765	0.7216
Maximum	1.0000	1.0000	1.0000
	SD maxT adjusted p-values		
Minimum	0.0014	–	0.0002
1st quartile	1.0000	–	1.0000
Median	1.0000	–	1.0000
Mean	0.9945	–	0.9983
3rd quartile	1.0000	–	1.0000
Maximum	1.0000	–	1.0000

Table 9.9. *Apo AI dataset: Gene descriptions from Entrez Gene database.* Description of the 4 genes found to be most significantly differentially expressed between Apo AI knock-out and control mice, based on the NCBI Entrez Gene database (*Mus musculus*, **`www.ncbi.nlm.nih.gov/entrez/query.fcgi?db=gene`**). A more detailed hyperlinked table, including information on chromosomal location and links to GenBank, Entrez Gene, NCBI Map Viewer, UniGene, PubMed, AmiGO, and KEGG, is provided on the website companion (Supplementary Table 9.1).

`Apo AI`	
Gene name:	`Apoa1`
Gene description:	Apolipoprotein A-I
Locations:	9 A2-A4; 9 27.0 cM
GeneID:	11806
Updated:	26-Jul-2006
`Apo CIII`	
Gene name:	`Apoc3`
Gene description:	Apolipoprotein C-III
Locations:	9 A5.2; 9 27.0 cM
GeneID:	11814
Updated:	26-Jul-2006
`Catechol-O-methyltransferase`	
Gene name:	`Comt`
Gene description:	Catechol-O-methyltransferase
Locations:	16 A3; 16 11.2 cM
GeneID:	12846
Updated:	26-Jul-2006
`Sterol desaturase`	
Gene name:	`Sc5d`
Gene description:	Sterol-C5-desaturase (fungal ERG3, delta-5-desaturase) homolog (S. cerevisae)
Location:	9 B
GeneID:	235293
Updated:	26-Jul-2006

9.3 Cancer microRNA study of Lu et al. (2005)

In addition to playing the important role of passing genetic messages from DNA to the protein-making machinery of the cell, *ribonucleic acids* (RNA) serve many other cellular functions. A new class of small, non-coding RNAs, known as *microRNAs* (miRNA), are currently the subject of intense study due to their provocative roles as gene regulators (miRBase, `microrna.sanger.ac.uk`; Wienholds and Plasterk (2005)). By binding to *messenger RNA* (mRNA), miRNAs regulate gene expression post-transcriptionally and affect the abundance of a wide range of proteins, in diverse biological processes.

The first known miRNAs, *lin-4* and *let-7*, were identified by forward genetic screens in the nematode *Caenorhabditis elegans* (*C. elegans*). These were initially termed *small temporal RNAs* (stRNA) for their involvement in developmental timing. By now, hundreds of miRNAs have been identified, in various multicellular organisms, including the fruitfly *Drosophila melanogaster* (*D. melanogaster*) and humans, and many are evolutionary conserved.

Although the biological functions of miRNAs are still largely unknown, miRNAs are predicted to regulate up to 30% of all protein-coding genes. Each mammalian miRNA is believed to regulate approximately 200 genes and many genes have several target sites for one or several different miRNAs. The large number of miRNA genes, their diverse expression patterns, and the abundance of miRNA targets, suggest the involvement of miRNAs in a variety of diseases, including cancers and viruses. More than half of the known human miRNA genes are located in genomic regions related to cancers, such as, fragile sites, minimal regions of loss of heterozigosity, minimal regions of amplification, and common breakpoint regions. miRNAs have also been implicated in several mammalian viruses, such as, the Epstein-Barr virus and the human immunodeficiency virus (HIV).

In their recent study, monitoring miRNA levels in cells derived from cancerous and non-cancerous tissues, Lu et al. (2005) made an astonishing discovery: predictors based on expression levels for the several hundred known mammalian miRNAs are better able to distinguish developmental lineage, differentiation state, and cancer state, than the best corresponding predictors based on genome-wide mRNA expression levels from the same cells. miRNA expression profiling may therefore be a valuable tool for the classification of poorly differentiated tumors.

The analysis in Lu et al. (2005) includes a comparison of miRNA expression measures between cancerous and non-cancerous tissues, using a permutation-based version of FWER-controlling marginal single-step Bonferroni Procedure 3.1, with modified two-sample t-statistics. For a test at nominal FWER level $\alpha = 0.05$, the authors reported that 59% of the miRNAs were significantly under-expressed in cancerous compared to non-cancerous tissues. Only a few miRNAs were over-expressed in cancerous tissues and none significantly so. Furthermore, miRNA expression measures were observed to vary greatly among the 19 different tissue types represented in the dataset (e.g., colon, pancreas, stomach). Tissue type should therefore be regarded as a confounding variable.

Motivated by the findings in Lu et al. (2005), we have undertaken further analyses of this publicly available miRNA dataset. We first consider the identification of differentially expressed miRNAs between cancerous and non-cancerous tissues. Our approach is based on tests for regression coefficients in (non-linear) logistic models relating cancer status to miRNA expression measures, while adjusting for the confounding tissue type variable. The second analysis concerns the identification of co-expressed miRNAs, i.e., pairs of miRNAs with correlated expression measures. Both testing problems are addressed using FWER-controlling joint single-step maxT Procedure 3.5, based on the general non-parametric bootstrap null shift and scale-transformed test statistics null distribution summarized in Procedure 2.3. This method identifies 90 (58% of the 155 studied) significantly differentially expressed miRNAs, as well as hundreds of significantly co-expressed pairs of miRNAs.

The reader is referred to Chapter 8 for a comparison of different test statistics null distributions in testing problems concerning correlation coefficients and regression coefficients (Pollard et al., 2005a).

9.3.1 Cancer miRNA dataset

Lu et al. (2005) measured expression levels for 217 known human miRNAs, by a bead-based flow cytometric profiling method, in cells from 46 cancerous and 140 non-cancerous tissues ($n = 186$ target samples in total). The pre-processed, $\log_2$-transformed data are available from the authors' website (`www.broad.mit.edu/cancer/pub/miGCM`: miRNA expression measures in file `miGCM_218.gct`; probe sequence information in file `supplementary_table_1.xls`; target sample information, such as cancer status and tissue type, in file `supplementary_table_2.xls`).

The analyses described below exclude cell lines and any miRNA with expression measures below a detection threshold of $\log_2 32 = 5$ in more than half of the $n = 186$ target samples.

The data for each of the $n = 186$ target samples consist of a binary *outcome/phenotype* Y_i for cancer status (1 for cancerous vs. 0 for non-cancerous tissues), a J-dimensional *covariate/genotype* vector $X_i = (X_i(j) : j = 1, \ldots, J)$ of real-valued expression measures for each of $J = 155$ miRNAs, and a 19-dimensional tissue type indicator vector W_i, $i = 1, \ldots, n$.

9.3.2 Multiple testing procedures

Differentially expressed miRNAs between cancerous and non-cancerous tissues: Tests for logistic regression coefficients

Our first analysis of the Lu et al. (2005) dataset concerns the identification of *differentially expressed* miRNAs between cancerous and non-cancerous tissues.

It involves fitting, for each miRNA, a *logistic regression model* relating cancer status Y to expression measure $X(j)$ and tissue type W. Specifically, the logistic regression model for the jth miRNA is

$$\text{logit}(\text{E}[Y|X(j), W]) \equiv \alpha(j) + \beta(j)X(j) + \gamma(j)W, \qquad j = 1, \ldots, J, \quad (9.3)$$

where $\text{logit}(z) \equiv \log(z/(1-z))$ is the *logit function*, $\alpha(j)$ a baseline effect parameter, $\beta(j)$ a main effect parameter for the expression measure $X(j)$ of the jth miRNA, and $\gamma(j)$ a miRNA-specific 19-dimensional parameter vector adjusting for tissue type W.

The parameter of interest for the identification of differentially expressed miRNAs is the J-vector $\beta = (\beta(j) : j = 1, \ldots, J)$ of *logistic regression coefficients* for the expression measures $X(j)$ of each miRNA, $j = 1, \ldots, J$. Thus, one considers the two-sided tests of the J null hypotheses $H_0(j) = \text{I}(\beta(j) = 0)$, of no association between the expression measures $X(j)$ and cancer status Y, vs. the alternative hypotheses $H_1(j) = \text{I}(\beta(j) \neq 0)$. Two-sided tests are used to identify both over- and under-expressed miRNAs in cancerous tissues.

The J null hypotheses are tested based on the following *t-statistics*,

$$T_n(j) \equiv \frac{\beta_n(j) - \beta_0(j)}{\sigma_n(j)}, \qquad j = 1, \ldots, J, \qquad (9.4)$$

where the null values $\beta_0(j)$ are zero and $\beta_n(j)$ are logistic regression parameter estimators with estimated standard errors $\sigma_n(j)$ (as implemented in the function `glm` from the R package `stats`, with the call `glm(Y ~ X(j) + W, family="binomial")`, using the binomial family and iteratively reweighted least squares (IWLS)).

In order to simultaneously test the $J = 155$ null hypotheses of no association between miRNA expression measures and cancer status, we apply *FWER-controlling joint single-step maxT Procedure 3.5*, with the general *non-parametric bootstrap null shift and scale-transformed test statistics null distribution* of Procedure 2.3 ($B = 5{,}000$ samples). Note that fitting the logistic regression model of Equation (9.3) allows the identification of differentially expressed miRNAs, while adjusting for the confounding variable tissue type.

Co-expressed miRNAs: Tests for correlation coefficients

A biological question of great interest in gene expression analysis is the identification of *co-expressed* miRNAs, i.e., pairs of miRNAs with correlated expression measures. Although some tests of association between expression measures and a binary outcome, such as cancer status, could be performed with standard multiple testing methods (e.g., MTPs based on a permutation data generating null distribution), correlation tests are a problem for which the general test statistics null distributions of Sections 2.3 and 2.4, and corresponding bootstrap estimators (Procedures 2.3 and 2.4), truly allow one to perform previously unavailable analyses.

Consider the $M \equiv J(J-1)/2 = 155 \times 154/2 = 11,935$ *correlation coefficients* for the expression measures of distinct pairs (j, j') of miRNAs,

$$\rho(j, j') \equiv \mathrm{Cor}[X(j), X(j')], \qquad j = 1, \ldots, J-1, \; j' = j+1, \ldots, J. \quad (9.5)$$

In order to identify pairs of significantly co-expressed miRNAs, consider the two-sided tests of the null hypotheses $H_0(j, j') = \mathrm{I}(\rho(j, j') = 0)$, of no association in expression measures, vs. the alternative hypotheses $H_1(j, j') = \mathrm{I}(\rho(j, j') \neq 0)$.

The M null hypotheses are tested based on the following *difference statistics*,

$$T_n(j, j') \equiv \sqrt{n}(\rho_n(j, j') - \rho_0(j, j')), \qquad j = 1, \ldots, J-1, \; j' = j+1, \ldots, J, \quad (9.6)$$

where the null values $\rho_0(j, j')$ are zero and $\rho_n(j, j')$ are empirical correlation coefficients, as defined in Equation (8.15).

In order to simultaneously test the $M = 11,935$ null hypotheses of no association in expression measures for pairs of miRNAs, we again apply *FWER-controlling joint single-step maxT Procedure 3.5*, with the general *non-parametric bootstrap null shift and scale-transformed test statistics null distribution* of Procedure 2.3 ($B = 5,000$ samples).

9.3.3 Results

Differentially expressed miRNAs between cancerous and non-cancerous tissues: Tests for logistic regression coefficients

The multiple testing analysis for the logistic regression model of Equation (9.3) suggests that a large fraction of miRNAs are significantly differentially expressed between cancerous and non-cancerous tissues. Bootstrap-based single-step maxT Procedure 3.5 yields 90 (58% of the 155 studied) and 53 (34% of the 155 studied) miRNAs with adjusted p-values less than nominal FWER level $\alpha = 0.05$ and 0.01, respectively (Table 9.10; Figure 9.13, Panel (a)). All 90 miRNAs that are significantly differentially expressed at level $\alpha = 0.05$ have negative test statistics ($T_n(j) < -3.8$), suggesting *under-expression* in cancerous compared to non-cancerous tissues.

Our findings are in agreement with the original publication of Lu et al. (2005), the main distinctions being that single-step maxT Procedure 3.5 takes into account the *joint distribution* of the test statistics and the logistic regression model of Equation (9.3) allows *adjusting* for the tissue type *confounding variable* when comparing expression measures between cancerous and non-cancerous tissues.

Five of the highly significant miRNAs listed in Table 9.10 are located in minimal deleted regions, minimal amplified regions, and breakpoint regions involved in human cancers (Calin et al., 2004). Specifically, `hsa-let-7d` and hsa-miR-23b have been associated with urothelial cancer; hsa-miR-22 with hepatocellular cancer; hsa-miR-99a with lung cancer; hsa-miR-100 with breast, cervical, lung, and ovarian cancers.

It would be of interest, as a follow-up analysis, to examine the target sequences of the differentially expressed miRNAs for the potential identification of common motifs.

Note that another approach for comparing mean miRNA expression measures in cancerous vs. non-cancerous tissues could be based on standard two-sample t-statistics. For such simple tests, data generating null distributions, such as permutation distributions, lead to proper Type I error control, provided (i) the two populations have the same covariance matrices or (ii) the two sample sizes are equal (Section 2.9; Pollard and van der Laan (2004)). Our multiple testing methodology, however, allows one to use more general and flexible models, such as the logistic regression model of Equation (9.3), which facilitates adjustment for covariates and also provides a simple predictor of cancer status.

Co-expressed miRNAs: Tests for correlation coefficients

The multiple testing analysis suggests that a large fraction of miRNA pairs have significantly correlated expression measures. Interestingly, bootstrap-based single-step maxT Procedure 3.5 yields 8,916 miRNA pairs (75% of all $M = 11,935$ pairs) with adjusted p-values less than nominal FWER level $\alpha = 0.05$ and 7,479 miRNA pairs (63% of all $M = 11,935$ pairs) with adjusted p-values approximately equal to zero (Table 9.11; Figure 9.13, Panel (b)). Correlation coefficients found to be significantly different from zero at nominal FWER level $\alpha = 0.05$ range from 0.26 to 0.99, with median value 0.55. Only 8% of all pairwise correlation coefficients are negative and none significantly so.

The 20 most significantly correlated pairs of miRNAs are listed in Table 9.11. Several of the identified pairs are composed of miRNAs in the same family (e.g., hsa-miR-10a and hsa-miR-10b). The two most significantly correlated miRNAs are a pair of paralogs, hsa-miR-17-5p (chromosome 17) and hsa-miR-106a (chromosome X), which belong to miRNA clusters believed to be up-regulated by the proto-oncogene c-MYC (O'Donnell et al., 2005). hsa-miR-19a, hsa-miR-19b, and hsa-miR-20 are also members of these paralogous miRNA clusters. Several other co-expressed miRNAs have been linked to cancer. In particular, hsa-miR-107 has been shown to increase cell growth in lung carcinomas (Cheng et al., 2005). hsa-miR-143 and hsa-miR-145, located within 1.7 kb on human chromosome 5, were found to be expressed at lower levels in cancerous and pre-cancerous tissues compared to normal colon tissues (Michael et al., 2003).

The fact that a majority of miRNA pairs are identified as being significantly co-expressed, even after adjusting for multiple tests (nominal FWER level 0.05), suggests a great deal of structure in miRNA expression. In order to focus on strongly and positively co-expressed pairs of miRNAs, one could perform one-sided tests of the null hypotheses $H_0(j,j') = \mathrm{I}(\rho(j,j') \leq \rho_0(j,j'))$ vs. the alternative hypotheses $H_1(j,j') = \mathrm{I}(\rho(j,j') > \rho_0(j,j'))$, with null values $\rho_0(j,j')$ greater than zero, e.g., $\rho_0(j,j') = 0.5$.

Co-expressed miRNAs: Cluster analysis

The above multiple testing analysis clearly suggests the existence of clusters of miRNAs with highly correlated expression measures. This leads us to perform a cluster analysis of the miRNA expression measures, in order to identify general expression patterns and groups of co-expressed miRNAs. We use the *hierarchical ordered partitioning and collapsing hybrid* (HOPACH) algorithm, with Pearson correlation distance (van der Laan and Pollard, 2003). HOPACH is implemented in the Bioconductor R package hopach (Pollard and van der Laan, 2005).

Figure 9.14 displays a pseudo-color image of the 155×155 matrix of pairwise correlation coefficients for the expression profiles of the $J = 155$ miRNAs, with rows and columns ordered according to the final level of the HOPACH tree. Similarly expressed miRNAs appear near each other and are visualized as blocks in the pseudo-color image. It would be of interest to investigate the biological and medical implications of the identified clusters of co-expressed miRNAs.

Summary

The analysis of this new miRNA dataset illustrates the flexibility and power of our proposed multiple testing methodology. Stepwise, augmentation, and empirical Bayes procedures could be used for more powerful analyses and control of a broader class of Type I error rates (Chapters 1–7). The large number of significant findings for both the differential expression and co-expression testing problems is likely linked to the reasonably large sample size ($n = 186$ target samples) relative to the number of tests ($M = 155$ regression coefficients and $M = 11,935$ correlation coefficients), as compared to similar studies of mRNA expression. This analysis of a rich dataset, using novel and rigorous statistical methods, highlights the potential for meaningful biological and medical discovery from high-throughput gene expression studies.

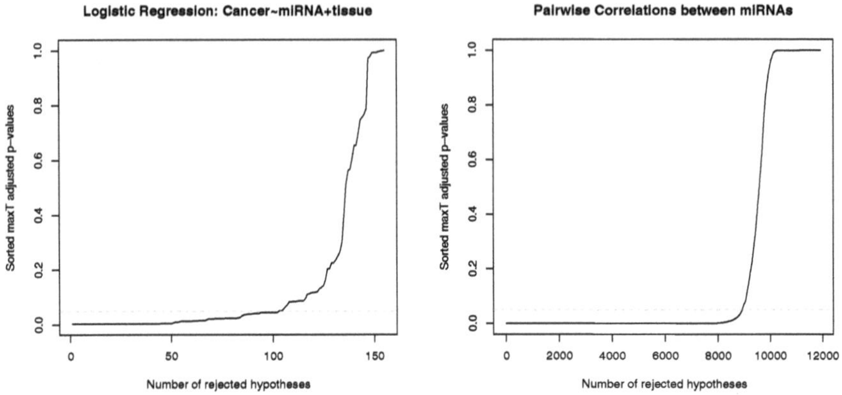

Panel (a): Differential expression Panel (b): Co-expression

Figure 9.13. *Cancer miRNA dataset, differential expression and co-expression: Single-step maxT adjusted p-values for tests for logistic regression coefficients and correlation coefficients.* Plots of sorted adjusted p-values for single-step maxT Procedure 3.5, based on the non-parametric bootstrap null shift and scale-transformed test statistics null distribution of Procedure 2.3 ($B = 5,000$ samples). Panel (a): Identification of differentially expressed miRNAs between cancerous and non-cancerous tissues, based on tests for logistic regression coefficients. Panel (b): Identification of pairs of co-expressed miRNAs, based on tests for correlation coefficients. The dashed horizontal line indicates the adjusted p-value cut-off for a test at nominal FWER level $\alpha = 0.05$.

Table 9.10. *Cancer miRNA dataset, differential expression: Tests for logistic regression coefficients.* The table reports the names, target sequences, adjusted p-values, and test statistics, for the 53 most significantly differentially expressed miRNAs between cancerous and non-cancerous tissues, according to FWER-controlling bootstrap-based single-step maxT Procedure 3.5. Adjusted p-values are computed under the non-parametric bootstrap null shift and scale-transformed test statistics null distribution of Procedure 2.3, based on $B = 5,000$ samples. miRNAs are sorted in decreasing order of their absolute test statistics $T_n(j)$. All 53 miRNAs have adjusted p-values less than 0.01 and negative test statistics, suggesting under-expression in cancerous compared to non-cancerous tissues. The target sequence is the reverse complement of the miRNA sequence and identifies potential binding sites for the miRNA.

Name	miRNA target sequence	Adjusted p-value	Test statistic
hsa-miR-98	UGAGGUAGUAAGUUGUAUUGUU	0.0038	-4.88
hsa-miR-28	AAGGAGCUCACAGUCUAUUGAG	0.0038	-4.79
hsa-miR-196	UAGGUAGUUUCAUGUUGUUGG	0.0038	-4.79
hsa-miR-30a	CUUUCAGUCGGAUGUUUGCAGC	0.0038	-4.78
hsa-miR-30e	UGUAAACAUCCUUGACUGGA	0.0038	-4.78
hsa-miR-99a#	AACCCGUAGAUCCGAUCUUGUG	0.0038	-4.77
hsa-miR-335	UCAAGAGCAAUAACGAAAAAUGU	0.0038	-4.72
hsa-let-7e	UGAGGUAGGAGGUUGUAUAGU	0.0038	-4.69
hsa-miR-23b#	AUCACAUUGCCAGGGAUUACCAC	0.0038	-4.67
hsa-miR-214	ACAGCAGGCACAGACAGGCAG	0.0038	-4.67
hsa-miR-99b	CACCCGUAGAACCGACCUUGCG	0.0038	-4.67
hsa-miR-30c	UGUAAACAUCCUACACUCUCAGC	0.0038	-4.66
hsa-miR-30b	UGUAAACAUCCUACACUCAGC	0.0038	-4.66
hsa-miR-338	UCCAGCAUCAGUGAUUUUGUUGA	0.0038	-4.65
hsa-miR-103	AGCAGCAUUGUACAGGGCUAUGA	0.0038	-4.64
hsa-miR-185	UGGAGAGAAAGGCAGUUC	0.0038	-4.63
hsa-miR-151*	UCGAGGAGCUCACAGUCUAGUA	0.0038	-4.62
hsa-miR-100#	AACCCGUAGAUCCGAACUUGUG	0.0038	-4.61
hsa-miR-20_(sub_1)	UAAAGUGCUUAUAGUGCAGGUAG	0.0038	-4.61
hsa-miR-129*	AAGCCCUUACCCCAAAAAGCAU	0.0038	-4.60
hsa-miR-22#	AAGCUGCCAGUUGAAGAACUGU	0.0038	-4.60
hsa-let-7d#	AGAGGUAGUAGGUUGCAUAGU	0.0038	-4.58
hsa-miR-107	AGCAGCAUUGUACAGGGCUAUCA	0.0038	-4.58
rno-miR-352	AGAGUAGUAGGUUGCAUAGUA	0.0038	-4.58
hsa-miR-197	UUCACCACCUUCUCCACCCAGC	0.0038	-4.57
hsa-miR-32	UAUUGCACAUUACUAAGUUGC	0.0038	-4.57
hsa-miR-342	UCUCACACAGAAAUCGCACCCGUC	0.0038	-4.56
hsa-miR-324-5p	CGCAUCCCCUAGGGCAUUGGUGU	0.0038	-4.51
hsa-miR-128b	UCACAGUGAACCGGUCUCUUUC	0.0038	-4.51
hsa-miR-126*	CAUUAUUACUUUUGGUACGCG	0.0038	-4.50
hsa-miR-19b	UGUGCAAAUCCAUGCAAAACUGA	0.0038	-4.49
hsa-miR-151_(sub_1)	ACUAGACUGAAGCUCCUUGAGG	0.0038	-4.49
hsa-miR-199a*	UACAGUAGUCUGCACAUUGGUU	0.0038	-4.48
hsa-let-7i	UGAGGUAGUAGUUUGUGCU	0.0038	-4.48
hsa-miR-10b	UACCCUGUAGAACCGAAUUUGU	0.0038	-4.47
miR-292-3p	AAGUGCCGCCAGGUUUUGAGUGU	0.0040	-4.46
hsa-miR-136	ACUCCAUUUGUUUUGAUGAUGGA	0.0042	-4.45
mmu-miR-10b	CCCUGUAGAACCGAAUUUGUGU	0.0042	-4.45
hsa-let-7f	UGAGGUAGUAGAUUGUAUAGUU	0.0042	-4.44
hsa-miR-302	UAAGUGCUUCCAUGUUUUGGUGA	0.0042	-4.43
mmu-let-7g	UGAGGUAGUAGUUUGUACAGU	0.0042	-4.43

Continued on next page ...

... continued from previous page

Name	**miRNA target sequence**	**Adjusted** ***p*****-value**	**Test statistic**
hsa-miR-10a	UACCCUGUAGAUCCGAAUUUGUG	0.0042	-4.42
hsa-miR-34b	AGGCAGUGUCAUUAGCUGAUUG	0.0042	-4.42
hsa-miR-92	UAUUGCACUUGUCCCGGCCUGU	0.0042	-4.42
hsa-miR-101	UACAGUACUGUGAUAACUGAAG	0.0044	-4.38
hsa-miR-16	UAGCAGCACGUAAAUAUUGGCG	0.0046	-4.37
mmu-miR-339	UCCCUGUCCUCCAGGAGCUCA	0.0046	-4.37
hsa-miR-19a	UGUGCAAAUCUAUGCAAAACUGA	0.0046	-4.37
hsa-miR-152	UCAGUGCAUGACAGAACUUGG	0.0052	-4.35
hsa-miR-23a	AUCACAUUGCCAGGGAUUUCC	0.0052	-4.34
hsa-miR-186	CAAAGAAUUCUCCUUUUGGGCUU	0.0072	-4.30
rno-miR-343	UCUCCCUCCGUGUGCCCAGU	0.0096	-4.29
hsa-miR-140	AGUGGUUUUACCCUAUGGUAG	0.0096	-4.28

[#]Located in minimal deleted regions, minimal amplified regions, and breakpoint regions involved in human cancers (Calin et al., 2004).

Table 9.11. *Cancer miRNA dataset, co-expression: Tests for correlation coefficients.* The table reports the names and correlation coefficients for the 20 most significantly co-expressed pairs of miRNAs, according to FWER-controlling bootstrap-based single-step maxT Procedure 3.5. Adjusted p-values are computed under the non-parametric bootstrap null shift and scale-transformed test statistics null distribution of Procedure 2.3, based on $B = 5,000$ samples. miRNA pairs are sorted in decreasing order of their absolute correlation coefficients $\rho_n(j, j')$. All 20 miRNA pairs have adjusted p-values approximately equal to zero and positive correlation coefficients.

Names		Correlation coefficient
hsa-miR-106a#	hsa-miR-17-5p#	0.99
mmu-miR-200b	hsa-miR-200b	0.99
mmu-miR-200b	hsa-miR-200c	0.99
hsa-miR-107†	hsa-miR-103	0.99
hsa-miR-200b	hsa-miR-200c	0.99
hsa-miR-145‡	hsa-miR-143‡	0.98
hsa-miR-199a_(sub_1)	mmu-miR-199b	0.98
hsa-miR-17-5p	hsa-miR-20_(sub_1)	0.97
hsa-miR-19a#	hsa-miR-19b#	0.97
hsa-miR-29a	hsa-miR-30a*	0.97
hsa-miR-181a	hsa-miR-181c	0.97
hsa-miR-199a_(sub_1)	hsa-miR-199a*	0.97
hsa-miR-29b_(sub_2)	hsa-miR-29c	0.97
hsa-miR-199a*	mmu-miR-199b	0.96
hsa-miR-200a	hsa-miR-141	0.96
hsa-miR-20_(sub_1)#	mmu-miR-106a	0.96
hsa-miR-106a	hsa-miR-20_(sub_1)#	0.96
hsa-miR-200a	hsa-miR-200a	0.96
hsa-miR-23b	hsa-miR-23a	0.96
hsa-miR-10a	hsa-miR-10b	0.96

Several pairs are composed of miRNAs in the same family (e.g., `hsa-miR-10a` and `hsa-miR-10b`).

Up-regulated by the proto-oncogene c-MYC (O'Donnell et al., 2005).

† Increases cell growth in lung carcinomas (Cheng et al., 2005).

‡ Expressed at lower levels in cancerous and pre-cancerous tissues compared to normal colon tissues (Michael et al., 2003).

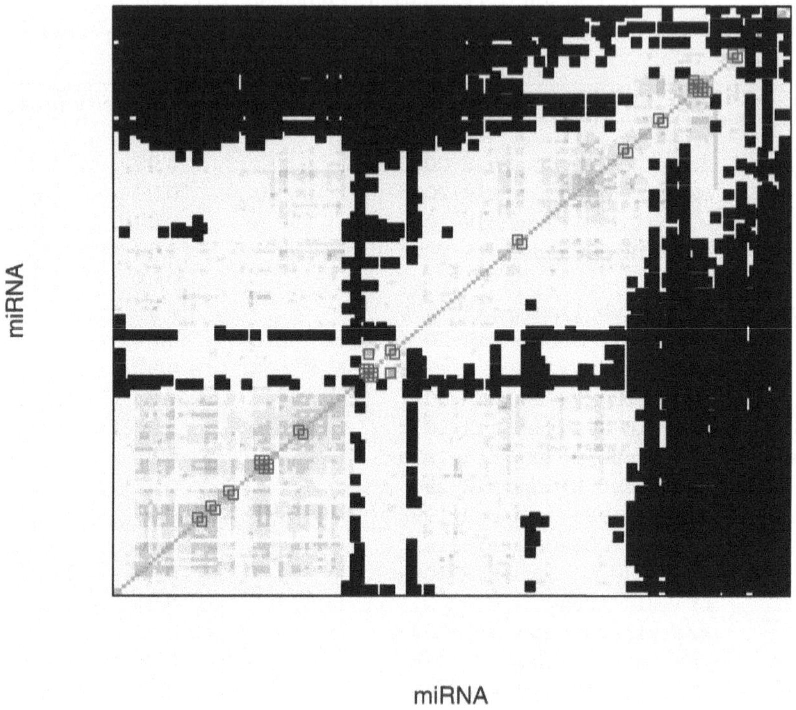

Figure 9.14. *Cancer miRNA dataset, co-expression: HOPACH clustering of miRNA expression profiles.* The figure provides a pseudo-color image of the 155×155 matrix of pairwise correlation coefficients for the expression profiles of the $J = 155$ miRNAs. Rows and columns are ordered according to the final level of the hierarchical tree of miRNA clusters produced by the HOPACH algorithm with Pearson correlation distance. Pairwise correlation coefficients not significantly different from zero are displayed in black. The remaining correlation coefficients are represented using a white (anti-correlated) to red (positively-correlated) color palette. Groups of co-expressed miRNAs appear as red blocks along the diagonal of the correlation matrix. The 20 most significantly correlated pairs of miRNAs from Table 9.11 are highlighted in blue. (Color plate p. 335)

10

Multiple Tests of Association with Biological Annotation Metadata

10.1 Introduction

10.1.1 Motivation

Experimental data, such as microarray gene expression measures, gain much in relevance from their association with *biological annotation metadata*, i.e., data on data, such as, GenBank sequences, Gene Ontology terms, KEGG pathways, and PubMed abstracts. A challenging and fascinating area of research for statisticians concerns the development of methods for relating experimental data to the wealth of metadata available publicly on the WWW. Tasks include accessing and pre-processing the data, making inference from these data, and summarizing and interpreting the results.

In this context, an important class of statistical problems involves testing for associations between known fixed features of a genome and unknown parameters of the distribution of variable features of this genome in a population of interest. Here, features of a genome are said to be *fixed*, if they remain constant among population units. In contrast, *variable* features are allowed to differ among population units. Fixed features typically consist of gene annotation metadata, that reflect current knowledge on gene properties, such as, nucleotide and protein sequences, regulation, and function. Variable features often consist of gene expression measures, that reflect cellular type/state/activity under particular conditions. The fixed and variable features define, respectively, *gene-annotation profiles* and *gene-parameter profiles*; the parameter of interest then corresponds to *measures of association between known gene-annotation profiles and unknown gene-parameter profiles.*

For instance, for the yeast *Saccharomyces cerevisiae* (in short, *S. cerevisiae*), one may be interested in detecting associations between the vector of mean transcript (i.e., mRNA) levels for all (approximately 6,500) genes under heat-shock conditions and *Gene Ontology* (GO) annotation for these genes. The reader is referred to the Gene Ontology Consortium website (`www.geneontology.org`) and to Section 10.3, below, for more information on gene ontologies, and to the Saccharomyces Genome Database (SGD) website (`www.yeastgenome.org`), for details on *S. cerevisiae*. In this example, the population of interest may consist of all heat-shocked yeast cells from well-defined cultures of a particular strain of *S. cerevisiae* (e.g., strain S288C). For each of the three gene ontologies (BP, CC, and MF, as described in Section 10.3.1), each gene is annotated with a fixed set of GO terms (i.e., this set is constant across population units for a given version of the GO Database). Thus, for a given GO term, one may define a gene-annotation profile as a known, fixed binary vector indicating for each gene whether it is annotated or not with the particular GO term. The transcript levels, however, vary among population units and the gene-parameter profile, i.e., the vector of genome-wide mean transcript levels in the population of heat-shocked yeast cells, is unknown and may be estimated, for example, from a microarray experiment involving a sample of yeast cells from the population. The association parameter of interest, between GO annotation and transcript levels, is then a vector of association measures (e.g., two-sample t-statistics) between the known binary gene-annotation profiles and the unknown continuous gene-parameter profile.

Similar inference questions arise in many other contexts and involve a variety of definitions for the gene-annotation profiles, the gene-parameter profiles, and the association parameters of interest. For example, in cancer microarray studies, one may seek associations between GO gene-annotation profiles and a gene-parameter profile of regression coefficients relating (censored) patient survival data to genome-wide transcript levels or DNA copy numbers. Furthermore, gene-annotation profiles need not be binary or even polychotomous, and may correspond to pathway membership, regulation by particular transcription factors, nucleotide sequences, and protein sequences.

Note that, for the sake of illustration, we focus on gene-level features. However, our proposed methodology is generic and may be applied to other types of features, such as those concerning gene isoforms and proteins. For instance, as in alternative splicing microarray analysis, one may collect data at the finer level of gene isoforms, where one gene may have multiple isoforms (Blanchette et al., 2005). In this context, *isoform-parameter profiles* may refer to the distribution of isoform microarray expression measures in a well-defined population, while *isoform-annotation profiles* may consist of intron/exon counts/lengths/nucleotide distributions. One may also consider protein-level features, where, for example, *protein-parameter profiles* correspond to antibody microarray expression measures and *protein-annotation profiles* refer to protein function, domain structure, and post-translational modification (e.g., Swiss-Prot, `www.expasy.org/sprot`).

10.1.2 Contrast with other approaches

Existing approaches for tests of association with biological annotation metadata focus primarily on relating microarray gene expression measures and GO annotation. Relevant articles and software packages include:

FatiGO from the BABELOMICS suite (Al-Shahrour et al. (2004, 2005); www.babelomics.org); GOstat (Beissbarth and Speed (2004); gostat.wehi.edu.au); Ontologizer (Grossmann et al. (2006); www.charite.de/ch/medgen/ontologizer); CSEPCT (McCarroll et al. (2004); genome3.ucsf.edu:8080/cgi-bin/ compareExp.cgi); GSEA-P (Mootha et al. (2003), Subramanian et al. (2005); www.broad.mit.edu/gsea/doc/doc_index.html); Tian et al. (2005).

Methods proposed thus far suffer from a number of limitations, related, to a large extent, to the absence of a clear and precise statement of the statistical inference question. As a result, the analyses often lack statistical rigor and tend to be ad hoc and dataset-specific.

One of our main contributions is the systematic and precise translation of a general class of biological questions into a corresponding class of multiple hypothesis testing problems. A key step in this process is the proper definition of the gene-annotation profiles, gene-parameter profiles, and association parameters of interest. This general formulation then allows us to apply the multiple testing methodology developed in Chapters 1–7, to control a broad class of Type I error rates, defined as generalized tail probabilities (gTP), $gTP(q, g) = \Pr(g(V_n, R_n) > q)$, for arbitrary functions $g(V_n, R_n)$ of the numbers of false positives V_n and rejected hypotheses R_n.

We wish to emphasize the crucial and often ignored distinction between: (i) *defining a parameter* of interest, measuring the association between gene-annotation and gene-parameter profiles, i.e., the statistical formulation of the biological question; (ii) *making inferences*, i.e., *estimating* and *testing hypotheses* concerning this parameter, based on a sample drawn from the population under consideration. Most methods proposed to date focus on (ii), without providing a clear statement of the question being answered in (i), that is, various estimation and testing procedures are proposed for an undefined parameter of interest.

Due to its general and rigorous statistical framework, our approach to multiple tests of association with biological annotation metadata differs in a number of important ways from current approaches, such as those developed for inference with Gene Ontology metadata and implemented in the software packages listed on the Gene Ontology Tools webpage (www.geneontology.org/GO.tools.shtml).

General gene-annotation profiles. Existing approaches typically consider binary gene-annotation profiles, e.g., vectors of indicators of GO term annotation. Our general definition of gene-annotation profiles allows consideration of arbitrary qualitative and quantitative fixed features of a genome, e.g., membership of genes to any number of pathways or clusters, intron/exon counts/lengths/nucleotide distributions, mean transcript levels.

General gene-parameter profiles. Existing approaches typically consider binary gene-parameter profiles, e.g., vectors of indicators of differential expression. Our general definition of gene-parameter profiles allows consideration of a much broader class of testing problems, concerning arbitrary qualitative and quantitative parameters, such as differences in mean expression levels or regression coefficients relating expression levels to clinical outcomes.

Estimated gene-parameter profiles. Existing approaches typically assume known gene-parameter profiles. For example, the list of differentially expressed genes from a microarray experiment is usually treated as known and fixed in subsequent analyses with GO, while in fact it corresponds to an unknown and estimated parameter. Distinguishing between the definition of a parameter and inference concerning this parameter, as in Section 10.2, provides a more rigorous and general formulation of the statistical question.

General tests of association. Common approaches to tests of association with GO annotation are typically limited to tests of independence in 2×2 contingency tables (e.g., based on the hypergeometric distribution, Fisher's exact test). As in Table 10.1, rows correspond to gene annotation with a given GO term (fixed binary gene-annotation profile) and columns to a gene property of interest, such as differential expression (treated as a fixed binary gene-parameter profile). The approach proposed in Section 10.2 allows consideration of a broader class of biological testing problems, while properly accounting for the fact that gene-parameter profiles are usually unknown and replaced by a random (i.e., data-driven) estimator.

10.1.3 Outline

This chapter proposes a general and formal statistical framework for multiple tests of association with biological annotation metadata, using the multiple testing methodology described in Chapters 1–7.

Section 10.2 presents the proposed statistical framework for multiple tests of association with biological annotation metadata and discusses in detail the main components of the inference problem, namely, the gene-annotation profiles, the gene-parameter profiles, and the association parameters. Multiple testing procedures (MTP) for tests of association between gene-annotation profiles and gene-parameter profiles are outlined. Section 10.3 gives an overview of the Gene Ontology (GO) and R software for accessing and analyzing GO annotation metadata (e.g., for assembling GO gene-annotation profiles). The proposed statistical and computational methods are illustrated in Section 10.4, using the acute lymphoblastic leukemia (ALL) microarray dataset of Chiaretti et al. (2004), with the aim of relating GO annotation to differential gene expression between B-cell ALL with the BCR/ABL fusion and cytogenetically normal NEG B-cell ALL. Finally, Section 10.5 summarizes our findings and outlines ongoing work.

10.2 Statistical framework for multiple tests of association with biological annotation metadata

Sections 10.2.1–10.2.3 introduce the main components of our approach to multiple tests of association with biological annotation metadata, namely, the gene-annotation profiles A, the gene-parameter profiles λ, and the association measures $\psi = \rho(A, \lambda)$ between gene-annotation and gene-parameter profiles. We stress that the choice of a suitable association parameter ψ is perhaps the most important and hardest aspect of the inference problem, as this parameter represents the statistical translation of the biological question of interest. Once the association parameter ψ is appropriately and precisely defined, one can rely on a variety of statistical methods to estimate and test hypotheses concerning this parameter. Section 10.2.4 describes how the multiple testing methodology of Chapters 1–7 may be used to detect associations between gene-annotation and gene-parameter profiles.

Note that, for the sake of illustration, we focus on gene-level features. However, as mentioned in Section 10.1.1, the methodology is generic and may be applied to other types of features, such as those concerning gene isoforms and proteins.

10.2.1 Gene-annotation profiles

Gene-annotation profiles refer to features of a genome that are assumed to be known and constant among units in a population of interest. Such features typically consist of gene annotation metadata, that reflect current knowledge on gene properties, such as, nucleotide and protein sequences, regulation, and function.

Specifically, let $A = (A(g, m) : g = 1, \ldots, G; m = 1, \ldots, M)$ denote a $G \times M$ *gene-annotation matrix*, providing data on M features for G genes in an organism of interest. Thus, row $A(g, \cdot) = (A(g, m) : m = 1, \ldots, M)$ denotes an M-dimensional gene-specific feature vector for the gth gene, $g = 1, \ldots, G$, and column $A(\cdot, m) = (A(g, m) : g = 1, \ldots, G)$ denotes a G-dimensional *gene-annotation profile* for the mth feature, $m = 1, \ldots, M$.

In many applications, the element $A(g, m)$ is a binary indicator, coding the YES/NO answer to the mth question, among a collection of M questions one may ask about gene g. For example, $A(g, m)$ could indicate whether gene g is annotated with a particular GO term m, among M terms in one of the three ontologies (BP, CC, or MF), i.e., whether gene g is an element of the node corresponding to the mth term in the GO directed acyclic graph (DAG). Other gene-annotation profiles of interest may refer to intron/exon counts/lengths/nucleotide distributions, gene pathway membership (e.g., from the Kyoto Encyclopedia of Genes and Genomes, KEGG, `www.genome.ad.jp/kegg`), or gene regulation by particular transcription factors. Regarding transcription regulation, one could use data from the Transcription Factor DataBase (TRANSFAC, `www.gene-regulation.com`) to generate gene-annotation profiles as follows. For a given transcription factor binding motif, a binary gene-annotation profile could consist of indicators for the presence or absence of the motif in the upstream control region of each gene. A continuous gene-annotation profile could be based on the position weight matrix of the binding motif.

Note that the aforementioned features are only *fixed in time* for a given version/release of the corresponding database(s), i.e., such biological data are constantly evolving as our knowledge of the roles of genes and proteins is accumulating and changing. The dynamic nature of biological annotation metadata is an important issue in terms of software design (Section 10.3.2; Gentleman et al. (2005b)).

Note also that the gene-annotation profiles are not restricted to be binary or even polychotomous and, in particular, could be continuous gene-parameter profiles, suitably estimated from previous studies.

The main point, regarding the formulation of the statistical inference question, is that gene-annotation profiles are *known* and *constant among population units*.

10.2.2 Gene-parameter profiles

Gene-parameter profiles are generally unknown and concern the distribution of variable features of a genome in a well-defined population. Gene-specific variables of interest reflect cellular type/state/activity under particular conditions and include microarray measures of transcript levels and comparative genomic hybridization (CGH) measures of DNA copy numbers.

Specifically, let $X = (X(j) : j = 1, \ldots, J)$ be a J-dimensional random vector, containing G *gene-specific random variables* $(X(g) : g = 1, \ldots, G)$. In addition to the G gene-specific variables, X may include various biological and clinical covariates (e.g., age, sex, treatment, timepoint) and outcomes (e.g., survival time, response to treatment, tumor class). Let P denote the data generating distribution for the random J-vector X and suppose that P belongs to a (possibly non-parametric) model $\mathcal{M}$.

Let the parameter mapping $\Lambda : \mathcal{M} \rightarrow \mathbb{R}^G$ define a G-dimensional *gene-parameter profile*, $\Lambda(P) = \lambda = (\lambda(g) : g = 1, \ldots, G) \in \mathbb{R}^G$, where each $\lambda(g) = \Lambda(P)(g) \in \mathbb{R}$ is a gene-specific real-valued parameter. For example, $\lambda(g)$ could be the mean expression measure $\mathrm{E}[X(g)]$ of gene g or a regression coefficient in a model relating an outcome component of X to the expression measure $X(g)$ of gene g, $g = 1, \ldots, G$.

While gene-annotation profiles are known and fixed, gene-parameter profiles are typically *unknown* and need to be *estimated*, e.g., from a microarray experiment involving a sample of population units. The sample $\mathcal{X}_n = \{X_i : i = 1, \ldots, n\}$ is assumed to consist of n independent and identically distributed (IID) copies of $X \sim P$, corresponding to n randomly sampled population units.

10.2.3 Association measures for gene-annotation and gene-parameter profiles

Let the parameter mapping $\Psi : \mathcal{M} \rightarrow I\!R^M$ specify an M-dimensional *association parameter vector*,

$$\Psi(P) = \psi = (\psi(m) : m = 1, \ldots, M) \equiv \rho(A, \Lambda(P)), \tag{10.1}$$

defined in terms of an *association measure* $\rho : I\!R^{G \times M} \times I\!R^G \rightarrow I\!R^M$, *known fixed gene-annotation profiles* A, and an *unknown gene-parameter profile* $\lambda = \Lambda(P)$.

The choice of a suitable association parameter is subject matter-dependent and requires careful consideration. For instance, for Gene Ontology annotation, it is desirable that the association parameter reflect the *structure of the GO directed acyclic graph* (Section 10.3.1). In principle, the dimension of the association parameter vector ψ could differ from the number M of features under consideration. In addition, one could accommodate several gene-parameter profiles λ.

The various quantities in the inference problem are summarized in Figure 10.1; examples of association parameters are given next and in Section 10.4.

Univariate association measures

In the simplest case, one could define the M association parameters univariately, i.e., define $\psi(m)$ based only on the mth gene-annotation profile $A(\cdot, m)$, $m = 1, \ldots, M$. Specifically, for the mth feature, let

$$\Psi(P)(m) = \psi(m) \equiv \rho_m(A(\cdot, m), \Lambda(P)), \tag{10.2}$$

where $\rho_m : I\!R^G \times I\!R^G \rightarrow I\!R$ provides a measure of association (e.g., correlation coefficient) between the G-dimensional gene-annotation profile $A(\cdot, m)$ and gene-parameter profile $\lambda = \Lambda(P)$. In many situations, the same association measure ρ_m may be used for each of the M features.

Continuous gene-annotation profiles and continuous gene-parameter profiles

For continuous gene-annotation and gene-parameter profiles, one may use as association measure the Pearson *correlation coefficient* between two G-vectors. That is,

$$\psi(m) = \frac{\sum_{g=1}^{G} (A(g, m) - \bar{A}(m))(\lambda(g) - \bar{\lambda})}{\sqrt{\sum_{g=1}^{G} (A(g, m) - \bar{A}(m))^2} \sqrt{\sum_{g=1}^{G} (\lambda(g) - \bar{\lambda})^2}}, \tag{10.3}$$

where $\bar{A}(m) \equiv \sum_g A(g, m)/G$ and $\bar{\lambda} \equiv \sum_g \lambda(g)/G$ denote, respectively, the averages of the G elements of the gene-annotation profile $A(\cdot, m)$ and gene-parameter profile λ.

Binary gene-annotation profiles and binary gene-parameter profiles

For binary gene-annotation and gene-parameter profiles, one may build 2×2 contingency Table 10.1 and use as association measure the χ^2-*statistic* (or corresponding p-value) for the test of independence of rows and columns. That is,

$$\psi(m) = \frac{G(g_{00}(m)g_{11}(m) - g_{01}(m)g_{10}(m))^2}{g_{0\cdot}(m)g_{\cdot 0}(m)g_{\cdot 1}(m)g_{1\cdot}(m)}, \tag{10.4}$$

where $g_{kk'}(m) \equiv \sum_g \mathrm{I}(A(g,m) = k)\,\mathrm{I}(\lambda(g) = k')$, $g_{k\cdot}(m) \equiv g_{k0}(m) + g_{k1}(m) = \sum_g \mathrm{I}(A(g,m) = k)$, and $g_{\cdot k'}(m) \equiv g_{0k'}(m) + g_{1k'}(m) = \sum_g \mathrm{I}(\lambda(g) = k')$, $k, k' \in \{0,1\}$. Note that in this context *the* χ^2-*statistic* $\psi(m)$ *is a parameter*, i.e., it is a function of the data generating distribution P, via the gene-parameter profile $\lambda = \Lambda(P)$, and is therefore *unknown* and *constant* among population units.

Binary gene-annotation profiles

For binary gene-annotation profiles, one may consider association parameter vectors of the form

$$\psi = A^\top \lambda. \tag{10.5}$$

That is, the association parameter for the mth feature is the *sum*,

$$\psi(m) = \sum_{g=1}^{G} A(g,m)\lambda(g) = \sum_{g=1}^{G} \mathrm{I}\,(A(g,m) = 1)\,\lambda(g),$$

of the parameters $\lambda(g)$ for genes g that have the property of interest, i.e., such that $A(g,m) = 1$. Such an association parameter is considered by Tian et al. (2005), to relate continuous microarray differential expression gene-parameter profiles to binary pathway gene-annotation profiles.

The following standardized association parameters, corresponding to association measures based on *two-sample t-statistics*, may also be considered,

$$\psi(m) = \frac{\bar{\lambda}_1(m) - \bar{\lambda}_0(m)}{\sqrt{\frac{v[\lambda]_1(m)}{A_1(m)} + \frac{v[\lambda]_0(m)}{A_0(m)}}}, \tag{10.6}$$

where, for the mth feature, $A_k(m) \equiv \sum_g \mathrm{I}\,(A(g,m) = k)$, $\bar{\lambda}_k(m) \equiv \sum_g \mathrm{I}\,(A(g,m) = k)\,\lambda(g)/A_k(m)$, and $v[\lambda]_k(m) \equiv \sum_g \mathrm{I}\,(A(g,m) = k)\,(\lambda(g) - \bar{\lambda}_k(m))^2/(A_k(m) - 1)$ denote, respectively, the numbers, averages, and variances of annotated ($k = 1$) and unannotated ($k = 0$) gene-parameters $\lambda(g)$.

In commonly-encountered combined GO annotation and microarray data analyses, a binary gene-parameter profile could indicate whether genes are differentially expressed in two populations of cells, a continuous gene-parameter profile could consist of coefficients for the regression of a (censored) clinical outcome on gene expression measures, and binary gene-annotation profiles could denote whether genes are annotated with particular GO terms (Section 10.4; Al-Shahrour et al. (2004, 2005); Beissbarth and Speed (2004); Grossmann et al. (2006)).

Multivariate association measures

More generally, the mth association parameter could be based on the entire gene-annotation matrix A or a subset of columns thereof, that is, $\Psi(P)(m) = \psi(m) \equiv \rho_m(A, \Lambda(P))$, for an association measure $\rho_m : \mathbb{R}^{G \times M} \times \mathbb{R}^G \rightarrow \mathbb{R}$.

Association parameters of interest include: linear combinations of association parameters for several features, partial correlation coefficients, χ^2-statistics for higher-dimensional contingency tables (e.g., with one dimension corresponding to a gene-parameter profile λ and other dimensions to several gene-annotation profiles $A(\cdot, m)$), and (contrasts of) regression coefficients of a gene-parameter profile λ on several gene-annotation profiles $A(\cdot, m)$.

In the case of Gene Ontology annotation, the association parameter ψ should preferably reflect the structure of the GO directed acyclic graph, by taking into account, for instance, annotation information for ancestor (i.e., less specific) or offspring (i.e., more specific) terms (Section 10.3.1). Specifically, let $\mathcal{P}(m)$ denote the set of (immediate) parents of a term m. As the genes annotated by the child term m are subsets of the genes annotated by the parent terms $\mathcal{P}(m)$, then $A(g, m) = 1$ implies $A(g, p) = 1$ for $p \in \mathcal{P}(m)$.

Following the causal inference literature (van der Laan, 2006; van der Laan and Robins, 2003), an association parameter of interest for GO term m is the *marginal causal effect parameter*, defined as

$$\psi(m) = \mathrm{E}[\mathrm{E}[\lambda | A(\cdot, m) = 1, A(\cdot, \mathcal{P}(m))]] - \mathrm{E}[\mathrm{E}[\lambda | A(\cdot, m) = 0, A(\cdot, \mathcal{P}(m))]], \tag{10.7}$$

where $A(\cdot, \mathcal{P}(m))$ denotes the submatrix of gene-annotation profiles for parent terms $\mathcal{P}(m)$ and the expected values are defined with respect to the empirical distribution of $\{(A(g, m), A(g, \mathcal{P}(m)), \lambda(g)) : g = 1, \ldots, G\}$.

In the special case of binary (differential expression) gene-parameter profiles, the so-called parent-child method of Grossmann et al. (2006) takes into account the structure of the GO DAG by testing for associations between gene-annotation and gene-parameter profiles using hypergeometric p-values computed conditionally on the annotation status of parent terms.

One could also consider Boolean combinations of annotation indicators for multiple features, that is, a transformed gene-annotation matrix whose columns are Boolean combinations of the columns of the original gene-annotation matrix. Such an approach would be particularly relevant in the context of transcription regulation, where individual features correspond to single transcription factor binding motifs and Boolean combinations to binding modules for multiple transcription factors.

10.2.4 Multiple hypothesis testing

Null and alternative hypotheses

Certain biological annotation metadata analyses may involve the *two-sided tests* of the M null hypotheses of no association between the gene-annotation profiles $A(\cdot, m)$ and a gene-parameter profile λ, i.e., tests of

$$H_0(m) = \mathrm{I}\left(\psi(m) = \psi_0(m)\right) \qquad \text{vs.} \qquad H_1(m) = \mathrm{I}\left(\psi(m) \neq \psi_0(m)\right). \quad (10.8)$$

Other analyses may call for the *one-sided tests* of

$$H_0(m) = \mathrm{I}\left(\psi(m) \leq \psi_0(m)\right) \qquad \text{vs.} \qquad H_1(m) = \mathrm{I}\left(\psi(m) > \psi_0(m)\right). \quad (10.9)$$

One-sided tests are appropriate if, for example, one is interested in determining whether differentially expressed genes are enriched in terms of GO annotation.

The M-vector $\psi_0 = (\psi_0(m) : m = 1, \ldots, M)$, of *null values* for the association parameter ψ, is determined by the biological question. For example, if $\psi(m) = \rho_m(A(\cdot, m), \lambda)$ is the Pearson correlation coefficient between the gene-annotation profile $A(\cdot, m)$ and the gene-parameter profile λ, then one may set $\psi_0(m) = 0$.

Note that in many situations, the same association measure ρ_m is used for each of the M features and one only has a single, common null value $\psi_0(m)$.

Test statistics

As in Sections 1.2.5 and 2.6, consider the general situation where, given a random sample $\mathcal{X}_n$ from the data generating distribution P, one has an *asymptotically linear estimator* $\psi_n = \hat{\Psi}(P_n)$ of the association parameter vector $\psi = \Psi(P)$, with M-dimensional vector *influence curve* $IC(X|P)$. Let $\hat{\Sigma}(P_n) = \sigma_n = (\sigma_n(m, m') : m, m' = 1, \ldots, M)$ denote a consistent estimator of the covariance matrix $\Sigma(P) = \sigma = (\sigma(m, m') : m, m' = 1, \ldots, M)$ of the vector influence curve $IC(X|P)$. For example, σ_n could be a bootstrap-based estimator of the covariance matrix σ or could be computed from an estimator $IC_n(X)$ of the influence curve $IC(X|P)$.

A broad range of association parameters ψ and corresponding estimators ψ_n satisfy the above conditions. In particular, suppose $\lambda_n = \hat{\Lambda}(P_n)$ is an asymptotically linear estimator of the gene-parameter profile $\lambda = \Lambda(P)$, based on a random sample $\mathcal{X}_n$ from P. Let $\psi_n \equiv \rho(A, \lambda_n)$ denote the corresponding *resubstitution*, or *plug-in*, *estimator* of the association parameter vector $\psi = \rho(A, \lambda)$. Then, if the function $\rho(A, \lambda)$ is differentiable with respect to λ, the resubstitution estimator ψ_n is also asymptotically linear.

One can therefore handle tests where the gene-parameter profiles λ are (functions of) means, variances, correlation coefficients, and regression coefficients, and where the association measures ρ are correlation coefficients, two-sample t-statistics, and χ^2-statistics. Examples are provided in Section 10.4, in the context of tests of association between differential gene expression in ALL and GO annotation.

Each null hypothesis $H_0(m)$ may then be tested using a (unstandardized) *difference statistic*,

$$T_n(m) \equiv \sqrt{n}\,(\psi_n(m) - \psi_0(m))\,, \tag{10.10}$$

or a (standardized) *t-statistic*,

$$T_n(m) \equiv \sqrt{n}\frac{\psi_n(m) - \psi_0(m)}{\sqrt{\sigma_n(m,m)}}. \tag{10.11}$$

Certain testing problems may call for other test statistics T_n, such as F-statistics, χ^2-statistics, and likelihood ratio statistics.

Let $Q_n = Q_n(P)$ denote the typically unknown (finite sample) joint distribution of the M-vector of test statistics $T_n = (T_n(m) : m = 1, \ldots, M)$, under the data generating distribution P.

Test statistics null distribution

As detailed in Chapter 2, a key feature of our proposed multiple testing procedures is the *test statistics null distribution* (rather than a data generating null distribution) used to obtain rejection regions (i.e., cut-offs) for the test statistics, confidence regions for the parameters of interest, and adjusted p-values. In practice, the true distribution $Q_n(P)$ of the test statistics T_n is unknown and replaced by a null distribution Q_0. The choice of a proper null distribution is crucial in order to ensure that (finite sample or asymptotic) control of the Type I error rate under the *assumed null distribution* does indeed provide the desired control under the *true distribution.* This issue is particularly relevant for large-scale testing problems, such as those involving gene annotation metadata, which concern high-dimensional multivariate distributions, with complex and unknown dependence structures among variables.

Section 2.2 provides a general characterization for a proper test statistics null distribution, in terms of *null domination* conditions for the joint distribution of the $\mathcal{H}_0$-specific test statistics $(T_n(m) : m \in \mathcal{H}_0)$ (Assumptions jtNDT, NDV, and ND$\boldsymbol{\Theta}$). This general characterization leads to the explicit proposal of two test statistics null distributions $Q_0 = Q_0(P)$: the asymptotic distribution of a vector of *null shift and scale-transformed test statistics* (Section 2.3) and the asymptotic distribution of a vector of *null quantile-transformed test statistics* (Section 2.4).

In practice, the test statistics null distribution $Q_0 = Q_0(P)$ is also unknown, as it depends on the unknown data generating distribution P. Resampling procedures are provided to conveniently obtain consistent estimators of the null distribution (e.g., bootstrap Procedures 2.3 and 2.4) and of the corresponding test statistic cut-offs, parameter confidence regions, and adjusted p-values.

We stress the generality of the aforementioned test statistics null distributions: Type I error control does not rely on restrictive assumptions such as subset pivotality and holds for general data generating distributions (with arbitrary dependence structures among variables), null hypotheses (defined in terms of submodels for the data generating distribution), and test statistics (e.g., t-statistics, χ^2-statistics, F-statistics).

Multiple testing procedures

Having identified a suitable test statistics null distribution Q_0 (or estimator thereof, Q_{0n}), there remains the main task of specifying *rejection regions* for the test statistics, confidence regions for the parameters of interest, and adjusted p-values.

One can apply the multiple testing methodology developed in Chapters 1–7 to control a broad class of Type I error rates, defined as generalized tail probabilities, $gTP(q,g) = \Pr(g(V_n, R_n) > q)$, and generalized expected values, $gEV(g) = \mathrm{E}[g(V_n, R_n)]$, for arbitrary functions $g(V_n, R_n)$ of the numbers of false positives V_n and rejected hypotheses R_n.

An overview of available MTPs is provided in Chapter 3. Core methodological Chapters 4–7 discuss the following main approaches for deriving rejection regions.

Chapter 4. *Joint single-step common-cut-off* and *common-quantile procedures* for controlling *general Type I error rates* $\Theta(F_{V_n})$, defined as arbitrary parameters of the distribution of the number of Type I errors V_n (Dudoit et al., 2004b; Pollard and van der Laan, 2004). Error rates of the form $\Theta(F_{V_n})$ include the generalized family-wise error rate (gFWER), $gFWER(k) = 1 - F_{V_n}(k) = \Pr(V_n > k)$, i.e., the chance of at least $(k+1)$ Type I errors.

Chapter 5. *Joint step-down common-cut-off (maxT)* and *common-quantile (minP) procedures* for controlling the *family-wise error rate* (FWER), $FWER = gFWER(0) = 1 - F_{V_n}(0) = \Pr(V_n > 0)$ (van der Laan et al., 2004a).

Chapter 6. (Marginal/joint single-step/stepwise) *augmentation multiple testing procedures* (AMTP) for controlling *generalized tail probability* error rates, $gTP(q,g) = \Pr(g(V_n, R_n) > q)$, for arbitrary functions $g(V_n, R_n)$ of the numbers of false positives V_n and rejected hypotheses R_n, based on an initial gFWER-controlling procedure (Dudoit et al., 2004a; van der Laan et al., 2004b). Error rates treated in detail include the generalized family-wise error rate, with $g(v,r) = v$, and tail probabilities for the proportion of false positives (TPPFP) among the rejected hypotheses, with $g(v,r) = v/r$.

Chapter 7. *Joint resampling-based empirical Bayes procedures* for controlling *generalized tail probability* error rates. The special case of TPPFP control is discussed in detail in van der Laan et al. (2005).

These multiple testing procedures are implemented in the Bioconductor R package multtest (Section 13.1; Pollard et al. (2005b); `www.bioconductor.org`).

10.3 The Gene Ontology

10.3.1 Overview of the Gene Ontology

The *Gene Ontology* (GO) Consortium (`www.geneontology.org`) provides *ontologies*, i.e., structured and controlled vocabularies, to describe gene products in terms of their associated biological processes, cellular components, and molecular functions. The ontologies specify terminologies and relationships among terms. They are organism-independent and can be applied even as our knowledge of the roles of genes and proteins is accumulating and changing.

The GO Consortium and other organizations supply *mappings* between GO terms and genes in various organisms.

Detailed documentation is available on the Gene Ontology Documentation webpage (`www.geneontology.org/GO.contents.doc.html`).

The three gene ontologies: BP, CC, and MF

The GO Consortium provides three ontologies, each consisting of a structured network of terms describing gene products.

Biological Process (BP or P). The Biological Process ontology refers to series of biological events that are accomplished by one or more ordered assemblies of molecular functions. Examples of broad BP terms are *cellular physiological process* (`GO:0050875`) and *signal transduction* (`GO:0007165`); examples of more specific BP terms are *pyrimidine base metabolism* (`GO:0006206`) and *alpha-glucoside transport* (`GO:0000017`).

Cellular Component (CC or C). The Cellular Component ontology refers to subcellular structures, with the proviso that the components be part of some larger object, which may be an anatomical structure (e.g., *rough endoplasmic reticulum* (`GO:0005791`), *nucleus* (`GO:0005634`)) or a gene product group (e.g., *ribosome* (`GO:0005840`)).

Molecular Function (MF or F). The Molecular Function ontology refers to tasks or activities performed by individual (or assembled complexes of) gene products. Examples of broad MF terms are *catalytic activity* (`GO:0003824`), *transporter activity* (`GO:0005215`), and *binding* (`GO:0005488`); examples of more specific MF terms are *adenylate cyclase activity* (`GO:0004016`) and *Toll binding* (`GO:0005121`).

A gene product may be used in one or more biological processes, may be associated with one or more cellular components, and may have one or more molecular functions.

Example 10.1. Gene product *ABL1_HUMAN*. The *Homo sapiens* gene product *Splice Isoform IA of Proto-oncogene tyrosine-protein kinase ABL1* (*ABL1_HUMAN*) can be described by the following terms in each of the three gene ontologies (AmiGO browser, Last updated: 2006-05-25, `www.godatabase.org/cgi-bin/amigo/go.cgi?view=details\&search_constraint=gp\&session_id=6973b1139030258\&gp=P00519`).

Biological Process:
regulation of progression through cell cycle (`GO:0000074`);
S-phase-specific transcription in mitotic cell cycle (`GO:0000115`);
mismatch repair (`GO:0006298`);
regulation of transcription, DNA-dependent (`GO:0006355`);
DNA damage response, signal transduction resulting in induction of apoptosis (`GO:0008630`).

Cellular Component:
nucleus (`GO:0005634`).

Molecular Function:
DNA binding (`GO:0003677`);
protein-tyrosine kinase activity (`GO:0004713`);
protein binding (`GO:0005515`).

GO directed acyclic graphs

For each of the three gene ontologies, GO terms are organized in a *directed acyclic graph* (DAG), where a *directed* graph has one-way edges and an *acyclic* graph has no path starting and ending at the same vertex. Each GO term is associated with a single vertex, or node, in the DAG. The words *term*, *node*, and *vertex*, may therefore be used interchangeably.

For a given GO term, an *ancestor* refers to a less specialized term; an *offspring* refers to a more specialized term. A *parent* is an immediate/direct ancestor of a term; a *child* is an immediate/direct offspring of a term. A *root* node has no parents (i.e., no incoming edges); a *leaf* node has no children (i.e., no outgoing edges). In a DAG, a child may have several parents.

GO terms must obey the so-called *true path rule*: if a (child) term describes a gene product, then all its immediate parent and more distant ancestor terms must also apply to the gene product.

The DAG structure of GO terms and corresponding true path rule are germane to the definition of a suitable association measure between gene-annotation profiles and gene-parameter profiles (Section 10.2.3). Furthermore, as discussed in Sections 10.3.2–10.3.5, in the context of Bioconductor annotation software, the true path rule is also relevant when assembling gene-annotation matrices.

GO software tools

Many software tools have been developed to deal with GO annotation metadata. The Gene Ontology Tools webpage (`www.geneontology.org/GO.tools.shtml`) provides a list of consortium and non-consortium software for searching and browsing the three gene ontologies, for annotating genes and gene products using GO, and for combined GO and gene expression microarray data analysis.

For instance, the AmiGO browser (`www.godatabase.org`) allows: searching for a GO term and viewing all gene products annotated with this term; searching for a gene product and viewing all its associated GO terms; and browsing the ontologies to view relationships among terms and gene products annotated with a given term.

The QuickGO browser (`www.ebi.ac.uk/ego`), developed by the European Bioinformatics Institute (EBI), also permits searches and graphical displays of the Gene Ontology by GO term, GO term identifier (ID), gene product, and other identifiers.

Software packages developed as part of the Bioconductor Project are discussed in Sections 10.3.2–10.3.5.

Example 10.2. GO term *protein-tyrosine kinase activity*. To get a sense of the information provided by the GO Consortium, consider the Molecular Function ontology and the GO term *protein-tyrosine kinase activity*, with GO term ID `GO:0004713`.

Go to the AmiGO browser (`www.godatabase.org`), enter the GO term ID `GO:0004713` in the `Search GO` box, select `Exact Match`, select `Terms`, and click on the `Submit Query` button. There are two main options for displaying information on a GO term: a "tree view" and a "graphical view". Click on the small tree-like icon (top-left corner of the table) to display the tree view with all ancestors (i.e., less specific terms) of the GO term *protein-tyrosine kinase activity*. Click on the `Graphical View` button to display the portion of the MF DAG corresponding to the GO term. Additional information may be obtained by clicking on the hyperlinked text `protein-tyrosine kinase activity`.

The GO term *protein-tyrosine kinase activity* has one (immediate) parent, *protein kinase activity* (`GO:0004672`), which itself has two parents, *kinase activity* (`GO:0016301`) and *phosphotransferase activity, alcohol group as acceptor* (`GO:0016773`). Altogether, the term *protein-tyrosine kinase activity* has 7 ancestors. According to the true path rule, any gene annotated with the GO term *protein-tyrosine kinase activity* should also be annotated with all of its less specific ancestor terms.

The portion of the MF DAG for the GO term *protein-tyrosine kinase activity* is displayed in Figures 10.2 and 10.3 using, respectively, the AmiGO and QuickGO browsers (note the different ordering of nodes in these two representations: for AmiGO, the offspring nodes are at the top of the graph, whereas for QuickGO, they are at the bottom of the graph).

GO gene-annotation matrices

For each of the three gene ontologies, one may define a $G \times M$ *binary gene-annotation matrix* A, indicating for each gene g whether it is annotated with each GO term m,

$$A(g,m) \equiv \begin{cases} 1, & \text{if gene } g \text{ is annotated with GO term } m \\ 0, & \text{otherwise} \end{cases}, \qquad (10.12)$$

$$g = 1, \ldots, G, \; m = 1, \ldots, M.$$

Section 10.3.5 provides sample R code for assembling GO gene-annotation matrices using Bioconductor annotation metadata packages.

As detailed in Section 10.2, detecting associations between GO annotation and other interesting features of a genome may be viewed as the multiple tests of the null hypotheses of no association between gene-annotation profiles $A(\cdot, m)$ and a gene-parameter profile $\lambda = \Lambda(P)$. The multiple testing methodology proposed in Chapters 1–7 is well-suited to handle the complex and unknown dependence structure among test statistics implied by the DAG structure of GO terms. The methods are illustrated in Section 10.4, for tests of association between differential gene expression in ALL and GO annotation.

10.3.2 Overview of R and Bioconductor software for GO annotation metadata analysis

As discussed in Gentleman et al. (2005b), the *Bioconductor Project* provides R packages for accessing and performing statistical inference with GO annotation metadata (`www.bioconductor.org`; `www.r-project.org`). The packages include:

- a general annotation software package: annotate;
- packages for graph theoretical analyses: e.g., graph, Rgraphviz;
- a GO-specific metadata package for navigating the three GO DAGs: GO;
- an Entrez Gene[1]-specific metadata package, providing bi-directional mappings between Entrez Gene IDs and GO term IDs: humanLLMappings (`www.ncbi.nlm.nih.gov/entrez/query.fcgi?db=gene`);
- various Affymetrix chip-specific metadata packages, providing bi-directional mappings between Affymetrix probe[2] IDs and GO term IDs: e.g., hgu95av2, hu6800 (`www.affymetrix.com`);
- a package for annotating and generating HTML reports for Affymetrix chip data: annaffy.

[1] N.B. The NCBI LocusLink database has been superseded by the Entrez Gene database.

[2] N.B. In the context of Affymetrix oligonucleotide chips, we use the shorter term *probe* to refer to a *probe-pair-set*, i.e., a collection of *perfect match* (PM) and *mismatch* (MM) *probe-pairs* that map to a particular gene.

Bioconductor metadata packages are updated regularly to reflect the evolving nature of biological annotation metadata; it is therefore crucial to keep track of *version* numbers. For information on Bioconductor software, please consult the book edited by Gentleman et al. (2005a) and the Documentation (`www.bioconductor.org/docs`) and Workshops (`www.bioconductor.org/workshops`) sections of the Bioconductor Project website, in addition to the standard R help facilities (e.g., `help` function, manuals, etc.).

The remainder of this section provides sample R code demonstrating Bioconductor software (results reported for R Release 2.2.1 and Bioconductor Release 1.7). In order to run through the examples, one needs to install and load the Bioconductor packages annotate, GO, and hgu95av2. The annotation metadata used in the examples correspond to the following package versions.

```
> library(annotate)
> library(GO)
> library(hgu95av2)
>
> packageDescription("annotate")$Version
[1] "1.8.0"
> packageDescription("GO")$Version
[1] "1.10.0"
> packageDescription("hgu95av2")$Version
[1] "1.10.0"
```

Accessing and analyzing annotation metadata from databases such as GenBank (`www.ncbi.nlm.nih.gov/Genbank`), GO (`www.geneontology.org`), and PubMed (`www.pubmed.gov`), presupposes the ability to perform the following essential bookkeeping task: *mapping between different identifiers* (ID) for a given gene/probe. Bioconductor annotation metadata packages consist of *environment* objects that provide *key-value mappings* between different sets of gene/probe identifiers.

For instance, in the annotation metadata package hgu95av2, for the Affymetrix chip series HG-U95Av2, the `hgu95av2PMID` environment provides mappings from Affymetrix probe IDs (keys) to PubMed IDs (values); similarly, the `hgu95av2GO` environment provides mappings from Affymetrix probe IDs (keys) to GO term IDs (values).

Example 10.3. Affymetrix probe ID `1635_at`. As of Version 1.10.0 of the hgu95av2 package, the Affymetrix probe with ID `1635_at` corresponds to the gene with symbol `ABL1` and long name `v-abl Abelson murine leukemia viral oncogene homolog 1`, located on the long arm of chromosome 9. This probe maps to one GenBank accession number, one Entrez Gene ID, 14 distinct GO term IDs, and 160 distinct PubMed IDs.

```
> probe <- "1635_at"
> get(probe, env=hgu95av2SYMBOL)
```

```
[1] "ABL1"
> get(probe, env=hgu95av2GENENAME)
[1] "v-abl Abelson murine leukemia viral oncogene homolog 1"
> get(probe, env=hgu95av2MAP)
[1] "9q34.1"
> get(probe, env=hgu95av2ACCNUM)
[1] "U07563"
> get(probe, env=hgu95av2LOCUSID )
[1] 25
> unique(names(get(probe, env=hgu95av2GO)))
 [1] "GO:0000074" "GO:0000115" "GO:0000166" "GO:0003677"
 [5] "GO:0004713" "GO:0005515" "GO:0005524" "GO:0005634"
 [9] "GO:0006298" "GO:0006355" "GO:0006468" "GO:0007242"
[13] "GO:0008630" "GO:0016740"
> length(get(probe, env=hgu95av2PMID))
[1] 160
```

The remainder of this section gives a brief overview of two main types of Bioconductor annotation metadata packages: the GO package (Section 10.3.3) and the hgu95av2 package for the Affymetrix chip series HG-U95Av2 (Section 10.3.4). Section 10.3.5 illustrates how these two packages may be used to assemble a GO gene-annotation matrix.

10.3.3 The annotation metadata package GO

The GO package provides environment objects containing key-value pairs for mappings between GO term IDs, GO terms, GO term ancestors, GO term parents, GO term children, GO term offspring, and Entrez Gene IDs. The `GO()` command lists all environments available in the GO package.

```
> GO()

Quality control information for  GO
Date built: Created: Fri Sep 30 03:02:24 2005

Mappings found for non-probe based rda files:
         GOALLLOCUSID found 9556
         GOBPANCESTOR found 9888
         GOBPCHILDREN found 4989
         GOBPOFFSPRING found 4989
         GOBPPARENTS found 9888
         GOCCANCESTOR found 1612
         GOCCCHILDREN found 578
         GOCCOFFSPRING found 578
         GOCCPARENTS found 1612
```

```
GOLOCUSID2GO found 70818
GOLOCUSID found 8017
GOMFANCESTOR found 7334
GOMFCHILDREN found 1403
GOMFOFFSPRING found 1403
GOMFPARENTS found 7334
GOOBSOLETE found 1032
GOTERM found 18834
```

For information on any of the GO environments, use the `help` function, e.g., `help(GOTERM)` or `?GOBPPARENTS`. For instance, the environment `GOTERM` provides mappings from *GO term IDs* (keys) to *GO terms* (values); the environments `GOBPPARENTS`, `GOCCPARENTS`, and `GOMFPARENTS`, provide ontology-specific mappings from *GO term IDs* (keys) to *GO term parent IDs* (values). The environments `GOALLLOCUSID`, `GOLOCUSID2GO`, and `GOLOCUSID`, provide mappings between *GO term IDs* and *Entrez Gene IDs* and are used in Section 10.3.5, below, to assemble an Entrez Gene ID-by-GO term ID gene-annotation matrix for the MF gene ontology.

Example 10.4. GO term ID `GO:0004713`. Let us use the GO package to obtain information on (all) ancestors, (immediate) parents, (immediate) children, and (all) offspring of the term corresponding to the GO term ID `GO:0004713`.

```
> ## List all GO IDs
> GOID <- ls(env = GOTERM)
> length(GOID)
[1] 18834
> GOID[1:10]
 [1] "GO:0000001" "GO:0000002" "GO:0000003" "GO:0000004"
 [5] "GO:0000006" "GO:0000007" "GO:0000009" "GO:0000010"
 [9] "GO:0000011" "GO:0000012"
>
> ## Get information on GO term corresponding to GO ID
> ## GO:0004713
> GOID <- "GO:0004713"
> term <- get(GOID,env=GOTERM)
> class(term)
[1] "GOTerms"
attr(,"package")
[1] "annotate"
> slotNames(term)
[1] "GOID"       "Term"       "Synonym"    "Secondary"
[5] "Definition" "Ontology"
> term
GOID = GO:0004713
```

```
Term = protein-tyrosine kinase activity
Synonym = protein tyrosine kinase activity
Definition = Catalysis of the reaction: ATP + a protein
     tyrosine = ADP + protein tyrosine phosphate.
Ontology = MF
>
> ## Get GO IDs of parents
> parents <- get(GOID,env=GOMFPARENTS)
> parents
        isa
"GO:0004672"
> mget(parents,env=GOTERM)
$"GO:0004672"
GOID = GO:0004672
Term = protein kinase activity
Definition = Catalysis of the transfer of a phosphate
     group, usually from ATP, to a protein substrate.
Ontology = MF


>
> ## Get GO IDs of ancestors
> ancestors <- get(GOID,env=GOMFANCESTOR)
> ancestors
[1] "all"        "GO:0003674" "GO:0003824" "GO:0016740"
[5] "GO:0016772" "GO:0016773" "GO:0016301" "GO:0004672"
>
> ## Get GO IDs of children
> children <- get(GOID,env=GOMFCHILDREN)
> children
[1] "GO:0004714" "GO:0004715" "GO:0004716"
>
> ## Get GO IDs of offspring
> offspring <- get(GOID,env=GOMFOFFSPRING)
> offspring
 [1] "GO:0004714" "GO:0004715" "GO:0004716" "GO:0005020"
 [5] "GO:0005021" "GO:0005023" "GO:0005010" "GO:0005011"
 [9] "GO:0005017" "GO:0005003" "GO:0005006" "GO:0005007"
[13] "GO:0005008" "GO:0005009" "GO:0008288" "GO:0005018"
[17] "GO:0005019" "GO:0005004" "GO:0005005" "GO:0008313"
[21] "GO:0004718"
```

As already noted in Example 10.2 and Figures 10.2 and 10.3, the term corresponding to the GO term ID `GO:0004713` is *protein-tyrosine kinase activity*, in the Molecular Function ontology. It has one (immediate) parent term, *protein kinase activity*.

10.3.4 Affymetrix chip-specific annotation metadata packages: The hgu95av2 package

The Bioconductor Project provides *Affymetrix chip-specific annotation metadata packages* for the main chip series for the human, mouse, rat, and other genomes (e.g., HG-U133, HG-U95, HU-6800, MG-U74, and RG-U34 series). These packages, built using the infrastructure package AnnBuilder, contain environment objects for mappings between Affymetrix probe IDs and other types of gene/probe identifiers.

Note that analogous packages are not supplied for two-color spotted microarrays, as there is no standard microarray design for this type of platform and specialized annotation metadata packages may have to be created for each microarray facility (e.g., using AnnBuilder). Once annotation metadata packages are available to provide mappings between different sets of gene/probe identifiers, the tools in annotate and related packages may be used in a similar manner for any type of microarray platform.

Consider the hgu95av2 package, for the Affymetrix chip series HG-U95Av2. This package provides the following environments.

```
> ? hgu95av2
> hgu95av2()

Quality control information for  hgu95av2
Date built: Created: Tue Oct  4 21:31:35 2005

Number of probes: 12625
Probe number missmatch: None
Probe missmatch: None
Mappings found for probe based rda files:
         hgu95av2ACCNUM found 12625 of 12625
         hgu95av2CHRLOC found 11673 of 12625
         hgu95av2CHR found 12145 of 12625
         hgu95av2ENZYME found 1886 of 12625
         hgu95av2GENENAME found 11418 of 12625
         hgu95av2GO found 9942 of 12625
         hgu95av2LOCUSID found 12203 of 12625
         hgu95av2MAP found 12109 of 12625
         hgu95av2OMIM found 9881 of 12625
         hgu95av2PATH found 3928 of 12625
         hgu95av2PMID found 12086 of 12625
         hgu95av2REFSEQ found 12008 of 12625
         hgu95av2SUMFUNC found 0 of 12625
         hgu95av2SYMBOL found 12159 of 12625
         hgu95av2UNIGENE found 12118 of 12625
Mappings found for non-probe based rda files:
         hgu95av2CHRLENGTHS found 25
```

```
hgu95av2ENZYME2PROBE found 643
hgu95av2GO2ALLPROBES found 5480
hgu95av2GO2PROBE found 3890
hgu95av2ORGANISM found 1
hgu95av2PATH2PROBE found 155
hgu95av2PFAM found 10439
hgu95av2PMID2PROBE found 98214
hgu95av2PROSITE found 8249
```

For information on any of these environments, use the **help** function, e.g., `help(hgu95av2GO)` or `?hgu95av2GO`. We focus on the three environments related to GO: `hgu95av2GO`, `hgu95av2GO2ALLPROBES`, and `hgu95av2GO2PROBE`.

The HG-U95Av2 chip contains 12,625 probes (keys in the `hgu95av2GO` environment), with the first 10 Affymetrix probe IDs listed below.

```
> ## List all Affymetrix IDs
> AffyID <- ls(env = hgu95av2GO)
> length(AffyID)
[1] 12625
> AffyID[1:10]
 [1] "1000_at"    "1001_at"    "1002_f_at" "1003_s_at" "1004_at"
 [6] "1005_at"    "1006_at"    "1007_s_at" "1008_f_at" "1009_at"
```

Probes-to-most specific GO terms mappings: The `hgu95av2GO` environment

The `hgu95av2GO` environment provides key-value pairs for the mappings from *Affymetrix probe IDs* (keys) to *GO term IDs* (values). Each Affymetrix probe ID is mapped to a list of one or more elements, where each element corresponds to a particular GO term and is itself a list with the following three elements.

- `"GOID"`: A GO term ID corresponding to the Affymetrix probe ID (key).
- `"Evidence"`: A code for the evidence supporting the association of the GO term to the Affymetrix probe.
- `"Ontology"`: An abbreviation for the name of the ontology to which the GO term belongs: BP (Biological Process), CC (Cellular Component), or MF (Molecular Function).

Note that only the *directly associated terms* or *most specific terms* (i.e., not their less specific ancestor terms) a probe is annotated with are returned as values in `hgu95av2GO`. The GO package may be used to obtain more information on the GO term IDs, e.g., GO term, (all) ancestors, (immediate) parents, (immediate) children, and (all) offspring (Section 10.3.3).

Example 10.5. GO terms directly associated with Affymetrix probe ID `1635_at`. Let us obtain GO annotation information for the probe with Affymetrix ID `1635_at`, corresponding to the `ABL1` gene. The code below shows that probe `1635_at` is directly annotated with 14 distinct GO terms (the same GO term ID may be returned multiple times with a different evidence code). As already noted in Example 10.1, one of these terms, with GO term ID `GO:0004713`, is *protein-tyrosine kinase activity*, in the Molecular Function ontology.

```
> probe <- "1635_at"
> probe2GO <- get(probe, env = hgu95av2GO)
> length(probe2GO)
[1] 14
> unique(names(probe2GO))
 [1] "GO:0000074" "GO:0000115" "GO:0000166" "GO:0003677"
 [5] "GO:0004713" "GO:0005515" "GO:0005524" "GO:0005634"
 [9] "GO:0006298" "GO:0006355" "GO:0006468" "GO:0007242"
[13] "GO:0008630" "GO:0016740"
> probe2GO[[5]]
$GOID
[1] "GO:0004713"

$Evidence
[1] "TAS"

$Ontology
[1] "MF"

> get(probe2GO[[5]]$GOID, env=GOTERM)
GOID = GO:0004713
Term = protein-tyrosine kinase activity
Synonym = protein tyrosine kinase activity
Definition = Catalysis of the reaction: ATP + a protein
     tyrosine = ADP + protein tyrosine phosphate.
Ontology = MF
```

The `hgu95av2GO` environment (and analogous environments for other chip series) may be used to assemble an Affymetrix probe ID-by-GO term ID gene-annotation matrix, row by row. This may entail, however, a number of data processing steps. Firstly, only the most specific terms a probe is annotated with are returned as values in `hgu95av2GO`. One therefore needs to add all (less specific) ancestor terms in order to comply with the true path rule. Secondly, several probes may correspond to the same gene, i.e., several Affymetrix probe IDs may map to the same Entrez Gene ID according to the environment `hgu95av2LOCUSID`. Thirdly, the `hgu95av2GO` environment returns GO terms for all three gene ontologies at once. One may need to separate terms according to membership in the BP, CC, and MF ontologies (e.g., using the `GOTERM` environment from the GO package).

Alternately, one may assemble an Affymetrix probe ID-by-GO term ID gene-annotation matrix, column by column, using the `hgu95av2GO2ALLPROBES` environment described below.

GO terms-to-directly annotated probes mappings: The `hgu95av2GO2PROBE` environment

The `hgu95av2GO2PROBE` environment provides key-value pairs for the mappings from *GO term IDs* (keys) to *Affymetrix probe IDs* (values). Values are vectors of length one or greater depending on whether a given GO term ID is mapped to one or more Affymetrix probe IDs. The value names are evidence codes for the GO term IDs.

Note that the probes a particular GO term is mapped to are only those associated *directly* with the GO term (vs. indirectly via its immediate children or more distant offspring). For a list of all probes associated directly or indirectly with a particular GO term, one may use the `hgu95av2GO2ALLPROBES` environment.

Example 10.6. Affymetrix probes directly associated with GO term ID `GO:0004713`. In the following example, 205 distinct Affymetrix probe IDs are associated directly with the GO term *protein-tyrosine kinase activity* (`GO:0004713`). The Affymetrix probe IDs include `1635_at`, corresponding to the `ABL1` gene.

```
> GOID <- "GO:0004713"
> GO2Probes <- get(GOID, env = hgu95av2GO2PROBE)
> length(unique(GO2Probes))
[1] 205
> GO2Probes[1:10]
        <NA>         <NA>         <NA>         <NA>         <NA>
   "1635_at"  "1636_g_at"  "1656_s_at"  "2040_s_at"  "2041_i_at"
         TAS          IEA          IEA          IEA          TAS
  "39730_at"    "1084_at" "35162_s_at"    "1564_at"     "854_at"
> is.element("1635_at", GO2Probes)
[1] TRUE
```

GO terms-to-all annotated probes mappings: The `hgu95av2GO2ALLPROBES` environment

The `hgu95av2GO2ALLPROBES` environment provides key-value pairs for the mappings from *GO term IDs* (keys) to *Affymetrix probe IDs* (values). Values are vectors of length one or greater depending on whether a given GO term ID is mapped to one or more Affymetrix probe IDs. The value names are evidence codes for the GO term IDs.

Note that, in accordance with the true path rule, the probes a particular GO term is mapped to are associated either *directly* with the GO term or *indirectly* via any of its immediate children or more distant offspring. The main difference between the **hgu95av2GO2PROBE** and **hgu95av2GO2ALLPROBES** environments is that the former considers only the GO term itself, whereas the latter considers the GO term and any of its offspring. Thus, the Affymetrix probe IDs returned by **hgu95av2GO2PROBE** are a subset of the probe IDs returned by **hgu95av2GO2ALLPROBES**.

Example 10.7. Affymetrix probes directly or indirectly associated with GO term ID GO:0004713. In the following example, 319 distinct Affymetrix probe IDs (some with multiple evidence codes) are associated either directly or indirectly with the GO term *protein-tyrosine kinase activity* (GO:0004713). This list of 319 Affymetrix probe IDs indeed includes the list of 205 probe IDs associated directly with the GO term ID GO:0004713.

```
> GOID <- "GO:0004713"
> GO2AllProbes <- get(GOID, env = hgu95av2GO2ALLPROBES)
> length(GO2AllProbes)
[1] 370
> length(unique(GO2AllProbes))
[1] 319
> sum(is.element(GO2Probes,GO2AllProbes))
[1] 205
```

The **hgu95av2GO2ALLPROBES** environment immediately yields an Affymetrix probe ID-by-GO term ID gene-annotation matrix, column by column. However, as with the **hgu95av2GO** environment, a number of data processing steps may be required, concerning, for example, uniqueness of Entrez Gene IDs and membership in the BP, CC, and MF ontologies.

10.3.5 Assembling a GO gene-annotation matrix

This section provides R code for assembling an Entrez Gene ID-by-GO term ID gene-annotation matrix A, column by column. Specifically, rows correspond to (unique) Entrez Gene IDs mapping to probes on the HG-U95Av2 chip and columns to terms in the Molecular Function ontology that map directly or indirectly to at least 10 Entrez Gene IDs for the HG-U95Av2 chip.

In practice, it may not be desirable to build the full $G \times M$ gene-annotation matrix, as this matrix could potentially be very large and sparse (padded with zeros). Rather, we assemble a (smaller) *gene-annotation list*, that provides, for each GO term ID, a list of Entrez Gene IDs annotated with the GO term.

Example 10.8. Entrez Gene ID-by-GO term ID gene-annotation matrix for MF ontology.

```
> ## List all Affymetrix IDs for HG-U95Av2 chip
> AffyID <- ls(env=hgu95av2GO)
> length(AffyID)
[1] 12625
>
> ## Get all unique Entrez Gene IDs for HG-U95Av2 chip
> LLID <- as.character(unique(unlist(mget(AffyID,
+   env=hgu95av2LOCUSID))))
> length(LLID)
[1] 9085
>
> ## Get MF GO IDs
> GOID <- ls(env=GOTERM)
> O <- unlist(lapply(mget(GOID, env=GOTERM),
+   function(z) z@Ontology))
> table(O)
O
  BP   CC   MF
9888 1612 7334
> MFID <- GOID[O=="MF"]
>
> ## For each MF GO ID, get all Entrez Gene IDs for genes
> ## annotated directly or indirectly with the GO term
> allMFLLID <- mget(MFID, env=GOALLLOCUSID)
>
> ## For each MF GO ID, get HG-U95Av2-specific Entrez Gene IDs
> ## for genes annotated directly or indirectly with the GO term
> MFLLID <- lapply(allMFLLID, function(z) intersect(z, LLID))
> numMFLLID <- unlist(lapply(MFLLID, length))
> summary(numMFLLID)
    Min.  1st Qu.   Median     Mean  3rd Qu.     Max.
   0.000    1.000    1.000    9.539    1.000 6762.000
>
> ## Retain only MF GO IDs that map to at least 10
> ## Entrez Gene IDs for the HG-U95Av2 chip
> MFAnnotList <- MFLLID[numMFLLID > 9]
> length(MFAnnotList)
[1] 466
> summary(unlist(lapply(MFAnnotList, length)))
   Min. 1st Qu.  Median    Mean 3rd Qu.    Max.
   10.0    16.0    27.5   132.2    70.0  6762.0
> MFAnnotList[1]
```

```
$"GO:0000146"
 [1] "4620"   "4621"   "4624"   "4625"   "4640"   "4643"   "4644"
 [8] "4646"   "4647"   "4650"   "58498"

>
> ## Get Entrez Gene IDs for probes annotated with GO ID
> ## GO:0004713
> is.element("GO:0004713",names(MFAnnotList))
[1] TRUE
> length(MFAnnotList["GO:0004713"][[1]])
[1] 180
```

10.4 Tests of association between GO annotation and differential gene expression in ALL

10.4.1 Acute lymphoblastic leukemia study of Chiaretti et al. (2004)

Our proposed approach to tests of association with biological annotation metadata is illustrated using the *acute lymphoblastic leukemia* (ALL) microarray dataset of Chiaretti et al. (2004) and Gene Ontology (GO) annotation metadata (Dudoit et al., 2006).

Bioconductor experimental data R package ALL

The ALL dataset is available in the Bioconductor experimental data R package ALL (Version 1.0.2, Bioconductor Release 1.7). The main object in this package is `ALL`, an instance of the class *exprSet*. The `exprs` slot of `ALL` provides a matrix of 12,625 *microarray expression measures* (Affymetrix chip series HG-U95Av2) for each of 128 ALL cell samples. The `phenoData` slot contains 21 *phenotypes* (i.e., covariates and outcomes) for each of the 128 cell samples. Phenotypes of interest include: `ALL$BT`, the type and stage of the cancer (i.e., B-cell ALL or T-cell ALL, of stage 1, 2, 3, or 4), and `ALL$mol.biol`, the molecular class of the cancer (i.e., BCR/ABL, NEG, ALL1/AF4, E2A/PBX1, p15/p16, or NUP-98).

The expression measures have been obtained using the three-step robust multichip average (RMA) pre-processing method, implemented in the Bioconductor R package affy (Bolstad et al., 2005), and have been subjected to a base 2 logarithmic transformation.

For greater detail on the ALL dataset, please consult the ALL package documentation.

The BCR/ABL fusion

A number of recent articles have investigated the prognostic relevance of the *BCR/ABL fusion* in adult ALL of the B-cell lineage (Gleissner et al., 2002).

The BCR/ABL fusion is the molecular analogue of the *Philadelphia chromosome*, one of the most frequent cytogenetic abnormalities in human leukemias. This *t(9;22) translocation* leads to a head-to-tail fusion of the **`v-abl Abelson murine leukemia viral oncogene homolog 1`** (`ABL1`) from chromosome 9 with the 5' half of the **`breakpoint cluster region`** (`BCR`) on chromosome 22 (Figure 10.4). The `ABL1` proto-oncogene encodes a cytoplasmic and nuclear protein tyrosine kinase that has been implicated in processes of cell differentiation, cell division, cell adhesion, and stress response. Although the BCR/ABL fusion protein, encoded by sequences from both the `ABL1` and `BCR` genes, has been extensively studied, the function of the normal product of the `BCR` gene is not clear. The `BCR/ABL` proto-oncogene has been found to be highly expressed in chronic myeloid leukemia (CML) and acute myeloid leukemia (AML) cells (Mukhopadhyay et al., 2002).

An interesting question is therefore the identification of genes that are differentially expressed between B-cell ALL with the BCR/ABL fusion and cytogenetically normal NEG B-cell ALL.

In order to address this quesion, we consider the expression measures of the $n = 79$ B-cell ALL cell samples (`ALL$BT` equal to `B`, `B1`, `B2`, `B3`, or `B4`), of the BCR/ABL or NEG molecular types (`ALL$mol.biol` equal to `BCR/ABL` or `NEG`).

Gene filtering

Many of the genes represented by the 12,625 probes are not expressed in B-cell lymphocytes. Accordingly, as in von Heydebreck et al. (2004), we only retain the 2,391 probes that meet the following two criteria: (i) fluorescence intensities greater than 100 (absolute scale) for at least 25% of the 79 cell samples; (ii) interquartile range (IQR) of the fluorescence intensities for the 79 cell samples greater than 0.5 (log base 2 scale).

Furthermore, different probes may correspond to the same gene, i.e., map to the same Entrez Gene ID, according to the environment `hgu95av2LOCUSID` from the `hgu95av2` package. In order to obtain one expression measure per gene, we choose to average the expression measures of multiple probes mapping to the same gene.

These various pre-processing steps lead to $G = 2,071$ genes with unique Entrez Gene IDs.

Reduced ALL dataset: Genotypes and phenotypes of interest

The combined genotypic and phenotypic data for the $n = 79$ B-cell ALL cell samples of the BCR/ABL and NEG molecular types may be summarized by the set $\mathcal{XY}_n \equiv \{(X_i, Y_i) : i = 1, \ldots, n\}$, of n pairs of G-dimensional gene expression profiles $X_i = (X_i(g) : g = 1, \ldots, G)$, $G = 2,071$, and cancer class labels $Y_i \in \{NEG, BCR/ABL\}$. Among the $n = 79$ B-cell ALL cell samples, there are $n_{BCR/ABL} \equiv \sum_i \mathrm{I}(Y_i = BCR/ABL) = 37$ BCR/ABL and $n_{NEG} \equiv \sum_i \mathrm{I}(Y_i = NEG) = 42$ NEG samples.

10.4.2 Multiple hypothesis testing framework

Our primary question of interest is the identification of genes that are *differentially expressed* (DE) between BCR/ABL and NEG B-cell ALL. A subsequent question involves *relating differential gene expression to GO annotation.*

As detailed below, GO annotation metadata for the filtered list of $G = 2,071$ unique genes from the HG-U95Av2 chip may be summarized by binary gene-annotation profiles.

The gene-parameter profiles of interest concern differential gene expression between BCR/ABL and NEG B-cell ALL, i.e., the association between microarray gene expression measures and cancer class. Continuous gene-parameter profiles of unstandardized and standardized measures of differential expression are estimated, respectively, by (unstandardized) differences of empirical means and (standardized) two-sample t-statistics. Binary gene-parameter profiles, indicating whether genes are differentially expressed, are estimated by imposing cut-off rules on two-sample t-statistics or adjusted p-values.

The following association measures between GO gene-annotation profiles and DE gene-parameter profiles are considered: two-sample t-statistics for tests of association between binary GO gene-annotation profiles and continuous DE gene-parameter profiles; χ^2-statistics for tests of association between binary GO gene-annotation profiles and binary DE gene-parameter profiles.

Significant associations between differential gene expression and GO annotation are identified by applying FWER-controlling single-step maxT Procedure 3.5, based on the non-parametric bootstrap null shift and scale-transformed test statistics null distribution of Procedure 2.3.

Gene-annotation profiles

Gene Ontology annotation metadata for the HG-U95Av2 chip series are obtained as described in Sections 10.3.2–10.3.5, from the following Bioconductor R packages: the GO-specific metadata package GO (Version 1.10.0, Bioconductor Release 1.7) and the Affymetrix chip-specific metadata package hgu95av2 (Version 1.10.0, Bioconductor Release 1.7).

For each of the three gene ontologies, *binary gene-annotation matrices* A_{BP}, A_{CC}, and A_{MF}, are assembled for the GO terms annotating at least 10 of the $G = 2,071$ filtered genes (sample R code provided in Section 10.3.5). Specifically, for gene ontology $o \in \{BP, CC, MF\}$, $A_o = (A_o(g, m) : g = 1, \ldots, G;\ m = 1, \ldots, M_o)$ is a $G \times M_o$ matrix, with element $A_o(g, m)$ indicating whether gene g is annotated by GO term m and such that $\sum_g A_o(g,m) \geq 10$ for each term m. The numbers of terms considered in each gene ontology are $M_{BP} = 367$, $M_{CC} = 81$, and $M_{MF} = 185$.

Gene-parameter profiles

Definition of gene-parameter profiles

Consider a data structure $(X,Y) \sim P$, where $X = (X(g) : g = 1,\ldots,G)$ is a $G = 2,071$-dimensional vector of microarray gene expression measures and $Y \in \{NEG, BCR/ABL\}$ is a binary cancer class label. Let $\eta_k \equiv \Pr(Y = k)$ denote the proportion of cancers of class $k \in \{NEG, BCR/ABL\}$. Define conditional G-dimensional mean vectors and $G \times G$ covariance matrices for the expression measures of class $k \in \{NEG, BCR/ABL\}$ cancers by

$$\mu_k \equiv \mathrm{E}[X|Y = k] \qquad \text{and} \qquad \sigma_k \equiv \mathrm{Cov}[X|Y = k],$$

respectively.

Gene-parameter profiles, concerning differential gene expression between BCR/ABL and NEG B-cell ALL, may be specified in various ways. *Continuous* DE gene-parameter profiles may be defined in terms of the following *unstandardized* and *standardized* measures of differential gene expression between BCR/ABL and NEG B-cell ALL,

$$\lambda^d(g) \equiv \mu_{BCR/ABL}(g) - \mu_{NEG}(g) \tag{10.13}$$

and

$$\lambda^t(g) \equiv \frac{\mu_{BCR/ABL}(g) - \mu_{NEG}(g)}{\sqrt{\frac{\sigma_{BCR/ABL}(g,g)}{\eta_{BCR/ABL}} + \frac{\sigma_{NEG}(g,g)}{\eta_{NEG}}}}.$$

Absolute values of $\lambda^d(g)$ and $\lambda^t(g)$ may be used for measuring two-sided differential expression, i.e., either over- or under-expression in BCR/ABL compared to NEG B-cell ALL.

Binary DE gene-parameter profiles may be defined in terms of indicators for two-sided and one-sided differential expression.

$$\begin{aligned}
\lambda^{\neq}(g) &\equiv \mathrm{I}\left(\mu_{BCR/ABL}(g) \neq \mu_{NEG}(g)\right) \\
&= \mathrm{I}\left(\lambda^d(g) \neq 0\right) = \mathrm{I}\left(\lambda^t(g) \neq 0\right), \\
\lambda^{+}(g) &\equiv \mathrm{I}\left(\mu_{BCR/ABL}(g) > \mu_{NEG}(g)\right) \\
&= \mathrm{I}\left(\lambda^d(g) > 0\right) = \mathrm{I}\left(\lambda^t(g) > 0\right), \\
\lambda^{-}(g) &\equiv \mathrm{I}\left(\mu_{BCR/ABL}(g) < \mu_{NEG}(g)\right) \\
&= \mathrm{I}\left(\lambda^d(g) < 0\right) = \mathrm{I}\left(\lambda^t(g) < 0\right).
\end{aligned} \tag{10.14}$$

Estimation of gene-parameter profiles

The above DE gene-parameter profiles may be estimated as follows, based on the sample $\mathcal{XY}_n$ of gene expression measures for the $n = 79$ B-cell ALL cell samples of the BCR/ABL and NEG molecular types.

The *resubstitution estimators* of the continuous gene-parameter profiles of Equation (10.13) are based, respectively, on differences of empirical means and two-sample Welch t-statistics (up to the multiplier $1/\sqrt{n}$). That is,

$$\lambda_n^d(g) \equiv \mu_{BCR/ABL,n}(g) - \mu_{NEG,n}(g) \tag{10.15}$$

and

$$\lambda_n^t(g) \equiv \frac{1}{\sqrt{n}} \frac{\mu_{BCR/ABL,n}(g) - \mu_{NEG,n}(g)}{\sqrt{\frac{\sigma_{BCR/ABL,n}(g,g)}{n_{BCR/ABL}} + \frac{\sigma_{NEG,n}(g,g)}{n_{NEG}}}},$$

where $\mu_{k,n}(g) \equiv \sum_i \mathrm{I}(Y_i = k) X_i(g)/n_k$ and $\sigma_{k,n}(g,g) \equiv \sum_i \mathrm{I}(Y_i = k)(X_i(g) - \mu_{k,n}(g))^2/(n_k - 1)$ denote, respectively, the empirical means and variances of the gene expression measures for cancers of class $k \in \{NEG, BCR/ABL\}$.

Estimating the two-sided binary gene-parameter profile $\lambda^{\neq}$ of Equation (10.14) involves the *two-sided tests* of the G null hypotheses $H_0(g) = \mathrm{I}(\mu_{BCR/ABL}(g) = \mu_{NEG}(g))$, of no differences in mean gene expression measures between BCR/ABL and NEG B-cell ALL. Likewise, estimating the one-sided binary gene-parameter profiles λ^+ and λ^- involves, respectively, the *one-sided tests* of the G null hypotheses of no over-expression ($H_0(g) = \mathrm{I}(\mu_{BCR/ABL}(g) \leq \mu_{NEG}(g))$) and no under-expression ($H_0(g) = \mathrm{I}(\mu_{BCR/ABL}(g) \geq \mu_{NEG}(g))$) in BCR/ABL compared to NEG B-cell ALL. For single-step common-cut-off maxT Procedure 3.5, adjusted p-values produce the same gene ranking as the test statistics defined in Equation (10.15). Simple and naive estimators of the three sets of differentially expressed genes (i.e., false null hypotheses), represented by the gene-parameter profiles $\lambda^{\neq}$, λ^+, and λ^-, are therefore given, respectively, by the sets of genes with the largest γG values of $|\lambda_n^t(g)|$, $\lambda_n^t(g)$, and $-\lambda_n^t(g)$. That is,

$$\lambda_{n,\gamma G}^{\neq}(g) \equiv \mathrm{I}\left(\sum_{g'=1}^{G} \mathrm{I}\left(|\lambda_n^t(g)| \geq |\lambda_n^t(g')|\right) \geq (1-\gamma)G \right), \tag{10.16}$$

$$\lambda_{n,\gamma G}^{+}(g) \equiv \mathrm{I}\left(\sum_{g'=1}^{G} \mathrm{I}\left(\lambda_n^t(g) \geq \lambda_n^t(g')\right) \geq (1-\gamma)G \right),$$

$$\lambda_{n,\gamma G}^{-}(g) \equiv \mathrm{I}\left(\sum_{g'=1}^{G} \mathrm{I}\left(-\lambda_n^t(g) \geq -\lambda_n^t(g')\right) \geq (1-\gamma)G \right).$$

Analogous estimators may also be based on other test statistics, such as unstandardized difference statistics λ_n^d.

More sophisticated estimators, that translate the proportion γ of rejected hypotheses into a Type I error rate such as the gFWER, TPPFP, or FDR, could be based on adjusted p-values for the multiple tests of the G null hypotheses $H_0(g)$. For example, one could estimate $\lambda^{\neq}$ by

$$\lambda^{\neq}_{n,\alpha}(g) \equiv \mathrm{I}\left(\widetilde{P}^{\neq}_{0n}(g) \leq \alpha\right), \tag{10.17}$$

where $\widetilde{P}^{\neq}_{0n}(g)$ are adjusted p-values for a suitably chosen multiple testing procedure, such as, FWER-controlling single-step maxT Procedure 3.5 or TPPFP-controlling augmentation multiple testing Procedure 3.26. One-sided binary gene-parameter profiles λ^+ and λ^- could be estimated likewise.

Association measures for gene-annotation and gene-parameter profiles

The association between continuous DE gene-parameter profiles, as in Equation (10.13), and binary GO gene-annotation profiles may be measured by *two-sample Welch t-statistics* (or corresponding p-values). Specifically, given a continuous G-vector x and a binary G-vector y, define the following association measure,

$$\rho^t(x,y) \equiv \frac{\bar{x}_1 - \bar{x}_0}{\sqrt{\frac{v[x]_1}{y_1} + \frac{v[x]_0}{y_0}}}, \tag{10.18}$$

where $y_k \equiv \sum_g \mathrm{I}(y(g) = k)$, $\bar{x}_k \equiv \sum_g \mathrm{I}(y(g) = k)\, x(g)/y_k$, and $v[x]_k \equiv \sum_g \mathrm{I}(y(g) = k)\,(x(g) - \bar{x}_k)^2/(y_k - 1)$, $k \in \{0,1\}$.

The association between binary DE gene-parameter profiles, as in Equation (10.14), and binary GO gene-annotation profiles may be measured by χ^2*-statistics* (or corresponding p-values) for the test of independence of rows and columns in a 2×2 contingency table, such as Table 10.1. Specifically, given binary G-vectors x and y, define the following association measure,

$$\rho^{\chi}(x,y) \equiv \frac{G(g_{00}g_{11} - g_{01}g_{10})^2}{(g_{00} + g_{01})(g_{00} + g_{10})(g_{11} + g_{01})(g_{11} + g_{10})}, \tag{10.19}$$

where $g_{kk'} \equiv \sum_g \mathrm{I}(x(g) = k)\,\mathrm{I}(y(g) = k')$, $k, k' \in \{0,1\}$.

Given an association measure[3] $\rho : I\!R^{G\times M} \times I\!R^{G} \rightarrow I\!R^{M}$, a $G \times M$ GO gene-annotation matrix A, and a G-dimensional DE gene-parameter profile $\lambda = \Lambda(P)$, the M-dimensional *association parameter* vector $\psi = \Psi(P)$ of primary interest is defined as

$$\psi \equiv \rho(A, \lambda). \tag{10.20}$$

[3] N.B. For ease of notation, ρ^t and ρ^{χ}, defined in Equations (10.18) and (10.19) as real-valued association measures, may also refer loosely to $I\!R^M$-valued association measures, defined as $\rho^t(X,y) \equiv (\rho^t(X(\cdot,m),y) : m = 1,\ldots,M)$ and $\rho^{\chi}(X,y) \equiv (\rho^{\chi}(X(\cdot,m),y) : m = 1,\ldots,M)$ for $X \in I\!R^{G\times M}$ and $y \in I\!R^{G}$.

The corresponding *resubstitution estimator* $\psi_n = \hat{\Psi}(P_n)$ is simply obtained by replacing the gene-parameter profile λ by an estimator thereof $\lambda_n = \hat{\Lambda}(P_n)$, that is,

$$\psi_n \equiv \rho(A, \lambda_n). \tag{10.21}$$

Null and alternative hypotheses

For the t-statistic-based association measure ρ^t of Equation (10.18), the identification of GO terms m that are significantly (positively or negatively) associated with BCR/ABL vs. NEG differential gene expression involves the *two-sided tests* of the M null hypotheses $H_0(m) = \mathrm{I}(\psi(m) = \psi_0(m))$ against the alternative hypotheses $H_1(m) = \mathrm{I}(\psi(m) \neq \psi_0(m))$, with null values $\psi_0(m) = 0$. In some contexts, one may be interested in identifying positive (negative) associations, i.e., in the *one-sided tests* of the M null hypotheses $H_0(m) = \mathrm{I}(\psi(m) \leq \psi_0(m))$ $(H_0(m) = \mathrm{I}(\psi(m) \geq \psi_0(m)))$ against the alternative hypotheses $H_1(m) = \mathrm{I}(\psi(m) > \psi_0(m))$ $(H_1(m) = \mathrm{I}(\psi(m) < \psi_0(m)))$.

For the χ^2-statistic-based association measure ρ^{χ} of Equation (10.19), the identification of GO terms m that are significantly (positively or negatively) associated with BCR/ABL vs. NEG differential gene expression involves the *one-sided tests* of the M null hypotheses $H_0(m) = \mathrm{I}(\psi(m) \leq \psi_0(m))$ against the alternative hypotheses $H_1(m) = \mathrm{I}(\psi(m) > \psi_0(m))$. A natural choice for the null values is the mean of the $\chi^2(1)$-distribution, $\psi_0(m) = 1$.

Test statistics

One-sided and two-sided tests of null hypotheses concerning any of the association parameters defined above may be based on (unstandardized) *difference statistics* $T_n(m)$, defined as in Equation (10.10).

For one-sided tests, large values of the test statistics $T_n(m)$ provide evidence against the corresponding null hypotheses $H_0(m)$, that is, rejection regions are of the form $\mathcal{C}_n(m) = (c_n(m), +\infty)$. For two-sided tests, large values of the absolute test statistics $|T_n(m)|$ provide evidence against the corresponding null hypotheses $H_0(m)$.

Multiple testing procedures

For the purpose of illustration, we focus on control of the *family-wise error rate*, using *single-step maxT Procedure 3.5*, based on the *non-parametric bootstrap null shift-transformed test statistics null distribution* of Section 2.3 (null shift values $\lambda_0(m) = 0$ and no scaling). The main steps are outlined in Procedures 2.3, 4.21 ($gFWER(0)$ special case), 8.1, and 8.2.

Let $O_n(m)$ denote the indices for the ordered adjusted p-values, so that $\widetilde{P}_{0n}(O_n(1)) \leq \cdots \leq \widetilde{P}_{0n}(O_n(M))$. GO terms with adjusted p-values less than or equal to α are declared significantly associated with differential gene expression at nominal FWER level α. That is, the list of GO terms found to be associated with differential gene expression is

$$\mathcal{R}_n(\alpha) \equiv \left\{m : \widetilde{P}_{0n}(m) \leq \alpha\right\} = \{O_n(1), \ldots, O_n(R_n(\alpha))\},$$

where $R_n(\alpha) \equiv |\mathcal{R}_n(\alpha)|$ denotes the number of identified GO terms.

Summary of testing scenarios

This section summarizes our approach for identifying GO terms associated with BCR/ABL vs. NEG differential gene expression. For each of the three gene ontologies (i.e., BP, CC, and MF), we consider the following three types of testing scenarios, each corresponding to a different association parameter $\psi = \rho(A, \lambda)$ for GO annotation and BCR/ABL vs. NEG differential gene expression. Scenarios $\mathsf{MT}[t, t]$ and $\mathsf{MT}[d, t]$ are very similar and correspond, respectively, to *continuous* gene-parameter profiles of *standardized* and *unstandardized* measures of differential gene expression. In contrast, Scenario $\mathsf{MT}[\neq, \chi]$ corresponds to a *binary* gene-parameter profile of differential gene expression indicators.

Scenario $\mathsf{MT}[t, t]$: Association parameter $\psi^{t,t} = \rho^t(A, |\lambda^t|)$, for standardized continuous DE gene-parameter profile λ^t. Consider the two-sided tests of

$$H_0^{t,t}(m) \equiv \mathrm{I}\left(\psi^{t,t}(m) = \psi_0^{t,t}(m)\right)$$

vs.

$$H_1^{t,t}(m) \equiv \mathrm{I}\left(\psi^{t,t}(m) \neq \psi_0^{t,t}(m)\right),$$

where the association parameter vector of interest is defined as $\psi^{t,t} \equiv \rho^t(A, |\lambda^t|)$, based on Equations (10.13) and (10.18), and the null values are $\psi_0^{t,t}(m) \equiv 0$. The continuous DE gene-parameter profile λ^t is estimated by λ_n^t, as in Equation (10.15), and the association parameter $\psi^{t,t}$ is estimated by the resubstitution estimator $\psi_n^{t,t} \equiv \rho^t(A, |\lambda_n^t|)$, as in Equation (10.21). The test statistics are defined as (unstandardized) difference statistics,

$$T_n^{t,t}(m) \equiv \sqrt{n}(\psi_n^{t,t}(m) - \psi_0^{t,t}(m)),$$

and the null hypotheses $H_0^{t,t}(m)$ are rejected for large absolute values of $T_n^{t,t}(m)$.

Scenario $\mathsf{MT}[d, t]$: Association parameter $\psi^{d,t} = \rho^t(A, |\lambda^d|)$, for unstandardized continuous DE gene-parameter profile λ^d. Consider the two-sided tests of

$$H_0^{d,t}(m) \equiv \mathrm{I}\left(\psi^{d,t}(m) = \psi_0^{d,t}(m)\right)$$

vs.

$$H_1^{d,t}(m) \equiv \mathrm{I}\left(\psi^{d,t}(m) \neq \psi_0^{d,t}(m)\right),$$

where the association parameter vector of interest is defined as $\psi^{d,t} \equiv \rho^t(A, |\lambda^d|)$, based on Equations (10.13) and (10.18), and the null values are $\psi_0^{d,t}(m) \equiv 0$. The continuous DE gene-parameter profile λ^d is estimated by λ_n^d, as in Equation (10.15), and the association parameter $\psi^{d,t}$ is estimated by the resubstitution estimator $\psi_n^{d,t} \equiv \rho^t(A, |\lambda_n^d|)$, as in Equation (10.21). The test statistics are defined as (unstandardized) difference statistics,

$$T_n^{d,t}(m) \equiv \sqrt{n}(\psi_n^{d,t}(m) - \psi_0^{d,t}(m)),$$

and the null hypotheses $H_0^{d,t}(m)$ are rejected for large absolute values of $T_n^{d,t}(m)$.

Scenario MT[$\neq, \chi$]: Association parameter $\psi^{\neq,\chi} = \rho^\chi(A, \lambda^{\neq})$, for binary DE gene-parameter profile $\lambda^{\neq}$. Consider the one-sided tests of

$$H_0^{\neq,\chi}(m) \equiv \mathrm{I}\left(\psi^{\neq,\chi}(m) \leq \psi_0^{\neq,\chi}(m)\right)$$

vs.

$$H_1^{\neq,\chi}(m) \equiv \mathrm{I}\left(\psi^{\neq,\chi}(m) > \psi_0^{\neq,\chi}(m)\right),$$

where the association parameter vector of interest is defined as $\psi^{\neq,\chi} \equiv \rho^\chi(A, \lambda^{\neq})$, based on Equations (10.14) and (10.19), and the null values are $\psi_0^{\neq,\chi}(m) \equiv 1$ (the mean of the $\chi^2(1)$-distribution). The following two types of estimators $\lambda_n^{\neq}$ are considered for the binary DE gene-parameter profile $\lambda^{\neq}$: $\lambda_{n,\gamma G}^{\neq}$, with numbers of DE genes $\gamma G = 20, 50, 100$ (Equation (10.16)); $\lambda_{n,\alpha}^{\neq}$, defined in terms of adjusted p-values for FWER-controlling permutation-based single-step maxT Procedure 3.5 ($B = 1,000$ permutations of the cancer class labels) and nominal FWER level $\alpha = 0.05$ (Equation (10.17)). Given an estimator $\lambda_n^{\neq}$ of $\lambda^{\neq}$, the association parameter $\psi^{\neq,\chi}$ is estimated by the resubstitution estimator $\psi_n^{\neq,\chi} \equiv \rho^\chi(A, \lambda_n^{\neq})$, as in Equation (10.21). The test statistics are defined as (unstandardized) difference statistics,

$$T_n^{\neq,\chi}(m) \equiv \sqrt{n}(\psi_n^{\neq,\chi}(m) - \psi_0^{\neq,\chi}(m)),$$

and the null hypotheses $H_0^{\neq,\chi}(m)$ are rejected for large values of $T_n^{\neq,\chi}(m)$.

For each of the three testing scenarios, the null shift-transformed test statistics null distribution Q_0 is estimated as in Procedure 2.3, with $B = 5,000$ non-parametric bootstrap samples of the data $\mathcal{XY}_n$ and $Z_n^B(m,b) = T_n^B(m,b) - \mathrm{E}[T_n^B(m,\cdot)]$ (i.e., null shift values $\lambda_0(m) = 0$ and no scaling). For one-sided testing Scenario MT[$\neq, \chi$], bootstrap-based single-step maxT adjusted p-values $\widetilde{P}_{0n}(m)$ are computed as in Procedures 4.21 and 8.2. For two-sided testing Scenarios MT[t,t] and MT[d,t], adjusted p-values are computed based on absolute values of $Z_n^B(m,b)$ and $T_n(m)$.

In what follows, the G-dimensional gene-parameter profiles λ correspond to the $G = 2,071$ genes with unique Entrez Gene IDs, obtained as described in Section 10.4.1. For each of the three gene ontologies, binary gene-annotation matrices are assembled for the GO terms annotating at least 10 of the $G = 2,071$ genes of interest: $G = 2,071 \times M_{BP} = 367$ gene-annotation matrix A_{BP} for the BP ontology, $G = 2,071 \times M_{CC} = 81$ gene-annotation matrix A_{CC} for the CC ontology, and $G = 2,071 \times M_{MF} = 185$ gene-annotation matrix A_{MF} for the MF ontology.

10.4.3 Results

Differentially expressed genes between BCR/ABL and NEG B-cell ALL

In order to identify differentially expressed genes between BCR/ABL and NEG B-cell ALL, two-sided tests of the G null hypotheses $H_0(g) = \mathrm{I}(\mu_{BCR/ABL}(g) = \mu_{NEG}(g))$ are performed using the two-sample t-statistics $\lambda_n^t(g)$ of Equation (10.15) and FWER-controlling bootstrap-based single-step maxT Procedure 8.2. Adjusted p-values $\widetilde{P}_{0n}^{\neq}(g)$ are obtained using the `MTP` function from the `multtest` package (Version 1.8.0, Bioconductor Release 1.7), with $B = 5,000$ non-parametric bootstrap samples and other arguments set to their default values.

Figure 10.5 displays a normal quantile-quantile plot of the test statistics $\lambda_n^t(g)$ (Panel (a)) and a plot of the sorted bootstrap-based single-step maxT adjusted p-values $\widetilde{P}_{0n}^{\neq}(g)$ (Panel (b)). A handful of genes stand out in terms of their large absolute test statistics and small adjusted p-values.

For control of the FWER at nominal level $\alpha = 0.05$, Procedure 8.2 identifies 16 differentially expressed genes, i.e., 16 genes with $\widetilde{P}_{0n}^{\neq}(g) \leq \alpha$. Table 10.2 provides the test statistics, adjusted p-values, and various identifiers for these 16 genes. A more detailed hyperlinked table is posted on the website companion (Supplementary Table 10.1; `www.stat.berkeley.edu/~sandrine/MTBook/BAM/BAM.html`).

Only 2 of the 16 identified genes have a negative test statistic (`MX1` and `TPD52L2`), suggesting that most DE genes tend to be *over-expressed* in cell samples with the BCR/ABL fusion.

The gene showing the most over-expression in BCR/ABL cell samples, as measured by the t-statistics λ_n^t, is the `ABL1` gene (`v-abl Abelson murine leukemia viral oncogene homolog 1`), located on the long arm of chromosome 9 (9q34.1). As mentioned in Section 10.4.1, the BCR/ABL phenotype is indeed defined in terms of the `ABL1` gene.

Furthermore, many of the DE genes seem to be related to apoptosis or oncogenesis. For example, the `Kruppel-like factor 9` (`KLF9`) gene encodes a transcription factor that binds GC-box elements in gene promoter regions. The Krüppel-like factor (KLF) family is comprised of highly related zinc-finger proteins, that are important components of the eukaryotic cellular transcriptional machinery and that take part in a wide range of cellular functions (e.g., cell proliferation, apoptosis, differentiation, and neoplastic transformation). In particular, KLFs have been linked to various cancers (Kaczynski et al., 2003). The intron-less gene `AHNAK nucleoprotein (desmoyokin)` (`AHNAK`), located on the long arm of chromosome 11 (11q12.2), encodes an unusually large protein ($\approxeq$ 700 kiloDalton (kDa)) that is typically repressed in cell lines derived from human neuroblastomas and several other types of tumors (Shtivelman et al., 1992). Yet another example, the `caspase 8, apoptosis-related cysteine peptidase` (`CASP8`) gene, encodes a key enzyme at the top of the apoptotic cascade and has been linked to neuroblastoma (Banelli et al., 2002). Likewise, other genes listed in Table 10.2, including `MX1`, `FYN`, `ACTN1`, `FHL1`, and `TRAM2`, appear to be related to the molecular biology of cancer.

Our results are in general agreement with those of von Heydebreck et al. (2004), slight differences being due, most likely, to our preliminary gene filtering, which involves averaging the expression measures of multiple probes mapping to the same Entrez Gene ID.

For greater detail, the interested reader is invited to consult Supplementary Table 10.1 on the website companion and follow links to PubMed and other databases. Further exploration of the DE genes may be performed in R using the Bioconductor packages annotate and annaffy.

GO terms associated with differential gene expression between BCR/ABL and NEG B-cell ALL

Figure 10.6 displays, for each of the three gene ontologies and each of the three testing scenarios, plots of the sorted adjusted p-values, $\widetilde{P}_{0n}(O_n(1)) \leq \cdots \leq \widetilde{P}_{0n}(O_n(M))$, for FWER-controlling bootstrap-based single-step maxT Procedure 8.2 ($B = 5,000$ bootstrap samples). The smaller the adjusted p-values, the less conservative the procedure, and the longer the list $\mathcal{R}_n(\alpha) = \{m : \widetilde{P}_{0n}(m) \leq \alpha\}$ of identified GO terms at any given nominal Type I error level α.

Table 10.3 summarizes the results in terms of the numbers $R_n(\alpha) = |\mathcal{R}_n(\alpha)|$ of GO terms found to be significantly associated with BCR/ABL vs. NEG differential gene expression for different nominal FWER levels α.

In general, adjusted p-values tend to be quite large, with only a handful of GO terms identified as being significantly associated with BCR/ABL vs. NEG differential gene expression for nominal FWER level $\alpha \in \{0.05, 0.10, 0.20\}$. The adjusted p-values for Scenarios $\mathsf{MT}[t,t]$ and $\mathsf{MT}[d,t]$ (red and blue plotting symbols), corresponding, respectively, to standardized and unstandardized continuous DE gene-parameter profiles, are similar. For the BP and MF gene ontologies, Scenario $\mathsf{MT}[t,t]$ seems to be slightly more conservative than Scenario $\mathsf{MT}[d,t]$; however, this does not hold for the CC ontology. Scenario $\mathsf{MT}[\neq,\chi]$, with four different estimators of the binary DE gene-parameter profile $\lambda^{\neq}$, tends to be more conservative than either Scenario

$\mathsf{MT}[t,t]$ or $\mathsf{MT}[d,t]$. Furthermore, the choice of parameter γG, for the number of genes called differentially expressed, can have a substantial impact on the adjusted p-values for Scenario $\mathsf{MT}[\neq,\chi:\gamma G]$. There are some indications, especially for the CC ontology, that greater values of the parameter γG lead to greater numbers of identified GO terms. Note that for Scenario $\mathsf{MT}[\neq,\chi]$, the p-value-based estimator $\lambda^{\neq}_{n,\alpha}$, with $\alpha = 0.05$, and the naive estimator $\lambda^{\neq}_{n,\gamma G}$, with $\gamma G = 20$, yield very similar results (green and purple plotting symbols). Indeed, when applied to the entire dataset for the $n = 79$ cell samples, permutation-based single-step maxT Procedure 3.5 identifies 20 genes as being differentially expressed between BCR/ABL and NEG B-cell ALL at nominal FWER level $\alpha = 0.05$. In other words, $\lambda^{\neq}_{n,0.05}$ and $\lambda^{\neq}_{n,20}$ yield the same estimate of the binary gene-parameter profile $\lambda^{\neq}$ for the set of DE genes. Minor discrepancies between the results of Scenarios $\mathsf{MT}[\neq,\chi:\alpha=0.05]$ and $\mathsf{MT}[\neq,\chi:\gamma G=20]$ are due to the fact that while the estimators $\lambda^{\neq}_{n,0.05}$ and $\lambda^{\neq}_{n,20}$ coincide on the original sample, they may differ on bootstrap samples of these data.

Next, the three testing scenarios are compared in terms of the contents of the lists $\mathcal{R}_n(\alpha)$ of identified GO terms. Specifically, let $\mathcal{O}_n(r) \equiv \{O_n(1),\ldots,O_n(r)\}$ denote the set of indices corresponding to the r smallest adjusted p-values for a given gene ontology and testing scenario. Measures of agreement between testing scenarios are provided by the numbers of common GO terms among sets of ordered GO terms $\mathcal{O}_n(r)$ of various cardinality r, i.e., by the cardinality of intersections of sets $\mathcal{O}_n(r)$ for different testing scenarios. Figure 10.7 displays plots of numbers of common GO terms for pairs of testing scenarios. As expected, there is substantial overlap between the GO terms identified by Scenarios $\mathsf{MT}[t,t]$ and $\mathsf{MT}[d,t]$ for continuous DE gene-parameter profiles (blue plotting symbols in top panels). This suggests that, for the ALL dataset, standardized (λ^t) and unstandardized (λ^d) continuous measures of differential gene expression have similar properties. In contrast, there is much less overlap between the GO terms identified by Scenario $\mathsf{MT}[\neq,\chi]$, for binary DE gene-parameter profiles, and either Scenario $\mathsf{MT}[t,t]$ or $\mathsf{MT}[d,t]$. For example, for the MF gene ontology, among the top 10 GO terms $\mathcal{O}_n(10)$ identified by each testing scenario, 6 are common to Scenarios $\mathsf{MT}[t,t]$ and $\mathsf{MT}[d,t]$, whereas at most 3 are common to Scenarios $\mathsf{MT}[t,t]$ and $\mathsf{MT}[\neq,\chi]$. Again, note the near perfect agreement between Scenarios $\mathsf{MT}[\neq,\chi:\alpha=0.05]$ and $\mathsf{MT}[\neq,\chi:\gamma G=20]$ (purple plotting symbols in lower panels). Figure 10.7 again illustrates the lack of robustness of Scenario $\mathsf{MT}[\neq,\chi:\gamma G]$ to the choice of parameter γG.

Moreover, examine graphical summaries of the joint distributions of the estimated continuous DE gene-parameter profile λ^t_n and the gene-annotation profiles $A(\cdot,m)$ for the top two GO terms $m \in \{O_n(1),O_n(2)\}$ identified according to each testing scenario. Figure 10.8 displays conditional boxplots of λ^t_n given $A(\cdot,m)$, that is, boxplots of the unannotated and annotated estimated gene-parameter profiles, $(\lambda^t_n(g):A(g,m)=0)$ and $(\lambda^t_n(g):$

$A(g, m) = 1$), respectively. Although the boxplots reveal clear differences (non-overlapping notches) between unannotated and annotated profiles for some of the GO terms (e.g., MF term GO:0003735), the differences can be subtle for other terms (e.g., MF term GO:0003924). Not surprisingly, the most extreme differences are seen for Scenarios $\mathsf{MT}[t, t]$ and $\mathsf{MT}[d, t]$, and, to a lesser extent, for Scenario $\mathsf{MT}[\neq, \chi : \alpha = 0.05]$ for the CC ontology. The boxplots again illustrate differences between Scenario $\mathsf{MT}[\neq, \chi]$ and either Scenario $\mathsf{MT}[t, t]$ or $\mathsf{MT}[d, t]$.

Tables 10.4, 10.5, and 10.6 report various p-value-based measures of association between the estimated DE gene-parameter profiles λ_n^t and $\lambda_{n,\alpha}^{\neq}$ and the gene-annotation profiles $A(\cdot, m)$ for the top two GO terms $m \in \{O_n(1), O_n(2)\}$ identified according to each testing scenario, in the BP, CC, and MF gene ontologies, respectively. The transformation to the $[0, 1]$ p-value scale allows a more direct comparison of the various testing scenarios. The tables again highlight the differences between Scenario $\mathsf{MT}[\neq, \chi]$, for binary DE gene-parameter profiles, and either Scenario $\mathsf{MT}[t, t]$ or $\mathsf{MT}[d, t]$, for continuous DE gene-parameter profiles. As expected, Scenarios $\mathsf{MT}[t, t]$ and $\mathsf{MT}[d, t]$ tend to identify GO terms with small p-values $P_{0n}^{t,t}(m)$ for t-tests of association between estimated continuous gene-parameter profiles λ_n^t and gene-annotation profiles $A(\cdot, m)$. In contrast, and also as expected, Scenario $\mathsf{MT}[\neq, \chi]$ tends to identify GO terms with small p-values $P_{0n}^{\neq,\chi}(m)$ for χ^2-tests of association between estimated binary gene-parameter profiles $\lambda_{n,\alpha}^{\neq}$ and gene-annotation profiles $A(\cdot, m)$. Furthermore, the tables corroborate our earlier observation that Scenario $\mathsf{MT}[\neq, \chi]$ tends to be more conservative than either Scenario $\mathsf{MT}[t, t]$ or $\mathsf{MT}[d, t]$. Indeed, some of the GO terms with small p-values $P_{0n}^{t,t}(m)$ for continuous gene-parameter profiles have very large p-values $P_{0n}^{\neq,\chi}(m)$ for binary gene-parameter profiles (e.g., MF term GO:0003735 in Table 10.6). Such terms are likely to be identified by Scenarios $\mathsf{MT}[t, t]$ and $\mathsf{MT}[d, t]$, but missed by Scenario $\mathsf{MT}[\neq, \chi]$. The converse phenomenon is not as striking. However, one should keep in mind that Scenario $\mathsf{MT}[\neq, \chi]$ depends on the choice of estimator for the binary DE gene-parameter profile $\lambda^{\neq}$, i.e., on parameters such as α and γG. In particular, with certain values of α (or γG), binary Scenario $\mathsf{MT}[\neq, \chi]$ may become more similar to either continuous Scenario $\mathsf{MT}[t, t]$ or $\mathsf{MT}[d, t]$. Column $A_1(m)$ in Tables 10.4–10.6 suggests that, compared to Scenario $\mathsf{MT}[\neq, \chi]$, Scenarios $\mathsf{MT}[t, t]$ and $\mathsf{MT}[d, t]$ tend to identify GO terms annotating a greater number of genes (this observation also holds for the top 20 terms identified according to each testing scenario; data not shown).

Figure 10.9 displays a scatterplot matrix of the 50 smallest adjusted p-values, based on Scenario $\mathsf{MT}[t, t]$, for each of the three gene ontologies. The plots indicate that more terms tend to be identified in the BP ontology compared to either the CC or MF ontologies, and fewer terms tend to be identified in the MF ontology compared to either the BP or CC ontologies. Note that comparisons based on adjusted p-values take into account differences in the numbers of tested hypotheses, $M_{BP} = 367$, $M_{CC} = 81$, and $M_{MF} = 185$, for each ontology.

Tables 10.7, 10.8, and 10.9 list the 20 GO terms with the smallest adjusted p-values for Scenario $\mathsf{MT}[t,t]$, applied to the BP, CC, and MF gene ontologies, respectively. Figures 10.10, 10.11, and 10.12 display portions of the directed acyclic graphs for the top 20 GO terms in each ontology. The figures suggest that GO terms associated with BCR/ABL vs. NEG differential gene expression tend to concentrate in certain branches of the DAGs, i.e., differential expression is associated with related properties of gene products. Although it is known that many of the effects of the BCR/ABL fusion are mediated by tyrosine kinase activity, the MF GO term *protein-tyrosine kinase activity* (`GO:0004713`) does not appear to be significantly associated with differential gene expression between BCR/ABL and NEG B-cell ALL (adjusted p-value of 0.8890 for Scenario $\mathsf{MT}[t,t]$).

For illustration purposes, we further investigate two of the GO terms from Tables 10.7 and 10.9: GO term *anti-apoptosis* (`GO:0006916`), with ninth smallest adjusted p-value for Scenario $\mathsf{MT}[t,t]$ applied to the BP gene ontology, and GO term *structural constituent of ribosome* (`GO:0003735`), with the smallest adjusted p-value for Scenario $\mathsf{MT}[t,t]$ applied to the MF gene ontology. Tables 10.10 and 10.11 list genes directly or indirectly annotated with GO terms `GO:0006916` and `GO:0003735`, respectively. Figure 10.13 displays mean-difference plots of the average expression measures in BCR/ABL and NEG cell samples for genes annotated with GO terms `GO:0006916` and `GO:0003735`.

Panel (a) in Figure 10.13 indicates that genes annotated with BP GO term *anti-apoptosis* (`GO:0006916`) tend to be over-expressed in BCR/ABL compared to NEG cell samples. Among these 21 genes, only `SOCS2` is significantly differentially expressed between BCR/ABL and NEG B-cell ALL (nominal FWER level $\alpha = 0.05$, Table 10.2). However, a brief survey of the literature reveals that several of the genes in Table 10.10 interact with the `BCR/ABL` proto-oncogene. For instance, Kirchner et al. (2003) investigate mechanisms for the `BCR/ABL`-mediated activation of the transcription factor NF-κB/Rel encoded by the `NFKB1` gene. Their findings suggest that NF-κB/Rel may be a potential target for molecular therapies of leukemia. Mukhopadhyay et al. (2002) demonstrate that ectopic expression of `BCR/ABL` interferes with the `tumor necrosis factor` (TNF) signaling pathway through the down-regulation of TNF receptors. The `TNF` gene encodes a multifunctional proinflammatory cytokine involved in the regulation of a wide spectrum of biological processes, including cell proliferation, differentiation, apoptosis, lipid metabolism, and coagulation. The `TNF` gene has been implicated in a variety of diseases, including autoimmune diseases, insulin resistance, and cancer.

As seen in Table 10.11, 22 of the 24 genes annotated with MF GO term *structural constituent of ribosome* (`GO:0003735`) code for ribosomal proteins. Although none of the 24 annotated genes is identified as being significantly differentially expressed between BCR/ABL and NEG B-cell ALL (nominal

FWER level $\alpha = 0.05$, Table 10.2), Panel (b) in Figure 10.13 suggests that these genes tend to be under-expressed in BCR/ABL cell samples.

10.5 Discussion

We have proposed a general and formal statistical framework for multiple tests of association with biological annotation metadata. A key component of our approach is the systematic and precise translation of a generic biological question into a corresponding multiple hypothesis testing problem, concerning association measures between known gene-annotation profiles and unknown gene-parameter profiles. This general and rigorous formulation of the statistical inference question allows us to apply the multiple testing methodology developed in Chapters 1–7, to control a broad class of Type I error rates, in testing problems involving general data generating distributions (with arbitrary dependence structures among variables), null hypotheses, and test statistics.

The flexibility of our approach was illustrated using the ALL microarray dataset of Chiaretti et al. (2004), with the aim of relating GO annotation to differential gene expression between BCR/ABL and NEG B-cell ALL. This analysis demonstrates the importance of selecting a suitable DE gene-parameter profile λ and measure ρ for the association between this gene-parameter profile and GO gene-annotation profiles A. Indeed, for the ALL dataset, the choice of gene-parameter profile for measuring differential expression between BCR/ABL and NEG B-cell ALL has a large impact on the list of identified GO terms. Testing scenarios based on binary DE gene-parameter profiles (Scenario $\mathsf{MT}[\neq, \chi]$) tended to be more conservative than scenarios based on continuous DE gene-parameter profiles (Scenarios $\mathsf{MT}[t, t]$ and $\mathsf{MT}[d, t]$), with little overlap between the lists of identified GO terms. Furthermore, testing scenarios based on binary gene-parameter profiles were sensitive to the somewhat arbitrary DE/non-DE gene dichotomization, that is, Scenario $\mathsf{MT}[\neq, \chi : \gamma G]$ lacked robustness with respect to the choice of parameter γG for the number of genes called differentially expressed according to the estimator $\lambda^{\neq}_{n,\gamma G}$. In contrast, continuous gene-parameter profiles based on standardized and unstandardized measures of differential gene expression lead to very similar results (Scenarios $\mathsf{MT}[t, t]$ and $\mathsf{MT}[d, t]$).

Our analysis of the ALL microarray dataset clearly shows the limitations of binary gene-parameter profiles of differential expression indicators, which are still the norm for combined GO annotation and microarray data analyses. Our proposed statistical framework, with general definitions for the gene-annotation and gene-parameter profiles, allows consideration of a much broader class of inference problems, that extend beyond GO annotation and microarray data analysis. Gene-annotation profiles may be continuous or polychotomous and may correspond, for example, to intron/exon counts/lengths/nucleotide distributions, gene pathway membership, or gene regulation by particular transcription factors. Likewise, gene-parameter profiles may be continuous or polychotomous and may correspond, for example, to regression coefficients relating possibly censored biological and clinical outcomes to genome-wide transcript levels, DNA copy numbers, and other covariates.

This first application of our proposed methodology only considered control of the family-wise error rate using single-step common-cut-off maxT Procedure 3.5, based on the non-parametric bootstrap null shift and scale-transformed test statistics null distribution of Procedure 2.3. Adjusted p-values tended to be quite large, with only a handful of GO terms identified as being significantly associated with BCR/ABL vs. NEG differential gene expression. Joint augmentation and empirical Bayes procedures could be used for control of a broader and more biologically relevant class of Type I error rates, defined as tail probabilities, $gTP(q,g) = \Pr(g(V_n, R_n) > q)$, for arbitrary functions $g(V_n, R_n)$ of the numbers of false positives V_n and rejected hypotheses R_n (Chapters 6 and 7; Dudoit et al. (2004a); van der Laan et al. (2004b, 2005)). Error rates based on the proportion V_n/R_n of false positives (e.g., TPPFP and FDR) are especially appealing for large-scale testing problems, compared to error rates based on the number V_n of false positives (e.g., gFWER), as they do not increase exponentially with the number M of tested hypotheses. More powerful analyses may also be achieved with the new null quantile-transformed test statistics null distribution of van der Laan and Hubbard (2006). The multiple testing methodology developed in Chapters 1–7 is particularly well-suited to handle the variety of parameters of interest and the complex and unknown dependence structures among test statistics (e.g., implied by the DAG structure of GO terms) that are likely to be encountered in these and other high-dimensional inference problems in biomedical and genomic research.

Ongoing efforts include consideration of more general and biologically pertinent multivariate association measures ρ. For instance, for GO annotation metadata, the association parameter for a given GO term could take into account the structure of the DAG by considering the gene-annotation profiles of offspring or ancestor terms. We are also interested in developing better numerical and graphical approaches for representing and interpreting the multiple testing results, e.g., the lists of GO terms and associated adjusted p-values. Finally, we are planning on implementing the proposed methods in an R package to be released as part of the Bioconductor Project.

Table 10.1. *Binary gene-annotation and gene-parameter profiles.* Given a binary gene-annotation profile $A(\cdot, m)$ and a binary gene-parameter profile λ, one may build a 2×2 contingency table, with rows corresponding to the gene-annotation profile and columns to the gene-parameter profile. Cell counts are defined as $g_{kk'}(m) = \sum_g \mathrm{I}(A(g,m) = k)\,\mathrm{I}(\lambda(g) = k')$, $k, k' \in \{0, 1\}$. For example, for tests of association between GO annotation and differential gene expression, $g_{11}(m)$ could correspond to the number of genes that are annotated with GO term m and differentially expressed.

Gene-annotation profile, $A(\cdot, m)$	**Gene-parameter profile, λ**: **1**	**0**	
1	$g_{11}(m) = \sum_{g=1}^{G} A(g,m)\lambda(g)$	$g_{10}(m) = \sum_{g=1}^{G} A(g,m)(1-\lambda(g))$	$A_1(m) = \sum_{g=1}^{G} A(g,m)$
0	$g_{01}(m) = \sum_{g=1}^{G}(1-A(g,m))\lambda(g)$	$g_{00}(m) = \sum_{g=1}^{G}(1-A(g,m))(1-\lambda(g))$	$A_0(m) = \sum_{g=1}^{G}(1-A(g,m))$
	$G\bar{\lambda} = \sum_{g=1}^{G} \lambda(g)$	$G(1-\bar{\lambda}) = \sum_{g=1}^{G}(1-\lambda(g))$	G

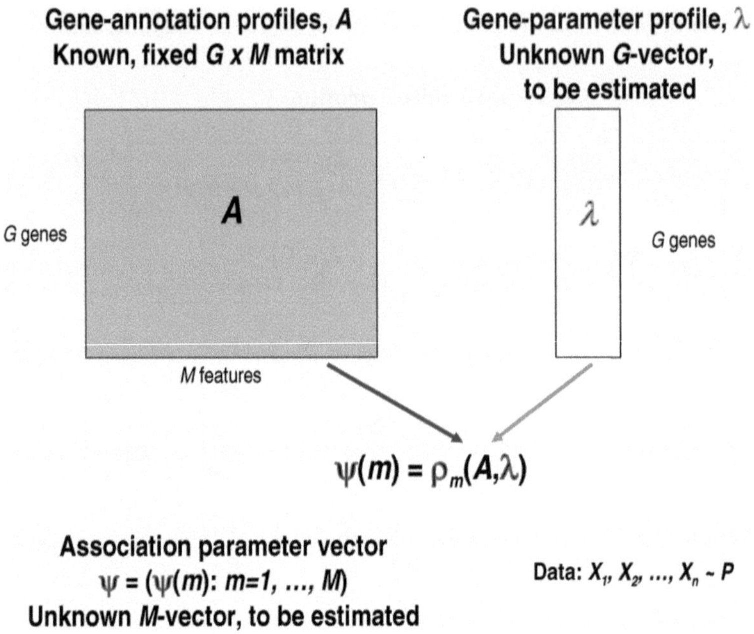

Figure 10.1. *Parameters for tests of association with biological annotation metadata.* This figure represents the main ingredients involved in multiple tests of association with biological annotation metadata: the gene-annotation profiles, the gene-parameter profile, and the association parameters.

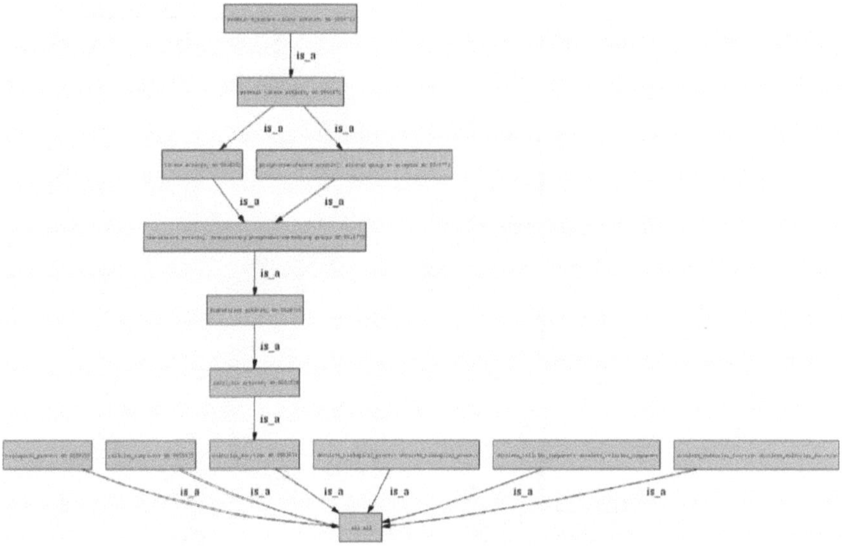

Figure 10.2. *DAG for MF GO term* `GO:0004713`*, AmiGO.* Portion of the directed acyclic graph for the GO term *protein-tyrosine kinase activity* (`GO:0004713`), in the Molecular Function ontology. This display, obtained using the AmiGO browser (Last updated 2006-02-14, **`www.godatabase.org`**), shows the nodes corresponding to all (less specific) ancestors of the term *protein-tyrosine kinase activity.*

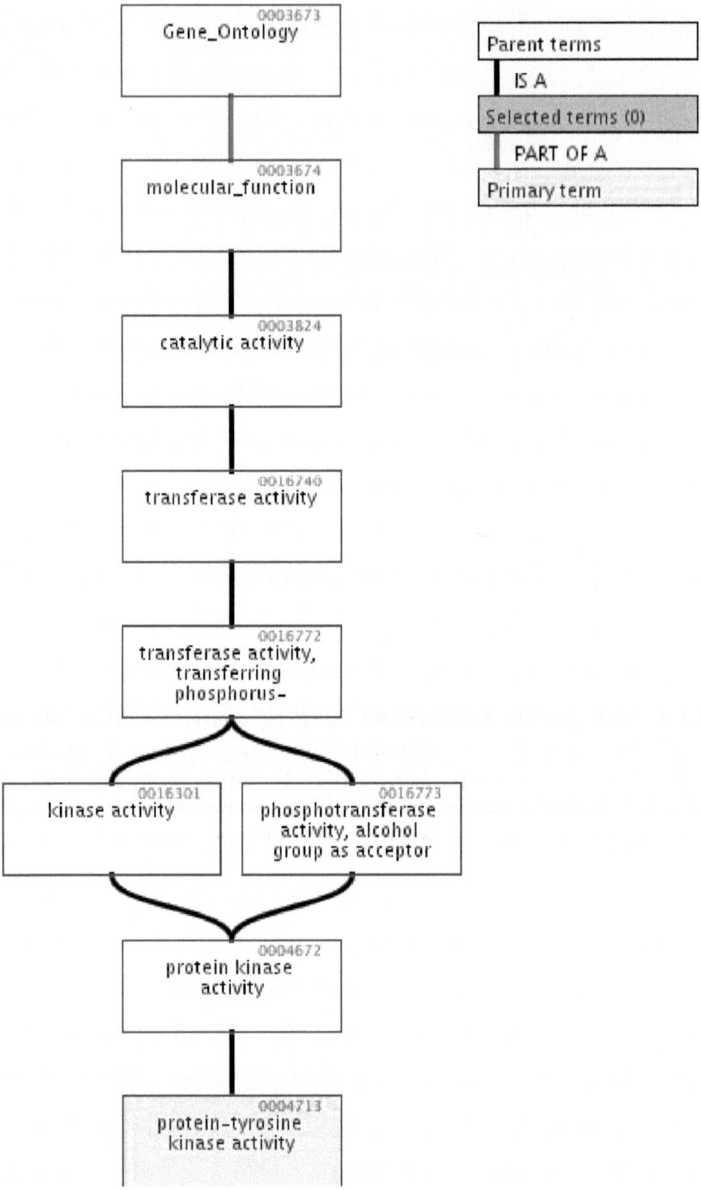

Figure 10.3. *DAG for MF GO term* `GO:0004713`*, QuickGO.* Portion of the directed acyclic graph for the GO term *protein-tyrosine kinase activity* (`GO:0004713`), in the Molecular Function ontology. This display, obtained using the EBI QuickGO browser (Last updated 2001-03-30 04:29:44.0, `www.ebi.ac.uk/ego`), shows the nodes corresponding to all (less specific) ancestors of the term *protein-tyrosine kinase activity*.

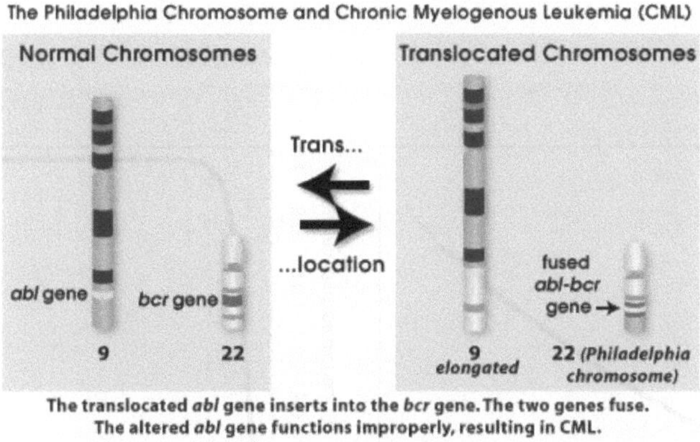

Panel (a): t(9;22) translocation

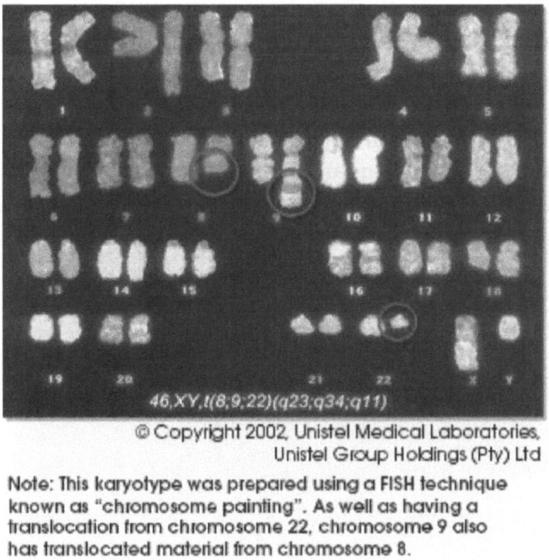

Panel (b): Karyotype

Figure 10.4. *The Philadelphia chromosome and the BCR/ABL fusion.* The BCR/ABL fusion is the molecular analogue of the Philadelphia chromosome. This t(9;22) translocation leads to a head-to-tail fusion of the `v-abl Abelson murine leukemia viral oncogene homolog 1` (`ABL1`) from chromosome 9 with the 5' half of the `breakpoint cluster region` (`BCR`) on chromosome 22. (Figure obtained from the Genetic Science Learning Center, The University of Utah, `gslc.genetics.utah.edu/units/disorders/karyotype/reciprocal.cfm`.) (Color plate p. 336)

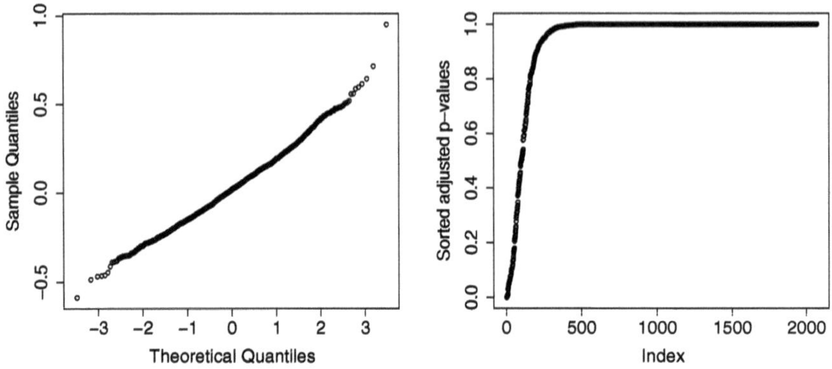

Panel (a): Test statistics Panel (b): Adjusted p-values

Figure 10.5. *Differentially expressed genes between BCR/ABL and NEG B-cell ALL.* Panel (a): Normal quantile-quantile plot of two-sample t-statistics $\lambda_n^t(g)$. Panel (b): Plot of sorted bootstrap-based single-step maxT adjusted p-values $\widetilde{P}_{0n}^{\neq}(g)$.

Table 10.2. *Differentially expressed genes between BCR/ABL and NEG B-cell ALL.* This table provides the Affymetrix probe IDs, Entrez Gene IDs (**hgu95av2LOCUSID** environment in hgu95av2 package), gene symbols (**hgu95av2SYMBOL** environment), gene names (**hgu95av2GENENAME** environment), test statistics $\lambda_n^t(g)$ (Equation (10.15)), and adjusted p-values $\widetilde{P}_{0n}^{\neq}(g)$, for the 16 genes found to be significantly differentially expressed between BCR/ABL and NEG B-cell ALL, at nominal FWER level $\alpha = 0.05$, according to the bootstrap-based single-step maxT procedure, with two-sample t-statistics $\lambda_n^t(g)$ and $B = 5,000$ bootstrap samples. A more detailed hyperlinked table, including information on gene function, chromosomal location, links to GenBank, Entrez Gene, NCBI Map Viewer, UniGene, PubMed, AmiGO, and KEGG, is provided on the website companion (Supplementary Table 10.1).

Probe ID	**Entrez Gene ID**	**Symbol**	$\lambda_n^t(g)$	$\widetilde{P}_{0n}^{\neq}(g)$
1635_at	25	ABL1	8.44	0
v-abl Abelson murine leukemia viral oncogene homolog 1				
40202_at	687	KLF9	6.33	0
Kruppel-like factor 9				
37027_at	79026	AHNAK	5.71	0.0014
AHNAK nucleoprotein (desmoyokin)				
39837_s_at	168544	ZNF467	5.45	0.0034
zinc finger protein 467				
33774_at	841	CASP8	5.29	0.0042
caspase 8, apoptosis-related cysteine peptidase				
37014_at	4599	MX1	-5.23	0.0050
myxovirus (influenza virus) resistance 1, interferon-inducible protein p78 (mouse)				
2039_s_at	2534	FYN	5.21	0.0050
FYN oncogene related to SRC, FGR, YES				
39329_at	87	ACTN1	4.97	0.0096
actinin, alpha 1				
32542_at	2273	FHL1	4.96	0.0102
four and a half LIM domains 1				
40051_at	9697	TRAM2	4.59	0.0268
translocation associated membrane protein 2				
38032_at	9900	SV2A	4.54	0.0308
synaptic vesicle glycoprotein 2A				
39319_at	3937	LCP2	4.50	0.0346
lymphocyte cytosolic protein 2 (SH2 domain containing leukocyte protein of 76 kDa)				
33232_at	1396	CRIP1	4.46	0.0368
cysteine-rich protein 1 (intestinal)				
36591_at	7277	TUBA1	4.37	0.0444
tubulin, alpha 1 (testis specific)				
38994_at	8835	SOCS2	4.35	0.0466
suppressor of cytokine signaling 2				
40076_at	7165	TPD52L2	-4.33	0.0480
tumor protein D52-like 2				

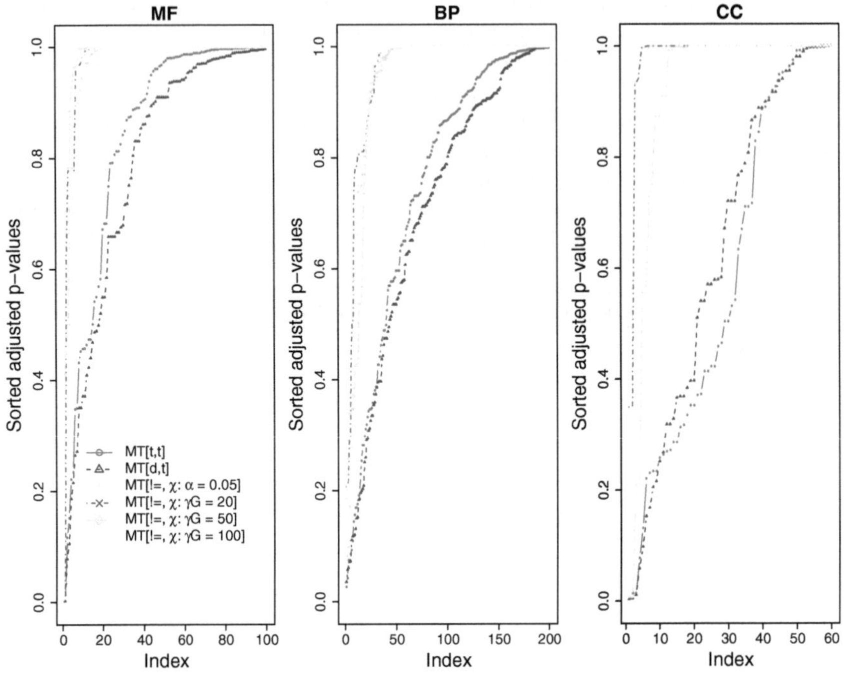

Figure 10.6. *GO terms associated with differential gene expression between BCR/ABL and NEG B-cell ALL, adjusted p-values.* Plots of sorted bootstrap-based single-step maxT adjusted p-values $\widetilde{P}_{0n}(m)$, for each of the three gene ontologies and each of the three testing scenarios. (Color plate p. 337)

Table 10.3. *GO terms associated with differential gene expression between BCR/ABL and NEG B-cell ALL.* This table reports, for each of the three gene ontologies and each of the three testing scenarios, the numbers $R_n(\alpha) = |\mathcal{R}_n(\alpha)|$ of GO terms found to be significantly associated with BCR/ABL vs. NEG differential gene expression for different nominal FWER levels α.

	Nominal FWER level, α								
	BP 0.05	BP 0.10	BP 0.20	CC 0.05	CC 0.10	CC 0.20	MF 0.05	MF 0.10	MF 0.20
MT[t,t]	2	6	14	3	4	5	1	1	3
MT[d,t]	1	5	16	3	5	7	1	2	4
MT[$\neq,\chi:\alpha=0.05$]	0	3	5	0	0	0	1	1	1
MT[$\neq,\chi:\gamma G=20$]	0	0	0	0	0	0	1	1	1
MT[$\neq,\chi:\gamma G=50$]	0	0	1	2	2	2	0	0	0
MT[$\neq,\chi:\gamma G=100$]	0	0	2	1	1	2	0	0	0

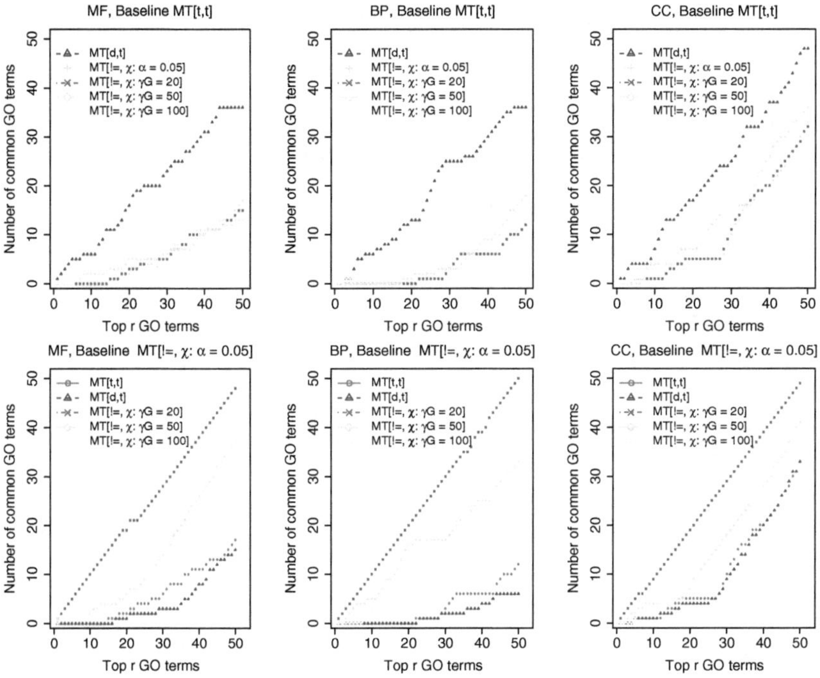

Figure 10.7. *GO terms associated with differential gene expression between BCR/ABL and NEG B-cell ALL, common terms between testing scenarios.* Plots of numbers of common GO terms among sets of ordered GO terms $\mathcal{O}_n(r)$ of various cardinality r for pairs of testing scenarios. Scenario $\mathsf{MT}[t,t]$ is used as the baseline in the top panels and Scenario $\mathsf{MT}[\neq, \chi : \alpha = 0.05]$, with adjusted p-value-based estimator $\lambda^{\neq}_{n,\alpha}$, $\alpha = 0.05$, for the binary DE gene-parameter profile $\lambda^{\neq}$, is used as the baseline in the bottom panels. For example, the blue curve in the top-left panel is a plot of $\left|\mathcal{O}^{d,t}_n(r) \cap \mathcal{O}^{t,t}_n(r)\right|$ vs. r for the MF gene ontology, i.e., of the overlap between the r most significant MF GO terms according to Scenarios $\mathsf{MT}[d,t]$ and $\mathsf{MT}[t,t]$. (Color plate p. 338)

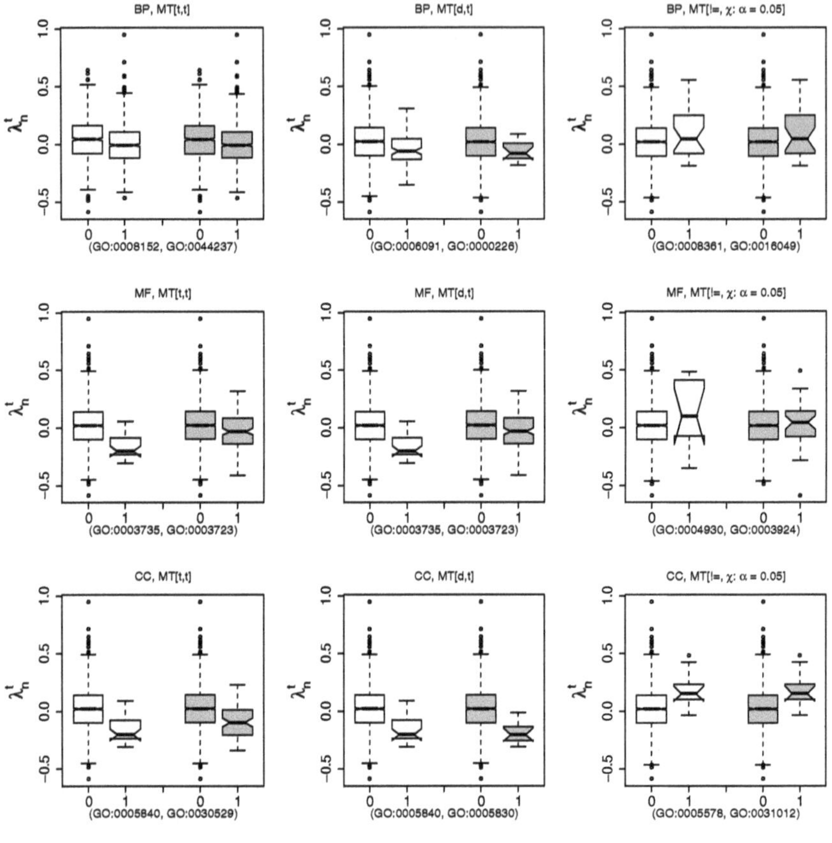

Figure 10.8. *GO terms associated with differential gene expression between BCR/ABL and NEG B-cell ALL, conditional distribution of* λ_n^t *given* A. Conditional boxplots of the estimated continuous DE gene-parameter profile λ_n^t given the gene-annotation profiles $A(\cdot, m)$ for the top two GO terms $m \in \{O_n(1), O_n(2)\}$ identified according to each of the three testing scenarios. Rows correspond to gene ontologies and columns to testing scenarios. In each panel, the white and gray boxplots correspond, respectively, to the GO terms with the smallest and second smallest adjusted p-values; boxplots for unannotated and annotated estimated gene-parameter profiles, $(\lambda_n^t(g) : A(g,m) = 0)$ and $(\lambda_n^t(g) : A(g,m) = 1)$, are labeled as 0 and 1, respectively. Non-overlapping notches (informally) represent large differences in medians.

Table 10.4. *GO terms associated with differential gene expression between BCR/ABL and NEG B-cell ALL, top two BP GO terms.* This table provides association measures between the estimated DE gene-parameter profiles λ_n^t and $\lambda_{n,\alpha}^{\neq}$ and the gene-annotation profiles $A(\cdot, m)$ for the top two BP GO terms $m \in \{O_n(1), O_n(2)\}$ identified according to each of the three testing scenarios. $A_1(m) = \sum_g A(g,m)$: Number of genes directly or indirectly annotated with GO term m (out of $G = 2,071$ genes, `GOALLLOCUSID` environment in GO package). $P_{0n}^{t,t}(m)$: Nominal unadjusted p-value for the two-sample t-test comparing the unannotated and annotated estimated continuous DE gene-parameter profiles, $(\lambda_n^t(g) : A(g,m) = 0)$ and $(\lambda_n^t(g) : A(g,m) = 1)$, respectively (`t.test` function from the R package `stats`, with default argument values). $P_{0n}^{\neq,\chi}(m)$: Unadjusted p-value for the χ^2-test of independence between the estimated binary DE gene-parameter profile $\lambda_{n,\alpha}^{\neq}$, $\alpha = 0.05$, and the gene-annotation profile $A(\cdot, m)$ (`chisq.test` function from the R package `stats`, with arguments `simulate.p.value = TRUE`, `correct=FALSE`). $\widetilde{P}_{0n}(m)$: Bootstrap-based single-step maxT adjusted p-value, according to which the top two GO terms are identified for each testing scenario.

BP

Scenario	**GO term**	$A_1(m)$	$P_{0n}^{t,t}(m)$	$P_{0n}^{\neq,\chi}(m)$	$\widetilde{P}_{0n}(m)$
MT[t,t]	GO:0008152	1076	0	0.1704	0.0262
	GO:0044237	1045	0	0.1824	0.0428
MT[d,t]	GO:0006091	98	0	0.6172	0.0366
	GO:0000226	14	0.0018	1	0.0582
MT[$\neq,\chi : \alpha = 0.05$]	GO:0008361	27	0.0553	0.0035	0.0828
	GO:0016049	27	0.0553	0.0010	0.0828
MT[$\neq,\chi : \gamma G = 20$]	GO:0008361	27	0.0553	0.0020	0.2078
	GO:0016049	27	0.0553	0.0020	0.2078
MT[$\neq,\chi : \gamma G = 50$]	GO:0048522	87	0.0356	0.0120	0.1860
	GO:0048518	96	0.0439	0.0145	0.2338
MT[$\neq,\chi : \gamma G = 100$]	GO:0050793	24	0.0854	0.0175	0.1458
	GO:0007155	59	0.0006	0.1109	0.1980

Table 10.5. *GO terms associated with differential gene expression between BCR/ABL and NEG B-cell ALL, top two CC GO terms.* Details in Table 10.4 caption.

CC

Scenario	GO term	$A_1(m)$	$P^{t,t}_{0n}(m)$	$P^{\neq,\chi}_{0n}(m)$	$\widetilde{P}_{0n}(m)$
MT$[t,t]$	GO:0005840	25	0	1	0.0056
	GO:0030529	77	0	0.6387	0.0138
MT$[d,t]$	GO:0005840	25	0	1	0.0040
	GO:0005830	11	0	1	0.0052
MT$[\neq,\chi:\alpha=0.05]$	GO:0005578	10	0.0167	0.0775	0.4940
	GO:0031012	10	0.0167	0.0815	0.4940
MT$[\neq,\chi:\gamma G=20]$	GO:0005578	10	0.0167	0.1069	0.3500
	GO:0031012	10	0.0167	0.0975	0.3500
MT$[\neq,\chi:\gamma G=50]$	GO:0005576	54	0.0009	1	0.0078
	GO:0005615	31	0.0480	0.2509	0.0078
MT$[\neq,\chi:\gamma G=100]$	GO:0005576	54	0.0009	1	0.0488
	GO:0005615	31	0.0480	0.2439	0.1280

Table 10.6. *GO terms associated with differential gene expression between BCR/ABL and NEG B-cell ALL, top two MF GO terms.* Details in Table 10.4 caption.

MF

Scenario	GO term	$A_1(m)$	$P^{t,t}_{0n}(m)$	$P^{\neq,\chi}_{0n}(m)$	$\widetilde{P}_{0n}(m)$
MT$[t,t]$	GO:0003735	24	0	1	0.0024
	GO:0003723	143	0	0.4068	0.1168
MT$[d,t]$	GO:0003735	24	0	1	0.0022
	GO:0003723	143	0	0.3968	0.0784
MT$[\neq,\chi:\alpha=0.05]$	GO:0004930	10	0.2241	0.0065	0.0366
	GO:0003924	34	0.6501	0.0395	0.7046
MT$[\neq,\chi:\gamma G=20]$	GO:0004930	10	0.2241	0.0025	0.0168
	GO:0003924	34	0.6501	0.0495	0.6210
MT$[\neq,\chi:\gamma G=50]$	GO:0004930	10	0.2241	0.0040	0.4108
	GO:0030246	22	0.8582	0.1919	0.4794
MT$[\neq,\chi:\gamma G=100]$	GO:0005509	69	0.0004	0.1399	0.3140
	GO:0004930	10	0.2241	0.0025	0.3262

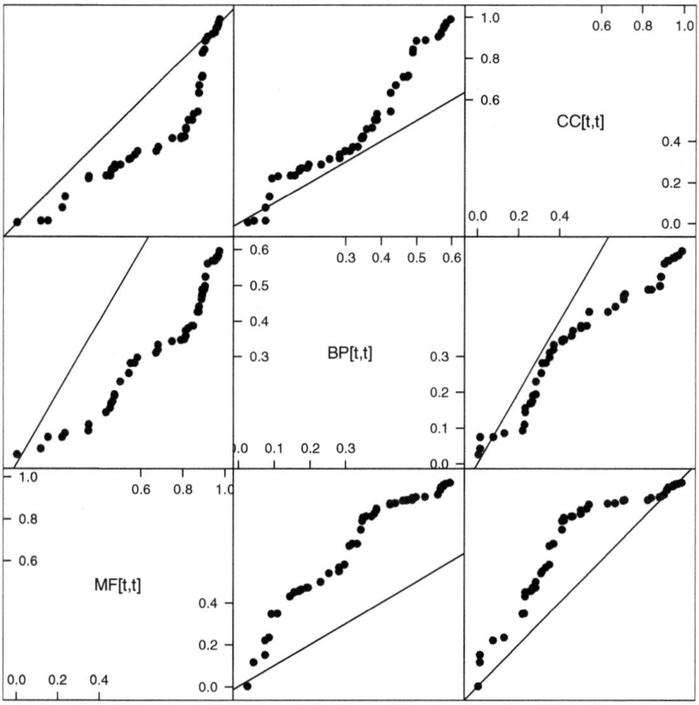

Figure 10.9. *GO terms associated with differential gene expression between BCR/ABL and NEG B-cell ALL, comparison of adjusted p-values for the three gene ontologies.* Scatterplot matrix of the 50 smallest adjusted p-values for each of the three gene ontologies, based on Scenario $\mathsf{MT}[t,t]$. The identity line is drawn for reference.

Table 10.7. *GO terms associated with differential gene expression between BCR/ABL and NEG B-cell ALL, top 20 BP GO terms.* This table lists the 20 GO terms with the smallest adjusted p-values for Scenario $\mathsf{MT}[t,t]$ applied to the BP gene ontology. $A_1(m) = \sum_g A(g,m)$: Number of genes directly or indirectly annotated with GO term m (out of $G = 2,071$ genes, `GOALLLOCUSID` environment in GO package). $\widetilde{P}_{0n}(m)$: Bootstrap-based single-step maxT adjusted p-value for Scenario $\mathsf{MT}[t,t]$.

BP, Scenario $\mathsf{MT}[t,t]$

GO term ID	GO term	$A_1(m)$	$\widetilde{P}_{0n}(m)$
GO:008152	*metabolism*	1076	0.0262
GO:044237	*cellular metabolism*	1045	0.0428
GO:009058	*biosynthesis*	187	0.0750
GO:044238	*primary metabolism*	1002	0.0750
GO:044249	*cellular biosynthesis*	169	0.0862
GO:006091	*generation of precursor metabolites and energy*	98	0.0928
GO:019882	*antigen presentation*	15	0.1098
GO:030333	*antigen processing*	14	0.1444
GO:006916	*anti-apoptosis*	21	0.1564
GO:043066	*negative regulation of apoptosis*	26	0.1692
GO:043069	*negative regulation of programmed cell death*	26	0.1692
GO:007154	*cell communication*	390	0.1754
GO:006457	*protein folding*	52	0.1910
GO:007165	*signal transduction*	351	0.1946
GO:000226	*microtubule cytoskeleton organization and biogenesis*	14	0.2302
GO:006082	*organic acid metabolism*	65	0.2538
GO:006163	*purine nucleotide metabolism*	29	0.2820
GO:007155	*cell adhesion*	59	0.2822
GO:007028	*cytoplasm organization and biogenesis*	10	0.2976
GO:019752	*carboxylic acid metabolism*	63	0.3108

Table 10.8. *GO terms associated with differential gene expression between BCR/ABL and NEG B-cell ALL, top 20 CC GO terms.* Details in Table 10.7 caption.

CC, Scenario MT$[t, t]$

GO term ID	**GO term**	$A_1(m)$	$\widetilde{P}_{0n}(m)$
GO:0005840	*ribosome*	25	0.0056
GO:0030529	*ribonucleoprotein complex*	77	0.0138
GO:0005830	*cytosolic ribosome (sensu Eukaryota)*	11	0.0144
GO:0043234	*protein complex*	334	0.0778
GO:0005886	*plasma membrane*	200	0.1316
GO:0005829	*cytosol*	78	0.2204
GO:0005737	*cytoplasm*	578	0.2304
GO:0005887	*integral to plasma membrane*	125	0.2338
GO:0031226	*intrinsic to plasma membrane*	125	0.2338
GO:0019866	*inner membrane*	37	0.2574
GO:0005743	*mitochondrial inner membrane*	28	0.2636
GO:0005746	*mitochondrial electron transport chain*	11	0.2692
GO:0000502	*proteasome complex (sensu Eukaryota)*	26	0.2714
GO:0000323	*lytic vacuole*	28	0.2866
GO:0005764	*lysosome*	28	0.2866
GO:0005576	*extracellular region*	54	0.3130
GO:0005773	*vacuole*	29	0.3172
GO:0005622	*intracellular*	1152	0.3350
GO:0043228	*non-membrane-bound organelle*	218	0.3524
GO:0043232	*intracellular non-membrane-bound organelle*	218	0.3524

Table 10.9. *GO terms associated with differential gene expression between BCR/ABL and NEG B-cell ALL, top 20 MF GO terms.* Details in Table 10.7 caption.

MF, Scenario MT[t, t]

GO term ID	GO term	$A_1(m)$	$\widetilde{P}_{0n}(m)$
GO:0003735	*structural constituent of ribosome*	24	0.0024
GO:0003723	*RNA binding*	143	0.1168
GO:0048037	*cofactor binding*	11	0.1518
GO:0051082	*unfolded protein binding*	47	0.2210
GO:0016853	*isomerase activity*	28	0.2348
GO:0016491	*oxidoreductase activity*	89	0.3476
GO:0005509	*calcium ion binding*	69	0.3496
GO:0015399	*primary active transporter activity*	57	0.4314
GO:0004872	*receptor activity*	101	0.4518
GO:0004871	*signal transducer activity*	242	0.4566
GO:0016765	*transferase activity, transferring alkyl or aryl (other than methyl) groups*	10	0.4570
GO:0016860	*intramolecular oxidoreductase activity*	13	0.4636
GO:0016614	*oxidoreductase activity, acting on CH-OH group of donors*	18	0.4734
GO:0016616	*oxidoreductase activity, acting on the CH-OH group of donors, NAD or NADP as acceptor*	18	0.4734
GO:0043169	*cation binding*	230	0.5002
GO:0005489	*electron transporter activity*	47	0.5420
GO:0005386	*carrier activity*	73	0.5502
GO:0004888	*transmembrane receptor activity*	59	0.5690
GO:0003824	*catalytic activity*	635	0.5826
GO:0003676	*nucleic acid binding*	449	0.6718

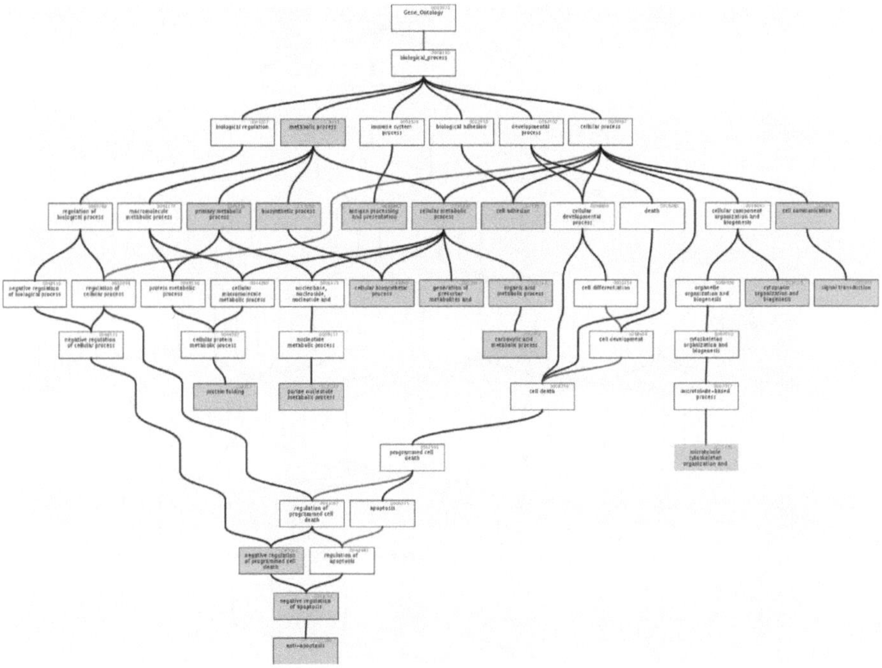

Figure 10.10. *GO terms associated with differential gene expression between BCR/ABL and NEG B-cell ALL, DAG for top 20 BP GO terms.* Portion of the directed acyclic graph for the 20 GO terms with the smallest adjusted p-values for Scenario $\mathsf{MT}[t, t]$ applied to the BP gene ontology. Nodes for the top 20 terms are shaded in lavender; black and red edges indicate, respectively, "is a" and "part of a" relationships among terms. The figure was produced using the QuickGO browser. According to QuickGO, the GO term IDs `GO:019882` and `GO:0030333` listed in Table 10.7 correspond to the same term, *antigen processing and presentation.*

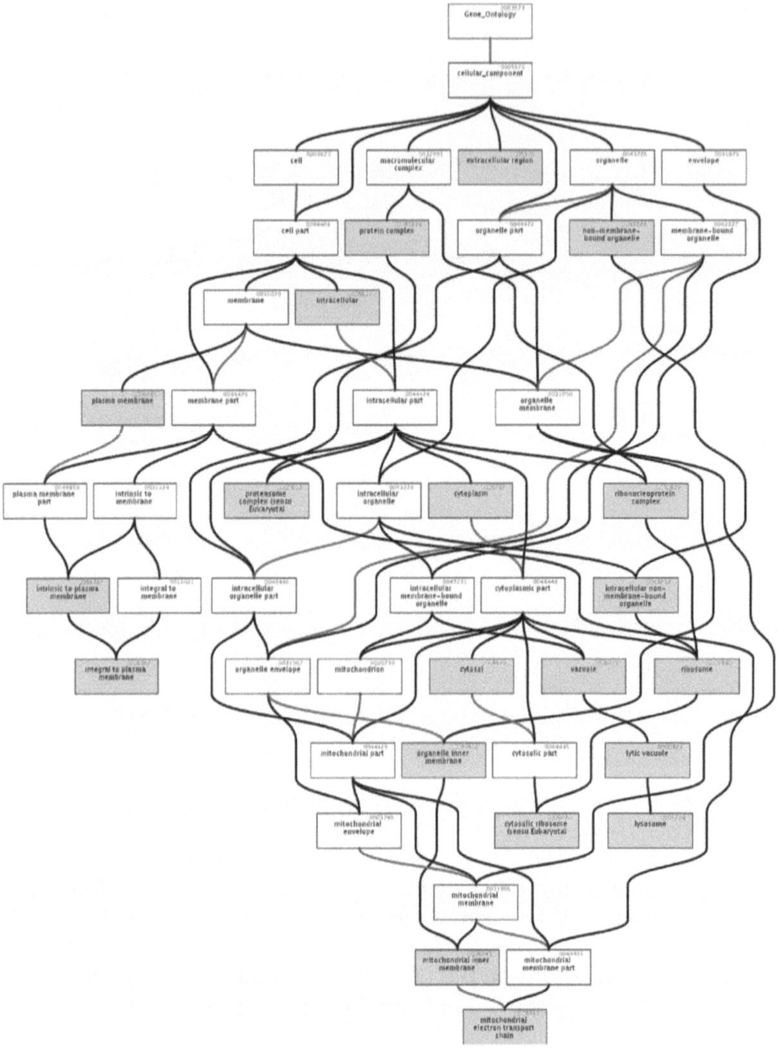

Figure 10.11. *GO terms associated with differential gene expression between BCR/ABL and NEG B-cell ALL, DAG for top 20 CC GO terms.* Portion of the directed acyclic graph for the 20 GO terms with the smallest adjusted p-values for Scenario $\mathsf{MT}[t, t]$ applied to the CC gene ontology. Nodes for the top 20 terms are shaded in lavender; black and red edges indicate, respectively, "is a" and "part of a" relationships among terms. The figure was produced using the QuickGO browser.

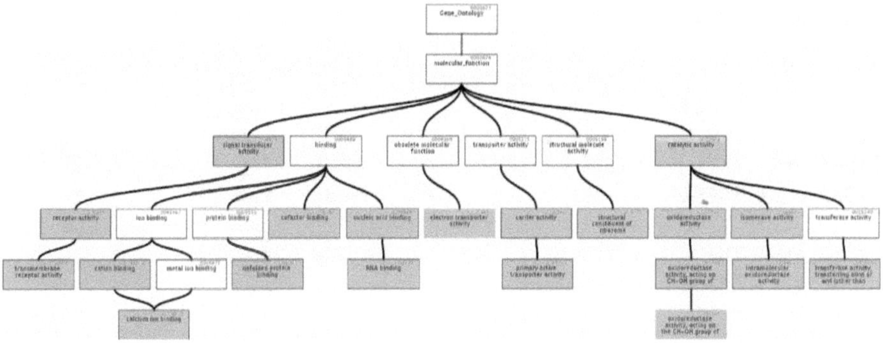

Figure 10.12. *GO terms associated with differential gene expression between BCR/ABL and NEG B-cell ALL, DAG for top 20 MF GO terms.* Portion of the directed acyclic graph for the 20 GO terms with the smallest adjusted p-values for Scenario $\mathsf{MT}[t,t]$ applied to the MF gene ontology. Nodes for the top 20 terms are shaded in lavender; black and red edges indicate, respectively, "is a" and "part of a" relationships among terms. The figure was produced using the QuickGO browser.

Table 10.10. *GO terms associated with differential gene expression between BCR/ABL and NEG B-cell ALL, BP GO term `GO:0006916`.* This table lists genes directly or indirectly annotated with GO term *anti-apoptosis* (out of $G = 2,071$ genes, `GOALLLOCUSID` environment in GO package). The term *anti-apoptosis* (`GO:0006916`) has the ninth smallest adjusted p-value for Scenario $\mathsf{MT}[t, t]$ applied to the BP gene ontology (Table 10.7).

BP `GO:0006916`

Probe ID	Symbol	Name
1237_at	IER3	immediate early response 3
1295_at	RELA	v-rel reticuloendotheliosis viral oncogene homolog A, nuclear factor of kappa light polypeptide gene enhancer in B-cells 3, p65 (avian)
1377_at	NFKB1	nuclear factor of kappa light polypeptide gene enhancer in B-cells 1 (p105)
1564_at	AKT1	v-akt murine thymoma viral oncogene homolog 1
1830_s_at	TGFB1	transforming growth factor, beta 1 (Camurati-Engelmann disease)
1852_at	TNF	tumor necrosis factor (TNF superfamily, member 2)
1997_s_at	BAX	BCL2-associated X protein
277_at	MCL1	myeloid cell leukemia sequence 1 (BCL2-related)
31536_at	RTN4	reticulon 4
32060_at	BNIP2	BCL2/adenovirus E1B 19 kDa interacting protein 2
33284_at	MPO	myeloperoxidase
36578_at	BIRC2	baculoviral IAP repeat-containing 2
38578_at	TNFRSF7	tumor necrosis factor receptor superfamily, member 7
38771_at	HDAC1	histone deacetylase 1
38994_at	SOCS2	suppressor of cytokine signaling 2
39097_at	SON	SON DNA binding protein
39378_at	BECN1	beclin 1 (coiled-coil, myosin-like BCL2 interacting protein)
39436_at	BNIP3L	BCL2/adenovirus E1B 19 kDa interacting protein 3-like
40570_at	FOXO1A	forkhead box O1A (rhabdomyosarcoma)
595_at	TNFAIP3	tumor necrosis factor, alpha-induced protein 3
641_at	PSEN1	presenilin 1 (Alzheimer disease 3)

Table 10.11. *GO terms associated with differential gene expression between BCR/ABL and NEG B-cell ALL, MF GO term* `GO:0003735`. This table lists genes directly or indirectly annotated with GO term *structural constituent of ribosome* (out of $G = 2,071$ genes, `GOALLLOCUSID` environment in GO package). The term *structural constituent of ribosome* (`GO:0003735`) has the smallest adjusted p-value for Scenario $\mathsf{MT}[t,t]$ applied to the MF gene ontology (Table 10.9).

MF `GO:0003735`

Probe ID	Symbol	Name
2016_s_at	RPL10	ribosomal protein L10
31511_at	RPS9	ribosomal protein S9
31546_at	RPL18	ribosomal protein L18
31955_at	FAU	Finkel-Biskis-Reilly murine sarcoma virus (FBR-MuSV) ubiquitously expressed (fox derived)
32221_at	MRPS18B	mitochondrial ribosomal protein S18B
32315_at	RPS24	ribosomal protein S24
32394_s_at	RPL23	ribosomal protein L23
32433_at	RPL15	ribosomal protein L15
32437_at	RPS5	ribosomal protein S5
33117_r_at	RPS12	ribosomal protein S12
33485_at	RPL4	ribosomal protein L4
33614_at	RPL18A	ribosomal protein L18a
33661_at	RPL5	ribosomal protein L5
33668_at	RPL12	ribosomal protein L12
33674_at	RPL29	ribosomal protein L29
34316_at	RPS15A	ribosomal protein S15a
36358_at	RPL9	ribosomal protein L9
36572_r_at	ARL6IP	ADP-ribosylation factor-like 6 interacting protein
36786_at	RPL10A	ribosomal protein L10a
39856_at	RPL36AL	ribosomal protein L36a-like
39916_r_at	RPS15	ribosomal protein S15
41152_f_at	RPL36A	ribosomal protein L36a
41214_at	RPS4Y1	ribosomal protein S4, Y-linked 1
41746_at	NHP2L1	NHP2 non-histone chromosome protein 2-like 1 (S. cerevisiae)

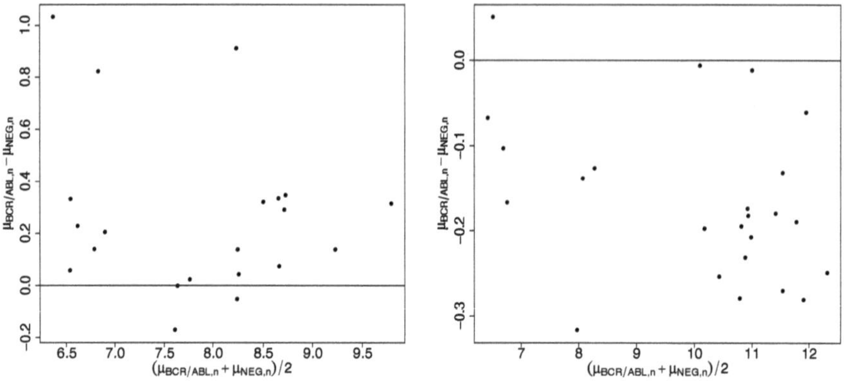

Panel (a): BP `GO:0006916` Panel (b): MF `GO:0003735`

Figure 10.13. *GO terms associated with differential gene expression between BCR/ABL and NEG B-cell ALL, BP GO term* `GO:0006916` *and MF GO term* `GO:0003735`*.* This figure displays mean-difference plots of average expression measures in BCR/ABL and NEG cell samples, i.e., plots of $\mu_{BCR/ABL,n}(g) - \mu_{NEG,n}(g)$ vs. $(\mu_{BCR/ABL,n}(g) + \mu_{NEG,n}(g))/2$, for genes directly or indirectly annotated with GO terms `GO:0006916` (Panel (a)) and `GO:0003735` (Panel (b)). The term *anti-apoptosis* (`GO:0006916`) has the ninth smallest adjusted p-value for Scenario $\mathsf{MT}[t,t]$ applied to the BP gene ontology (Tables 10.7 and 10.10) and the term *structural constituent of ribosome* (`GO:0003735`) has the smallest adjusted p-value for Scenario $\mathsf{MT}[t,t]$ applied to the MF gene ontology (Tables 10.9 and 10.11).

11

HIV-1 Sequence Variation and Viral Replication Capacity

11.1 Introduction

Amino acid sequence variation can have a substantial impact on the secondary and tertiary *structure* of a *protein*, thereby affecting its *function* and, ultimately, a variety of *phenotypes*.

This chapter analyzes the HIV-1 dataset of Segal et al. (2004), with the aim of relating HIV-1 sequence variation to viral replication capacity in AIDS patients. Section 11.2 provides a brief presentation of the HIV-1 dataset. Section 11.3 describes multiple testing procedures (MTP) for identifying protease and reverse transcriptase codons that are significantly associated with viral replication capacity. Specifically, the null hypotheses of no genotype-phenotype associations are tested using two-sample t-statistics comparing differences in mean replication capacity for viruses with mutant and wild-type codons. The following multiple testing procedures are applied: FWER-controlling single-step maxT Procedure 3.5, gFWER-controlling augmentation Procedure 3.20, TPPFP-controlling augmentation Procedure 3.26, and TPPFP-controlling resampling-based empirical Bayes Procedure 7.1. Section 11.4 illustrates the software implementation of the MTPs using the SAS macros of Section 13.2. Finally, Section 11.5 discusses the results of the MTPs and their biological implications.

11.2 HIV-1 dataset of Segal et al. (2004)

11.2.1 HIV-1 sequence variation and viral replication capacity

Studying *genomic sequence variation* for the *human immunodeficiency virus type 1* (HIV-1) genome could potentially give important insight into *genotype-phenotype associations* for the *acquired immune deficiency syndrome* (AIDS).

In this context, a key *phenotype* is the *replication capacity* (RC) of HIV-1, as it reflects the severity of the disease. A measure of replication capacity may be obtained by monitoring viral replication in an ideal environment, with many cellular targets, no exogenous or endogenous inhibitors, and no immune system responses against the virus (Barbour et al., 2002; Segal et al., 2004).

Genotypes of interest correspond to *codons* in the protease and reverse transcriptase regions of the viral strand. The *protease* (PR) enzyme affects the reproductive cycle of the virus by breaking protein peptide bonds during replication. The *reverse transcriptase* (RT) enzyme synthesizes double-stranded DNA from the virus' single-stranded RNA genome, thereby facilitating integration into the host's chromosome. Because the PR and RT regions are essential to viral replication, many antiretrovirals (protease inhibitors and reverse transcriptase inhibitors) have been developed to target these specific genomic locations. Studying PR and RT genotypic variation involves sequencing the corresponding HIV-1 genomic regions and determining the amino acids encoded by each codon, i.e., each nucleotide triplet. Figure 11.1 provides a diagram of the HIV-1 lifecycle and modes of action of protease and reverse transcriptase inhibitors.

11.2.2 HIV-1 dataset

The HIV-1 dataset comprises $n = 317$ records, linking viral replication capacity with protease and reverse transcriptase sequence data, from individuals participating in studies at the San Francisco General Hospital and the Gladstone Institute of Virology and Immunology (Segal et al., 2004).

Protease codon positions 4 to 99 (i.e., `pr4`–`pr99`) and reverse transcriptase codon positions 38 to 223 (i.e., `rt38`–`rt223`) on the viral strand are considered in the present analysis. In order to obtain accurate PR and RT codon genotypes for each patient, mutations were confirmed using data curated by the Stanford University HIV Drug Resistance Database (HIVdb, `hivdb.stanford.edu`). Note that, in the remainder of this chapter, the term *codon* is used loosely to refer to one of twenty *amino acids* (each encoded by possibly multiple nucleotide triplets), rather than one of $4^3 = 64$ triplets of nucleotides.

The data for each of the $n = 317$ patients consist of a replication capacity measure Y_i and an $M = 282$-dimensional covariate vector $X_i = (X_i(m) : m = 1, \ldots, M)$ of codon genotypes in the PR and RT HIV-1 regions, $i = 1, \ldots, n$.

Specifically, the *outcome/phenotype* Y of interest is the natural logarithm of a continuous measure of replication capacity, ranging from 0.261 to 151.

The M *covariates/genotypes* each correspond to one of the $M = 96 + 186 = 282$ codon positions in the PR and RT regions, with the observed number of codons ranging from 1 to a maximum of 10 at any given location. A wide majority of patients typically exhibit one particular codon at each position. Codons are therefore recoded as *binary* covariates, with value of 0 corresponding to the *wild-type* codon, i.e., the most common codon among the $n = 317$ patients, and value of 1 for *mutant* codons, i.e., all other codons.

Thus, X is a binary covariate M-vector, where $X(m) \in \{0,1\}$ denotes the binary codon genotype at position m, $m = 1, \ldots, M$.

11.3 Multiple testing procedures

The aim is to identify protease and reverse transcriptase codons significantly associated with viral replication capacity. Specifically, we wish to test for each of the $M = 282$ codon positions whether viral replication capacity Y is associated with the corresponding binary codon genotype $X(m) \in \{0,1\}$, $m = 1, \ldots, M$. For the mth codon position (i.e., mth hypothesis), the parameter of interest is the *difference in mean replication capacity* $\psi(m)$ for viruses with mutant and wild-type codons, that is,

$$\psi(m) \equiv \mathrm{E}[Y|X(m) = 1] - \mathrm{E}[Y|X(m) = 0], \qquad m = 1, \ldots, M. \tag{11.1}$$

We consider two-sided tests of the null hypotheses $H_0(m) = \mathrm{I}\left(\psi(m) = 0\right)$ of no differences in mean RC vs. the alternative hypotheses $H_1(m) = \mathrm{I}\left(\psi(m) \neq 0\right)$ of different mean RC, for each codon position m.

The tests are based on *two-sample pooled-variance t-statistics*,

$$\begin{aligned} T_n(m) &\equiv \frac{\psi_n(m) - \psi_0(m)}{\sigma_n(m)} \qquad (11.2)\\ &= \frac{\bar{Y}_{1,n}(m) - \bar{Y}_{0,n}(m) - 0}{\sigma_{p,n}(m)\sqrt{\frac{1}{n_0(m)} + \frac{1}{n_1(m)}}}, \\ \sigma^2_{p,n}(m) &\equiv \frac{(n_0(m) - 1)\sigma^2_{0,n}(m) + (n_1(m) - 1)\sigma^2_{1,n}(m)}{n_0(m) + n_1(m) - 2}, \end{aligned}$$

where the null values $\psi_0(m)$ are zero, $n_k(m) \equiv \sum_i \mathrm{I}\left(X_i(m) = k\right)$ denotes the number of patients with codon genotype $X(m) = k \in \{0,1\}$ at position m, and $\bar{Y}_{k,n}(m) \equiv \sum_i \mathrm{I}\left(X_i(m) = k\right) Y_i / n_k(m)$ and $\sigma^2_{k,n}(m) \equiv \sum_i \mathrm{I}\left(X_i(m) = k\right) (Y_i - \bar{Y}_{k,n}(m))^2 / (n_k(m) - 1)$ denote, respectively, the sample means and sample variances for the RC of patients with codon genotype $X(m) = k \in \{0,1\}$ at position m. The pooled-variance estimators are denoted by $\sigma^2_{p,n}(m)$. The estimators of the differences in mean replication capacity, $\psi(m)$, are simply the corresponding differences in sample means, $\psi_n(m) \equiv \bar{Y}_{1,n}(m) - \bar{Y}_{0,n}(m)$. The estimated standard errors of these estimators are $\sigma_n(m) \equiv \sigma_{p,n}(m)\sqrt{1/n_0(m) + 1/n_1(m)}$. The null hypothesis $H_0(m)$ is rejected, i.e., the corresponding codon position is declared significantly associated with RC, for large absolute values of the test statistic $T_n(m)$. Note that the above two-sample pooled-variance t-statistics correspond to t-statistics for the univariate regression of the outcome Y on the binary covariates $X(m)$ (cf. one-way ANOVA).

11.3.1 Multiple testing analysis, Part I

In the main analysis, a variety of bootstrap-based multiple testing procedures, controlling the FWER, gFWER, and TPPFP, are applied to assess the statistical significance of the genotype-phenotype associations. The results of each MTP are reported in terms of adjusted p-values, so that codon positions with adjusted p-values less than or equal to a user-supplied α are declared significantly associated with the RC phenotype at nominal Type I error level α (Section 1.2.12).

Specifically, the *non-parametric bootstrap* version of Procedure 2.3 is applied to obtain an estimate Q_{0n} of the *null shift and scale-transformed test statistics null distribution* Q_0 of Section 2.3 (with $B = 7,500$ bootstrap samples). This estimated null distribution is used to compute unadjusted p-values and adjusted p-values for the following procedures.

- *FWER control*: Joint single-step (common-cut-off) maxT Procedure 3.5.
- *gFWER control*: $gFWER(k)$-controlling augmentation multiple testing Procedure 3.20, based on FWER-controlling joint single-step maxT Procedure 3.5, for an allowed number $k \in \{5, 10, 50\}$ of false positives.
- *TPPFP control*: $TPPFP(q)$-controlling augmentation multiple testing Procedure 3.26, based on FWER-controlling joint single-step maxT Procedure 3.5, for an allowed proportion $q \in \{0.10, 0.20, 0.50\}$ of false positives.

These three multiple testing procedures are implemented as described in Section 11.4, below, using SAS macros. Note that all procedures are also available in the Bioconductor R package multtest, discussed in detail in Section 13.1 (Pollard et al. (2005b); `www.bioconductor.org`).

11.3.2 Multiple testing analysis, Part II

The simulation studies in Dudoit et al. (2004a) and van der Laan et al. (2005) suggest that, although augmentation multiple testing procedures (AMTP) compare favorably to TPPFP-controlling marginal procedures, they become more conservative as the number of tested hypotheses increases. Motivated by these observations, van der Laan et al. (2005) have developed a TPPFP-controlling resampling-based empirical Bayes procedure, which, as does the augmentation method, provides asymptotic Type I error control for general data generating distributions, but is less conservative for finite samples (Chapter 7).

In a secondary analysis of the HIV-1 dataset, we compare results for *TPPFP-controlling augmentation multiple testing Procedure 3.26* and *TPPFP-controlling resampling-based empirical Bayes Procedure 7.1*. This analysis was performed in R, using prototype code; the new empirical Bayes approach will be available in the multtest package, in a forthcoming Bioconductor software release.

Both TPPFP-controlling procedures are based on the *non-parametric bootstrap null shift and scale-transformed test statistics null distribution* Q_{0n} of Procedure 2.3, with $B = 10,000$ bootstrap samples.

For empirical Bayes Procedure 7.1, $B_0 = 50$ pairs $\{(T_{0n}^b, \mathcal{H}_{0n}^b) : b = 1, \ldots, B_0\}$, of null test statistics and random guessed sets of true null hypotheses, are generated as follows. The reader is referred to Chapter 7 for details.

1. The M-vectors of *null test statistics* T_{0n}^b have the non-parametric bootstrap null shift and scale-transformed test statistics null distribution Q_{0n} of Procedure 2.3, i.e., $\{T_{0n}^b : b = 1, \ldots, B_0\}$ correspond to B_0 columns from the $M \times B$ matrix $\mathbf{Z}_n^B$ of null shift and scale-transformed bootstrap test statistics ($B = 10,000$).
2. The *random guessed sets of true null hypotheses* $\mathcal{H}_{0n}^b$ have a distribution $Q_{0n}^{\mathcal{H}}$ that corresponds to M independent Bernoulli random variables with parameters $\pi_{0n}(T_n(m))$. That is, generate binary random M-vectors $H_{0n}^b = (H_{0n}^b(m) : m = 1, \ldots, M)$ of null hypotheses as
$$H_{0n}^b(m) \overset{\perp}{\sim} \text{Bernoulli}\left(\pi_{0n}(T_n(m))\right), \qquad m = 1, \ldots, M, \tag{11.3}$$
and define
$$\mathcal{H}_{0n}^b \equiv \left\{m : H_{0n}^b(m) = 1\right\}. \tag{11.4}$$
Here, $\pi_{0n}(t)$ is an estimated true null hypothesis posterior probability function, such as the estimated local q-value function
$$\pi_{0n}(t) = \min\left\{1, \frac{\pi_{0n} f_{0n}(t)}{f_n(t)}\right\}, \tag{11.5}$$
corresponding to the marginal non-parametric mixture model of Section 7.2.4. In the HIV-1 dataset application, the estimated marginal null density f_{0n} is simply a standard normal density. The estimated marginal density f_n is obtained by applying a Gaussian kernel density estimator to the $M \times B$ pooled elements of the matrix $\mathbf{T}_n^B$ of bootstrap test statistics, before shifting and scaling (Procedure 2.3, with $B = 10,000$). The kernel density estimator is implemented using the R function `density` (`stats` package), with default argument values.
3. Null test statistics T_{0n}^b and guessed sets $\mathcal{H}_{0n}^b$ are generated *independently*, given the empirical distribution P_n, for each $b = 1, \ldots, B_0$.

11.4 Software implementation in SAS

The multiple testing procedures implemented in the SAS macros of Section 13.2 are applied to perform Part I of the analysis of the HIV-1 dataset (SAS, Version 9, `www.sas.com`).

1. Read in the data in the form of a SAS dataset `[y:X]`, with first column corresponding to an outcome Y (here, natural logarithm of viral replication capacity measure) and M subsequent columns to an M-dimensional covariate vector $X = (X(m) : m = 1, \ldots, M)$ (here, binary codon genotypes).
2. Define the following parameters or macro variables.
   - `&row`: the number of rows for the dataset `[y:X]`, here $n = 317$ patients.
   - `&col`: the number of columns for the dataset `[y:X]`, here $M+1 = 283$.
   - `&boots`: the number of bootstrap samples for estimating the null shift and scale-transformed test statistics null distribution, here $B = 7,500$.
   - `&k`: the allowed number of false positives for the gFWER-controlling augmentation procedure, here $k \in \{5, 10, 50\}$.
   - `&q`: the allowed proportion of false positives for the TPPFP-controlling augmentation procedure, here $q \in \{0.10, 0.20, 0.50\}$.
   - `&nt`: the number of hypotheses, here $M = 282$.
3. Apply the `%lmt` macro to compute $M = 282$ codon-specific t-statistics, T_n.
4. Apply the `%boot` and `%bootnull` macros to obtain a bootstrap estimate Q_{0n} of the null shift and scale-transformed test statistics null distribution.
5. Compute adjusted p-values for FWER-controlling single-step maxT Procedure 3.5 using the `%ssmaxT` macro.
6. Given adjusted p-values for the initial FWER-controlling MTP, apply the `%gfwer` and `%tppfp` macros to obtain, respectively, adjusted p-values for gFWER- and TPPFP-controlling augmentation Procedures 3.20 and 3.26.

Note that the Type I error rate parameters k and q and the number of bootstrap samples B are user-supplied, in contrast to the other macro variables which are predetermined by the dataset and null hypotheses under consideration. SAS code for the analysis of the HIV-1 dataset reported in Section 11.5.1 is provided in Appendix C and on the book's website (`www.stat.berkeley.edu/~sandrine/MTBook/SAS/SAS.html`). The total computing time for the analysis was approximately 34.5 hours (Dell GX270, Intel Pentium 4, 2.8 GHz, 512 MB RAM).

11.5 Results

11.5.1 Multiple testing analysis, Part I

Sorted adjusted p-values are plotted in Figure 11.2 for FWER-controlling single-step maxT Procedure 3.5, $gFWER(k)$-controlling augmentation Procedure 3.20 (allowed number of false positives, $k \in \{10, 50\}$), and $TPPFP(q)$-controlling augmentation Procedure 3.26 (allowed proportion of false positives, $q \in \{0.10, 0.20, 0.50\}$).

Table 11.1 reports the 13 smallest sorted adjusted p-values and corresponding t-statistics for multiple testing procedures controlling the FWER, gFWER ($k = 5$), and TPPFP ($q = 0.10$).

The MTPs identify several codon positions as significantly associated with viral replication capacity. For a nominal Type I error level $\alpha = 0.05$, FWER-controlling single-step maxT Procedure 3.5 identifies 7 codon positions (out of $M = 282$), while gFWER- and TPPFP-controlling augmentation Procedures 3.20 and 3.26 identify, respectively, 12 ($= 5 + 7$) and 7 codon positions.

Adjusted p-values for the gFWER-controlling AMTP, with $k = 5$ allowed false positives, are simply k-shifted versions of the single-step maxT adjusted p-values. Adjusted p-values for the TPPFP-controlling AMTP, with an allowed proportion $q = 0.10$ of false positives, are mq-shifted versions (up to a ceiling integer transformation) of the single-step maxT adjusted p-values. In particular, the 13 smallest adjusted p-values for the TPPFP-controlling AMTP correspond to single-step maxT adjusted p-values with the following ranks: 1, 2, 3, 4, 5, 6, 7, 8, 9, 9, 10, 11, and 12.

11.5.2 Multiple testing analysis, Part II

Table 11.2 presents results from two TPPFP-controlling MTPs, augmentation multiple testing Procedure 3.26 and resampling-based empirical Bayes Procedure 7.1. The table reports, for each MTP, the number of rejected hypotheses, i.e., the number of codon positions (out of $M = 282$) found to be associated with replication capacity, for an allowed proportion $q = 0.05$ of false positives and nominal Type I error levels $\alpha = 0.05, 0.10$.

As expected from the simulation studies in van der Laan et al. (2005), empirical Bayes Procedure 7.1 rejects a greater number of null hypotheses than augmentation Procedure 3.26, at both nominal Type I error levels α.

11.5.3 Biological interpretation

Among the identified protease (`pr32`, `pr34`, `pr43`, `pr46`, `pr47`, `pr54`, `pr55`, `pr82`, and `pr90`) and reverse transcriptase (`rt41`, `rt184`, and `rt215`) codon positions, several have been singled out in previous research as related to replication capacity and/or antiretroviral resistance (Birkner et al., 2005c; Segal et al., 2004; Shafer et al., 2001).

It is interesting to note that the 13 codon positions with the smallest adjusted p-values in Table 11.1 all have negative t-statistics, suggesting that mutant codons (recoded as 1) are associated with *decreased viral replication capacity*. Indeed, although mutations may allow the virus to become resistant to antiretroviral therapies, this gain may come at the cost of a decrease in its replication capacity.

The specific mutations observed in the present study are consistent with those found in the literature. For example, `Vpr32I`, `Mpr46I`, `Ipr54V/L/T`, `Vpr82A/T/F/S`, and `Lpr90M`, correspond to protease codon positions for which mutations increase the resistance to various protease inhibitors. Here, the nomenclature Vpr32I refers to protease codon position pr32, with wild-type amino acid V (valine, Val) and mutant amino acid I (isoleucine, Ile). Mutations in several of the identified codons also have an impact on the replication capacity of the virus. Mutation at reverse transcriptase codon position rt41 (Mrt41L) increases azidothymidine (AZT) resistance when present with Trt215Y/F, i.e., with a mutated Y or F amino acid at position rt215. In addition, mutation Mrt184V/I partially suppresses Trt215Y/Y-mediated AZT resistance (Shafer et al., 2001). AZT, also known as Zidovudine, is a nucleoside reverse transcriptase inhibitor. It affects HIV's ability to replicate by producing faulty reverse transcriptase and, hence, inhibiting the transcription of RNA to DNA.

11.6 Discussion

As illustrated in the above HIV-1 data analysis, the multiple testing methodology of Chapters 1–7 provides simple and flexible procedures for identifying specific codons, or regions of the viral strand, that are significantly associated with replication capacity.

Our results are consistent with previous research and other analyses of this HIV-1 dataset. The reader is referred to articles by Birkner et al. (2005c) and Segal et al. (2004) for alternative statistical analyses and biological discussion of a related HIV-1 dataset. In particular, Birkner et al. (2005c) apply a loss-based *deletion/substitution/addition* (D/S/A) algorithm to build predictors for viral replication capacity Y, based on binary codon genotypes X (Sinisi and van der Laan, 2004). Such methods accommodate a large number of covariates and possibly higher-order interactions among these covariates.

Recall that a crude binary (wild-type vs. mutant) coding is used for codon genotypes; a more sensitive analysis may be achieved by using individual amino acids or grouping amino acids based on their biochemical properties.

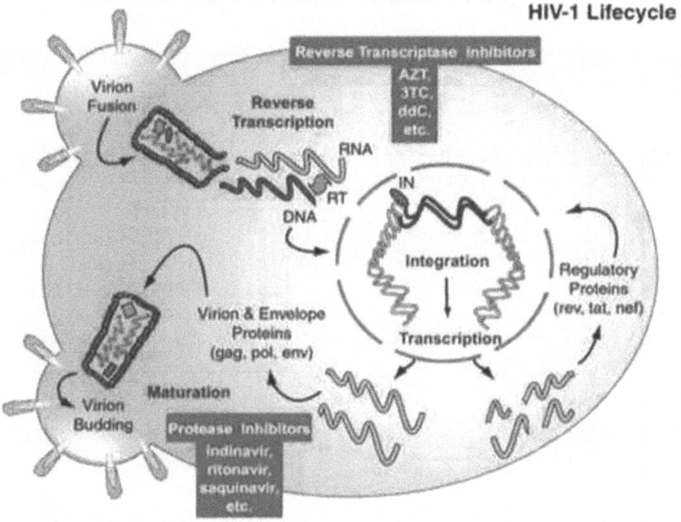

Figure 11.1. *HIV-1 lifecycle.* Diagram of the HIV-1 lifecycle and modes of action of protease and reverse transcriptase inhibitors. (Color plate p. 339)

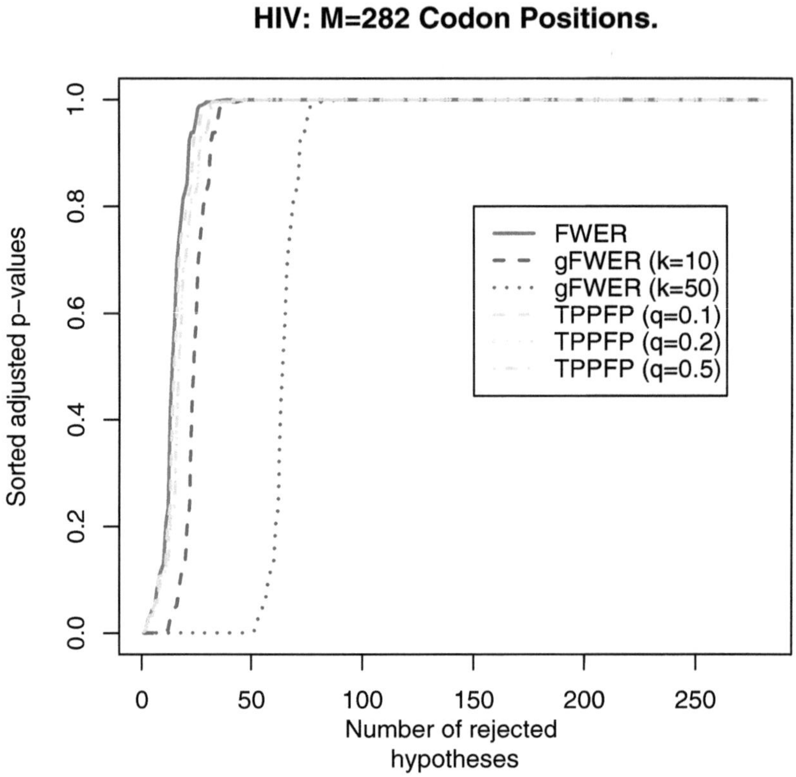

Figure 11.2. *HIV-1 dataset: Multiple testing analysis, Part I.* Sorted adjusted p-values for FWER-controlling single-step maxT Procedure 3.5, $gFWER(k)$-controlling augmentation Procedure 3.20 (allowed number of false positives, $k \in \{10, 50\}$), and $TPPFP(q)$-controlling augmentation Procedure 3.26 (allowed proportion of false positives, $q \in \{0.10, 0.20, 0.50\}$). (Color plate p. 340)

Table 11.1. *HIV-1 dataset: Multiple testing analysis, Part I.* Sorted adjusted p-values and t-statistics for multiple testing procedures controlling the FWER, gFWER, and TPPFP: FWER-controlling single-step maxT Procedure 3.5 (SS maxT); gFWER-controlling augmentation Procedure 3.20, with allowed number $k = 5$ of false positives (gFWER AMTP); TPPFP-controlling augmentation Procedure 3.26, with allowed proportion $q = 0.10$ of false positives (TPPFP AMTP). Results are displayed for the top 13 codon positions, where codon positions are sorted in increasing order of their SS maxT adjusted p-values. For a nominal Type I error level $\alpha = 0.05$, SS maxT, gFWER AMTP, and TPPFP AMTP identify, respectively, 7, 12, and 7 codon positions (out of $M = 282$) as associated with replication capacity.

Codon	**t-statistic**	**Adjusted p-values**		
position		SS maxT	gFWER AMTP	TPPFP AMTP
pr32	-9.755	0.0001	0	0.0001
pr47	-9.579	0.0013	0	0.0013
pr34	-8.843	0.0087	0	0.0087
pr55	-8.150	0.0104	0	0.0104
pr90	-6.237	0.0396	0	0.0396
rt184	-6.162	0.0431	0.0001	0.0431
pr43	-6.118	<u>0.0444</u>	0.0013	<u>0.0444</u>
pr54	-5.539	0.0780	0.0087	0.0780
rt41	-5.225	0.0978	0.0104	0.0978
pr46	-5.224	0.0980	0.0396	0.0978
pr82	-4.521	0.1678	0.0431	0.0980
rt215	-4.479	0.1740	<u>0.0444</u>	0.1678
rt121	-4.070	0.2380	0.0780	0.1740

Table 11.2. *HIV-1 dataset: Multiple testing analysis, Part II.* Comparison of TPPFP-controlling augmentation multiple testing Procedure 3.26 (TPPFP AMTP) and resampling-based empirical Bayes Procedure 7.1 (TPPFP EBayes). The table reports, for each MTP, the number of rejected hypotheses, i.e., the number of codon positions (out of $M = 282$) found to be associated with replication capacity, for an allowed proportion $q = 0.05$ of false positives and nominal Type I error levels $\alpha = 0.05, 0.10$.

	Number of rejected hypotheses	
α	TPPFP EBayes	TPPFP AMTP
0.05	11	5
0.10	13	8

12

Genetic Mapping of Complex Human Traits Using Single Nucleotide Polymorphisms: The ObeLinks Project

12.1 Introduction

12.1.1 Motivation

A central question in genetic mapping is to relate *genotypes* at multiple genetic markers to *phenotypes*, i.e., relate DNA variation to biological and clinical outcomes. A *genetic marker* is simply a genomic region, i.e., a segment of DNA, that varies among individuals. Genetic markers provide convenient landmarks on chromosomes and form the basis of *genetic maps* against which new genes are mapped.

In recent years, *single nucleotide polymorphisms* (SNP) have become the genetic markers of choice in genome-wide and candidate gene/pathway approaches for the genetic mapping of complex traits, such as, diabetes, multiple sclerosis, and obesity. As the name suggests, SNPs consist of a single base of DNA, where the most frequent nucleotide may be replaced in some individuals by another nucleotide. An example of a SNP corresponds to the alteration of the DNA segment `AAGGTTA` to `ATGGTTA`, where the second nucleotide `A` in the first sequence is replaced by a `T` in the second sequence. SNPs are thought to occur, on average, in more than 1% of the human genome. Because only 3–5% of our genomic DNA codes for the production of proteins, most SNPs are found in non-coding sequences (i.e., introns). SNPs located within coding sequences (i.e., exons) are of particular interest because they are more likely to alter the biological function of a protein. High-throughput genotyping technologies, such as microarrays, allow these binary markers to be readily and simultaneously assayed at thousands of loci spanning the genome. For more information on SNPs, please refer to the NCBI primer, *SNPs: Variations on a Theme* (`www.ncbi.nlm.nih.gov/About/primer/snps.html`), and Entrez SNP database (dbSNP, `www.ncbi.nlm.nih.gov/projects/SNP`; Maglott et al. (2005)).

Current genetic mapping efforts are focused on the identification of susceptibility genes for common *complex traits*, such as, diabetes, multiple sclerosis, and obesity, to name only a few. Such traits exhibit two levels of complexity: phenotypic complexity and etiological complexity (Hauser and Boehnke, 1998). *Phenotypic complexity* is due, for instance, to different clinical forms of a disease, diagnostic uncertainties (e.g., late age of onset), or variations in the definition of a disease. *Etiological complexity* arises when the trait of interest is influenced by several interacting genes, genotypes are incompletely penetrant (not all genetically predisposed individuals have the trait), or there are sporadic cases (the trait may occur in individuals who are not genetically predisposed). Etiological complexity is further increased by interactions between environmental and genetic factors. Important study design questions in the genetic mapping of complex traits include not only the selection of suitable genetic markers, but also the identification of relevant phenotypes to monitor, in order to successfully quantify the combined effects of multiple genetic and environmental factors.

The detection of *gene interaction* effects on phenotypes is of particular interest in the genetic mapping of complex traits (Hoh and Ott, 2003, 2004). For this purpose, single-locus genotypes can be recoded into multilocus composite genotypes, by exploiting the haplotype block structure of the human genome (Daly et al., 2001; Gabriel et al., 2002; Wang and Dudoit, 2004) or by computational approaches, based, for example, on Galois lattices (Birkner et al., 2007).

The biological question of detecting *genotype-phenotype associations* can be restated as a multiple hypothesis testing problem: the simultaneous test of the null hypotheses of no association between multilocus composite genotypes and one or more (possibly censored, qualitative or quantitative) phenotypes. To date, efforts to identify significant gene-gene and/or gene-environment interactions have only seen limited success, for reasons that include the paucity of suitable analysis methods and the lack of power of existing studies.

12.1.2 Outline

This chapter analyzes SNP data from the ObeLinks Project, with the aim of identifying genotypic combinations associated with human obesity-related phenotypes (`www.obelinks.org`; Birkner et al. (2007)).

Section 12.2 provides a brief presentation of the ObeLinks Project. Section 12.3 describes multiple testing procedures (MTP) for identifying multilocus composite SNP genotypes that are significantly associated with obesity-related phenotypes, such as body mass index and the metabolic variables glycemia and insulinemia. Specifically, the null hypotheses of no genotype-phenotype associations are tested using t-statistics comparing the phenotypes of patients possessing/not possessing the multilocus composite SNP genotypes corresponding to nodes in a Galois lattice. The following four multiple testing procedures are applied: FWER-controlling single-step maxT Procedure 3.5, gFWER-controlling augmentation Procedure 3.20,

TPPFP-controlling augmentation Procedure 3.26, and FDR-controlling step-up Benjamini and Hochberg (1995) Procedure 3.22. Section 12.4 presents the results of the MTPs and outlines their biological significance. Finally, Section 12.5 summarizes our findings and ongoing efforts.

12.2 The ObeLinks Project

12.2.1 ObeLinks dataset

The goal of the *ObeLinks Project* is to identify genotypic combinations associated with human obesity (`www.obelinks.org`; Birkner et al. (2007)). Obesity is a *multifactorial disease*, which, by definition, could be caused or influenced by multiple *genetic* and *environmental factors*, via *gene-gene* and/or *gene-environment interactions* (Human Obesity Gene Map, `obesitygene.pbrc.edu`, Pérusse et al. (2005); Clément (2005); Clément and Ferré (2003); Gloyn et al. (2003); Swarbrick and Vaisse (2003)). The condition is becoming increasingly prevalent in North America and Europe and has been linked to other causes of morbidity, such as, type 2 diabetes, hypertension, cardiovascular diseases, and several cancers.

This chapter reports analyses performed on data from $n = 386$ morbidly obese patients (i.e., with body mass index exceeding 40) participating in the ObeLinks Project. For each patient, 22 common SNPs were genotyped and 29 phenotypes were measured. The data are provided on the book's website (`www.stat.berkeley.edu/~sandrine/MTBook/ObeLinks/ObeLinks.html`). As detailed in Tables 12.1 and 12.2 and Figure 12.1, the 29 obesity-related phenotypes are divided into five categories and include body mass index (BMI), glycemia, and insulinemia. The 22 SNPs are classified into three sets, based on pathway membership and potential significance for obesity (Table 12.3): insulin resistance (`OB-IR`), signaling (`OB-Signaling`), and thermogenesis (`OB-ThermoG`). Note that, although biologically-founded, this classification is somewhat subjective, as SNPs could in principle be involved in any combination of the aforementioned three processes. Furthermore, different SNPs may belong to the same gene.

The SNP genotype data are *unphased*, i.e., the parental origins of the two alleles at a given locus are generally unknown. The most frequent SNP allele is denoted by `0` and referred to as the *wild-type* or *major allele*; the least frequent allele is denoted by `1` and referred to as the *mutant* or *minor allele*. SNP genotypes are coded as `(wt=00)` for the homozygous wild-type genotype, `(ht=01)` for the heterozygous genotype, and `(hm=11)` for the homozygous mutant genotype.

Three *penetrance* models are considered for the genotype-phenotype association. With the *codominance* model, the three genotypes, `(wt=00)`, `(ht=01)`, and `(hm=11)`, are treated as having distinct effects, or penetrances, on the phenotype of interest. The *dominance* model only distinguishes between two genotypes, `(wt=00)` and `(ht=01 or hm=11)`, i.e., `(ht=01)` and `(hm=11)` are assumed to have the same effect on the phenotype. The *recessive* model only distinguishes between two genotypes, `(wt=00 or ht=01)` and `(hm=11)`, i.e., `(wt=00)` and `(ht=01)` are assumed to have the same effect on the phenotype.

Each of the three sets of SNPs (namely, `OB-IR`, `OB-Signaling`, and `OB-ThermoG`) is analyzed separately, under each of the three penetrance models, thus yielding nine sets of SNP genotypes. Note that considering different modes of inheritance, in addition to multiple SNPs and multiple phenotypes, further compounds the multiple testing problem.

To assist in the detection of gene interaction effects on phenotypes, single-locus SNP genotypes are recoded into *multilocus composite genotypes* using *Galois lattices* (Section 12.2.2; Table 12.4; Figure 12.2). Each node in the Hass diagram representation of a Galois lattice corresponds to a set of patients with a particular multilocus composite SNP genotype. For example, node N_4 in the Hass diagram of Figure 12.2 corresponds to patients with multilocus genotype `(SNP 1 = wt, SNP 2 = hm, SNP 4 = wt)`. Based on Galois lattices, one can recode SNP genotypes as *binary* vectors, with elements corresponding to nodes in a lattice and indicating whether a particular patient has the multilocus composite genotype corresponding to a given node.

The data for each of the $n = 386$ patients consist of an $L = 29$-dimensional vector of qualitative or quantitative *outcomes/phenotypes*, $Y_i = (Y_i(l) : l = 1, \ldots, L)$, and, for each SNP genotype set, an M-dimensional vector of binary *covariates/genotypes*, $X_i = (X_i(m) : m = 1, \ldots, M)$, where

$$X_i(m) \equiv \begin{cases} 1, & \text{if individual } i \text{ has the multilocus composite SNP genotype} \\ & \text{corresponding to the } m\text{th node in the Galois lattice,} \\ 0, & \text{otherwise} \end{cases}$$
$$m = 1, \ldots, M, \; i = 1, \ldots, n. \tag{12.1}$$

That is, the genotype data structure implied by the Galois lattice for a given SNP genotype set is an $n \times M$ binary matrix, with rows corresponding to the $n = 386$ patients and columns to the M nodes in the lattice. Table 12.5 provides the number of nodes M for Galois lattices corresponding to each of the nine sets of SNP genotypes. Note that a Galois lattice for n objects could have up to 2^n nodes. The numbers of nodes in Table 12.5 are therefore remarkably small compared to their theoretical maximum of 2^{386}.

The present chapter reports results for three quantitative phenotypes, $(Y(l) : l \in \{\text{BMI}, \text{glycemia}, \text{insulinemia}\})$, from the set of 29 phenotypes, and for the `OB-IR Codominant` SNP genotype set. Six-number summaries of the phenotype distributions are provided in Table 12.2; boxplots and density plots are displayed in Figure 12.1. The genotype data for the $n = 386$ patients, for the 6 SNPs in the `OB-IR Codominant` set, led to an $M = 428$-node Galois lattice, where each node corresponds to a set of patients with a particular multilocus composite SNP genotype. The coverage of the $M = 428$ nodes in the `OB-IR Codominant` Galois lattice ranges from 2 to 342 patients, as indicated by the following six-number summary of the node coverage distribution

```
Min. 1st Qu.  Median     Mean 3rd Qu.     Max.
    2.00    6.75   17.50   35.62   41.00  342.00
```

One is interested in testing, for each node in the Galois lattice, the null hypothesis of no *association* between the *multilocus composite SNP genotype* (now binary) and a *phenotype*.

12.2.2 Galois lattices

This section provides a brief introduction to Galois lattices; the reader is referred to Godin et al. (1995a,b), Wille (1982, 1992), and references therein for further detail. The numerous applications of Galois lattices include, information retrieval, software engineering, and knowledge discovery, representation, and management.

Galois lattices are based on the notion of formal context. A *formal context* is a triple $(\mathcal{O}, \mathcal{D}, \mathcal{I})$, where $\mathcal{O}$ and $\mathcal{D}$ are finite sets and $\mathcal{I}$ is a *binary relation* between $\mathcal{O}$ and $\mathcal{D}$, i.e., a set $\mathcal{I} \subseteq \mathcal{O} \times \mathcal{D}$ of ordered pairs (o, d) with $o \in \mathcal{O}$ and $d \in \mathcal{D}$. Let $\mathcal{P}(\mathcal{O}) \equiv \{O : O \subseteq \mathcal{O}\}$ and $\mathcal{P}(\mathcal{D}) \equiv \{D : D \subseteq \mathcal{D}\}$ denote, respectively, the power sets of $\mathcal{O}$ and $\mathcal{D}$, and represent $(o, d) \in \mathcal{I}$ by $o\mathcal{I}d$. Depending on the application, elements of the set $\mathcal{O}$ may be referred to as *objects*, *documents*, *instances*, *items*, or *transactions*, and elements of the set $\mathcal{D}$ as *descriptions*, *attributes*, *features*, *properties*, or *terms*.

Define the following *derivation operators*

$$f_{\mathcal{D}} : \mathcal{P}(\mathcal{O}) \rightarrow \mathcal{P}(\mathcal{D}), \qquad f_{\mathcal{D}}(O) \equiv \{d \in \mathcal{D} : o\mathcal{I}d, \forall\, o \in O\}, \tag{12.2}$$

and

$$f_{\mathcal{O}} : \mathcal{P}(\mathcal{D}) \rightarrow \mathcal{P}(\mathcal{O}), \qquad f_{\mathcal{O}}(D) \equiv \{o \in \mathcal{O} : o\mathcal{I}d, \forall\, d \in D\}.$$

That is, $f_{\mathcal{D}}(O)$ is the set of descriptions shared by all objects in O and $f_{\mathcal{O}}(D)$ is the set of objects possessing all the descriptions in D. In particular, $f_{\mathcal{D}}(\emptyset) = \mathcal{D}$ and $f_{\mathcal{O}}(\emptyset) = \mathcal{O}$.

A pair of subsets $(O, D) \in \mathcal{P}(\mathcal{O}) \times \mathcal{P}(\mathcal{D})$ is said to be *complete* with respect to the binary relation $\mathcal{I}$ or a *concept* of the context $(\mathcal{O}, \mathcal{D}, \mathcal{I})$, if $f_{\mathcal{D}}(O) = D$ and $f_{\mathcal{O}}(D) = O$. A *partial order* $\preceq$ on concepts from the context $(\mathcal{O}, \mathcal{D}, \mathcal{I})$ is defined as follows: $(O_1, D_1) \preceq (O_2, D_2)$ i.f.f. $O_2 \subseteq O_1$ or, equivalently, $(O_1, D_1) \preceq (O_2, D_2)$ i.f.f. $D_1 \subseteq D_2$. The set

$$\mathcal{G}(\mathcal{O}, \mathcal{D}, \mathcal{I}) \equiv \{(O, D) \in \mathcal{P}(\mathcal{O}) \times \mathcal{P}(\mathcal{D}) : f_{\mathcal{D}}(O) = D,\ f_{\mathcal{O}}(D) = O\}, \tag{12.3}$$

of all concepts (i.e., complete pairs) from the context $(\mathcal{O}, \mathcal{D}, \mathcal{I})$, with partial order $\preceq$, forms a complete lattice called the *concept lattice* or *Galois lattice* of the context.

The graph representation of a Galois lattice $\mathcal{G}(\mathcal{O}, \mathcal{D}, \mathcal{I})$, commonly-termed *Hass diagram*, organizes concepts according to the partial order $\preceq$ as follows. Each *node* N corresponds to a complete pair of subsets $N = (O, D) \in \mathcal{G}(\mathcal{O}, \mathcal{D}, \mathcal{I})$. The set of objects $O \subseteq \mathcal{O}$, represented by a node $N = (O, D)$, is referred to as the *coverage* or *extent* of the node; the set of descriptions $D \subseteq \mathcal{D}$, that are common to all objects in O, is referred to as the *description* or *intent* of the node.

A *subsumption relation* between nodes is defined in terms of the aforementioned partial order $\preceq$, whereby, $N_1 \preceq N_2$ i.f.f. $O_2 \subseteq O_1$ and $D_1 \subseteq D_2$. If $N_1 \preceq N_2$, one says that N_1 is a *parent* of N_2 and N_2 is a *child* of N_1. In other words, a parent node has larger coverage and smaller description, i.e., is less specific/more general, than a child node. The *root* node $(\mathcal{O}, f_{\mathcal{D}}(\mathcal{O}))$ has coverage the set $\mathcal{O}$ of all objects and has no parents. The *leaf* nodes $(\{o\}, f_{\mathcal{D}}(\{o\}))$ have coverage of single objects $\{o\}$ and have no children. The Hass diagram for the Galois lattice has an *edge* between nodes N_1 and N_2 if $N_1 \preceq N_2$ and there is no other node N_3 such that $N_1 \preceq N_3 \preceq N_2$.

Galois lattices provide very interesting structures for organizing objects according to their descriptions. In contrast to classical conceptual hierarchies, the graph corresponding to a Galois lattice is not restricted to be a tree. Furthermore, a Galois lattice represents all existing relationships among objects. For example, if two objects o_1 and o_2 have a common description d, there exists a concept in the Galois lattice covering these objects and described by d. That is, if $o_1 \mathcal{I} d$ and $o_2 \mathcal{I} d$, there exists $N = (O, D) \in \mathcal{G}(\mathcal{O}, \mathcal{D}, \mathcal{I})$ with $\{o_1, o_2\} \subseteq O$ and $d \in D$. Thus, a Galois lattice for n objects may comprise up to 2^n concepts.

One distinguishes between the following two families of algorithms for building Galois lattices. *Batch* or *non-incremental* algorithms typically assemble a table representing the binary relation $\mathcal{I}$ for entire sets $\mathcal{O}$ and $\mathcal{D}$ and identify all concepts and partial order relations between these concepts before building the lattice (Bordat, 1986; Chein, 1969; Ganter, 1984). For instance, a batch algorithm may start with the root node $(\mathcal{O}, f_{\mathcal{D}}(\mathcal{O}))$ and iterate the child generation process down to all leaf nodes $(\{o\}, f_{\mathcal{D}}(\{o\}))$. In contrast, *incremental* algorithms allow the lattice to be expanded, without having to be regenerated from scratch, whenever new objects are considered (Carpineto and Romano, 1996; Godin et al., 1995b; Norris, 1978). Existing algorithms have exponential complexity (w.r.t. the numbers of objects $|\mathcal{O}|$ and descriptions $|\mathcal{D}|$) and lead to the same lattice. Many techniques have been developed with the aim of reducing computational complexity (Kuznetsov and Obiedkov, 2002; Mephu Nguifo and Njiwoua, 1998) and allowing consideration of more complex object descriptions, such as relational descriptions (Bisson, 1990; Bournaud et al., 2000).

In the application of Galois lattices to the genetic mapping ObeLinks Project, the set of objects $\mathcal{O}$ corresponds to patients and the set of descriptions $\mathcal{D}$ to multilocus SNP genotypes. For instance, the formal context in Table 12.4 and the Hass diagram in Figure 12.2 correspond to the Galois lattice for a simple artificial example with $n = 4$ patients genotyped at 5 SNPs. The set of objects $\mathcal{O} = \{o_1, o_2, o_3, o_4\}$ corresponds to the $n = 4$ patients and the set of descriptions $\mathcal{D} = \{\texttt{(SNP 1 = wt)}, \texttt{(SNP 1 = ht)}, \ldots, \texttt{(SNP 5 = hm)}\}$ to the 5×3 possible unphased genotypes for the 5 SNPs. The Hass diagram has $M = 6$ nodes (in addition to the 4 leaf nodes), each corresponding to a subset of patients with a particular multilocus composite SNP genotype. For example, node $N_2 = (O_2, D_2)$ has coverage $O_2 = \{o_1, o_2, o_3\}$ of three patients and description $D_2 = \{$`(SNP 1 = wt)`, `(SNP 2 = hm)`$\}$; node $N_4 = (O_4, D_4)$ has coverage $O_4 = \{o_1, o_2\}$ of two patients and description $D_4 = \{$`(SNP 1 = wt)`, `(SNP 2 = hm)`, `(SNP 4 = wt)`$\}$. Node N_2 is a parent of the more specific node N_4, that is, $N_2 \preceq N_4$.

In the ObeLinks Project, Galois lattices for SNP genotypes are built using an incremental concept formation algorithm of Godin et al. (1995b). This algorithm is based on the following idea: to add a new object to the lattice, one only needs to compare its description with the descriptions of existing concepts, as these provide a summarization of all the objects already classified. In order to reduce the number of concepts to explore, two strategies are used: considering concepts in ascending order of their coverage and exploiting the partial order between concepts. Another advantage of the Godin et al. (1995b) algorithm is that it builds the Hass diagram at the same time as it generates the complete pairs comprising the Galois lattice. Although this algorithm has quadratic complexity in the number of objects $n = |\mathcal{O}|$, an efficiently computed hash function makes it well-suited to the ObeLinks Project, where contexts are small and sparse, i.e., lead to few nodes and nodes with small coverage (Kuznetsov and Obiedkov, 2002). In particular, datasets related to the ObeLinks Project tend to have small description sets $\mathcal{D}$ (i.e., small numbers of SNP genotypes) and high similarity among objects, leading to relatively small numbers of nodes.

Note that a Galois lattice accounts for each combination of SNP genotypes found in the data. Although this property has the advantage that an individual is never misclassified (i.e., there is at least one node covering each individual), it can lead to many nodes having small coverage, i.e., representing a small number of individuals.

In what follows, we may (loosely) use the following terms interchangeably: formal context/Galois lattice/Hass diagram and concept/node.

12.3 Multiple testing procedures

In order to identify genes associated with obesity-related phenotypes, we test, for each of the $M = 428$ nodes in the `OB-IR Codominant` Galois lattice, the null hypothesis of no association between the corresponding multilocus composite SNP genotype, $(X(m) : m = 1, \ldots, M)$, and one of the three quantitative phenotypes of interest, $(Y(l) : l \in \{\text{BMI}, \text{glycemia}, \text{insulinemia}\})$. Thus, the parameter of interest for the mth node (i.e., mth hypothesis) and lth phenotype is the *difference in mean phenotype* $\psi(m; l)$ for individuals possessing $(X(m) = 1)$/not possessing $(X(m) = 0)$ the multilocus composite SNP genotype corresponding to node m. That is,

$$\psi(m;l) \equiv \mathrm{E}[Y(l)|X(m)=1] - \mathrm{E}[Y(l)|X(m)=0], \qquad (12.4)$$
$$m = 1,\ldots,M, \; l \in \{\text{BMI, glycemia, insulinemia}\}.$$

We consider one-sided tests of the null hypotheses $H_0(m;l) = \mathrm{I}(\psi(m;l) \leq 0)$ of no elevated phenotype vs. the alternative hypotheses $H_1(m;l) = \mathrm{I}(\psi(m;l) > 0)$ of elevated phenotype for individuals with the mth genotype.

The tests are based on *two-sample pooled-variance t-statistics*,

$$T_n(m;l) \equiv \frac{\psi_n(m;l) - \psi_0(m;l)}{\sigma_n(m;l)} \qquad (12.5)$$
$$= \frac{\bar{Y}_{1,n}(m;l) - \bar{Y}_{0,n}(m;l) - 0}{\sigma_{p,n}(m;l)\sqrt{\frac{1}{n_0(m)} + \frac{1}{n_1(m)}}},$$
$$\sigma^2_{p,n}(m;l) \equiv \frac{(n_0(m)-1)\sigma^2_{0,n}(m;l) + (n_1(m)-1)\sigma^2_{1,n}(m;l)}{n_0(m)+n_1(m)-2},$$

where the null values $\psi_0(m;l)$ are zero, $n_k(m) \equiv \sum_i \mathrm{I}(X_i(m) = k)$ denotes the number of patients possessing $(k = 1)$/not possessing $(k = 0)$ the multilocus composite SNP genotype corresponding to node m, and $\bar{Y}_{k,n}(m;l) \equiv \sum_i \mathrm{I}(X_i(m) = k)\, Y_i(l)/n_k(m)$ and $\sigma^2_{k,n}(m;l) \equiv \sum_i \mathrm{I}(X_i(m) = k)\,(Y_i(l) - \bar{Y}_{k,n}(m;l))^2/(n_k(m) - 1)$ denote, respectively, the sample means and sample variances for the lth phenotype of individuals possessing $(k = 1)$/not possessing $(k = 0)$ the mth genotype. The pooled-variance estimators are denoted by $\sigma^2_{p,n}(m;l)$. The estimators of the differences in mean phenotypes, $\psi(m;l)$, are simply the corresponding differences in sample means, $\psi_n(m;l) \equiv \bar{Y}_{1,n}(m;l) - \bar{Y}_{0,n}(m;l)$. The estimated standard errors of these estimators are $\sigma_n(m;l) \equiv \sigma_{p,n}(m;l)\sqrt{1/n_0(m) + 1/n_1(m)}$. The null hypothesis $H_0(m;l)$ is rejected, i.e., the corresponding multilocus composite SNP genotype is declared significantly associated with the lth phenotype, for large values of the test statistic $T_n(m;l)$. Note that the above two-sample pooled-variance t-statistics correspond to t-statistics for the univariate regression of the outcome $Y(l)$ on the binary covariates $X(m)$ (cf. one-way ANOVA).

A variety of bootstrap-based multiple testing procedures, controlling the FWER, gFWER, TPPFP, and FDR, are applied to assess the statistical significance of the genotype-phenotype associations. The results of each MTP are reported in terms of adjusted p-values, so that multilocus composite SNP genotypes (i.e., nodes) with adjusted p-values less than or equal to a user-supplied α are declared significantly associated with the phenotype at nominal Type I error level α (Section 1.2.12).

Specifically, the *non-parametric bootstrap* version of Procedure 2.3 is applied to obtain an estimate Q_{0n} of the *null shift and scale-transformed test statistics null distribution* Q_0 of Section 2.3 (with $B = 7{,}500$ bootstrap samples). This estimated null distribution is used to compute unadjusted p-values and adjusted p-values for the following procedures.

- *FWER control*: Joint single-step (common-cut-off) maxT Procedure 3.5.
- *gFWER control*: $gFWER(k)$-controlling augmentation multiple testing Procedure 3.20, based on FWER-controlling joint single-step maxT Procedure 3.5, for an allowed number $k \in \{10, 50\}$ of false positives.
- *TPPFP control*: $TPPFP(q)$-controlling augmentation multiple testing Procedure 3.26, based on FWER-controlling joint single-step maxT Procedure 3.5, for an allowed proportion $q \in \{0.10, 0.20, 0.50\}$ of false positives.
- *FDR control*: Marginal step-up Benjamini and Hochberg (1995) Procedure 3.22.

Note that all procedures are implemented in the Bioconductor R package multtest, discussed in detail in Section 13.1 (Pollard et al. (2005b); `www.bioconductor.org`).

12.4 Results

12.4.1 Body mass index

Body mass index (BMI) is a simple function of height (in meters, m) and weight (in kilograms, kg), defined as

$$BMI \equiv \frac{\text{Weight}}{\text{Height}^2} \text{ kg/m}^2. \tag{12.6}$$

This index is not gender-specific and represents a commonly-used indirect measure of fat mass, overweightedness, and obesity. Individuals are classified as slim, normal, overweight, obese, and morbidly obese, if, respectively, $BMI < 18$, $BMI \in [18, 25)$, $BMI \in [25, 30)$, $BMI \in [30, 40)$, and $BMI \geq 40$. As indicated in Table 12.2, patients considered in the ObeLinks Project are morbidly obese, with BMI ranging from 40.00 to 87.24 and with an average BMI value of 47.55, i.e., $Y_i(\text{BMI}) \in [40.00, 87.24]$, $i = 1, \ldots, n$, and $\bar{Y}_n(\text{BMI}) = \sum_i Y_i(\text{BMI})/n = 47.55$.

The results of the multiple testing procedures for BMI are reported in Table 12.6 and Figure 12.3. For a nominal Type I error level $\alpha = 0.05$, FWER-controlling single-step maxT Procedure 3.5 and FDR-controlling step-up Benjamini and Hochberg (1995) Procedure 3.22 identify, respectively, 6 and 9 nodes (out of $M = 428$) as significantly associated with BMI.

Nodes `n126` and `n125` correspond to multilocus composite SNP genotypes with mutations in two genes known to be associated with obesity-related phenotypes: `FABP2` homozygous mutant and `ABCC8.IVS15-3TC` homozygous mutant (Hani et al., 1997; Ordovas, 2001; Pérusse et al., 2005; Prochazka et al., 1993). Furthermore, all 10 nodes listed in Table 12.6 exhibit the `ABCC8.IVS15-3TC=hm` genotype; other SNP genotypes are homozygous wild-type.

`FABP2`. As detailed in the Entrez Gene database (Table 12.9), the `FABP2` gene codes for an intra-cellular fatty acid binding protein (FABP). FABPs are divided into at least three distinct types: hepatic, intestinal, and cardiac FABPs. They form 14–15 kiloDalton (kDa) proteins and are thought to participate in the uptake, intra-cellular metabolism, and/or transport of long-chain fatty acids. FABPs may also be involved in the modulation of cell growth and proliferation. The intestinal fatty acid binding protein 2 (FABP2) gene contains 4 exons and its product is an abundant cytosolic protein in epithelial cells of the small intestine. This gene has a common polymorphism at codon 54 that corresponds to an alanine-encoding allele and a threonine-encoding allele (Table 12.3). The Thr-54 protein is associated with increased fat oxidation and insulin resistance.

ABCC8,IVS15-3TC. As detailed in the Entrez Gene database (Table 12.9), the protein encoded by the ABCC8 gene is a member of the super-family of ATP-binding cassette (ABC) transporters. ABC proteins are divided into 7 distinct sub-families (ABC1, MDR/TAP, MRP, ALD, OABP, GCN20, White) and transport various molecules across extra- and intra-cellular membranes. The ABCC8 protein product is a member of the MRP sub-family, which is involved in multidrug resistance. This protein functions as a modulator of ATP-sensitive potassium channels and insulin release. Mutations and deficiencies in the ABCC8 protein product have been observed in patients with hyperinsulinemic hypoglycemia of infancy, an autosomal recessive disorder of unregulated and high insulin secretion. Mutations have also been associated with non-insulin-dependent diabetes mellitus (NIDDM), or type 2 diabetes, an autosomal dominant disease of defective insulin secretion. The gene alias SUR1 encodes the sulfonylurea receptor-1, a pancreatic regulatory subunit, which binds a widely-used class of insulin-secreting drugs and which has been associated with hyperinsulinemia.

To date, mutations FABP2=hm and ABCC8,IVS15-3TC=hm have only been studied singly in terms of their association with obesity-related phenotypes such as BMI. Our ongoing research efforts include investigating the interaction between the two genes FABP2 and ABCC8, as they both seem to have an important relationship with insulin sensitivity.

12.4.2 Glucose metabolism

The ObeLinks Project monitors various measures of glucose homeostasis, such as, glycemia, insulinemia, leptin levels, and triglyceride levels. Initially, glucose and insulin levels are studied separately, although one could use markers of insulin sensitivity derived from combined fasting glucose and insulin levels: e.g., homeostasis model assessment (HOMA) and quantitative insulin-sensitivity check index (QUICKI) (Hoffman et al., 2004).

Glycemia

Glycemia refers to the concentration of glucose in the blood system. An increase in blood glucose levels increases glycemia and, therefore, leads to an increase in insulin levels to reestablish a balanced glycemia. Obesity worsens insulin sensitivity, eventually exhausting pancreatic production of insulin and causing hyperglycemia and diabetes.

The results of the multiple testing procedures for glycemia are reported in Table 12.7 and Figure 12.4. For a nominal Type I error level $\alpha = 0.05$, FWER-controlling single-step maxT Procedure 3.5 and FDR-controlling step-up Benjamini and Hochberg (1995) Procedure 3.22 both identify 3 nodes (out of $M = 428$) as significantly associated with glycemia. Note, however, that the node coverages are much smaller than in Table 12.6 for BMI; the reported associations should therefore be interpreted cautiously.

Node **n99** exhibits the same two homozygous mutant genotypes discussed in Section 12.4.1, above, for tests of association with BMI: FABP2=hm and ABCC8,IVS15-3TC=hm. Nodes **n317** and **n255** correspond to 5-locus composite SNP genotypes that are identical, except for the occurrence of FABP2=hm at **n317** and ABCC8,IVS15-3TC=hm at **n255**. As discussed for BMI, the two mutations FABP2=hm and ABCC8,IVS15-3TC=hm have interesting biological implications. Nodes **n356**, **n461**, **n354**, and **n378**, all correspond to multilocus composite genotypes with homozygous mutant genotypes for SNPs FABP2 and ENPP1. Most of the other SNP genotypes are homozygous wild-type.

ENPP1. As detailed in the Entrez Gene database (Table 12.9), the ENPP1 gene is a member of the ectonucleotide pyrophosphatase/phosphodiesterase (ENPP) family. The encoded protein is a type II transmembrane glycoprotein, comprising two identical disulfide-bonded subunits. This protein has broad specificity and cleaves a variety of substrates, including phosphodiester and pyrophosphate bonds of nucleotides and nucleotide sugars. The ENPP1 protein product may function to hydrolyze nucleoside 5' triphosphates to their corresponding monophosphates and may also hydrolyze diadenosine polyphosphates. Mutations in the ENPP1 gene have been associated with idiopathic infantile arterial calcification, ossification of the posterior longitudinal ligament (OPLL) of the spine, and insulin resistance. Most recently, Meyre et al. (2005) have identified variants of ENPP1 that are associated with childhood and adult obesity and increase the risk of glucose intolerance and type 2 diabetes.

Insulinemia

Insulinemia refers to the concentration of insulin in the blood system. *Insulin* is a hormone secreted by islet cells within the pancreas in response to increases in blood glucose levels. Most cells have insulin receptors which bind insulin circulating in the blood stream. When insulin is attached to the surface of a cell, the cell activates other receptors designed to absorb glucose into the interior of the cell. Individuals suffering from *type 1 diabetes*, or *insulin-dependent diabetes mellitus* (IDDM), are unable to produce insulin and require insulin injections in order to metabolize glucose. In the more common form of *type 2 diabetes*, or *non-insulin dependent diabetes mellitus* (NIDDM), the pancreas does produce insulin, however, cells respond sluggishly to insulin and are therefore hampered in their glucose utilization. Obesity and diabetes are known to be associated with some of the same gene variations. In particular, obese subjects typically exhibit decreased insulin sensitivity.

The results of the multiple testing procedures for insulinemia are reported in Table 12.8 and Figure 12.5. Neither FWER-controlling single-step maxT Procedure 3.5 nor FDR-controlling step-up Benjamini and Hochberg (1995) Procedure 3.22 identifies any nodes as significantly associated with insulinemia, for nominal Type I error levels $\alpha \leq 0.10$.

Most nodes exhibit homozygous wild-type SNP genotypes; the only node with a homozygous mutant genotype is node n258, with FABP2=hm. Nonetheless, a potentially interesting biological finding concerns the occurrence of heterozygous genotypes IRS1=ht and PPI=ht for node n261 with the smallest SS maxT and SU BH adjusted p-values (with the caveat of lack of statistical significance and small node coverage).

IRS1. As detailed in the Entrez Gene database (Table 12.9), the IRS1 gene codes for a pivotal insulin receptor substrate protein and is an important element in insulin signaling pathways. Mutations in the IRS1 gene have been associated with various phenotypes related to type 2 diabetes. IRS1 protein signaling has been shown to be an important mechanism for insulin resistance in obese individuals. Understanding regulation and signaling by IRS1 in cell growth, metabolism, and survival, could reveal new strategies to prevent and/or cure diabetes and other metabolic diseases.

PPI. As detailed in the Entrez Gene database (Table 12.9), the protein encoded by the PPI gene (also known as PPP1R1A, I-1, and IPP-1) is a protein phosphatase 1, regulatory (inhibitor) subunit 1A. Phosphatase inhibitor-1 plays an important role in the regulation of glycogen metabolism, through the inhibition of protein phosphatase-1 (PP1) activity. The PP1 protein has been implicated in the regulation of cell growth. Glycogen metabolism is controlled predominantly by the coordinated action of two enzymes, glycogen synthase and glycogen phosphorylase, both of which are regulated by phosphorylation and allosteric modulators. Insulin promotes the net dephosphorylation of both glycogen synthase and glycogen phosphorylase, through the inhibition of protein kinases and the activation of protein phosphatases. Among the protein kinases, glycogen synthase kinase-3 (GSK-3) is thought to be an important target for insulin in its stimulation of glycogen synthase activity. Among the protein phosphatases, protein phosphatase-1 was also found to be a target of insulin.

The protein product of PPI is located near the tyrosine-phosphorylated IRS1 peptide in insulin signaling pathways. Thus, node n261 could be representing a possible insulin-stimulated interaction involving PPI and IRS1 near the plasma membrane in the cytosol.

12.5 Discussion

One of the key questions facing genetic epidemiologists is the determination of genetic and environmental factors predisposing to the development of complex traits, such as obesity. Challenges regarding the identification of gene-gene and gene-environment interactions include the limited sample sizes of current studies, the large numbers of hypotheses to be tested, and the replication of findings. Overcoming these challenges requires the collection of suitable data on phenotypes, genotypes, and environmental variables, and the development of sound computational and statistical methods. For nutrition-related conditions, well-controlled intervention studies, involving large samples, are currently underway in Europe and the United States of America. European projects, such as NUGENOB (`www.nugenob.com`) and Diogenes (`www.diogenes-eu.org`), are producing large databases, similar in structure to the ObeLinks database, which could potentially be combined for more powerful analyses.

Various approaches have been taken to identify genes associated with complex traits, including the multifactor dimensionality reduction (MDR) method of Hahn et al. (2003) and the two-stage method of Herbert et al. (2006), which consists of a SNP screening step followed by a moderate number of family-based association tests (FBAT).

Here, we have applied the multiple testing methodology developed in Chapters 1–7 to SNP genotype data from the ObeLinks Project, with the aim of identifying genotypic combinations associated with various obesity-related phenotypes. Galois lattices were constructed to recode single-locus SNP genotypes into multilocus composite genotypes. The following multiple testing procedures were used to control four different Type I error rates: FWER-controlling single-step maxT Procedure 3.5; gFWER-controlling augmentation Procedure 3.20, based on single-step maxT Procedure 3.5; TPPFP-controlling augmentation Procedure 3.26, based on single-step maxT Procedure 3.5; FDR-controlling step-up Benjamini and Hochberg (1995) Procedure 3.22. The first three joint common-cut-off procedures produce the same rankings of nodes (i.e., multilocus composite SNP genotypes), based on the ordering of the corresponding test statistics, while the fourth marginal common-quantile procedure ranks nodes based on the ordering of the corresponding unadjusted p-values. The two types of rankings are in principle different, unless the test statistics have identical marginal null distributions (e.g., standard normal marginal test statistics null distributions for each node).

The multiple testing procedures identified a number of multilocus composite genotypes defined by SNPs located in genes known to be associated with insulin resistance (e.g., type 2 diabetes). The identified SNPs tend to describe neighboring nodes of the Hass diagram for the `OB-IR Codominant` Galois lattice, thereby suggesting that the corresponding genes belong to the same biological pathway. In particular, our methods suggest a possible novel interaction between the two genes FABP2 and ABCC8 in their effect on BMI and glycemia. Although SNPs in the `OB-IR` set were selected based on their location in genes known to be involved in insulin signaling or insulin resistance, the identified multilocus composite genotypes may point to more specific and/or novel interactions between genes/pathways.

Note that while the test of hypotheses corresponding to individual nodes in a SNP Galois lattice reveals multilocus genotype combinations associated with a particular phenotype, further examination of the identified nodes may be necessary. For instance, certain SNP genotypes may be present in a node description simply because most individuals in the sample possess these genotypes (e.g., for glycemia, the homozygous wild-type genotype `ABCC8,Thr759Thr=wt` occurs in the descriptions of all Table 12.7 nodes). Post-processing or follow-up analyses include the test of hypotheses concerning Boolean combinations of nodes, in order to zero in on relevant SNP combinations. Alternately, one could pre-process the Galois lattice to exclude common genotypes. It would also be of interest to investigate methods that account for population SNP frequencies when building the Galois lattice. In addition, a number of nodes in the SNP Galois lattices have very small coverage, i.e., correspond to rare genotype combinations among the $n = 386$ patients. Filtering nodes according to coverage may prove beneficial.

It should be noted that some patients have extreme glycemia values. Indeed, as shown in Table 12.2 and Figure 12.1, one patient has a glycemia value of 21 mmol (i.e., nearly 4 g/l), which corresponds to uncontrolled diabetes. It would be of interest, from a biological and medical point of view, to perform the multiple testing analyses on a subset of non-diabetic patients. Furthermore, the use of indices such as HOMA or QUICKI may be more appropriate to test for associations between SNP genotypes and insulin sensitivity, because these measures take into account both insulin and glucose plasma levels. It may also be appropriate to consider two-sided rather than one-sided tests of association between genotypes and phenotypes.

Although a sample size of $n = 386$ patients is rather small for the purpose of identifying significant gene-gene interaction effects on phenotypes, note that the sample of $n = 386$ individuals is only a subset of patients followed as part of the ObeLinks Project. In addition, we only considered the **`OB-IR Codominant`** SNP genotype set, related to insulin signaling or insulin resistance. The main goal of the analyses presented in this chapter was to establish a proof of principle for the use of the proposed multiple testing methods. Power calculations and sample size determination in the context of multiple testing remain open questions, which are complicated by the typically complex and unknown dependence structure among variables (here, SNP genotypes). Initial efforts to address these important practical issues include the simulation-based method of Li et al. (2005), motivated by the identification of differentially expressed genes in microarray experiments.

In addition to analyzing other SNP genotype sets and phenotypes, our ongoing projects include adapting the proposed multiple testing methodology to the challenging question of identifying gene-environment interactions.

Table 12.1. *ObeLinks dataset: Phenotypes.* The table lists the 29 obesity-related phenotypes monitored in the ObeLinks Project. The phenotypes are divided into five categories. Here, "Max" refers to the maximum value of a variable during life span, "Min" to the minimum value reached after a diet, and "20" to the value at age 20. "Zscore" is a BMI standard deviation score adjusting for age (Rolland-Cachera et al., 1991). "Age WeightMax" and "Age WeightMin" refer, respectively, to the age at maximum and minimum weight. "PluriM" is an indicator for a metabolic syndrome, such as, insulin resistance (i.e., type 2 diabetes), dyslipidaemia, hypertension, or microalbuminuria.

Phenotype	**Type of variable**
Sex	
Sex	Binary
Weight	
Weight	Quantitative
WeightMax-WeightMin	Quantitative
WeightMax-Weight20	Quantitative
BMI	Quantitative
BMI20	Quantitative
BMIMax	Quantitative
BMIMin	Quantitative
Zscore	Quantitative
ZscoreMax	Quantitative
Age WeightMax	Quantitative
Age WeightMin	Quantitative
Waist-to-hip ratio	Quantitative
Antecedents	
Precocious obesity	Binary
Family obesity	Binary
Childhood weight gain	Binary
Complications	
PluriM	Binary
Arterial hypertension	Binary
Systolic blood pressure	Quantitative
Diastolic blood pressure	Quantitative
Diabetes	Binary
Diabetic status	Polychotomous
Biological tests	
Glycemia (fasting)	Quantitative
Insulinemia (fasting)	Quantitative
Triglyceride (fasting)	Quantitative
Leptin (fasting)	Quantitative
Cholesterol HDL	Quantitative
Cholesterol TT (Total)	Quantitative
Lipoprotein a (Lpa)	Quantitative

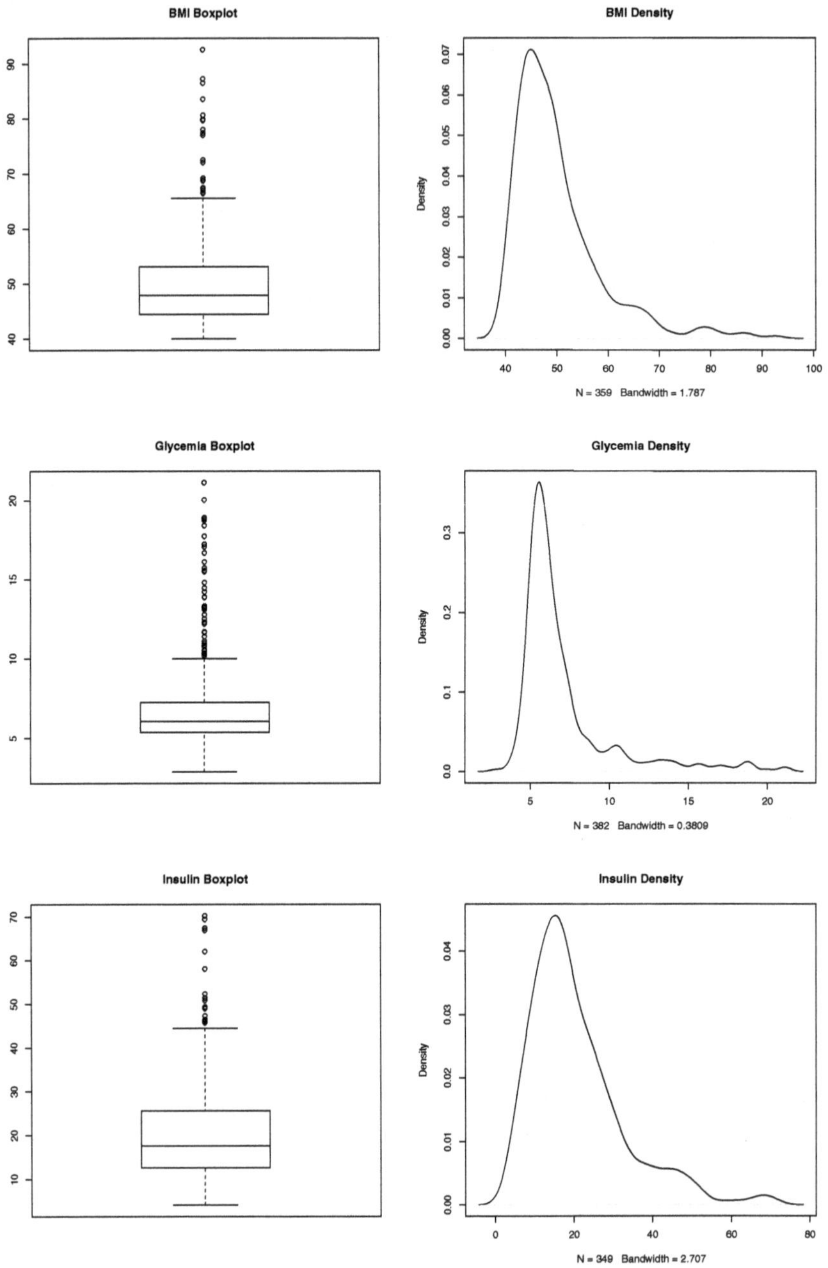

Figure 12.1. *ObeLinks dataset: Phenotype distributions.* Boxplots and density plots of the distributions of body mass index (BMI), glycemia, and insulinemia, for $n = 386$ patients participating in the ObeLinks Project.

Table 12.2. *ObeLinks dataset: Phenotype distributions.* Six-number summaries of the distributions of body mass index (BMI), glycemia, and insulinemia, for $n = 386$ patients participating in the ObeLinks Project.

	BMI	**Glycemia**	**Insulinemia**
Minimum	40.00	2.880	4.00
1st quartile	42.64	5.380	12.60
Median	45.14	6.050	17.60
Mean	47.55	7.057	20.62
3rd quartile	49.60	7.242	25.60
Maximum	87.24	21.100	70.20
# NA's	1.00	4.00	37.00

Table 12.3. *ObeLinks dataset: SNP sets.* The table provides the following information for each of the 22 SNPs genotyped in the ObeLinks Project: the short gene name (Symbol); the UniGene identifier (UG ID; **www.ncbi.nlm.nih.gov/entrez/query.fcgi?db=unigene**); the Entrez Gene identifier (GeneID; **www.ncbi.nlm.nih.gov/entrez/query.fcgi?db=gene**); the genomic location of the SNP (Location); the mutation (SNP); the gene name (Name); a description of the SNP and its associated variant in the protein product (Variant).

Symbol	UG ID	GeneID	Location	SNP	Name
			OB-IR **SNP set**		
FABP2	Hs.282265	2169	4q28-q31	G → A	Fatty acid binding protein 2, intestinal
	Variant: Ala54Thr				
IRS1	Hs.471508	3667	2q36	G → A	Insulin receptor substrate 1
	Variant: Gly972Arg				
ENPP1	Hs.527295	5167	6q22-q23	A → C	Ectonucleotide pyrophosphatase/ phosphodiesterase 1
	Variant: Lys121Gln				
PPI or PPP1R1A	Hs.505662	5502	12q13.2	G → A	Protein phosphatase 1, regulatory (inhibitor) subunit 1A
	Variant: Glu145Glu (silent)				
ABCC8	Hs.54470	6833	11p15.1	C → T	ATP-binding cassette, sub-family C (CFTR/MRP), member 8
	Variant: Mutation in Exon 18, ACC → ACT, Thr759Thr (silent) (ABCC8,Thr759Thr)				
ABCC8	Hs.54470	6833	11p15.1	T → C	ATP-binding cassette, sub-family C (CFTR/MRP), member 8
	Variant: Mutation in Intron 15, -3bp from Exon 16, alternative splicing acceptor site (ABCC8,IVS15-3TC)				
			OB-Signaling **SNP set**		
DRD2	Hs.73893	1813	11q23	G → A	Dopamine receptor D2
	Variant: Val154Ile				
ESR1	Hs.208124	2099	6q25.1	T → C	Estrogen receptor 1
	Variant: Mutation in Intron 1, causing absence of PvuII restriction enzyme site				
LEP or OB	Hs.194236	3952	7q31.3	A → G	Leptin (obesity homolog, mouse)
	Variant: Mutation +19bp in non-coding Exon 1				
LEPR or OBR	Hs.23581	3953	1p31	Ins/Del	Leptin receptor
	Variant: 45bp insertion or deletion in 3'UTR				
LEPR or OBR	Hs.23581	3953	1p31	A → G	Leptin receptor
	Variant: Mutation in Exon 12, Gln223Arg				
LPL	Hs.180878	4023	8p22	T → G	Lipoprotein lipase
	Variant: Mutation in Intron 8, causing absence of Hind III restriction enzyme site				
VDR	Hs.524368	7421	12q13.11	G → A	Vitamin D (1,25- dihydroxyvitamin D3) receptor
	Variant: Mutation in 3'UTR of Intron 8, causing absence of BSM1 restriction enzyme site				
GAL	Hs.278959	51083	11q13.2	(AC)n	Galanin
	Variant: AC repeat in 3'UTR of preprogalanin gene, +10–15bp from stop codon				
			OB-ThermoG **SNP set**		
ADRB2	Hs.591251	154	5q31-q32	A → G	Adrenergic, beta-2-, receptor, surface
	Variant: Arg16Gly				
ADRB2	Hs.591251	154	5q31-q32	G → C	Adrenergic, beta-2-, receptor, surface
	Variant: Gln27Glu				
ADRB3	Hs.2549	155	8p12-p11.2	A → G	Adrenergic, beta-3-, receptor
	Variant: Trp64Arg				

Continued on next page ...

... continued from previous page

Symbol	UG ID	GeneID	Location	SNP	Name
UCP1	Hs.249211	7350	4q28-q31	A → G	Uncoupling protein 1 (mitochondrial, proton carrier)
	Variant: Mutation -3,826bp in 5'UTR				
UCP2	Hs.80658	7351	11q13		Uncoupling protein 2 (mitochondrial, proton carrier)
	Variant: 45bp insertion or deletion in 3'UTR of Exon 8, +158bp from stop codon				
UCP3	Hs.101337	7352	11q13	T → C	Uncoupling protein 3 (mitochondrial, proton carrier)
	Variant: Mutation in Exon 3, Tyr99Tyr (silent)				
UCP3	Hs.101337	7352	11q13	C → T	Uncoupling protein 3 (mitochondrial, proton carrier)
	Variant: Mutation in Exon 5, Tyr210Tyr (silent)				
UCP3	Hs.101337	7352	11q13	C → T	Uncoupling protein 3 (mitochondrial, proton carrier)
	Variant: Mutation in 5'UTR of Exon 5, -36bp upstream of splice site				

SNP sets. The 22 SNPs are classified into the following three sets, based on pathway membership and potential significance for obesity.

- `OB-IR`: 6 SNPs located in the coding or non-coding sequence of a gene for a protein involved in insulin signaling or the pathophysiology of insulin resistance (e.g., lipid transport).
- `OB-Signaling`: 8 SNPs located in the coding or non-coding sequence of a gene for an adiposity signal protein or its receptor or, more generally, for a protein involved in signal message transmission.
- `OB-ThermoG`: 8 SNPs located in the coding or non-coding sequence of a gene related to the thermogenesis process.

UTR refers to the untranslated region of a gene transcript, i.e., of a mRNA sequence. The 5'UTR is the portion of a mRNA sequence from its 5' end to the first codon used in translation. The 3'UTR is the portion of a mRNA sequence from its 3' end to the last codon used in translation.

+$x$bp refers to a locus x base-pairs (bp) past the transcription initiation site, i.e., bases are counted (positively) in the 5' to 3' direction of transcription.

Lys121Gln means substitution at position 121 of the wild-type amino acid lysine (Lys, K) by the mutant amino acid glutamine (Gln, Q).

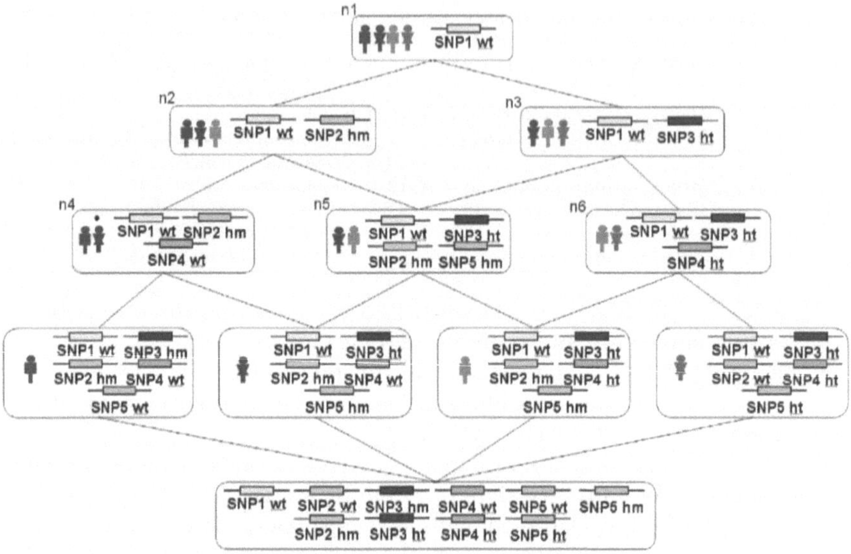

Figure 12.2. *Galois lattice for SNP genotypes.* The Hass diagram represents the Galois lattice for a simple artificial example with $n = 4$ patients genotyped at 5 SNPs. The set of objects $\mathcal{O} = \{o_1, o_2, o_3, o_4\}$ corresponds to the $n = 4$ patients and the set of descriptions $\mathcal{D} = \{$`(SNP 1 = wt)`, `(SNP 1 = ht)`$, \ldots,$ `(SNP 5 = hm)`$\}$ to the 5×3 possible unphased genotypes for the 5 SNPs. Table 12.4 represents the formal context $(\mathcal{O}, \mathcal{D}, \mathcal{I})$, i.e., the binary relation $\mathcal{I}$ between the sets $\mathcal{O}$ and $\mathcal{D}$, underlying the Galois lattice. The Hass diagram has $M = 6$ nodes (in addition to the 4 leaf nodes), each corresponding to a subset of patients with a particular multilocus composite SNP genotype. For example, node $N_2 = (O_2, D_2)$ has coverage $O_2 = \{o_1, o_2, o_3\}$ of three patients and description $D_2 = \{$`(SNP 1 = wt)`, `(SNP 2 = hm)`$\}$. The leaf nodes (shaded in yellow) represent the set of all SNP genotypes (i.e., descriptions) observed in the sample of $n = 4$ individuals, in this case, 10 out of a possible 5×3 genotypes. (Color plate p. 341)

Table 12.4. *Galois lattice for SNP genotypes.* The table represents the formal context $(\mathcal{O}, \mathcal{D}, \mathcal{I})$ for a simple artificial example with $n = 4$ patients genotyped at 5 SNPs. The set of objects $\mathcal{O} = \{o_1, o_2, o_3, o_4\}$ corresponds to the $n = 4$ patients and the set of descriptions $\mathcal{D} = \{$`(SNP 1 = wt)`, `(SNP 1 = ht)`, ..., `(SNP 5 = hm)`$\}$ to the 5×3 possible unphased genotypes for the 5 SNPs. The binary relation $\mathcal{I}$ can be represented by a binary matrix, with rows corresponding to descriptions and columns to objects. Entry (d, o) is one if $(o, d) \in \mathcal{I}$, i.e., $o\mathcal{I}d$, and zero otherwise. For instance, column 1 indicates that patient o_1 has multilocus composite SNP genotype `(SNP 1 = wt, SNP 2 = hm, SNP 3 = hm, SNP 4 = wt, SNP 5 = wt)`. The Hass diagram for the SNP genotype Galois lattice implied by the formal context $(\mathcal{O}, \mathcal{D}, \mathcal{I})$ is displayed in Figure 12.2.

		Objects, $\mathcal{O}$			
	Binary relation, $\mathcal{I}$	o_1	o_2	o_3	o_4
Descriptions, $\mathcal{D}$	`(SNP 1 = wt)`	1	1	1	1
	`(SNP 1 = ht)`	0	0	0	0
	`(SNP 1 = hm)`	0	0	0	0
	`(SNP 2 = wt)`	0	0	0	1
	`(SNP 2 = ht)`	0	0	0	0
	`(SNP 2 = hm)`	1	1	1	0
	`(SNP 3 = wt)`	0	0	0	0
	`(SNP 3 = ht)`	0	1	1	1
	`(SNP 3 = hm)`	1	0	0	0
	`(SNP 4 = wt)`	1	1	0	0
	`(SNP 4 = ht)`	0	0	1	1
	`(SNP 4 = hm)`	0	0	0	0
	`(SNP 5 = wt)`	1	0	0	0
	`(SNP 5 = ht)`	0	0	0	1
	`(SNP 5 = hm)`	0	1	1	0

Table 12.5. *ObeLinks dataset: Galois lattices for SNP genotype sets.* The table reports the number of nodes M (excluding leaf nodes) for the Galois lattices corresponding to each of the nine SNP genotype sets in the ObeLinks Project (Section 12.2.1). A genotype set is defined in terms of one of three sets of SNPs (namely, `OB-IR`, `OB-Signaling`, and `OB-ThermoG`; Table 12.3) and one of three penetrance models (namely, `Codominant`, `Dominant`, and `Recessive`).

SNP genotype set		Number of nodes, M
SNP set	**Penetrance model**	
OB-IR	Codominant	428
	Dominant	282
	Recessive	156
OB-Signaling	Codominant	356
	Dominant	665
	Recessive	862
OB-ThermoG	Codominant	9367
	Dominant	8113
	Recessive	3486

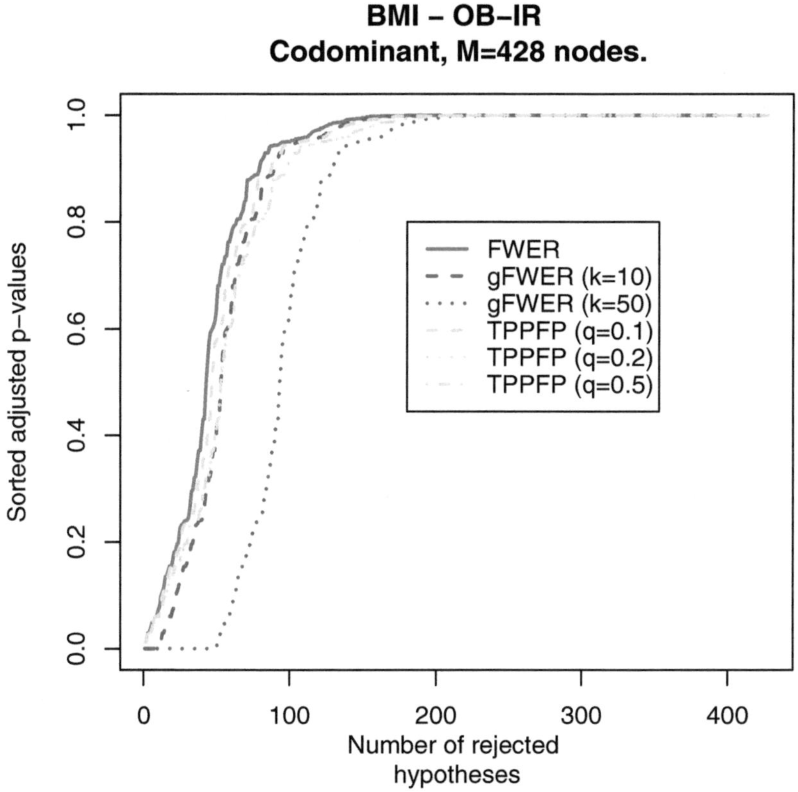

Figure 12.3. *ObeLinks dataset: BMI phenotype,* `OB-IR Codominant` *SNP genotype set.* Sorted adjusted p-values for FWER-controlling single-step maxT Procedure 3.5, $gFWER(k)$-controlling augmentation Procedure 3.20 (allowed number of false positives, $k \in \{10, 50\}$), and $TPPFP(q)$-controlling augmentation Procedure 3.26 (allowed proportion of false positives, $q \in \{0.10, 0.20, 0.50\}$). (Color plate p. 342)

Table 12.6. *ObeLinks dataset: BMI phenotype,* `OB-IR Codominant` *SNP genotype set.* The table reports adjusted p-values, node ID, coverage, and description (i.e., multilocus composite SNP genotype), for the 10 nodes, out of $M = 428$ nodes in the `OB-IR Codominant` Galois lattice, with the largest t-statistics for tests of association with the BMI phenotype. Adjusted p-values are listed for FWER-controlling single-step maxT Procedure 3.5 (SS maxT) and FDR-controlling step-up Benjamini and Hochberg (1995) Procedure 3.22 (SU BH). For a nominal Type I error level $\alpha = 0.05$, SS maxT and SU BH identify, respectively, 6 and 9 nodes as significantly associated with BMI.

Node ID	Coverage	Adjusted p-values SS maxT	Adjusted p-values SU BH
n104	78	0.0119	0.0214
(ENPP1=wt, ABCC8,IVS15-3TC=hm)			
n105	63	0.0129	0.0214
(IRS1=wt, ENPP1=wt, ABCC8,IVS15-3TC=hm)			
n121	72	0.0299	0.0428
(ENPP1=wt, ABCC8,Thr759Thr=wt, ABCC8,IVS15-3TC=hm)			
n126	30	0.0299	0.0428
(FABP2=hm, IRS1=wt, ENPP1=wt, ABCC8,Thr759Thr=wt, ABCC8,IVS15-3TC=hm)			
n123	57	0.0379	0.0428
(IRS1=wt, ENPP1=wt, ABCC8,Thr759Thr=wt, ABCC8,IVS15-3TC=hm)			
n92	40	<u>0.0499</u>	0.0476
(ENPP1=wt, PPI=wt, ABCC8,IVS15-3TC=hm)			
n102	90	0.0549	0.0476
(ABCC8,IVS15-3TC=hm)			
n125	38	0.0609	0.0476
(FABP2=hm, ENPP1=wt, ABCC8,Thr759Thr=wt, ABCC8,IVS15-3TC=hm)			
n47	35	0.0709	<u>0.0476</u>
(IRS1=wt, ENPP1=wt, PPI=wt, ABCC8,IVS15-3TC=hm)			
n103	75	0.0789	0.0514
(IRS1=wt, ABCC8,IVS15-3TC=hm)			

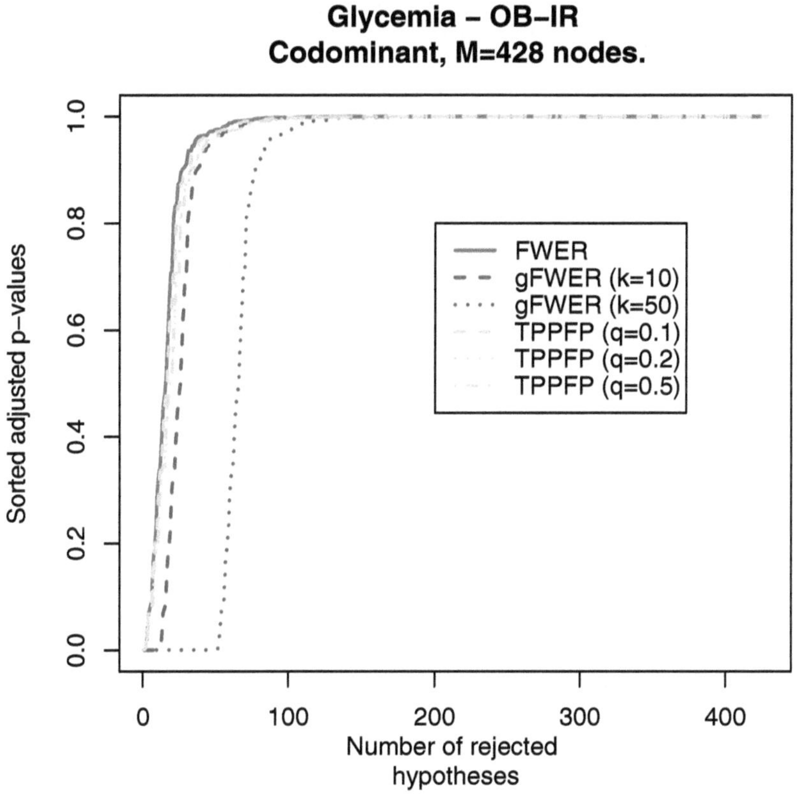

Figure 12.4. *ObeLinks dataset: Glycemia phenotype,* `OB-IR Codominant` *SNP genotype set.* Sorted adjusted p-values for FWER-controlling single-step maxT Procedure 3.5, $gFWER(k)$-controlling augmentation Procedure 3.20 (allowed number of false positives, $k \in \{10, 50\}$), and $TPPFP(q)$-controlling augmentation Procedure 3.26 (allowed proportion of false positives, $q \in \{0.10, 0.20, 0.50\}$). (Color plate p. 343)

Table 12.7. *ObeLinks dataset: Glycemia phenotype,* ***OB-IR Codominant*** *SNP genotype set.* The table reports adjusted p-values, node ID, coverage, and description (i.e., multilocus composite SNP genotype), for the 10 nodes, out of $M = 428$ nodes in the `OB-IR Codominant` Galois lattice, with the largest t-statistics for tests of association with the glycemia phenotype. Adjusted p-values are listed for FWER-controlling single-step maxT Procedure 3.5 (SS maxT) and FDR-controlling step-up Benjamini and Hochberg (1995) Procedure 3.22 (SU BH). For a nominal Type I error level $\alpha = 0.05$, SS maxT and SU BH both identify 3 nodes as significantly associated with glycemia.

Node ID	Coverage	Adjusted p-values	
		SS maxT	SU BH
n356	2	0.0008	0.0214
(FABP2=hm, IRS1=wt, ENPP1=hm, PPI=wt, ABCC8,Thr759Thr=wt, ABCC8,IVS15-3TC=ht)			
n99	3	0.0037	0.0214
(FABP2=hm, IRS1=ht, ENPP1=wt, PPI=wt, ABCC8,Thr759Thr=wt, ABCC8,IVS15-3TC=hm)			
n461	3	<u>0.0217</u>	<u>0.0428</u>
(FABP2=hm, IRS1=wt, ENPP1=hm, ABCC8,Thr759Thr=wt, ABCC8,IVS15-3TC=ht)			
n355	4	0.0689	0.1284
(IRS1=wt, ENPP1=hm, PPI=wt, ABCC8,Thr759Thr=wt, ABCC8,IVS15-3TC=ht)			
n354	4	0.0781	0.1284
(FABP2=hm, IRS1=wt, ENPP1=hm, PPI=wt, ABCC8,Thr759Thr=wt)			
n317	6	0.0904	0.1355
(FABP2=hm, IRS1=ht, ENPP1=wt, PPI=wt, ABCC8,Thr759Thr=wt)			
n255	5	0.1641	0.2690
(IRS1=ht, ENPP1=wt, PPI=wt, ABCC8,Thr759Thr=wt, ABCC8,IVS15-3TC=hm)			
n392	6	0.1895	0.2889
(ENPP1=hm, ABCC8,Thr759Thr=wt, ABCC8,IVS15-3TC=ht)			
n460	5	0.2241	0.3281
(IRS1=wt, ENPP1=hm, ABCC8,Thr759Thr=wt, ABCC8,IVS15-3TC=ht)			
n378	7	0.2987	0.4494
(FABP2=hm, IRS1=wt, ENPP1=hm, ABCC8,Thr759Thr=wt)			

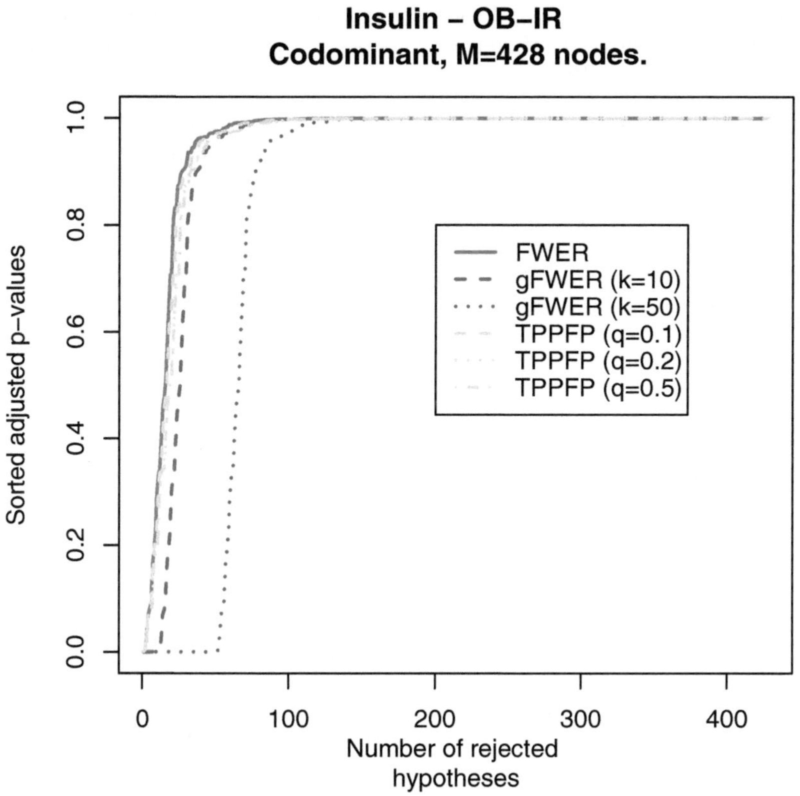

Figure 12.5. *ObeLinks dataset: Insulinemia phenotype,* `OB-IR Codominant` *SNP genotype set.* Sorted adjusted p-values for FWER-controlling single-step maxT Procedure 3.5, $gFWER(k)$-controlling augmentation Procedure 3.20 (allowed number of false positives, $k \in \{10, 50\}$), and $TPPFP(q)$-controlling augmentation Procedure 3.26 (allowed proportion of false positives, $q \in \{0.10, 0.20, 0.50\}$). (Color plate p. 344)

Table 12.8. *ObeLinks dataset: Insulinemia phenotype, `OB-IR Codominant` SNP genotype set.* The table reports adjusted p-values, node ID, coverage, and description (i.e., multilocus composite SNP genotype), for the 10 nodes, out of $M = 428$ nodes in the **`OB-IR Codominant`** Galois lattice, with the largest t-statistics for tests of association with the insulinemia phenotype. Adjusted p-values are listed for FWER-controlling single-step maxT Procedure 3.5 (SS maxT) and FDR-controlling step-up Benjamini and Hochberg (1995) Procedure 3.22 (SU BH). Neither SS maxT nor SU BH identifies any nodes as significantly associated with insulinemia, for nominal Type I error levels $\alpha \leq 0.10$.

Node ID	Coverage	Adjusted p-values	
		SS maxT	SU BH
n261	3	0.1362	0.4423
(IRS1=ht, ENPP1=wt, PPI=ht, ABCC8,Thr759Thr=wt)			
n29	148	0.1484	0.4423
(ENPP1=wt, PPI=wt)			
n70	28	0.1671	0.4423
(ENPP1=wt, PPI=wt, ABCC8,Thr759Thr=wt, ABCC8,IVS15-3TC=wt)			
n68	34	0.2264	0.4470
(ENPP1=wt, PPI=wt, ABCC8,IVS15-3TC=wt)			
n38	26	0.2892	0.4470
(IRS1=wt, ENPP1=wt, PPI=wt, ABCC8,Thr759Thr=wt,ABCC8,IVS15-3TC=wt)			
n31	132	0.3121	0.4470
(ENPP1=wt, PPI=wt, ABCC8,Thr759Thr=wt)			
n258	14	0.3509	0.4470
(FABP2=hm, IRS1=ht, ENPP1=wt)			
n27	176	0.3747	0.4470
(PPI=wt)			
n13	12	0.3911	0.4470
(FABP2=ht, IRS1=wt, ENPP1=wt, PPI=wt, ABCC8,Thr759Thr=wt,ABCC8,IVS15-3TC=wt)			
n15	132	0.4449	0.4470
(FABP2=ht, IRS1=wt, ENPP1=wt, PPI=wt)			

Table 12.9. *ObeLinks dataset: Gene descriptions from Entrez Gene database.* This table provides information from the NCBI Entrez Gene database (*Homo sapiens*, `www.ncbi.nlm.nih.gov/entrez/query.fcgi?db=gene`), for some of the genes defining the multilocus composite SNP genotypes that are found to be associated with obesity-related phenotypes in the ObeLinks dataset.

Gene name:	FABP2
Gene description:	Fatty acid binding protein 2, intestinal
Location:	4q28-q31
GeneID:	2169
Updated:	26-Jul-2006
Gene name:	IRS1
Gene description:	Insulin receptor substrate 1
Location:	2q36
GeneID:	3667
Updated:	26-Jul-2006
Gene name:	ENPP1
Gene description:	Ectonucleotide pyrophosphatase/phosphodiesterase 1
Location:	6q22-q23
GeneID:	5167
Other designations:	Plasma-cell membrane glycoprotein 1 (PC-1)
Updated:	07-Jul-2006
Gene name:	PPP1R1A
Gene description:	Protein phosphatase 1, regulatory (inhibitor) subunit 1A
Location:	12q13.2
GeneID:	5502
Other designations:	Inhibitor-1 (I-1); Protein phosphatase inhibitor-1 (PPI-1)
Updated:	26-Jul-2006
Gene name:	ABCC8
Gene description:	ATP-binding cassette, sub-family C (CFTR/MRP), member 8
Location:	11p15.1
GeneID:	6833
Other designations:	Sulfonylurea receptor (hyperinsulinemia) (SUR)
Updated:	24-Jul-2006

13

Software Implementation

13.1 R package multtest

13.1.1 Introduction

The multiple testing procedures (MTP) proposed in Chapters 1–7 are implemented in the latest version of the *R package* multtest, released as part of the *Bioconductor Project*, an open-source software project for the analysis of biomedical and genomic data (Gentleman et al. (2004); Pollard et al. (2005b); R Development Core Team (2006); multtest package, Version 1.10.0, Bioconductor Release 1.8, `www.bioconductor.org`; R Release 2.3.0, `www.r-project.org`). Note that although the multtest package emphasizes microarray data analysis, the MTPs implemented in multtest are applicable to any type of multiple testing problem (e.g., from astronomy, marketing). The multtest package may also be downloaded from the Comprehensive R Archive Network (CRAN, `cran.r-project.org/mirrors.html`). Please consult the Bioconductor Project website regularly for the latest package version and documentation.

Version 1.10.0 of multtest provides MTPs for tests concerning means, differences in means, and regression parameters in linear and Cox proportional hazards models. Procedures are available for controlling the following Type I error rates: family-wise error rate (FWER), generalized family-wise error rate (gFWER), tail probabilities for the proportion of false positives (TPPFP) among the rejected hypotheses, and false discovery rate (FDR). In addition to standard marginal MTPs (e.g., Bonferroni procedure), the package implements FWER-controlling single-step and step-down (common-cut-off) maxT and (common-quantile) minP procedures, that take into account the joint distribution of the test statistics. Augmentation multiple testing procedures (AMTP) are provided for controlling the gFWER and TPPFP, based on any initial FWER-controlling procedure. The results of a multiple testing procedure can be summarized using rejection regions for the test statistics, confidence regions for the parameters of interest, and adjusted p-values. A key ingredient of the MTPs implemented in multtest is the test statistics null distribution used to derive rejection regions and corresponding confidence regions and adjusted p-values. Both bootstrap and permutation estimators of the null distribution are available. The multiple testing procedures currently implemented in multtest are listed in Table 13.1.

The S4 class/method object-oriented programming approach is adopted to summarize the results of a MTP. The modular design of multtest allows interested users to readily extend the package's functionality.

Various testing scenarios are illustrated in Section 9.2, using the Apo AI microarray dataset of Callow et al. (2000), to identify differentially expressed genes between mice with the Apo AI gene knocked-out and inbred control mice (experimental data R package ApoAI, `www.stat.berkeley.edu/~sandrine/MTBook/ApoAI/ApoAI.html`). Pollard et al. (2005b) apply multtest functions to the acute lymphoblastic leukemia (ALL) microarray dataset of Chiaretti et al. (2004), with the aim of identifying genes whose expression measures are associated with possibly censored biological and clinical outcomes, such as, cancer cellular class (ALL B-cell vs. ALL T-cell), cancer molecular class (BCR/ABL, NEG, ALL1/AF4, E2A/PBX1, p15/p16, NUP-98), and time to relapse (Bioconductor experimental data R package ALL, Version 1.0.2, Bioconductor Release 1.7).

Ongoing efforts involve expanding the class of MTPs implemented in multtest, enhancing software design and the user interface, and increasing computational efficiency. In particular, we are planning on implementing the general gTP-controlling joint augmentation and resampling-based empirical Bayes procedures of Chapters 6 and 7. We also intend to provide additional resampling-based estimators for both the null shift and scale-transformed (Section 2.3) and null quantile-transformed (Section 2.4) test statistics null distributions (e.g., parametric bootstrap, Bayesian bootstrap).

13.1.2 Overview

Early versions of the multtest package (Version $< 1.5.0$) focused on FWER-controlling permutation-based step-down maxT and minP procedures. More recent versions (Version $\geq 1.5.0$) include the following new features: an expanded class of tests, such as tests for regression parameters in linear and Cox proportional hazards models; control of a wider selection of Type I error rates (e.g., gFWER, TPPFP); bootstrap estimation of the test statistics null distribution; augmentation multiple testing procedures; and confidence regions for the parameter vector of interest.

Because of their generality and novelty, this section emphasizes MTPs that rely on the non-parametric bootstrap null shift and scale-transformed test statistics null distribution of Section 2.3 and that are available through the package's main *user-level function* `MTP`. Note that MTPs based on a permutation null distribution are also applicable in certain testing scenarios (Section 2.9, Chapter 9). In particular, FWER-controlling permutation-based step-down maxT and minP MTPs are implemented in the functions `mt.maxT` and `mt.minP`, respectively, and can also be applied directly by invoking the `MTP` function.

As detailed in Section 1.2, one needs to specify the following main ingredients when applying a multiple testing procedure.

Data, $\mathcal{X}_n = \{X_i : i = 1, \ldots, n\}$.
Test statistics, T_n, for each of the null hypotheses, e.g., one-sample t-statistics, robust rank-based F-statistics, t-statistics for regression coefficients in Cox proportional hazards model.
Type I error rate, $\Theta(F_{V_n,R_n})$, providing an appropriate measure of false positives for the testing problem under consideration, e.g., TPPFP, with an allowed proportion $q = 0.10$ of false positives.
Test statistics null distribution, Q_0 (or estimator thereof, Q_{0n}), e.g., non-parametric bootstrap null shift and scale-transformed test statistics null distribution of Procedure 2.3.
Multiple testing procedure, $\mathcal{R}_n = \mathcal{R}(T_n, Q_{0n}, \alpha)$, for controlling the Type I error rate $\Theta(F_{V_n,R_n})$ at a nominal level α, e.g., FWER-controlling single-step maxT Procedure 3.5.

Accordingly, the multtest package has adopted a modular and extensible approach to the implementation of MTPs, with the following four main types of functions.

Functions for computing the test statistics, T_n. These are *internal functions* (e.g., `meanX`, `coxY`), i.e., functions that are generally not called directly by the user. As shown in Section 13.1.3, below, the type of test statistic is specified by the argument `test` of the main user-level function `MTP`. Advanced users, interested in extending the class of tests available in multtest, can simply add their own test statistic functions to the existing library of such internal functions. Section 13.1.5 provides a brief discussion of the function closure approach for specifying test statistics.
Functions for obtaining the test statistics null distribution, Q_0, or an estimator thereof, Q_{0n}. The main function currently available is the internal function `boot.resample`, implementing the non-parametric bootstrap null shift and scale-transformed test statistics null distribution of Procedure 2.3.
Functions for implementing the multiple testing procedure, $\mathcal{R}_n = \mathcal{R}(T_n, Q_{0n}, \alpha)$. The main user-level function is the wrapper function `MTP`, which returns rejection regions for the test statistics, confidence regions for the parameters of interest, and adjusted p-values, for MTPs controlling a variety of Type I error rates. In particular, the `MTP` function implements the FWER-controlling single-step and step-down maxT and minP procedures (Procedures 3.5, 3.6, 3.11, and 3.12). The functions `fwer2gfwer`, `fwer2tppfp`, and `fwer2fdr` provide, respectively, gFWER-, TPPFP-, and FDR-controlling augmentation multiple testing procedures,

based on adjusted p-values from any FWER-controlling procedure, and can be called via the `typeone` argument to `MTP` (Procedures 3.20 and 6.2, Procedures 3.26 and 6.4, Theorems 6.6 and 6.7).

Functions for numerical and graphical summaries of a MTP. As described in Section 13.1.4, below, a number of summary methods are available to operate on objects of class *MTP*, output from the main `MTP` function.

Note that the multtest package also provides several simple FWER-controlling marginal MTPs, such as, single-step Bonferroni Procedure 3.1, step-down Holm Procedure 3.7, step-up Hochberg Procedure 3.13, single-step Šidák Procedure 3.3, and step-down Šidák-like Procedure 3.9. It also implements FDR-controlling marginal step-up Benjamini and Hochberg (1995) Procedure 3.22 and Benjamini and Yekutieli (2001) Procedure 3.23. These MTPs are available through the `mt.rawp2adjp` function, which takes as input a vector of unadjusted p-values and returns the corresponding adjusted p-values.

We stress that all bootstrap-based MTPs implemented in multtest can be applied using the main user-level function `MTP`. Therefore, most users only need to be familiar with this function. Other functions are provided primarily for the benefit of more advanced users, interested in extending the package's functionality (Section 13.1.5).

The multiple testing procedures currently implemented in multtest are listed in Table 13.1. For greater detail on multtest functions, the reader is referred to the book chapter by Pollard et al. (2005b) and to the package documentation, in the form of helpfiles (e.g., ?MTP) and vignettes (e.g., `openVignette("multtest")`).

13.1.3 `MTP` function for resampling-based multiple testing procedures

The main *user-level function* for resampling-based multiple testing is `MTP`. Its arguments and values (i.e., input and output) are described in detail below.

Input

The *arguments* of `MTP` may be listed using the function **args**.

```
> library("multtest")
> packageDescription("multtest")$Version
[1] "1.10.0"
> args(MTP)
function (X, W = NULL, Y = NULL, Z = NULL, Z.incl = NULL,
    Z.test = NULL, na.rm = TRUE, test = "t.twosamp.unequalvar",
    robust = FALSE, standardize = TRUE,
    alternative = "two.sided", psi0 = 0, typeone = "fwer",
```

```
k = 0, q = 0.1, fdr.method = "conservative", alpha = 0.05,
smooth.null = FALSE, nulldist = "boot", B = 1000,
method = "ss.maxT", get.cr = FALSE, get.cutoff = FALSE,
get.adjp = TRUE, keep.nulldist = TRUE, seed = NULL)
```

Data. The data components (`X`, `W`, `Y`, `Z`, `Z.incl`, and `Z.test`) are the first six arguments to the `MTP` function. Only `X` is required.

- `X`: The data `X` consist of a J-dimensional random vector, observed on each of n sampling units (e.g., patients, mice, cell lines). These data can be stored in a $J \times n$ *matrix*, *data.frame*, or *exprs* slot of an object of class *exprSet*.
- `W`: One may supply a $J \times n$ *matrix* `W` of non-negative weights corresponding to each element in the data `X`. One may also specify an n-dimensional *vector* `W` of observation-level weights (where all J variables are weighted equally for a given observation i) or a J-dimensional *vector* `W` of variable-level weights (where all n observations are weighted equally for a given variable j).
- `Y`: The argument `Y` corresponds to a possibly censored continuous or polychotomous outcome, obtained, for example, from the *phenoData* slot of an object of class *exprSet*.
- `Z`, `Z.incl`, and `Z.test`: Additional covariates, measured on each sampling unit, may be supplied using the argument `Z`. When the tests concern parameters in regression models with covariates from `Z` (e.g., values `lm.XvsZ`, `lm.YvsXZ`, and `coxph.YvsXZ`, for the `MTP` argument `test`, as described below), the arguments `Z.incl` and `Z.test` specify, respectively, which covariates (e.g., which columns of `Z`, including `Z.test`) should be included in the model and which regression parameter is to be tested (only when `test="lm.XvsZ"`).
- `na.rm`: The argument `na.rm` controls the treatment of missing values (`NA`). By default `na.rm=TRUE`, so that an observation with a missing value for the jth variable, $j = 1, \ldots, J$, is excluded from the computation of any test statistic based on this variable.

Test statistics. The test statistics should be chosen based on the parameters of interest (e.g., location, scale, or regression parameters) and the hypotheses to be tested. In the current implementation of multtest, the following test statistics are available through the argument `test`, with default value `t.twosamp.unequalvar`, for two-sample Welch t-statistics.

- `t.onesamp`: One-sample t-statistics for tests of means.
- `t.twosamp.equalvar`: Two-sample equal-variance (or pooled-variance) t-statistics for tests of equality of means.
- `t.twosamp.unequalvar`: Two-sample unequal-variance t-statistics for tests of equality of means (also known as two-sample Welch t-statistics).
- `t.pair`: Two-sample paired t-statistics for tests of equality of means.

- `f`: Multi-sample F-statistics for tests of equality of means in a one-way design.
- `f.block`: Multi-sample F-statistics for tests of equality of means in a two-way or block design.
- `lm.XvsZ`: t-statistics for tests of regression coefficients in linear models with outcome `X[j,]` ($j = 1, \ldots, J$) and covariate of interest `Z.test`, with possibly additional covariates `Z.incl` from `Z` (in the case of no covariates, one recovers the one-sample t-statistics, `t.onesamp`).
- `lm.YvsXZ`: t-statistics for tests of regression coefficients in linear models with outcome `Y` and covariate of interest `X[j,]` ($j = 1, \ldots, J$), with possibly additional covariates `Z.incl` from `Z`.
- `coxph.YvsXZ`: t-statistics for tests of regression coefficients in Cox proportional hazards survival models with outcome `Y` and covariate of interest `X[j,]` ($j = 1, \ldots, J$), with possibly additional covariates `Z.incl` from `Z`.

Robust, rank-based versions of the above test statistics can be implemented by setting the argument `robust` to `TRUE` (by default, `robust=FALSE`).
Both standardized and unstandardized difference statistics are available through the argument `standardize` (by default, `standardize=TRUE`; Section 1.2.5; Pollard and van der Laan (2004)).
The type of alternative hypothesis is specified via the `alternative` argument: default value of `two.sided`, for two-sided test, and values of `less` or `greater`, for one-sided tests.
The (common) null value for the parameters of interest is specified through the `psi0` argument (by default, `psi0=0`).

Type I error rate. By default, the `MTP` function implements FWER-controlling single-step maxT Procedure 3.5 (argument `typeone="fwer"`). Augmentation multiple testing procedures, controlling other Type I error rates, such as the gFWER, TPPFP, and FDR, can be specified through the argument `typeone`. Related arguments include `k` and `q`, for the allowed number and proportion of false positives for control of $gFWER(k)$ and $TPPFP(q)$, respectively, and `fdr.method`, for the type of FDR-controlling TPPFP-based procedure (i.e., `"conservative"` or `"restricted"` methods, corresponding, respectively, to Bound 1 and Bound 2 in Theorems 6.6 and 6.7). The nominal Type I error level of the test is determined by the argument `alpha` (by default, `alpha=0.05`). Testing can be performed for a range of nominal Type I error levels by specifying a vector of levels `alpha`.

Test statistics null distribution. The test statistics null distribution is estimated by default using the non-parametric version of bootstrap Procedure 2.3, for the null shift and scale-transformed null distribution of Section 2.3 (argument `nulldist="boot"`). The bootstrap procedure is implemented in the internal function `boot.resample`, which calls C to compute test statistics for each bootstrap sample. The null shift and scale values,

λ_0 and τ_0, are determined by the type of test statistic (e.g., $\lambda_0 = 0$ and $\tau_0 = 1$ for t-statistics). Permutation null distributions are available by setting `nulldist="perm"`. The number of resampling steps is specified by the argument `B` (by default, `B=1000`). A kernel density smoothed test statistics null distribution may be used by setting the argument `smooth.null=TRUE` (by default, `smooth.null=FALSE`).

Multiple testing procedure. Several methods are available in multtest for controlling a given Type I error rate and can be specified via the argument `method` of the `MTP` function (Table 13.1).

- *FWER-controlling procedures.* The `MTP` function implements the FWER-controlling single-step and step-down (common-cut-off) maxT and (common-quantile) minP procedures (Procedures 3.5, 3.6, 3.11, and 3.12). The default MTP is the single-step maxT procedure (`method="ss.maxT"`), as it requires the least computation. These four main MTPs are implemented in the internal functions `ss.maxT`, `ss.minP`, `sd.maxT`, and `sd.minP`.
- *gFWER-, TPPFP-, and FDR-controlling procedures.* As detailed in Chapter 6, any FWER-controlling MTP can be trivially augmented to control additional Type I error rates, such as the gFWER and TPPFP. Two FDR-controlling procedures can then be derived from such TPPFP-controlling AMTPs (Section 6.4). AMTPs are implemented in the functions `fwer2gfwer`, `fwer2tppfp`, and `fwer2fdr`, which take as input FWER adjusted p-values and return augmentation adjusted p-values for control of the gFWER, TPPFP, and FDR, respectively. Note that the aforementioned AMTPs can be applied directly via the `typeone` argument of the main function `MTP`.

Output control. Various arguments are available to specify which combination of the following quantities should be returned: confidence regions for the parameters of interest (argument `get.cr`); cut-offs for the test statistics (argument `get.cutoff`); adjusted p-values (argument `get.adjp`); test statistics null distribution (argument `keep.nulldist`). Note that parameter estimators and confidence regions only apply to the test of single-parameter null hypotheses using t-statistics (i.e., not the F-tests). In addition, in the current implementation of `MTP`, parameter confidence regions and test statistic cut-offs are only provided when `typeone="fwer"`, so that `get.cr` and `get.cutoff` should be set to `FALSE` when using the error rates gFWER, TPPFP, or FDR. The seed for the random number generator used for bootstrap resampling may be set with the argument `seed`.

Output

The S4 class/method object-oriented programming approach is adopted to summarize the results of a MTP (Section 13.1.5). The output of the `MTP` function is an instance of the *class MTP*, with the following *slots*.

```
> slotNames("MTP")
 [1] "statistic" "estimate"  "sampsize"  "rawp"      "adjp"
 [6] "conf.reg"  "cutoff"    "reject"    "nulldist"  "call"
[11] "seed"
```

A brief description of the class is given next; for greater detail, consult class?MTP.

MTP results.

- statistic: The *numeric* M-vector of test statistics, specified by the MTP arguments test, robust, standardize, alternative, and psi0. In many testing problems, $M = J =$ nrow(X).
- estimate: For the test of single-parameter null hypotheses using t-statistics (i.e., not the F-tests), the *numeric* M-vector of parameter estimates.
- sampsize: The sample size, i.e., $n =$ ncol(X).
- rawp: The *numeric* M-vector of unadjusted p-values.
- adjp: The *numeric* M-vector of adjusted p-values (computed only if get.adjp=TRUE).
- conf.reg: For the test of single-parameter null hypotheses using t-statistics (i.e., not the F-tests), the *numeric* $M \times 2 \times$ length(alpha) *array* of lower and upper simultaneous confidence limits for the parameter vector, for each value of the nominal Type I error level alpha (computed only if get.cr=TRUE).
- cutoff: The *numeric* $M \times$ length(alpha) *matrix* of cut-offs for the test statistics, for each value of the nominal Type I error level alpha (computed only if get.cutoff=TRUE).
- reject: The $M \times$ length(alpha) *matrix* of rejection indicators (TRUE for a rejected null hypothesis), for each value of the nominal Type I error level alpha.

Test statistics null distribution. The nulldist slot contains the $M \times B$ *matrix* for the estimated test statistics null distribution (returned only if keep.nulldist=TRUE). This option is not currently available for permutation null distributions (i.e., for nulldist="perm"). By default (i.e., for nulldist="boot"), the elements of nulldist are the null shift and scale-transformed bootstrap test statistics, as defined in Procedure 2.3.

Reproducibility. The last two slots of an *MTP* object provide information on a specific call of the MTP function and can be used for reproducibility in other calls of MTP. The slot call contains the MTP function call and seed is an *integer* specifying the state of the random number generator used to create the resampled datasets. The seed argument is currently used only for the bootstrap null distribution (i.e., for nulldist="boot").

13.1.4 Numerical and graphical summaries of a multiple testing procedure

The following *methods* are defined to operate on *MTP* instances and summarize the results of a MTP. For greater detail, consult `methods?MTP`.

`[`: The null hypothesis subsetting method `[` operates selectively on each slot of an *MTP* instance to retain only the data related to the specified hypotheses.

`as.list`: The `as.list` method converts an object of class *MTP* to an object of class *list*, with an element for each *MTP* slot.

`plot`: The `plot` method produces the following graphical summaries of the results of a MTP. The type of display may be specified via the `which` argument.

1. Scatterplot of number of rejected hypotheses vs. nominal Type I error level.
2. Plot of ordered adjusted p-values; can be viewed as a plot of nominal Type I error level vs. number of rejected hypotheses.
3. Scatterplot of adjusted p-values vs. test statistics (also known as "volcano plot").
4. Plot of unordered adjusted p-values.
5. Plot of confidence regions for user-specified parameters, by default the 10 parameters corresponding to the 10 smallest adjusted p-values (argument `top`).
6. Plot of test statistics and cut-offs (for each value of `alpha`) for user-specified hypotheses, by default the 10 hypotheses corresponding to the 10 smallest adjusted p-values (argument `top`).

The argument `logscale` allows one to use the negative decimal logarithm of the adjusted p-values in the second, third, and fourth graphical displays (by default, `logscale=FALSE`). The arguments `caption` and `sub.caption` may be used to change the titles and subtitles for each of the plots (the default subtitle is the `MTP` function call). Note that some of these plots are implemented in the older function `mt.plot`.

`print`: The `print` method returns a description of an object of class *MTP*, including: the sample size n, the number M of tested hypotheses, the type of test performed (value of argument `test`), the Type I error rate (value of argument `typeone`), the nominal level of the test (value of argument `alpha`), the name of the MTP (value of argument `method`), and the `MTP` function call. In addition, this method produces a table with the class, mode, length, and dimension of each slot of the *MTP* instance.

`summary`: The `summary` method provides numerical summaries of the results of a MTP and returns a *list* with the following three components.

- `rejections`: A *data.frame* with the number(s) of rejected hypotheses for the nominal Type I error level(s) specified by the `alpha` argument of the function `MTP`.

- `index`: An *integer* M-vector of indices for ordering the hypotheses according to first `adjp`, then `rawp`, and finally the absolute value of `statistic` (not printed in the summary).
- `summaries`: When applicable (i.e., when the corresponding quantities are returned by `MTP`), a table with six-number summaries of the distributions of the adjusted p-values, unadjusted p-values, test statistics, and parameter estimates.

`update`: The `update` method for the *MTP* class provides a mechanism for rerunning the `MTP` function with different choices for the following arguments: `alternative`, `typeone`, `k`, `q`, `fdr.method`, `alpha`, `smooth.null`, `method`, `get.cr`, `get.cutoff`, `get.adjp`, and `keep.nulldist`. When `evaluate` is `TRUE`, a new object of class *MTP* is returned. Else, the updated function call is returned. The *MTP* object passed to the `update` method must have a non-empty `nulldist` slot (i.e., must have been created by a call to `MTP` with `keep.nulldist=TRUE`).

Examples of the above numerical and graphical summaries of a MTP are provided in Section 9.2, for the analysis of the Apo AI microarray dataset of Callow et al. (2000).

13.1.5 Software design

The following features of the programming approach employed in multtest may be of interest to users interested in extending the functionality of the package.

Function closures

The use of *function closures*, as in the Bioconductor package genefilter, allows uniform data input for all MTPs and facilitates the extension of the package's functionality, by implementing, for example, new types of test statistics.

Specifically, a function closure is defined for each value of the `MTP` argument `test`. The function closure consists of a *function* for computing the test statistic (with only two arguments, a data vector `x` and a corresponding weight vector `w`, with default value of `NULL`) and its enclosing *environment*, with bindings for relevant additional arguments, such as, null values `psi0`, outcomes `Y`, and covariates `Z`.

Existing internal test statistic functions are located in the file `R/statistics.R`.

Thus, new test statistics can be added to multtest by simply defining a new function closure and a corresponding new value for the `MTP` argument `test`.

Class/method object-oriented programming

Like many other Bioconductor packages, multtest has adopted the *S4 class/method object-oriented programming* approach of Chambers (1998).

In particular, a new *MTP* class and associated methods are provided to represent and operate on the results of multiple testing procedures.

Calls to C

Because resampling procedures, such as bootstrap Procedure 2.3, for estimating the null shift and scale-transformed test statistics null distribution, are computationally intensive, care must be taken to ensure that the resampling steps are not prohibitively slow.

The use of function closures for the test statistics prevents, however, programming the entire procedure in C. In the current implementation, we have chosen to define the function closure and compute the observed test statistics in R and then call C to apply the closure to each bootstrap sample (using the R random number generator). This approach puts the for loops over the B bootstrap samples and M null hypotheses in the compiled code, thus speeding up this computationally costly step of the MTP.

13.2 SAS macros

This section, based on Birkner et al. (2005b), discusses the software implementation in SAS of the following three MTPs (SAS, Version 9, `www.sas.com`): FWER-controlling single-step maxT Procedure 3.5, gFWER-controlling augmentation multiple testing Procedure 3.20, and TPPFP-controlling augmentation multiple testing Procedure 3.26.

SAS macros were written to compute various components of a MTP, including: test statistics T_n, bootstrap estimators Q_{0n} of the null shift and scale-transformed test statistics null distribution Q_0 (Procedure 2.3), and adjusted p-values $\widetilde{P}_{0n}(m)$.

`%lmt`: The `%lmt` macro takes as input a SAS dataset of the form `[y:X]` (e.g., `resample.hivdata` for the HIV-1 dataset of Chapter 11), with rows corresponding to n observations and with the first column referring to an outcome Y and the remaining columns to an M-dimensional covariate vector $X = (X(m) : m = 1, \ldots, M)$. This macro uses `PROC REG` to compute t-statistics $T_n(m)$, for the univariate linear regression of the outcome Y on each of the M covariates $X(m)$, $m = 1, \ldots, M$. The test statistics T_n are stored in the dataset `tstats`.

`%boot`: The `%boot` macro generates B (macro variable `&boots`) bootstrap samples from the original dataset and computes M-vectors of test statistics $T_n^{\#}$ for each of these B bootstrap samples using the `%lmt` macro. Specifically, the rows of the `[y:X]` dataset are sampled at random, with replacement using `PROC SURVEYSELECT`. Note that `PROC SURVEYSELECT` uses the bootstrap iteration index as the seed and the `method=urs` command for "unrestricted random sampling", i.e., sampling at random, with replacement. The bootstrap test statistics $T_n^\#$ are stored in the dataset `tstatsB`.

`%bootnull`: The `%bootnull` macro reads in the B bootstrap M-vectors of test statistics $T_n^\#$ produced by the `%boot` macro and stored in the `tstatsB` dataset. `PROC IML` is used to compute B M-vectors of corresponding null-transformed test statistics $Z_n^\#$, which are then stored in the dataset `Qo`. The empirical distribution of these new M-vectors of test statistics $Z_n^\#$ yields a bootstrap estimator Q_{0n} of the null distribution Q_0 (Procedure 2.3).

`%ssmaxT`: The `%ssmaxT` macro takes as input the M-vector of test statistics T_n for the original sample (i.e., `tstats` dataset from `%lmt` macro) and B M-vectors of null-transformed test statistics $Z_n^\#$, corresponding to a bootstrap estimator Q_{0n} of the null distribution Q_0 (i.e., `Qo` dataset of dimension BM from `%bootnull` macro). For each of the B bootstrap samples, the maximum of the M null-transformed test statistics $Z_n^\#$ is computed using `PROC IML` (i.e., row maxima of the $B \times M$ matrix obtained from the `Qo` dataset). Single-step maxT adjusted p-values are computed for each of the M null hypotheses as the proportions of the B maxima that are greater than or equal to the corresponding test statistics for the original sample (Procedure 3.5). The adjusted p-values are stored in the dataset `fwer`. Note that the current implementation of `%ssmaxT` provides adjusted p-values for two-sided tests only (i.e., based on the absolute values of the test statistics).

`%gfwer`: Given an allowed number k of false positives and adjusted p-values for an arbitrary initial FWER-controlling MTP (e.g., `fwer` dataset from `%ssmaxT` macro), the `%gfwer` macro uses `PROC IML` to compute adjusted p-values for gFWER-controlling augmentation multiple testing Procedure 3.20.

`%tppfp`: Given an allowed proportion q of false positives and adjusted p-values for an arbitrary initial FWER-controlling MTP (e.g., `fwer` dataset from `%ssmaxT` macro), the `%tppfp` macro uses `PROC IML` to compute adjusted p-values for TPPFP-controlling augmentation multiple testing Procedure 3.26.

Note that the macros `%fwer`, `%gfwer`, and `%tppfp` return adjusted p-values as datasets, as opposed to matrices, to facilitate the identification and labeling of null hypotheses.

The above macros are applied in Chapter 11 to the HIV-1 dataset of Segal et al. (2004), with the aim of relating HIV-1 sequence variation to viral replication capacity (Birkner et al., 2005b).

Code is provided in Appendix C and on the book's website (`www.stat.berkeley.edu/~sandrine/MTBook/SAS/SAS.html`). The user could easily modify and extend this collection of macros to adapt to other data structures and/or testing problems.

Table 13.1. *multtest package: Multiple testing procedures implemented in the R package multtest.* The table provides a list of multiple testing procedures implemented in the R package multtest, Version 1.10.0, Bioconductor Release 1.8, `www.bioconductor.org`. The procedures are summarized in Tables A.2–A.9. The main user-level function `MTP` implements bootstrap-based MTPs, while the functions `mt.maxT` and `mt.minP` implement permutation-based MTPs. Unadjusted p-values may be obtained using either `MTP`, `mt.maxT`, or `mt.minP`, and supplied to `mt.rawp2adjp` for simple marginal MTPs.

MTP	Function(s)
FWER	
Single-step Bonferroni Procedure 3.1	`mt.rawp2adjp` via `MTP`, `mt.maxT`, or `mt.minP`
Step-down Holm Procedure 3.7	`mt.rawp2adjp` via `MTP`, `mt.maxT`, or `mt.minP`
Step-up Hochberg Procedure 3.13	`mt.rawp2adjp` via `MTP`, `mt.maxT`, or `mt.minP`
Single-step Šidák Procedure 3.3	`mt.rawp2adjp` via `MTP`, `mt.maxT`, or `mt.minP`
Step-down Šidák-like Procedure 3.9	`mt.rawp2adjp` via `MTP`, `mt.maxT`, or `mt.minP`
Single-step maxT Procedure 3.5	`MTP` and `ss.maxT`
Single-step minP Procedure 3.6	`MTP` and `ss.minP`
Step-down maxT Procedure 3.11	`MTP`, `sd.maxT`, and `mt.maxT`
Step-down minP Procedure 3.12	`MTP`, `sd.minP`, and `mt.minP`
gFWER	
Augmentation Procedure 3.20	`MTP` and `fwer2gfwer`
TPPFP	
Augmentation Procedure 3.26	`MTP` and `fwer2tppfp`
FDR	
Step-up Benjamini and Hochberg Procedure 3.22	`mt.rawp2adjp` via `MTP`, `mt.maxT`, or `mt.minP`
Step-up Benjamini and Yekutieli Procedure 3.23	`mt.rawp2adjp` via `MTP`, `mt.maxT`, or `mt.minP`
TPPFP-based procedure, Theorems 6.6 and 6.7	`fwer2fdr`

A

Summary of Multiple Testing Procedures

Table A.1. *Definitions and notation.* This table summarizes basic definitions, notation, and usage for the main components of a multiple testing procedure. The reader is referred to Section 1.2 for details.

Distributions, models, and random variables (Section 1.2.2)

Data generating distribution: P
Model: $\mathcal{M}$, $P \in \mathcal{M}$
Submodel: $\mathcal{M}(m) \subseteq \mathcal{M}$
Random variable/vector: $X \sim P$
Realization of random variable X: x

Marginal distributions
Scalar random variable $X \in I\!R$
Cumulative distribution function (CDF):
$F_X(x) \equiv \Pr(X \leq x)$
$F_X^{-1}(\alpha) \equiv \inf\{x \in I\!R : F_X(x) \geq \alpha\}$
Survivor function:
$\bar{F}_X(x) \equiv \Pr(X > x) = 1 - F_X(x)$
$\bar{F}_X^{-1}(\alpha) \equiv \inf\left\{x \in I\!R : \bar{F}_X(x) \leq \alpha\right\} = F_X^{-1}(1-\alpha)$
Probability density function (PDF):
$f_X(x) \equiv \frac{d}{dx} F_X(x)$
δ-quantile: $F_X^{-1}(\delta)$

Joint distributions
Random vector $X = (X(j) : j = 1, \ldots, J) \in I\!R^J$
Joint CDF:
$F_X(x) \equiv \Pr(X \in (-\infty, x]) = \Pr(\cap_j \{X(j) \leq x(j)\})$,
where $(-\infty, x] \equiv \prod_j (-\infty, x(j)]$

Joint survivor function:
$\bar{F}_X(x) \equiv \Pr(X \in (x, +\infty)) = \Pr(\cap_j \{X(j) > x(j)\})$,
where $(x, +\infty) \equiv \prod_j (x(j), +\infty)$
Joint PDF:
$f_X(x) \equiv \frac{d}{dx} F_X(x) = \frac{\partial}{\partial x(1)} \cdots \frac{\partial}{\partial x(J)} F_X(x)$

Conditional distributions
Random variables X and Y
Conditional CDF:
$F_{X|Y}(x|Y) \equiv \Pr(X \in (-\infty, x]|Y)$
Conditional survivor function:
$\bar{F}_{X|Y}(x|Y) \equiv \Pr(X \in (x, +\infty)|Y)$
Conditional PDF:
$f_{X|Y}(x|Y) \equiv \frac{d}{dx} F_{X|Y}(x|Y)$

X and Y are independent: $X \perp Y$
X and Y are independent and identically distributed (IID): $X, Y \stackrel{IID}{\sim} P$
Standard Gaussian or normal N(0, 1) distribution:
PDF $\phi(x) = \frac{1}{\sqrt{2\pi}} \exp\left(-\frac{1}{2}x^2\right)$,
CDF $\Phi(x) = \int_{-\infty}^{x} \phi(z)dz$,
survivor function $\bar{\Phi} = 1 - \Phi$

Parameters (Section 1.2.3)
Parameter: $\Psi(P) = \psi = (\psi(j) : j = 1, \ldots, J)$
Expected value: $\mathrm{E}_P[g(X)] \equiv \int g(x) dF_X(x) = \int g(x) f_X(x) dx$
Mean vector: $\psi = \Psi(P) = \mathrm{E}_P[X]$, $\psi(j) = \Psi(P)(j) = \mathrm{E}_P[X(j)]$
Covariance matrix:
$\sigma = \Sigma(P) = \mathrm{Cov}_P[X] \equiv \mathrm{E}_P[(X - \mathrm{E}_P[X])(X - \mathrm{E}_P[X])^\top]$,
$\sigma(j, j') = \mathrm{Cov}_P[X(j), X(j')] \equiv \mathrm{E}_P[(X(j) - \mathrm{E}_P[X(j)])(X(j') - \mathrm{E}_P[X(j')])]$,
$\sigma^2(j) = \sigma(j, j) = \mathrm{Var}_P[X(j)] \equiv \mathrm{E}_P[(X(j) - \mathrm{E}_P[X(j)])^2]$
Correlation matrix:
$\sigma^* = \Sigma^*(P) = \mathrm{Cor}_P[X]$,
$\sigma^*(j, j') = \mathrm{Cor}_P[X(j), X(j')] \equiv \sigma(j, j')/\sigma(j)\sigma(j')$

Null and alternative hypotheses (Section 1.2.4)
Null hypotheses: $H_0(m) \equiv \mathrm{I}(P \in \mathcal{M}(m))$, $m = 1, \ldots, M$
Alternative hypotheses: $H_1(m) \equiv \mathrm{I}(P \notin \mathcal{M}(m))$
E.g. $H_0(m) = \mathrm{I}(\psi(m) \leq \psi_0(m))$ and $H_1(m) = \mathrm{I}(\psi(m) > \psi_0(m))$
True null hypotheses: $\mathcal{H}_0 = \mathcal{H}_0(P) \equiv \{m : H_0(m) = 1\}$, $h_0 \equiv |\mathcal{H}_0|$
False null hypotheses: $\mathcal{H}_1 = \mathcal{H}_1(P) \equiv \{m : H_1(m) = 1\} = \mathcal{H}_0^c$, $h_1 \equiv |\mathcal{H}_1| = M - h_0$
Complete null hypothesis: $H_0^C \equiv \prod_{m=1}^{M} H_0(m) = \mathrm{I}\left(P \in \cap_{m=1}^{M} \mathcal{M}(m)\right)$

Data and empirical distributions (Section 1.2.2)
Random sample: $\mathcal{X}_n \equiv \{X_i : i = 1, \ldots, n\}$, $X_i \stackrel{IID}{\sim} P$, $i = 1, \ldots, n$

Empirical distribution: P_n
$\Pr_{P_n}(X \leq x) \equiv \frac{1}{n}\sum_i \mathrm{I}(x_i \leq x)$, $\Pr_{P_n}(X = x_i) \equiv \frac{1}{n}$, $i = 1, \ldots, n$

Estimators (Section 1.2.5)
Estimator: $\hat{\Psi}(P_n) = \psi_n = (\psi_n(j) : j = 1, \ldots, J)$
Empirical mean vector: $\psi_n = \hat{\Psi}(P_n) = \mathrm{E}_{P_n}[X] = \bar{X}_n \equiv \frac{1}{n}\sum_i X_i$,
$\psi_n(j) = \hat{\Psi}(P_n)(j) = \mathrm{E}_{P_n}[X(j)] = \bar{X}_n(j) \equiv \frac{1}{n}\sum_i X_i(j)$
Empirical covariance matrix:
$\sigma_n = \hat{\Sigma}(P_n) = \mathrm{Cov}_{P_n}[X]$,
$\sigma_n(j, j') = \mathrm{Cov}_{P_n}[X(j), X(j')] \equiv \frac{1}{n}\sum_i (X_i(j) - \bar{X}_n(j))(X_i(j') - \bar{X}_n(j'))$,
$\sigma_n^2(j) = \sigma_n(j, j) = \mathrm{Var}_{P_n}[X(j)] \equiv \frac{1}{n}\sum_i (X_i(j) - \bar{X}_n(j))^2$
Empirical correlation matrix:
$\sigma_n^* = \hat{\Sigma}^*(P_n) = \mathrm{Cor}_{P_n}[X]$,
$\sigma_n^*(j, j') = \mathrm{Cor}_{P_n}[X(j), X(j')] \equiv \sigma_n(j, j')/\sigma_n(j)\sigma_n(j')$

Test statistics (Sections 1.2.5, 2.6, and 2.7)
Test statistics: $T_n = (T_n(m) : 1, \ldots, M)$
Test statistics true distribution: $T_n \sim Q_n = Q_n(P)$
Difference statistics: $T_n(m) \equiv \sqrt{n}(\psi_n(m) - \psi_0(m))$
t-statistics: $T_n(m) \equiv \sqrt{n}\frac{\psi_n(m) - \psi_0(m)}{\sigma_n(m)}$
F-statistics: $T_n(m) \equiv \frac{\frac{1}{K-1}\sum_k n_k(\bar{X}_{k,n_k}(m) - \bar{X}_n(m))^2}{\frac{1}{n-K}\sum_k \sum_i (X_{k,i}(m) - \bar{X}_{k,n_k}(m))^2}$

Test statistics null distributions (Chapter 2)
Null shift and scale-transformed null distribution (Section 2.3)
$Z_n(m) \equiv \sqrt{\min\{1, \tau_0(m)/\mathrm{Var}[T_n(m)]\}}(T_n(m) - \mathrm{E}[T_n(m)]) + \lambda_0(m)$
Test statistics null distribution: $Q_0 = Q_0(P)$, where $Z_n \stackrel{\mathcal{L}}{\Rightarrow} Q_0$
Estimator of test statistics null distribution (Procedure 2.3): Q_{0n}
t-statistics: $\lambda_0(m) = 0$, $\tau_0(m) = 1$
F-statistics: $\lambda_0(m) = 1$, $\tau_0(m) = 2/(K-1)$

Null quantile-transformed null distribution (Section 2.4)
$Z_n(m) \equiv q_{0,m}^{-1} Q_{n,m}^{\Delta}(T_n(m))$
Test statistics null distribution: $Q_0 = Q_0(P)$, where $Z_n \stackrel{\mathcal{L}}{\Rightarrow} Q_0$
Estimator of test statistics null distribution (Procedure 2.4): Q_{0n}

Multiple testing procedures and rejection regions (Sections 1.2.6 and 1.2.7)
Rejection regions: $\mathcal{C}_n(m) = \mathcal{C}_n(m; \alpha) = \mathcal{C}(m; T_n, Q_{0n}, \alpha)$
Critical values/cut-offs: $c_n(m) = c_n(m; \alpha) = c(m; T_n, Q_{0n}, \alpha)$

Multiple testing procedure (MTP):

$$\begin{aligned}\mathcal{R}_n &= \mathcal{R}_n(\alpha) = \mathcal{R}(T_n, Q_{0n}, \alpha) \\ &\equiv \{m : T_n(m) \in \mathcal{C}_n(m)\} \\ &= \{m : \widetilde{P}_{0n}(m) \leq \alpha\}\end{aligned}$$

Augmentation multiple testing procedure (AMTP): $\mathcal{R}_n^+ \equiv \mathcal{R}_n \cup \mathcal{A}_n$
E.g. One-sided rejection regions: $\mathcal{C}_n(m) = (c_n(m), +\infty)$
Common cut-offs: $\gamma^{(M)} \in I\!R^M$, where $\gamma^{(M)}(m) \equiv \gamma \in I\!R$, $\forall\, m = 1, \ldots, M$
Common-quantile cut-offs: $q^{-1}(\delta) \equiv (Q_m^{-1}(\delta) : m = 1, \ldots, M)$, where $Q_m^{-1}(\delta)$ are the δ-quantiles of the marginal distributions Q_m of an M-variate distribution Q

Errors in multiple hypothesis testing (Section 1.2.8)
Number of rejected hypotheses: $R_n \equiv |\mathcal{R}_n| = \sum_{m=1}^{M} \mathrm{I}\left(T_n(m) \in \mathcal{C}_n(m)\right)$
Number of true negatives: $W_n \equiv |\mathcal{R}_n^c \cap \mathcal{H}_0| = \sum_{m \in \mathcal{H}_0} \mathrm{I}\left(T_n(m) \notin \mathcal{C}_n(m)\right)$
Number of Type I errors: $V_n \equiv |\mathcal{R}_n \cap \mathcal{H}_0| = \sum_{m \in \mathcal{H}_0} \mathrm{I}\left(T_n(m) \in \mathcal{C}_n(m)\right)$
Number of Type II errors: $U_n \equiv |\mathcal{R}_n^c \cap \mathcal{H}_1| = \sum_{m \in \mathcal{H}_1} \mathrm{I}\left(T_n(m) \notin \mathcal{C}_n(m)\right)$
Number of true positives: $S_n \equiv |\mathcal{R}_n \cap \mathcal{H}_1| = \sum_{m \in \mathcal{H}_1} \mathrm{I}\left(T_n(m) \in \mathcal{C}_n(m)\right)$

Type I error rates (Section 1.2.9)
Parameter of the joint distribution of the numbers of Type I errors and rejected hypotheses: $\Theta(F_{V_n,R_n})$
E.g. $gTP(q,g) \equiv \Pr(g(V_n, R_n) > q)$, $gEV(g) \equiv \mathrm{E}[g(V_n, R_n)]$

Power (Section 1.2.10)
Parameter of the joint distribution of the numbers of Type II errors and rejected hypotheses: $\Theta(F_{U_n,R_n})$
E.g. $AvgPwr \equiv 1 - \frac{1}{h_1}\mathrm{E}[U_n]$

Unadjusted p-values (Section 1.2.12)

$$P_{0n}(m) = P(T_n(m), Q_{0n,m}) \equiv \inf\left\{\alpha \in [0,1] : T_n(m) \in \mathcal{C}_n(m;\alpha)\right\},$$

where $\Pr_{Q_{0n,m}}(T_n(m) \in \mathcal{C}_n(m;\alpha)) \leq \alpha$
E.g. $p_{0n}(m) = \bar{Q}_{0n,m}(t_n(m)) = \Pr_{Q_{0n,m}}(T_n(m) > t_n(m))$

Adjusted p-values (Section 1.2.12)

$$\begin{aligned}\widetilde{P}_{0n}(m) &= \widetilde{P}(m; T_n, Q_{0n}) \\ &\equiv \inf\left\{\alpha \in [0,1] : m \in \mathcal{R}_n(\alpha)\right\} \\ &= \inf\left\{\alpha \in [0,1] : T_n(m) \in \mathcal{C}_n(m;\alpha)\right\}\end{aligned}$$

E.g. Bonferroni adjusted p-values, $\widetilde{P}_{0n}(m) = \min\left\{M P_{0n}(m), 1\right\}$

Confidence regions (Section 4.6)

$\mathcal{CR}_n = \mathcal{CR}(\mathcal{X}_n, \mathcal{Q}_{0n}, \alpha)$

E.g.

$$\mathcal{CR}_n = \left\{ \psi \in I\!R^M : \psi(m) \in \left[\psi_n(m) - u(m)\frac{\sigma_n(m)}{\sqrt{n}}, \right.\right.$$
$$\left.\left. \psi_n(m) - l(m)\frac{\sigma_n(m)}{\sqrt{n}} \right], \ \forall\, m = 1, \ldots, M \right\}$$

Table A.2. *Multiple hypothesis testing flowchart.*

Specify data generating distribution and parameters of interest

$P, \psi = (\psi(j) : j = 1, \ldots, J)$

⇓

Define null and alternative hypotheses

$H_0(m) = \mathrm{I}(P \in \mathcal{M}(m))$ and $H_1(m) = \mathrm{I}(P \notin \mathcal{M}(m))$

⇓

Specify test statistics

$T_n = (T_n(m) : m = 1, \ldots, M)$

⇓

Estimate test statistics null distribution

Q_{0n} (Procedures 2.3 and 2.4)

⇓

Select Type I error rate

$\Theta(F_{V_n,R_n})$

⇓

Apply MTP

FWER	$\Pr(V_n > 0)$	Single-step common-cut-off maxT Procedure 3.5 Single-step common-quantile minP Procedure 3.6 Step-down common-cut-off maxT Procedure 3.11 Step-down common-quantile minP Procedure 3.12 Resampling-based empirical Bayes Procedure 7.1
gFWER	$\Pr(V_n > k)$	Single-step common-cut-off $T(k+1)$ Procedure 3.18 Single-step common-quantile $P(k+1)$ Procedure 3.19 Augmentation multiple testing Procedure 3.20 Resampling-based empirical Bayes Procedure 7.1
General	$\Theta(F_{V_n})$	Single-step common-cut-off Procedure 4.2 Single-step common-quantile Procedure 4.1 Resampling-based empirical Bayes Procedure 7.1*
TPPFP	$\Pr(V_n/R_n > q)$	Augmentation multiple testing Procedure 3.26 Resampling-based empirical Bayes Procedure 7.1
gTP	$\Pr(g(V_n, R_n) > q)$	Augmentation multiple testing Procedure 6.9 Resampling-based empirical Bayes Procedure 7.1
FDR	$\mathrm{E}[V_n/R_n]$	TPPFP-based procedure, Theorems 6.6 and 6.7 Resampling-based empirical Bayes Procedure 7.1*
gEV	$\mathrm{E}[g(V_n, R_n)]$	gTP-based procedure, Theorem 6.12 Resampling-based empirical Bayes Procedure 7.1*
General	$\Theta(F_{g(V_n,R_n)})$	Resampling-based empirical Bayes Procedure 7.1*

⇓

Summarize results

Adjusted p-values, rejection regions, and confidence regions

$*$ Generalization of Procedure 7.1, as discussed in Section 7.8.

Table A.3. *Type I error rates.* Commonly-used Type I error rates, defined as parameters $\Theta(F_{V_n,R_n})$ of the joint distribution of the numbers of Type I errors V_n and rejected hypotheses R_n (Section 1.2.9).

Type I error rate	**Parameter $\Theta(F_{V_n,R_n})$**
Family-wise error rate	$FWER = \Pr(V_n > 0)$
Generalized family-wise error rate	$gFWER(k) = \Pr(V_n > k)$
Per-comparison error rate	$PCER = \mathrm{E}[V_n]/M$
Per-family error rate	$PFER = \mathrm{E}[V_n]$
Median-based per-family error rate	$mPFER = F_{V_n}^{-1}(1/2)$
Quantile number of false positives	$QNFP(\delta) = F_{V_n}^{-1}(\delta)$
Tail probability for the proportion of false positives	$TPPFP(q) = \Pr(V_n/R_n > q)$
False discovery rate	$FDR = \mathrm{E}[V_n/R_n]$
Proportion of expected false positives	$PEFP = \mathrm{E}[V_n]/\mathrm{E}[R_n]$
Quantile proportion of false positives	$QPFP(\delta) = F_{V_n/R_n}^{-1}(\delta)$
Generalized tail probability error rate	$gTP(q,g) = \Pr(g(V_n, R_n) > q)$
Generalized expected value error rate	$gEV(g) = \mathrm{E}[g(V_n, R_n)]$

Table A.4. *Multiple testing procedures.* Main properties of the multiple testing procedures presented in Chapters 1–7. "Distribution": whether the MTP takes into account the joint distribution of the test statistics ("Joint" vs. "Marginal"); "Step": stepwise nature of the MTP ("Single", "Down", or "Up"); "Cut-offs": whether the MTP is based on common cut-offs or common-quantile cut-offs ("Common-cut-off" vs. "Common-quantile"); "Control": conditions for Type I error control ("General" vs. "Restricted"). Type I error rates are defined in Table A.3; Tables A.6–A.9 provide further detail on the MTPs.

Short name	Distribution	Step	Cut-offs	Control
	FWER, Table A.6			
SS Bonferroni	Marginal	Single	Common-quantile	General
SD Holm	Marginal	Down	Common-quantile	General
SU Hochberg	Marginal	Up	Common-quantile	Restricted
SS Sidak	Marginal	Single	Common-quantile	Restricted
SD Sidak	Marginal	Down	Common-quantile	Restricted
SS maxT	Joint	Single	Common-cut-off	General
SS minP	Joint	Single	Common-quantile	General
SD maxT	Joint	Down	Common-cut-off	General
SD minP	Joint	Down	Common-quantile	General
FWER EBayes	Joint	Single	Any	General
	gFWER, Table A.7			
gFWER SS LR	Marginal	Single	Common-quantile	General
gFWER SD LR	Marginal	Down	Common-quantile	General
SS T(k+1)	Joint	Single	Common-cut-off	General
SS P(k+1)	Joint	Single	Common-quantile	General
gFWER AMTP	Any	Any	Any	General
gFWER EBayes	Joint	Single	Any	General
	TPPFP, Table A.8			
TPPFP Rest SD LR	Marginal	Down	Common-quantile	Restricted
TPPFP Gen SD LR	Marginal	Down	Common-quantile	General
TPPFP AMTP	Any	Any	Any	General
TPPFP EBayes	Joint	Single	Any	General
	FDR, Table A.9			
SU BH	Marginal	Up	Common-quantile	Restricted
SU BY	Marginal	Up	Common-quantile	General
TPPFP-based	Any	Any	Any	General
FDR EBayes	Joint	Single	Any	General

Table A.5. *Multiple testing procedures.* References for the multiple testing procedures proposed in Chapters 1–7. Type I error rates are defined in Table A.3; Tables A.6–A.9 provide further detail on the MTPs.

FWER, Table A.6

SS maxT: Single-step common-cut-off maxT procedure
Procedure 3.5, p. 118; Procedure 4.2, p. 165; Corollary 4.9, Equation (4.20), p. 173

SS minP: Single-step common-quantile minP procedure
Procedure 3.6, p. 118; Procedure 4.1, p. 164; Corollary 4.8, Equation (4.17), p. 172

SD maxT: Step-down common-cut-off maxT procedure
Procedure 3.11, p. 126; Procedure 5.1, p. 202; Proposition 5.5, Equation (5.22), p. 211

SD minP: Step-down common-quantile minP procedure
Procedure 3.12, p. 126; Procedure 5.6, p. 213; Proposition 5.11, Equation (5.39), p. 219

FWER EBayes: Resampling-based empirical Bayes procedure
Procedure 7.1, p. 298; Equations (7.21)–(7.30), p. 300

gFWER, Table A.7

SS T(k+1): Single-step common-cut-off $T(k+1)$ procedure
Procedure 3.18, p. 138; Procedure 4.2, p. 165; Corollary 4.9, Equation (4.20), p. 173

SS P(k+1): Single-step common-quantile $P(k+1)$ procedure
Procedure 3.19, p. 138; Procedure 4.1, p. 164; Corollary 4.8, Equation (4.17), p. 172

gFWER AMTP: Augmentation multiple testing procedure
Procedure 3.20, p. 139; Procedure 6.2, p. 242; Equation (6.24), p. 245; Procedure 6.9, p. 258; Theorem 6.11, p. 263

gFWER EBayes: Resampling-based empirical Bayes procedure
Procedure 7.1, p. 298; Equations (7.21)–(7.30), p. 300

$\Theta(F_{V_n})$

SS T: Single-step common-cut-off procedure
Procedure 3.18, p. 138; Procedure 4.2, p. 165; Proposition 4.5, Equation (4.14), p. 170

SS P: Single-step common-quantile procedure
Procedure 3.19, p. 138; Procedure 4.1, p. 164; Proposition 4.4, Equation (4.12), p. 169

EBayes: Resampling-based empirical Bayes procedure (generalized)
Procedure 7.1, p. 298; Section 7.8

TPPFP, Table A.8

TPPFP AMTP: Augmentation multiple testing procedure
Procedure 3.26, p. 153; Procedure 6.4, p. 246; Equation (6.35), p. 250; Procedure 6.9, p. 258; Theorem 6.11, p. 263
TPPFP EBayes: Resampling-based empirical Bayes procedure
Procedure 7.1, p. 298; Equations (7.21)–(7.30), p. 300

gTP

gTP AMTP: Augmentation multiple testing procedure
Procedure 6.9, p. 258; Theorem 6.11, p. 263
gTP EBayes: Resampling-based empirical Bayes procedure
Procedure 7.1, p. 298; Equations (7.21)–(7.30), p. 300

FDR, Table A.9

TPPFP-based: TPPFP-based procedure
Theorems 6.6 and 6.7, p. 253; Proposition 6.8, p. 255
FDR EBayes: Resampling-based empirical Bayes procedure (generalized)
Procedure 7.1, p. 298; Section 7.8

gEV

gTP-based: gTP-based procedure
Theorem 6.12, p. 270; Proposition 6.13, p. 271
gEV EBayes: Resampling-based empirical Bayes procedure (generalized)
Procedure 7.1, p. 298; Section 7.8

$\Theta(F_{g(V_n,R_n)})$

EBayes: Resampling-based empirical Bayes procedure (generalized)
Procedure 7.1, p. 298; Section 7.8

Table A.6. *FWER-controlling multiple testing procedures,* $\Theta(F_{V_n,R_n}) = 1 - F_{V_n}(0)$. Summary and properties of FWER-controlling multiple testing procedures, in terms of: whether they take into account the joint distribution of the test statistics ("Joint" vs. "Marginal"); their stepwise nature ("Single-step", "Step-down", or "Step-up"); whether they are based on common cut-offs or common-quantile cut-offs ("Common-cut-off" vs. "Common-quantile"); conditions for FWER control ("General" vs. "Restricted"); their adjusted p-values $\widetilde{P}_{0n}(O_n(m))$. For common-quantile (i.e., p-value-based) procedures, $O_n(m)$ denote the indices for the ordered unadjusted p-values $P_{0n}(O_n(m))$, so that $P_{0n}(O_n(1)) \leq \cdots \leq P_{0n}(O_n(M))$. For common-cut-off procedures, $O_n(m)$ denote the indices for the ordered test statistics $T_n(O_n(m))$, so that $T_n(O_n(1)) \geq \cdots \geq T_n(O_n(M))$.

1. **SS Bonferroni: Bonferroni (1936)**
   Proc. 3.1, p. 113
   Marginal/Single-step/Common-quantile
   General control
$$\widetilde{P}_{0n}(m) = \min\{M\, P_{0n}(m), 1\}$$
2. **SD Holm: Holm (1979)**
   Proc. 3.7, p. 121
   Marginal/Step-down/Common-quantile
   General control
$$\widetilde{P}_{0n}(O_n(m)) = \max_{h=1,\ldots,m} \{\min\{(M-h+1)\, P_{0n}(O_n(h)), 1\}\}$$
3. **SU Hochberg: Hochberg (1988)**
   Proc. 3.13, p. 129
   Marginal/Step-up/Common-quantile
   Restricted control: e.g., independence, Simes' Inequality (Equation (B.5))
$$\widetilde{P}_{0n}(O_n(m)) = \min_{h=m,\ldots,M} \{\min\{(M-h+1)\, P_{0n}(O_n(h)), 1\}\}$$
4. **SS Sidak: Šidák (1967)**
   Proc. 3.3, p. 115
   Marginal/Single-step/Common-quantile
   Restricted control: e.g., independence, Šidák's Inequality (Equations (B.3) and (B.4))
$$\widetilde{P}_{0n}(m) = 1 - (1 - P_{0n}(m))^M$$
5. **SD Sidak: Šidák-like, Holland and Copenhaver (1987)**
   Proc. 3.9, p. 123
   Marginal/Step-down/Common-quantile
   Restricted control: e.g., independence, Šidák's Inequality (Equations (B.3) and (B.4))

$$\widetilde{P}_{0n}(O_n(m)) = \max_{h=1,\ldots,m} \left\{1 - (1 - P_{0n}(O_n(h)))^{(M-h+1)}\right\}$$

6. **SS maxT: Single-step common-cut-off maxT, Dudoit et al. (2004b)**
   Proc. 3.5, p. 118; Proc. 4.2, p. 165; Cor. 4.9, p. 173
   Joint/Single-step/Common-cut-off
   General control

$$\widetilde{p}_{0n}(m) = \Pr_{Q_0}\left(\max_{m=1,\ldots,M} Z(m) \geq t_n(m)\right)$$

7. **SS minP: Single-step common-quantile minP, Dudoit et al. (2004b)**
   Proc. 3.6, p. 118; Proc. 4.1, p. 164; Cor. 4.8, p. 172
   Joint/Single-step/Common-quantile
   General control

$$\widetilde{p}_{0n}(m) = \Pr_{Q_0}\left(\min_{m=1,\ldots,M} P_0(m) \leq p_{0n}(m)\right)$$

8. **SD maxT: Step-down common-cut-off maxT, van der Laan et al. (2004a)**
   Proc. 3.11, p. 126; Proc. 5.1, p. 202; Prop. 5.5, p. 211
   Joint/Step-down/Common-cut-off
   General control

$$\widetilde{p}_{0n}(o_n(m)) = \max_{h=1,\ldots,m} \left\{\Pr_{Q_0}\left(\max_{l\in\{o_n(h),\ldots,o_n(M)\}} Z(l) \geq t_n(o_n(h))\right)\right\}$$

9. **SD minP: Step-down common-quantile minP, van der Laan et al. (2004a)**
   Proc. 3.12, p. 126; Proc. 5.6, p. 213; Prop. 5.11, p. 219
   Joint/Step-down/Common-quantile
   General control

$$\widetilde{p}_{0n}(o_n(m)) = \max_{h=1,\ldots,m} \left\{\Pr_{Q_0}\left(\min_{l\in\{o_n(h),\ldots,o_n(M)\}} P_0(l) \leq p_{0n}(o_n(h))\right)\right\}$$

10. **FWER EBayes: Resampling-based empirical Bayes, van der Laan et al. (2005)**
    Proc. 7.1, p. 298; Eqn. (7.21)–(7.30), p. 300
    Joint/Single-step/Common-cut-off or Common-quantile
    General control

$$\widetilde{p}_{0n}(o_n(m)) \approxeq \min_{h\in\{o_n(m),\ldots,o_n(M)\}} \frac{1}{B}\sum_{b=1}^{B} \mathrm{I}\left(V(c_n(h); \mathcal{H}_{0n}^b, T_{0n}^b) > 0\right)$$

    Common-cut-off: $c_n(h) = (t_n(h))^{(M)}$
    Common-quantile: $c_n(h) = q_{0n}^{-1}(1 - p_{0n}(h))$

Table A.7. *gFWER-controlling multiple testing procedures,* $\Theta(F_{V_n,R_n}) = 1 - F_{V_n}(k)$. Summary and properties of gFWER-controlling multiple testing procedures, in terms of: whether they take into account the joint distribution of the test statistics ("Joint" vs. "Marginal"); their stepwise nature ("Single-step", "Step-down", or "Step-up"); whether they are based on common cut-offs or common-quantile cut-offs ("Common-cut-off" vs. "Common-quantile"); conditions for gFWER control ("General" vs. "Restricted"); their adjusted p-values $\widetilde{P}_{0n}(O_n(m))$. For common-quantile (i.e., p-value-based) procedures, $O_n(m)$ denote the indices for the ordered unadjusted p-values $P_{0n}(O_n(m))$, so that $P_{0n}(O_n(1)) \leq \cdots \leq P_{0n}(O_n(M))$. For common-cut-off procedures, $O_n(m)$ denote the indices for the ordered test statistics $T_n(O_n(m))$, so that $T_n(O_n(1)) \geq \cdots \geq T_n(O_n(M))$. In the case of augmentation multiple testing procedures, the adjusted p-values for the initial FWER-controlling MTP and the AMTP are denoted by $\widetilde{P}_{0n}(m)$ and $\widetilde{P}^{+}_{0n}(m)$, respectively. The indices $O_n(m)$ are such that $\widetilde{P}_{0n}(O_n(1)) \leq \cdots \leq \widetilde{P}_{0n}(O_n(M))$.

1. **gFWER SS LR: Lehmann and Romano (2005)**
   Proc. 3.15, p. 134
   Marginal/Single-step/Common-quantile
   General control

$$\widetilde{P}_{0n}(m) = \min\left\{\frac{M}{k+1} P_{0n}(m), 1\right\}$$

   $k = 0$: Bonferroni Proc. 3.1

2. **gFWER SD LR: Lehmann and Romano (2005)**
   Proc. 3.17, p. 136
   Marginal/Step-down/Common-quantile
   General control

$$\widetilde{P}_{0n}(O_n(m)) = \begin{cases} \min\left\{\frac{M}{k+1} P_{0n}(O_n(m)), 1\right\}, & \text{if } m \leq k \\ \max_{h=1,\ldots,m-k}\left\{\min\left\{\frac{M-h+1}{k+1} P_{0n}(O_n(h+k)), 1\right\}\right\}, & \text{if } m > k \end{cases}$$

   $k = 0$: Holm Proc. 3.7

3. **SS T(k+1): Single-step common-cut-off $T(k+1)$, Dudoit et al. (2004b)**
   Proc. 3.18, p. 138; Proc. 4.2, p. 165; Cor. 4.9, p. 173
   Joint/Single-step/Common-cut-off
   General control

$$\widetilde{p}_{0n}(m) = \Pr_{Q_0}\left(Z^{\circ}(k+1) \geq t_n(m)\right)$$

   $k = 0$: Single-step maxT Proc. 3.5

4. **SS P(k+1): Single-step common-quantile $P(k+1)$, Dudoit et al. (2004b)**
   Proc. 3.19, p. 138; Proc. 4.1, p. 164; Cor. 4.8, p. 172
   Joint/Single-step/Common-quantile
   General control

$$\widetilde{p}_{0n}(m) = \Pr_{Q_0}\left(P_0^\circ(k+1) \leq p_{0n}(m)\right)$$

   $k = 0$: Single-step minP Proc. 3.6

5. **gFWER AMTP: Augmentation, van der Laan et al. (2004b)**
   Proc. 3.20, p. 139; Proc. 6.2, p. 242; Eqn. (6.24), p. 245
   Marginal or Joint/Single-step, Step-down, or Step-up/Common-cut-off or Common-quantile
   General control

$$\widetilde{P}_{0n}^{+}(O_n(m)) = \begin{cases} 0, & \text{if } m \leq k \\ \widetilde{P}_{0n}(O_n(m-k)), & \text{if } m > k \end{cases}$$

6. **gFWER EBayes: Resampling-based empirical Bayes, van der Laan et al. (2005)**
   Proc. 7.1, p. 298; Eqn. (7.21)–(7.30), p. 300
   Joint/Single-step/Common-cut-off or Common-quantile
   General control

$$\widetilde{p}_{0n}(o_n(m)) \approxeq \min_{h \in \{o_n(m),\ldots,o_n(M)\}} \frac{1}{B}\sum_{b=1}^{B} \mathrm{I}\left(V(c_n(h); \mathcal{H}_{0n}^b, T_{0n}^b) > k\right)$$

   Common-cut-off: $c_n(h) = (t_n(h))^{(M)}$
   Common-quantile: $c_n(h) = q_{0n}^{-1}(1 - p_{0n}(h))$

Table A.8. *TPPFP-controlling multiple testing procedures,* $\Theta(F_{V_n,R_n}) = \Pr(V_n/R_n > q)$. Summary and properties of TPPFP-controlling multiple testing procedures, in terms of: whether they take into account the joint distribution of the test statistics ("Joint" vs. "Marginal"); their stepwise nature ("Single-step", "Step-down", or "Step-up"); whether they are based on common cut-offs or common-quantile cut-offs ("Common-cut-off" vs. "Common-quantile"); conditions for TPPFP control ("General" vs. "Restricted"); their adjusted p-values $\widetilde{P}_{0n}(O_n(m))$. For common-quantile (i.e., p-value-based) procedures, $O_n(m)$ denote the indices for the ordered unadjusted p-values $P_{0n}(O_n(m))$, so that $P_{0n}(O_n(1)) \leq \cdots \leq P_{0n}(O_n(M))$. For common-cut-off procedures, $O_n(m)$ denote the indices for the ordered test statistics $T_n(O_n(m))$, so that $T_n(O_n(1)) \geq \cdots \geq T_n(O_n(M))$. In the case of augmentation multiple testing procedures, the adjusted p-values for the initial FWER-controlling MTP and the AMTP are denoted by $\widetilde{P}_{0n}(m)$ and $\widetilde{P}^+_{0n}(m)$, respectively. The indices $O_n(m)$ are such that $\widetilde{P}_{0n}(O_n(1)) \leq \cdots \leq \widetilde{P}_{0n}(O_n(M))$.

1. **TPPFP Rest SD LR: Lehmann and Romano (2005)**
   Proc. 3.24, p. 151
   Marginal/Step-down/Common-quantile
   Restricted control: e.g., Assumption LR.TPPFP1 or LR.TPPFP2

$$\widetilde{P}_{0n}(O_n(m)) = \max_{h=1,\ldots,m} \left\{\min\left\{b(h)\, P_{0n}(O_n(h)), 1\right\}\right\}$$

$$b(m) = \frac{M + \lfloor qm \rfloor + 1 - m}{\lfloor qm \rfloor + 1}$$

2. **TPPFP Gen SD LR: Lehmann and Romano (2005)**
   Proc. 3.25, p. 152
   Marginal/Step-down/Common-quantile
   General control

$$\widetilde{P}_{0n}(O_n(m)) = \max_{h=1,\ldots,m} \left\{\min\left\{b(h)\, P_{0n}(O_n(h)), 1\right\}\right\}$$

$$b(m) = C(\lfloor qM \rfloor + 1)\frac{M + \lfloor qm \rfloor + 1 - m}{\lfloor qm \rfloor + 1},$$

   where $C(M) \equiv \sum_{m=1}^{M} 1/m$

3. **TPPFP AMTP: Augmentation, van der Laan et al. (2004b)**
   Proc. 3.26, p. 153; Proc. 6.4, p. 246; Eqn. (6.35), p. 250
   Marginal or Joint/Single-step, Step-down, or Step-up/Common-cut-off or Common-quantile
   General control

$$\widetilde{P}^+_{0n}(O_n(m)) = \widetilde{P}_{0n}(O_n(\lceil (1-q)m \rceil))$$

4. **TPPFP EBayes: Resampling-based empirical Bayes, van der Laan et al. (2005)**
   Proc. 7.1, p. 298; Eqn. (7.21)–(7.30), p. 300
   Joint/Single-step/Common-cut-off or Common-quantile
   General control

$$\widetilde{p}_{0n}(o_n(m)) \approxeq \min_{h \in \{o_n(m),\ldots,o_n(M)\}} \frac{1}{B} \sum_{b=1}^{B} \mathrm{I}\left(\frac{V(c_n(h); \mathcal{H}_{0n}^b, T_{0n}^b)}{R(c_n(h); \mathcal{H}_{0n}^b, T_{0n}^b, T_n)} > q \right)$$

   Common-cut-off: $c_n(h) = (t_n(h))^{(M)}$
   Common-quantile: $c_n(h) = q_{0n}^{-1}(1 - p_{0n}(h))$

Table A.9. *FDR-controlling multiple testing procedures, $\Theta(F_{V_n,R_n}) = \mathrm{E}[V_n/R_n]$.* Summary and properties of FDR-controlling multiple testing procedures, in terms of: whether they take into account the joint distribution of the test statistics ("Joint" vs. "Marginal"); their stepwise nature ("Single-step", "Step-down", or "Step-up"); whether they are based on common cut-offs or common-quantile cut-offs ("Common-cut-off" vs. "Common-quantile"); conditions for FDR control ("General" vs. "Restricted"); their adjusted p-values $\widetilde{P}_{0n}(O_n(m))$. For common-quantile (i.e., p-value-based) procedures, $O_n(m)$ denote the indices for the ordered unadjusted p-values $P_{0n}(O_n(m))$, so that $P_{0n}(O_n(1)) \leq \cdots \leq P_{0n}(O_n(M))$. For common-cut-off procedures, $O_n(m)$ denote the indices for the ordered test statistics $T_n(O_n(m))$, so that $T_n(O_n(1)) \geq \cdots \geq T_n(O_n(M))$.

1. **SU BH: Benjamini and Hochberg (1995)**
   Proc. 3.22, p. 146
   Marginal/Step-up/Common-quantile
   Restricted control: e.g., independence, positive regression dependence

$$\widetilde{P}_{0n}(O_n(m)) = \min_{h=m,\ldots,M} \left\{ \min \left\{ \frac{M}{h} P_{0n}(O_n(h)), 1 \right\} \right\}$$

2. **SU BY: Benjamini and Yekutieli (2001)**
   Proc. 3.23, p. 147
   Marginal/Step-up/Common-quantile
   General control

$$\widetilde{P}_{0n}(O_n(m)) = \min_{h=m,\ldots,M} \left\{ \min \left\{ C(M)\frac{M}{h} P_{0n}(O_n(h)), 1 \right\} \right\},$$

   where $C(M) \equiv \sum_{m=1}^{M} 1/m$

3. **TPPFP-based: TPPFP-based, van der Laan et al. (2004b)**
   Thm. 6.6 and 6.7, p. 253; Prop. 6.8, p. 255
   Marginal or Joint/Single-step, Step-down, or Step-up/Common-cut-off or Common-quantile
   General control

$$\widetilde{P}_{0n}(m) = \inf \left\{ \alpha \in [0,1] : \widetilde{P}_{0n}^{TPPTP(q(\alpha))}(m) \leq \alpha^{TPPFP}(\alpha) \right\}$$

   Bound 1: $q(\alpha) + \alpha^{TPPFP}(\alpha) = \alpha$
   Bound 2: $q(\alpha)(1 - \alpha^{TPPFP}(\alpha)) + \alpha^{TPPFP}(\alpha) = \alpha$

4. **FDR EBayes: Resampling-based empirical Bayes**
   Proc. 7.1, p. 298; Sect. 7.8 (generalized)
   Joint/Single-step/Common-cut-off or Common-quantile
   General control

B

Miscellaneous Mathematical and Statistical Results

B.1 Probability inequalities

Boole's Inequality. Given a collection of M events, $B_1, \ldots, B_M$, then

$$\Pr\left(\bigcup_{m=1}^{M} B_m\right) \leq \sum_{m=1}^{M} \Pr(B_m), \tag{B.1}$$

with equality for disjoint events, i.e., for $B_m \cap B_{m'} = \emptyset$, $\forall\ m \neq m'$. Boole's Inequality is also known as *Bonferroni's Inequality.*

Markov's Inequality. Given a non-negative random variable X and a constant $a > 0$, then

$$\Pr(X \geq a) \leq \frac{\mathrm{E}[X]}{a}. \tag{B.2}$$

Šidák's Inequality. [Section 3.2.2, p. 115; Šidák (1967)] Consider a random M-vector $Z = (Z(m) : m = 1, \ldots, M)$, with joint distribution Q_0, and an M-vector of constants $c = (c(m) : m = 1, \ldots, M) \in I\!R^M$. Then, *Šidák's Inequality* states that

$$\Pr_{Q_0}\left(\bigcap_{m=1}^{M} \{Z(m) \leq c(m)\}\right) \geq \prod_{m=1}^{M} \Pr_{Q_0}(Z(m) \leq c(m)). \tag{B.3}$$

For a random M-vector of unadjusted p-values $P_0 = (P_0(m) : m = 1, \ldots, M)$, defined as in Equation (1.45) based on Q_0, and an M-vector of constants $c = (c(m) : m = 1, \ldots, M) \in [0,1]^M$, the p-value version of Šidák's Inequality states that

$$\Pr_{Q_0}\left(\bigcap_{m=1}^{M} \{P_0(m) > c(m)\}\right) \geq \prod_{m=1}^{M} \Pr_{Q_0}(P_0(m) > c(m)). \tag{B.4}$$

Simes' Inequality. [Section 3.2.4, p. 128; Simes (1986)] Consider a random M-vector $Z = (Z(m) : m = 1, \ldots, M)$ with joint distribution Q_0, unadjusted p-values $P_0 = (P_0(m) : m = 1, \ldots, M)$ defined as in Equation (1.45) based on Q_0, and ordered unadjusted p-values $P_0^\circ(m)$ such that $P_0^\circ(1) \leq \cdots \leq P_0^\circ(M)$. Then, *Simes' Inequality* states that

$$\Pr_{Q_0}\left(\bigcap_{m=1}^{M}\left\{P_0^\circ(m) > \frac{m}{M}\alpha\right\}\right) \geq 1 - \alpha \tag{B.5}$$

or, equivalently,

$$\Pr_{Q_0}\left(\bigcup_{m=1}^{M}\left\{P_0^\circ(m) \leq \frac{m}{M}\alpha\right\}\right) \leq \alpha.$$

B.2 Convergence results

Consider J-dimensional random vectors $\{X_n\}$ and X, with respective joint cumulative distribution functions $\{F_n\}$ and F, defined as $F_n(x) \equiv \Pr(X_n \in (-\infty, x]) = \Pr(\cap_j\{X_n(j) \leq x(j)\})$ and $F(x) \equiv \Pr(X \in (-\infty, x]) = \Pr(\cap_j\{X(j) \leq x(j)\})$ for $x \in I\!R^J$ and corresponding J-dimensional rectangle $(-\infty, x] \equiv \prod_j(-\infty, x(j)] = (-\infty, x(1)] \times \cdots \times (-\infty, x(J)] = \{y \in I\!R^J : y(j) \leq x(j),\ j = 1, \ldots, J\}$.

Theorem B.1. [Weak convergence] *The following are equivalent definitions of* weak convergence *of the sequence $\{X_n\}$ to X.*
(i) For every bounded continuous function $\ell : I\!R^J \rightarrow I\!R$,

$$\lim_{n\rightarrow\infty} \mathrm{E}[\ell(X_n)] = \mathrm{E}[\ell(X)]. \tag{B.6}$$

(ii) For every continuity point x of F, that is, for every $x \in I\!R^J$ such that the J-dimensional rectangle $(-\infty, x]$ has null boundary probability $\Pr(X \in \partial(-\infty, x])$ *under F,*

$$\lim_{n\rightarrow\infty} F_n(x) = F(x). \tag{B.7}$$

[Theorem 29.1, p. 390, Billingsley (1986)]

Weak convergence may also be referred to as *convergence in distribution* or *convergence in law* and may be denoted by $X_n \stackrel{\mathcal{L}}{\Rightarrow} X$, $X_n \Rightarrow X$, or $F_n \Rightarrow F$.

Proposition B.2. [Weak convergence for discrete distributions] *Consider discrete integer-valued random variables $\{X_n\}$ and X, with respective cumulative distribution functions $\{F_n\}$ and F. If the sequence $\{X_n\}$ converges weakly to X, then $\lim_n F_n(x) = F(x)$ for every $x \in I\!R$.*

Proof of Proposition B.2. Because X is a discrete integer-valued random variable, its CDF F has a countable number of discontinuity points, $\mathcal{D}_F \equiv \{x \in I\!R : \lim_{x' \to x} F(x') \neq F(x)\} = \{x \in I\!R : \Pr(X = x) > 0\} \subseteq Z\!\!\!Z$. If $x \notin \mathcal{D}_F$, then $\lim_n F_n(x) = F(x)$, by definition of weak convergence in Equation (B.7). If $x \in \mathcal{D}_F$, and since $\mathcal{D}_F$ is countable and CDFs are right-continuous, there exists $\delta > 0$ such that $x + \delta \notin \mathcal{D}_F$, $F_n(x + \delta) = F_n(x)$, and $F(x + \delta) = F(x)$. Thus, by definition of weak convergence in Equation (B.7), $\lim_n F_n(x) = \lim_n F_n(x + \delta) = F(x + \delta) = F(x)$.

□

Theorem B.3. [Continuous Mapping Theorem] *Consider a function* $\ell : I\!R^J \to I\!R^K$, *with discontinuity set* $\mathcal{D}_\ell \equiv \{x \in I\!R^J : \ell \text{ is discontinuous at } x\}$. *If* $X_n \stackrel{\mathcal{L}}{\Rightarrow} X$ *and* $\Pr(X \in \mathcal{D}_\ell) = 0$, *then*

$$\ell(X_n) \stackrel{\mathcal{L}}{\Rightarrow} \ell(X). \tag{B.8}$$

[Theorem 29.2, p. 391, Billingsley (1986)]

Theorem B.4. [Central Limit Theorem] *Suppose* $\{X_n\}$ *is a sequence of independent and identically distributed* J*-dimensional random vectors, with* J*-dimensional mean vector* μ *and* $J \times J$ *covariance matrix* σ. *Then,*

$$\sqrt{n}(\bar{X}_n - \mu) \stackrel{\mathcal{L}}{\Rightarrow} N(0, \sigma), \tag{B.9}$$

where $\bar{X}_n \equiv \sum_i X_i/n$ *denotes the empirical mean vector.*
[Theorem 2.9.1, p. 51, Mardia et al. (1979)]

B.3 Properties of floor and ceiling functions

Floor. The *floor function* $\lfloor\ \rfloor : I\!R \to Z\!\!\!Z$ is defined as

$$\lfloor x \rfloor \equiv \sup\{n \in Z\!\!\!Z : n \leq x\}, \tag{B.10}$$

that is, $\lfloor x \rfloor$ is the greatest integer less than or equal to x.

For any $x \in I\!R$ and $n \in Z\!\!\!Z$, the floor function satisfies the following properties.

- $\lfloor x \rfloor \leq x < \lfloor x \rfloor + 1$ or, equivalently, $x - 1 < \lfloor x \rfloor \leq x$.
- $\lfloor x + n \rfloor = \lfloor x \rfloor + n$.
- $\lfloor x \rfloor \geq n \iff x \geq n$.
- $\lfloor x \rfloor < n \iff x < n$ or, equivalently, $\lfloor x \rfloor \leq n - 1 \iff x < n$.

Ceiling. The *ceiling function* $\lceil\ \rceil : I\!R \to Z\!\!\!Z$ is defined as

$$\lceil x \rceil \equiv \inf\{n \in Z\!\!\!Z : n \geq x\}, \tag{B.11}$$

that is, $\lceil x \rceil$ is the least integer greater than or equal to x.

For any $x \in \mathbb{R}$ and $n \in \mathbb{Z}$, the ceiling function satisfies the following properties.

- $\lceil x \rceil - 1 < x \leq \lceil x \rceil$ or, equivalently, $x \leq \lceil x \rceil < x + 1$.
- $\lceil x + n \rceil = \lceil x \rceil + n$.
- $\lceil x - 1 \rceil \geq n \iff x > n$.
- $\lceil x \rceil < n \iff x \leq n - 1$ or, equivalently, $\lceil x \rceil \leq n \iff x \leq n$.
- $\lceil x \rceil = -\lfloor -x \rfloor$.

C

SAS Code

This appendix provides SAS code, written by Merrill D. Birkner, for the multiple testing analyses discussed in Chapter 11 and Section 13.2 (Birkner et al. (2005b); SAS, Version 9, `www.sas.com`).

The code may be downloaded as a text file from the book's website (`www.stat.berkeley.edu/~sandrine/MTBook/SAS/SAS.html`).

```
/*****************************************************************/
/* Read in HIV-1 SAS dataset, [y:X] */

options nonotes;

/* number of patients */
%let row = 317;

/* first column corresponds to RC outcome, */
/* remaining columns to PR and RT codons */
%let col = 283;

/* number of bootstrap samples */
%let boots = 7500;

/* allowed number of false positives for gFWER AMTP */
%let k = 5;

/* allowed proportion of false positives for TPPFP AMTP */
%let q = 0.05;

/* number of tested hypotheses */
%let nt= 282;

libname resample "c:\hivexample";
```

```
/*****************************************************************/
/* lmt: t-statistics for univariate linear regression model */

%macro lmt;
  %do a = 2 %to &col;
    proc reg data=resample.hivdata noprint outest=outest&a tableout;
        model VAR1=VAR&a;
    data outest&a(rename=(VAR&a=t)); set outest&a; where _type_='T';
         keep VAR&a;
    proc append base=tstats data=outest&a;
  %end;
%mend;
%lmt;
proc print data=tstats; title "t-statistics for data";
run;

/*****************************************************************/
/* boot and bootnull: Non-parametric bootstrap null */
/* shift-transformed test statistics null distribution */

%macro boot;
%do j=1 %to &boots;
proc surveyselect noprint data=resample.hivdata out=datanew
        seed=&j
        method=urs
        rep=&row
        sampsize=1
        stats;
%do a = 2 %to &col;
proc reg data=datanew noprint outest=outest&a tableout;
        model VAR1=VAR&a;
data outest&a(rename=(VAR&a=t)); set outest&a; where _type_='T';
     keep VAR&a;
proc append base=tstatsB data=outest&a;
%end;
%end;
%mend;
%boot;
quit;

/**********************/
%macro bootnull;
proc iml;
        use tstatsB;
        read all var {t} into x;
        close tstatsB;
        nt=&nt;
        bt=&boots;
        tB=shape(x,bt,nt);
```

```
        b=tB[:,];
        bb=J(bt,nt,0);
        do i=1 to bt;
                bb[i,]=(tB[i,] - b);
        end;
        bb2=shape(bb,bt*nt,1);
        create Qo from bb2;
        append from bb2;
        close Qo;
        quit;
%mend;
%bootnull;

/*****************************************************************/
/* ssmaxT: FWER-controlling single-step maxT procedure */

%macro ssmaxT;
proc iml;
        use Qo;
        read all into bb;
        close Qo;
        nt=&nt;
        bt=&boots;
        Qo=shape(bb,bt,nt);
        mx=J(1,bt,0);
        do i=1 to bt;
                mb=abs(Qo[i,]);
                mx[,i]=max(mb);
        end;
        c=mx;
        brank=rank(mx);
        mx[brank]=c;
        use tstats;
        read all var {t} into t;
        close tstats;
        print t;
        pval=J(1,nt,0);
        do j=1 to nt;
                tmp=J(1,bt,0);
                do i=1 to bt;
                        if abs(t[j]) < mx[i] then tmp[i]=1;
                end;
                st=sum(tmp);
                pval[j]=st/bt;
        end;
        create fwer from pval;
        append from pval;
        close fwer;
        quit;
```

```
proc print data=fwer;
run;
%mend;
%ssmaxT;

/****************************************************************/
/* gfwer: gFWER-controlling augmentation multiple testing */
/* procedure */

%macro gfwer;
proc iml;
        use fwer;
        read all into pval;
        close fwer;
        k=&k;
        nt=&nt;
        gp=J(1,nt,0);
        j=k+1;
        c=pval;
        brank=rank(pval);
        pval[brank]=c;
        do i=1 to (nt-k);
                gp[j]=pval[i];
                j=j+1;
        end;
        gg=gp[brank];
        gpval=shape(gg,1,nt);
        create gfwer from gpval;
        append from gpval;
        close gfwer;
        quit;
proc print data=gfwer;
run;
%mend;
%gfwer;

/****************************************************************/
/* tppfp: TPPFP-controlling augmentation multiple testing */
/* procedure */

%macro tppfp;
proc iml;
        use fwer;
        read all into pval;
        close fwer;
        q=&q;
        nt=&nt;
        c=pval;
        brank=rank(pval);
```

```
        pval[brank]=c;
        tp=J(1,nt,0);
        do i=1 to nt;
                m=ceil(i*(1-q));
                tp[i]=pval[m];
        end;
        tt=tp[brank];
        tpval=shape(tt,1,nt);
        create tppfp from tpval;
        append from tpval;
        close tppfp;
        quit;
proc print data=tppfp;
run;
%mend;
%tppfp;

/*******************************************************************/
```

References

F. Abramovich, Y. Benjamini, D. Donoho, and I. Johnstone. Adapting to unknown sparsity by controlling the false discovery rate. Technical Report 2000–19, Department of Statistics, Stanford University, Stanford, CA 94305, 2000.

F. Al-Shahrour, R. Díaz-Uriarte, and J. Dopazo. FatiGO: A web tool for finding significant associations of Gene Ontology terms with groups of genes. *Bioinformatics*, 20(4):578–580, 2004. Available at `fatigo.bioinfo.cipf.es`.

F. Al-Shahrour, P. Minguez, J. M. Vaquerizas, L. Conde, and J. Dopazo. BABELOMICS: A suite of web tools for functional annotation and analysis of groups of genes in high-throughput experiments. *Nucleic Acids Research*, 33:W460–W464, 2005. Available at `www.babelomics.org`.

A. A. Alizadeh, M. B. Eisen, R. E. Davis, C. Ma, I. S. Lossos, A. Rosenwald, J. C. Boldrick, H. Sabet, T. Tran, X. Yu, J. I. Powell, L. Yang, G. E. Marti, T. Moore, J. Hudson Jr, L. Lu, D. B. Lewis, R. Tibshirani, G. Sherlock, W. C. Chan, T. C. Greiner, D. D. Weisenburger, J. O. Armitage, R. Warnke, R. Levy, W. Wilson, M. R. Grever, J. C. Byrd, D. Botstein, P. O. Brown, and L. M. Staudt. Distinct types of diffuse large B-cell lymphoma identified by gene expression profiling. *Nature*, 403(6769):503–511, 2000.

B. Banelli, I. Casciano, M. Croce, A. Di Vinci, I. Gelvi, G. Pagnan, C. Brignole, G. Allemanni, S. Ferrini, M. Ponzoni, and M. Romani. Expression and methylation of CASP8 in neuroblastoma: Identification of a promoter region. *Nature Medicine*, 8(12):1333–1335, 2002.

J. D. Barbour, T. Wrin, R. M. Grant, J. N. Martin, M. R. Segal, C. J. Petropoulos, and S. G. Deeks. Evolution of phenotypic drug susceptibility and viral replication capacity during long-term virologic failure of protease inhibitor therapy in human immunodeficiency virus-infected adults. *Journal of Virology*, 76(21):11104–11112, 2002.

A. Barrier, P.-Y. Boëlle, A. Lemoine, C. Tse, D. Brault, F. Chiappini, F. Lacaine, S. Houry, M. Huguier, A. Flahault, and S. Dudoit. Gene expression profiling of nonneoplastic mucosa may predict clinical outcome of colon cancer patients. *Diseases of the Colon and Rectum*, 48(12):2238–2248, 2005a.

A. Barrier, A. Lemoine, P.-Y. Boëlle, C. Tse, D. Brault, F. Chiappini, J. Breittschneider, F. Lacaine, S. Houry, M. Huguier, M. J. van der Laan, T. P. Speed, B. Debuire, A. Flahault, and S. Dudoit. Colon cancer prognosis prediction by gene expression profiling. *Oncogene*, 24(40):6155–6164, 2005b.

A. Barrier, N. Olaya, F. Chiappini, F. Roser, O. Scatton, C. Artus, B. Franc, S. Dudoit, A. Flahault, B. Debuire, D. Azoulay, and A. Lemoine A. Ischemic preconditioning modulates the expression of several genes, leading to the overproduction of IL-1Ra, iNOS, and Bcl-2 in a human model of liver ischemia-reperfusion. *The FASEB Journal*, 19(12):1617–1626, 2005c.

A. Barrier, P.-Y. Boëlle, F. Roser, J. Gregg, C. Tse, D. Brault, F. Lacaine, S. Houry, M. Huguier, B. Franc, A. Flahault, A. Lemoine, and S. Dudoit. Stage II colon cancer prognosis prediction by tumor gene expression profiling. *Journal of Clinical Oncology*, 24(29):4685–4691, 2006.

T. Beissbarth and T. P. Speed. GOstat: Find statistically overrepresented Gene Ontologies within a group of genes. *Bioinformatics*, 20(9):1464–1465, 2004. Available at `gostat.wehi.edu.au`.

Y. Benjamini and H. Braun. John W. Tukey's contributions to multiple comparisons. *Annals of Statistics*, 30(6):1576–1594, 2002.

Y. Benjamini and Y. Hochberg. Controlling the false discovery rate: A practical and powerful approach to multiple testing. *Journal of the Royal Statistical Society, Series B*, 57:289–300, 1995.

Y. Benjamini and D. Yekutieli. The control of the false discovery rate in multiple testing under dependency. *Annals of Statistics*, 29(4):1165–1188, 2001.

Y. Benjamini and D. Yekutieli. Quantitative trait loci analysis using the false discovery rate. *Genetics*, 171(2):783–790, 2005.

R. Beran. Balanced simultaneous confidence sets. *Journal of the American Statistical Association*, 83:679–686, 1988.

P. Billingsley. *Probability and Measure*. Probability and Mathematical Statistics. Wiley, New York, 2nd edition, 1986.

M. D. Birkner, A. E. Hubbard, and M. J. van der Laan. Application of a multiple testing procedure controlling the proportion of false positives to protein and bacterial data. Technical Report 186, Division of Biostatistics, University of California, Berkeley, Berkeley, CA 94720-7360, 2005a. Available at `www.bepress.com/ucbbiostat/paper186`.

M. D. Birkner, K. S. Pollard, M. J. van der Laan, and S. Dudoit. Multiple testing procedures and applications to genomics. Technical Report 168, Division of Biostatistics, University of California, Berkeley, Berkeley, CA 94720-7360, 2005b. Available at `www.bepress.com/ucbbiostat/paper168`.

M. D. Birkner, S. E. Sinisi, and M. J. van der Laan. Multiple testing and data adaptive regression: An application to HIV-1 sequence data. *Statistical Applications in Genetics and Molecular Biology*, 4(1):Article 8, 2005c. Available at `www.bepress.com/sagmb/vol4/iss1/art8`.

M. D. Birkner, A. E. Hubbard, M. J. van der Laan, C. F. Skibola, C. M. Hegedus, and M. T. Smith. Issues of processing and multiple testing of SELDI-TOF MS proteomic data. *Statistical Applications in Genetics and Molecular Biology*, 5(1):Article 11, 2006. Available at `www.bepress.com/sagmb/vol5/iss1/art11`.

M. D. Birkner, M. Courtine, M. J. van der Laan, K. Clément, J.-D. Zucker, and S. Dudoit. Statistical methods for detecting genotype/phenotype associations in the ObeLinks Project. Technical report, Division of Biostatistics, University of California, Berkeley, Berkeley, CA 94720-7360, 2007. (In preparation).

G. Bisson. KBG: A knowledge based generalizer. In B. W. Porter and R. J. Mooney, editors, *Machine Learning: Proceedings of the Seventh International Conference (1990), Austin, Texas, June 21–23, 1990*, pages 9–15, Palo Alto, CA, 1990. Morgan Kaufmann.

M. Blanchette, R. E. Green, S. E. Brenner, and D. C. Rio. Global analysis of positive and negative pre-mRNA splicing regulators in *Drosophila. Genes & Development*, 19(11):1306–1314, 2005.

J. C. Boldrick, A. A. Alizadeh, M. Diehn, S. Dudoit, C. L. Liu, C. E. Belcher, D. Botstein, L. M. Staudt, P. O. Brown, and D. A. Relman. Stereotyped and specific gene expression programs in human innate immune responses to bacteria. *Proc. Natl. Acad. Sci.*, 99(2):972–977, 2002.

B. M. Bolstad, R. A. Irizarry, L. Gautier, and Z. Wu. Preprocessing high-density oligonucleotide arrays. In R. C. Gentleman, V. J. Carey, W. Huber, R. A. Irizarry, and S. Dudoit, editors, *Bioinformatics and Computational Biology Solutions Using R and Bioconductor*, chapter 2, pages 13–32. Springer, New York, 2005. Available at `www.bioconductor.org/pubs/docs/mogr`.

C. E. Bonferroni. Teoria statistica delle classi e calcolo delle probabilità. *Pubblicazioni del R Istituto Superiore di Scienze Economiche e Commerciali di Firenze*, pages 3–62, 1936.

J. P. Bordat. Calcul pratique du treillis de Galois d'une correspondance. *Mathématique des Sciences Humaines*, 96(4):31–47, 1986.

I. Bournaud, M. Courtine, and J.-D. Zucker. KIDS: An iterative algorithm to organize relational knowledge. In R. Dieng and O. Corby, editors, *Knowledge Engineering and Knowledge Management. Methods, Models, and Tools: 12th International Conference, EKAW 2000, Juan-les-Pins, France, October 2–6, 2000, Proceedings*, volume 1937 of *Lecture Notes in Computer Science*, pages 217–232, Berlin/Heidelberg, 2000. Springer. Available at `www.springerlink.com/content/472a404urcgxjqn9/?p=db5cfd25a3c54db7a2dcf7c2e1cfd3fc&pi=15`.

G. A. Calin, C. Sevignani, C. D. Dumitru, T. Hyslop, E. Noch, S. Yendamuri, M. Shimizu, S. Rattan, F. Bullrich, M. Negrini, and C. M. Croce. Human microRNA genes are frequently located at fragile sites and genomic regions involved in cancers. *Proc. Natl. Acad. Sci.*, 101(9):2999–3004, 2004.

M. J. Callow, S. Dudoit, E. L. Gong, T. P. Speed, and E. M. Rubin. Microarray expression profiling identifies genes with altered expression in HDL deficient mice. *Genome Research*, 10(12):2022–2029, 2000.

C. Carpineto and G. Romano. A lattice conceptual clustering system and its application to browsing retrieval. *Machine Learning*, 24(2):95–122, 1996.

S. Cawley, S. Bekiranov, H. H. Ng, P. Kapranov, E. A. Sekinger, D. Kampa, A. Piccolboni, V. Sementchenko, J. Cheng, A. J. Williams, R. Wheeler, B. Wong, J. Drenkow, M. Yamanaka, S. Patel, S. Brubaker, H. Tammana, G. Helt, K. Struhl, and T. R. Gingeras. Unbiased mapping of transcription factor binding sites along human chromosomes 21 and 22 points to widespread regulation of noncoding RNAs. *Cell*, 116:499–509, 2004.

J. M. Chambers. *Programming with Data.* Springer, New York, 1998.

M. Chein. Algorithmes de recherche des sous-matrices premières d'une matrice. *Bull. Math. Soc. Sci. Math. RS Roumanie*, 13(61):21–25, 1969.

A. M. Cheng, M. W. Byrom, J. Shelton, and L. P. Ford. Antisense inhibition of human miRNAs and indications for an involvement of miRNA in cell growth and apoptosis. *Nucleic Acids Research*, 33(4):1290–1297, 2005.

F. Chiappini, A. Barrier, R. Saffroy, M.-C. Domart, N. Dagues, D. Azoulay, M. Sebagh, B. Franc, S. Chevalier, B. Debuire, S. Dudoit, and A. Lemoine. Exploration of global gene expression in human liver steatosis by high-density oligonucleotide microarray. *Laboratory Investigation*, 86(2):154–165, 2006.

S. Chiaretti, X. Li, R. Gentleman, A. Vitale, M. Vignetti, F. Mandelli, J. Ritz, and R. Foa. Gene expression profile of adult T-cell acute lymphocytic leukemia identifies distinct subsets of patients with different response to therapy and survival. *Blood*, 103(7):2771–2778, 2004.

K. Clément. Genetics of human obesity. *Proc. Nutr. Soc.*, 64(2):133–142, 2005.

K. Clément and P. Ferré. Genetics and the pathophysiology of obesity. *Pediatric Research*, 53(5):721–725, 2003.

M. J. Daly, J. D. Rioux, S. F. Schaffner, T. J. Hudson, and E. S. Lander. High-resolution haplotype structure in the human genome. *Nature Genetics*, 29 (2):229–232, 2001.

T. Z. DeSantis, C. E. Stone, S. R. Murray, J. P. Moberg, and G. L. Andersen. Rapid quantification and taxonomic classification of environmental DNA from both prokaryotic and eukaryotic origins using a microarray. *FEMS Microbiology Letters*, 245(2):271–278, 2005.

S. Dudoit and Y. H. Yang. Bioconductor R packages for exploratory analysis and normalization of cDNA microarray data. In G. Parmigiani, E. S. Garrett, R. A. Irizarry, and S. L. Zeger, editors, *The Analysis of Gene Expression Data: Methods and Software*, pages 73–101. Springer, New York, 2003.

S. Dudoit, Y. H. Yang, M. J. Callow, and T. P. Speed. Statistical methods for identifying differentially expressed genes in replicated cDNA microarray experiments. *Statistica Sinica*, 12(1):111–139, 2002.

S. Dudoit, J. P. Shaffer, and J. C. Boldrick. Multiple hypothesis testing in microarray experiments. *Statistical Science*, 18(1):71–103, 2003.

S. Dudoit, M. J. van der Laan, and M. D. Birkner. Multiple testing procedures for controlling tail probability error rates. Technical Report 166, Division of Biostatistics, University of California, Berkeley, Berkeley, CA 94720-7360, 2004a. Available at `www.bepress.com/ucbbiostat/paper166`.

S. Dudoit, M. J. van der Laan, and K. S. Pollard. Multiple testing. Part I. Single-step procedures for control of general Type I error rates. *Statistical Applications in Genetics and Molecular Biology*, 3(1):Article 13, 2004b. Available at `www.bepress.com/sagmb/vol3/iss1/art13`.

S. Dudoit, S. Keleş, and M. J. van der Laan. Multiple tests of association with biological annotation metadata. Technical Report 202, Division of Biostatistics, University of California, Berkeley, Berkeley, CA 94720-7360, 2006. Available at `www.bepress.com/ucbbiostat/paper202`.

O. J. Dunn. Estimation of the means of dependent variables. *Annals of Mathematical Statistics*, 29:1095–111, 1958.

B. Efron. Local false discovery rates. Technical report, Department of Statistics, Stanford University, Stanford, CA 94305, 2005. Available at `www-stat.stanford.edu/~brad/papers`.

B. Efron, J. D. Storey, and R. Tibshirani. Microarrays, empirical Bayes methods, and false discovery rates. Technical Report 2001–218, Department of Statistics, Stanford University, Stanford, CA 94305, 2001a.

B. Efron, R. Tibshirani, J. D. Storey, and V. Tusher. Empirical Bayes analysis of a microarray experiment. *Journal of the American Statistical Association*, 96:1151–1160, 2001b.

H. Finner. Stepwise multiple test procedures and control of directional errors. *Annals of Statistics*, 27:274–289, 1999.

S. B. Gabriel, S. F. Schaffner, H. Nguyen, J. M. Moore, J. Roy, B. Blumenstiel, J. Higgins, M. DeFelice, A. Lochner, M. Faggart, S. N. Liu-Cordero, C. Rotimi, A. Adeyemo, R. Cooper, R. Ward, E. S. Lander, M. J. Daly, and D. Altshuler. The structure of haplotype blocks in the human genome. *Science*, 296:2225–2229, 2002.

B. Ganter. Two basic algorithms in concept analysis. Technical Report 831, Technische Hochschule, Darmstadt, Germany, 1984.

Y. Ge, S. Dudoit, and T. P. Speed. Resampling-based multiple testing for microarray data analysis. *TEST*, 12(1):1–44, 2003.

C. R. Genovese and L. Wasserman. A stochastic process approach to false discovery control. *Annals of Statistics*, 32(3):1035–1061, 2004a.

C. R. Genovese and L. Wasserman. Exceedance control of the false discovery proportion. Technical Report 807, Department of Statistics, Carnegie Mellon University, Pittsburgh, PA 15213, July 2004b. Available at `www.stat.cmu.edu/tr/tr807/tr807.html`.

C. R. Genovese, N. A. Lazar, and T. E. Nichols. Thresholding of statistical maps in functional neuroimaging using the false discovery rate. *NeuroImage*, 15(4):870–878, 2002.

R. C. Gentleman, V. J. Carey, D. J. Bates, B. Bolstad, M. Dettling, S. Dudoit, B. Ellis, L. Gautier, Y. Ge, J. Gentry, K. Hornik, T. Hothorn, W. Huber, S. Iacus, R. A. Irizarry, F. Leisch, C. Li, M. Maechler, A. J. Rossini, G. Sawitzki, C. Smith, G. K. Smyth, L. Tierney, Y. H. Yang, and J. Zhang. Bioconductor: Open software development for computational biology and bioinformatics. *Genome Biology*, 5(10):R80, 2004. Available at `genomebiology.com/2004/5/10/R80`,`www.bioconductor.org`.

R. C. Gentleman, V. J. Carey, W. Huber, R. A. Irizarry, and S. Dudoit, editors. *Bioinformatics and Computational Biology Solutions Using R and Bioconductor*. Statistics for Biology and Health. Springer, New York, 2005a. Available at `www.bioconductor.org/pubs/docs/mogr`.

R. C. Gentleman, V. J. Carey, and J. Zhang. Meta-data resources and tools in Bioconductor. In R. C. Gentleman, V. J. Carey, W. Huber, R. A. Irizarry, and S. Dudoit, editors, *Bioinformatics and Computational Biology Solutions Using R and Bioconductor*, chapter 7, pages 111–133. Springer, New York, 2005b. Available at `www.bioconductor.org/pubs/docs/mogr`.

R. D. Gill. Non- and semi-parametric maximum likelihood estimators and the von Mises method. I. *Scandinavian Journal of Statistics*, 16(2):97–128, 1989. (With a discussion by J. A. Wellner and J. Præstgaard and a reply by the author).

R. D. Gill, M. J. van der Laan, and J. A. Wellner. Inefficient estimators of the bivariate survival function for three models. *Annales de l'Institut Henri Poincaré. Probabilités et Statistiques*, 31(3):545–597, 1995.

B. Gleissner, N. Gokbuget, C. R. Bartram, B. Janssen, H. Rieder, J. W. Janssen, C. Fonatsch, A. Heyll, D. Voliotis, J. Beck, T. Lipp, G. Munzert, J. Maurer, D. Hoelzer, E. Thiel, and German Multicenter Trials of Adult Acute Lymphoblastic Leukemia Study Group. Leading prognostic relevance of the BCR-ABL translocation in adult acute B-lineage lymphoblastic leukemia: A prospective study of the German Multicenter Trial Group and confirmed polymerase chain reaction analysis. *Blood*, 99(5):1536–1543, 2002.

A. L. Gloyn, M. N. Weedon, K. R. Owen, M. J. Turnerand B. A. Knight, G. Hitman, M. Walker, J. C. Levy, M. Sampson, S. Halford, M. I. McCarthy, A. T. Hattersley, and T. M. Frayling. Large-scale association studies of variants in genes encoding the pancreatic β-cell K_{ATP} channel subunits Kir6.2 (*KCNJ11*) and SUR1 (*ABCC8*) confirm that the *KCNJ11* E23K variant is associated with type 2 diabetes. *Diabetes*, 52(2):568–572, 2003.

R. Godin, G. Mineau, R. Missaoui, and H. Mili. Méthodes de classification conceptuelle basées sur les treillis de Galois et applications. *Revue d'Intelligence Artificielle*, 9(2):105–137, 1995a.

R. Godin, R. Missaoui, and H. Alaoui. Incremental concept formation algorithms based on Galois (concept) lattices. *Computational Intelligence*, 11 (2):216–267, 1995b.

T. R. Golub, D. K. Slonim, P. Tamayo, C. Huard, M. Gaasenbeek, J. P. Mesirov, H. Coller, M.L. Loh, J. R. Downing, M. A. Caligiuri, C. D. Bloomfield, and E. S. Lander. Molecular classification of cancer: Class discovery and class prediction by gene expression monitoring. *Science*, 286(5439): 531–537, 1999.

V. Goss Tusher, R. Tibshirani, and G. Chu. Significance analysis of microarrays applied to the ionizing radiation response. *Proc. Natl. Acad. Sci.*, 98: 5116–5121, 2001.

S. Grossmann, S. Bauer, P. N. Robinson, and M. Vingron. An improved statistic for detecting over-represented gene ontology annotations in gene sets. In A. Apostolico, C. Guerra, S. Istrail, P. Pevzner, and M. Waterman, editors, *Research in Computational Molecular Biology: 10th Annual International Conference, RECOMB 2006, Venice, Italy, April 2–5, 2006, Proceedings*, volume 3909 of *Lecture Notes in Computer Science*, pages 85–98, Berlin/Heidelberg, 2006. Springer. Available at `www.springerlink.com/content/w83h54235ku1v142/?p=a2cb65a9ec484e519ad1ff83b41c707a&pi=8`.

L. W. Hahn, M. D. Ritchie, and J. H. Moore. Multifactor dimensionality reduction software for detecting gene-gene and gene-environment interactions. *Bioinformatics*, 19(3):376–382, 2003.

E. H. Hani, K. Clément, G. Velho, N. Vionnet, J. Hager, A. Philippi, C. Dina, H. Inoue, M. A. Permutt, A. Basdevant, M. North, F. Demenais, B. Guy-Grand, and P. Froguel. Genetic studies of the sulfonylurea receptor gene locus in NIDDM and in morbid obesity among French Caucasians. *Diabetes*, 46(4):688–694, 1997.

E. R. Hauser and M. Boehnke. Genetic linkage analysis of complex genetic traits by using affected sibling pairs. *Biometrics*, 54:1238–1246, 1998.

A. Herbert, N. P. Gerry, M. B. McQueen, I. M. Heid, A. Pfeufer, T. Illig, H.-E. Wichmann, T. Meitinger, D. Hunter, F. B. Hu, G. Colditz, A. Hinney, J. Hebebrand, K. Koberwitz, X. Zhu, R. Cooper, K. Ardlie, H. Lyon, J. N. Hirschhorn, N. M. Laird, M. E. Lenburg, C. Lange, and M. F. Christman. A common genetic variant is associated with adult and childhood obesity. *Science*, 312(5771):279–283, 2006.

A. von Heydebreck, W. Huber, and R. Gentleman. Differential expression with the Bioconductor Project. Technical Report 7, Bioconductor Project Working Papers, 2004. Available at `www.bepress.com/bioconductor/paper7`.

Y. Hochberg. A sharper Bonferroni procedure for multiple tests of significance. *Biometrika*, 75:800–802, 1988.

Y. Hochberg and A. C. Tamhane. *Multiple Comparison Procedures*. Probability and Mathematical Statistics. Wiley, New York, 1987.

R. P. Hoffman, P. Vicini, and C. Cobelli. Pubertal changes in HOMA and QUICKI: Relationship to hepatic and peripheral insulin sensitivity. *Pediatric Diabetes*, 5(3):122–125, 2004.

J. Hoh and J. Ott. Mathematical multi-locus approaches to localizing complex human trait genes. *Nature Reviews Genetics*, 4(9):701–709, 2003.

J. Hoh and J. Ott. Genetic dissection of diseases: Design and methods. *Current Opinion in Genetics & Development*, 14(3):229–232, 2004.

B. Holland and M. D. Copenhaver. An improved sequentially rejective Bonferroni test procedure. *Biometrics*, 43(2):417–423, 1987.

S. Holm. A simple sequentially rejective multiple test procedure. *Scandinavian Journal of Statistics*, 6:65–70, 1979.

G. Hommel. Tests of the overall hypothesis for arbitrary dependence structures. *Biometrical Journal*, 25:423–430, 1983.

G. Hommel. A stagewise rejective multiple test procedure based on a modified Bonferroni test. *Biometrika*, 75:383–386, 1988.

G. Hommel and T. Hoffman. Controlled uncertainty. In P. Bauer, G. Hommel, and E. Sonnemann, editors, *Multiple Hypotheses Testing*, pages 154–161. Springer, 1988.

J. C. Hsu. *Multiple Comparisons: Theory and Methods.* Chapman & Hall/CRC, London, New York, 1996. Available at `www.stat.ohio-state.edu/~jch`.

Y. Huang and J. C. Hsu. Hochberg's step-up method: Cutting corners off Holm's step-down method. Technical Report 761, Department of Statistics, Ohio State University, Columbus, OH 43210-1247, 2005.

K. Jogdeo. Association and probability inequalities. *Annals of Statistics*, 5: 495–504, 1977.

J. Kaczynski, T. Cook, and R. Urrutia. Sp1- and Krüppel-like transcription factors. *Genome Biology*, 4(2):206, 2003.

T. Kamae, U. Krengel, and G. L. O'Brien. Stochastic inequalities on partially ordered spaces. *Annals of Probability*, 5(6):899–912, 1977.

S. Keleş, M. J. van der Laan, S. Dudoit, and S. E. Cawley. Multiple testing methods for ChIP-Chip high density oligonucleotide array data. *Journal of Computational Biology*, 13(3):579–613, 2006. Available at `www.liebertonline.com/doi/abs/10.1089/cmb.2006.13.579?prevSearch=allfield%3A%28dudoit%29`.

D. Kirchner, J. Duyster, O. Ottmann, R. M. Schmid, L. Bergmann, and G. Munzert. Mechanisms of Bcr-Abl-mediated NF-κB/Rel activation. *Experimental Hematology*, 31(6):504–511, 2003.

E. L. Korn, J. F. Troendle, L. M. McShane, and R. Simon. Controlling the number of false discoveries: Application to high-dimensional genomic data. *Journal of Statistical Planning and Inference*, 124(2):379–398, 2004.

S. O. Kuznetsov and S. A. Obiedkov. Comparing performance of algorithms for generating concept lattices. *Journal of Experimental and Theoretical Artificial Intelligence*, 14(2–3):189–216, 2002.

M. J. van der Laan. Statistical inference for variable importance. *International Journal of Biostatistics*, 2(1):Article 2, 2006. Available at `www.bepress.com/ijb/vol2/iss1/2`.

M. J. van der Laan and A. E. Hubbard. Quantile-function based null distribution in resampling based multiple testing. *Statistical Applications in Genetics and Molecular Biology*, 5(1):Article 14, 2006. Available at `www.bepress.com/sagmb/vol5/iss1/art14`.

M. J. van der Laan and K. S. Pollard. A new algorithm for hybrid hierarchical clustering with visualization and the bootstrap. *Journal of Statistical Planning and Inference*, 117(2):275–303, 2003.

M. J. van der Laan and J. M. Robins. *Unified Methods for Censored Longitudinal Data and Causality.* Springer, New York, 2003.

M. J. van der Laan, S. Dudoit, and K. S. Pollard. Multiple testing. Part II. Step-down procedures for control of the family-wise error rate. *Statistical Applications in Genetics and Molecular Biology*, 3(1):Article 14, 2004a. Available at `www.bepress.com/sagmb/vol3/iss1/art14`.

M. J. van der Laan, S. Dudoit, and K. S. Pollard. Augmentation procedures for control of the generalized family-wise error rate and tail probabilities for the proportion of false positives. *Statistical Applications in Genetics and Molecular Biology*, 3(1):Article 15, 2004b. Available at `www.bepress.com/sagmb/vol3/iss1/art15`.

M. J. van der Laan, M. D. Birkner, and A. E. Hubbard. Empirical Bayes and resampling based multiple testing procedure controlling tail probability of the proportion of false positives. *Statistical Applications in Genetics and Molecular Biology*, 4(1):Article 29, 2005. Available at `www.bepress.com/sagmb/vol4/iss1/art29`.

E. L. Lehmann. *Testing Statistical Hypotheses.* Wiley, New York, 2nd edition, 1986.

E. L. Lehmann and J. P. Romano. Generalizations of the familywise error rate. *Annals of Statistics*, 33(3):1138–1154, 2005.

S. S. Li, J. Bigler, J. W. Lampe, J. D. Potter, and Z. Feng. FDR-controlling testing procedures and sample size determination for microarrays. *Statistics in Medicine*, 24(15):2267–2280, 2005.

J. Lu, G. Getz, E. A. Miska, E. Alvarez-Saavedra, J. Lamb, D. Peck, A. Sweet-Cordero, B. L. Ebert, R. H. Mak, A. A. Ferrando, J. R. Downing, T. jacks, H. R. Horvitz, and T. R. Golub. MicroRNA expression profiles classify human cancers. *Nature*, 435(9):834–838, 2005. Available at `www.broad.mit.edu/cancer/pub/miGCM`.

D. Maglott, J. Ostell, K. D. Pruitt, and T. Tatusova. Entrez Gene: Gene-centered information at NCBI. *Nucleic Acids Research*, 33:D54–D58, 2005. Available at `nar.oxfordjournals.org/cgi/content/full/33/suppl_1/D54`.

E. Manduchi, G. R. Grant, S. E. McKenzie, G. C. Overton, S. Surrey, and C. J. Stoeckert. Generation of patterns from gene expression data by assigning confidence to differentially expressed genes. *Bioinformatics*, 16:685–698, 2000.

K. V. Mardia, J. T. Kent, and J. M. Bibby. *Multivariate Analysis.* Academic Press, London, New York, 1979.

S. A. McCarroll, C. T. Murphy, S. Zou, S. D. Pletcher, C-S Chin, Y. N. Jan, C. Kenyon, C. I. Bargmann, and H. Li. Comparing genomic expression patterns across species identifies shared transcriptional profile in aging. *Nature Genetics*, 36(2):197–204, 2004.

E. Mephu Nguifo and P. Njiwoua. Using lattice-based framework as a tool for feature extraction. In C. Nédellec and C. Rouveirol, editors, *Machine Learning: ECML-98: 10th European Conference on Machine Learning, Chemnitz, Germany, April 21–23, 1998, Proceedings*, volume 1398 of *Lecture Notes in Computer Science*, pages 304–309, Berlin/Heidelberg, 1998. Springer.

D. Meyre, N. Bouatia-Naji, A. Tounian, C. Samson, C. Lecoeur, V. Vatin, M. Ghoussaini, C. Wachter, S. Hercberg, G. Charpentier, W. Patsch, F. Pattou, M.-A. Charles, P. Tounian, K. Clément, B. Jouret, J. Weill, B. A. Maddux, I. D. Goldfine, A. Walley, P. Boutin, C. Dina, and P. Froguel. Variants of ENPP1 are associated with childhood and adult obesity and increase the risk of glucose intolerance and type 2 diabetes. *Nature Genetics*, 37(8): 863–867, 2005. Available at `www.nature.com/ng/journal/v37/n8/abs/ng1604.html;jsessionid=741AF577246F06B4483ED78D3CAD091A`.

M. Z. Michael, S. M. O'Connor, N. G. van Holst Pellekaan, G. P. Young, and R. J. James. Reduced accumulation of specific microRNAs in colorectal neoplasia. *Molecular Cancer Research*, 1(12):882–891, 2003.

V. K. Mootha, C. M. Lindgren, K-F. Eriksson, A. Subramanian, S. Sihag, J. Lehar, P. Puigserver, E. Carlsson, M. Ridderstråle, E. Laurila, N. Houstis, M. J. Daly, N. Patterson, J. P. Mesirov, T. R. Golub, P. Tamayo, B. Spiegelman, E. S. Lander, J. H. Hirschhorn, D. Altshuler, and L. C. Groop. PGC–1α-responsive genes involved in oxidative phosphorylation are coordinately downregulated in human diabetes. *Nature Genetics*, 34(3):267–273, 2003.

A. Mukhopadhyay, S. Shishodia, J. Suttles, K. Brittingham, B. Lamothe, R. Nimmanapalli, K. N. Bhalla, and B. B. Aggarwal. Ectopic expression of protein-tyrosine kinase Bcr-Abl suppresses tumor necrosis factor (TNF)-induced NF-κB activation and IκBα phosphorylation. Relationship with down-regulation of TNF receptors. *Journal of Biological Chemistry*, 277 (34):30622–30628, 2002.

M. A. Newton, C. M. Kendziorski, C. S. Richmond, F. R . Blattner, and K. W. Tsui. On differential variability of expression ratios: Improving statistical inference about gene expression changes from microarray data. *Journal of Computational Biology*, 8:37–52, 2001.

E. M. Norris. An algorithm for computing the maximal rectangles in a binary relation. *Revue Roumaine de Mathématiques Pures et Appliquées*, 23(2): 243–250, 1978.

K. A. O'Donnell, E. A. Wentzel, K. I. Zeller, C. V. Dang, and J. T. Mendell. c-Myc-regulated microRNAs modulate E2F1 expression. *Nature*, 435(7043): 839–843, 2005.

J. M. Ordovas. Genetics, postprandial lipemia and obesity. *Nutrition, Metabolism & Cardiovascular Diseases*, 11(2):118–133, 2001.

A. Packer, editor. *The Chipping Forecast II*, volume 32(4s) of *Nature Genetics*. Nature Publishing Group, December 2002. Available at `www.nature.com/ng/journal/v32/n4s`. (Supplement).

A. Packer and M. Axton, editors. *The Chipping Forecast III*, volume 37(6s) of *Nature Genetics*. Nature Publishing Group, June 2005. Available at `www.nature.com/ng/journal/v37/n6s`. (Supplement).

L. Pérusse, T. Rankinen, A. Zuberi, Y. C. Chagnon, S. J. Weisnagel, G. Argyropoulos, B. Walts, E. E. Snyder, and C. Bouchard. The human obesity gene map: The 2004 update. *Obesity Research*, 13(3):381–490, 2005.

E. Phimister and B. Cohen, editors. *The Chipping Forecast*, volume 21(1s) of *Nature Genetics*. Nature of America, January 1999. Available at `www.nature.com/ng/journal/v21/n1s`. (Supplement).

K. S. Pollard and M. J. van der Laan. Choice of a null distribution in resampling-based multiple testing. *Journal of Statistical Planning and Inference*, 125(1–2):85–100, 2004.

K. S. Pollard and M. J. van der Laan. Cluster analysis of genomic data. In R. C. Gentleman, V. J. Carey, W. Huber, R. A. Irizarry, and S. Dudoit, editors, *Bioinformatics and Computational Biology Solutions Using R and Bioconductor*, chapter 13, pages 209–228. Springer, New York, 2005. Available at `www.bioconductor.org/pubs/docs/mogr`, `www.bepress.com/ucbbiostat/paper167`.

K. S. Pollard, M. D. Birkner, M. J. van der Laan, and S. Dudoit. Test statistics null distributions in multiple testing: Simulation studies and applications to genomics. *Journal de la Société Française de Statistique*, 146(1–2):77–115, 2005a. Available at `www.stat.berkeley.edu/~sandrine/Docs/Papers/SFdS05/SFdS.html`. Numéro double spécial *Statistique et Biopuces*.

K. S. Pollard, S. Dudoit, and M. J. van der Laan. Multiple testing procedures: The multtest package and applications to genomics. In R. C. Gentleman, V. J. Carey, W. Huber, R. A. Irizarry, and S. Dudoit, editors, *Bioinformatics and Computational Biology Solutions Using R and Bioconductor*, chapter 15, pages 249–271. Springer, New York, 2005b. Available at `www.bioconductor.org/pubs/docs/mogr`, `www.bepress.com/ucbbiostat/paper164`.

M. Prochazka, S. Lillioja, J. F. Tait, W. C. Knowler, D. M. Mott, M. Spraul, P. H. Bennett, and C. Bogardus. Linkage of chromosomal markers on 4q with a putative gene determining maximal insulin action in Pima Indians. *Diabetes*, 42(4):514–519, 1993.

M. L. Puri and P. K. Sen. *Nonparametric Methods in Multivariate Analysis*. Wiley, New York, 1971.

R Development Core Team. *R: A Language and Environment for Statistical Computing*. R Foundation for Statistical Computing, Vienna, Austria, 2006. Available at `www.r-project.org`.

P. H. Ramsey. Power differences between pairwise multiple comparisons. *Journal of the American Statistical Association*, 73:479–485, 1978.

A. Reiner, D. Yekutieli, and Y. Benjamini. Identifying differentially expressed genes using false discovery rate controlling procedures. *Bioinformatics*, 19 (3):368–375, 2003.

M. F. Rolland-Cachera, T. J. Cole, M. Sempe, J. Tichet, C. Rossignol, and A. Charraud. Body Mass Index variations: Centiles from birth to 87 years. *European Journal of Clinical Nutrition*, 45(1):13–21, 1991.

D. M. Rom. A sequentially rejective test procedure based on a modified Bonferroni inequality. *Biometrika*, 77:663–665, 1990.

J. P. Romano and M. Wolf. Control of generalized error rates in multiple testing. Technical Report 2005–12, Department of Statistics, Stanford University, Stanford, CA 94305, 2005.

D. Rubin, M. J. van der Laan, and S. Dudoit. A method to increase the power of multiple testing procedures through sample splitting. *Statistical Applications in Genetics and Molecular Biology*, 5(1):Article 19, 2006. Available at `www.bepress.com/sagmb/vol5/iss1/art19`.

S. K. Sarkar. Some results on false discovery rate in stepwise multiple testing procedures. *Annals of Statistics*, 30(1):239–257, 2002.

S. K. Sarkar. Generalizing Simes' test and Hochberg's stepup procedure. Technical report, Fox School of Business and Management, Temple University, Philadelphia, PA 19122, August 2005.

S. K. Sarkar. Probability inequalities for ordered MTP2 random variables: A proof of the Simes conjecture. *Annals of Statistics*, 26(2):494–504, 1998.

S. K. Sarkar and C-K. Chang. The Simes method for multiple hypothesis testing with positively dependent test statistics. *Journal of the American Statistical Association*, 92:1601–1608, 1997.

S. R. Searle. *Linear Models*. Wiley, New York, 1971.

P. Seeger. A note on a method for the analysis of significances en masse. *Technometrics*, 10(3):586–593, 1968.

M. R. Segal, J. D. Barbour, and R. M. Grant. Relating HIV-1 sequence variation to replication capacity via trees and forests. *Statistical Applications in Genetics and Molecular Biology*, 3(1):Article 2, 2004. Available at `www.bepress.com/sagmb/vol3/iss1/art2`.

R. W. Shafer, K. M. Dupnik, M. A. Winters, and S. H. Eshleman. A guide to HIV-1 reverse transcriptase and protease sequencing for drug resistance studies. In *HIV Sequencing Compendium*, pages 83–133. Theoretical Biology and Biophysics Group, Los Alamos National Laboratory, 2001.

J. P. Shaffer. Multiplicity, directional (Type III) errors, and the null hypothesis. *Psychological Methods*, 7:356–369, 2002.

J. P. Shaffer. Modified sequentially rejective multiple test procedures. *Journal of the American Statistical Association*, 81:826–831, 1986.

J. P. Shaffer. Multiple hypothesis testing. *Annu. Rev. Psychol.*, 46:561–584, 1995.

E. Shtivelman, F. E. Cohen, and J. M. Bishop. A human gene (AHNAK) encoding an unusually large protein with a 1.2-μm polyionic rod structure. *Proc. Natl. Acad. Sci.*, 89(12):5472–5476, 1992.

Z. Šidák. Rectangular confidence regions for the means of multivariate normal distributions. *Journal of the American Statistical Association*, 62:626–633, 1967.

R. J. Simes. An improved Bonferroni procedure for multiple tests of significance. *Biometrika*, 73:751–754, 1986.

S. E. Sinisi and M. J. van der Laan. Deletion/substitution/addition algorithm in learning with applications in genomics. *Statistical Applications in Genetics and Molecular Biology*, 3(1):Article 18, 2004. Available at www.bepress.com/sagmb/vol3/iss1/art18.

B. Sorić. Statistical "discoveries" and effect-size estimation. *Journal of the American Statistical Association*, 84(406):608–610, 1989.

T. P. Speed, editor. *Statistical Analysis of Gene Expression Microarray Data.* Chapman & Hall/CRC, Boca Raton, FL, 2003.

J. D. Storey. The positive false discovery rate: A Bayesian interpretation and the q-value. *Annals of Statistics*, 31(6):2013–2035, 2003.

J. D. Storey. A direct approach to false discovery rates. *Journal of the Royal Statistical Society, Series B*, 64(3):479–498, 2002.

J. D. Storey and R. Tibshirani. SAM thresholding and false discovery rates for detecting differential gene expression in DNA microarrays. In G. Parmigiani, E. S. Garrett, R. A. Irizarry, and S. L. Zeger, editors, *The Analysis of Gene Expression Data: Methods and Software*, pages 272–290. Springer, New York, 2003.

J. D. Storey and R. Tibshirani. Estimating false discovery rates under dependence, with applications to DNA microarrays. Technical Report 2001–28, Department of Statistics, Stanford University, Stanford, CA 94305, 2001.

J. D. Storey, J. E. Taylor, and D. O. Siegmund. Strong control, conservative point estimation and simultaneous conservative consistency of false discovery rates: A unified approach. *Journal of the Royal Statistical Society, Series B*, 66(1):187–205, 2004.

A. Subramanian, P. Tamayo, V. K. Mootha, S. Mukherjee, B. L. Ebert, M. A. Gillette, A. Paulovich, S. L. Pomeroy, T. R. Golub, E. S. Lander, and J. P. Mesirov. Gene set enrichment analysis: A knowledge-based approach for interpreting genome-wide expression profiles. *Proc. Natl. Acad. Sci.*, 102 (43):15545–15550, 2005. Available at www.broad.mit.edu/gsea/doc/doc_index.html.

M. M. Swarbrick and C. Vaisse. Emerging trends in the search for genetic variants predisposing to human obesity. *Current Opinion in Clinical Nutrition and Metabolic Care*, 6(4):369–375, 2003.

L. Tian, S. A. Greenberg, S. W. Kong, J. Altschuler, I. S. Kohane, and P. J. Park. Discovering statistically significant pathways in expression profiling studies. *Proc. Natl. Acad. Sci.*, 102(38):13544–13549, 2005.

J. F. Troendle. A stepwise resampling method of multiple hypothesis testing. *Journal of the American Statistical Association*, 90(429):370–378, 1995.

J. F. Troendle. A permutational step-up method of testing multiple outcomes. *Biometrics*, 52:846–859, 1996.

M. Tsunoda, J. Tenhunen, C. Tilgmann, H. Arai, and K. Imai. Reduced membrane-bound catechol-O-methyltransferase in the liver of spontaneously hypertensive rats. *Hypertension Research*, 26(11):923–927, 2003.

A. W. van der Vaart and J. Wellner. *Weak Convergence and Empirical Processes.* Springer, New York, 1996.

Y. Wang and S. Dudoit. Quantification and visualization of LD patterns and identification of haplotype blocks. Technical Report 150, Division of Biostatistics, University of California, Berkeley, Berkeley, CA 94720-7360, 2004. Available at `www.bepress.com/ucbbiostat/paper150`.

P. H. Westfall. Resampling-based multiple testing for microarray data analysis. *TEST*, 12(1), 2003. (Discussion).

P. H. Westfall and S. S. Young. *Resampling-Based Multiple Testing: Examples and Methods for P-Value Adjustment.* Wiley, New York, 1993.

E. Wienholds and R. H. A. Plasterk. MicroRNA function in animal development. *Federation of European Biochemical Societies Letters*, 579(26): 5911–5922, 2005.

R. Wille. Restructuring lattice theory: An approach based on hierarchies of concepts. In I. Rival, editor, *Ordered Sets*, volume 83, pages 445–470. D. Reidel, Dordrecht-Boston, 1982.

R. Wille. Concept lattices and conceptual knowledge systems. *Computers & Mathematics with Application*, 23(6–9):493–515, 1992.

Y. H. Yang and A. C. Paquet. Preprocessing two-color spotted arrays. In R. C. Gentleman, V. J. Carey, W. Huber, R. A. Irizarry, and S. Dudoit, editors, *Bioinformatics and Computational Biology Solutions Using R and Bioconductor*, chapter 4, pages 49–69. Springer, New York, 2005. Available at `www.bioconductor.org/pubs/docs/mogr`.

Y. H. Yang, S. Dudoit, P. Luu, and T. P. Speed. Normalization for cDNA microarray data. In M. L. Bittner, Y. Chen, A. N. Dorsel, and E. R. Dougherty, editors, *Microarrays: Optical Technologies and Informatics*, volume 4266 of *Proceedings of SPIE*, pages 141–152, Bellingham, WA, May 2001. SPIE-International Society for Optical Engineering.

Y. H. Yang, M. J. Buckley, S. Dudoit, and T. P. Speed. Comparison of methods for image analysis on cDNA microarray data. *Journal of Computational and Graphical Statistics*, 11(1):108–136, 2002.

D. Yekutieli and Y. Benjamini. Resampling-based false discovery rate controlling multiple test procedures for correlated test statistics. *Journal of Statistical Planning and Inference*, 82:171–196, 1999.

Z. Yu and M. J. van der Laan. Construction of counterfactuals and the G-computation formula. Technical Report 122, Division of Biostatistics, University of California, Berkeley, Berkeley, CA 94720-7360, 2002. Available at `www.bepress.com/ucbbiostat/paper122`.

Author Index

Subject Index

Springer Series in Statistics

(continued from p. II)

Knottnerus: Sample Survey Theory: Some Pythagorean Perspectives
Konishi: Information Criteria and Statistical Modeling
Küchler/Sørensen: Exponential Families of Stochastic Processes
Kutoyants: Statistical Inference for Ergodic Diffusion Processes
Lahiri: Resampling Methods for Dependent Data
Lavallée: Indirect Sampling
Le Cam: Asymptotic Methods in Statistical Decision Theory
Le Cam/Yang: Asymptotics in Statistics: Some Basic Concepts, 2nd edition
Le/Zidek: Statistical Analysis of Environmental Space-Time Processes
Liu: Monte Carlo Strategies in Scientific Computing
Manski: Partial Identification of Probability Distributions
Marshall/Olkin: Life Distributions: Structure of Nonparametric, Semiparametric and Parametric Families
Mielke/Berry: Permutation Methods: A Distance Function Approach, 2nd edition
Molenberghs/Verbeke: Models for Discrete Longitudinal Data
Mukerjee/Wu: A Modern Theory of Factorial Designs
Nelsen: An Introduction to Copulas, 2nd edition
Pan/Fang: Growth Curve Models and Statistical Diagnostics
Politis/Romano/Wolf: Subsampling
Ramsay/Silverman: Applied Functional Data Analysis: Methods and Case Studies
Ramsay/Silverman: Functional Data Analysis, 2nd edition
Reinsel: Elements of Multivariate Time Series Analysis, 2nd edition
Rosenbaum: Observational Studies, 2nd edition
Rosenblatt: Gaussian and Non-Gaussian Linear Time Series and Random Fields
Särndal/Swensson/Wretman: Model Assisted Survey Sampling
Santner/Williams/Notz: The Design and Analysis of Computer Experiments
Schervish: Theory of Statistics
Shao/Tu: The Jackknife and Bootstrap
Simonoff: Smoothing Methods in Statistics
Song: Correlated Data Analysis: Modeling, Analytics, and Applications
Sprott: Statistical Inference in Science
Stein: Interpolation of Spatial Data: Some Theory for Kriging
Taniguchi/Kakizawa: Asymptotic Theory for Statistical Inference for Time Series
Tanner: Tools for Statistical Inference: Methods for the Exploration of Posterior Distributions and Likelihood Functions, 3rd edition
Tillé: Sampling Algorithms
Tsaitis: Semiparametric Theory and Missing Data
van der Laan/Robins: Unified Methods for Censored Longitudinal Data and Causality
van der Vaart/Wellner: Weak Convergence and Empirical Processes: With Applications to Statistics
Verbeke/Molenberghs: Linear Mixed Models for Longitudinal Data
Weerahandi: Exact Statistical Methods for Data Analysis

MIX
Papier aus verantwortungsvollen Quellen
Paper from responsible sources
FSC® C105338

If you have any concerns about our products,
you can contact us on
ProductSafety@springernature.com

In case Publisher is established outside the EU,
the EU authorized representative is:
Springer Nature Customer Service Center GmbH
Europaplatz 3, 69115 Heidelberg, Germany

Printed by Libri Plureos GmbH
in Hamburg, Germany